Experimental Methods in the Physical Sciences

VOLUME 30

LASER ABLATION AND DESORPTION

EXPERIMENTAL METHODS IN THE PHYSICAL SCIENCES

Robert Celotta and Thomas Lucatorto, *Editors in Chief*

Founding Editors

L. MARTON
C. MARTON

Volume 30

Laser Ablation and Desorption

Edited by

John C. Miller
Chemical and Biological Physics Section
Oak Ridge National Laboratory, Oak Ridge, Tennessee

and

Richard F. Haglund, Jr.
Department of Physics and Astronomy and
W. M. Keck Foundation Free-Electron Laser Center
Vanderbilt University, Nashville, Tennessee

ACADEMIC PRESS

San Diego   New York   Boston   London   Sydney   Tokyo   Toronto

This book is printed on acid-free paper. ∞

Copyright © 1998 by ACADEMIC PRESS

All Rights Reserved.
No part of this publication may be reproduced or transmitted in any form or by any means, electronic or mechanical, including photocopy, recording, or any information storage and retrieval system, without permission in writing from the Publisher.
The appearance of the code at the bottom of the first page of a chapter in this book indicates the Publisher's consent that copies of the chapter may be made for personal or internal use of specific clients. This consent is given on the condition, however, that the copier pay the stated per-copy fee through the Copyright Clearance Center, Inc. (222 Rosewood Drive, Danvers, Massachussetts 01923) for copying beyond that permitted by Sections 107 or 108 of the U.S. Copyright Law. This consent does not extend to other kinds of copying, such as copying for general distribution, for advertising or promotional purposes, for creating new collective works, or for resale. Copy fees for pre-1998 chapters are as shown on the title pages. If no fee code appears on the title page, the copy fee is the same as for current chapters.
1079-4042/98 $25.00

Academic Press

525 B Street, Suite 1900, San Diego, CA 92101-4495, USA
1300 Boylston Street, Chestnut Hill, MA 02167, USA
http://www.apnet.com

United Kingdom Edition published by
Academic Press Limited
24-28 Oval Road, London NW1 7DX

International Standard Serial Number: 1079-4042/98

International Standard Book Number: 0-12-475975-0

PRINTED IN THE UNITED STATES OF AMERICA
97 98 99 00 01 IP 9 8 7 6 5 4 3 2 1

CONTENTS

Contributors — xi
Preface — xiii
Volumes in Series — xv

1. Introduction to Laser Desorption and Ablation
John C. Miller

1.1. Introduction	1
1.2. Historical Development	2
1.3. Annotated Bibliography	4
References	12

2. Mechanisms of Laser-Induced Desorption and Ablation
Richard F. Haglund, Jr.

2.1. Introduction	15
2.2. Photoexcitation and Material Response at Surfaces	20
2.3. Competition Between Localized and Delocalized Excitations	38
2.4. Laser-Induced Desorption from Metals, Semiconductors, and Insulators	58
2.5. Laser Ablation	76
2.6. Novel Experimental Tools for Studying Laser Ablation	107
2.7. Modeling Laser-Induced Desorption and Ablation	116
2.8. Conclusion and Outlook	123
References	126

3. Low Fluence Laser Desorption and Plume Formation from Wide Bandgap Crystalline Materials
J. Thomas Dickinson

3.1.	Introduction	139
3.2.	Photoelectron Emission	140
3.3.	Photodesorption of Neutral Species	140
3.4.	Photodesorption of Positive Ions	142
3.5.	Low Fluence Plume Formation	152
3.6.	Conclusions	169
	References	170

4. Lasers, Optics, and Thermal Considerations in Ablation Experiments
Costas P. Grigoropoulos

4.1.	Definition of Laser Intensity and Fluence Variables	173
4.2.	Thermal Considerations	175
4.3.	Lasers Used in Ablation	187
4.4.	Beam Delivery	191
4.5.	Temperature of the Target	201
	References	219

5. Plume Formation and Characterization in Laser-Surface Interactions
Roger Kelly, Antonio Miotello, Aldo Mele, and Anna Giardini Guidoni

5.1.	Introduction	225
5.2.	Emission Mechanisms and Plume Formation	232

5.3. Plume Characterization	245
5.4. General Conclusions	281
References	286

6. Surface Characterization
MARIKA SCHLEBERGER, SYLVIA SPELLER, and WERNER HEILAND

6.1. Introduction	291
6.2. LEED and RHEED	292
6.3. Scanning Tunneling Microscopy (STM) and Atomic Force Microscopy (AFM)	299
6.4. Ion Scattering Spectrometry	306
6.5. Structural Analysis Using Ion Beams	312
6.6. Secondary Ion Mass Spectrometry (SIMS)	315
6.7. XPS and AES	318
6.8. Summary	327
References	327

7. Surface Modification with Lasers
ZANE BALL and ROLAND SAUERBREY

7.1. Introduction	333
7.2. Physical and Chemical Effects of Laser Irradiation	340
7.3. Surface Morphology Modification	353
7.4. Surface Modification of Polymers	359
7.5. Surface Modification of Metals and Semiconductors	362
7.6. Excimer Laser Lithography	366
7.7. Conclusion	369
References	369

8. Chemical Analysis by Laser Ablation
RICHARD E. RUSSO and XIANGLEI MAO

Introduction	375
8.1. Direct Solid-Phase Chemical Analysis	375
8.2. Laser-Induced Plasmas	376
8.3. Laser Ablation—Inductively Coupled Plasma	384
8.4. Correlation of ICP Emission to Ablated Mass	390
8.5. Ablation in Noble Gases	393
8.6. Quantitative Analysis	397
8.7. Particles	401
8.8. Conclusion	403
References	405

9. Matrix-Assisted Laser Desorption and Ionization
JAMES A. CARROLL and RONALD C. BEAVIS

9.1. Introduction	413
9.2. Protein-Doped Matrix Crystals	417
9.3. Protein Ion Sources	423
9.4. Applications to Polymer Analysis	430
References	439

10. Physical Mechanisms Governing the Ablation of Biological Tissue
GLENN EDWARDS

10.1. Introduction	449
10.2. Gedanken Experiments	450

10.3. Laser-Tissue Interactions	450
10.4. Ultraviolet and Visible Laser Ablation	456
10.5. IR-Laser Ablation	457
10.6. Toward Clinical Applications	468
10.7. Future Research	468
10.8. Concluding Remarks	469
References	470

11. Growth and Doping of Compound Semiconductor Films by Pulsed Laser Ablation
Douglas H. Lowndes

11.1. Introduction	475
11.2. Characteristics of Laser Ablation Important for Film Growth	477
11.3. Growth of Compound Semiconductor Films by Pulsed Laser Ablation	517
References	557

12. Laser Ablation in Optical Components and Thin Films
Michael Reichling

12.1. Introduction	573
12.2. Experimental Techniques	576
12.3. Laser Ablation from Bulk Optical Materials	581
12.4. Laser Ablation from Thin Film Optical Materials	607
References	616

13. The Future of Laser Ablation
RICHARD F. HAGLUND, JR.

13.1. Introduction	625
13.2. Material Cutting and Joining	627
13.3. Cluster-Beam Generation by Laser Ablation	628
13.4. Laser Plasma X-ray Generation	630
13.5. Lasers in the Semiconductor Industry	631
13.6. High-Speed Color Printing	634
13.7. Laser Ablation in Microfabrication	636
13.8. Conclusions	638
References	639

INDEX 641

CONTRIBUTORS

Numbers in parentheses indicate the pages on which the authors' contributions begin.

ZANE BALL (333), *Institut für Optik und Quantenelektronik, Friedrich-Schiller-Universität, D-07743 Jena, Germany* currently with *Intel Corporation, Hillsboro, OR, 97124*

RONALD C. BEAVIS (413), *Department of Pharmacology, New York University Medical Center, New York, New York, 10016*

JAMES A. CARROLL (413), *Monsanto Company*

J. THOMAS DICKINSON (139), *Washington State University, Pullman, Washington, 99164*

GLENN EDWARDS (449), *Department of Physics and Astronomy and W. M. Keck Foundation Free-Electron Laser Center, Vanderbilt University, Nashville, Tennessee, 37212*

COSTAS P. GRIGOROPOULOS (173), *Department of Mechanical Engineering, University of California, Berkeley, Berkeley, California, 94720*

ANNA GIARDINI GUIDONI (225), *Dipartimento di Chimica, Università di Roma, La Sapienza, I-00185 Roma, Italy*

RICHARD F. HAGLUND, JR. (15, 625), *Department of Physics and Astronomy and W. M. Keck Foundation Free-Electron Laser Center, Vanderbilt University, Nashville, Tennessee, 37212*

WERNER HEILAND (291), *FB Physik, Universität Osnabrück, D-49069 Osnabrück, Germany*

ROGER KELLY (225), *Dipartimento di Fisica, Università di Trento, I-38050 Povo, Italy*

DOUGLAS H. LOWNDES (475), *Solid State Division, Oak Ridge National Laboratory, Oak Ridge, Tennessee, 37831 and Department of Materials Science and Engineering, The University of Tennessee, Knoxville, Tennessee, 37996*

XIANGLEI MAO (375), *Lawrence Berkeley National Laboratory, Berkeley, California, 94720*

ALDO MELE (225), *Dipartimento di Chimica, Università di Roma, La Sapienza, I-00185 Roma, Italy*

JOHN C. MILLER (1), *Chemical and Biological Physics Section, Oak Ridge National Laboratory, Oak Ridge, Tennessee, 37831*

ANTONIO MIOTELLO (225), *Dipartimento di Fisica, Università di Trento, I-38050 Povo, Italy*

MICHAEL REICHLING (573), *Fachbereich Physik, Freie Universität Berlin, 14195 Berlin, Germany*

RICHARD E. RUSSO (375), *Lawrence Berkeley National Laboratory, Berkeley, California, 94720*

ROLAND SAUERBREY (333), *Institut für Optik und Quantenelektronik, Friedrich-Schiller-Universität, D-07743 Jena, Germany*

MARIKA SCHLEBERGER (291), *FB Physik, Universität Osnabrück, D-49069 Osnabrück, Germany*

SYLVIA SPELLER (291), *FB Physik, Universität Osnabrück, D-49069 Osnabrück, Germany*

PREFACE

This book introduces the subject of laser ablation and laser-induced desorption for scientists and engineers. It is designed to provide an overview of fundamental experimental and theoretical tools, models and techniques, as well as an introduction to the most important applications. As used in this book, the terms "ablation" and "desorption" refer to laser-induced particle removal from surfaces under the two extremes of massive and negligible rates of surface erosion, respectively. In the case of ablation, removal rates of a substantial fraction of a monolayer per laser pulse lead to gas-dynamic effects among the ejected particles in the atmosphere above the surface, as well as to collective effects and substantial disturbance of the surface at mesoscopic scale lengths. In desorption, on the other hand, ejection of individual atoms, molecules or ions is essentially an isolated event, with little residual damage to the surface. In between these two extremes, of course, there exists a continuum of possible states of surface and plume conditions.

Laser-surface interactions lead into a tangle of scientific questions revolving around the multifaceted aspects of photon-matter coupling at high photon flux across a broad spectrum of laser wavelengths and pulse durations, and under environmental conditions ranging from ultrahigh vacuum to substantial ambient-gas pressures. These questions include the nature of atomic and molecular binding in and on surfaces; the correlation between bulk and surface structure and dynamics; the constraints imposed by broken symmetry, defects, impurities, and adsorbed atoms and molecules; and the mechanisms of electronic or vibrational excitation and relaxation which govern the dynamics of bond-breaking. These complexities justify Wolfgang Pauli's celebrated dictum that "Die Oberfläche ist eine Erfindung des Teufels!" ("The surface is an invention of the Devil!") However, they are slowly yielding their secrets to spectroscopic and microscopic investigations, many extending now to the shortest distance and time scales presently attainable. Laser ablation, on the other hand, with its complex panoply of gas-dynamic and collective, mesoscale surface modification, is still largely unexplored at the fundamental level. No doubt this circumstance will change as the demand for better control of laser-driven surface processing grows.

The first section of this book is devoted to the fundamental experimental and theoretical aspects of laser desorption and ablation. It begins with an outline of the historical development and phenomenology of laser ablation, including an annotated bibliography. This is followed by a chapter on mechanisms and models of laser-induced desorption and ablation, emphasizing the relationship between selective electronic and vibrational excitation and relaxation, on one hand, and mesoscopic thermal, gas-dynamic and thermo-mechanical processes, on the other. Next, the initial stages in the formation of the laser ablation plasma and the associated plume are discussed, in the context of ultraviolet laser interactions with a simple insulator. Experimental basics of laser-induced desorption and ablation are described in three chapters covering: laser sources and laser diagnostics for the regime where thermal effects are dominant; tools and techniques for characterizing the gas dynamics of atoms, ions, and molecules in the plume; and a survey of the most important analytical techniques for characterizing surfaces before, during, and after desorption or ablation.

The second section of this book is devoted to a survey of applications which make use of laser desorption and ablation. An important new analytical application is laser desorption and ionization which makes it possible to volatilize biological and other macromolecules for accurate mass analysis. A related chapter on laser ablation for surface analysis surveys applications of mass spectrometry and laser-induced fluorescence diagnostics for real-time materials processing applications. A number of books on pulsed laser deposition of thin films now exist; our chapter on laser-assisted thin-film deposition surveys the generic techniques and describes specific applications to compound semiconductor film growth. Laser ablation is rapidly becoming part of the armamentarium of many medical and dental specialties, as described in an overview of basic principles and current applications in general surgery, dermatology, opthalmology, vascular surgery, and dentistry. The chapter on surface modification describes techniques of lithography and alterations in surface topography and electronic properties by etching and low-level irradiation. The role of laser ablation in damage to thin films and optical components, a major problem in the optics industry, is treated in the penultimate chapter. The final chapter looks briefly to future applications of laser ablation based on some of the newest developments, such as laser-ablation cluster-beam sources for materials synthesis and laser-driven X-ray sources for coherent X-ray imaging, holography, and lithography.

This volume is structured so as to lead the reader in a straightforward way through the fundamentals of laser-surface interactions. However, each chapter is also self-contained, and includes references to other chapters as necessary, so that the interested reader may begin with the topic of greatest interest and follow the threads to other aspects of the subject as needed.

In addition to the authors of the various chapters, we express our appreciation to our colleagues and the editors of journals who have given permission to reproduce previously published results.

John C. Miller
Richard F. Haglund, Jr.

VOLUMES IN SERIES
EXPERIMENTAL METHODS IN THE PHYSICAL SCIENCES

(formerly Methods of Experimental Physics)

Editors in Chief

Robert Celotta and Thomas Lucatorto

Volume 1. Classical Methods
Edited by Immanuel Estermann

Volume 2. Electronic Methods, Second Edition (in two parts)
Edited by E. Bleuler and R. O. Haxby

Volume 3. Molecular Physics, Second Edition (in two parts)
Edited by Dudley Williams

Volume 4. Atomic and Electron Physics—Part A: Atomic Sources and Detectors; Part B: Free Atoms
Edited by Vernon W. Hughes and Howard L. Schultz

Volume 5. Nuclear Physics (in two parts)
Edited by Luke C. L. Yuan and Chien-Shiung Wu

Volume 6. Solid State Physics—Part A: Preparation, Structure, Mechanical and Thermal Properties; Part B: Electrical, Magnetic and Optical Properties
Edited by K. Lark-Horovitz and Vivian A. Johnson

Volume 7. Atomic and Electron Physics—Atomic Interactions (in two parts)
Edited by Benjamin Bederson and Wade L. Fite

Volume 8. Problems and Solutions for Students
Edited by L. Marton and W. F. Hornyak

Volume 9. Plasma Physics (in two parts)
Edited by Hans R. Griem and Ralph H. Lovberg

Volume 10. Physical Principles of Far-Infrared Radiation
By L. C. Robinson

Volume 11. Solid State Physics
Edited by R. V. Coleman

Volume 12. Astrophysics—Part A: Optical and Infrared Astronomy
Edited by N. Carleton

Part B: Radio Telescopes; Part C: Radio Observations
Edited by M. L. Meeks

Volume 13. Spectroscopy (in two parts)
Edited by Dudley Williams

Volume 14. Vacuum Physics and Technology
Edited by G. L. Weissler and R. W. Carlson

Volume 15. Quantum Electronics (in two parts)
Edited by C. L. Tang

Volume 16. Polymers—Part A: Molecular Structure and Dynamics; Part B: Crystal Structure and Morphology; Part C: Physical Properties
Edited by R. A. Fava

Volume 17. Accelerators in Atomic Physics
Edited by P. Richard

Volume 18. Fluid Dynamics (in two parts)
Edited by R. J. Emrich

Volume 19. Ultrasonics
Edited by Peter D. Edmonds

Volume 20. Biophysics
Edited by Gerald Ehrenstein and Harold Lecar

Volume 21. Solid State: Nuclear Methods
Edited by J. N. Mundy, S. J. Rothman, M. J. Fluss, and L. C. Smedskjaer

Volume 22. Solid State Physics: Surfaces
Edited by Robert L. Park and Max G. Lagally

Volume 23. Neutron Scattering (in three parts)
Edited by K. Sk"ld and D. L. Price

Volume 24. Geophysics—Part A: Laboratory Measurements; Part B: Field Measurements
Edited by C. G. Sammis and T. L. Henyey

Volume 25. Geometrical and Instrumental Optics
Edited by Daniel Malacara

Volume 26. Physical Optics and Light Measurements
Edited by Daniel Malacara

Volume 27. Scanning Tunneling Microscopy
Edited by Joseph Stroscio and William Kaiser

Volume 28. Statistical Methods for Physical Science
Edited by John L. Stanford and Stephen B. Vardeman

Volume 29. Atomic, Molecular, and Optical Physics—Part A: Charged Particles; Part B: Atoms and Molecules; Part C: Electromagnetic Radiation
Edited by F. B. Dunning and Randall G. Hulet

Volume 30. Laser Ablation and Desorption
Edited by John C. Miller and Richard F. Haglund, Jr.

1. INTRODUCTION TO LASER DESORPTION AND ABLATION

John C. Miller

Chemical and Biological Physics Section
Oak Ridge National Laboratory
Oak Ridge, Tennessee

1.1 Introduction

At the ripe age of 35 years, the laser has become a mature technological device with many varied applications. This was not always true of course. For many years, the laser was viewed as "an answer in search of a question." That is, it was seen as an elegant device, but one with no real, useful application outside of fundamental scientific research. In the last two decades, however, numerous laser applications have moved from the laboratory to the industrial workplace or the commercial market.

Lasers are unique energy sources characterized by their spectral purity, spatial and temporal coherence, and high average and peak intensity. Each of these characteristics has led to applications that take advantage of these laser qualities. For instance, spatial coherence, which allows a highly collimated laser beam, has spawned remote-sensing, range-finding, and target designation applications. Other applications based on coherence include interferometry and holography. Likewise the property of monochromaticity (spectral purity) enables chemical and physical sensing techniques based on high-resolution spectroscopy. Many other unique uses of lasers include communications, information storage and manipulation, and entertainment. All of these "high-tech" applications have come to define everyday life in the late twentieth century.

One property of lasers, however, that of high intensity, did not immediately lead to "delicate" applications but rather to those requiring "brute force." That is, the laser was used in a macroscopic way either for material removal or heating. The first realistic applications involved cutting, drilling, and welding, and the laser was little more advanced than a saw, a drill, or a torch. In a humorous vein, A. L. Schawlow proposed and demonstrated the first "laser eraser" in 1965 [1]. The different absorbencies of paper and ink allowed the laser to selectively remove typewriter print without damaging the underlying paper. Another early application used a laser to generate a plasma at the surface of a solid, and the resulting spectral emission could be used for elemental analysis [2]. Vastly more expensive than traditional tools, however, the laser only slowly found niche uses where its advantages made up for the added cost and complexity. Ready's book [3] provides much detail about the applications of high powered lasers up to 1971. The present volume gives comprehensive reviews of the major applications of intense pulsed lasers since that classic reference.

1.2 Historical Development

The laser age began with the birth of the first laser in 1960. T. H. Maiman [4] demonstrated the optical pumping of a ruby rod and the emission of coherent radiation. The 1960s were very important times for research involving laser-material interactions and the stage was set in that decade for virtually all of the later applications. In particular, many aspects of laser ablation were first studied during this decade. For instance, Brech and Cross [2] collected and spectrally dispersed the emitted light from ruby-laser ablated metals. This work formed the basis for the technique of laser microprobe emission spectroscopy for the elemental analysis of solid materials. Linlor [5] made measurements of the energy of ejected ions by time-of-flight determinations. This work, along with that of Honig and Woolston [6], was the first example of laser mass spectrometry, which was eventually introduced as a commercial instrument by Leybold-Heraeus in 1978. Muray [7] was the first of many to investigate laser photoemission of electrons. The study of ablation plumes by photography was initiated by Ready [8] in 1963. Other important papers appearing in 1963 were by Rosan *et al.* [9] for the first ablation studies of biological material and Howe [10] who used rotationally and vibrationally resolved molecular emission bands to measure temperatures of ablation plumes. Berkowitz and Chupka [11] were the first to observe clusters in an ablation plume. They observed carbon, magnesium, and boron cluster ions after postionization of ablated neutrals.

Later, Basov and Krokhin [12] made the first suggestion of laser fusion, and as higher power lasers were used, vacuum ultraviolet [13] and X-ray emissions [14] were detected. Higher power also led to the observation of multiply-charged ions [15] and to two- [16] and three-photon [17] photoemission. Measurable neutron fluxes from laser-heated targets were first reported in 1968 [18]. Finally, of great importance in terms of modern applications of ablation, the first laser deposition of thin films was demonstrated by Smith and Turner [19] in 1965. Unfortunately these early films were of poor quality and the technique could not compete with other established techniques. It was not until twenty years later that laser produced films were competitive. More details of these early studies can be found in the book by Ready [3]. Clearly the 1960s were a period of exploring many different aspects of laser ablation and coming to a first stage of experimental and theoretical understanding.

During the 1970s and early 1980s the development and understanding of laser desorption and ablation processes were incremental and steady. In particular, the field was driven by advances in laser technology. As lasers got brighter, with shorter pulse lengths and correspondingly higher peak powers, and more stable in terms of shot-to-shot reproducibility, ablation studies moved into new regimes. The availability of new lasers, especially excimer lasers, and a broader wavelength capability led to new coupling possibilities. And, of course, the application of ablation studies to ever more types of materials vastly increased the diversity of data available. Concurrently the sophistication of ablation diagnostics also improved dramatically, led by advances in electronics and computers. However, the application of laser ablation to problems of practical import advanced very little in this period. The principal uses continued to be in emission and mass spectrometry for chemical analysis.

This situation changed dramatically in the late 1980s. Zaitsev-Zotov et al. [20] first produced superconducting thin films by laser ablation in 1983. But the technique went virtually unnoticed until Venkatesan and coworkers [21, 22] demonstrated in 1987 the growth of the newly discovered high-temperature superconductors by laser ablation of bulk Y-Ba-Cu material followed by annealing in air or oxygen. Amazingly the stoichiometry of the thin films was virtually identical to that of the bulk. The ablation technique offered several advantages of simplicity, versatility, and experimental ease over traditional methods of sputtering or coevaporation. These results produced a "feeding frenzy" as research groups around the world further refined and extended the technique. The recent book edited by Chrisey and Hubler [23] comprehensively reviews the last decade of research in pulsed laser deposition (PLD) of thin films.

Several other applications of laser ablation "came of age" in the late 1980s. Particularly spectacular has been the growth of laser-based medical procedures. Laser surgery has matured tremendously, and many techniques have been approved for general clinical use. Laser-based ophthalmology is now widely available, and laser reshaping of soft tissue of the throat for the treatment of sleep apnea and control of snoring is an established technique. Laser ablation is a useful tool in the dermatology field for the removal of birthmarks or tattoos and most recently for cosmetic smoothing of wrinkled skin. When coupled with fiber optic delivery and viewing systems, laser surgery is increasingly being used for internal arthroscopic cutting and for arterial angioplasty. Dental applications are being studied as well.

Although the PLD and medical applications are by far the most important in terms of economic impact, laser ablation has found several other important new uses. For instance, even though laser microanalysis by laser ablation coupled to mass spectrometry has been in use for two decades, the late 1980s saw the development of a new application to biological molecules. In matrix-assisted laser desorption/ionization (MALDI), the laser couples to an organic matrix such as nicotinic acid and pumps energy into the system. Large biological molecules dissolved in the matrix are desorbed and ionized in such a way that they are carried intact into the gas phase with little or no fragmentation. This technique has revolutionized the identification and study of large molecular weight biomolecules and polymers and has even been used to sequence genome fragments. Another application of laser ablation to another research field is its use with supersonic jet sources to volatilize, cool, and condense solids into clusters, which may then be studied by various spectroscopic techniques. Finally, extremely high power laser ablation has paved the way to the generation of high-energy plasmas, which serve as the source for bright X-ray sources and even coherent devices.

Each of these modern applications of laser ablation is described in more detail in the following chapters of the present volume. As a paradigm for the evolving sophistication of laser ablation, the laser eraser described by Schawlow thirty years ago has now become a tool for graffiti removal and, more delicately, for art restoration [24, 25].

1.3 Annotated Bibliography

1.3.1 Journals

In the one or two decades after the demonstration of the first laser, most papers on laser ablation and desorption appeared in physics journals. As the emphasis shifted to applications, many more journals published such papers. It is not useful to simply list these journals as it would be a very long list. However, a few review articles have appeared, and some journals have published special issues on the subjects of interest here. They are as follows:

J. T. Cheung, and H. Sankur, "Growth of Thin Films by Laser-Induced Evaporation," *CRC Critical Reviews in Solid State and Material Sciences* **15**, 63, 1988. A comprehensive review.

R. Srinivasan, and B. Braren, "Ultraviolet Laser Ablation of Organic Polymers," *Chem. Rev.* **89**, 1303, 1989. An excellent review.

P. Kelly, ed., "Laser-Induced Material Modification," *Optical Engineering*, vol. 28 (10), 1989. A special issue with twelve articles; several are relevant to laser ablation and desorption.

F. Hillencamp, R. C. Beavis, and B. Chait, "Matrix-Assisted Laser Desorption/Ionization Mass Spectrometry of Biopolymers," *Anal. Chem.* **63**, 1193A, 1991. The first review of this subject by three of its first practitioners.

G. K. Hubler, ed., "Pulsed Laser Ablation," *MRS Bulletin*, vol. 17 (2), 1992. A special issue with four overview articles on PLD.

R. Russo, "Laser Ablation for Spectrochemical Analysis," *Appl. Spectrosc.* **49**, 14A, 1995. A good overview of laser ablation, with emphasis on its use as a source for inductively coupled plasma (ICP) spectroscopy.

"Applications of Laser Technology in Materials Processing," *Opt. and Quant. Electron.* **27** (12), (1995). A special issue devoted to the industrial use of lasers. Several chapters cover different aspects of ablation.

1.3.2 Books

The following chronological list is not meant to be exhaustive, but it does include most of the major books relevant to laser ablation/desorption. In some cases, books are listed where only one or two chapters are relevant. Clearly it is not possible to list all such examples. Some foreign language or limited-access books have been omitted.

J. F. Ready, *Effects of High Power Laser Radiation*, Academic Press, New York, 1971. The "classic" reference. Definitive account of the early state of the subject at that time. Still a useful overview.

C. A. Anderson, ed., *Microprobe Analysis*, John Wiley & Sons, New York, 1973. Four chapters (12–15) provide a background summary of laser microprobe analysis and applications to geology, biological samples, and metals. Dated and of mostly historical interest.

T. P. Hughes, *Plasmas and Laser Light*, John Wiley & Sons, New York, 1975. An excellent summary of the theory of ablation as well as experimental results up to the publication date.

G. Bekefi, *Principles of Laser Plasmas*, John Wiley & Sons, New York, 1976. A good account of theoretical approaches with some, now dated, experimental results.

J. F. Ready, *Industrial Applications of Lasers*, Academic Press, New York, 1978. Of mostly historical interest as industrial processing has changed dramatically since the 1970s.

R. F. Wood, *Laser Damage in Optical Materials*, Adam Hilger, Bristol, 1986. A good overview of this field.

E. H. Piepmeier, ed., *Analytical Applications of Lasers*, John Wiley & Sons, New York, 1986. Chapter 19 is on laser ablation for atomic spectroscopy.

L. J. Radziemski, R. W. Solarz, and J. A. Paisner, eds., *Laser Spectroscopy and Its Applications*, Marcel Dekker, New York, 1987. Chapter 5 contains a good account of laser plasma generation and analysis.

L. J. Radziemski, and D. A. Cremers, eds., *Laser-Induced Plasmas and Applications*, Marcel Dekker, New York, 1989. An excellent source covering laser-produced plasmas in both gaseous and solid samples. Applications to chemical analysis, fusion, semiconductor fabrication, and X-ray generation are described in detail.

R. M. Wood, ed., *Selected Papers on Laser Damage in Optical Materials*, SPIE Milestone Series vol. MS24, 1990.

D. M. Lubman, ed., *Lasers and Mass Spectrometry*, Oxford University Press, Oxford, 1990. With an emphasis on chemical analysis, over half of the 23 chapters are relevant to laser ablation/desorption of solids or surfaces for subsequent examination by mass spectrometry.

A. Vertes, R. Gijbels, and F. Adams, *Laser Ionization Mass Analysis*, John Wiley & Sons, New York, 1993. Similar to the previous listing.

J. C. Miller, ed., *Laser Ablation Principles and Applications*, Springer Series in Material Science, vol. 28, Springer-Verlag, Berlin, 1994. Laser ablation is covered in detail with chapters on historical development, theory, optical surface damage, superconducting thin films, polymer ablation, mass spectrometry, and chemical analysis.

B. Chrisey, and G. K. Hubler, eds., *Pulsed Laser Deposition of Thin Films*, John Wiley & Sons, New York, 1994. A comprehensive overview of the subject. Thirteen chapters cover fundamentals and experimental aspects of PLD. The latter twelve chapters offer a detailed literature survey of PLD studies of specific material types such as oxides, ferrites, biomaterials.

M. von Allmen, and A. Blatter, *Laser Beam Interactions With Materials: Physical Principles and Applications*, Springer Series in Material Science, vol. 2, 2nd ed, Springer-Verlag, New York, 1995. A mostly theoretical account of laser-matter interactions with chapters on heating, melting and solidification, and evaporation and plasma formation.

D. Bäuerle, *Laser Processing and Chemistry*, 2nd ed, Springer-Verlag, Berlin, 1996. An up-to-date revision of an excellent book originally entitled *Chemical Processing With Lasers*. Thirty chapters cover a variety of topics very comprehensively.

1.3.3 Conference Series

The newest developments in any field first surface at conferences, and the proceedings of such conferences provide a useful snapshot of the current state of the subject. Conference series further allow one to follow the historical evolution of the hot topics over many years. The useful lifetime of such proceedings are, of course, limited because more complete accounts of the research are usually rapidly published in primary journals. Furthermore, conference proceedings often have a limited distribution and hence are sometimes difficult to obtain. Nonetheless the following list may prove useful.

Annual Symposium on Optical Materials for High Power Lasers. Sometimes referred to as "the Boulder Damage Conference," this meeting is the oldest and longest running conference in this field for which published proceedings are available. Initially published by the American Society for

Testing Materials (ASTM), later proceedings were published by the National Bureau of Standards (NBS), The National Institute of Standards and Technology (NIST), or The Society of Photo-Optical Instrumentation Engineers (SPIE). Somewhat confusingly the proceedings title is "Laser Induced Damage in Optical Materials" rather than the conference name.

A. J. Glass, and A. H. Guenter, eds., *Damage in Laser Glass*, ASTM Spec. Tech. Pub. 469, ASTM, Philadelphia, 1969.

A. J. Glass, and A. H. Geunther, eds., *Damage in Laser Materials*, Nat. Bur. Stand. (U.S.) Spec. Publ. 341, 1970.

A. J. Glass, and A. H. Geunther, eds., *Damage in Laser Materials: 1971*, Nat. Bur. Stand. (U.S.) Spec. Publ. 356, 1971.

A. J. Glass, and A. H. Geunther, eds., *Laser-Induced Damage in Optical Materials: 1972*, Nat. Bur. Stand. (U.S.) Spec. Publ. 372, 1972.

A. J. Glass, and A. H. Geunther, eds., *Laser-Induced Damage in Optical Materials: 1973*, Nat. Bur. Stand. (U.S.) Spec. Publ. 387, 1973.

A. J. Glass, and A. H. Geunther, eds., Laser-Induced Damage in Optical Materials: A Conference Report, *Appl. Opt.* **13**, 74, 1974.

A. J. Glass, and A. H. Geunther, eds., *Laser-Induced Damage in Optical Materials: 1974*, Nat. Bur. Stand. (U.S.) Spec. Publ. 414, 1974.

A. J. Glass, and A. H. Geunther, eds., Laser-Induced Damage in Optical Materials: The ASTM Symposium, *Appl. Opt.* **14**, 698, 1975.

A. J. Glass, and A. H. Geunther, eds., *Laser-Induced Damage in Optical Materials: 1975*, Nat. Bur. Stand. (U.S.) Spec. Publ. 435, 1975.

A. J. Glass, and A. H. Geunther, eds., Laser-Induced Damage in Optical Materials: 7th ASTM Symposium, *Appl. Opt.* **15**, 1510, 1976.

A. J. Glass, and A. H. Geunther, eds., *Laser-Induced Damage in Optical Materials: 1976*, Nat. p Bur. Stand. (U.S.) Spec. Publ. 462, 1976.

A. J. Glass, and A. H. Geunther, eds., Laser-Induced Damage in Optical Materials: 8th ASTM Symposium, *Appl. Opt.* **16**, 1214, 1977.

A. J. Glass, and A. H. Geunther, eds., *Laser-Induced Damage in Optical Materials: 1977*, Nat. Bur. Stand. (U.S.) Spec. Publ. 509, 1977.

A. J. Glass, and A. H. Geunther, eds., Laser-Induced Damage in Optical Materials: 9th ASTM Symposium, *Appl. Opt.* **17**, 2386, 1978.

A. J. Glass, and A. H. Geunther, eds., *Laser-Induced Damage in Optical Materials: 1978*, Nat. Bur. Stand. (U.S.) Spec. Publ. 541, 1978.

A. J. Glass, and A. H. Geunther, eds., Laser-Induced Damage in Optical Materials: 10th ASTM Symposium, *Appl. Opt.* **18**, 2212, 1979.

H. E. Bennett, A. J. Glass, A. H. Geunther, and B. E. Newnam, eds., *Laser-Induced Damage in Optical Materials: 1979*, Nat. Bur. Stand. (U.S.) Spec. Publ. 568, 1979.

H. E. Bennett, A. J. Glass, A. H. Geunther, and B. E. Newnam, eds., Laser-

Induced Damage in Optical Materials: 11th ASTM Symposium, *Appl. Opt.* **19**, 2212, 1980.

H. E. Bennett, A. J. Glass, A. H. Geunther, and B. E. Newnam, eds., *Laser-Induced Damage in Optical Materials: 1980*, Nat. Bur. Stand. (U.S.) Spec. Publ. 620, 1981.

H. E. Bennett, A. J. Glass, A. H. Geunther, and B. E. Newnam, eds., Laser-Induced Damage in Optical Materials: 12th ASTM Symposium, *Appl. Opt.* **20**, 3003, 1981.

H. E. Bennett, A. H. Geunther, D. Milam, and B. E. Newnam, eds., *Laser-Induced Damage in Optical Materials: 1981*, Nat. Bur. Stand. (U.S.) Spec. Publ. 638, 1983.

H. E. Bennett, A. H. Geunther, D. Milam, and B. E. Newnam, eds., Laser-Induced Damage in Optical Materials: 13th ASTM Symposium, *Appl. Opt.* **22**, 3276, 1983.

H. E. Bennett, A. H. Geunther, D. Milam, and B. E. Newnam, eds., *Laser-Induced Damage in Optical Materials: 1982*, Nat. Bur. Stand. (U.S.) Spec. Publ. 669, 1984.

H. E. Bennett, A. H. Geunther, D. Milam, and B. E. Newnam, eds., Laser-Induced Damage in Optical Materials: 14th ASTM Symposium, *Appl. Opt.* **23**, 3782, 1984.

H. E. Bennett, A. H. Geunther, D. Milam, and B. E. Newnam, eds., *Laser-Induced Damage in Optical Materials: 1983*, Nat. Bur. Stand. (U.S.) Spec. Publ. 688, 1985.

H. E. Bennett, A. H. Geunther, D. Milam, and B. E. Newnam, eds., Laser-Induced Damage in Optical Materials: 15th ASTM Symposium, *Appl. Opt.* **25**, 258, 1986.

H. E. Bennett, A. H. Geunther, D. Milam, and B. E. Newnam, eds., *Laser-Induced Damage in Optical Materials: 1984*, Nat. Bur. Stand. (U.S.) Spec. Publ. 727, 1986.

H. E. Bennett, A. H. Geunther, D. Milam, and B. E. Newnam, eds., Laser-Induced Damage in Optical Materials: 16th ASTM Symposium, *Appl. Opt.* **26**, 813, 1987.

H. E. Bennett, A. H. Geunther, D. Milam, and B. E. Newnam, eds., *Laser-Induced Damage in Optical Materials: 1985*, Nat. Bur. Stand. (U.S.) Spec. Publ. 746, 1987.

H. E. Bennett, A. H. Geunther, D. Milam, and B. E. Newnam, eds., *Laser-Induced Damage in Optical Materials: 1986*, Nat. Inst. Stand. Tech. (U.S.) Spec. Publ. 752, 1987.

H. E. Bennett, A. H. Geunther, D. Milam, and B. E. Newnam, and M. J. Soileau, eds., *Laser-Induced Damage in Optical Materials: 1987*, Nat. Inst. Stand. Tech. (U.S.) Spec. Publ.756, 1988.

H. E. Bennett, L. L. Chase, A. H. Geunther, B. E. Newnam, and M. J. Soileau, eds., *Laser-Induced Damage in Optical Materials: 1989*, Nat. Inst. Stand. Tech. (U.S.) Spec. Publ. 801, ASTM STP 1117 and SPIE vol. 1438, 1989.

H. E. Bennett, L. L. Chase, A. H. Geunther, B. E. Newnam, and M. J. Soileau, eds., *Laser-Induced Damage in Optical Materials: 1990*, ASTM STP 1141 and SPIE vol. 1441, 1991.

H. E. Bennett, L. L. Chase, A. H. Geunther, B. E. Newnam, and M. J. Soileau, eds., *Laser-Induced Damage in Optical Materials: 1991*, SPIE vol. 1624, 1992.

H. E. Bennett, L. L. Chase, A. H. Geunther, B. E. Newnam, and M. J. Soileau, eds., *Laser-Induced Damage in Optical Materials: 1992*, SPIE vol. 1848, 1993.

H. E. Bennett, L. L. Chase, A. H. Geunther, B. E. Newnam, and M. J. Soileau, eds., *Laser-Induced Damage in Optical Materials: 1993*, SPIE vol. 2114, 1994.

Desorption Induced by Electronic Transitions (DIET). Earlier conferences emphasized electron and ion beam excitation, but more recent conferences describe many laser-based studies.

N. H. Tolk, M. M. Traum, J. C. Tully, and T. E. Madey, eds., *Desorption Induced by Electronic Transitions—DIET I*, Springer Series in Chemical Physics, vol. 24, Springer-Verlag, Berlin, 1983.

W. Brenig and D. Menzil, eds., *Desorption Induced by Electronic Transitions—DIET II*, Springer Series in Surface Science, vol. 4, Springer, Heidelberg, 1985.

R. H. Stulen, and M. L. Knotek, eds., *Desorption Induced by Electronic Transitions—DIET III*, Springer Series in Surface Science, vol. 13, Springer, Heidelberg, 1988.

G. Betz, and P. Varga, eds. *Desorption Induced by Electronic Transitions—DIET IV*, Springer Series in Surface Science, vol. 19, Springer-Verlag, Hiedelberg, 1990.

A. R. Burns, E. B. Stechel, and D. R. Jennison, eds., *Desorption Induced by Electronic Transitions—DIET V*, Springer Series in Surface Science, vol 31, Springer-Verlag, Berlin, 1993.

M. Szymonski, and Z. Postawa, Nucl. Instr. and Methods, *Phys. Res. B.* 101, 1995.

International Conference on Laser Ablation (COLA) Begun as a workshop in 1991, this was the first conference to focus solely on laser ablation. The series has emphasized fundamental understanding and has attempted to cover the broad applications in material science, analytical chemistry, biomedical sciences, thin films, X-ray generation, pulsed laser deposition, and so on.

J. C. Miller, and R. F. Haglund, Jr., eds., *Laser Ablation Mechanisms and Applications*, Lecture Notes in Physics, vol. 389, Springer-Verlag, Berlin, 1991.

J. C. Miller, and D. B. Geohegan, eds., *Laser Ablation Mechanisms and Applications II*, American Institute of Physics Conference Proceedings, vol. 288, American Institute of Physics, New York, 1994.

E. Fogarassy, D. B. Geohegan, and M. Stuke, eds., *Laser Ablation Mechanisms and Applications III*, European Materials Research Society Symposia Proceedings, Elsevier, New York, 1996; see also *Applied Surface Science*, vols. 96–98, 1996.

R. F. Haglund, Jr., D. B. Geohegan, K. Murakami, and R. E. Russo, eds., *Laser Ablation Mechanisms and Applications IV*, Elsevier, New York, 1998; see also *Applied Surface Science*, vols, xxx, 1998.

Materials Research Society Symposia. In their two annual meetings, the MRS co-locates a number of focused symposia, each of which usually publishes a proceedings volume. Although many such proceedings may have a few papers relevant to laser ablation/desorption, the following symposia have limited themselves to this topic or have a significant number of related papers.

D. C. Paine, and J. C. Bravman, eds., *Laser Ablation for Material Synthesis*, Materials Research Society Symposium Proceedings, vol. 191, Materials Research Society, Pittsburgh, 1990.

B. Braren, J. J. Dubowski, and D. P. Norton, eds., *Laser Ablation in Materials Processing: Fundamentals and Applications*, Materials Research Society Proceedings, vol. 285, Materials Research Society, Pittsburgh, 1993.

H. A. Atwater, J. T. Dickinson, D. H. Lowndes, and A. Polman, eds., *Film Synthesis and Growth Using Energetic Beams*, Materials Research Society Proceedings, vol. 388, Materials Research Society, Pittsburgh, 1995.

D. C. Jacobson, D. E. Luzzi, T. F. Heinz, and M. Iwaki, eds., *Beam-Solid Interactions for Materials Synthesis and Characterization*, Materials Research Society Proceedings, vol. 354, 1995.

R. Singh, D. Norton, L. Laude, J. Narayan, and J. Cheung, eds., *Advanced Laser Processing of Materials—Fundamentals and Applications*, Material Research Society Proceedings, vol. 397, 1996.

European Materials Research Society. Similar to the American society, the E-MRS sponsors focused symposia. In 1993, the COLA conference (see above) was co-located with the E-MRS meeting.

I. W. Boyd, and E. F. Krimmel, eds., *Photon, Beam and Plasma Assisted Processing*, North Holland, Amsterdam, 1989.

E. Fogarassy, and S. Lazare, eds., *Laser Ablation of Electronic Materials: Basic*

Mechanisms and Applications, European Materials Research Society Monographs, North Holland, Amsterdam, 1992.

International Symposium on Resonance Ionization Spectroscopy. Although resonance ionization spectroscopy (RIS) is primarily a gas-phase technique, it has often been used to analyze solids or surfaces following laser or electron beam vaporization. The conference series has thus always included a session on solids and a number of relevant papers may be found in each of the proceedings. In more recent years, MALDI has also been a frequent topic at these meetings.

G. S. Hurst, and M. G. Payne, eds., *Resonance Ionization Spectroscopy 1984*, The Institute of Physics Conference Series, vol. 71, The Institute of Physics, Bristol, 1984.

G. S. Hurst and C. Gray Morgan, eds., *Resonance Ionization Spectroscopy 1986*, The Institute of Physics Conference Series, vol. 84, The Institute of Physics, Bristol, 1987.

T. B. Lucatorto, and J. E. Parks, eds., *Resonance Ionization Spectroscopy 1988*, The Institute of Physics Conference Series, vol. 94, The Institute of Physics, Bristol, 1989.

J. E. Parks, and N. Omenetto, eds., *Resonance Ionization Spectroscopy 1990*, The Institute of Physics Conference Series, vol. 114, The Institute of Physics, Bristol, 1991.

C. M. Miller, and J. E. Parks, eds., *Resonance Ionization Spectroscopy 1992*, The Institute of Physics Conference Series, vol. 128, The Institute of Physics, Bristol, 1992.

H.-J. Kluge, J. E. Parks, and K. Wendt, eds., *Resonance Ionization Spectroscopy 1994*, American Institute of Physics Proceedings, vol. 329, American Institute of Physics, New York, 1995.

N. Winograd, and J. E. Parks, eds., *Resonance Ionization Spectroscopy 1966*, American Institute of Physics Proceedings, vol. 388, American Institute of Physics, New York, 1997.

NATO Advanced Study Institutes. Each is a one-time NATO-sponsored conference.

L. D. Laude, D. Bäerle, and M. Wautelet, eds., *Interfaces Under Laser Irradiation*, Martinus Nijhoff Publishers, Dordrecht, 1987.

L. D. Laude, ed., *Excimer Lasers*, Kluwer Academic Publishers, Dordrecht, 1994.

International Laser Science Conference (ILS). Later called the Interdisciplinary Laser Science Conference, this meeting is sponsored by the Laser Science Division of the American Physical Society. Only the first four meetings published proceedings, but each conference had one or more sessions devoted to laser-condensed matter interaction, including ablation, PLD, and so on.

W. C. Stwalley, and M. Lapp, eds., *Advances in Laser Science—I*, American Institute of Physics Conference Series, vol. 146, American Institute of Physics, New York, 1986.

M. Lapp, W. C. Stwalley, and G. A. Kenney-Wallace, eds., *Advances in Laser Science—II*, American Institute of Physics Conference Series, vol. 160, American Institute of Physics, New York, 1987.

A. C. Tam, J. L. Gole, and W. C. Stwalley, eds., *Advances in Laser Science—III*, American Institute of Physics Conference Series, vol. 172, American Institute of Physics, New York, 1988.

J. L. Gole, D. F. Heller, M. Lapp, and W. C Stwalley, eds., *Advances in Laser Science—IV*, American Institute of Physics Conference Series, vol. 191, American Institute of Physics, New York, 1989.

Acknowledgments

Oak Ridge National Laboratory, managed by Lockheed Martin Energy Research Corporation for the U.S. Department of Energy under contract number DE-AC05-96OR22464.

The submitted manuscript has been authored by a contractor of the U.S. Government under contract No. DE-AC05-96OR22464. Accordingly, the U.S. Government retains a nonexclusive, royalty-free license to publish or reproduce the published form of this contribution, or allow others to do so, for U.S. Government purposes.

References

1. Schawlow, A. L. Lasers. *Science* **149**, 13 (1965).
2. Brech, F., and Cross, L. Optical micromission stimulated by a ruby maser. *Appl. Spectrosc.* **16**, 59 (1962).
3. Ready, J. F. *Effects of High-Power Laser Radiation.* Academic Press, New York, 1971.
4. Maiman, T. H. Stimulated optical radiation. *Nature* **187**, 493 (1960).
5. Linlor, W. I. Plasmas produced by laser bursts. *Bull. Am. Phys. Soc.* **7**, 440 (1962).
6. Honig, R. E. and Woolston, J. R. Laser-induced emission of electrons, ions, and neutral atoms from solid surfaces. *Appl. Phys. Lett.* **2**, 138 (1963).
7. Muray, J. J. Photoelectric effect induced by high-intensity laser light beam from quartz and borosilicate glass. *Bull. Am. Phys. Soc.* **8**, 77 (1963).
8. Ready, J. F. Development of plume of material vaporized by giant-pulse laser. *Appl. Phys. Lett.* **3**, 11 (1963).
9. Rosan, R. C., Healy, M. K., and McNary, Jr., W. F. Spectroscopic ultramicroanalysis with a laser. *Science* **142**, 236 (1963).

REFERENCES

10. Howe, J. A. Observations on the maser-induced graphite jet. *J. Chem. Phys.* **39**, 1362 (1963).
11. Berkowitz, J., and Chupka, W. A. Mass spectrometric study of vapor ejected from graphite and other solids by focused laser beams. *J. Chem. Phys.* **40**, 2735 (1964).
12. Basov, N. G., and Krokhin, O. N. Conditions for heating up of a plasma by the radiation from an optical generator. *Zh. Eksp. Teor. Fiz.* **46**, 171 (1964) [Sov. Phys. JETP **19**, 123 (1964)].
13. Ehler, A. W., and Weissler, G. L. Vacuum ultraviolet radiation from plasmas formed by a laser on metal surfaces. *Appl. Phys. Lett.* **8**, 89 (1966).
14. Langer, P., Tonon, G., Floux, F., and Ducauze, A. Laser-induced emission of electrons, ions, and X-rays from solid targets. *IEEE* **QE2**, 499 (1966).
15. Archibold, E., and Hughes, T. P. Electron temperature in a laser-heated plasma. *Nature* **204**, 670 (1964).
16. Sonneburg, H., Heffner, H., and Spicer, W. Two-photon photoelectric effect in Cs_3Sb. *Appl. Phys. Lett.* **5**, 95 (1964).
17. Logothetis, E. M., and Hartman, P. L. Three-photon photoelectric effect in gold. *Phys. Rev. Lett.* **18**, 581 (1967).
18. Basov, N. G., Kruikov, P. G., Zakharov, S. D., Senatskii, Y. V., and Tchekalin, S. V. Experiments on the observation of neutron emission at the focus of the high-power laser radiation on a lithium deuteride surface. *IEEE J. Quant. Electron.* **QE4**, 864 (1968).
19. Smith, H. M., and Turner, Vacuum, A. F. Deposited thin films using a ruby laser. *Appl. Optics* **4**, 147 (1965).
20. Zaitsev-Zotov, S. V., Martynyuk, A. N., and Protasov, N. E. Superconductivity of $BaPb_{1-x}Bi_xO_3$ films prepared by laser evaporation method. *Sov. Phys. Solid State* **25**, 100 (1983).
21. Dijkkamp, D., Venkatesan, T., Wu, X. D., Shaheen, S. A., Jiswari, N., Min-Lee, Y. H., McLean, W. L., and Croft, M. Preparation of Y-Ba-Cu oxide superconductor thin films using pulsed laser evaporation from high T_c bulk material. *Appl. Phys. Lett.* **51**, 619 (1987).
22. Wu, X. D., Dijkkamp, D., Ogale, S. B., Inam, A., Chase, E. W., Miceli, P. F., Chang, C. C., Tarascon, J. M., and Venkatesan, T. Epitaxial ordering of oxide superconductor thin films on (100) $SrTiO_3$ prepared by pulsed laser evaporation. *Appl. Phys. Lett.* **51**, 861 (1987).
23. Chrisey, D. B., and Hubler, G. K., eds. *Pulsed Laser Deposition of Thin Films*. John Wiley & Sons, New York, 1994.
24. Matthews, D. L. Graffiti removal by laser. *Sci. and Tech. Rev.* April 1996.
25. Fotakis, C. Lasers for art's sake! *Opt. and Phot. News* **30** (1995).

2. MECHANISMS OF LASER-INDUCED DESORPTION AND ABLATION

Richard F. Haglund, Jr.

Department of Physics and Astronomy
and W. M. Keck Foundation Free-Electron Laser Center
Vanderbilt University
Nashville, Tennessee

2.1 Introduction

Laser-induced desorption and ablation result from the conversion of an initial electronic or vibrational photoexcitation into kinetic energy of nuclear motion, leading to the ejection of atoms, ions, molecules, and even clusters from a surface. The widely accepted term for beam-induced ejection of particles from surfaces is *sputtering*, and in generic descriptions we shall use the term *laser sputtering*. We define *laser-induced desorption* as particle ejection without any detectable mesoscopic modification of surface composition or structure; with a particle yield that is a linear function of the density of electronic or vibrational excitation; and without any significant gas-dynamic effects in the steam of particles leaving the surface. *Laser ablation*, in contrast, is a sputtering process in which material removal rates typically exceed one-tenth monolayer per pulse; the surface is structurally or compositionally modified at mesoscopic length scales; and particle yields are superlinear functions of the density of excitation. The formation of an ablation plume—a weakly ionized, low-to-moderate density expanding gas cloud—adds to laser ablation the complications of plasma-surface interactions, gas dynamics, and laser-induced photochemistry.

Laser-induced desorption and laser ablation are not entirely distinct phenomena, however. Laser-induced desorption initiated by low-fluence "conditioning" of a surface may lead to material modifications that affect subsequent laser ablation. Moreover, laser ablation need not involve massive, catastrophic destruction of a surface; indeed, much of pulsed-laser film deposition relies on laser ablation as a reasonably well controlled and repeatable method of injecting target material into the gas phase. It is, therefore, probably more correct to view desorption and ablation as two points on a continuum of phenomena seen in laser interactions with material surfaces, beginning with desorption and perhaps ending at avalanche ionization and massive thermomechanical damage of surfaces, as suggested by Fig. 1.

It is useful to begin this survey from an historical perspective. A generation ago, laser ablation in the nanosecond regime—using the ruby and Nd:YAG

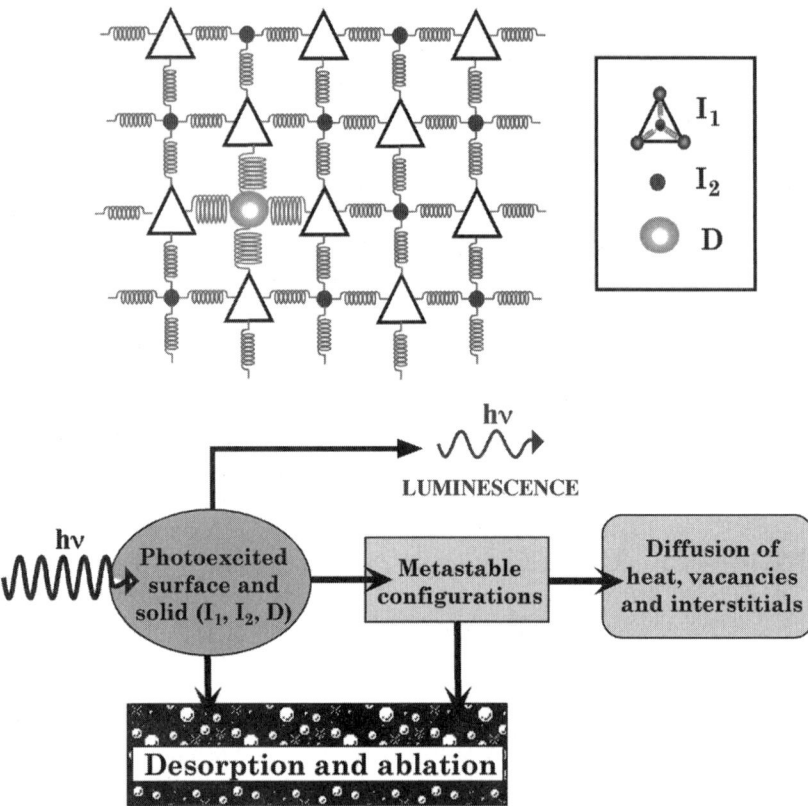

FIG. 1. Schematic of pathways leading from photoexcitation to laser-induced desorption and ablation. Laser light can excite intrinsic molecular (I_1) or atomic (I_2) sites or defect sites (D) by coupling to their respective electronic or vibrational modes; energy is dissipated as heat to the lattice by electron-lattice (electron-phonon) coupling. The horizontal arrow represent loss channels competing with desorption or ablation.

lasers then commonly available in the visible and near-infrared—was widely understood to result from rapid heating of the surface layer followed by vaporization at the melting temperature of the surface [1]. In the 1970s, studies of damage in optical components and thin films confronted problems arising from imperfect materials [2] and flawed growth and polishing processes (for example, the notorious problems of columnar growth and residual polishing contaminants in thin optical films). The materials problems were particularly critical for the high peak power (TW class) lasers constructed for fusion experiments. The view that laser sputtering mechanisms were primarily thermal seemed reasonable given that typical electron-hole thermalization times even in insulators were much shorter than the typical laser pulse duration.

In the 1980s, the formidable armamentarium of tools and techniques from ultrahigh-vacuum surface science and laser spectroscopy was increasingly applied to fundamental questions about laser-induced desorption and ablation. It became evident rather quickly that purely thermal descriptions of laser sputtering were inadequate to explain the observed species and velocity distributions, particularly from the growing set of insulating and semiconducting materials that were being studied both for scientific and technological reasons. These newer mechanistic studies, which provided the details of the velocity and species distributions of the *ejecta* from laser sputtering experiments, showed that specific, identifiable electronic mechanisms played a significant, and in some cases dominant, role in all aspects of laser-induced desorption and ablation. For example, color centers or other electronic defects created by the initial laser irradiation changed the absorption characteristics—and thus the deposition of laser energy—for photons late in a given laser pulse or in subsequent laser pulses. The development of high-power, high-repetition-rate lasers forced the recognition of long-term conditioning and incubation effects even over time scales long compared to thermal relaxation times. Diffusion of electronic defects to the surface weakens atomic bonds, resulting in preferential ejection or evaporation of certain atomic or ionic species. Thermomechanical defects in particular may develop or relax with long time constants.

The current decade has seen exciting developments in laser sources and surface microscopy that are radically altering our understanding of laser sputtering, often driven by or giving birth to technological innovations in laser processing and analysis of materials. Ultrashort pulse solid-state lasers in the wavelength range from 400 to 1100 nm clearly generate nonequilibrium states of the material, which relax along complex electronic and vibrational pathways. Infrared free-electron lasers capable of direct vibrational excitation of more complex materials with strong absorption bands in the "molecular fingerprint" region (roughly 2–10 µm) also are shedding light on the meaning of thermal ablation. The development of greater range in both pulse duration and wavelength of lasers used in desorption and ablation experiments highlights the inadequacy of the old vocabulary of "thermal" and "nonthermal" processes. Instead, we are being forced to adopt instead a more dynamical vocabulary describing excitation and relaxation processes with their typical time scales.

In this chapter, we review the mechanisms of laser-induced desorption and ablation induced by both *electronic* and *vibrational* processes, over wavelengths ranging from ultraviolet to infrared, and spanning laser-pulse durations from femtoseconds to nanoseconds. The conceptual framework within which we discuss the phenomenology of laser sputtering involves five distinctive phases:

1. Absorption of laser light by excitation of electronic or vibrational modes of the solid
2. Competition between localized and delocalized modes for the absorbed photon energy
3. Mechanisms of laser-induced desorption from metals, semiconductors, and insulators
4. Phenomenology of laser ablation as a function of wavelength or pulse duration
5. Emission of photoelectrons, ions, and neutral species into the incipient ablation plume

The first and second phases are governed by the optical physics of the irradiated solid and by well-understood physical conservation laws. The third phase is much better understood for collisionally induced sputtering than for the electronic and vibrational processes described in this chapter. The fourth and fifth phases have no analog in sputtering by either ions or energetic electrons, and they are particularly critical to understanding the plasma physics of the plume. The gas dynamics of the plume requires a more extensive discussion, which is provided in Chapter 5.

The kinetics and dynamics of this conversion depend critically on the mechanism of light absorption, electron-lattice interactions characteristic of the laser-irradiated solid, such as scattering of free electrons by phonons; phonon emission; localized lattice rearrangements and configuration changes, such as self-trapping of holes and excitons; defect formation and defect reactions; and surface decomposition due to electronic interactions of defects with lattice ions. Accompanying or following the electron-lattice interaction are secondary electronic and vibrational processes such as photoabsorption by free electrons; successive (multiple-photon) excitations of self-trapped excitons and electronic defects; transient changes in optical absorption as the surface layer or near-surface bulk decomposes; photoemission and ionization of neutral species by incident laser light; generation of a plasma and neutral gas cloud (which may also interact with the surface); and photoacoustic or photothermal effects that generate mechanical disturbances in the bulk.

Because laser photons have a penetration depth much greater than the electron and soft X-ray probes typically used in surface science, laser sputtering necessarily involves processes both in the near-surface bulk and at the surface. This chapter emphasizes the relationship between sputtering phenomenology and the vast literature on the physics of photo-excited solids. The initial sections discuss the absorption of laser light, electronic and vibrational excitation, and the competition between localized and delocalized relaxation processes that determine whether the excitation will lead to desorption. We treat laser-induced desorption in relationship to the microscopic properties of surfaces, as exemplified by desorption of adsorbed molecules on metal surfaces, and of surface atoms and ions in metals, semiconductors, and insulators. Laser ablation, on the other hand, is viewed as exemplifying the role of the mesoscopic properties of materials, as seen in ablation experiments on metal, semiconductor, and insulator surfaces across a wide spectrum of wavelengths and a range of pulse durations from femtoseconds to nanoseconds. The initial stages of plume formation are considered as a prelude to the extensive discussion of gas dynamics in ablation plumes found in Chapter 5. We conclude with brief discussions of experimental tools and techniques that appear to promise the greatest opportunities for understanding the fundamental mechanisms of laser-induced desorption, and with a discussion of the most important classes of physical models that have been applied to laser sputtering.

Various aspects of this complex and variegated subject have been treated in recent reviews, including laser ablation of insulators [3]; comparisons of laser ablation and heavy particle sputtering [4]; electronic processes in the laser-induced desorption and ablation of insulators and semiconductors [5]; and modification of solids by high-intensity lasers [6]. Bulk and surface electronic processes in insulating and semiconducting solids have been reviewed by Itoh [7]. All of these reviews remain useful, though some are dated in light of the newer developments. Because of the well-documented recent history, this chapter concentrates on fundamental mechanisms and on those techniques and results that are illustrative, rather than comprehensive. Examples have been chosen to highlight the application of experimental techniques from surface and optical science so that the conceptual richness of information available from the experimental toolkit is evident.

The most important conference on the fundamental mechanisms of desorption and ablation is the Conference on Laser Ablation; its proceedings [8, 9, 10] are significant snapshots of the field as it has been developing in recent years. A much older Symposium on Laser-Induced Damage to Optical Materials, the so-called "Boulder Damage Symposium," has focused principally on the practical problems of damage to laser optics; the proceedings of this conference are published annually by the National Institute of Standards and Technology. A useful introduction to laser ablation as it relates to surface damage to optics in high power laser systems is provided in an excellent review by Chase [11]. A somewhat broader survey, still useful in the proper context, is the older book by Wood [12]. An important related conference series is the biennial-to-triennial DIET conference on Desorption Induced by Electronic Transitions; the proceedings of this series have often carried material of direct relevance to laser-induced desorption.

This chapter has a dual emphasis, partly tutorial, partly a survey of the most recent illustrative experimental results and techniques. Sections 2.2 and 2.3 are designed for readers who want an introduction to the fundamental, relevant ideas about the interactions of light with solids, and about the relationship between electronic response and those structural properties relevant to desorption and ablation. Readers who are already familiar with the optical physics of solids and with the concepts of localization and delocalization as they relate to bond-breaking, desorption, and ablation state may want to go directly to Section 2.4 to pick up with experimental studies of laser-induced desorption and ablation. In any case, readers of either background or experience level will find plenty of references to the relevant literature.

2.2 Photoexcitation and Material Response at Surfaces

The alteration of material surfaces by light has been observed for centuries in such common phenomena as the solarization of glass due to ultraviolet light (first remarked by Michael Faraday [13], of course!) and the discoloration, checking, and cracking of painted surfaces exposed to Florida or California sunshine. Nevertheless, large-scale light-induced modification of surfaces on a large scale had to await the advent of the laser, which had sufficient flux density to cause massive changes in surface binding properties in a short period of time. In this section, we consider the basic parameters that describe the laser excitation; the material response of the surface and near-surface bulk; the relationship of that response to the structural properties of the material; and the excitation mechanisms through which the incident photon energy is eventually converted into atoms, ions, molecules, and clusters expelled energetically from the surface. We conclude with a brief discussion of nonlinear phenomena—such as self-focusing—that may affect laser-surface interactions under special circumstances.

2.2.1 Absorption of Laser Light at Surfaces

Lasers deposit energy in irradiated surfaces and the near-surface region of the bulk material down to a penetration depth that is characteristic of the laser frequency (or wavelength) and the material. The energy may be deposited either by exciting free electrons or by exciting electronic or vibrational transitions in atoms, ions molecules, or optically active defects. The mechanism, density, and lifetime of the induced excitation depend on the electronic structure, composition, surface topography, and defect populations of the irradiated solid as well as on the laser frequency and pulse duration. The details of surface alteration by laser ablation—such as large-scale material removal—will also be influenced by surface morphology, by thermodynamic or hydrodynamic surface roughening and instabilities, and by the ambient atmosphere. Some nonlinear effects in laser-surface interactions depend on the intensity and intensity distribution of the incident laser beam as well as on its frequency. Finally, the amount of light absorbed at the surface may be influenced by laser interactions with the ejected material and, in the case of thin irradiated samples, by internal reflections and phase changes at internal boundaries. In this section, we consider the mechanisms of light absorption in metals, semiconductors, and insulators, and the density of excitation induced by photons.

In metals, both conduction- and valence-band electrons may participate in laser excitations, the former through a free-electron-like optical response and the latter an *interband* response that has a threshold corresponding to the energy separation between the valence and conduction bands. The nearly free conduction-band electrons absorb photons by direct heating of the electron gas well out into the infrared. The dielectric function of the free electrons at a wavelength ω is

$$\varepsilon(\omega) = 1 - \frac{\omega_p^2}{\omega^2 + i\omega\Gamma} = 1 - \frac{\omega_p^2}{\omega^2 + \Gamma^2} + i\frac{\omega_p^2 \Gamma}{\omega(\omega^2 + \Gamma^2)}$$

$$\approx 1 - \frac{\omega_p^2}{\omega^2 + \Gamma^2} + i\frac{\omega_p^2 \Gamma}{\omega^3} \quad (2.1)$$

where the *plasma frequency* is $\omega_p^2 = \sqrt{n_e e^2/\varepsilon_0 m_e}$ and the relaxation frequency is related to the Fermi velocity and the electron mean free path by $\Gamma = v_F/\ell$. In metals, the skin depth for optical excitations—that is, the depth within which the laser intensity falls off to $2/e$ of its initial value—is typically on the order of tens of nanometers. The laser-induced excitation is dissipated by means of collisions between excited electrons and the lattice. The mean free paths at room temperature are also on the order of tens of nanometers, which means that initial electronic excitations are completely thermalized in a few picoseconds at most. Table I summarizes relevant information on skin depths, mean free paths, Fermi velocities, and interband transition energies for a few metals.

The optical response of insulators and semiconductors involves both electronic and ionic contributions to the dielectric function; the variation of the

TABLE I. Skin depth d, mean free path ℓ, Fermi velocity v_F and interband transition energy E_{IB} for simple (s–p) and noble metals.

Metal	Na	Al	Cu	Ag	Au	Hg
δ (2 eV) in nm	38	13	30	24	31	255
δ (3 eV) in nm	42	13	30	29	37	141
δ (4 eV) in nm	48	13	29	82	27	115
ℓ (nm)	34	16	42	52	42	11
v_F ($\cdot 10^{-6}$ m·s^{-1})	1.07	2.02	1.57	1.39	1.39	
E_{IB} (eV)	2.1	1.5	2.1	3.9	2.4	2–4

dielectric function with wavelength is illustrated schematically in Fig. 2. The dielectric function of nonmetallic solids is complex, with the real (refractive) part dominating away from vibrational or electronic resonances and the imaginary (absorptive) component in the ascendancy near resonances. The dielectric constant at high frequencies below the ultraviolet electronic absorption edge can be estimated as

$$\varepsilon_{opt} - 1 = \frac{Ne^2}{m_e \varepsilon_o \omega_e^2} \qquad (2.2)$$

where the frequency ω_e corresponds to the average excitation energy of the electrons. The static dielectric constant is related to the optical (high-frequency) dielectric constant by

$$\varepsilon_s - \varepsilon_{opt} = \left[\frac{\varepsilon_{opt} + 2}{3}\right]^2 \cdot \frac{N(e_T^*)^2}{M\varepsilon_o \omega_v^2} \qquad (2.3)$$

The quantity e_T is the transverse charge and can be directly related to the atomic

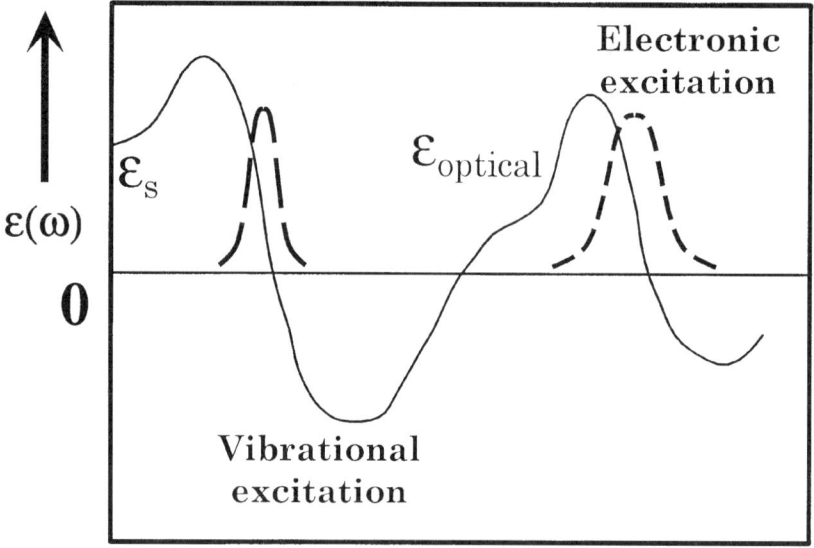

FIG. 2. A generic dielectric function for a nonmetallic solid showing the locations of electronic and vibrational resonances. The static (s) or long-wavelength limit is in the infrared, the optical limit in the visible and ultraviolet. The solid curve indicates the real part, the dashed lines the imaginary part, of the dielectric function.

potentials in a bond-orbital model of a dielectric solid; it is essentially the *dynamical* effective charge that participates in an optical excitation.

There are virtually no free (conduction-band) electrons at room temperature in insulators and only a very small number in semiconductors. Thus optical absorption in laser-irradiated semiconductors and insulators at moderate to high laser intensities (in the range of $MW \cdot cm^{-2}$ to $GW \cdot cm^{-2}$) leads to the creation of electron-hole pairs rather than electron heating. Only at extremely high intensities (generally in excess of $TW \cdot cm^{-2}$) is it possible to generate enough conduction-band electrons in insulators to generate significant free-electron heating. The relaxation mechanism by which the electrons and holes recombine and the response of the lattice to the creation and motion of free charges and electron-hole pairs depend in a complex way on both the ratio of the laser photon energy $h\nu$ to the bulk bandgap energy E_G, and on the electronic structure and binding configuration of the solid and its incorporated defects, as suggested in the schematic band-structure diagram in Fig. 3.

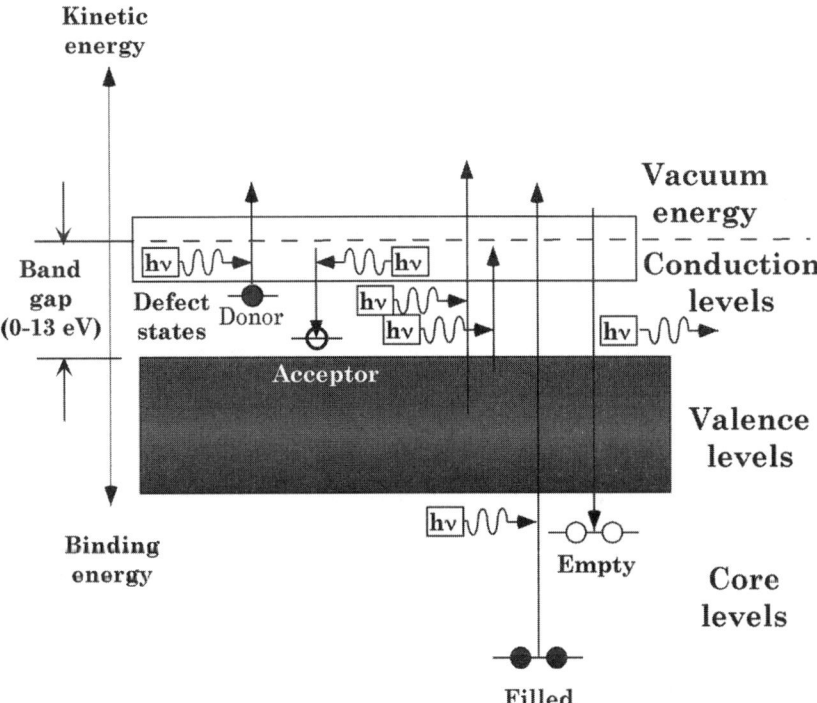

FIG. 3. Schematic of the band structure of metallic and nonmetallic solids, showing the energetic configuration of the valence and conduction band states as well as defects. In metals, the band gap vanishes and the valence and conduction levels overlap.

When hν is greater than or equal to E_G, as is often true for laser-irradiated semiconductors, the laser-solid interaction is dominated by one-photon transitions. If hν is less than E_G, on the other hand, the laser may induce either one-photon transitions between surface or defect states and the conduction band, or *multiphoton* or *multiple-photon* transitions between valence and conduction bands. Optical absorption coefficients of nonmetallic solids for photons with energies greater than the bulk bandgap typically exceeds 10^5 cm^{-1} for direct transitions near the band edge. Visible and ultraviolet lasers create electron-hole pairs in nonmetals by optical transitions from a filled state (for example, in the valence band, near the Fermi level in the conduction band, or in a defect state) to an empty state (which might be in the conduction band or be an empty defect or surface state). Infrared laser photons may be absorbed in harmonic phonon modes of the solid or in specific vibrational modes corresponding to natural frequencies of constituent molecular groups. Typical two-photon absorption coefficients for transparent solids [14] are in the range of 10^{-11}–10^{-12} cm $\cdot$ W^{-1}; three-photon absorption coefficients are correspondingly smaller. Resonant absorption processes, of course, can produce greatly enhanced nonlinear aborption coefficients.

Multiphoton band-to-band transitions also produce electron-hole pairs and excitons, just as do single-photon transitions. However, because the volume density of electronic excitation generated by multiphoton transitions is extremely low, the number of electron-hole pairs and the desorption yields are comparable to those produced in one-photon band-to-band transitions excited by much weaker light sources, such as electron storage rings. Multiphoton band-to-band transitions can be resonantly enhanced by defect states, greatly increasing the density of electronic excitation and the number of electron-hole pairs [15]. The probability for this resonant enhancement depends on the electronic structure of the solid, the precise character of the defects involved in the excitation, and the laser wavelength. Once produced by multiphoton band-to-band transitions near the surface, defects may subsequently be excited by single-photon or multiphoton transitions.

The most common terms for describing the laser light are the *fluence* and the *intensity*; the latter is properly called *irradiance*, although intensity is in much more common usage. The fluence is the energy density per unit area and is usually stated in J $\cdot$ cm^{-2}; the intensity measures the laser power per unit area and is commonly quoted in W $\cdot$ cm^{-2}. Although one often reads in the literature that the desorption or ablation behavior depends on fluence, the situation is generally more complicated in laser-sputtering experiments, most of which show a convoluted dependence on both pulse energy and pulse duration. The intensity is, in fact, the more fundamental quantity because it is the intensity that when multiplied by a cross section, translates into a probability per unit time for creating some particular reaction product.

The number and energy of absorbed photons establish the temperature rise in a laser-irradiated metal, whereas the number of valence-to-conduction-band excitations establishes the initial density of electron-hole pairs in an insulator or semiconductor. When coupled to the relevant mechanisms for relaxation, this in turn determines the energy and the density of electrons and holes available for radiative and nonradiative recombination, including heating, Auger transitions, and the like. The intensity of the laser light as a function of depth, $I(z)$, is given by

$$I(z) = I_o \cdot \exp\left\{-\left(\sum_i \alpha_i(\omega) + \sum_j \beta_j(\omega)I + \sum_k \gamma_k(\omega)I^2 \ldots\right)z\right\} \quad (2.4)$$

where I_o is the incident intensity, and α_i, β_j, and γ_k are, respectively, the one-, two-, and three-photon absorption coefficients corresponding to the absorbing species $\{i, j, k\}$; the ellipsis in Eq. (2.4) refers to higher order absorption processes. The absorption coefficients are simply related to the number density of each particular absorber and the corresponding cross sections; for example, $\alpha = N\sigma^{(1)}$ gives the linear absorption coefficient in terms of the one-photon cross section and number density of one-photon absorbing species. Table II gives measured multiphoton absorption coefficients for some nonmetallic solids [16].

The density of electronic or vibrational excitation has a major role in determining the response of a solid to photon irradiation. The number of excitations per second of a specific site in a solid is given by $N^{(n)} = \sigma^{(n)} \cdot (I/h\nu)^n$ for multiphoton excitation of order n, where σ is the excitation cross section and I is the irradiance. For typical single- and multiphoton absorption cross sections, the number of band-to-band excitations for a nanosecond pulse with a fluence of $1 \text{ J} \cdot \text{cm}^{-1}$ is of order 10^{-2} for single-photon excitation, but of order 10^{-7}–10^{-10}

TABLE II. Multiphoton absorption coefficients for some common semiconductors and insulators from various sources [16].

Material	Bandgap (eV)	β (cm·W^{-1}) at 248 nm	β (cm·W^{-1}) at 266 nm	γ (cm^2·W^{-1}) 1064 nm	γ (cm^2·W^{-1}) 532 nm	γ (cm^2·W^{-1}) 266 nm
Fused silica	7.8	$4.5 \cdot 10^{-11}$	$5 \cdot 10^{-11}$	$2.73 \cdot 10^{-16}$	$1.6 \cdot 10^{-16}$	$5.8 \cdot 10^{-16}$
BaF$_2$	9.1	$1.1 \cdot 10^{-10}$	$6 \cdot 10^{-11}$	$2.85 \cdot 10^{-16}$	$1.6 \cdot 10^{-16}$	$2.3 \cdot 10^{-15}$
CaF$_2$	10.0	$8.3 \cdot 10^{-12}$	—	$1.90 \cdot 10^{-16}$	—	—
SrF$_2$	9.6	$1.1 \cdot 10^{-11}$	—	$1.76 \cdot 10^{-16}$	—	—
MgF$_2$	11.8	$\leq 1.3 \cdot 10^{-12}$	~0	$0.92 \cdot 10^{-16}$	$0.92 \cdot 10^{-16}$	$1 \cdot 10^{-15}$
LiF	11.6	$\leq 1.3 \cdot 10^{-12}$	~0	$1.05 \cdot 10^{-16}$	$0.92 \cdot 10^{-16}$	$2.1 \cdot 10^{-16}$
NaCl	8.5	—	3.5	$3.3 \cdot 10^{-16}$	—	—
KBr	6.0	—	2.2	$6.9 \cdot 10^{-16}$	$1.0 \cdot 10^{-15}$	—
LiNbO$_3$	4.0	—	—	$1.0 \cdot 10^{-15}$	$9.2 \cdot 10^{-14}$	—
BBO	6.2	—	$9 \cdot 10^{-10}$	$2.3 \cdot 10^{-15}$	$4.4 \cdot 10^{-15}$	$2.1 \cdot 10^{-17}$
Al$_2$O$_3$	7.3	—	$9 \cdot 10^{-11}$	$2.7 \cdot 10^{-16}$	$2.9 \cdot 10^{-15}$	$5.5 \cdot 10^{-16}$

for two- and three-photon excitations. Single-photon band-to-band transitions in the bulk produce electron-hole pairs with a spatial density $\pi = \sigma I n \tau / \hbar \omega$, where τ is the lifetime of electron-hole pairs and n is the number of atoms per unit volume. The energy of the electron-hole pairs, ultimately dissipated to phonons by recombination, heats the lattice at a rate of $\sigma In/hv$ per unit time and per unit volume. When electronic transitions are limited to surface states, heat is generated only at the surface and transported into the bulk; thus heating effects in this case may be insignificant. When defects are excited, electronic excitation is transferred to local lattice modes and the global temperature rise is, once again, not a dominant effect. When the photon flux is high, cascade excitation of surface defects by multiple-photon excitation can cause emission of atoms from defect sites.

The density of electron-hole pairs for absorption including optical processes up to third-order can then be written down as follows:

$$n_{eh} = n_o(1 - R)(1 - \exp[-(\alpha + \beta I + \gamma I^2)z])\zeta(I) \qquad (2.5)$$

where $\zeta(I)$ is the efficiency for producing electron-hole pairs including competition from all other relaxation channels, n_o is the number of absorbers, and R is the surface reflectivity. From this equation, we can directly calculate the local density of electronic excitation assuming that we know $\zeta(I)$. For example, for Ge at room temperature irradiated by 0.75 eV photons, α is of order 25 cm^{-1}; for a laser pulse of 1 J · cm^{-2} fluence, this absorption coefficient translates to an excitation probability of $\sim 10^{-3}$ per atom within the absorption length of the Ge. Multiphoton absorption coefficients in highly transparent materials may be comparable to linear absorption coefficients for sufficiently high intensity, as with picosecond lasers. Either single-photon or multiphoton absorption characteristics may change by formation of transient defects during a laser pulse, or by the accumulation of defects with successive laser pulses.

Where a multiphoton or multiple-photon transition is required to initiate the desorption or ablation process, it is possible to estimate the probabilities for finding k photons in the same volume simultaneously [17]. The mean photon number in a cubical cell with a linear dimension L and index of refraction n illuminated by a laser of intensity I at a wavelength λ is

$$m = \frac{\text{laser intensity}}{\text{energy/photon}} \cdot (\text{transit time}) \cdot (\text{area}) = \frac{I}{hc/\lambda} \cdot \frac{L}{c/n} \cdot L^2 = \frac{nIL^3}{hc^2} \qquad (2.6)$$

where m is the mean number of photons in the designated volume and n is the index of refraction. Assuming that laser photons obey Poisson statistics, the probability of k photons *simultaneously* being in a volume having a mean photon number m is

$$P_k = \frac{m^k}{k!} e^{-m} \cong \frac{m^k}{k!} \propto I^k \qquad (m \ll 1) \qquad (2.7)$$

where m is given by Eq. (2.6). Probabilities for *resonant* multiphoton transitions, of course, will be substantially higher.

Thus, for example, for a XeCl laser (308 nm ~ 4 eV) irradiating a volume of KCl 2 nm on a side, an *e-h* pair can be created by exciting an electron from the valence into the conduction band, a two-photon transition requiring $k = 2$. For typical excimer laser pulses of 100 mJ in 10 ns, focused into a 1-mm-square spot, we find $P_2 \approx 3 \cdot 10^{-7}$. On the other hand, for a chirped-pulse amplifier laser system, producing 1 mJ of 400-nm light in a 200 fs second-harmonic pulse, it takes a three-photon excitation to make a band-to-band transition. In this case, however, the higher intensity more than compensates for the substantially less probable three-photon transition because $P_3 \approx 0.1$. Even at the highest intensities seen in ablation experiments, the spatial density of electronic excitation remains low in a transparent dielectric, not much greater than that found by one-photon excitation in the deep ultraviolet using current synchrotron-light sources. In semiconductors, on the other hand, electron-hole pairs can be created by many lasers via one-photon transitions. For instance, a 100 mJ pulse from a Q-switched Nd:YAG laser incident on a silicon surface creates electron-hole pairs with probability $P_1 \approx 10^{-3}$ during a pulse lasting 10 ns.

2.2.2 Electronic, vibrational and structural properties of solids

The absorption of laser light by solids can be described with remarkable accuracy by semi-classical models of the harmonic oscillator with appropriate quantum-mechanical corrections where necessary. The natural frequencies of the electron modes of excitation in solids are in the ultraviolet; the corresponding frequencies of ion vibrations lie in the infrared due to the substantially greater mass of the ions. In a monatomic lattice, the quantized lattice vibrations—or *phonons*—are acoustical disturbances in which the ions move in phase with one another. In a lattice with two or more different ion species per unit cell, on the other hand, there are two species of normal modes, in which the ions move either in or out of phase with one another, as depicted schematically in Fig. 4. These out-of-phase modes are the so-called *optical* phonons, which are infrared active and can be directly excited by laser light. Excitation of the *acoustic* phonon modes, on the other hand, is associated with the melting of the lattice.

The physical mechanisms of bonding and bond-breaking in nonmetallic solids, as well as of the linear [18] and nonlinear [19] optical response, can be described in a bond-orbital picture [20, 21] of electronic structure for solids whose wave functions are amenable to a tight-binding description. The physical justification for the use of such a model is that the events that are of interest in laser-induced desorption and ablation—photon absorption, transformation, and lattice motion—are essentially localized events and thus not easily describable

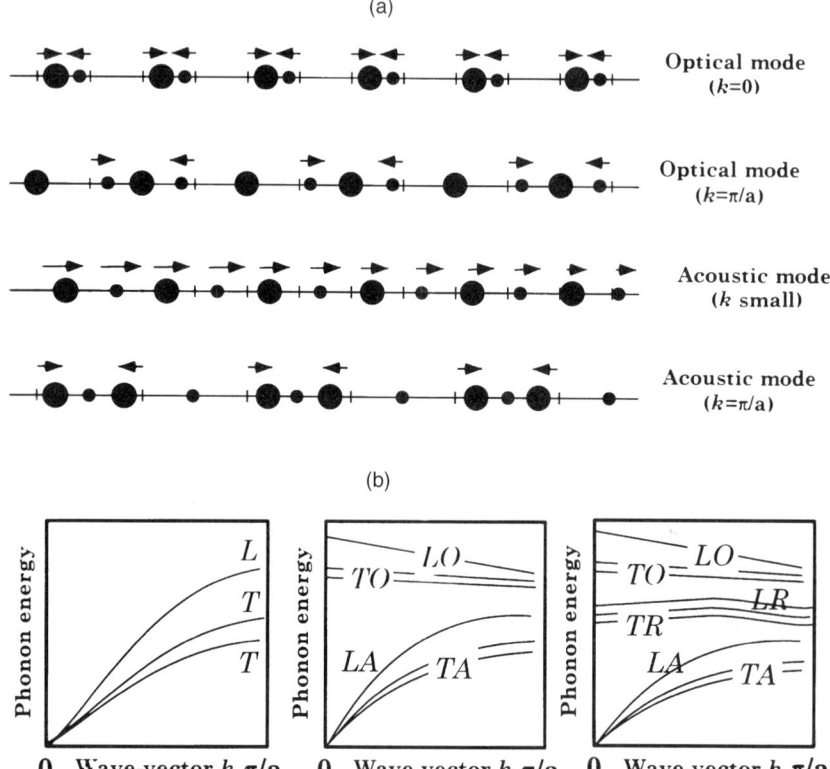

FIG. 4. Schematic of phonon modes showing (a) lattice vibrations corresponding to acoustic and optical modes; and (b) dispersion curves for solids with monatomic, diatomic, and triatomic unit cells. Note that the optical phonon modes—the ones to which the electrons couple most easily—do not occur in monatomic solids.

in terms of the delocalized picture of the band theory of solids. Nevertheless, the local properties of a solid, whether crystalline or amorphous, are determined by the global band structure of the solid, and the bond-orbital theory incorporates the band structure through approximate local potentials that are rooted in exact band structure calculations.

Imagine isolated atoms of two species, a and c, joined via s-p bonding to produce hybrid orbitals, and eventually bonds and bands, as illustrated schematically in Fig. 5. Creating the hybrid orbitals requires a transfer of charge ΔZ and a promotion energy

$$E_{pro} = \left(1 + \frac{\Delta Z}{4}\right)V_1^c + \left(1 - \frac{\Delta Z}{4}\right)V_1^a \tag{2.8}$$

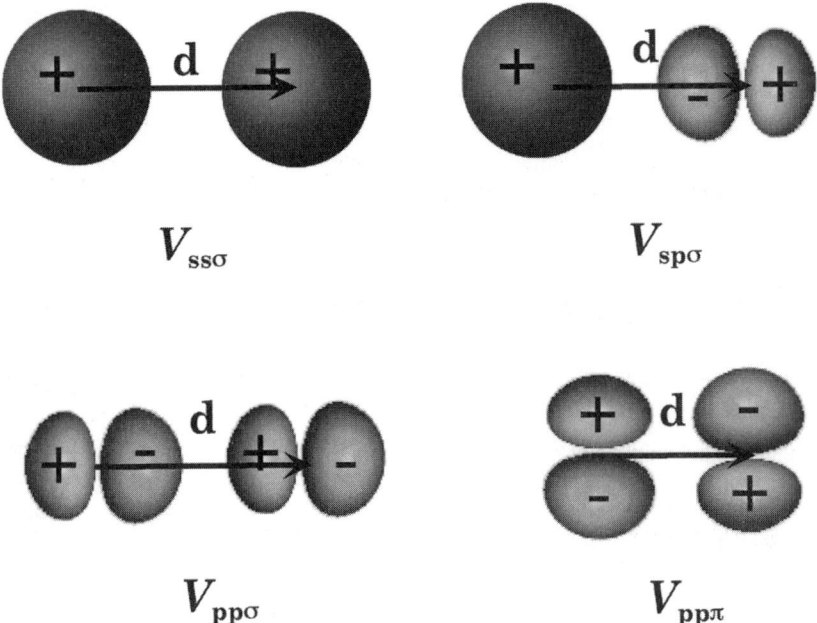

FIG. 5. Schematic of atomic orbitals that are used in constructing the sp_3 bond orbitals from which the bond-orbital potentials are calculated.

where the metallization energy is defined by $V_1 = (\varepsilon_p - \varepsilon_s)/4$ for either the anion a or cation c. The electrons from atoms a and c are hybridized to create a bonding orbital and an anti-bonding hybrid orbital separated by twice the hybridization energy V_3. The hybrid covalent energy V_2 is the matrix element between neighboring hybrids pointing into the bond, constructed from the matrix elements V_{ijk} of the atomic orbitals ($i, j = s, p$) that make up the bonding states ($k = \sigma, \pi$):

$$V_2 = \tfrac{1}{4}(-V_{ss\sigma} + 2\sqrt{3}\, V_{sp\sigma} + 3 V_{pp\pi}) \tag{2.9}$$

The energy needed to transform the hybrid orbitals into bond-orbital eigenstates of the full Hamiltonian is

$$E_{\text{bond}} = \left(1 + \frac{\Delta Z}{4}\right)[\sqrt{V_2^2 + V_3^2} - V_3] + \left(1 - \frac{\Delta Z}{4}\right)[\sqrt{V_2^2 + V_3^2} + V_3] \tag{2.10}$$

The cohesive energy is found by subtracting from the bond-formation energy the promotion energy needed to construct the hybrid orbitals and the "orbital overlap potential" arising from kinetic, Coulomb, exchange and correlation energies, $V_o(d)$, which yields

$$E_{coh} = -E_{pro} - V_o(d) + E_{bond}(d) \qquad (2.11)$$

This calculation of the cohesive energy can be generalized and extended to more ionically bonded solids (such as silica) by computing the binding energy as a function of the moments of the density of states [22]. The pressure and the bulk modulus are, respectively, the first and second derivatives of the cohesive energy with respect to bond length d.

The cohesive energy can be strongly affected if the laser pulse generates a sufficiently dense electron-hole plasma. This is unlikely in insulators, but might be expected, for example, in semiconductors irradiated with photons just above the bandgap energy. In this case, theoretical or computational results may need to be corrected phenomenologically or from first principles for the effects of an electron-hole plasma [23]. The specific correction probably depends on laser pulse duration because the pulse duration determines whether or not the electron-hole plasma is in equilibrium with the rest of the solid during irradiation, or whether the plasma is essentially created instantaneously.

In a heteropolar lattice, the transverse oscillations of an electromagnetic wave induce contrary motions of the positively and negatively charged sublattices, the so-called *optical phonon* modes. The optical modes of the solid are characterized experimentally by measuring reflectance at normal incidence (for transverse-optical or TO modes) and at nonnormal incidence (for the longitudinal optical or LO modes). This produces a localized polarization density and generates a *transverse charge* e_T^* defined by

$$e_T^* = \left[\frac{Mr_o^3 \varepsilon_1}{3e^2} \cdot (\Omega_{LO}^2 - \Omega_{TO}^2)\right]^{1/2} = Z^* + \frac{8}{3}\alpha_p(1 - \alpha_p^2) \qquad (2.12)$$

where Ω_{LO}^2 and Ω_{TO}^2 are the squared phonon frequencies of the longitudinal (LO) and transverse (TO) optical phonons; M is twice the reduced mass; r_o is the atomic radius; and ε_1 is the dielectric constant. While the *bond charge* $z^* = 4\alpha_p - \Delta Z$ is the effective *static* or structural charge, the transverse charge e_T^* is the relevant *dynamical* quantity. Note also that, by means of the bond-orbital formalism, the mesoscopic dielectric function can be related to the microscopic potentials embodied in the polarity parameter.

Electron-lattice coupling is a measure of the degree to which ion motion follows electronic motion in a solid. Although electron-lattice coupling clearly plays a definitive role in desorption and ablation of nonmetals, no quantitative criterion has yet been established for either regime. However, coupling strength is also a qualitative indicator of the relative dominance of radiative and nonradiative processes that compete for the energy poured into a laser-irradiated solid. The electron-lattice coupling strength is conventionally defined to be [24]

$$\alpha = \frac{1}{2}\left[\frac{1}{\varepsilon_\infty} - \frac{1}{\varepsilon_o}\right]\left[\frac{e^2}{r_p}\right]\left[\frac{1}{\hbar\Omega_{LO}}\right] = \frac{\text{average deformation energy}}{\text{average phonon energy}} \quad (2.13)$$

where the dielectric constants e_o and ε_∞ are, respectively, due to ionic and electronic screening; Ω_{LO} is the longitudinal optical phonon frequency, and the polaron radius r_p is defined by $r_p^2 = (\hbar^2/2m_p^*)(\hbar\Omega_{LO})$, with m_p the effective polaron mass. Qualitatively, this parameter measures the number of phonons (vibrational quanta) "dressing" an electron moving through a deformable lattice. The coupling constant α is large in ionic solids, small in covalently bonded polar materials, and vanishes for metals and homopolar semiconductors. The polaron radius, it turns out, varies only weakly with ionicity, over a range 10–15 Å; it is thus a minimally helpful index of the probability of laser-induced desorption. However, the coupling constant provides a definite bound on desorption probability: if α is less than 1, energy localization due to a lattice deformation is unlikely to compete favorably with transfer to delocalized LO phonon modes.

Figure 6 shows a map of materials from various solid "families" located by bandgap energy and electron-lattice coupling strength α. Data on electron-lattice coupling strength are taken from several sources and are uncertain to the extent that the effective polaron masses are not all well known [25]. The shaded vertical bar is located at a coupling strength $\alpha = 0.5$. To the left of that line, in line with the criterion proposed above, desorption is not likely except at defect sites; to the right of it, desorption from perfect lattice sites is a possibility, though the probability must be determined from other criteria. The horizontal bar coincides with the bandgap energy equal to the output from the ArF laser (193 nm, or 6.4eV), the shortest wavelength laser that is commonly used in desorption experiments. Desorption can be induced in materials above this line only by multiphoton transitions or by excitation of intra-gap defects; below the line, desorption can be initiated even from perfect lattice sites by either single- or multiple-photon excitations.

To summarize, we can say that the phenomenology of laser-induced desorption and laser ablation depends in an essential way on both the density of electronic excitation and on the electron-lattice coupling strength, and on the way in which these two parameters interact to produce energy localization and lattice instabilities. The density of electronic excitation depends on the bandgap energy and on the laser used to induce the lattice instability. Nanosecond lasers typically produce only low densities of excitation in wide bandgap solids, but very high densities in semiconductors. Picosecond and femtosecond lasers, on the other hand, may produce high densities of excitation even in wide bandgap dielectrics. Thus we can expect quite different ablation mechanisms to operate in the low- and high-density regimes even for the same material. The electron-lattice coupling strength is a function of the ionicity of the material. Because

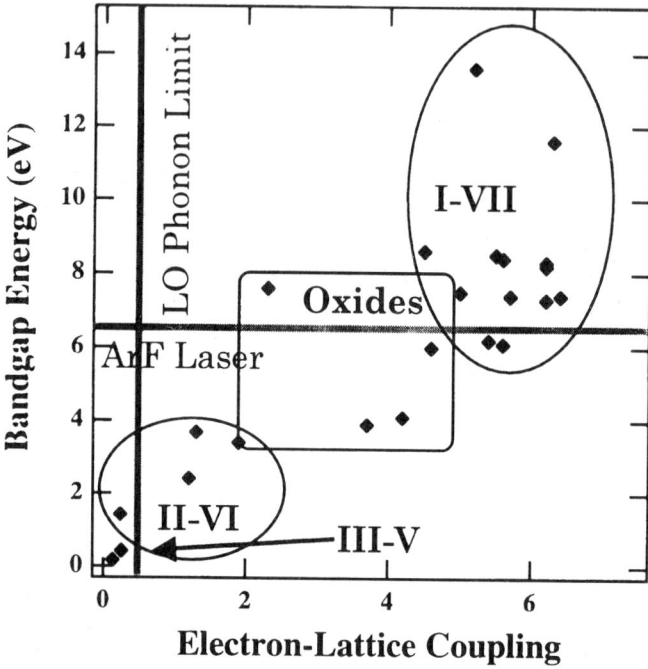

FIG. 6. Distribution of typical nonmetallic solids with respect to bandgap energy and electron-lattice coupling strength. For solids to the left of the LO phonon limit, electron-lattice coupling is so weak that no lattice distortions can lead directly to desorption without the help of defects. The line marked "ArF laser" corresponds to the highest single-photon laser transition typically available in the laboratory; for solids with larger bandgaps, band-to-band transitions are made only by multiphoton or multiple-photon transitions.

many wide-bandgap materials are also highly ionic, we expect most transparent insulators to have large values of α, whereas covalent semiconductors have small values of α.

2.2.3. Elementary Excitations in Solids

In laser-induced desorption and ablation, we are dealing with interactions between light and matter that lead to energy localization, bond-breaking and, ultimately, atomic and ionic motion. The interaction of light with solids comes through electronic excitation and thereafter to electron-phonon coupling. In this section, we briefly review the characteristics of the principal elementary excitations excited by the coupling of light to a solid: plasmons, polaritons, polarons, and excitons. These excitations provide the conceptual basis on which to form a first-principles picture of optical excitation and relaxation, based on the coupling of these excitations to phonons. Of course, the optical response of solids is also affected by the cohesive energy, bandgap energy, and the structural type of the irradiated material.

In *metals*, the dominant response is that of the free-electron gas, and the excitation quanta are generically referred to as *plasmons*. The plasmon is a collective oscillation of the electron gas, and is also important under certain circumstances, even in semiconductors where an electron-hole plasma can be formed. The plasmon is the quantum of the oscillation of the electron gas against the background ions in a metal, whereas the *plasmon polariton* is the quantum of the transverse wave field that couples photons and plasmons in metals. Although these excitations are often referred to simply as plasmons, this is, in fact, inaccurate terminology because the excitation really couples to the electromagnetic field and is therefore distinct from the (longitudinal) plasmons excited by free particles in such techniques as HEELS (high-energy electron loss spectroscopy) [26].

The *polaron* is an elementary excitation or quasi-particle consisting of a charge carrier and its accompanying charge cloud; in the case of an electron moving in a lattice, for example, the polaron comprises the electron dressed by the hole cloud dragged along as the crystal tries to preserve charge neutrality. If one assumes that the electron is spread out uniformly over a radius r, the energy change from the polarization of the surrounding lattice is approximately given by [21]

$$\Delta E = -\frac{e^2}{8\pi\varepsilon_o r} \cdot \left[\frac{1}{\varepsilon_{opt}} - \frac{1}{\varepsilon_s}\right] \qquad (2.14)$$

Competing with this energy-lowering interaction are the kinetic energy of the free charge carrier, and the energy gained by coupling into delocalized band states; this last term is approximately half the bandwidth. In weak-coupling materials such as the compound semiconductors, scattering is more probable than small-polaron formation or trapping, whereas the reverse is true for strong-coupling solids. Scattering of free carriers is important to desorption and ablation in two ways: First, inelastic scattering slows carriers to the point where they may be captured, thus creating electronic defects, which may lead to desorption or ablation. Second, scattering of laser-excited free carriers by phonons may produce rapid local heating; thermalization times required to bring energetic electrons to the bottom of the conduction band are of order 1 ps.

Polaron size is strongly correlated with electron-lattice coupling strength. One usually speaks of "large" polarons with radii of order 10–15 Å in weak-coupling solids, and of "small" polarons with radii a factor of two or three smaller in strong-coupling solids. The small polaron appears to be a precursor to the lattice distortion known as the self-trapped exciton, already implicated in laser-induced desorption from strong-coupling solids such as the alkali halides and alkaline-earth fluorides. Early studies of the self-trapped exciton in KBr [27] had shown a transient absorption feature associated with the self-trapped exciton; however, the source of this transient absorption had not been identified. In a series of important studies, the dynamics of small polaron formation in alkali iodides have now been measured with femtosecond laser techniques. In this experiment [28], samples of KI and RbI at temperatures in the 7–80°K range were irradiated by ultraviolet laser pulses with wavelength 302.5 nm and durations of 200 fs. In previous models, it had been assumed that the free hole created by laser excitation was self-trapped to form the self-trapped hole (STH) or V_K center, an X_2^- molecule localized on the site of a single halogen ion; it was assumed that this center then captured an electron to form the self-trapped exciton. The short-time-delay measurements show unequivocally, however, that the small-polaron state is the precursor to both the single-center STH and the two-center STH; the V_K center apparently stabilizes the polaron, leading to the possibility of forming the self-trapped exciton.

The *exciton* is an elementary excitation of a crystal that was first described theoretically in the 1930s [29]. It may be viewed as a hydrogenic atom comprising an electron and hole, bound together by their mutual Coulomb attraction, modified by the dielectric constant of the medium and by the effective masses of the electron and hole. Tightly bound excitons, roughly those confined to the vicinity of a single lattice site, are called *Frenkel* excitons; the larger excitons found in materials with larger dielectric constants and wider bands (and thus smaller effective masses) are generally called *Wannier* excitons. The energy spectrum of the exciton is

$$E_n = E_{gap} - \frac{1}{32\varepsilon_o^2\varepsilon_r^2} \cdot \frac{\mu e^4}{\hbar^2 n^2} \qquad (2.15)$$

where ε_r is the relative dielectric constant, μ is the reduced mass of the electron-hole pair, and n is the principal quantum number. In general, the effects of energy levels for principal quantum numbers greater than 1 are seen only at low temperatures. It is clear from this expression that the exciton states are typically just below the bulk band edge; hence they can be excited with slightly lower energy photons than the bulk bandgap. Excitons are associated with photon-stimulated and laser-induced desorption in wide bandgap materials; more will be said about this in Section 2.5. The self-trapped exciton represents the extreme case of a lattice-localized excitation, one which efficiently converts the absorbed photon energy into the motion of lattice ions.

The exciton may be formed whenever the basic conservation laws of energy and momentum permit the development of this bound state, that is, for

$$\hbar\omega = E_c(\mathbf{k}) - E_v(\mathbf{k}) \quad \text{and} \quad \nabla_\mathbf{k}[E_c(\mathbf{k}) - E_v(\mathbf{k})] = 0 \qquad (2.16)$$

where the subscripts c and v refer to conduction and valence bands, respectively.

This may occur in either direct-gap or indirect-gap solids; in the latter case, the formation of the exciton requires a phonon assist. Typical binding energies of excitons in a variety of solids are shown in Table III. Photons with an energy hν above the energy E_{exc} of the first exciton peak produce both electron-hole pairs and excitons. Those between the first exciton peak and the bandgap energy E_{gap} produce excitons, whereas those above E_{gap} produce electron-hole pairs. Surplus kinetic energy, (h$\nu - E_{exc}$) for excitons and (h$\nu - E_{gap}$) for electron-hole pairs, is transferred to phonons by electron-lattice interactions within a picosecond or so. Thus absorption of a photon in the visible or ultraviolet typically produces both electronic and vibrational excitation in the surface and near-surface regions of an irradiated solid. Absorption of low-energy infrared photons, on the other hand, generally does not excite the system out of the electronic ground state, so that excitons are not formed and the energy is transferred directly into lattice excitation.

In insulators with relatively small dielectric constants, excitons tend to be strongly bound. In most semiconductors, on the other hand, excitons are converted immediately to electron-hole pairs, except at very low temperatures. Moreover, at higher temperatures, excitons created near a direct bandgap in an indirect-gap semiconductor are energetically unstable against decay into a free electron and a free hole in most cases. Therefore, dense electronic excitation of a semiconductor produces an electron-hole plasma, whereas dense electronic excitation of an insulator yields a mixture of electron-hole pairs and excitons. Electron-hole pairs are converted to excitons on a time scale of order $\tau_c \sim K/(8\pi e\mu n)$, where K is the dielectric constant; e, the electronic charge; μ, the mobility; and n the concentration [30]. Where exciton-formation times are longer than the electron-hole recombination time, excitons will predominate over free

TABLE III. Bulk exciton binding energies for semiconductors and insulators.

Material	Bandgap (eV)	Dielectric constant	Exciton type	Exciton binding energy (meV)	Self-trapped exciton (eV)
Si	1.17	11.7	Wannier	14.7	—
Ge	0.66	15.8	Wannier	4.15	—
GaAs	1.42	13.13	Wannier	4.2	—
GaP	2.26	9.0	Wannier	3.5	—
InP	1.35	12.37	Wannier	4.0	—
InSb	0.17	17.7	Wannier	0.4	—
CdS	2.42	5.2	Wannier	29	—
CdSe	1.70	10.0	Wannier	15	—
KCl	7.6	2.20	Frenkel	400	2.6
KBr	6.0	2.39	Frenkel	400	2.3
KI	6.3	2.68	Frenkel	480	1.7
a-SiO$_2$	7.8	2.24	Frenkel	Not known	Not measured

TABLE IV. Exciton formation times for semiconductors and insulators. Mobilities are given in units of cm$^2 \cdot$V^{-1}s^{-1}, carrier densities in cm^{-3}, and formation times in ps.

Material	Bandgap (eV)	Dielectric constant	Electron mobility	Hole mobility	Carrier density	Formation time τ_c (ps)
Diamond	5.6	5.5	1800	1200	10^{14}	16.9
Si	1.17	11.7	1350	480	10^{14}	47.9
Ge	0.66	15.8	3600	1800	10^{14}	24.3
SiC	3.0	10.2	100	15	10^{14}	564
AlAs	2.20	10.1	280	0	10^{14}	300
GaAs	1.42	13.13	8000	300	10^{14}	9.08
GaP	2.26	9.0	200	140	10^{14}	249
InP	1.35	12.37	4500	100	10^{14}	15.2
CdS	2.42	5.2	400	15	10^{14}	71.9
PbS	0.41	17.0	550	600	10^{14}	114
PbTe	0.31	30.0	1700	800	10^{14}	36.8
KBr	6.0	2.2	100	0	10^{14}	122

electron-hole pairs while just the reverse is true in semiconductors. Values of τ_c for a number of materials are given in Table IV.

2.2.4 Nonlinear Response of Optically Excited Solids

Nonlinear optical effects originate in dynamical changes in a semiconductor or insulator produced by an applied optical field. The ith component of the polarization P induced in a medium by the optical field can be expanded up to third order in a power series in the applied electric field $E = E_o e^{i\omega t}$ as follows:

$$P_i = \sum_j \chi_{ij}^{(1)} E_j + \sum_{j,k} \chi_{ijk}^{(2)} E_j E_k + \sum_{j,k,l} \chi_{ijkl}^{(3)} E_j E_k E_l \qquad (2.17)$$

where the summation indices refer to Cartesian coordinates in the material and the polarization direction of the applied optical field. The susceptibilities $\chi^{(q)}$ are functions of the optical frequency and can, in principle, be calculated from the materials properties of the irradiated medium. The second-order susceptibility vanishes in any centrosymmetric medium. Second-harmonic generation can be used as a surface diagnostic because the broken symmetry at the surface permits the generation of the second harmonic even on centrosymmetric materials. So far, this possibility has not been exploited significantly in desorption or ablation experiments [31].

The first-order susceptibility is related to the linear index of refraction n_o and the linear (Beer's law) absorption coefficient α through the following equations:

$$n_o = \Re r[1 + \chi^{(1)}] \qquad \alpha = \frac{\omega}{n_o c} \cdot \Im m[\chi^{(1)}] \qquad (2.18)$$

For a material without a preferred axis of symmetry, the third-order susceptibility of the composite has an analogous relationship to the nonlinear index of refraction and to the nonlinear absorption coefficient, as follows [32]:

$$n_2 = \frac{12\pi}{n_o} \Re e[\chi^{(3)}] \qquad \beta = \frac{96\pi^2 \omega}{n_o^2 c^2} \Im m[\chi^{(3)}] \qquad (2.19)$$

The third-order susceptibility $\chi^{(3)}$ is, in general, a fourth-rank tensor with 81 components; however, because of the symmetry considerations applicable to most media, the actual number of components is usually substantially fewer. For example, materials with cubic symmetry have only four independent components of $\chi^{(3)}$.

Given the high intensities available from current lasers, nonlinear effects are a matter for obvious consideration in dealing with mechanisms of laser-induced desorption and laser ablation. Not all variation with intensity can properly be ascribed to multiphoton excitation, however. Whereas the signature of a multiphoton process of order k is the dependence of the yield on I^k, the converse does not hold true.

A particularly catastrophic effect of the third-order nonlinearity is self-focusing, a manifestation of the nonlinear index of refraction [33]. For a Gaussian beam propagating in a medium with an index of refraction $n(I) = n_o + n_2 I$, a dielectric medium will act as a lens with a focal length that varies with distance from the axis of the beam, as well as with distance from the focal plane of any real lens involved in focusing the beam. For sufficiently strong self-focusing, a laser beam will break up into filaments when the self-focusing just compensates for the tendency of the laser beam to self-diffract. The diameter of such a self-trapped filament can be shown to be

$$d = \frac{0.61\lambda}{\sqrt{2 n_o n_2 I}} \qquad (2.20)$$

The laser power in such a self-trapped filament in effect defines the critical power for self-focusing to occur, and is given by

$$P_{crit} = \frac{\pi}{4} d^2 I = \frac{\pi (0.61\lambda)^2}{8 n_o n_2} \qquad (2.21)$$

independent of the filament diameter. Table V shows the critical power for self-focusing for a variety of materials. Note that the critical powers are well within the range of even nanosecond lasers, meaning that this particular nonlinear effect cannot be neglected with impunity in any laser ablation experiment—or its interpretation [34]! For a beam undergoing self-focusing, the distance to the focus can be estimated by a simple application of Fermat's principle, and turns

TABLE V. Critical self-focusing powers P_{crit} and self-focusing distances z_f (in m) for various dielectrics at a wavelength of 1.064 μm. The two self-focusing distances correspond to the case of a typical ns-Nd:YAG laser with a peak power of 10 MW, and a chirped-pulse-amplifier system with a peak power of 0.1 TW, both with beam waists of 100 μm.

Material	Bandgap (eV)	Refractive index	Nonlinear index (cm^2·W^{-1})	P_{crit} (W)	z_f ($P = 10$ MW)	z_f ($P = 0.1$ TW)
CS$_2$	3.9 eV	1.700	2.6·10^{-18}	3.7·10^4	3.2·10^{-3}	3.2·10^{-5}
Fused silica	7.8 eV	1.458	2.7·10^{-20}	4.2·10^6	2.9·10^{-2}	2.9·10^{-4}
BK-7 glass	5.6	1.517	3.4·10^{-20}	3.2·10^6	2.6·10^{-2}	2.6·10^{-4}
Al$_2$O$_3$	7.3	1.768	2.5·10^{-20}	3.7·10^6	3.3·10^{-2}	3.3·10^{-4}
KDP	3.8	1.468	2.9·10^{-20}	3.9·10^6	2.8·10^{-2}	2.9·10^{-4}
BaF$_2$	9.2	1.475	2.9·10^{-20}	3.9·10^6	2.9·10^{-2}	2.9·10^{-4}
BBO	6.2	1.600	2.3·10^{-20}	4.5·10^6	3.3·10^{-2}	3.3·10^{-4}
CaCO$_3$	5.9	1.590	2.3·10^{-20}	4.5·10^6	3.3·10^{-2}	3.3·10^{-4}
LiNbO$_3$	3.9	2.200	1.0·10^{-19}	7.5·10^5	1.9·10^{-2}	1.9·10^{-4}
KBr	6.0	1.530	6.1·10^{-20}	1.8·10^6	2.0·10^{-2}	2.0·10^{-4}
Te glass	3.6	2.000	1.7·10^{-19}	4.9·10^5	1.4·10^{-2}	1.4·10^{-4}

out to be

$$z_{sf} = \frac{2n_o}{0.61} \cdot \frac{w_o^2}{\lambda} \cdot \frac{1}{\sqrt{P/P_{crit}}} \quad (2.22)$$

where w_o is the beam-waist size (radius). For a fused silica surface irradiated by an excimer laser with a 10 ns, 100 mJ pulse, this distance is of the order of 1 cm; for liquid carbon-disulfide, it is approximately 1 mm. The self-focusing distance will vary substantially with both wavelength and peak laser power. Table V presents the self-focusing distance z_{sf} for typical nanosecond and femtosecond laser peak powers at the Nd:YAG laser wavelength 1.064 μm. For nanosecond lasers, the self-focusing distance is of order centimeters and therefore much longer than typical ablation depth in transparent optical materials. On the other hand, with the terawatt peak powers obtainable from chirped-pulse amplifier systems, self-focusing may occur at distances much shorter than the ablation depth and therefore must be carefully taken into account.

2.3 Competition Between Localized and Delocalized Excitations

Following photoexcitation of electronic or vibrational modes, whether they belong to the perfect solid or to defects, the second stage in laser-induced desorption or ablation is marked by a competition between localized and delocalized relaxation processes for the absorbed energy, as suggested schematically in Fig. 7. Desorption of an atom or molecule can occur if the localization process wins out by retaining the absorbed energy at a single site for longer than a few vibrational periods—long enough for the bond to be broken and the atom or molecule to begin to move away from its equilibrium position. Radiative recombination, because of its long lifetime (typically a nanosecond or more), is rarely the major channel of energy dissipation; instead, desorption and ablation processes can be viewed as one of several possible nonradiative transitions involving the following elementary energy-transformation processes:

1. Electronic energy to vibrational energy by electron-phonon scattering
2. Electronic excitation energy to configurational energy
3. Electronic energy from one configuration coordinate to another
4. Vibrational energy from one degree of freedom to another

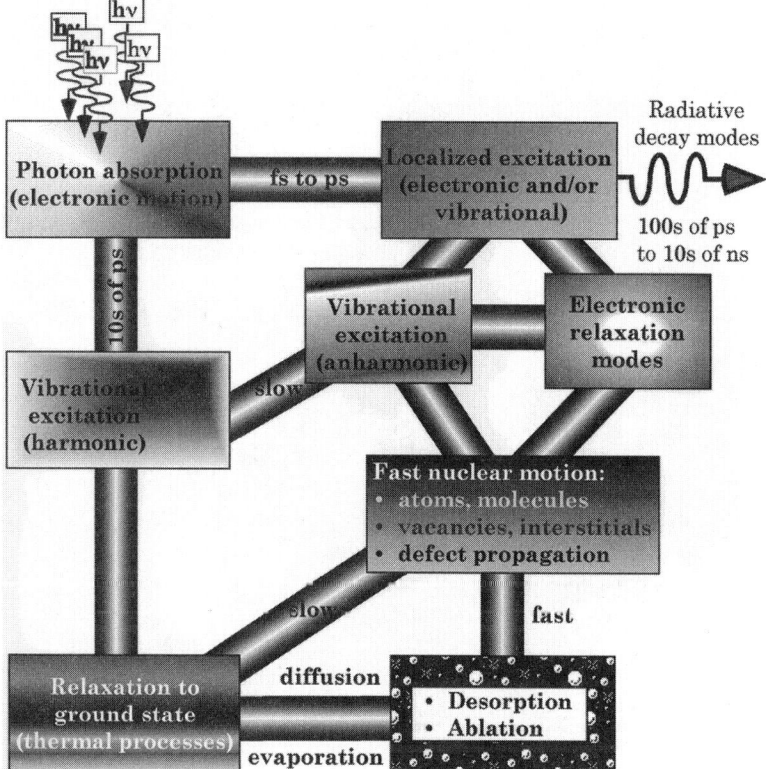

FIG. 7. Energy-flow pathways showing competition between localized and delocalized relaxation mechanisms.

If the localization condition is satisfied, any of these processes may lead to desorption or ablation; however, if localization does not occur at the surface, the radiationless relaxation process may simply return the system to its ground state.

The first of these processes is the only mechanism relevant to electronic relaxation in metals, except in such particular cases as excitation of surface plasmons or small metallic clusters; it also plays a critical role in laser-excited semiconductor. An electronic excitation in an ionic insulator generally takes the form of creating a hole on the anion; if that excitation leads to the creation of an exciton, electronic excitation energy is converted into *configurational energy* of the exciton, and the evolution of such a new configuration is followed on a *reaction coordinate*. That configurational energy can be carried forward as the new state evolves, and under some circumstances it may produce a new configuration; for example, an exciton might be converted into a vacancy-interstitial defect pair. Finally, it is also possible for vibrational energy in one degree of freedom, such as vibration in a particular bond, to be converted into a different vibrational mode.

2.3.1 Localization, Delocalization, and Relaxation Times

The process of laser-induced desorption or ablation begins with the excitation to a potential energy surface on which the photoexcited system will move. As the excited system begins to evolve on this surface, it follows a route determined by the competition between localized excitations—such as trapping at defects, or self-trapping by lattice distortion—and delocalized mechanisms of energy loss, such as electron-phonon scattering. It turns out that the relative roles of localized and delocalized excitations can be represented schematically using bond-orbital parameters

$$\alpha_p = \frac{V_3}{[V_2^2 + V_3^2]^{1/2}} \quad \text{and} \quad \alpha_m = \frac{V_1}{[V_2^2 + V_3^2]^{1/2}} \tag{2.23}$$

which are defined respectively as the *polarity* and the *metallicity*. (The Phillips ionicity parameter [35] is essentially the square of the polarity defined in this way.) Fig. 8 shows where in the space defined by these two parameters the most common solid types are to be found. From this diagram, note that as either the polarity or metallicity increases, we move from the open tetrahedral structure of the Group IV elements toward the cubic, close-packed structure of the metals and ionic solids. The direction of increasing metallicity corresponds to increased electron-lattice scattering as the primary nonradiative loss mechanism, whereas the direction of increasing polarity (ionicity) indicates increasing strength for electron-lattice coupling. Hence, this diagram also gives a qualitative indication of the relative strength of the localizing versus delocalizing mechanisms for

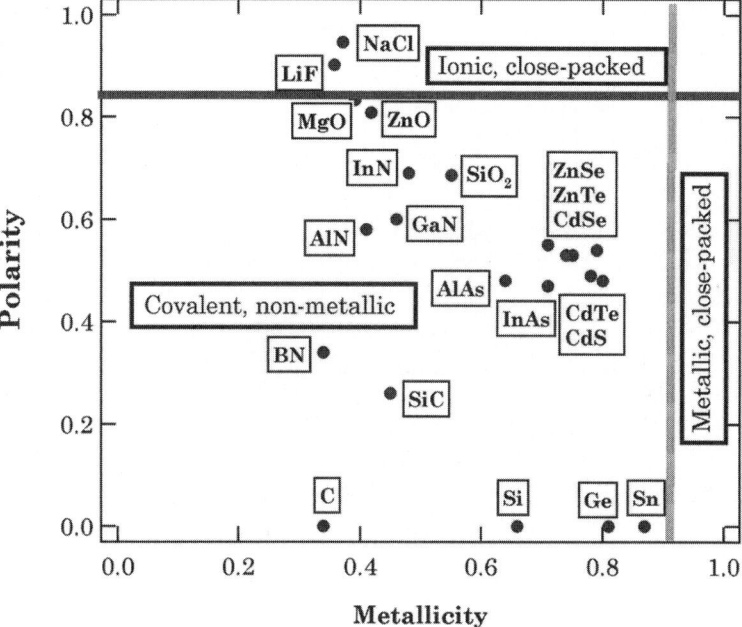

FIG. 8. Distribution of metallic and nonmetallic solids as functions of the bond-orbital polarity and metallicity as defined in Eq. (2.23).

absorbed energy and therefore of the processes that can lead to desorption and ablation.

At the origin of the metallicity axis, photoexcitation generally produces delocalized electrons in the conduction band for which scattering is the major energy-loss process; particularly for homopolar bonds (e.g., Si or Ge), there is very little opportunity for localization. In the case of the better conductors (metallicity approaching 1), there is basically no way for electrons to be localized in the metal unless an impurity on the surface or in the near-surface bulk provides a localization site. Along the polarity axis, on the other hand, one proceeds from essentially no lattice-induced localization to self-trapping, a variation that reflects the strength of the electron-lattice coupling for the solid.

In evaluating the competition between localizing and delocalizing relaxation processes following photoexcitation, there are excitation, relaxation and transport processes whose characteristic time constants need to be considered. The competition between these characteristic time constants varies strongly with material properties, laser wavelength, and pulse duration, and is known with some certainty for only a handful of cases—leaving substantial challenges open for future experiments. (See Tables VI and VII.)

TABLE VI. Localization characteristics as a function of electron-lattice-coupling strength α for various materials. "X" stands for any of F, Cl, or Br, but not I. Adapted from Reference [5].

Coupling	Localization characteristics	Materials
Vanishing ($\alpha = 0$)	Anderson localization in bulk and possible on the surface, otherwise primarily at defects	Si, Ge
Weak ($0 < a \leq 0.9$)	Two-hole localization possible by negative-U interaction at defects, or in dense electron-hole plasma	III-V and II-VI semiconductors, ZnO, TlX, CuX, AgBr
Intermediate ($0.5 < \alpha \leq 2$)	Polaronic response, localization by lattice deformation possible depending on degree of ionicity.	MgO, CaO, c-SiO$_2$, Al$_2$O$_3$, LiNbO$_3$, KNbO$_3$, LiI, NaI, KI, RbI, CsI, possibly high-T_c oxides
Strong ($2 < \alpha \leq 4$)	Either excitons or holes are self-trapped in the bulk, possibly also on the surface	LiX, NaX, KX, RbX, AgCl, MgF$_2$, CaF$_2$, SrF$_2$, BaF$_2$, BaFCl, RbMgF$_3$, a-SiO$_2$

2.3.2 Along the Metallicity (Covalency) Axis: Electron-Phonon Scattering

The metallicity axis represents the degree to which absorbed photon energy is dissipated in delocalized modes. It is marked principally by the efficiency of electron-phonon coupling in metals and elemental semiconductors. At high

TABLE VII. Possible configuration coordinates for desorption arranged according to electron-lattice coupling strength. Materials corresponding to various electron-lattice coupling strengths are given in Table V. Q_1 is the initial excitation; Q_2 is the lattice-localized excitation; Q_3 is the predissociative state; and Q_4 is the final-state configuration.

Coupling	Q_1	Q_2	Q_3	Q_4
Vanishing ($\alpha = 0$)	e-h pair (~1 eV)	Anderson localization	Jahn-Teller distortion	Neutral atoms, energetic ions
Weak ($0 < \alpha \leq 0.9$)	e-h pair (1–3 eV)	Trapped charge carriers at defects	Two holes localized on one site	Neutral atoms and energetic ions
Intermediate ($0.5 < \alpha \leq 2$)	e-h pair (3–10 eV)	Large to small polarons	E' center? F_2 center?	Neutral atoms, ions, clusters
Strong ($2 < \alpha \leq 4$)	e-h pair (8–12 eV)	Self-trapped exciton	F-centers and H-centers	Neutral atoms, small ion fraction

electron-hole-plasma densities, materials lying on or near this axis also experience a variety of screened Coulomb interactions which can, under some circumstances, lead to desorption and ablation. In metals and elemental semiconductors—that is, for low polarities and high metallicity in Fig. 8—polarons are not formed, and photoexcited electrons relax primarily by electron-phonon scattering, gradually returning to equilibrium with the ionic lattice. Electron-electron scattering is relatively infrequent in metals [36] because of the combined effects of the Pauli exclusion principle and the screening of the Coulomb interaction by the electron gas. Therefore, electron-phonon scattering represents the principal loss mechanism. The electron thermalization time in many different metals has been measured using femtosecond laser pulses by a number of techniques: femtosecond photoemission from W films [37]; thermal modulation measurements of the transmission through thin Nb films [38]; time-of-flight heat transport measurements in thin Au films [39]; and transient second-harmonic generation of Cu films [40]. In all cases, the relaxation times are less than a few picoseconds. Because both the electrons and the phonons are delocalized, the only way to localize energy to produce desorption or ablation from a perfect metal surface is to turn off the laser pulse before thermal diffusion out of the laser focal zone begins to heat the surrounding material. Experiments with ultrashort laser pulses, to be presented in Section 2.5, confirm that the primary process of laser ablation in pure metallic films appears to be thermal vaporization.

As one moves toward lower metallicity—for example, toward the elemental semiconductors—but still at low polarity, the competition between the radiative and nonradiative processes begins to be dominated by the density of the laser-generated electron-hole plasma. Because hole mobilities tend to be an order of magnitude less than electron mobilities in homopolar semiconductors, some possibility for exciton formation exists. However, the recombination rates are so high at normal temperatures that melting and vaporization dominate in most realistic circumstances. The electron-hole plasma in semiconductors can be de-excited by Auger recombination and radiative and nonradiative recombination at defect sites. The cross section for the former is inversely proportional to the density n^3 of electron-hole pairs, and hence it dominates at extremely high density of excitation. Recombination can occur at defect sites by capture of an electron followed by hole capture, and vice versa [41]. The electron is first captured into an excited state close to the continuum; the excitation energy is subsequently dissipated by phonon emission. The second particle is de-excited nonradiatively to the lowest excited state and returns to the ground state following photon or phonon emission.

Where localized excitations are involved in laser-induced desorption or laser ablation in metals, the circumstances generally involve resonant processes (such as the surface plasmon resonance), spatial localization induced by small metal particles, and morphological localization—for example, at asperities, step edges, or other surface features—induced by hydrodynamical or thermodynamical roughening. Examples of these unusual localization processes as they occur in desorption and ablation from metals are presented in Sections 2.4–2.5.

2.3.3 Along the Polarity Axis

The polarity axis represents the degree to which absorbed photon energy is stored in distorted lattice configurations. The lattice localization phenomena encountered as we move along the polarity axis exhibits a remarkable degree of complexity, ranging from polaron and exciton formation to lattice distortions that are so strong that they "self-trap" excitons. The electronic and vibrational energy transformation processes numbers (2) through (5) are all represented, often in competition with one another. For a heteropolar solid, as we move away from the origin of Fig. 8, the initial and final states in an electronic excitation tend to have progressively different configuration coordinates. Figure 9 shows how the initial electronic excitation in a solid is dissipated as it moves through different configurations enroute either to the ground state or to desorption and eventually to ablation. The configuration coordinates appropriate to strong and weak electron-lattice coupling may be viewed as typical of alkali halides or alkaline-earth fluorides and compound semiconductors, respectively. The strength of the electron-lattice coupling is basically determined by the degree of spatial separation along the configuration coordinate between the initial and final excited state.

The localized electron-lattice interactions have spatially compact wave functions that can evolve toward desorption and ablation. In nonmetallic solids, electron-hole pairs created by photon absorption relax to form a localized and excited electronic and vibrational state. The conversion of the electronic energy of the absorbed photon to nuclear motion, nonradiative electron-hole recombination or a radiative transition is determined by the character of electron-lattice interactions. In solids with the strongest electron-lattice coupling, such as the alkali halides and alkaline-earth fluorides, localization can occur even in perfectly crystalline material and at low densities of electronic excitation because the Coulomb forces guarantee that ions will follow electronic motion. In strong-coupling solids, luminescence with a large bandwidth from self-trapped excitons is also observed [24, 42, 43].

In general, a charge carrier, electron-hole pair, or exciton created by photoexcitation of a solid creates a localized distortion due to the coupling with the lattice. Two different types of such lattice distortion—polaronic and excitonic—are illustrated schematically in Fig. 10. The charge carrier or exciton can reduce its energy by delocalizing in a band state or localizing through lattice

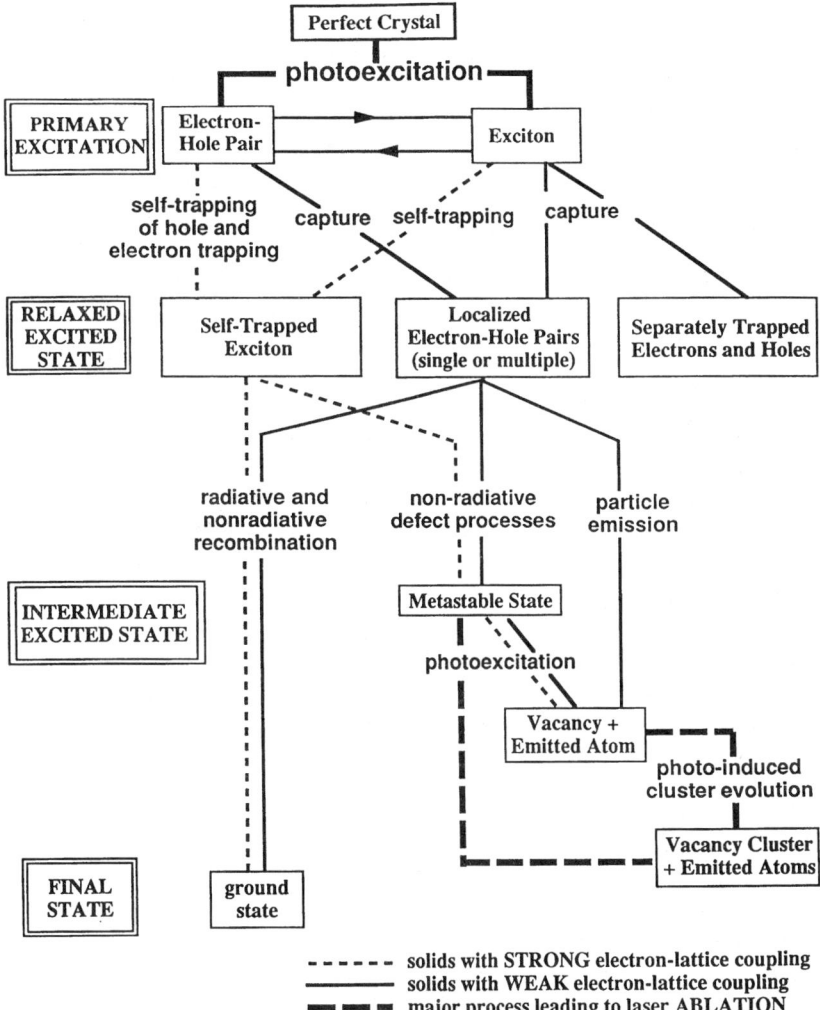

FIG. 9. Flow diagram showing the pathways leading from photoexcitation to photodesorption for the limiting cases of strong- and weak-coupling solids. There are four stages, corresponding generically to the four different configuration coordinates: initial excitation, relaxed excited state, predissociative state, and final state.

polarization. The criterion for localization is determined by the relative values of the bandwidth $2 \cdot B$, and the lattice relaxation energy E_{LR}. Self-trapping occurs if $B < E_{LR}$ [44]. Solids for which this is the case are said to have strong electron-lattice coupling. Weak electron-lattice coupling is characteristic of solids for

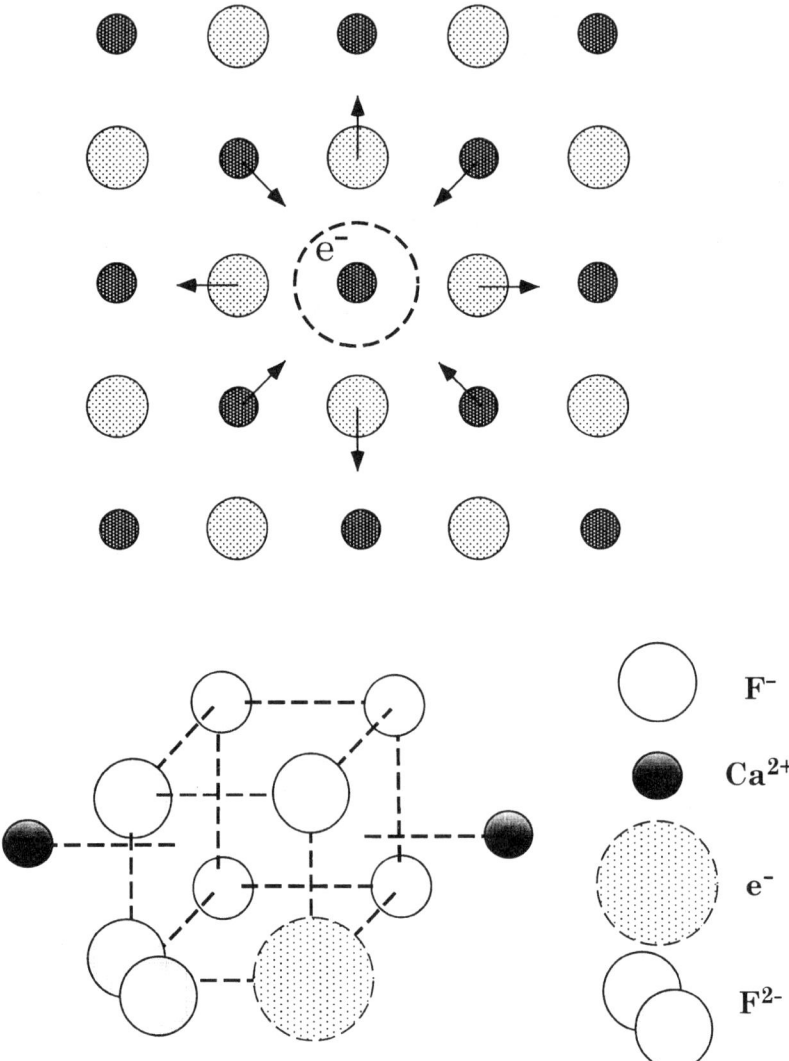

FIG. 10. Schematic diagram of lattice-localized excitations in a nonmetallic solid: (a) a small polaron trapped at a positive ion in a diatomic insulator and (b) a self-trapped exciton in the fluorite structure in one of the four possible nearest neighbor positions.

which $B > E_{LR}$. If more than one carrier can be localized at a single site, as often occurs in semiconductors, it is necessary to incorporate into the localization criterion the on-site Coulomb repulsion energy U [45]. If E_{LR} dominates over B and U, two-carrier localization can occur; if U and E_{LR} dominate B, localization of a carrier results; alternatively if B dominates the others, the carriers remain free. The criterion for two-carrier localization on defects, $E_{LR} > U$, was suggested first by Anderson, who pointed out that because of the diamagnetism of many materials (certainly most optical ones), charge carriers may well under appropriate circumstances localize at a single site with paired spins [46]. In the language of chemistry, the self-trapped hole is said to be localized because it is bound in a covalent configuration occupying the highest anti-bonding orbital of the combined species [47]. In many insulating solids, self-trapped holes are precursors of the self-trapped exciton, which has significant photochemical consequences for the evolution of the laser-irradiated material.

A self-trapped exciton (STE) is formed when a self-trapped hole traps an electron or as the result of free-exciton relaxation. The STE has a chemical interpretation analogous to the self-trapped hole, but unlike the self-trapped hole, has the possibility of relaxing to a number of excited states because of the extra electronic excitation energy of STEs [48]. The most well known of these relaxed (but metastable) states is a Frenkel pair (close vacancy-interstitial pair) [49]. The energy gain by the self-trapping of an exciton is larger than that for a hole, as much as one or two electron-volts [50]. Experimental evidence suggests that STEs may play an important role in laser-induced desorption in the alkali halides, alkaline-earth fluorides, and amorphous SiO_2 (fused silica).

2.3.4. The "In-Between" Case: Moderate Polarity and Moderate Metallicity

Solids with moderate polarity and moderate metallicity exhibit a broad range of behaviors not typical of either highly ionic or highly metallic solids. In solids with weak electron-lattice coupling, neither holes nor excitons are self-trapped and thus another mechanism of energy localization must be operative to produce laser-induced desorption. Experimentally it is observed that the localization of energy required for desorption tends to occur only at defects or at very high densities of electronic excitation in the perfect lattice [51]. In solids with weak electron-lattice coupling, luminescence also competes effectively with the motion of nuclei in the solid, and a sharp free-exciton luminescence band is often observed, particularly at low temperatures. Excitons in insulators with weak electron-lattice coupling de-excite through both radiative and nonradiative recombination channels. Delocalized electron-phonon interactions in heteropolar solids include the Fröhlich or polaron interaction with transverse optical phonons, the piezoelectric interaction with acoustic phonons, and the deformation potential.

In weak-coupling solids, defect-induced or plasma-generated processes are often the primary mechanisms that lead to desorption or ablation. One of the most intriguing localization mechanisms which can arise in weak-coupling solids is called the *negative-U* interaction, illustrated schematically in Fig. 11 for a dangling bond in an amorphous semiconductor. The electronic contribution to the total correlation energy is $U_{electronic} = I - A$, where I is the ionization potential and A is the electron affinity. The elastic restoring force exerted by the lattice on a doubly occupied dangling bond can be written as $U_{elastic} = -f^2/K$, where f is the defect force on a singly occupied, neutral dangling bond. The total correlation energy $U = U_{electronic} - U_{elastic}$ can be negative under a variety of circumstances, depending on the defect and host.

Two-hole localization by the Anderson negative-U interaction on a defect site has been observed in the bulk of semiconductors, including the V^+ center in Si [52] and a localized state of the DX center in GaAs [53]. For defects on surfaces (another kind of defect site!), one possibility for localization of two holes is that

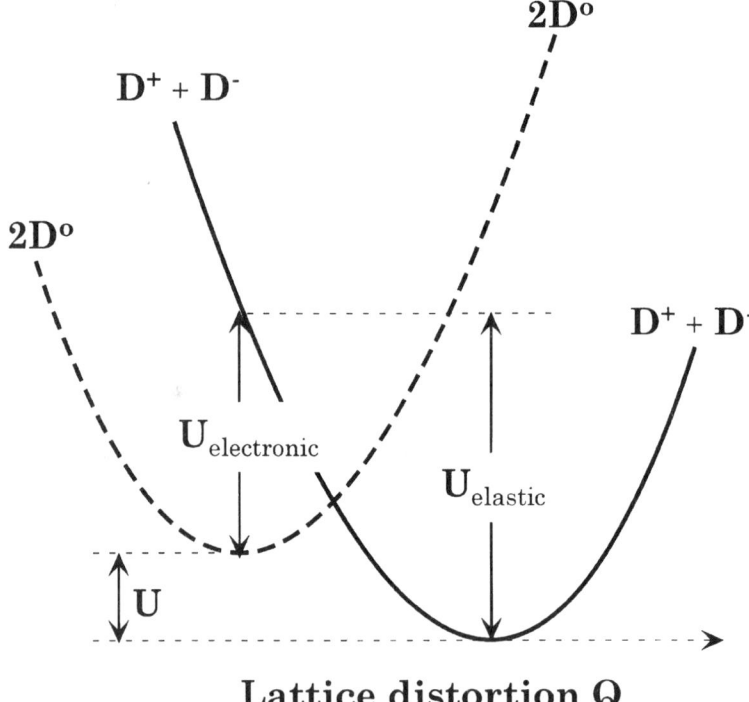

FIG. 11. Potential energy surfaces for a negative-U interaction, showing the lowering of the total correlation energy because of the greater depth of the elastic-distortion potential for two carriers localized at a single site. The configuration coordinate is the lattice distortion. D^o represents a singly occupied dangling bond, while D^+ and D^- are unoccupied and doubly occupied dangling bonds, respectively.

in which an atom on the surface is displaced outward. If the total correlation energy is driven negative by this lattice distortion, Anderson localization could then lead to desorption. Defect-mediated two-hole localization with accompanying lattice distortion is typical of weak-coupling solids, where it is normally initiated by carrier trapping. However, two-hole localized states can also be produced by sequential (or cascade) excitation of a defect state.

Two-hole localization can also occur at very high densities of excitation in a dense electron-hole plasma. For example, a two-hole localization model [54] was first proposed to explain the laser sputtering of Zn atoms from ZnO which exhibited a distinct ablation threshold and a two-temperature Maxwell-Boltzmann velocity distribution. The two-component velocity distribution shows that one is not dealing with an equilibrium process such as melting. Itoh and Nakayama proposed a phonon-assisted localization of two holes at a single surface site based on Anderson's idea that the Coulomb repulsion might be less than the energy gain from lattice distortion caused by the electron-hole plasma. The existence of a fluence threshold for ablation was ascribed to the need to reach a critical density of the electron-hole plasma to screen the effective charge of the two holes until they can move within the effective interaction distance of the Anderson negative-U potential. The screened Coulomb potential is given by

$$U_{sc} = \frac{e^2}{\varepsilon(n) \cdot r_c} \cdot \exp[-2k(n)r_c] \qquad (2.24)$$

where n is the density of the electron-hole plasma, $\varepsilon(n)$ is the dielectric constant, and r_c is the screening radius which is of order one or two lattice constants in a dense e-h plasma. The probability for particle emission can be calculated from reasonable assumptions about the density of the laser-induced electron-hole plasma [55]. An alternative to the idea of localization by screening in the e-h plasma is that the Coulomb barrier to two-hole localization is overcome if the density of electron-hole pairs is sufficient for the Fermi energy of the degenerate holes to exceed the potential barrier [56]. This is consistent with the general notion of the negative-U potential originating in a displacement of the Fermi energy in the gap.

2.3.5 Energy Transfer and Localization Mechanisms: Case Studies

With these general remarks about localization and delocalization of absorbed photon energy, we are in a position to categorize the specific mechanisms of energy transfer and storage after photoexcitation, and to apply these generic ideas to a couple of illustrative cases. The first thing that happens after an electronic or vibrational excitation is localized is *conversion of electronic energy to vibrational energy by cooling transitions*, as illustrated in Fig. 12. Cooling transitions can occur in either the ground state or in an excited state. In

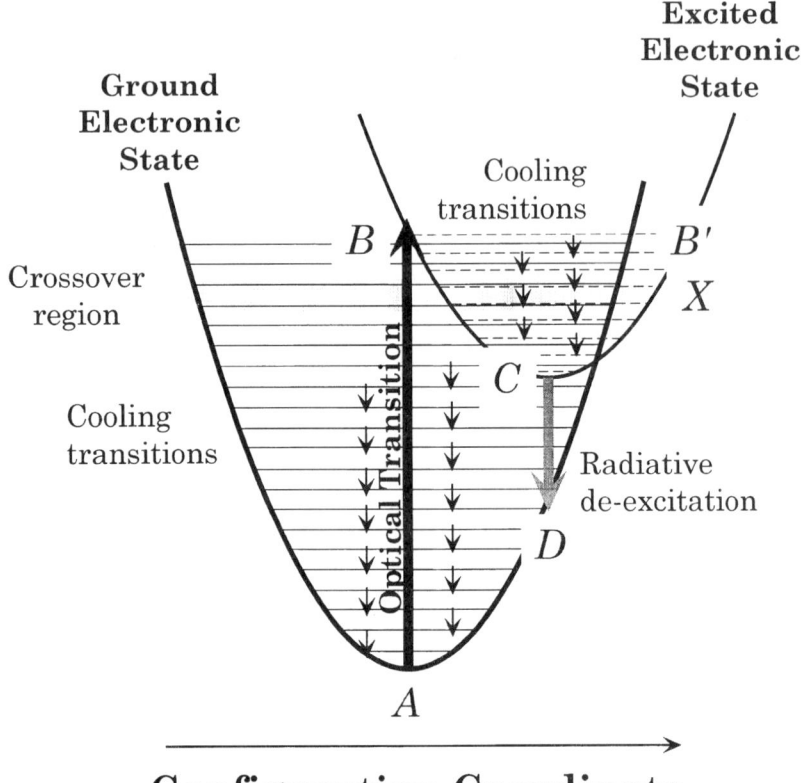

FIG. 12. Schematic diagram of cooling transitions in both excited and ground electronic states. Note that a nonradiative transition can take place at the level marked "crossover." If the excited system can relax to the bottom of the band at the point C, a radiative transition to a vibrationally excited level of the electronic ground state is most likely. Adapted from [42].

either case, the energy liberated appears in the form of heat within the laser-irradiated zone of the solid in a time constant of some tens to hundreds of vibrational periods, even if there is no localization. That heat is transferred to the surrounding material on a much slower time scale, namely, the characteristic diffusion time. Once the cooling transitions have brought the system to the lowest vibrational state of the excited system, the lattice relaxes to a configuration at a minimum of the adiabatic potential energy surface (APES) on a time scale less than the inverse of the characteristic lattice frequency.

In following the energy flow from this point, it is necessary to distinguish two types of relaxed configuration as shown in Fig. 13: (a) with the minimum of the relaxed excited state potential surface situated above the ground state, and (b) with the minimum of the relaxed excited state potential-energy surface located below the ground state APES. In case (a), both radiative and nonradiative transitions lead to the ground state, a situation typical in weak-coupling solids. Because the energy separation between the relaxed excited state and the ground state is large, the nonradiative transition involves a multiphonon process, which in turn requires strong electron-lattice coupling to compete with the radiative transition. In case (b), because of the large lattice relaxation energy E_{LR} and the configuration change at the APES minimum, the electronic excitation energy is largely transferred to configurational energy.

In solids with coupling strengths intermediate between cases 13(a) and 13(b), both radiative and nonradiative transitions can relax the system to its ground state before desorption occurs. Figure 14 illustrates such a case: A photon with energy $\hbar\omega$ is absorbed in the ground state at the equilibrium configuration O; this initiates a Franck-Condon transition to a vibrational level A in the excited-state vibronic manifold, still at the configuration coordinate O. As the localized phonon wave packet associated with state A oscillates along the APES of the excited state, it begins to lose energy by cooling transitions. The *Dexter-Klick-Russell* (*DKR*) rule [57] states that if state A lies above the crossover region, transitions to high-lying vibronic states in the electronic ground-state manifold will occur at a rate comparable to the characteristic lattice-vibration frequency.

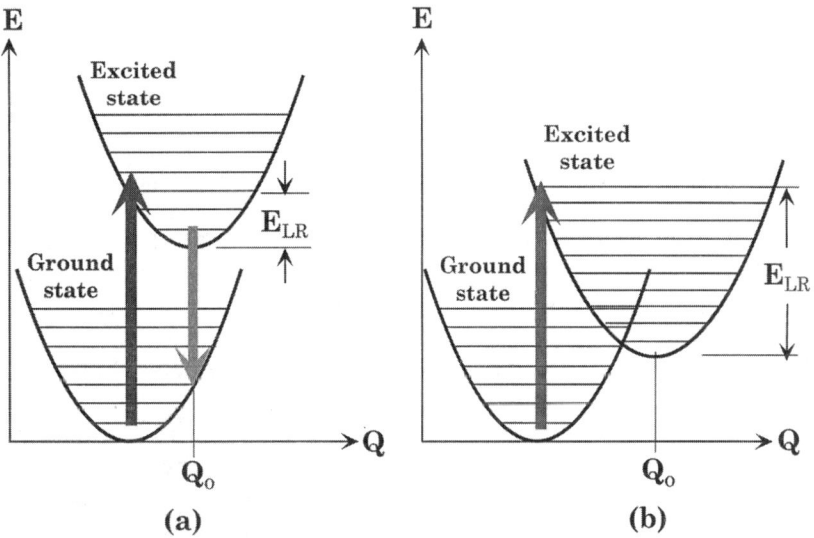

FIG. 13. Two different configurations for the ground state and relaxed excited states, shown for two different electron-lattice coupling strengths: (a) weak coupling and (b) strong coupling. Solid vertical arrows represent radiative transitions.

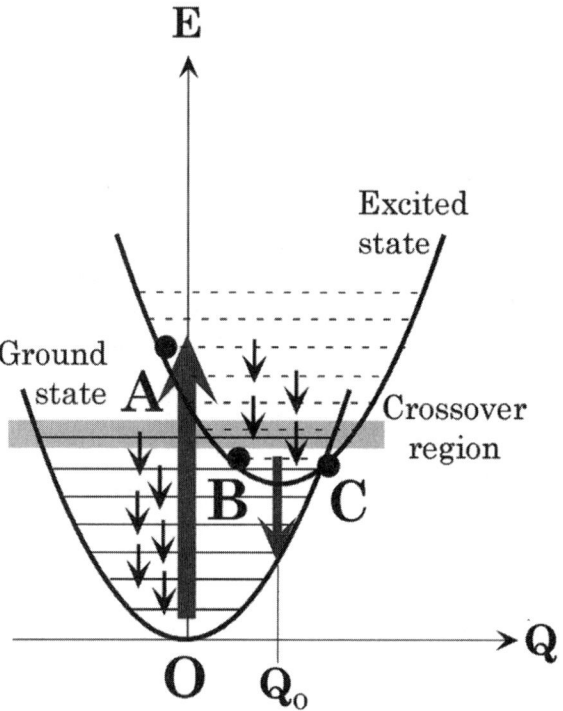

FIG. 14. A radiationless Dexter-Klick-Russell transition, in which the cross-over from one configuration to another (point B) and subsequent rapid cooling transitions prevent the formation of a pre-dissociative state.

Otherwise, the cooling transitions continue to state B whence a radiative transition to the ground state will occur. If we define the ratio $\Lambda = (E_A - E_B)/\hbar\omega$, the DKR rule implies that relaxation by luminescence is highly probable for $0 \leq \Lambda \leq 1/4$, less so for $1/4 \leq \Lambda \leq 1/2$, and vanishes for $\Lambda > 1/2$ [42]. Since Λ is related to electron-lattice coupling strength, this also suggests that ion motion leading to desorption is most likely for $\Lambda > 1/2$.

These so-called DKR processes are especially effective in de-excitation following capture of a charge carrier into a high-lying excited state, and may explain observations of fast nonradiative transitions from laser-generated electron-hole pairs to the lowest excited state of the self-trapped exciton [58]. Also, because the radiative transition probability from state B is enhanced by cooling transitions from the initial excited state A, the kinetics of the competing nonradiative transitions to the ground state may also depend on photon energy as well as on the detailed configuration of the excited levels.

In the previous examples, radiative transitions in weak-coupling solids and non-radiative DKR transitions in intermediate coupling solids lead back to the ground state, desorption is inhibited. However, an electronic transition from the excited electronic state to a vibrationally excited level of the ground state may also initiate desorption via the mechanism illustrated in Fig. 15. Here the excited-vibrational-state wave function of a localized intrinsic mode Q overlaps the wave function of a reaction mode, characterized by a different configuration coordinate Q_R. The energy in the intrinsic mode can be efficiently transferred to the reaction mode, inducing a configuration change and possibly desorption. Where the vibrational energy in the Q mode surpasses the potential barrier for the reaction, the process is often called *phonon kicking* and is believed to cause recombination-induced defect migration [59] and desorption from rare-gas

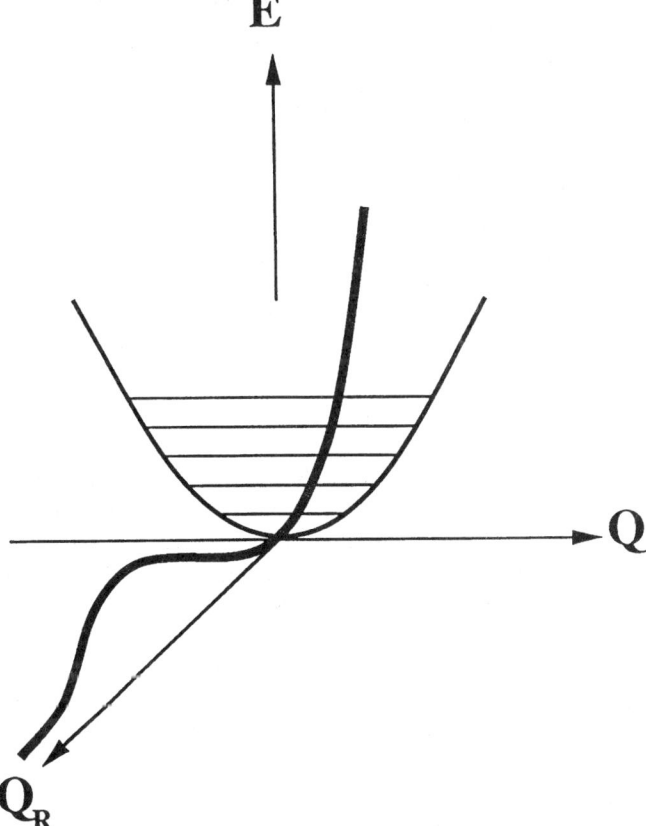

FIG. 15. Vibrational to vibrational energy transfer process in which the overlap between vibrational states in two different configurations permits the crossing of an activation barrier from one coordinate to the other.

solids [60]. If a configurational change fails to occur, the localized vibrational energy is ultimately converted to lattice phonons, that is, to heat. In infrared laser excitation of materials, where the system remains in the electronic ground state, this kind of phonon kicking may be especially important.

Nonradiative de-excitation can also occur by conversion of electronic energy to other electronic degrees of freedom if this second electronic system transfers its energy to vibrational modes or configurations. The Auger transition is an important example of this process. The localized electronic excitation energy is transferred to free electrons, which in turn are de-excited by emitting phonons [61]. This process dominates when the density of electron-hole pairs is high, as it can be in laser-excited semiconductors. In an electron-hole plasma, each charge carrier is surrounded by carriers of opposite charge, reducing the Coulomb repulsion between carriers of like sign by a screening factor $\exp(-kr)$, where k, the screening constant, varies as $n^{1/6}$, and r is the distance between the carriers [62]. The electron-hole plasma may exhibit ambipolar diffusion [63] in which the local density of the electrons and holes are strongly correlated.

Up to this point, we have been dealing with perfect solids, whereas in reality, all solids have a variety of intrinsic and extrinsic defects. The energy levels of defects that donate or accept charge generally lie in the bandgap, creating unique opportunities for localizing and releasing energy. Defects play three important roles in laser-induced desorption and ablation:

1. They may act as recombination centers for electrons and holes, thus removing electron-hole pairs from the potential drivers for desorption or ablation.
2. By weakening local bonds, they may serve as "nucleation centers" for rapid multiplication of desorption or ablation sites.
3. They may store absorbed energy at a lattice site for long periods of time in localized modes, thereby contributing to the possibility of bond-breaking.

In an electron-hole plasma of concentration n, each defect acts as a recombination center with a characteristic lifetime of $1/(\sigma_t v n)$, where σ_t is the carrier trapping cross section; v the velocity of the charge carrier; and n the concentration of defects. With $n = 10^{19}$ cm^{-3}, the time constant can be as short as 1 ns. Following photoexcitation, an electron or a hole is first captured into a high-lying excited state; relaxation follows by nonradiative transitions through a sequence of highly excited states. Here, we emphasize two points. First, the capture cross section is governed by the rate of de-excitation to lower levels because delocalization of trapped carriers from highly excited states is very probable. Capture is in fact the outcome of a competition between the delocalization and the transition to low-lying excited states. Second, de-excitation to the lowest excited state is usually much faster than the transition from the lowest excited state, to the ground state. In the lowest excited state, the lattice is relaxed to a configuration in which the adiabatic potential energy surface (APES) has a minimum, as shown schematically in Fig. 16. From this point, the detailed dynamical evolution of the system depends on the strength of the electron-lattice coupling.

Successive capture of an electron and a hole (or vice versa) leads to recombination of an electron-hole pair. Although this process is not a major loss channel for excitation energy when defect concentration is low and excitation density is extremely high, it is during the process of recombination that the electronic energy is imparted to local modes, which ultimately lead to particle emission. Just as in the case of charge-carrier capture, the relaxed excited defect state will be formed after successive capture of both carriers. The energy of an electron-hole pair is partially stored in this excited state. We shall see later that the recombination channel can be the source of desorption and even ablation.

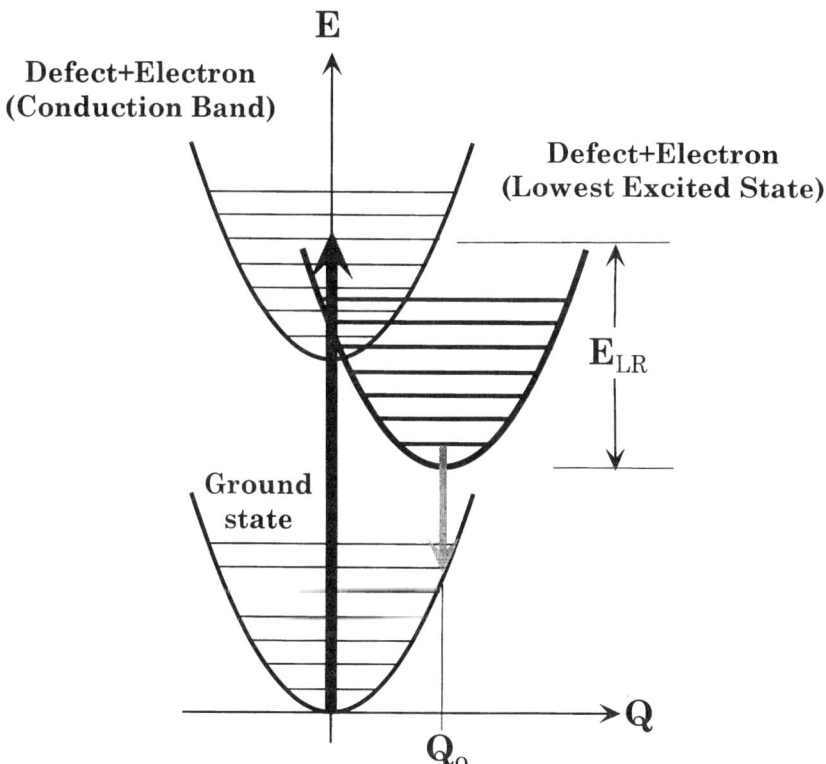

FIG. 16. Energetics involved in the creation of a relaxed excited state at a defect site in a compound semiconductor.

When the APES of an excited state has a minimum in a configuration substantially different from that for the minimum of the APES of the ground state, as depicted in Fig. 13(b), laser excitation results in configurational change (process 3). This change is represented by the configuration coordinate Q in Figure 13(b); physically, it could correspond to a rearrangement of the electron-hole pair into a self-trapped hole plus an electron, or into a self-trapped exciton. Unlike phonon kicking—where the electronic energy first goes to vibrational energy of a local mode and then to the reaction mode—the electronic energy in this case is converted directly to the reaction mode. Desorption induced by excitation of a substrate-absorbate system to an anti-bonding excited state, as in the Menzel-Gomer-Redhead mechanism, is one example of this process in photo-induced desorption.

Formation of a metastable excited state by electronic excitation of defects in semiconductors—as in Fig. 16—is another well-documented example of this type of energy transfer [64]. When an antisite defect in compound semiconductors—for example, an As atom occupying a Ga site of GaAs—is elecronically excited, the defect is converted to a metastable state with a configuration significantly different from the ground state [65]. Configurational change initiated by electronic excitation forces a change in lattice conformation to minimize the total system energy. However, such a conformational change reduces the rigidity of the lattice because the bonding energies are reduced by the electronic excitation energy [66]. The concept is an extension of the early suggestion by Van Vechten *et al.* [67]. Localized holes are especially effective in bond weakening and defect migration. The excitation of an electron from a bonding to an anti-bonding state in a tetrahedrally coordinated semiconductor softens the dispersion curve of the transverse acoustic phonons, as pointed out by Wautelet and Laude [68].

Example: Desorption induced by self-trapped exciton formation. The best understood example of the conversion of electronic energy to vibrational and configurational energies is the self-trapped exiton (STE) and its related defects in alkali halides. The STE in alkali halides consists of a halogen molecular ion X_2^- in an off-center configuration, and an electron; it is the precursor of a metastable state consisting of an F-center and an H-center (Frenkel pair on the halogen sublattice). The sequence of configurational changes that produces the Frenkel pair following photoexcitation is illustrated in Fig. 17. The initial optical excitation forms an electron-hole pair, which in turn leads, first by self-trapping and then through a series of nonradiative and cooling transitions, to the STE. From the STE, the Frenkel pair may be generated either by a simple thermal hopping over the barrier [69], or by a dynamical transfer from the STE to the Frenkel-pair configuration via a series of cooling transitions on the APES to the lowest state of the STE. That this latter possibility is realized has now been confirmed experimentally using ultrafast laser techniques [70].

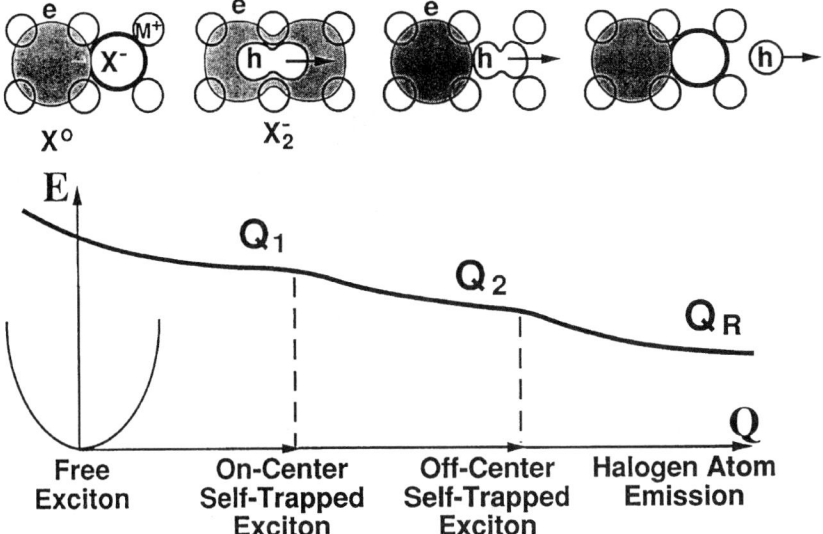

FIG. 17. Schematic of configuration coordinates for desorption initiated by the formation of a self-trapped exciton in a strong-coupling solid. (Compare Table VII.)

Example: Desorption of atoms from a compound semiconductor. Thus far, we have not considered the possible role of surface states in the localization of absorbed photon energy. Laser excitation from bulk valence-band states to the conduction band produces electron-hole pairs with three-dimensional wave functions, whereas excitations from a filled surface state to the conduction band, or from a bulk valence-band state to an empty surface state, produce excited states with two-dimensional wave functions. In the special cases where the energy gap E_{gap}^{surf} for transitions involving surface states is smaller than the bandgap energy E_{gap} for direct transitions in the bulk, surface-state transitions can be excited by photons whose energy is too small to excite band-to-band transitions in the bulk. For example, in GaP, there is an empty surface state 0.7 eV below the bulk band edge [71]; laser irradiation of GaP at the wavelength appropriate for initiating transitions to this state produces essentially no bulk absorption or heating, since the absorption coefficient of bulk GaP below the bulk band edge is 5000 cm^{-1}. The role of two-dimensional electron-hole pairs in desorption and ablation, demonstrated only recently [72, 73], can thus be studied without being masked by bulk thermodynamic effects. The electron-hole pairs produced by these surface-state transitions are initially constrained to the two-dimensional surface, but may be converted to three-dimensional electron-hole pairs by bulk phonon scattering.

If a tunable laser is used selectively to excite the surface state no bulk excitations occur. In this case, a hole is created on the surface site. This hole is not sufficient by itself to weaken the bond to the point of desorption, as shown by bond-orbital calculations [5] and illustrated by the transition from the ground state to the lower excited-state APES in Fig. 18. However, if a second hole is created in the already distorted bond following a second photoexcitation, desorption will result due to reversal of the Coulomb field surrounding the weakened bond. This picture is consistent with recent experiments on compound semiconductors, which are described in greater detail in Section 2.4.

2.4 Laser-Induced Desorption from Metals, Semiconductors, and Insulators

The removal of an isolated atom, molecule, or ion from a perfect, atomically smooth surface by electron, ion, atom, or photon impact is a fundamental problem in surface science. Laser-induced desorption is a low-yield sputtering process in which atoms, molecules, or ions are ejected from isolated surface sites without producing detectable residual disorder at the mesoscale, without appreciable heating of the volume surrounding the absorption site, and without either plasma generation or collisional effects. In fact, laser-induced desorption may have more in common with other fundamental studies of desorption—such as those initiated by electron and synchrotron light sources—than with laser ablation [74].

When laser light is absorbed at a surface, an electronic (Franck-Condon) excitation takes place; only afterwards does the atomic motion that constitutes the lattice response begin. If the initial electronic or vibrational excitation is ultimately localized so that there is energy enough to break a bond on a particular site, desorption is *dynamically* possible. Desorption clearly must originate at surface sites; bond-breaking deeper in the interior of the solid produces localized but internal defects. Laser-induced desorption may occur at perfect surface sites, steps, or terraces [75], or surface sites adjacent to point defects (for example, color centers), at sites disturbed by subsurface excitations or defects, at molecular adsorbate sites and in the vicinity of structural defects (such as dislocations or grain boundaries) [76]. Many well-known excitation and relaxation processes that generate interstitial defects—such as exciton decay, which produces interstitial-vacancy pairs—also lead to desorption when the defects are created near the surface.

In this section, we first consider desorption of adsorbed molecules from metal surfaces to introduce the concepts of excitation, relaxation, and localization; next comes a look at two unique situations where localization and desorption in

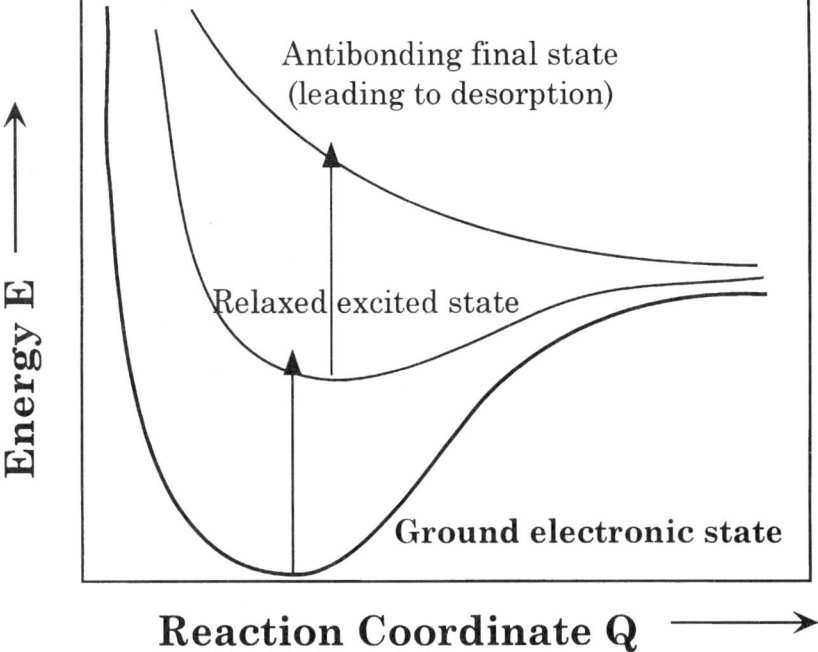

FIG. 18. Potential energy surfaces for excitation to an anti-bonding state from the relaxed excited state formed by two-hole localization in a compound semiconductor. Such potential energy surfaces are relevant to the case of desorption by resonant excitation of a surface state in GaP.

metals are possible. There follows in turn an examination of desorption induced by electronic processes in solids with strong and weak electron-lattice coupling, illustrating the competition between metallicity (delocalization) and polarity (localization). Finally, we discuss the role of surface defects and surface state impurities in desorption, with Ga-atom desorption from gas-dosed GaP and surface-state-mediated desorption from Al_2O_3 as experimental examples.

2.4.1 Bond-Breaking at the Microscale

The characteristic length scale associated with absorption of a single photon, and with the localization of energy that eventually leads to laser-induced desorption (not necessarily at the same site at which the absorption occurred!) is of the order $d_{micro} \cong \tau_{vib} \cdot v_s$, where τ_{vib} is a typical phonon vibrational period in the material, and v_s is the speed of sound. This definition is based on the ideas that desorption is a localized event and that the lattice instability must be initiated before the incident photon energy is dissipated into delocalized (harmonic) vibrational modes of the solid. The length scale derived from this notion—ranging from 0.1 to 10 nm—is commensurate with screening lengths for electron-hole plasmas in semiconductors and relaxation lengths for polarons and self-trapped excitons in ionic insulators [77]. If we define a characteristic microscopic time scale for desorption, $t_{micro} \equiv a/v_s$, we find that it varies from about 200 fs to 3 ps—just the time scale of ultrashort pulse lasers which produce interesting results in desorption and ablation.

Desorption induced by electronic transitions (DIET) was first observed by Menzel and Gomer [78] and by Redhead [79] in experiments on electron bombardment of atoms chemisorbed on metal surfaces in ultrahigh vacuum. The Menzel-Gomer-Redhead (MGR) model provides a generic microscopic vocabulary which is useful for describing desorption processes. The semi-classical MGR model, depicted schematically in Fig. 19(a), begins with an electronic excitation in the adsorbed atom as a Franck-Condon transition from the ground-state potential energy surface (PES) V_g to an excited-state PES V_e so rapidly that only the electronic configuration of the molecule is altered during the transition. As the adsorbate subsequently evolves along the nuclear coordinate axis on the excited-state PES, potential energy is converted into kinetic energy until the adsorbate is de-excited and returns to the ground state. If the molecule remains on the excited-state PES longer than some critical time τ_c, it has a total positive kinetic energy when the metal system returns to the ground state and the molecule desorbs. The probability of desorption is then roughly $p \approx \exp[-\tau_c/\tau]$. The emphasis in the MGR picture is on the excitation and desorption of a molecule or atom based on the internal excitation imparted to it by electron or photon impact. A variation of this mechanism has been discussed by Antoniewicz [80] in which the excited state instead of being excited directly to an antibonding state is actually attracted toward the surface following excitation and "bounced" from the repulsive hard-core surface potential before desorbing. Similar generic pictures can be applied, with appropriate modifications, to a number of the cases we examine in the following paragraphs.

A quite different kind of desorption mechanism was discovered experimentally and explained theoretically by Knotek and Feibelman [81]. It is illustrated schematically in Fig. 19(b) for the case in which a core-hole excitation is created on either ion in a maximally valent covalent oxide, such as TiO_2. The core hole can be filled by an electron from a neighboring Ti ion in an *inter*-atomic Auger process, creating a pair of holes in the valence band thereby upsetting the Coulomb balance in the lattice and causing the ejection of an energetic Ti ion. The key idea here is that in covalent solids, the two-hole state may be localized because the hole-hole interaction energy is greater than the width of the valence band. Thus, even though valence charge carriers tend to be strongly shared in covalent solids, the two-hole state can be localized. The screening length and screening time of the core hole is obviously critical to the dynamics of desorption. Because laser photons have insufficient energy to cause core-hole transitions, the principal application in laser-induced desorption will be for cases in which the initial hole is in a relatively well localized state, such as those in the d-bands of heavier elements or the deeper valence levels of oxygen. However, the Knotek-Feibelman mechanism provides a useful conceptual analog for a range of desorption processes in which the internal excitation of the desorbed entity is transformed by interatomic interactions. Analogous mechanisms have been proposed for the desorption of Si^+ ions from Si surfaces, and for a variety of other circumstances in which energetic ions are desorbed from insulating solids.

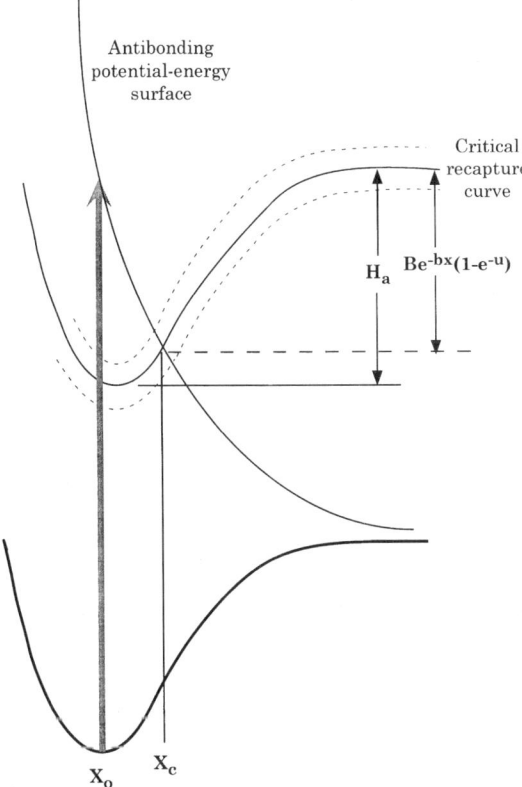

FIG. 19. Comparison of two different models for desorption initiated by electronic transitions (DIET). (a) The Menzel-Gomer-Redhead model, in which the system is excited directly to an anti-bonding state by a Franck-Condon transition from the valence band; (b) the Knotek-Feibelman model for TiO_2 in which the two-hole state is localized when the hole-hole repulsion exceeds the valence bandwidth, leading to desorption following inter-atomic Auger decay.

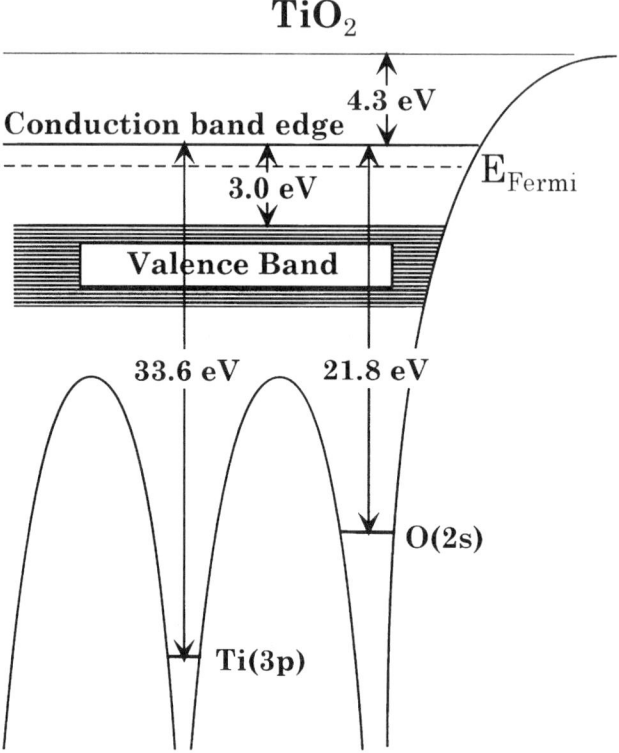

Fig. 19. *Continued*

2.4.2 Desorption from Metallic Surfaces and Clusters

On the perfect surface of a metal, the dissipation of initial electronic or vibrational excitations from absorbed photons occurs on a time scale of a few femtoseconds. This means that desorption cannot occur from a pure metallic surface—unless some imperfection is available to localize the energy. This localization site might be a structural (mechanical) defect, oxide layers (particularly likely if the surface is located in air), or an adsorbed atom or molecule that disturbs the otherwise perfect periodicity of the metallic surface. In this section, we consider three cases that illustrate a range of possibilities: desorption of molecules from atomically flat crystalline surfaces; desorption of metal atoms from atomic clusters; and desorption of metal atoms and ions from thin films through resonant excitation of the surface plasmon. Each case illustrates particular features of the problem of energy absorption, localization, and atomic motion.

We begin with what in some respects is the simplest case: an adsorbed diatomic molecule on a well-characterized metal surface. This should be the classic case to which the MGR model might be expected to apply. The excited-state lifetimes of molecular adsorbates on metallic surfaces are very short (~2–3 fs) because of the rapid interchange of energy with the electronic bath states in the metal. Hence desorption probabilities in this circumstance tend to be very low—as observed in experiments in which adsorbed molecules are excited by nanosecond or picosecond laser pulses. Nevertheless, even under these conditions, a tantalizing variety of electronic effects were observed in early studies of the phenomena of laser-induced desorption, including competition between thermal and non-thermal desorption channels and hot-electron effects seen in translational velocity distributions, dramatic alterations of the rotational Boltzmann distributions, and inverted spin-orbit populations of the scattered molecules [82].

When the local temporal as well as spatial density of electronic excitation is very high, repeated excitations of the molecule while it is on the excited-state potential energy surface are possible, and desorption efficiency can be increased. Femtosecond laser-induced desorption of NO chemisorbed on the (111) surface of Pd [83], of O_2 adsorbed on Pd(111) [84], and of CO adsorbed on the (111) surface of Cu [85], are illustrative of the general principle; a number of other studies consider the mechanisms in greater detail [86, 87]. When irradiated by 620nm light at pulse durations between 100–200 fs, the molecules leave the surface with kinetic energies much higher than predicted surface temperatures. Ho *et al.* have found similar behavior for O_2 desorption from Pt(111) [88]. Even though the initial laser excitation is clearly coupled into the metal substrate, significant vibrational population in $v = 0$, 1, and 2 levels of the desorbed molecules is observed [89].

A salient characteristic of these experiments is the highly nonlinear dependence of the molecular yield on laser intensity, illustrated in Fig. 20(a). This high yield—as much as 108 times as efficient as single-photon desorption—apparently results from repeated excitations of the adsorbed molecule while on the excited-state PES, as shown by model calculations [90]. The desorption yields of NO shown in panels (a) through (c) were measured with pairs of pulses, with energies of 1.7 and 1.8 mJ, 1.6 and 2.0 mJ, and 1.6 and 3.0 mJ, respectively. The laser energy absorbed in the substrate heats the electron gas near the surface of the metal to peak electron temperatures of several thousand degrees. Subsequently this electronic energy transfers to the lattice ions and adsorbate degrees of freedom on a time scale of picoseconds; because the adsorbate degrees of freedom are ill matched to those of the metal lattice, it is likely that this transfer is electronic in nature. Indeed, one of the most important results is that the process of electron-hole interactions in the metal substrate is not arbitrarily fast, but apparently has a lifetime of order 650 fs.

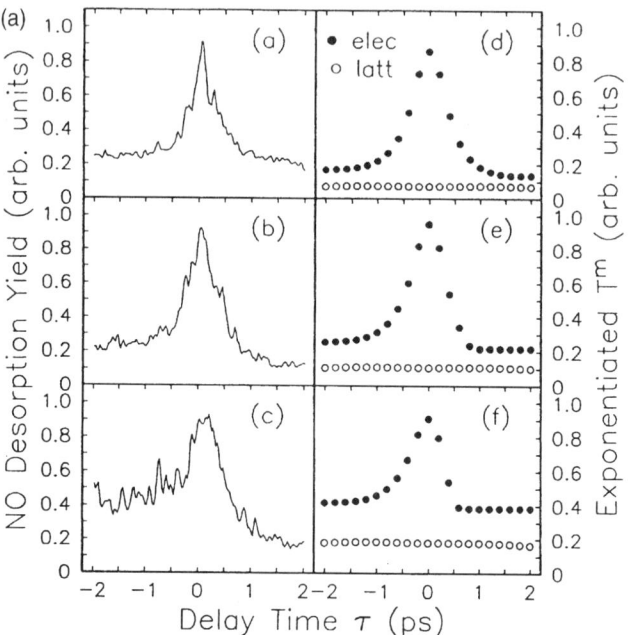

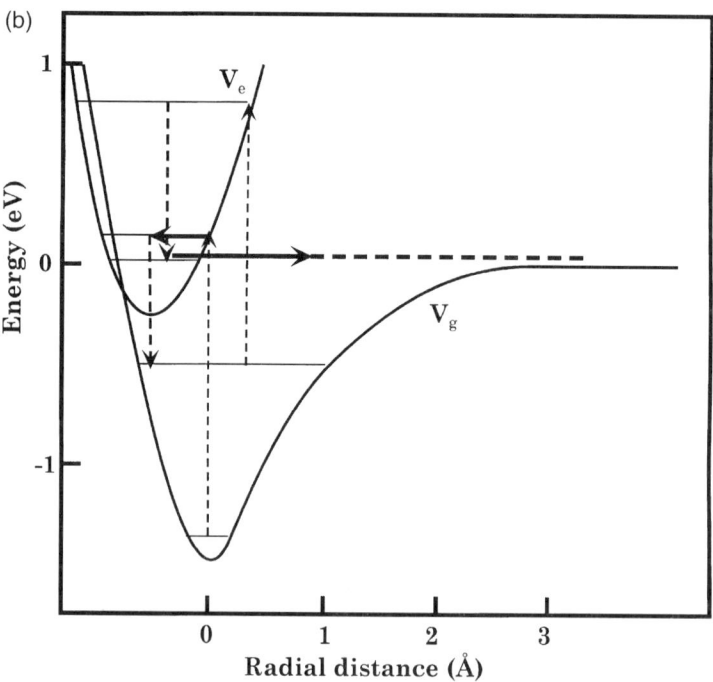

The high local density of excitation in the metal defeats the effect of the electronic collisions which would otherwise de-excite the adsorbed molecule before it gains sufficient kinetic energy to desorb from the metal surface [91]. The desorption mechanism, pictured in Fig. 20(b), thus involves a sequence of quasi-thermal but nonadiabatic excitation/de-excitation steps nicknamed DIMET (desorption induced by multiple electronic transitions). The initial excitation, simulated in the calculated excited state, pulls the adsorbate molecule in closer to the surface, as predicted by Antoniewicz [80]. If the photon flux is high enough, the molecule is re-excited each time it decays until it finally acquires an energy above the vacuum level and is able to desorb. This kind of quasithermal but nonadiabatic process in which energy is transferred from excited *electronic* states of the substrate formally violates the Born-Oppenheimer approximation in its conventional formulation, but is substantiated through an alternative but compatible theoretical description invoking electronic friction [92].

Another possibly athermal desorption phenomenon in metals is connected with the surface-plasmon resonance. Early evidence of plasmon-mediated desorption includes emission of atoms and ions from thin Al films deposited on a glass prism, excited by an totally internally reflected laser beam [93]. In these experiments, an attenuated-total-reflection geometry was used in which a thin (27 nm) Al film was illuminated at various angles from the back side by the second harmonic of a Nd:YAG laser, and desorption yields were monitored by nonresonant, multiphoton ionization time-of-flight mass spectrometry (see Section 2.8) using an excimer laser operated at a wavelength of 308 nm. The geometry (Fig. 21(a)) was arranged so that desorption was always monitored from a fresh spot on the surface of the film. Fig. 21(b) shows the comparison between the desorption yields of neutral Al atoms as a function of incident angle between the Nd:YAG laser beam and the aluminum film, and the yields obtained by illuminating the front side of the film with the Nd:YAG laser directly. The linear relationship between the desorption yield and the laser intensity (fluence) is strongly suggestive of an electronically mediated mechanism, although the fluence threshold ($210 \text{ mJ} \cdot \text{cm}^{-2}$) is rather high, corresponding to about $20 \text{ MW} \cdot \text{cm}^{-2}$. The early studies have been extended recently to Au, Al, and Ag films. Careful studies have ruled out a number of other possible effects, including shock waves, vibrational excitation leading to desorption, and

---

FIG. 20. Desorption of NO from a metallic surface induced by multiple electronic transitions from a femtosecond laser. (a) Measured NO desorption yields for different pulse energies, and calculated electron and lattice temperatures, as described in the text. (b) Schematic model for desorption induced by multiple electronic transitions. From Budde *et al.*, *Phys. Rev. Lett.* **66**, 3024 (1991) and Misewich *et al.*, *Phys. Rev. Lett.* **68**, 3737–3740 (1992).

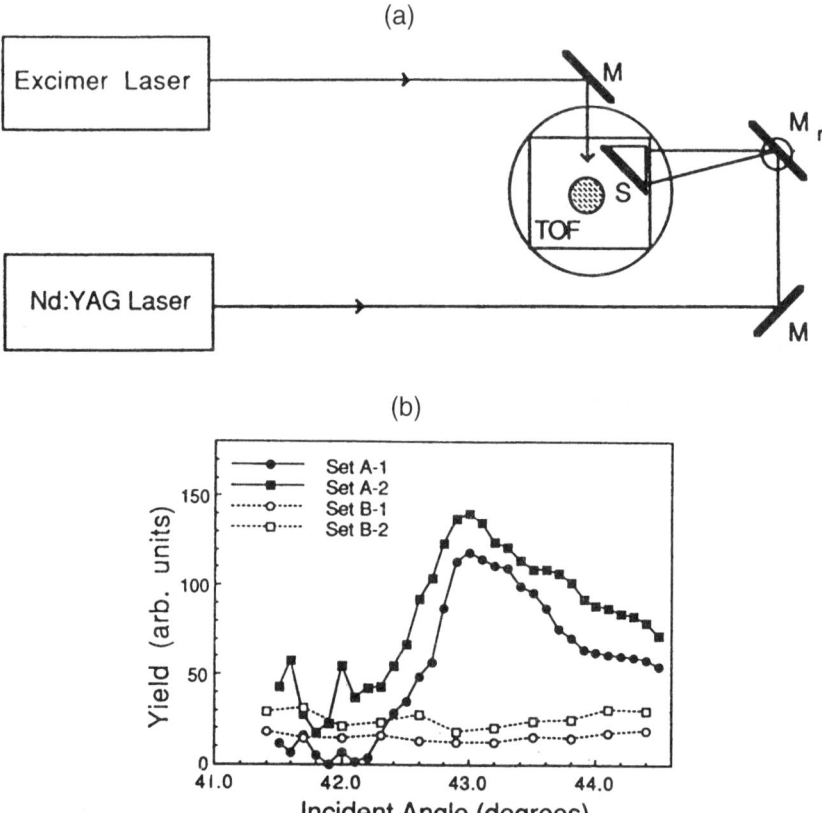

FIG. 21. Desorption induced by excitation of the surface plasmon resonance. (a) Experimental geometry; (b) comparison of the desorption yield of Al desorbed by a Nd:YAG laser under four different conditions: A-1, back-side illumination at a fluence of 0.43 J·cm^{-2}; A-2, same illumination conditions at 15% lower fluence; B-1, front-side illumination at a fluence of 0.43 J·cm^{-2}; B-2, same conditions as B-1 but at 15% lower fluence. From Lee et al., Phys. Rev. B **39**, 8012 (1989).

laser-induced thermal desorption. Because the desorption yield is consistent with the energy density at the film-vacuum interface (rather than with the total energy absorbed in the film), it is concluded that the plasmon excitation is not damped by coupling to phonon modes [94].

Perhaps the most dramatic demonstration of the critical connection between spatial localization and desorption has been shown in the desorption of Na atoms from Na clusters (diameter 5–100 nm) adsorbed on alkali-halide surfaces at by continuous laser irradiation at intensities in the mW·cm^{-2} range [95]. In these experiments, an atomic beam of Na or Ag was used to deposit atoms on a LiF crystal surface maintained in ultrahigh vacuum (pressure less than 10^{-10} mbar), mounted on a manipulator stage that could be cooled to liquid nitrogen temperatures and heated to approximately 700°K [96]. Following nucleation of the clusters at defect sites on the surface at typical area densities of 10^8–10^9 cm^{-2}, the surface was irradiated by a continuous-wave Ar or Kr ion laser at intensitives ranging from 40 mW·cm^{-2} to 160 W·cm^{-2}. Absorption spectra of the clusters were measured by sending light from a xenon high-pressure lamp through a monochromator and detecting the transmitted radiation with a photodiode. Desorbing neutral atoms were detected by a quadrupole mass spectrometer tuned to the mass of the sodium atom; the mass spectrometer signal disappears when the laser is turned off, as shown in Fig. 22(a). The most striking result of these experiments, however, is that the desorption yield is maximized at the wavelength corresponding to the surface plasmon resonance, as shown in Fig. 22(b). Once again, the desorption itself appears compatible with a Menzel-Gomer-Redhead mechanism; however, unlike the DIMET case, the energy localization in time and space needed for desorption is not supplied by multiple-photon excitation, but by the small spatial extent of the electron gas in the small metal clusters. A theoretical study has shown that the experimental results are consistent with a picture in which the surface plasmon enhances the electromagnetic field preferentially at the surface of the cluster, as in the surface photoelectric effect. Indeed, the ratio of the radial to incident electric fields for a cluster of diameter R has the form familiar from Mie theory: $E(R, \omega) = 3E_o/(\varepsilon + 2)$, which is maximal at the surface plasmon resonance. Desorption is caused by field-induced excitation from the valence band states into the antibonding states of the conduction band, reaching its maximum yield when the electric field is largest [97].

In conclusion, what we learn from the experiments on metals is that there are two possible desorption mechanisms, which are linked to free-electron heating and to interband transitions, respectively. Interband transitions, such as those involved in the surface-plasmon experiments and the cluster experiments, can cause desorption of the bare metal atoms as long as there is some localization mechanism for storing the energy for a few vibrational periods. Recent femtosecond measurements of the relaxation time of the surface plasmon resonance in small copper [36] and gold [98] nanocrystallites confirms that these relaxation times are on the order of a few picoseconds, and that they are lengthened slightly for optical transitions near the surface plasmon resonance.

2.4.3 Laser-Induced Desorption: The Case of Weak-Coupling Solids

The most sensitive measurements of laser-induced desorption from semiconductors have come from the application of resonant multiphoton ionization techniques to observe the desorption of the metal atoms from elemental and

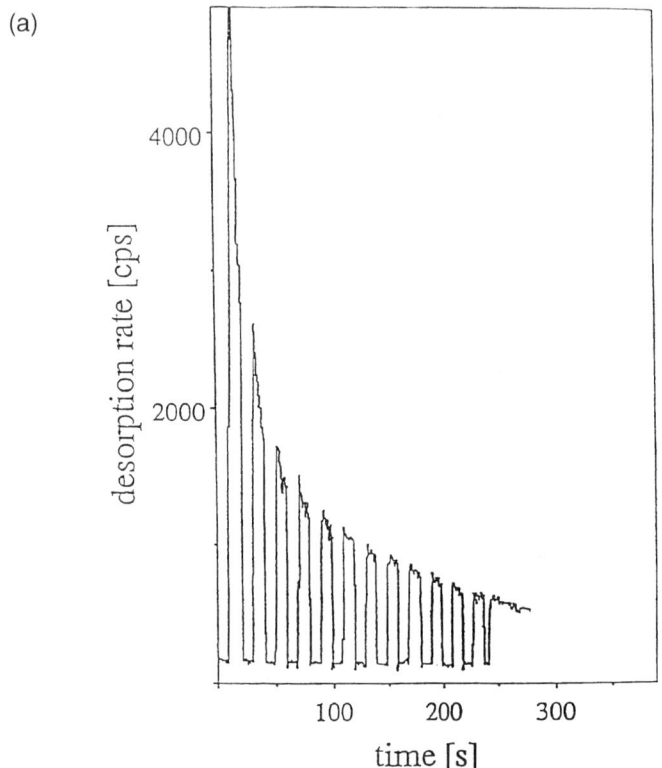

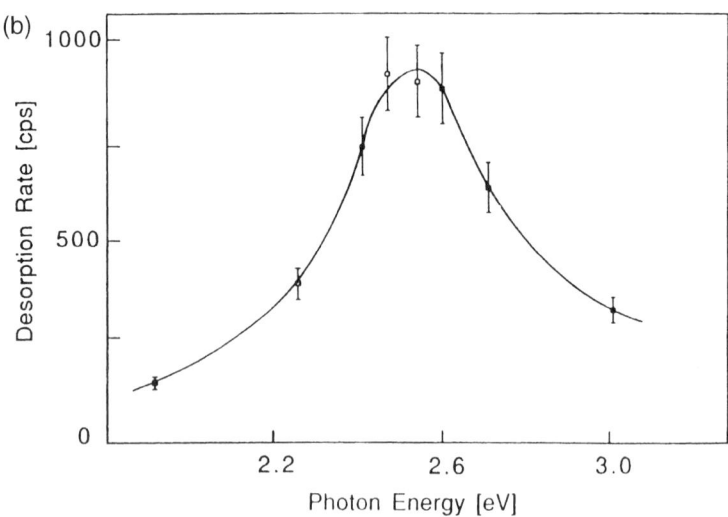

compound semiconductor surfaces [99]. These studies have revealed that in the case of compound semiconductors, the dynamical interaction of surface electron-hole pairs with surface defects—such as adatoms, steps, and weakly bound atoms near-surface vacancies—is responsible for desorption.

The earliest of these experiments involved clean GaP surfaces in ultrahigh vacuum on which an occupied surface state 0.7 eV below the conduction-band edge at room temperature was resonantly excited by a tunable, nanosecond excimer-pumped dye laser. Laser-induced Ga^o emission occurred at fluences much lower than the ablation threshold [100]; emission yields were found to be enhanced by Ar^+ irradiation and reduced by annealing, indicating that the emission was defect-related. Three distinctive types of yield versus fluence relations were observed. One component of the Ga atom yield, enhanced most noticably by the introduction of defects by Ar^+ irradiation, was quickly reduced by repeated irradiation with tens of laser pulses at the same fluence; it was identified with desorption from adatom or top sites. In addition, measurements of emission yield as a function of laser shot-number reveal another defect type from which is eliminated slowly (hundreds to thousands of pulses) upon repetitive laser irradiation; this emission was ascribed to Ga atoms eliminated from kink sites on steps. In both instances, no detectable change in mesoscale surface topography was detectable by LEED measurements, even after thousands of shots, indicating that the sputtering was in fact desorption. Yields of both components were lower for laser photons above the bandgap energy, suggesting that electron-hole pairs confined in surface states were particularly efficient in causing the Ga^o desorption [101].

In a later experiment [102], the (110) and (111) surfaces of sputter-cleaned GaAs single crystals were subjected to laser irradiation at a variety of different wavelengths. The yield of Ga neutral atoms was measured far below the laser ablation threshold as a function of the scaling parameter $\hbar\omega - E_{gap}(T)$, where the function $E_{gap}(T)$ was computed from well-known empirical formulas. The yield of Ga^o versus this scaling parameter is plotted in Fig. 23(a). When compared to the surface band structure shown in Fig. 23(b), the two sharp jumps in the yield at 0.4 and 1.10 eV are well correlated with the transitions T_1 and T_2 from occupied surface states to an unoccupied surface band and the bulk conduction band, respectively. When the temperature-dependent desorption yield of Ga^o was measured for laser pulses polarized parallel and perpendicular to the [1$\bar{1}$0] direction on the GaAs (110) surface, it was found that photons

---

FIG. 22. Desorption of Na atoms from the surface of Na clusters with a mean particle radius of 50 nm. (a) Desorption rate as a function of time while irradiated with an Ar^+ laser ($\lambda = 514$ nm) at an intensity of 113 W·cm^{-2}. (b) Desorption signal as a function of wavelength for the same cluster size, but with a constant laser intensity at each wavelength of 22 W·cm^{-2}. From Hoheisel et al., Appl. Phys. A **51**, 271 (1990).

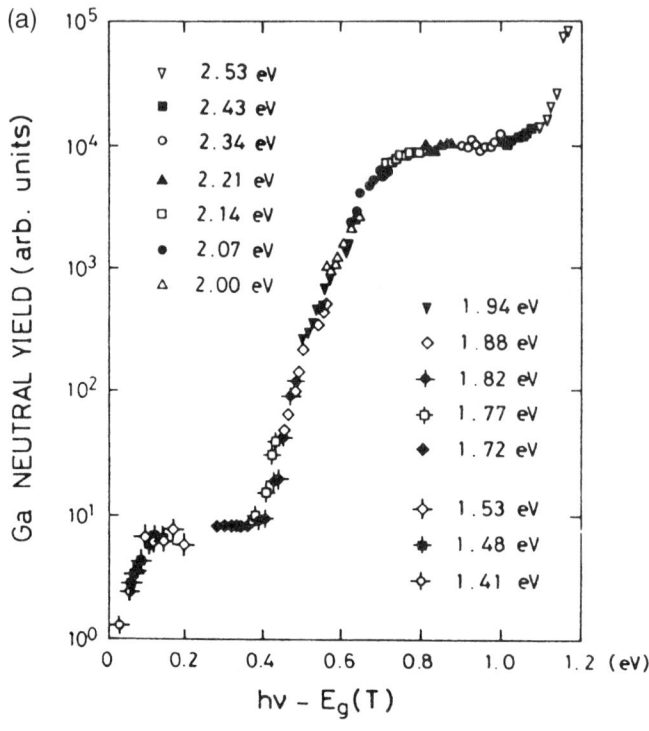

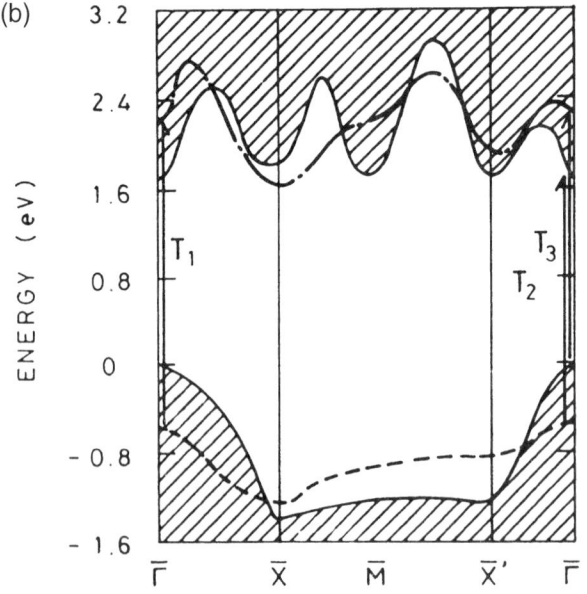

polarized parallel to that direction were much more effective in desorbing Ga° than those polarized perpendicular to that direction. This is also consistent with theoretical predictions [103] that the dipole moments of these states are strongly aligned in the T_2 surface state; however, the strong overlap of the T_2 and T_3 transitions prevented that correlation from being established unequivocally.

2.4.4 Laser-induced Desorption: The Case of Strong-Coupling Solids

In solids with strong electron-lattice coupling, self-trapping of excitons and possibly of holes provides a means of localizing the energy required to lift an electron from the valence into the conduction band and initiate an electronic transition into a relaxed excited state. Furthermore, the self-trapped excitons in the bulk are known to be converted to a variety of configurations, including closely coupled vacancy-interstitial pairs. Not much information on self-trapping near surfaces has been obtained, except that conversion from a self-trapped excitons to an emitted atom is energetically feasible. The signature for desorption should be a yield that is proportional to the density of electronic excitation, that is, to the nth power of intensity for an n-photon electronic transition from valence to conduction band. Subsequent electronic excitation leads to a metastable state from which the recovery to the ground state proceeds via a thermally activated process. Desorption of chemisorbed species from metal surfaces belong to this category; the relaxed configuration is the one from which an atom is ultimately ejected.

The first experimental evidence for photon-stimulated desorption in alkali halides was carried out using a lamp to erode the surface of KI mounted in an electron microscope [104]. Since these early experiments, it has been demonstrated that in alkali halides, ground-state atoms and ions are desorbed with a yield that is linear in the density of excitation whether that excitation is created by conventional ultraviolet sources [105], by multiphoton excitation [106], or by electron-beam irradiation [107]. The first experimental results hinting that an electronic, rather than a thermal, mechanism might be at work in laser-induced desorption and ablation of wide bandgap materials were obtained by Schmid *et al.* using four-photon excitation from a ruby laser ($\hbar\omega = 1.79$ eV) to eject Cl atoms from a KCl surface [106]. Both the yield of Cl ions, which exhibited a fourth-power dependence on intensity (Fig. 24(a)), and the apparently

---

FIG. 23. (a) Desorption yield of Ga° from the GaAs(110) 1 × 1 surface as a function of $\hbar\omega - E_g(T)$, the temperature-dependent bandgap energy, scaled such that yields at two different photon energies for the same value of $\hbar\omega - E_g(T)$ are equal. (b) Schematic band structure of the GaAs(110) 1 × 1 surface. The electronic transitions T_1 and T_2 are from an occupied surface band to unoccupied states in the surface and conduction bands, respectively. From Kanasaki *et al.*, *Phys. Rev. Lett.* **70**, 2495 (1993).

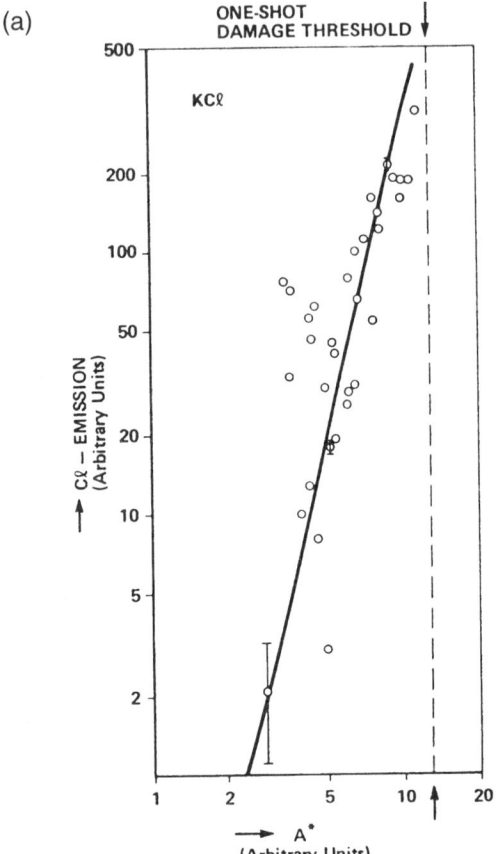

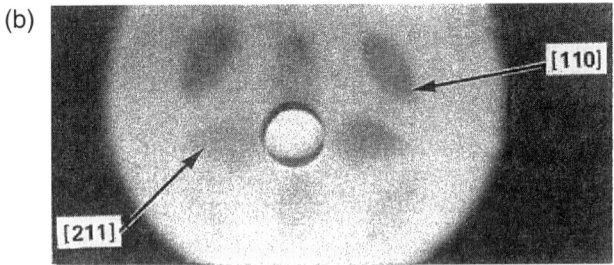

FIG. 24. Desorption of Cl neutrals from a potassium chloride single-crystal surface in high vacuum, employing a ruby laser ($\lambda = 694$ nm). (a) Yield as a function of laser intensity, showing the dependence on the fourth power of the laser intensity. (b) Spatial distribution of the Cl emission, showing the correlation with directions in the crystal surface. From Schmid et al., Phys. Rev. Lett. **35**, 1382 (1975).

directional emission pattern of the Cl atoms and ions (Fig. 24(b)) suggested a nonthermal origin, as foreshadowed in earlier studies (not using a laser) by Townsend [108].

Although all the details of the desorption mechanism are not yet completely understood, it is known that self-trapped excitons created by photoexcitation of alkali halides are transformed to a Frenkel pair (*F-center-H-center* pair) in the halogen sublattice. It is also generally accepted that halogen atom emission is the primary process; this is believed to occur by the dissociation of the *H* center (the halogen molecular ion X_2^-) into a neutral halogen atom X_o and a halogen ion X^- in the near-surface bulk. In the desorption regime, the surface is slowly decomposed by this process, gradually becoming enriched in neutral metal atoms which desorb thermally. Recent calculations using embedded-cluster Hartree-Fock techniques show that excitation of an *F* center would be sufficient to overcome the activation barrier (about 2 eV) for desorption of a Na atom; the *H* center approaching the surface is oriented along the $\langle 100 \rangle$ direction perpendicular to the surface and emits a neutral halogen atom, leaving behind a halogen ion on a lattice site [109].

Following up on the language developed in Section 2.3, Fig. 17 depicted configurational changes induced after formation of an exciton en route to emission of an atom from an alkali halide surface [110]. Evidently several configuration coordinates are involved in this process: the halogen-halogen bond length Q_1 representing the configuration for forming a halogen molecular ion X_2^-; the distance between the X_2^- molecule and the surface, described by the translational coordinate Q_2; and the decomposition mode Q_3 of X_2^- by dissociation of the *F*-center-*H*-center complex near the surface. The APES has the antibonding character typical of the MGR picture, although it involves several configurational coordinates. The APES describes emission of halogen atoms leaving halogen vacancies, resulting in a stoichiometry change [111]. An analogous configuration coordinate diagram describes laser-induced desorption in alkaline earth fluorides, in which the relaxation of excitons to self-trapped excitons and *F-H* pairs is similar, but not identical, to that in alkali halides; in this case, the STE appears to have an on-site configuration, so there is no *F-H*-center pair coordinate to be considered. Although the existence of self-trapped excitons in SiO_2 is now well established from the photothreshold for production of the 2.9 eV luminescence band [112], the fate of self-trapped excitons on a clean SiO_2 surface is not yet clear. Because self-trapping arises partly from weakening the bond on which a hole is localized, it is conceivable that an exciton near the surface leads to emission of an atom.

The basic conclusions about the mechanisms of laser-induced desorption in the strong-coupling solids have been confirmed by elegant studies of mass-selected cluster beams of positive and negative ions with the structure $(M_{n\pm1} \cdot X_n)^\pm$, where M = K, Cs and X = Br, I formed by laser photofragmentation [113].

Mass-selected beams of clusters created in a laser-ablation rare-gas jet source were introduced into an ultrahigh-vacuum chamber where they were irradiated by unfocused laser beams providing 5.0 (KrF) and 6.4 (ArF) eV photons. Neutral, positively and negatively charged clusters could be selected. The results are notable for the high efficiency of a single observed primary process, namely, the emission of a single neutral halogen atom in all cases. The one-photon processes have the form

$$(M_{n+1}X_n)^+ + \hbar v \rightarrow (M_{n+1}X_{n-1})^+ + X^o \quad \text{and}$$
$$(M_{n+1}X_n)^- + \hbar v \rightarrow (M_n X_n)^- + X^o \quad (2.25)$$

The high efficiency of the process argues for a direct excitation leading to the ejection of the halogen atom; the linear dependence of the fluence for the validates the idea that the desorption yield is linearly proportional to the density of excitation.

Figure 25 shows how the cross sections for halogen-atom photoemission from nanoclusters of CsI and KBr varies with wavelength and number of atoms in each cluster. The higher cross sections for 6.4 eV photon irradiation indicate that we are dealing with a one-electron one-hole excitation in the alkali halide

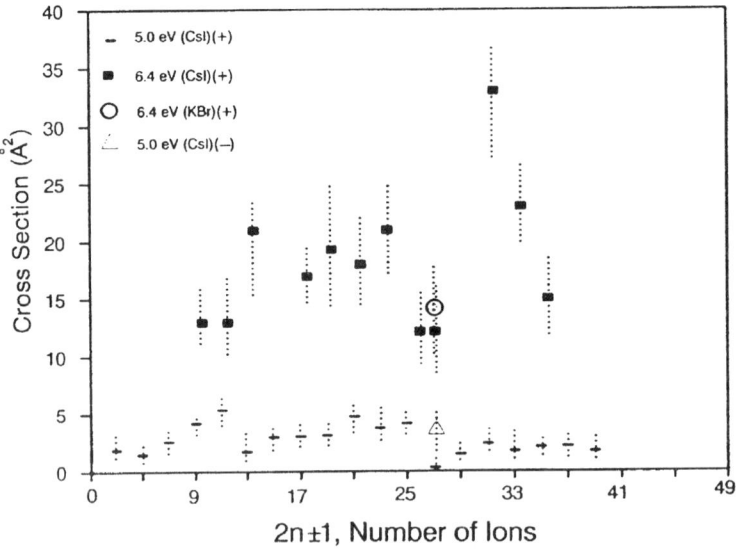

FIG. 25. Wavelength variation of the halogen-atom photoemission cross sections for cesium iodide and potassium bromide clusters illuminated at 6.4 eV and 5.0 eV, as a function of number of ions in the cluster. The cross sections at 5.0 eV are all approximately a factor 20 smaller than for the higher-energy photon. From Li et al., Phys. Rev. Lett. **68**, 3420 (1992).

system; there are also some suggestions of some "magic number" effects so often found in cluster phenomena.

2.4.5 The Role of Defects in Laser-Induced Desorption

Defects—such as interstitials, vacancies, and anti-site defects—play several important roles in laser-induced desorption. First, they change the electronic structure of the surface and near-surface bulk material by generating states in the bandgap, and hence they alter the local optical response of the solid. Second, defects represent sites that have different charge, binding, and electronic structure than perfect sites in the bulk lattice. Third, they can alter the vibrational response of the solid by generating localized modes. In any case, the alterations can have a significant effect on desorption.

Optical transitions involving mid-gap defect levels dominate the optical response for photons of sub-gap energies, as noted in the voluminous work on optical excitation of defects in the bulk [114]. Optical transitions between localized defect orbitals are often observed. The cross sections for these transitions depend on the transition matrix element between the defect ground state and excited state; for allowed transitions, they are of order 10^{-17} cm^2. Transitions from a defect-localized orbital to a continuum is also observed; in this case, optical absorption cross sections are smaller. Moreover, the vibrational frequencies and the lifetimes of defect states may be much enhanced compared to those of ordinary atoms at nondefect sites [115]. This comes about because the defect sites do not couple efficiently to the harmonic bath of phonons in the solid; hence, vibrational excitation in these states may be slow to be randomized by cooling transitions [116].

Not much information is available on the optical transitions of defects on surfaces. Excitation of defects on surfaces, including surface states, appears to be associated with desorption in gallium phosphide [117]. When an electron is trapped in an excited state of a defect that has previously trapped a hole (or vice versa), the same excited state may eventuate during the electron-hole recombination process as that produced by electronic excitation of a defect. Laser excitation of these defects leading to desorption may be a particularly effective way of studying the two-dimensional character of surface defects in insulators.

A good example of the role of surface defects, including surface states, in desorption is furnished by recent studies of laser-induced desorption of Al and AlO ions from sapphire [118, 119]. The existence of excitons on Al_2O_3 surfaces remains a matter of some disput, but these experiments may provide some positive evidence in the case. Clean and water-covered surfaces were characterized by low-energy electron diffraction (LEED), electron energy-loss spectroscopy (EELS) and temperature-programmed desorption (TPD). The reflection EELS spectra show the existence of a state about 4 eV below the bulk conduction-band minimum, which is shifted to 3.6 eV and broadened when dosed with saturation coverage of H_2O (Fig. 26(a)). This state has been identified tentatively by a number of experimenters as an intrinsic surface state.

The time-of-flight spectrum for Al atoms desorbed by 355-nm laser light, detected in a quadrupole mass spectrometer, is shown in Fig. 26(b). The main argument in favor of the involvement of an energetic surface process is the fact that the best-fit Boltzmann distribution corresponds to a very hot atomic emission process, with an ensemble-average energy of $\langle E_k \rangle = 7.0$ eV, far too energetic to be accounted for by a thermal vaporization process. However, one should not assume from this that a surface exciton is involved, since the dosing with water evidently changes the surface condition from the clean 2×1 surface reconstruction. Clearly, the crystalline sapphire surface will deserve further study.

Another common way of studying surface states is by dosing the surface with adsorbates to probe the electronic structure in the vicinity of sites where chemisorption or physisorption occurs. GaP(110) surfaces in ultrahigh vacuum were dosed with submonolayer quantities of oxygen [121] and bromine [120], and desorption of Ga° was initiated by light from an excimer-pumped nanosecond dye laser tuned to excite the occupied surface state at an excitation energy of 2.07 eV, at a fluence of $0.8 \text{J} \cdot \text{cm}^{-2}$, well below the ablation threshold. Figure 27 shows the relative yields as a function of number of laser pulses on a given site for clean and oxygen-dosed GaP surfaces. The adsorbed oxygen is evidently tying up Ga°, probably at step sites by forming a Ga-O-P bridging bond because O is unlikely to bond to Ga at a top site. Similar results are obtained for dosing with Br on GaAs. The fact that adsorbates weaken the Ga at step sites implies that repeated cycles of gas dosing and laser-induced desorption could actually eliminate this particular type of defect.

These measurements show that laser-induced desorption at high sensitivity opens the way for studying the dynamics of the surface states as they respond to optical excitation. That this information can be obtained with tunable nanosecond lasers is also important because it shows that where there is no heating of the substrate, information about electronic transitions on the surface can be obtained readily even with long pulse durations because there is essentially no communication with the bulk substrate. Possible technological implications of this work for the atom-scale planarization of compound semiconductor surfaces over large areas have been explored [122].

2.5 Laser Ablation

Laser ablation offers sharp contrasts to laser-induced desorption in almost every respect: where the latter involves only individual atomic sites, the former is a collective phenomenon that touches many atomic sites; ablation is a

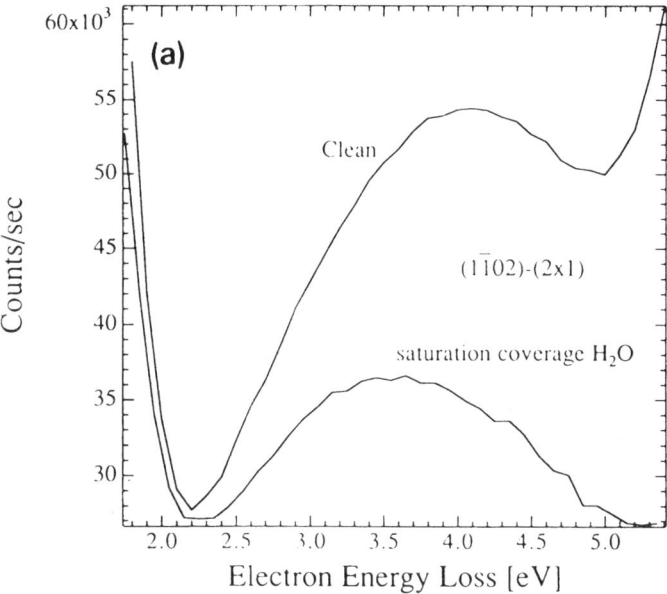
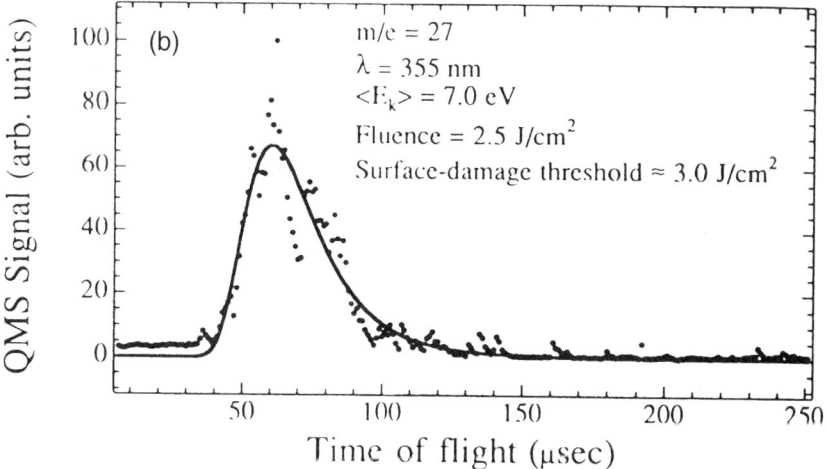

FIG. 26. Studies of the role of defects in desorption of Al and AlO ions from clean and water-covered Al_2O_3 (1102) surfaces in ultrahigh vacuum. (a) Reflection electron-energy-loss spectra for the clean (2 × 1) reconstructed surface and the water-covered surface, showing the shift in the surface state. (b) Time-of-flight spectra for aluminum ions, together with a modified Maxwell–Boltzmann fit to the data, for irradiation with the 355 nm third harmonic of a Nd:YAG laser at a fluence of 2.5 J·cm^{-2}. From Schildbach and Hamza, *Phys. Rev. B* **45**, 6197 (1992), and *Surf. Sci.* **282**, 306 (1993).

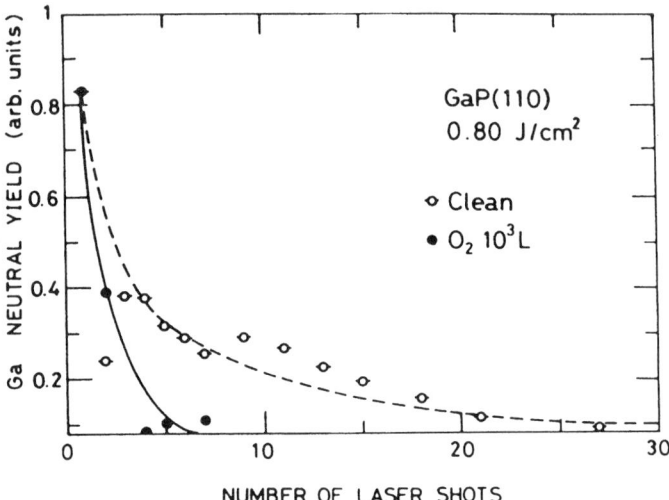

FIG. 27. Comparison of Ga° yields for clean and oxygen dosed GaP(11) surfaces at the indicated fluence. The desorption laser was tuned to 600 nm; it had a pulse duration of 28 ns and the focal-plane fluence was 0.8 J·cm^{-2}. From Kanasaki et al., Surf. Sci. Lett. **257**, L642 (1991).

threshold phenomenon accompanied by the formation of a dense plume of gas, and gas dynamics plays an important role. Because of the collective nature of the ablation process, we shall be looking for excitations that have a collective character rather than the single- or perhaps few-particle excitations we considered under the heading of laser-induced desorption. Most importantly, laser ablation is a mesoscale phenomenon, involving properties of the material on a scale large enough that bulk properties—elasticity, compressibility, and so on—begin to play a role. However, these properties must be seen in the context of the particular excitation and relaxation mechanisms that initiate ablation, together with the appropriate time and distance scales. Laser ablation refers to a high-yield photon sputtering process exhibiting all of the following characteristics: (1) material removal, not infrequently nonstoichiometric, at rates ranging from a fraction of a monolayer to a few monolayers per pulse; (2) a total yield of ejected particles that is proportional to a high power of the density of electronic or vibrational excitation; and (3) a threshold photon fluence or intensity below which only laser-induced desorption, without destruction of the surface, is observed. For sufficiently large removal rates, typically a tenth monolayer or more per pulse [123], formation of a plume above the surface is observed. This plume exhibits extremely complex gas-dynamic [124] and/or plasma behavior, and its presence differentiates laser sputtering from sputtering by energetic particle beams. The formation and dynamics of the laser-ablation plume is described in much greater detail in Chapter 5.

In addition to the nanosecond Nd:YAG, excimer and CO_2 lasers that have been most widely used in ablation experiments in the past, two new types of lasers have appeared on the scene which bring new capabilities to the study of laser ablation phenomena: the near infrared variable-pulse-duration chirped pulse amplifier systems (CPA) with pulse durations from 200 fs to 2–5 ns [125]; and picosecond infrared free-electron lasers covering wavelengths in the 2–10 μm range [126]. Here we review a few of the most interesting results in laser ablation, emphasizing those features that highlight the role of mesoscale parameters in ablation phenomenology.

2.5.1 Characteristics of Bond-Breaking at the Mesoscale

As we have seen, desorption is a localized event triggered by a lattice instability that must be initiated before the incident photon energy is dissipated into nonlocal vibrational modes of the solid. As depicted schematically in Fig. 28, desorption is associated with a characteristic length scale $d_{micro} = \tau_{vibr} \cdot v_s$, where τ_{vibr} is the inverse of a typical phonon frequency and v_s is the sound speed in the material. This is essentially the length scale over which electron-phonon coupling heats the surrounding material following absorption of a laser photon. Laser ablation, in contrast, is associated with a characteristic *mesoscopic* length scale of order $d_{meso} \cong \tau_{laser} \cdot v_s$, where τ_{laser} is the duration of the laser pulse; for ultrashort laser pulses, the relevant length scale may be defined as $d_{meso} \cong \tau_{ion} \cdot v_s$, where τ_{ion} is a lattice equilibration time. This choice reflects the idea that the excitation, thermalization, and lattice instability that drive ablation are "pumped" by the laser pulse for its entire duration. This length ranges up to a few tens of microns in some cases, consistent with the idea that ablation implies the excitation and removal of a substantial fraction of a monolayer, a volume of order $30 \cdot 10^4$ μm^3, or few tens of microns in linear dimensions. These size scales are consistent with typical diffusion lengths associated with the thermal response to laser ablation because diffusion constants have a range dictated by density and sound speed in the material. This length scale is also one on which the thermal and mechanical properties of materials can be said to be meaningful; the bond-orbital theory makes a straightforward connection between the microscopic and macroscopic length scales.

2.5.2 Laser Ablation of Metals

Laser ablation of metals furnishes important illustrations of the basic mechanisms of energy deposition, localization, and transformation at the mesoscale. In metals, we are in the lower right corner of the diagram of Fig. 8, with maximum metallicity and vanishing ionicity—that is, essentially no charge transfer in the bonds at all, and with electrons providing both the binding and the principal means of dissipating incident laser energy through the thermal conductivity. Thus no energy localization—except possibly at defect sites—is possible: all laser ablation of metals must occur through vaporization, perhaps, in the case of ultrafast lasers, without a melt or liquid phase.

Laser ablation of metals was studied soon after the development of the laser; many of the most important concepts were already developed by the early 1970s [127]. Irradiation by laser light produces an extremely high density of excitation in metals within a skin depth (~50–100 nm) of the surface. At the same time, electron-lattice coupling is essentially zero: the almost complete long-range dielectric screening of the ions by the electrons precludes loss mechanisms other than electron scattering from lattice ions until the electron energy is completely

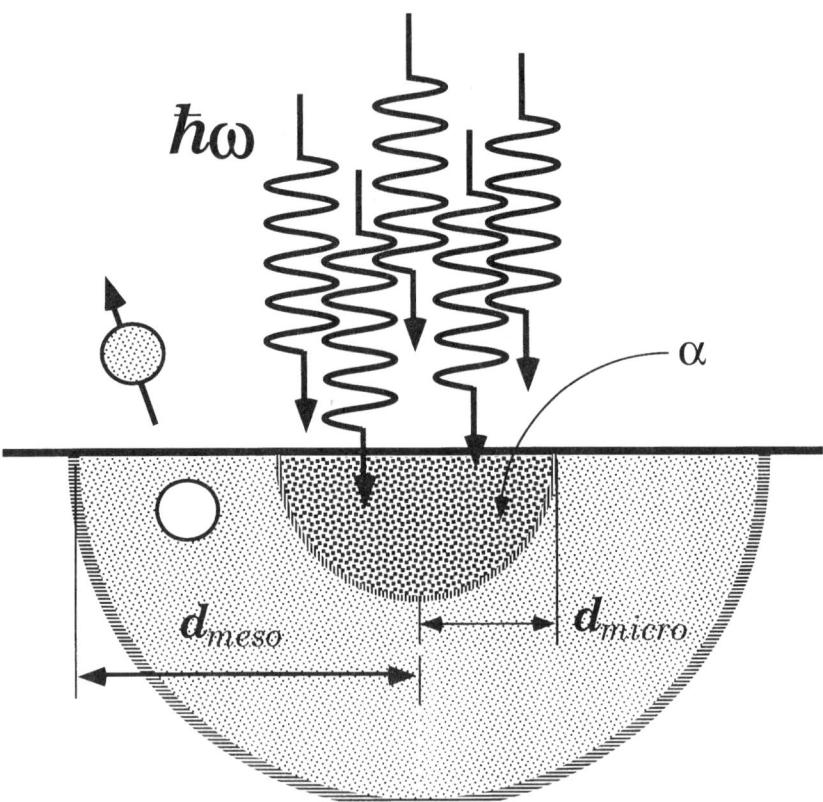

FIG. 28. Microscale versus mesoscale regimes in laser-induced desorption and ablation. The characteristic scale lengths d_{meso} and d_{micro} are discussed in the text.

thermalized. Electron-phonon collision times are on the order of 10–20 fs, but the collisions are relatively inefficient because of the momentum mismatch between electrons and phonons. Hence the energy tends to be thermalized on a time scale of up to a few picoseconds [128]. With the exception of the experiments described in Section 2.4, it appears that laser-induced desorption in the absence of defects does not occur in metals; this is primarily because the metallic bond has no mechanism for localizing energy. The relaxation of the initial photoexcitation in metals is described by an *electron-phonon coupling* constant, the quantity G appearing in the coupled equations for a one-dimensional, two-temperature model of a laser-heated system coupled to a heat bath by diffusion [129]:

$$C_e(T_e) \cdot \frac{\partial T_e}{\partial t} = \nabla \cdot (K\nabla T_e) - G(T_e - T_i) - S(r, t) \quad \text{and}$$

$$C_i(T_i) \cdot \frac{\partial T_i}{\partial t} = G(T_e - T_i)$$

(2.26)

Here the subscripts e and l refer to the electron and lattice, respectively, the T_j are temperatures, the C_j are heat capacities, and $S(r, t)$ is the source (laser) term. This equation has three interesting limits, depending on the ratio of cooling times for the electrons (τ_e) and ions (τ_i) to the duration of the laser pulse τ_L.

In a recent experiment [130], a commercial femtosecond Ti : sapphire system with a chirped-pulse amplifier provided laser pulses at 780 and 390 nm with pulse energies up to 100 mJ, at pulse durations ranging from 200 fs to 400 ps. Pulse durations of 3–5 ns were also obtained by seeding a conventional regenerative amplifier with pulses from the femtosecond oscillator. Pulse durations were measured by standard second-order autocorrelation techniques, streak cameras, or fast photodiodes. Metal targets in vacuum were irradiated by 10^4 laser pulses with durations in the fs, ps, and ns regimes. The results are shown in Fig. 29.

According to Eq. (2.26), in the limit in which $\tau_L \leq \tau_e \leq \tau_i$, the electrons are not in thermal equilibrium with the lattice when the laser pulse ends, and electron-phonon interactions occurs until equilibrium is reached. Ablation occurs when the energy deposited in the delocalized phonon modes exceeds the critical thermodynamic temperature T_{tc} at which a phase transformation from solid or liquid to the gas phase occurs instantaneously on the time scale of the laser pulse. After the laser pulse turns off, the electrons cool off exponentially until the electron and ion temperatures are in equilibrium. If the laser fluence satisfies the inequality $F_a \geq (\rho\Omega_{evap}/\alpha) \cdot \exp(\alpha z)$, the condition for instantaneous evaporation without passing through a melting phase is satisfied (Fig. 29(a)). The quantity in parentheses can be identified as the threshold fluence for ablation. However, for irradiation with a great many pulses, hydrodynamic instabilities in the melted phase begin to show up in the form of substructure in the bottom of the laser ablation pits.

For picosecond pulses, $\tau_e \leq \tau_L \leq \tau_i$. In this case, the electrons are already beginning to thermalize during the laser pulse, and the morphology of the ablated surface is determined by a competition among melting, vaporization, and solidification. Fig. 29(b) shows how the vaporization process generates a recoil pressure and a wavelike structure radiating away from the ablation zone; the structures in the ablation zone arise from the rapid melting and solidification in competition with hydrodynamic instabilities in the melt front. In this time regime, the electron temperature becomes quasi-stationary, and the ion temperature is increasing linearly with time. Remarkably the electron cooling time again determines the attained lattice temperature, and the value of the threshold

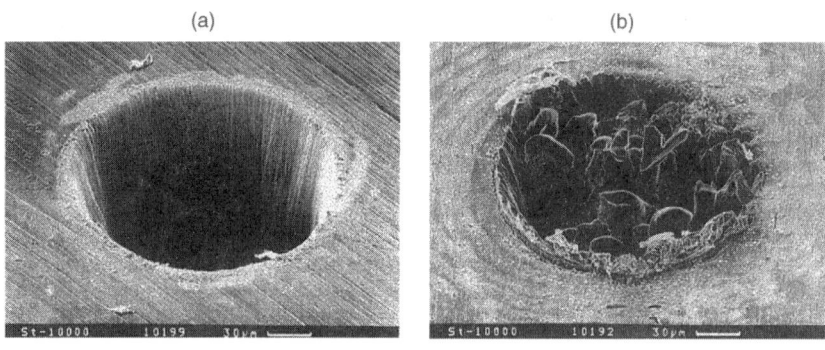

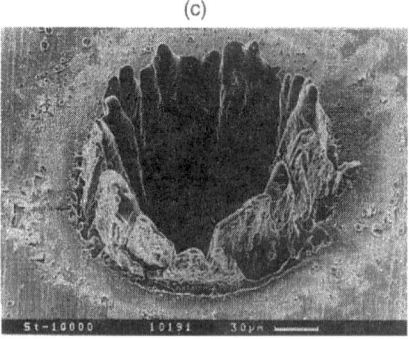

FIG. 29. Electron micrographs showing laser ablation of stainless steel from a 100 μm-thick foil for various conditions. The CPA laser wavelength was 780 nm in all cases. (a) Pulse duration 200 fs, fluence 0.5 J·cm^{-2}, energy 120 μJ per pulse. (b) Pulse duration 80 ps, fluence 3.7 J·cm^{-2}, energy 900 μJ per pulse. (c) Pulse duration 3.3 ns, fluence 4.2 J·cm^{-2}, pulse energy 1 mJ. From Chichkov et al., Appl. Phys. A **63**, 109 (1996). The length bar is 30 μm in all micrographs.

fluence for "strong evaporation" is the same as for femtosecond pulses. This should not obscure the fact that the physics is quite different from the femtosecond time scale.

Finally, in the limit of nanosecond pulses, there is a relatively large melt front, and sputtered material appears in both vapor and liquid phases. In addition, a substantial thermal wave is propagated into the material adjacent to the volume where the laser light is being absorbed. The combination of these two effects produces both the corona around the etch pit seen in Fig. 29(c) and the appearance of the columnar material, which appears to have been "frozen" in the midst of being ejected from the surface. The major correction to this treatment is including gas-dynamic effects at the vapor-liquid interface, which has been demonstrated in a number of papers [129].

An open question is whether or not ablation measurements can actually tell us anything about such fundamental quantities as the electron-phonon coupling constant G in Eq. (2.26). Corkum *et al.* measured the surface melting thresholds using a variable-pulse-duration CO_2 laser ($\lambda = 9.3$ μm) for pulse durations ranging from 100 ns to 2.5 ps, from which they estimated both the relaxation time τ_c and the electron-phonon coupling constants of bulk copper and molybdenum [131]. As criteria for surface melting, the experimenters used the appearance of a damaged spot following multiple-shot irradiation (evidence of gross material removal) and the appearance of a visible spark (plasma formation), which fit the criteria for a laser-ablation threshold. The threshold fluence for damage was found to be independent of pulse duration below about 1 ns. The use of the coupled rate equations (2.26) to extract the electron-phonon coupling constant G gives values of G much smaller than those measured for thin films: 10^{10} W·cm^{-3}·K^{-1} for Cu and $2 \cdot 10^{10}$ W·cm^{-3}·K^{-1} for Mo. There has been some debate about the origin of these differences [131]; it has been suggested that G depends on the precise morphology of a given sample. However, the reduced value of G suggests that ablation may be associated with reduced coupling efficiency to the phonon bath.

In most discussions about the mechanisms of laser ablation, it is assumed that, whatever difficulties there may be in handling ablation from nonmetallic surfaces, at least in the case of metals ablated by *nanosecond* pulsed lasers, the mechanism is simple vaporization at the melt front. More sophisticated measurements, however, show that the competition between vaporization and hydrodynamic sputtering is more complex than it appears in most modeling. In two recent studies of Au sputtering by a nanosecond-pulsed KrF laser ($\lambda = 248$ nm) [132], the Au surface developed a striking surface topography typical of a Rayleigh-Taylor instability with a periodicity around 5 μm, and micron-size droplets developed on the surface as a result of the competition between inertial and surface-tension forces. See Fig. 30(a). However, the expected thermal behavior of ablated Au atoms—with a Maxwell-Boltzmann distribution at a

(a)

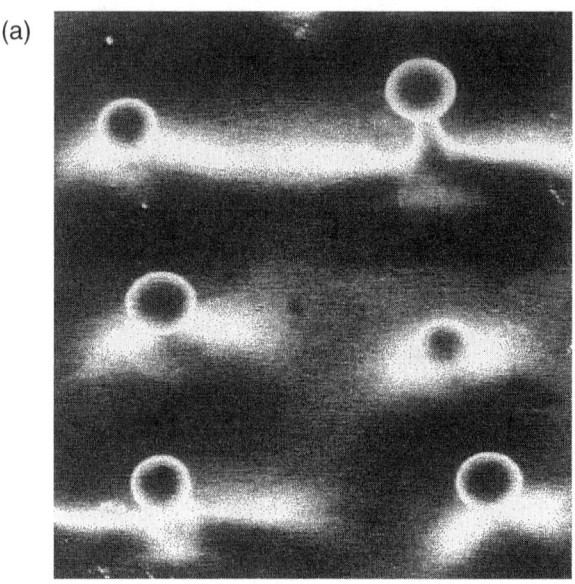

4 μm

(b)

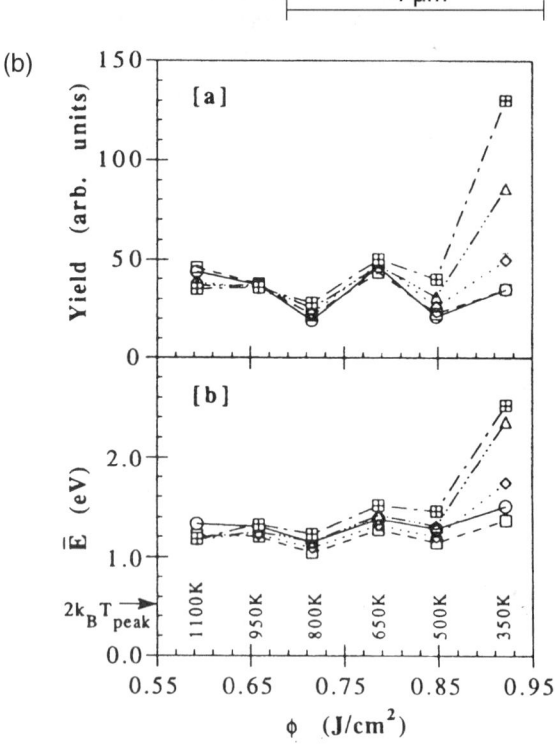

mean temperature of the surface melting and with a Knudsen-layer cosine angular distribution—was not observed. Instead, the direction of emission was strongly forward peaked even at pressures such that there could have been no collisional effects, and at densities such that there was no significant ion yield. The ablated Au atoms had a Maxwell-Boltzmann velocity distribution; however, mean translational energies ranged from 4.1 eV (for a fresh spot) to 6 eV (for a site conditioned by several hundred shots), far above any reasonable value for a thermal process. See Fig. 30(b).

This result is at variance with an earlier, equally painstaking, study [133] of excimer laser ablation of Cu, which appeared to show that the ablated Cu^o had translational velocities consistent with the surface melting temperature (2600°K) provided that suitable corrections for laser heating of the ablation plume were made. By simultaneously monitoring Cu^o, Cu^+, and Cu_2 in the plume by laser-induced fluorescence, it was shown that the very hot Cu^+ was created by multiphoton ionization of Cu^o in the plume as well as by electron-impact ionization of the Cu_2. Even though surface-second-harmonic generation on silver appear to corroborate the idea that laser interactions at ns time scales are thermal, this discrepancy continues to bedevil ablation experiments [134]. In the Cu ablation experiments, neither the spatial and temporal characteristics of the laser beam, nor the surface itself, were well characterized. Another attempt to separate thermal, topographical and electronic effects in nanosecond KrF-laser ablation of Au employed atomic-force microscopy and careful spatial beam shaping and characterization [135]. The result? Once again, the measured Maxwell-Boltzmann distributions correspond to temperatures much higher than surface temperatures calculated from standard heat-transfer models. Hence the thermal model may still need refinement.

A number of experiments have provided evidence for laser ablation by excitation of the surface plasmon resonance or related surface roughness-enhancement. For example, Shea and Compton reported positive ion distributions from roughened Ag with both thermal (~0.5 eV) and nonthermal (typical energies 3.6eV) ion groups [136]. Quantum mechanical calculations [137] suggested that the evaporative decay by ion emission of a surface plasmon should produce an ion yield increasing with the third power of the laser intensity, in general agreement with the experiments; the relative ion-production

---

FIG. 30. Nanosecond ablation experiments on well-characterized Au surfaces using a KrF laser ($\lambda = 248$ nm). (a) Micron-size droplets generated at crests of a Rayleigh-Taylor instability in the Au surface following irradiation at 248 nm. (b) Average yields [upper graph] and mean translational energies [lower graph] of Au atoms at six different combinations of target temperature and fluence chosen such that the surface temperature was 2700°K in all cases. From Bennett et al., J. Appl. Phys. **77**, 849 (1995); Phys. Rev. Lett. **76**, 1659 (1996).

efficiency—roughly 10^{-8} ions per incident photon—was also in reasonable agreement with these calculations. However, the experiments on rough surfaces by Bennett et al. [135] described earlier, although qualitatively consistent with some kind of plasmon excitation, still show ion energies far too high to be consistent with the model for ion desorption put forward by Ritchie et al. Helvajian and Welle [138] observed Ag ions ablated by 351 nm and 248 nm laser light from clean, high-purity, single-crystal surfaces with kinetic energies approaching 9 ± 1 eV, corresponding to temperatures far above the critical thermodynamic temperature T_{tc}. Although only a slight wavelength dependence was evident in the ablation threshold, the ion yield showed an intensity dependence of I^{13} for 248-nm laser irradiation and I^{18} for 351 nm irradiation. Similar high-energy-ion ejection occurring from Al and W(100) surfaces was observed by the same group, and it was specifically noted that the observed broadening of the kinetic-energy distribution was not due either to space-charge or to plasma effects [139]. On tungsten and aluminum surfaces, the threshold for ion ejection was lowered by adsorption of oxygen, hydrogen, and fluorine, with adsorbates evidently serving to localize the photoexcitation. This mechanism is likely to be especially effective at surface defect sites, such as vacancies or steps.

In laser ablation from metals, the key element is the competition between thermal diffusion and vaporization; the relevant physical parameters are the diffusion length $L_{diff} = \sqrt{2\kappa\tau_{Laser}}$ and the absorption depth. Relatively few ultrashort-pulse ablation experiments on metals have measured the characteristics of the ablation plume. However, a measurement of 0.5-ps KrF laser ablation of thin (0.1, 0.3, 0.6, and 1.0 μm) nickel films, in which the Ni^{58} atom yield was measured by nonresonant multiphoton mass spectrometry, showed results consistent with the electron micrographs [140]. The fluence threshold was dramatically reduced by comparison with nanosecond laser measurements on similar films, and the ablation threshold was independent of film thickness, suggesting that minimal diffusion of heat to the region surrounding the absorption zone. Although some photoemission is competing for the incident laser energy, the evaporative process clearly began after the laser pulse was turned off.

2.5.3 Phase Transitions and Shearing Constants: Laser Ablation in Semiconductors

In early experiments in which a nanosecond-pulse nitrogen-pumped dye laser was used to irradiate GaP, it was found that at wavelengths below the bulk band edge (at which the *bulk* optical absorption was too small to induce significant heating), there was a significant yield of P atoms as monitored in a quadrupole mass spectrometer. The yield was highly nonlinear as a function of fluence, and the fluence threshold increased by almost an order of magnitude as laser wavelengths shifted toward the red end of the spectrum [141]. Later it was found that low-energy electron-diffraction (LEED) patterns were changed from those corresponding to the clean (17 × 17) surface to a (1 × 1) structure as the fluence increased; again, this change required higher fluences for wavelengths below the bulk band edge. The yield at threshold increases more rapidly with laser fluence in compound semiconductors than in alkali halides. The ablation threshold was determined from the fluence dependence of the yield and by observation of a modified LEED pattern [142]. The extent of damage to the ablated surface was monitored by changes in stoichiometry—increasing metallization of the surface—deduced from scanning electron microscopy (SEM) and from Auger electron spectroscopy (AES). The ablation threshold was found to depend only slightly on the number of laser pulses, so that incubation effects did not appear to be important.

The tendency for the compound-semiconductor surface to metallize under laser ablation seems to be quite general. Long et al. oberved surface decomposition in GaAs by 5-ns high-repetition-rate (6 kHz) pulses from a copper-vapor laser ($\lambda = 510$ nm) irradiation at fluences as low as 1 mJ·cm^{-2} [143]. Even though this wavelength is above the bandgap for GaAs, the electronic character of the process is assured because the calculated temperature rise due to photon absorption is less than 13 K. Time-resolved photoemission measurements and electron micrographs show the growth of Ga islands with diameters as large as 100 Å, marking the surface decomposition, as seen in the photo inset of Fig. 31. Experiments by Brewer have shown that CdTe surface composition can be modified reversibly by excimer (KrF, $\lambda = 248$ nm) laser irradiation [144]. The experiments were carried out in high vacuum, with removal rates ranged from less than 1 Å to approximately 100 Å/pulse. Above a fluence threshold of 40 mJ·cm^{-2}, a metal-enriched surface is observed; below this threshold, stoichiometry was restored by laser irradiation. The velocity distributions of the Cd, Te, and Te$_2$ desorption products are all thermal in this regime. The experimenters explain their results as the competition between two thermally activated processes, under the assumption that the kinetics of Cd desorption is linked to the kinetics of Te$_2$ formation.

The question of the ablation threshold thus looms large in explaining the metallization of the surface, whether the threshold is viewed as the trigger for an electronic mechanism or as the energy needed to overcome an activation barrier. Measured ablation thresholds of both elemental and compound semiconductors show a decreasing threshold with increasing ionicity. In view of the close correlation between ionicity and bond strength [145], the data suggest that the bond strength or cohesive energy plays a role in ablation. Indeed, Ichige et al. [146] suggested that ablation was triggered by a phase change brought on by dimerization of the tetrahedral structure of the compound semiconductors to produce a nearly cubic structure much more vulnerable to dissociation. Most of

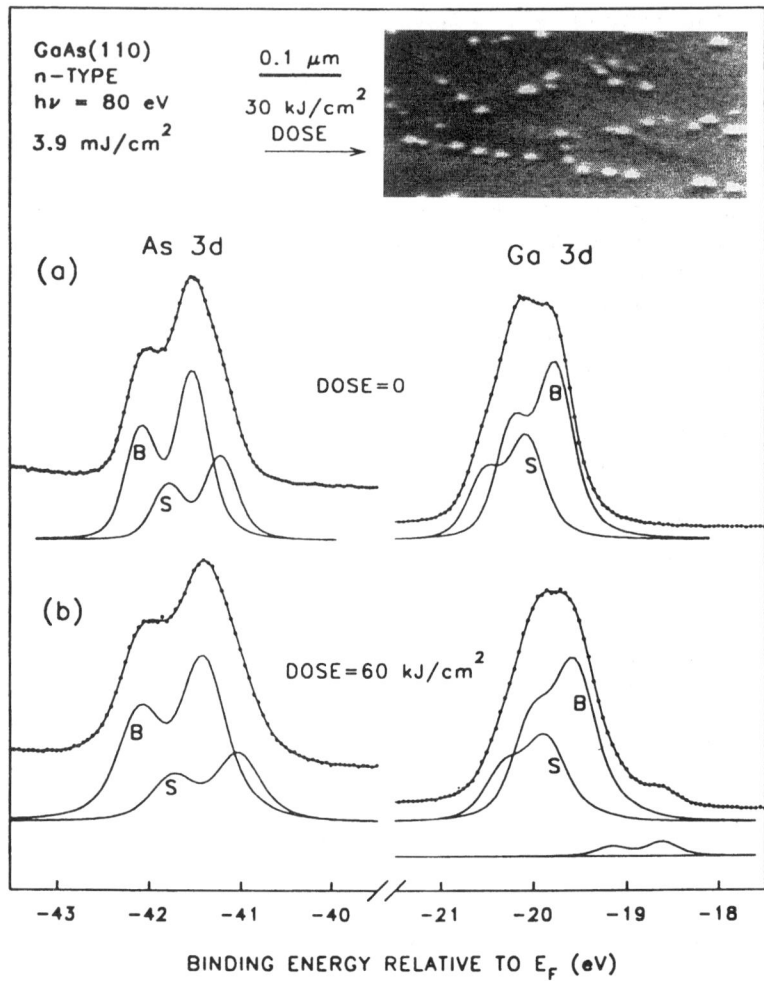

FIG. 31. Photoemission spectra [(a) and (b)] and electron micrograph showing the evolution of Ga islands on the surface of GaAs(11) during laser irradiation with a high-repetition-rate Cu-vapor laser. The spectra measured before (a) and after (b) the laser irradiation comprise bulk (B) and surface (S) core-level features. Evidence for laser-induced surface decomposition is the small shoulder in the Ga-3d difference spectrum at lower right. From Long *et al.*, *Phys. Rev. Lett.* **68**, 1014 (1992).

the measurements of threshold fluences were made with a nitrogen laser operating at 337 nm, well above the bandgap energy for any of the species studied; therefore band-to-band transitions and high density of electronic excitation can be inferred. One might think that, if this is a structural phase change, the thresholds should scale with the ionicity or polarity of the bond. When the ablation thresholds are plotted against the value of the polarity α_p, basically proportional to the bond charge, however, the trend to lower thresholds with greater polarity is confirmed, although the scaling remains unconvincing [147, 148].

Realistic scaling is more likely to come from identifying the forces on the lattice constituents introduced by electron-lattice (electron-phonon) coupling. On the microscopic length scale, the most probable distortion of the lattice is not lengthening of the bonds but shearing in which the rigid hybrids are twisted about a central axis, leaving the volume unchanged. The bond-orbital theory can be used to compute mesoscale thermoelastic properties of use in ablation studies. When two hybrid bonds have been distorted by an angle ϑ, the change in the hybrid covalent energy for small misalignment can be easily estimated to be $\delta V_2 = -\lambda V_2 \vartheta^2$, where the parameter λ—an angular force constant—can be obtained from the atomic wave functions. By equating the energy change $\delta \varepsilon_{mis} = 2\lambda V_2 \alpha_c^3 \Delta^2$ arising from the misalignment of the hybrid bond orbitals to the mesoscopic elastic shear energy $\delta \varepsilon_{shear} = (a^3/32)(c_{11} - c_{12})\Delta^2$, and solving for the angular force constant in terms of the atomic orbitals, we obtain

$$(c_{11} - c_{12}) = \frac{\sqrt{3}\,\lambda \alpha_c^3 V_2}{d^3} = \frac{\sqrt{3}\,\alpha_c^3 V_2}{d^3} \cdot \frac{3 V_{pp\pi} - 3 V_{pp\sigma} + \sqrt{3}\, V_{sp\sigma}}{V_{ss\sigma} - 2\sqrt{3}\, V_{sp\sigma} - 3 V_{pp\pi}} \qquad (2.27)$$

The shear constant derived in this way has a mesoscopic meaning because it stems from the mechanics of the solid viewed as an elastic medium [149]. However, it also is clearly linked to bond parameters at the microscopic scale, as evidenced by the appearance of the interatomic potentials V_{ijk}.

In Fig. 32, the threshold data are plotted proportional to the shearing constant, scaling simply as the cube of the covalency in accordance with Eq. (2.6). In this case, the threshold data seem to scale almost linearly, with the exception of ZnO and CdSe. Interestingly these "outlier" specimens have wurtzite rather than zincblende structures, and so would be expected to have quite different mechanical properties (including shearing coefficients) than those of the remaining materials. Also, if one looks at the three points for GaN, GaP, and GaAs, one sees quite exact scaling with the shearing constant. Similarly good scaling has been reported with cohesive energy, an agreement that is not surprising. Although these results are only qualitative correlations, there is certainly reason to pursue the idea that within families of related materials, the ablation threshold is related to the shear constant and/or the cohesive energy.

In the ultrashort-pulse ablation domain, recent experiments on Si and AlN show a remarkable persistence of thermal phenomena, even at pulse durations of 250–500 fs [130, 150]. As shown in Fig. 33, even for a single laser pulse at 390 nm, there are clear signs of melting and refreezing in the laser focal spot. In panel (b) of Fig. 33, one sees signs of thermal waves propagating outward from the laser spot, and in panels (c) and (d), there is clear evidence of redeposition of

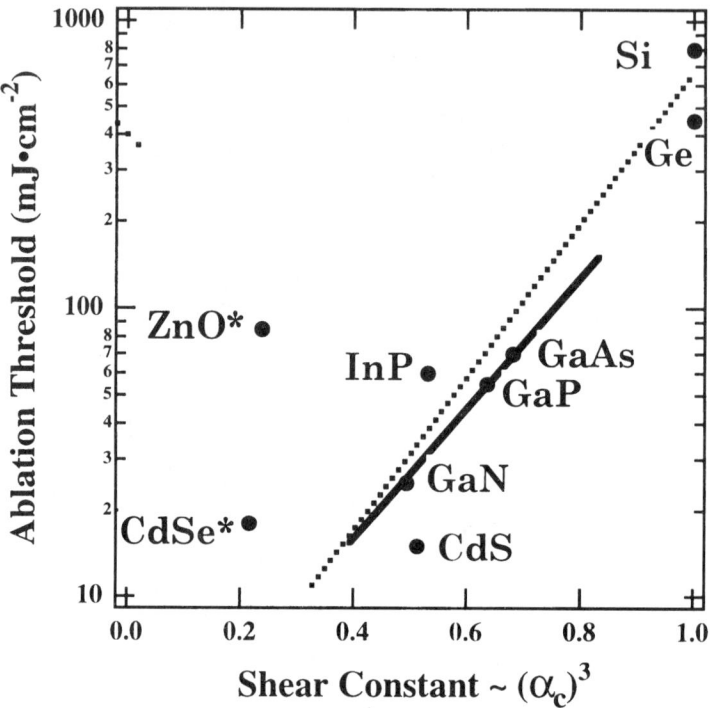

FIG. 32. Scaling of laser ablation thresholds in compound and elemental semiconductors as a function of the shearing strength calculated from the bond-orbital model. Data from Ichige et al., *Nucl. Instrum. Methods B* **33**, 820–823 (1988).

vaporized material around the ablation crater. Molecular dynamics (MD) simulation seems to confirm that there are long-lived excitations that do not decay during the laser pulse and remain free to work their mischief after the pulse is over [151]. A three-dimensional simulation on a lattice 50×100 Å $\times 100$ Å in size was run over a time duration extending as long as 20 ps after the incident pulse, assuming pulse durations of 200 fs, 1 ps, and 20 ps. However, these results should be treated as interesting but still preliminary (as indeed the authors note) because not all the parameters (for example, relaxation rates for excited atoms, focal spot diameter of 25 Å), are necessarily physical. The calculations show that as pulse width decreases below 1 ps, increasing numbers of energetic ions are produced, which initiate collisional sputtering in the bulk, as well as triggering a rapid surface expansion in a kind of "Coulomb explosion." It is not known presently from experiment whether there is a substantial ion fraction in the plume; this would be an important test case. But simulations at a pulse duration of 25 ps do predict the large fraction of redeposited atoms

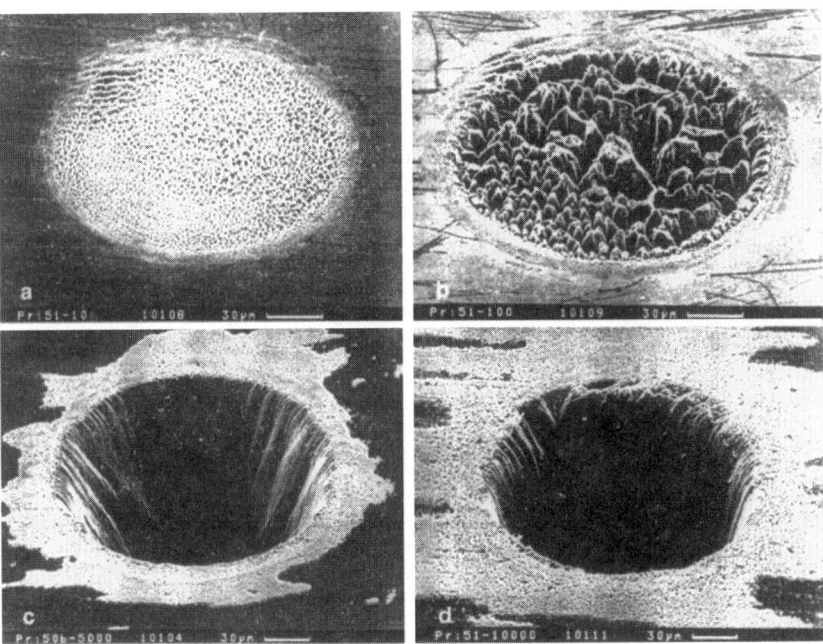

FIG. 33. Ablation of a 0.3 mm-thick Si target by a CPA laser system operating at a laser wavelength of 390 nm, pulse duration 250 fs, 0.5 mJ pulse energy and 2.5 J·cm^{-2} fluence, for varying numbers of shots as follows. (a) 10; (b) 100; (c) 1000; and (d) 5000 shots. From Chichkov et al., *Appl. Phys. A* **63**, 109 (1996).

seen in the electron micrographs, and refinement of the calculations can be expected to bring additional insight.

Some clues to ultrafast mesoscale disruption of semiconductor surfaces by laser ablation can probably be gleaned from experimental [152] and theoretical [153, 154] studies of the melting of diamond-structure semiconductors at intensities in the 1 TW·cm^{-2} range. The coupling of light into the solid proceeds via the transverse optical phonons; these have the most efficient momentum transfer with the electrons. The studies suggest (1) that the transverse-acoustic (TA) phonons (which are related to shear waves in a solid) "soften" as the electron-hole plasma density increases, vanishing at relative ionization fraction of approximately 0.1; and (2) that the longitudinal optical (LO) phonons exacerbate the instability initiated by the TA phonons. In a typical nanosecond pulsed laser ablation experiment, intensities are some four orders of magnitude lower, so that the electron-hole plasma is much less dense, while the longer duration of the laser pulse implies that heat is actually transferred to the lattice during the laser pulse, unlike the case of ultrafast (100 fs) melting where equilibration begins only after the laser pulse is over. Depending on the density of electrons in the conduction band, there may be significant heating due to electron-lattice collisions and the generation of elastic waves in semiconductors, and possibly even in wide-bandgap insulators [155].

Getting the ablation threshold right is only half the problem, however; we are still faced with finding a mechanism that can explain the intensity dependence of the ablation yields. Hattori *et al.* have proposed a *phonon*-assisted multihole localization mechanism to explain the superlinear dependence of the yield of GaP [99]. In Hattori's model, an excited defect state is created either by sequential trapping of an electron-hole pair or by defect excitation. Localization of two holes on these surface defect sites can occur either by cascade excitation or by negative-U localization. The defect-excited state is relaxed to form a metastable state, which is excited again, during the same laser pulse, to form a different excited state. In each cycle, a new hole is added or the existing hole is excited into a higher bonding orbital. A relaxed excited state of a defect can absorb photons by inducing electron transitions, hole transitions, and the perturbed band-to-band transitions. Of these three, the latter two are most effective in weakening the bonds; the second one promotes holes to inner bonding state, and the third produces a new hole.

Two-hole localization by these two processes differs from the *Knotek-Feibelman* (KF) two-hole localization induced by the e-hh Auger process initiated by core hole excitation [81]. The KF two-hole localized state is unstable against separation of the two holes because of the Coulomb repulsion, but in the negative-U two-hole localized state, the Coulomb repulsion energy is compensated by the lattice relaxation, so that the instability may affect the lattice spacing but not the distance between two holes.

2.5.4 Laser Ablation of Dielectrics with Strong Electron-Lattice Coupling

The technological motivation for studying laser ablation in wide-bandgap dielectrics with large electron-lattice coupling strength has been the problem of surface damage to optical components. Surfaces damages at fluences (intensities) much lower than bulk damage thresholds and multiple-shot and single-shot damage thresholds seem to be quite different in most cases. To complicate the picture further, there is a strong dependence on pulse duration. In these transparent solids, a great deal of evidence suggests that the critical step leading to ablation is the alteration of the one-photon absorption coefficient of the material, either during the damaging pulse or by previous irradiation history at fluences below the threshold for laser ablation. In this section, we consider laser damage and ablation in strong-coupling materials that support the formation of the self-trapped exciton. This will show how the microscopic effects due to strong electron-lattice coupling manifest themselves at the mesoscale.

One of the great challenges in studying ablation and damage in wide-bandgap solids is to distinguish between effects that are intrinsic to a pure material and those that arise from extrinsic defects. In typical commercial optical materials, the details of the surface treatment—for example, polishing and cleaving—play an enormous role in the absorption characteristics of the material. Thus one might well expect to see substantial variations in ablation behavior depending on whether the sample is polished or cleaved, in ultrahigh vacuum or air, and so on. In laser ablation, as opposed to laser-induced desorption, we should also expect intrinsic factors—such as the excitation and decay of the self-trapped exciton— to influence the mesoscale phenomenology only indirectly, for example, through changes in the optical absorption coefficient.

From the very beginning of studies of laser damage in wide-bandgap materials, two models have been advanced for explaining intrinsic laser-induced damage (that is, damage not caused by defects, impurities, or mechanical imperfections): multiphoton free-carrier generation and heating, and multiphoton free-carrier generation followed by impact ionization. In either model, mesoscale properties are key to macroscopic damage: the absorption coefficient determines what happens to the incident photon energy, and the course of laser ablation is determined by the mesoscopic thermal, structural, and mechanical response of the material. Even when the ultimate response is thermomechanical, the actual route taken to the damage will have electronic antecedents that underlie and may well determine the evolution of damage.

We begin a survey of this complicated question by looking at recent studies of the pulse-width dependence of laser damage to optical-quality samples of CaF_2 and SiO_2 ultrashort-pulse damage, made with a CPA laser system at the Lawrence Livermore National Laboratory operating at wavelengths of 1053 and 526 nm and covering pulse durations from 270 fs to 1 ns. The benchmark measurement for ablation in this experiment is the threshold for multiple-shot ("many-on-one") laser-induced surface damage, using the appearance of visible damage in a Nomarski microscope [156]. Although this seems like a very simple diagnostic, in the hands of skilled practitioners, it has a long history of making reproducible assessments, though the thresholds measured in this way tend to be 20%–30% higher than the thresholds measured by detection of ions, photoelectrons, or neutral atoms. In any case, the laser fluence at the damage threshold generally follows the $\sqrt{\tau_L}$ scaling law expected from thermal diffusion models down to pulse durations of about 10–20 ps (Fig. 34) for both SiO_2 and CaF_2. In fact, the scaling law works just as it does for metals in the limiting case of Eq. (2.26) where the pulse duration is longer than thermal relaxation times.

Stuart et al. argue that the melting and boiling observed in their dielectric surfaces are the signature of a thermal mechanism, whereas the fracture observed at pulse durations below 10 ps is evidence of an electronic excitation mechanism at work. However, this is probably an artifact of the "many-on-one"

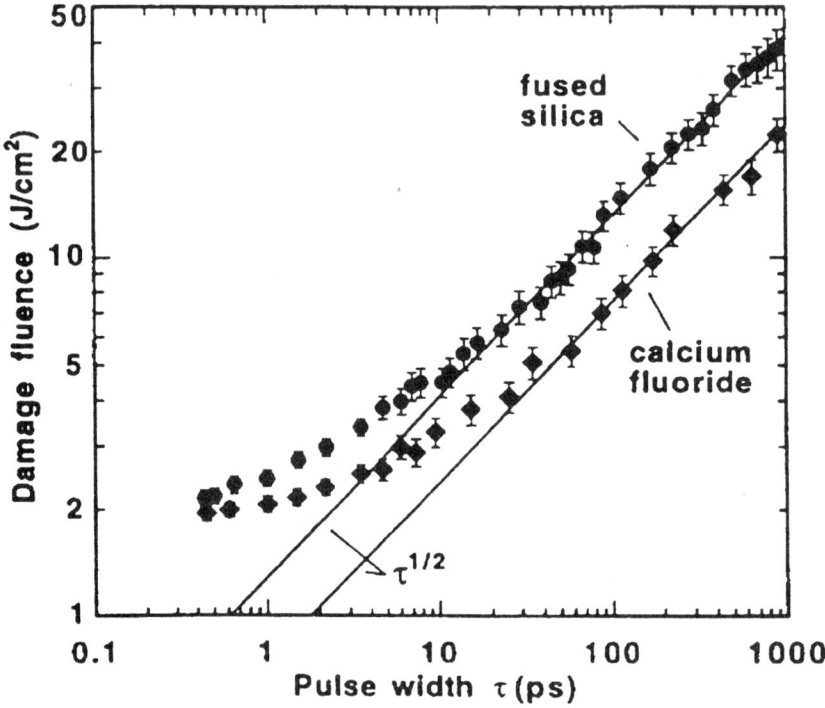

FIG. 34. Threshold fluence for the appearance of surface damage in a Nomarski micrograph, as a function of pulse duration for a CPA laser system operating at 1053 nm, for fused silica (upper curve) and calcium fluoride (lower curve). The solid lines follow a $\sqrt{\tau_L}$ scaling law. From Stuart et al., Phys. Rev. Lett. **74**, 2248 (1995).

experimental protocol. In fact, nanosecond pulsed laser ablation of CaF_2 has been extensively studied at a wide range of visible and ultraviolet wavelengths, and there fracture is a common occurrence at the damage threshold *for a single shot* on a virgin spot, provided that the probe used is more sensitive than visible microscopy. For example, sensitive measurements of Ca^0 emission from CaF_2 irradiated by 7-ns, 532 nm laser pulses generally showed a highly nonlinear increase near an intensity (0.5 GW·cm^{-2}) where transmission measurements through the irradiated spot also showed a decrease [157]. Although these experiments were complicated by hot spots in the beam, the damage morphology in all cases showed stress cracking and exfoliation of triangular shards of CaF_2 following normal cleavage planes of the crystal. Similar results were also observed for KrF laser ($\lambda = 248$ nm) laser irradiation at intensities approaching 1 GW·cm^{-2}—fracture along cleavage planes as observed just above the step edge (Fig. 35(a)). The curvature of the small platelets of CaF_2 was measured in

(a)

(b)

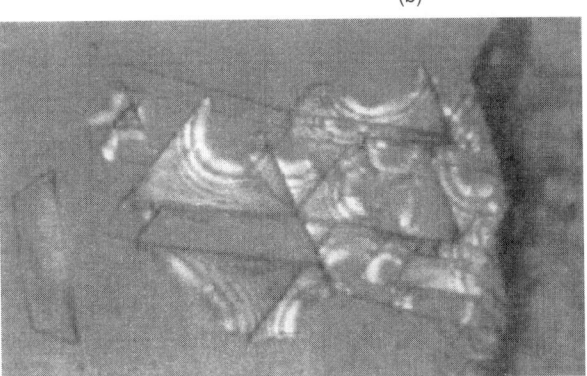 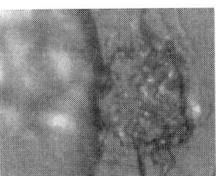

FIG. 35. Scanning electron micrographs of surface damage from laser ablation of CaF$_2$ using a KrF laser ($\lambda = 248$ nm). (a) Damage induced on a polished surface at a fluence of 2.8 J·cm^{-2}. (b) Damage induced where the laser spot overlaps a step edge, with evidence of surface melting; the out-of-focus region above the step (inset) shows damage morphology like that in (a). From Gogoll et al., Appl. Surf. Sci. **96–98**, 332 (1996).

a number of cases by looking at the interference fringes in a Newton's ring experiment. The curvature is consistent with a thermomechanical mechanism of fracture, with the ejection of platelets brought about by differential heat transport between the top and bottom surfaces of the platelet. However, when the observing microscope is focused at the step edge (Fig. 35(b))—where there is already a high density of dislocations and other defects—there is clear evidence of boiling and melting. This is, in fact, consistent with conjecture by the Livermore group that otherwise imperceptible defects, such as color centers or microscopic melted regions, may build up before the multiple shot damage becomes observable by the Nomarski microscope. Thus the experimental evidence acquired with longer pulse lasers across a wavelength range from ultraviolet to near-infrared appears to tell the same story.

This picture of damage caused by free-electron laser heating and thermomechanical stress is bolstered by exquisite laser-damage experiments and modeling done on *bulk* heating in alkali halide crystals at Washington State University in the 1980s [159]. Extremely high-purity alkali halide crystals were irradiated by the harmonics from a 60-ps Nd:YAG laser. As a "thermometer," the group used the recombination radiation from self-trapped excitons, which is extremely sensitive to temperature. In these relatively long-pulse experiments, it appears that multiphoton generation of free carriers is the major factor leading to thermal damage. In NaCl, for example, a free-electron concentration of 10^{18} cm^{-3} in NaCl can be generated by multiphoton band-to-band transitions at a flux of $5 \cdot 10^{29}$ cm$^{2} \cdot$s^{-1}. Rapidly heated by single-photon absorption, these electrons heat the crystal by electron-phonon scattering to a few hundred degrees, eventually producing mechanical stresses that far exceed the yield strength of the crystal.

Now consider the observations for pulse durations shorter than 10 ps. For pulses shorter than 10 ps, there is a marked leveling off in the rate of change in the threshold fluence with pulse duration; this change in the slope of the threshold is accompanied by a change in the appearance of the damaged surface, from melting to fracture. This evident change in physical appearance is ascribed to electron-impact ionization of the irradiated material because the probability of creating the initial electron avalanche increases dramatically at the higher intensities. This conjecture seems to be substantiated by a theoretical study of free-electron heating in SiO_2, using Monte Carlo techniques to integrate the Boltzmann transport equations in an intense radiation field [160]. The theory on which the computations are based treats the laser-electron interaction quantum mechanically at visible wavelengths, while giving a classical treatment in terms of dielectric breakdown appropriate to wavelengths longer than 2 μm. The calculations confirm the dominant role of lattice heating by multiphoton absorption for intensities typical of the 60-ps laser used in the Washington State experiments. The calculations also show, however, that at intensities of order 0.3 TW·cm^{-2}, the competition between multiphoton free-carrier absorption and impact ionization begins to tilt toward the latter; that shift in balance goes up dramatically at higher intensities. Because the calculations also incorporate realistic scattering rates and predict experimentally observable indicators of laser damage, at least at long pulse durations, they enhance the credibility of the conjecture about impact ionization as the source of damage (subject only to the caveat raised by Arnold and Cartier [155] that both classical and quantum mechanical constructions are suspect for 1 μm, thus making it possible only to bound the relative probabilities for either case).

However, the reduction in surface damage or ablation threshold with pulse duration observed in the Livermore experiments conflicts with single-shot measurements of plasma formation thresholds in the bulk of SiO_2 by Du et al. who also used a CPA laser system to study laser-induced breakdown (LIB) in solid fused silica [161]. This experiment detected plasma emission during a single CPA laser pulse by measuring transmission through the focal region of the lens. As shown in Fig. 36, the fluence threshold for LIB follows the perennial $\tau^{1/2}$ curve from the nanosecond regime down to about 10 ps; below that, the LIB threshold actually increases. To complicate matters further, Du et al. use an electron avalanche theory to substantiate their experimental finding.

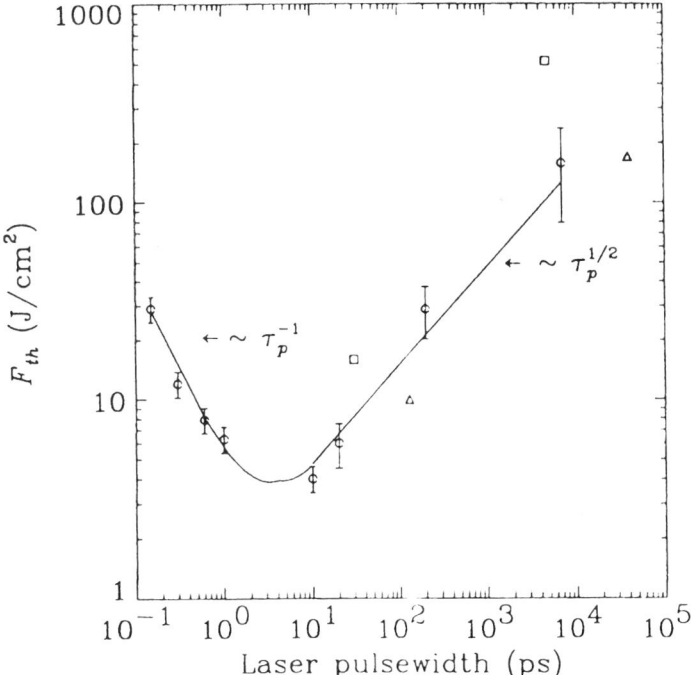

FIG. 36. Fluence threshold for plasma formation in bulk SiO_2 as a function of pulse duration, measured by light transmission through the focal spot of a Ti:sapphire CPA laser operating at its fundamental wavelength of 780 nm. From Du et al., *Appl. Phys. Lett.* **64**, 3071 (1994).

The theoretical underpinning of their expression for ionization rate per unit length, however, is held by Arnold and Cartier [155] to be incorrect for the specific case of SiO_2.

So now one is faced with the problem of sorting out the relation of surface to bulk damage and the relation of both to laser ablation, as well as the even more complex issue of single-shot versus multiple-shot measurements. Two points suggest themselves in thinking about these apparently contradictory results. One is that looking for plasma breakdown in the solid is clearly a "higher temperature" phenomenon than that found in the STERL technique, which is sensitive primarily to the local temperature rise. The second lesson is that surface damage—laser ablation—is likely to be quite different than that found in the bulk. The surface introduces a number of important complications, such as the role of surface defects and photoemission, that remain to be elucidated for these ultrashort laser pulses.

Recent studies of ablation phenomenology as a function of pulse duration in CaF_2 with a CPA laser suggest that for a small number of laser shots, an interpretation of surface damage based on multiphoton absorption by self-trapped excitons and heating, rather than on impact ionization, may be quite credible after all. In these experiments, 200- and 580-fs pulses at a wavelength of 790 nm produce photothermal or thermomechanical fracture (Fig. 37(a) and 37(b)), whereas for pulse durations between 1 and 5 ps, the laser-produced damage appears thermal, with distinct signs of thermal stress cracking in the ablation zone (Fig. 37(c)) [162]. These results can be understood in the following way. In calcium fluoride, "on-center" self-trapped excitons are generated by four-photon excitation at 790 nm on a time scale of approximately 1 ps; given the intensities generated by the CPA laser ($\sim 10^{12}$–10^{13} W·cm^{-2}), generating substantial numbers of STEs is not a problem. For subpicosecond pulses, then, the STEs are formed *after* the laser pulse is over, and all of the energy stored in the STEs will go into bond-weakening interactions via electron-lattice coupling. The lattice deformations so generated could, in principle, induce severe fracture in the remaining material. For pulse durations between 1 and 5 ps, on the other hand, electron-hole pairs created during the laser pulse by multiphoton excitation are converted to STEs with lifetimes of a few µs. The STEs alter the absorption coefficient of the calcium fluoride, however, because the electrons can be excited into the conduction band by one- or two-photon transitions. If a sufficient number of STEs can be created in the first picosecond of the laser pulse—a hypothesis that should be testable using reasonable kinetics calculations—rapid heating of the crystal will follow this change of optical absorption characteristics, leading to melting and other thermal effects.

The mechanism that produces the superlinear yield-fluence relation in laser ablation is almost certainly related to the formation of defects that contribute to the superlinear ablation yield either by providing real intermediate states to

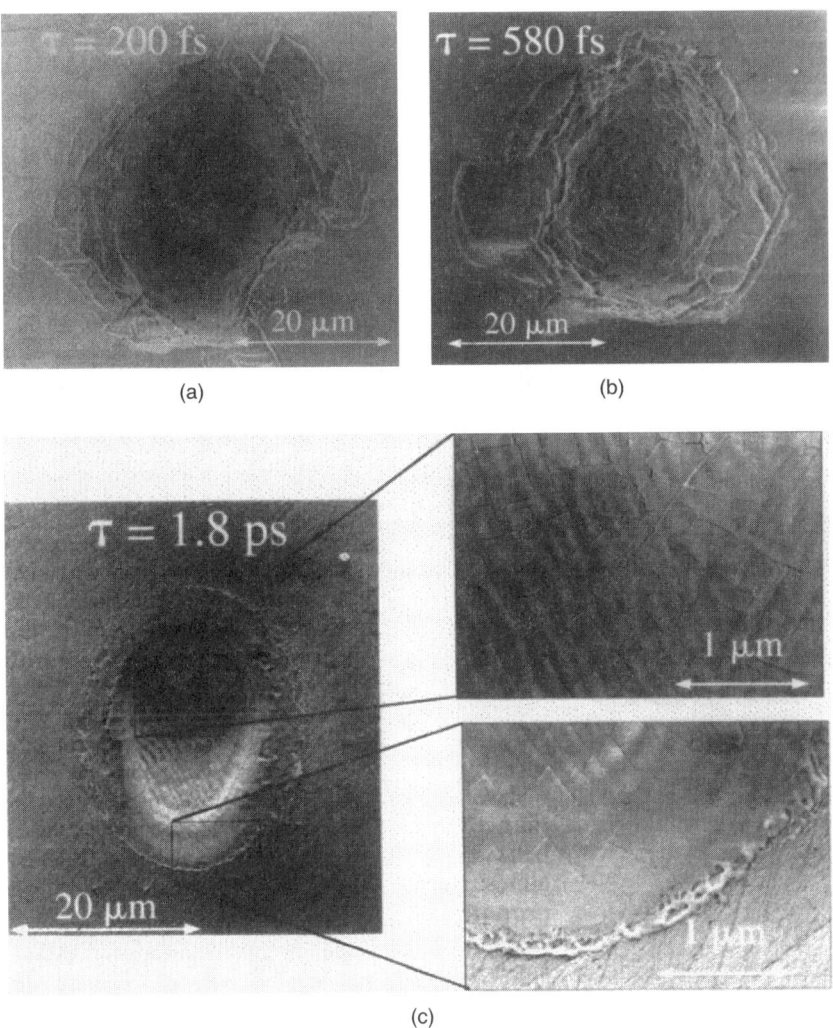

FIG. 37. Scanning-electron micrographs of polished CaF$_2$(111) irradiated by a Ti:sapphire at the fundamental wavelength of 790 nm; total exposure 50 shots, fluence of 7 J·cm^{-2} in all cases. Pulse duration (a) 200 fs; (b) 580 fs; (c) 1.8 ps. Note ripple formation in (c). From Ashkenasi et al., Appl. Phys. A **63**, 103 (1996).

enhance the multiphoton absorption probability [163] or by weakening the bonds of neighboring atoms to the point where these are ejected by one-photon excitation. The observed differences between alkali halides and alkaline earth fluorides with respect to the accumulation of metal layers under laser irradiation probably arise from differences in vapor pressure of the metallic component. The issue of "conditioning" or "incubation" at all pulse durations deserves further study. In studies of SiO_2, the Livermore group has found that only 2% below the fluence threshold for damage for 400-fs pulses at 1053 nm, it is possible to take 10^4 shots on the same spot without damage, suggesting that there are no long-term incubation effects in the material [164].

Among the most vexing questions in laser ablation of strong-coupling materials are the effects of the laser-ablation plasma in shielding the surface from the laser light and also in producing damage by energetic charged-particle sputtering. These questions are discussed in much greater detail in Chapter 5 but it is appropriate to examine some experimental evidence here. Double-pulse experiments have been carried out in Al_2O_3 [165] and in $LiNbO_3$ [166], the former at 80 ps pulse duration, the latter at 500 fs. The experiment on the $Al_2O_3(1120)$ surface used 80 fs pulses from the fundamental ($\lambda = 1064$ nm) of a Nd:YAG laser to irradiate the surface. If significant surface modifications occur only above threshold, using two ablating pulses with a variable time delay is one way to ascertain the lifetimes of material-dependent surface excitations which contribute to ablation, and of plasma shielding which cuts off the flux of ablating photons from the surface. In the experiments on Al_2O_3, the 80-ps fundamental of a Nd:YAG laser (1064 nm) was split into pump and probe beams of roughly equal intensity, separated by time delays ranging from 170 ps to 10 ns. The threshold for ablation was measured by monitoring the signal from neutral and charged particles arriving at a quadrupole mass spectrometer. At time delays exceeding 1 ns, the two pulses were found to behave as independent pulses, that is, each pulse separately had to exceed the ablation threshold. The onset of pulse-overlap effects near 200 ps was found to be consistent with surface exciton decay, a resonant absorption process due to formation of an intermediate reaction center, or the decay of a laser-induced plasma. In the experiments on $LiNbO_3$, illustrated in Fig. 38, a KrF laser with a pulse duration of 500 fs permitted accurate measurements of pulse-overlap effects for ultraviolet excitation (248 nm) of wide bandgap materials. Below a time delay of 50 ps, the ablation rate (measured in terms of etch depth per laser pulse) increased with decreasing delay, indicating that each pulse separately was absorbed and produced ablation. For pulse delays longer that 50 ps, on the other hand, the ablation rate is equal to that of the first pulse by itself, indicating that the ablation plasma has blocked the arrival of the second pulse.

An additional cut through the complex parameter space of laser ablation phenomenology is supplied by the recent generation of free-electron lasers and the experiments that can be done using them. The relative importance of thermal and nonthermal processes in laser ablation has been one of the enduring controversies in the study of laser-surface interactions. From first principles, it is understood that, in visible and ultraviolet laser-surface interactions, the initial

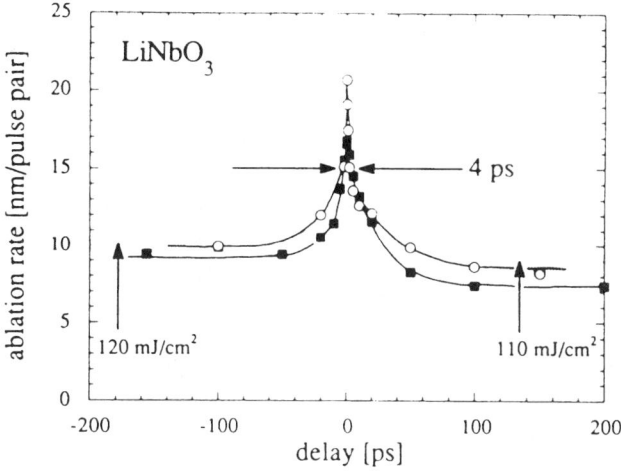

FIG. 38. Ablation rate, measured in nm etched per pulse pair, for double-pulse irradiation of LiNbO$_3$ at 248 nm, pulse duration 500 fs. The fluences in the two pulses are 95 and 110 mJ·cm^{-2}, respectively, for the bold rectangles, and 120 and 110 mJ·cm^{-2}, respectively, for the open circles. The data displayed represent averages over 500 shots (bold rectangles) and 200 shots (open circles). From Preuss et al., Appl. Phys. Lett. **62**, 3049 (1993).

excitation is electronic, and that this excitation is communicated by electron-phonon coupling to the surrounding lattice where it produces local heating and, under some circumstances, vaporization and melting. However, although the absorbed energy clearly must be thermalized, heating of the lattice via this indirect route seems, in most cases, not to produce the characteristic results of vaporization: the ejected atoms, ions and molecules, for example, often show velocity distributions appropriate to temperatures far above any physically reasonable surface temperature. Much more evidence points to a mechanism in which local heating of an electron gas eventually exceeds the yield strength of the material, producing surface debris resembling fracture.

Recently published results [167] on laser-induced surface modification and damage as a function of wavelength in calcium carbonate (calcite) show distinct changes in surface morphology, in ablation-plume emission, and in photoacoustic deflection for wavelengths in the near-ultraviolet, visible, and near- and mid-infrared. The infrared results were obtained with a tunable free-electron laser which directly excited vibrational modes of the OH defects and the fundamental $v_2 - v_4$ asymmetric stretch mode of the carbonate group. The morphology of surface damage is consistent with thermomechanical fracture at visible and near ultraviolet wavelengths, whereas in the infrared the mechanism is much more consistent with surface melting and vaporization, even for single shot damage. In addition, the plume spectra show that the absorbed photon energy exits through different channels for infrared versus visible laser light.

Calcite, a crystalline form of calcium carbonate ($CaCO_3$), occurs naturally in mineral form; its well-known crystal structure is planar, with each CO_3 group forms a triangular cluster in a plane is perpendicular to the optic axis and planes containing interspersed calcium ions.

The free-electron laser (FEL) used in these experiments [168] delivers a nominal 4 µs macropulse with energies from 20 to 60 mJ at a pulse repetition frequency of up to 30 Hz; the macropulse comprises about 10^4 micropulses 1 ps in duration, separated by intervals of 350 ps in a nearly perfect TEM_{00} mode. A broadband electro-optic switch (Pockels cell) was used to slice a short pulse train ($\tau \geq 100$ ns) out of the FEL macropulse [169]. Typical micropulse energies are of order 10–20 µJ. Because relaxation times following a picosecond excitation are much shorter than the interpulse spacing in $CaCO_3$, it is possible (though not obvious) that the net effect of a train of pulses switched out by the Pockels cell is simply a linear sum of the effects of the individual pulses.

The changes in surface topography induced by laser irradiation were studied by scanning electron microscopy (SEM). The ablation threshold in air for 1.064 µm light is ~ 5 J/cm^2, where the ablation threshold is defined as the minimum laser fluence for any visible change in the surface under the optical microscope. Cracks a few millimeters long formed due to the excitation of defects inside the material and the accompanying mechanical disturbance of the crystal; the cracks appear to form along cleavage planes and may be spawned at dislocations. The irradiated spots show a distinctive exfoliational topography, with platelets in irregular block-like shapes but no signs of heating or melting (Fig. 39 (a)). The surface topography of calcite samples irradiated at second harmonic (532 nm) and the third harmonic (355 nm) of the Nd:YAG laser was similar, even though the threshold fluences were about a factor of 2 lower than that for 1064 nm light. However, this result is consistent with the important role played by impurity absorption. Natural, optical-quality calcite contains chemical impurities such as Mn, Mg, and Sr [170]. The Mn^{2+} has an absorption band peaked at 550 nm corresponding to a ^6S to ^4G transition and at 362 nm corresponding to a ^6S to ^4D transition. Therefore, it is expected that the absorption of 532 nm or 355 nm laser light is enhanced at these defect centers; even at 1064 nm, these defects could be excited by multi-photon absorption.

At 7.0 µm, the nominal ablation threshold in air is about 1 J/cm^2 (pulse duration: 100 ns, spot radius: ~ 50 µm). According to calculations of thermal diffusion in a two-dimensional axisymmetric heat-transfer model, the melting threshold of calcite at 7 µm is 1.5 J·cm^{-2}. The lower value in the experimentally observed ablation threshold is probably due to enhanced absorption by impurities. In contrast to the exfoliational topography of the calcite irradiated at 1064 nm, the FEL-irradiated spots exhibited well-defined holes with carbonized walls and scattered droplets in the ablation crater, clear evidence of melting (Fig. 39 (c)). When the FEL irradiates the surface at 3.3 μm (Fig. 39 (b)), the shape of the ablated spot is similar to that produced by 7.0 μm light. However, the threshold fluence is much higher, a few J·cm^{-2}; this necessitates extending the macropulse duration to about a microsecond. Absorption at 3.3 μm is relatively smaller than at 7.0 μm, but impurities in the crystal such as OH and surface defects may absorb laser light, albeit with a much greater absorption length. As the surface becomes damaged, the crystal absorption coefficient may be enhanced due to laser-induced defects and consequently thermal ablation dominates. The requirement of long pulse width to reach ablation threshold and the evidence of explosive vaporization on the SEM photograph favor such a scenario. The optical transmission at this wavelength of a calcite sample

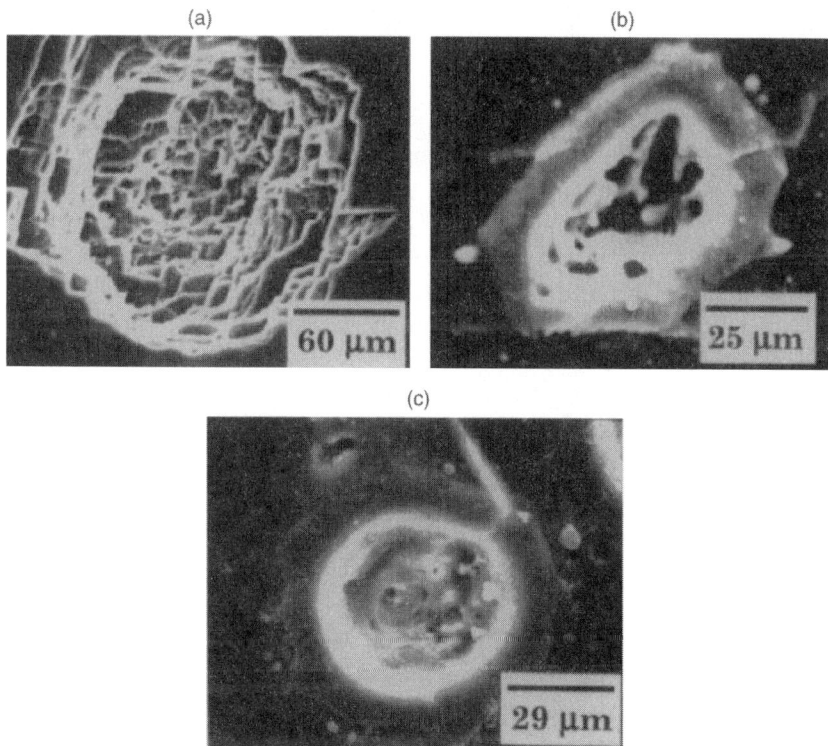

Fig. 39. Scanning electron micrographs of ablation damage natural to calcite crystal, all for single shot impact. (a) Nd:YAG laser, 532 nm, pulse duration 10 ns, fluence 10 J·cm^{-2}; (b) free-electron laser, 3.3 μm, pulse duration 100 ns, fluence 10 J·cm^{-2}; (c) free-electron laser, 7.0 μm, pulse duration 100 ns, fluence 2 J·cm^{-2}. From Park and Haglund, *Appl. Phys. A* **64**, 431 (1997).

(thickness = 3 mm) is only about 10%, proving the existence of absorbing impurities. However, questions about the effects of multiple pulses, and whether laser pulses clean up preexisting defects or introduce additional defects, have not been resolved yet. Moreover, multiple pulses (>100 shots) of 2.72 μm FEL irradiation did not produce melting but only fracture. The optical transmission of the same calcite sample at 2.72 μm is ~52%. Hence the density of absorbing defects or the absorption depth is thought to be related to the transition from fracture mechanism to melting. This idea is not new; it is rather well known from the thermomechanical ablation theory that acoustic effects in laser ablation or spallation are maximized when an "acoustic confinement" condition is met [171]. If the acoustic relaxation time, determined by the characteristic distance (optical absorption depth) divided by the speed of sound, is longer than the laser pulse width or the interval between consecutive pulses, an accumulation of mechanical stress inside the material occurs, eventually leading to fracture. In other words, as the optical absorption depth decreases due to enhanced absorption, the acoustic effect contributes less and thermal effects takes over, consistent with the observations at 3.3 μm.

The marked difference in ablation topography as a function of wavelength finds a parallel in the distinctive composition of the ablation plumes for single laser shots on fresh surface sites. First of all, the "color" of the plume shows an outstanding difference, manifesting a snowy-white color for 7.0 μm irradiation and bluish red color for 0.532 or 1.064 μm irradiation. The time-integrated spectral distribution of the plume emission is shown in Fig. 40(a). The plume emission spectrum is nearly identical for 0.532, 1.06, and 2.72 μm, all of which produced fracture, while the spectrum in Fig. 40(b) at 7.0 μm is distinct from the previous spectra. The plume excited by near-IR irradiation shows a rich structure in near-UV and blue spectrum, whereas the plume generated by 7.0 μm FEL light shows outstanding peaks only in green and orange. However, in the case of FEL irradiation, modulated nanosecond pulses could not produce a plasma plume strong enough for spectral analysis; hence a full macropulse (τ_L = 2.2 μs FWHM) was used on the sample. Therefore, not only is the mechanism of particle ejection varying with wavelength, but the composition and time scale of plume development also appear to vary markedly as one moves from near-infrared to mid-infrared radiation.

The plume produced by near-IR light consists of some calcium ions, CaO and CO molecules, and some neutral oxygen. It is amazing to see that identical composition of plume emission spectra was produced by widely different excitation sources, ranging from the Nd:YAG laser to the FEL, considering the wide gap in photon energy (1.17 eV versus 0.5 eV) and pulse width (nanosecond versus picosecond pulse trains). The calcite surface irradiated with 2.72 μm FEL beam shows the same fracture-like topography as in visible and near-ultraviolet irradiation, which suggests a relationship between the surface fracture and

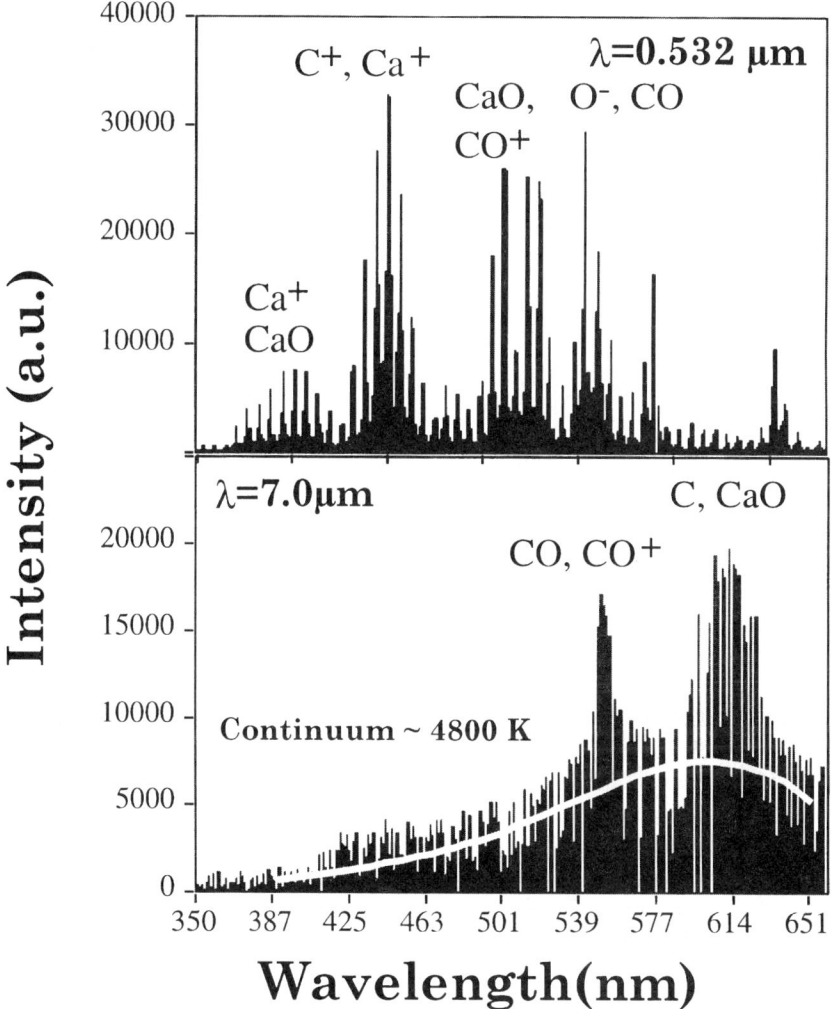

FIG. 40. Plume emission spectra from ablation of calcite by (a) ns Nd:YAG (532 nm), fluence of 10 J·cm^{-2}, pulse duration; (b) free-electron laser (7.0 μm), fluence of 2 J·cm^{-2}, pulse duration 100 ns. The smooth curve is the blackbody emission curve corresponding to 4800°K. From Park and Haglund, *Appl. Phys. A* **64**, 431 (1997).

plume composition. Therefore, a part of these species were perhaps emitted associated with fracture; the emission of electrons and particles associated with the plastic deformation of ionic crystals ("fractoemission") is well known [172].

In the case of the plume produced by resonant vibrational absorption, the overall increase of background continuum emission resembles black-body emission (Fig. 40(b)), consistent with the evidence of melting seen in SEM. This spectrum may result from the thermal emission from hot particulates and vapor. The thermal emission looks to be peaked around at 620 nm, corresponding, according to Wien's displacement law, to a temperature of about 4800 K, about three times the melting temperature (m.p. = 1612°K). The two outstanding emission peaks of the FEL-ablation plumes may be due to carbon neutrals (possibly carbon soot) and carbon monoxides. The lines at 606 nm region can be identified as CaO. The CaO could be also a product of thermal decomposition of calcite. The existence of carbon monoxide would result from thermal decomposition or oxidation of hot carbons. The carbonization of the laser-irradiated spot was also observed by SEM.

Photothermal beam-deflection measurements showed that the visible-ultraviolet excitation of calcite produced a much stronger acoustic shock wave, as measured by beam-deflection spectroscopy, than the infrared excitation at 7.0 μm; the shock-front velocities measured for visible-near-ultraviolet irradiation are consistent with calculations of an explosive or fracture event, while the much weaker shock front produced by the infrared free-electron laser at 7.0 μm is consistent with a gentler, thermal evaporation process.

Taken together, these results show that the mechanisms of particle ejection and plume formation are strongly dependent on the wavelength of the laser used to initiate ablation. This would be no surprise in experiments done with visible or ultraviolet lasers, but the differences observed for infrared wavelengths are arresting. Whether they result simply from differences in the absorption coefficient or whether there are deeper questions about the equilibration of energy when the initial excitation is thermal remains to be explored. In light of these results, however, it may be necessary to revise commonly held notions that the temporal duration of the laser pulse is the primary determinant of the ablation mechanism, and that, once the input laser energy is thermalized, ablation proceeds along the same route in configuration space, regardless of the mode of excitation.

The possibility of direct thermal coupling to wide bandgap materials with infrared lasers also highlights the possibility of ablation by photomechanical effects, a mechanism often described by the term *spallation*. In spallation experiments with Nd:YAG lasers on transparent materials such as glass, for example, it is generally necessary to use a strongly absorbing layer at one surface of the material in order to introduce the thermomechanical shock wave which drives through to the rear surface [173]. However, there are other contexts, including high-power infrared lasers and strongly absorbing media, where spallation can be important in the intrinsic material. This mechanism has recently been discussed in a number of contexts, including the testing of thin elastic coatings [174] and short-pulse ablation of biological tissues such as bone [175]. The theory of spallation for front-surface abalation in semitransparent materials has been reviewed recently by Dingus and Scammon [171].

2.6 Novel Experimental Tools for Studying Laser Ablation

Although the historic way of "detecting" laser damage or ablation has been to look at the topography and morphology of the ablated surface, it should be evident from the preceding discussions of laser ablation mechanisms that it is not possible to determine mechanisms, in most cases, without measuring essentially *all* of the following:

- Initial topography, composition, and defect or impurity content of the target surface
- Temporal and spatial profiles and, where necessary, spectral content, of the laser beam
- The number density, type, and internal state of the ejected particles
- The intensity or fluence marking the transition from below- to above-threshold conditions
- The gas-dynamic and plasma characteristics of the ablation plume
- The final state of the surface, including but not limited to topography

Although such a comprehensive set of measurements is not easy to produce, the great leap in understanding made in a few special cases where this has been done justifies the effort. Many of these points are covered elsewhere in the present volume, notably in Chapter 4 and Chapter 6.

The tools and techniques of mass spectrometry, laser-induced fluorescence, resonant (R) and nonresonant (NR) multiphoton-ionization (MPI) [176] mass spectrometry, and photoacoustic techniques are in routine use and are described in textbooks or substantial review articles. Most of these techniques are aimed at identifying and characterizing desorbed or ablated particles, and their use in this context been described in a number of reviews as well as in primary research literature.

In this section, we briefly introduce three measurement techniques that are less widely known or are just coming into use in experiments on laser-induced desorption or ablation. Some of them fall into the category of "ultra-sensitive" spectroscopy and are thus usable across the full range of laser sputtering phenomena. Again, our intention is to be illustrative rather than comprehensive. We make no apology for the fact that the choices may also be viewed as idiosyncratic!

2.6.1 Detection of Ablation by Surface-Plasmon Resonance Spectroscopy

Detection of the surface plasmon resonance in thin films (see Section 2.3) has been used to good effect in the detection of pulsed laser ablation at sensitivities in the monolayer regime. In experiments at Konstanz, a silver film approximately 500 Å thick is deposited on a prism using standard thermal or electron-beam evaporation techniques. The material to be studied can then be deposited or placed in contact with this silver film. When the film is illuminated from the rear by, for example, a He-Ne laser, the reflected light can be monitored using any kind of simple photodetector. When the angle of incidence of the laser is properly adjusted, the surface-plasmon resonance is excited and the detected intensity is drastically reduced (see Fig. 41(a)). The change in the angle at which resonant excitation occurs has been shown to be an extremely sensitive measure of changes to the material in contact with the silver reflective layer.

In one of its earliest applications to the study of laser ablation [177] this technique was used to study the competition between thermal and nonthermal mechanisms in the ablation of thin organic films. Isopropanol, acetone, and tetrafluoromethane vapors were condensed on the silver surface maintained at liquid nitrogen temperature (~ 77 K) in a vacuum chamber, to form films some 50 Å thick (about ten monolayers). The films were then irradiated by a KrF laser (248 nm) at varying fluence levels. The optical signal from the reflected HeNe laser beam was monitored by a fast p-i-n photodiode coupled to a wide-band amplifier, thus achieving nanosecond time resolution in addition to the high sensitivity of the reflected signal. Tetrafluoromethane is transparent in the ultraviolet but has a low enthalphy; it could, therefore, be expected to be thermally desorbed as the silver is heated by the KrF laser pulse. Isopropanol and acetone have similar enthalpies and other thermodynamic properties, but have optical absorption coefficients that differ by orders of magnitude.

Typical results of the technique are shown in Figs. 41(b) and 41(c), where the ablation temperature is presented as a function of laser fluence. The ablation temperature is influenced both by the energy deposited in the film and by the transient heating of the silver film which serves as a sensitive thermometer. In panel (a), it is apparent that the ablation temperature remains roughly constant for the tetrafluoromethane film; this result is consistent with a thermal mechanism in which ablation is controlled entirely by an activation barrier for evaporation, occurring as soon as that temperature is reached. For isopropanol [panel (b)] and acetone [panel (c)], on the other hand, the increase in ablation temperature with increasing fluence is a clear indication that thermal desorption alone is not the mechanism responsible for the ablation. In this case, the authors suggest that chemical transformations occur in the isopropanol and acetone layers, a conjecture that would have to be established from other measurements, of course. This technique has also been applied to the study of bubble nucleation and growth at surfaces, an important problem for laser ablation in liquid and vapor environments [178], as well as to the detection of planar and spherical shock waves from rear- and front-side ablation of calcite [179].

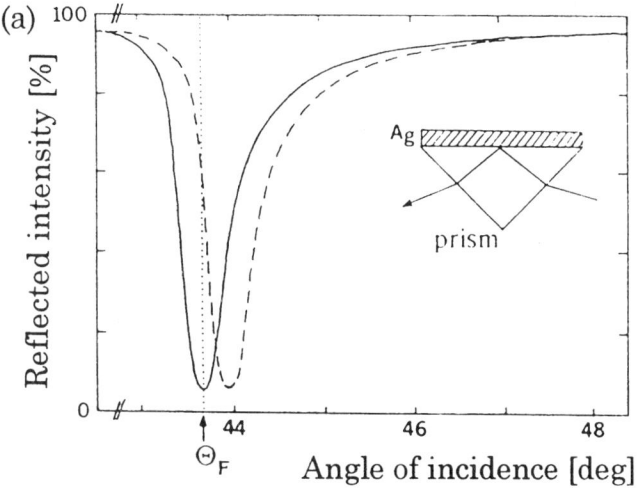

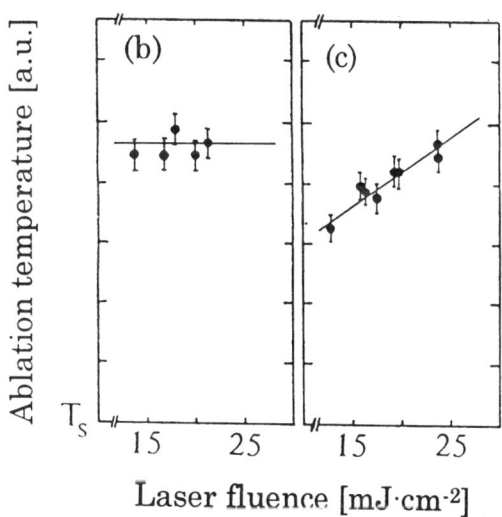

FIG. 41. Detection of laser ablation in thin organic films under KrF excimer laser irradiation using the surface plasmon resonance. (a) Experimental geometry and resonance shift for an adsorbed monolayer. (b) Temperature vs. fluence profile for atetrafluoromethane film, showing an ablation process controlled by thermal activation. (c) Temperature vs. fluence profile for isopropanol, showing the operation of an athermal mechanism. From Herminhaus and Leiderer, *Appl. Phys. Lett.* **58**, 352 (1991).

2.6.2 Scanning-Probe Microscopy

Many of the surface analysis tools described in Chapter 6 provide excellent mesoscopic information about reasonably well-ordered surfaces. However, in both desorption and ablation studies, it is also necessary to characterize the beginning stages of disorder on an atomic scale. Scanning-probe microscopy has become a well-established tool for measuring surface topography in virtually all areas of condensed-matter science within the last five years [180] and is ideally suited to making major contributions to our understanding of both laser-induced desorption and ablation. Scanning probe microscopy has recently been applied both to deliver beam to the surface and primarily to study surface topography at the site where laser-induced modification is occurring.

In *near-field scanning optical microscopy* (NSOM), a severely tapered optical fiber tip is produced by chemical etching or by stretching a CO_2-laser-heated fiber tip to the breaking point, then coating the tip with aluminum and chromium to achieve appropriate reflectivity. A constant distance between the fiber tip and the surface to be illuminated is maintained by feeding back the shearing-force on the tip as the tip is dithered near its mechanical resonance frequency. The tip thus becomes both the instrument of beam delivery and a scanning probe of the surface topography.

In one such experiment [181], ablation of holes of order 70 nm diameter and 40 nm deep were observed for single-pulse irradiation of anthracene crystals. Light from a commercial optical parametric oscillator pumped by a frequency-tripled Nd:YAG laser (355 nm) was coupled into the fiber, producing a transmitted far-field pulse energy of around 100 μJ at a wavelength of 450 nm and a pulse duration of 3–6 ns. The tip was maintained at a constant distance of 10 nm from the surface. After a single laser shot, nanometer-dimension holes were observed (Fig. 42), without evidence of debris deposited between holes; the holes, spaced about 750 nm apart, were observed in both the scanning-probe and electron microscope pictures. Scanning a cross section of the laser irradiated sample across the holes shows that the lateral resolution of this technique is of order 70 nm, confirming that indeed the amount of material removed puts the experiment within the ablation regime as defined here.

In a variation of the scanning-probe technique, laser irradiation of the Si(111) 7×7 surface with a KrF laser has been reported over a much larger area [182]. Earlier studies [7] have shown that the emission of atoms from semiconductors originates at defect sites, and that the defects that have the lowest binding energy are the adatom sites. It, therefore, is of interest to see whether such defect sites indeed lose adatoms preferentially under laser irradiation. In the currently accepted dimer-adatom-stacking-fault model of the 7×7 Si(111) surface, a unit cell contains 12 adatoms that have the longest bond length and lowest binding energy. If conjectures about the origin of the atomic emission at the most weakly bonded sites is correct, these sites should be the first from which desorption and ablation occur.

STM laser ablation experiments were performed in an ultrahigh vacuum chamber on atomically clean Si(111) surfaces (Fig. 43(a)), by irradiating them with dye laser pulses (2.48 eV) 28 ns in duration, with a laser spot 0.4 mm in diameter. During laser irradiation, the SiC tips were retracted so that the laser would not damage them during the ablation phase of the experiment. Si atoms were detected by resonance ionization mass spectrometry, making it possible to correlate the properties of the ejected atoms with the topographical information about the surface provided by the STM. As shown in Fig. 43(b), tunneling images of the Si(111) 7 × 7 clearly show a loss of adatoms following laser

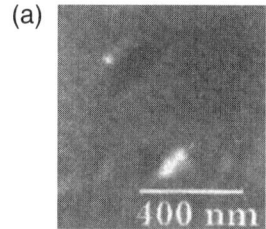

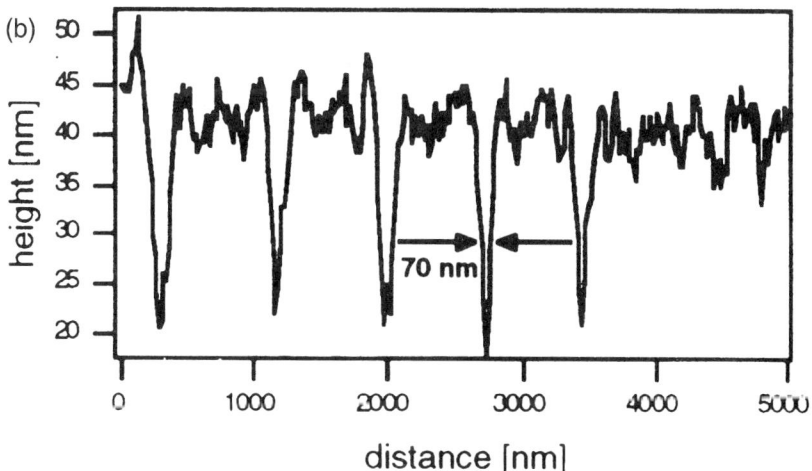

FIG. 42. (a) Shear-force image of an anthracene crystal with a central protusion and one desorption spot induced by a ns pulse from a Nd:YAG-pumped optical parametric oscillator at 45 nm, producing approximately 100 µJ energy through the metallized tip of a near-field scanning optical microscope. (b) Linear section through a series of desorption spots showing a lateral resolution of order 70 nm, made in the same way. From Zeisel *et al.*, *Appl. Phys. Lett.* **68**, 2491 (1996).

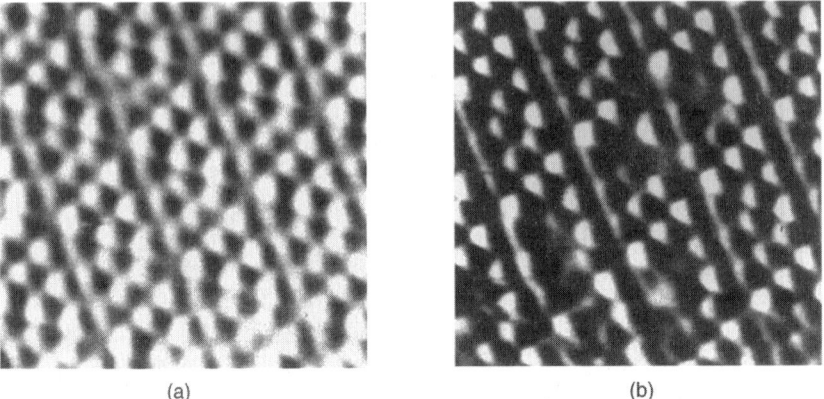

(a) (b)

FIG. 43. Tunneling images of the Si(111) 7 × 7 surface (a) before and (b) after bombardment by repeated pulses from a ns-excimer-pumped dye laser with photon energies of 2.48 eV and a fluence of 90 mJ·cm^{-2}. From Kanasaki *et al.*, *Solid State Comm.* **98**, 913 (1996).

irradiation at 300 mJ·cm^{-2}. Moreover, the fraction of unit cells that show vacancies after laser irradiation goes up for all sites, including isolated sites, dimer sites, and trimer or higher order binding sites. The fluence dependence of the Si atoms, measured by RMPI, shows a superlinear yield consistent with the rapidly increasing number of weakly bound atomic sites [179]. This is consistent with a model already proposed for ablation based on the picture of defect multiplication during laser irradiation.

Another experiment of this type has been performed on the technologically critical Si(100) surface, which is used as the substrate for many microelectronic devices and which reconstructs as a dimer layer. Following standard surface preparation and annealing, a clean, reconstructed Si(100) surface (Fig. 44(a)) was irradiated by a frequency doubled, pulsed Nd:YAG laser with a pulse duration of 7 ns. As shown in Fig. 44(b), the dimerized uppermost layer is selectively ablated below the melt fluence. While the atoms in the topmost layer continue to exhibit the (2 × 1) configuration as ablation proceeds, the next layer, as it is uncovered by laser irradiation, retains the (1 × 1) structure of the bulk-terminated solid. Then, as more and more of the upper layer is removed, the atoms in the decreasing number of remaining dimers move closer together.

2.6.3 Time-Resolved Holography, Interferometry and Plume Spectroscopy

Understanding the dynamics of plume formation, including the development of shock waves and plume chemistry, is central to understanding the initial stages of ablation. The plume chemistry is a matter of special concern in the

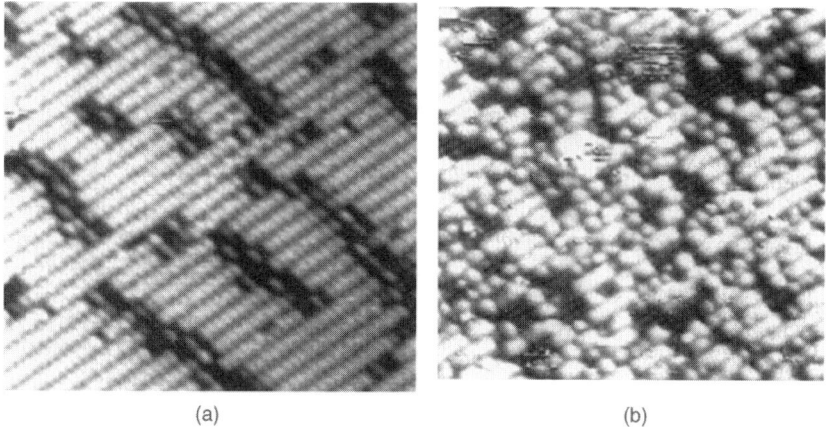

FIG. 44. Tunnelling images of the Si(100) 2 × 1 surface (a) before and (b) after irradiation by a single pulse from a Nd:YAG laser (($l = 1064$ nm) at a fluence of 150 mJ·cm^{-2}. The melt threshold for this surface is ~ 1 J·cm^{-2}, the ablation threshold ~ 1 J·cm^{-2}. From Xu et al., Phys. Rev. B **53**, R4245 (1996) and Phys. Rev. B **54**, 5180 (Erratum).

ablation of polymers and organic crystals (as in matrix-assisted laser desorption-ionization mass spectrometry of biomolecules). Here we cite two recent examples of useful diagnostic techniques.

The first is resonant holographic interferometry of ablation plumes [184]. This technique provides species-specific density contour information for the ground-state neutral species that dominate the composition of an ablation plume but are otherwise hard to detect. Unlike resonant absorption spectroscopy [185], or resonant absorption photography [186], the combination of a resonant excitation technique with holography can provide a time-resolved density profile for the neutral species as well. In a University of Michigan experiment, shown schematically in Fig. 45(a), two excimer lasers are used, a KrF ($\lambda = 248$ nm) laser for ablation and an XeCl laser ($\lambda = 308$ nm) to pump a dye laser. Synchronization of the two lasers is not a problem for the nanosecond pulse durations employed here. The ablation target is placed in one arm of a Mach Zehnder interferometer, with apparatus dimensions smaller than the coherence length of the dye laser (of order 5 cm). The beams from two separate laser shots are combined on a holographic plate, the first with the ablation laser firing, the second without to provide the reference flat background fringes. Figure 45(b) shows the results for ablation of an Al target in a background gas of 14 mtorr Ar gas, showing the density contours measured on the plate. For "production" use of the technique, of course, it would be desirable to replace the photographic plate with appropriate imaging optics and a CCD camera.

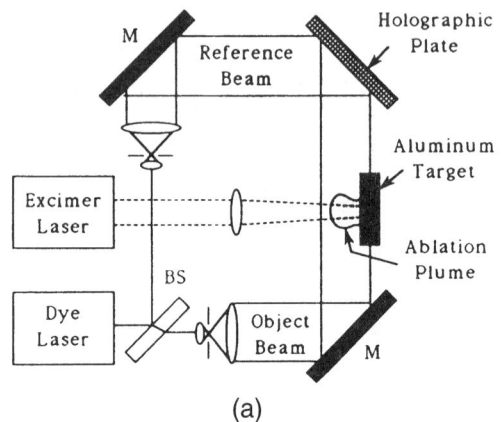

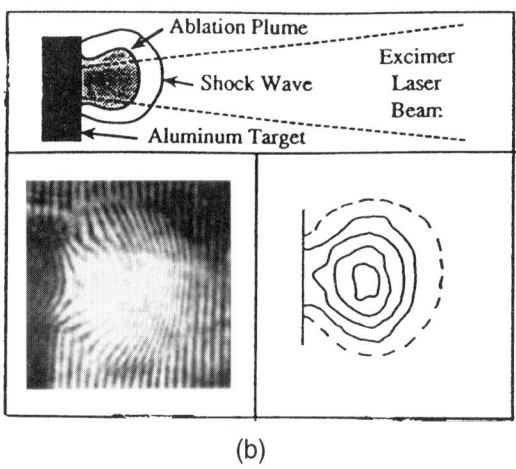

FIG. 45. Experiment using resonant time-resolved holographic interferometry to determine the spatial distribution of Al atoms ablated from an aluminum surface by a KrF excimer laser ($\lambda = 248$ nm, $\tau_p \sim 40$ ns) at fluences of 2–5 J·cm^{-2}. (a) Optical schematic of the experimental apparatus; the actual experiment is complicated by the vacuum chamber in which the ablation takes place. (b) Hologram and equiline density profiles of Al neutrals 4 μs after the arrival of the laser pulse, and in an atmosphere of 1 torr Ar gas. Each fringe shift represents a change in line-density of $1.4 \cdot 10^{14}$ cm^{-2}. From Lindley et al., Appl. Phys. Lett. **63**, 888 (1993).

The second example is measurement of the pressure and temperature of a thin polymer film during picosecond laser ablation by the use of coherent anti-Stokes Raman spectroscopy (CARS), a technique that provides a simultaneous sensitivity to chemical information in the plume [183]. The target was a film of poly(methylmethacrylate) (PMMA), a widely used photoresist and model material for many studies of laser-polymer interactions, doped with a near-IR-absorbing dye, IR-165, which absorbs at $\lambda = 308$ nm where PMMA is transparent. The 150-ps, Gaussian-profile ablating pulse from a Nd:YAG laser ($\lambda = 1064$ nm) was focused to a 100-μm spot on the surface of the film, generating a temperature gradient and a thermophysical shock wave because the laser pulse is shorter than the characteristic hydrodynamic relaxation time. Probe beams from synchronously pumped, narrow-band and a wide-band 50-ps dye lasers were used to probe a 50-μm-diameter near the center of the ablation spot.

Because the 808 cm^{-1} resonance in PMMA red-shifts by a known amount with temperature and blue-shifts a known amount with increasing pressure, a time-dependent measurement of the spectral shift can reveal both the temperature jump (which occurs on a picosecond time scale as the absorbing dye is heated) and the dynamics of the pressure wave (which occurs on a nanosecond, hydrodynamic time scale). Figure 46 shows the state diagram for the PMMA which is derived from the measurement of the CARS spectral shift. Note that the reason why this is possible is the vastly differing time scales for hydrodynamic versus thermal relaxation times (nanoseconds vs. microseconds). Also, because the CARS technique is based on a coherent process, the signal is much stronger than the incoherent background, leading to high sensitivity.

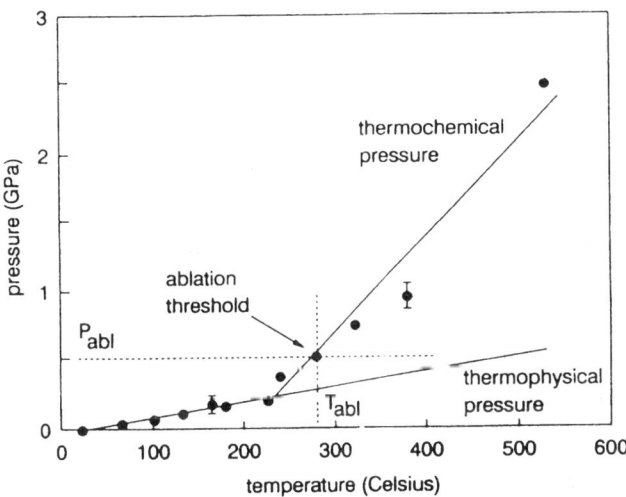

FIG. 46. Peak pressure vs. peak temperature for PMMA ablated by 150-ps laser pulses from a Nd:YAG laser ($l = 1064$ nm). This equation of state was calculated from the temperature-dependent redshifts and pressure-dependent blueshifts in the CARS spectrum of PMMA. From Hare et al., J. Appl. Phys. **77**(11), 5950–5960 (1993).

2.7 Modeling Laser-Induced Desorption and Ablation

The initial photon-surface interaction produces defects both by altering the electronic and geometrical structure of the surface and bulk, and by laser-induced particle emission. Models of laser-induced desorption—such as the MGF and KF models—can be used in special cases to calculate the detailed kinetics and dynamics of the desorption process. Because desorption is a localized event with the participation of only a few nearest-neighbor sites, it is possible, as in the DIMET process (Section 2.4), to simulate the process in some detail to build confidence in model interpretations.

Models for ablation pose a greater challenge because the ablation process and its relation to mesoscale physical properties are not well understood at present. In one of the earliest attempts to model semiconductor ablation, it was shown that weakly bonded atoms could be viewed as the nucleation sites for ablation because the weaker binding—for example, due to Jahn-Teller distortions at the surface of a solid—rendered these sites more susceptible to desorption. As these defects would grow in number, defects and defect clusters would rapidly evolve on the surface and in the sub-surface bulk. The changing optical, topological, and structural properties of the defect clusters then create the conditions for massive ejection of surface atoms, molecules, and clusters. This exceedingly generic mechanism, modeled using a Monte Carlo procedure, might operate in insulators and semiconductors alike [187]. The difference from one material to another would lie solely in the route taken to produce the defect clusters in the first place—but unravelling the precise mesoscale properties relevant to ablation is a major unmet challenge.

This section presents several recent studies that illustrate the range of theoretical and computational opportunities in desorption and ablation, as well as pointing out the range of data needed to help make the models more realistic.

2.7.1 Modeling Laser-Induced Desorption: Cluster Calculations

Because laser-induced desorption involves only the cooperative motion of a few atoms, a mesoscale simulation is often quite adequate to understand the basic driving forces. Calculations using finite clusters of atoms are appropriate to this task, particularly where surface states are involved, and have been used by several investigators. Among the important studies carried out thus far have been cluster calculations on laser-induced desorption from alkaline-earth fluorides by excitation of surface states [188], cluster calculations based on CNDO (complete neglect of differential overlap) in GaP [189], and studies of atomic emission from the (100) surface of NaCl [190]. In these more recent applications, cluster techniques are proving to be helpful in differentiating among competing models based on experimental observations.

One of the most important features of cluster-based calculations is the explicit recognition given to the role played by surface versus bulk phenomena. For example, the relaxation of two-dimensional excitons or electron-hole pairs may be quite different on the surface because of the broken symmetry which will influence the available relaxation modes. Localized phonon modes, which are well known to arise from the presence of defects in a perfect lattice [191], also may be created at surfaces by localized optical excitations, even when the excitation cannot be localized in three dimensions [192]. An excited F center, for example, can produce a local mode in two dimensions but not in three, according to these calculations: This agrees with other calculations carried out in conjunction with electron-stimulated desorption experiments on LiF, which showed the existence of a desorption threshold at the formation energy of the surface exciton.

In an application of the embedded-cluster Hartree-Fock computational technique, F and H centers were simulated near the (100) surface. Arguments based on the energetics required to remove halogen atoms from the surface (the same as the energy required to form an F center on the surface) seem to be insufficient to decide on the face of it whether or not the halogen atom can be emitted. However, a dynamical calculation in which the relaxation of the lattice is properly taken into account shows that, although an F center in the ground state is stable, an excited F center can provide enough energy to induce direct desorption of a neutral Na atom. The same calculations show that an exciton formed in the third layer down from the surface can also produce halogen-atom desorption, as shown in Fig. 47. In this calculation, several stages in the evolution of the desorbing atom are shown. Initially an exciton is created (Fig. 47(a)), and some distortion of the lattice is evident; as the halogen ions begin to move along the (110) direction, the lattice distorts and then relaxes in a configuration in which the F center has occupied the place of the original photoexcited Cl ion, and a neutral halogen atom has left the surface. Further testing of these calculations will have to include high-sensitivity measurements of the desorption of both halogen and alkali atoms from alkali halide crystal surfaces, a task complicated by the need for ultrasensitive detection methods for the halogen neutrals.

2.7.2 Modeling Laser Damage to Materials: Kinetics Calculations

Laser ablation clearly has something in common with the phenomenon of bulk optical damage in solids—but precisely what that something is remains an open question. Much of the early work in this area measures little more than the extrinsic damage caused by impurities or structural defects in the materials. However, in their pioneering work on intrinsic damage—defined as a microscopically visible, irreversible change in the material—Bräunlich, Kelly, and

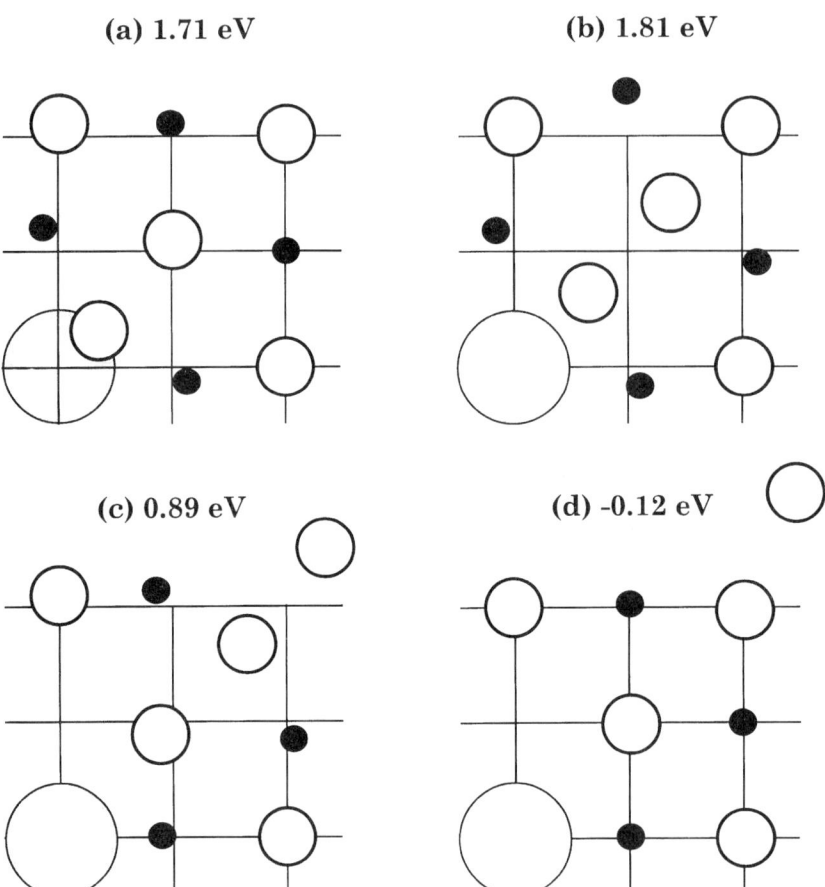

FIG. 47. Cluster calculation of the atomic configuration of a NaCl crystal following creation of a self-trapped exciton (STE) in the third layer from the surface. In (a), the initial position of the STE is shown, with halogen ions represented by small open circles, metal ions by small closed circles, and the STE by a large open circle; in (b), a halogen ion has begun to move at the saddle-point (predissociative) configuration of the potential energy surface; in (c) the halogen ion, now neutralized, has begun to move away from the surface (note that the Na surface ion has begun to relax); at (d), the desorption of the halogen atom is complete, and the charge balance in the crystal has been restored. An F center now sits at the site of the original excitation event. From Puchin *et al.*, *Phys. Rev. B* **47**(16), 10 760–10 768 (1993).

their collaborators [194] demonstrated that four-photon excitations in high-purity KBr damage the solid not by electron avalanche formation, but by multiphoton generation and subsequent heating of conduction-band electrons. Significantly the kind of kinetics calculations that they used to make this demonstration could probably be applied to map out the kinetics of desorption and ablation.

In their experiments, the Washington State group used luminescence from the self-trapped exciton in KBr as a thermometer to demonstrate that lattice heating by free-carrier generation and scattering could heat the lattice up to, but not beyond, the melting point [195]. Their experiment showed that the polaron-heating mode, which assumes strong conduction-electron coupling to the lattice, was insufficient to account for the observed increase in temperature, in particular for the saturation that sets in at high intensities as melting begins, as shown in Fig. 48. It turns out that the polaron model, which assumes that the heating mechanism is due to scattering of charge carriers from acoustic phonons, also saturates, but at a value much too high to reflect accurately the experimental data.

This work is notable for the extensive modeling of the electronic excitation and relaxation processes, which are incorporated into an elaborate rate-equation

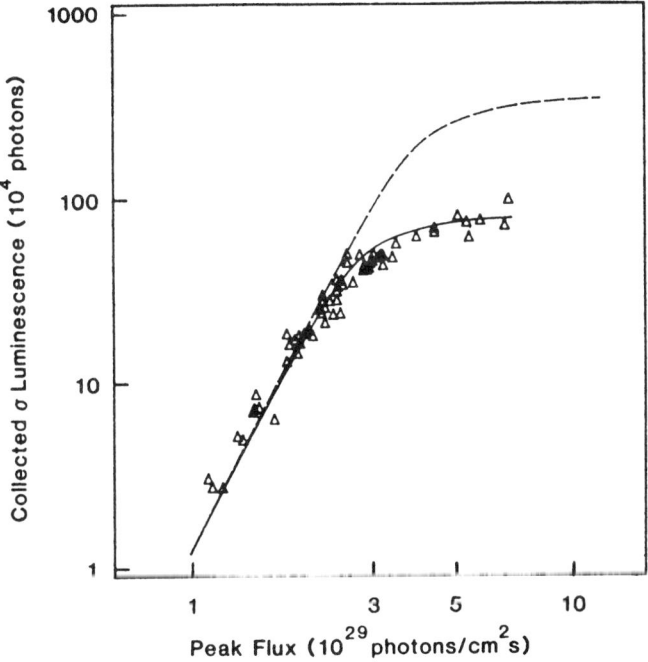

FIG. 48. Calculated (smooth curves) and measured (open triangles) self-trapped exciton luminescence as a function of laser flux for four-photon (bulk) excitation of KBr by 532-nm, 100-ps laser pulses. The STE luminescence is a measure of lattice temperature. The dashed curve assumes a polaron-heating model, the solid curve a free-carrier heating model, based on measured rate coefficients for the various excitation and loss channels in the experiment. From Shen et al., Phys. Rev. B **38**(5), 3494–3504 (1988).

model and solved to obtain the dynamics of the damage process. It should be pointed out that these kinetics calculations are impossible to carry out in detail without a knowledge of the relevant multiphoton excitation cross sections and without identifying the major energy loss (delocalization) channels. Moreover, as we have seen with the cluster calculations, it is likely that the surface will introduce an additional set of kinetics parameters that need to be measured by careful studies of laser-induced desorption.

2.7.3 Modeling Laser Ablation: Bond-Orbital Calculations

Thus far there have been almost no attempts to make use of the general tools of bond-orbital theory to describe laser-induced desorption or ablation, even though it seems a relatively natural one to use. Haglund and Itoh compared the potential energy surfaces for one-hole (in alkali halides) and two-hole (in compound semiconductors) bond disruption and found that indeed it required only a single hole on the halogen ion in an alkali halide crystal to put the system on an anti-bonding potential energy surface, whereas compound semiconductors required two holes in a covalent bond to achieve the same end [5].

The most sophisticated applications of bond-orbital techniques have been carried out by Stampfli and Bennemann, who have calculated the dynamics of ultrafast melting in semiconductors. In their model, the bond-orbital calculations are used to infer the softening and melting of the long-wavelength acoustic phonons due to excitation by a femtosecond laser pulse [196]. As we have pointed out earlier, the coupling of an electromagnetic wave with a solid occurs most efficiently through the transverse optical phonons. However, in a dense electron-hole plasma, the transverse acoustical phonon—responsible for shearing distortions—couples to the optical modes through the anharmonic component of the longitudinal optical phonon. When the cohesive energy is calculated from the bond-orbital parameters, it is possible to expand the cohesive energy in powers of the phonon amplitudes. At a density of the electron-hole plasma corresponding to excitation of 15% of the valence-band electrons into the conduction band, the acoustical phonon frequency goes to zero, indicating melting of the lattice and a structural transformation to a nearly cubic structure with metallic electronic properties. (See Fig. 49.) This kind of structural phase transformation, confirmed by numerous studies of the surface optical properties, has not yet been linked to ablation at this time scale. However, the possibility that it is the precursor to vaporization for sufficiently high temperature cannot be discounted; experimental evidence is not yet forthcoming.

2.7.4 Modeling Laser Ablation: Molecular Dynamics Calculations

Molecular dynamics has been used for years now to study the behavior of complex materials undergoing heavy-particle sputtering. In principle, molecular dynamics should also be interesting for laser sputtering: the computational demands are simple enough to handle relatively large systems, and simulations have been carried out for polymer ablation. However, the range of time constants found in laser desorption and ablation is so large that there are significant practical difficulties in carrying out the computations. Moreover, because molecular dynamics calculations are based on following the classical trajectories of the ions, quantum mechanical effects—for example, electronic excitation—are generally difficult to incorporate, although there have been important new developments in this area [197].

Garrison has recently proposed a "breathing sphere" model for laser-induced desorption from organic crystals, which may represent a way around some of these difficulties and provide some clues as to how to proceed with simpler crystalline systems [198]. For organic crystals, the systems are modeled as weakly interacting spheres, bound in the crystal by van der Waals forces. The vibrational modes of a given molecule are lumped together in a single degree of freedom and an adjustable relaxation time, which represents the rate at which the internal excitation is communicated to the surrounding volume. The internal excitation is simulated, in turn, as an excitation in a particular vibrational mode, or as the end-product of internal conversion following an electronic excitation by an ultraviolet or visible laser. Guest species—such as the analyte molecules in a matrix-assisted laser desorption-ionization problem—can be incorporated into the host matrix by permitting them to occupy the space of the equivalent number of host molecules; a second relaxation time allows the simulation of interactions between the host and guest molecules. Perhaps the most important features of the model are that pressure, temperature, and velocity distributions can be calculated realistically, permitting detailed comparisons with experimental results in an intuitively satisfying way.

The results of a calculation for a "typical" molecular solid containing analyte molecules, irradiated by an ultraviolet laser, are shown in Fig. 50; the internal parameters of the simulated organic solid are shown in Table VIII. The simulation assumes a Beers law exponential absorption profile as a function of depth, and a laser spot diameter much larger than the penetration depth so that a one-dimensional absorption profile is adequate. Almost 60,000 molecules are incorporated into the simulation, arranged in monolayers containing 560 molecules each. As light is absorbed by randomly selected host molecules, it is transferred to the surrounding molecules with a time constant of 10 ps. In this case, the heating rate is more rapid than the thermal diffusion time, so that a strong pressure wave builds up in the irradiated volume, driving a compressional wave into the unheated part of the solid and simultaneously accelerating the top layers in the direction of the surface normal. This is consistent with the phase explosion picture of plume development, which is discussed in greater detail in Chapter 5.

(a)

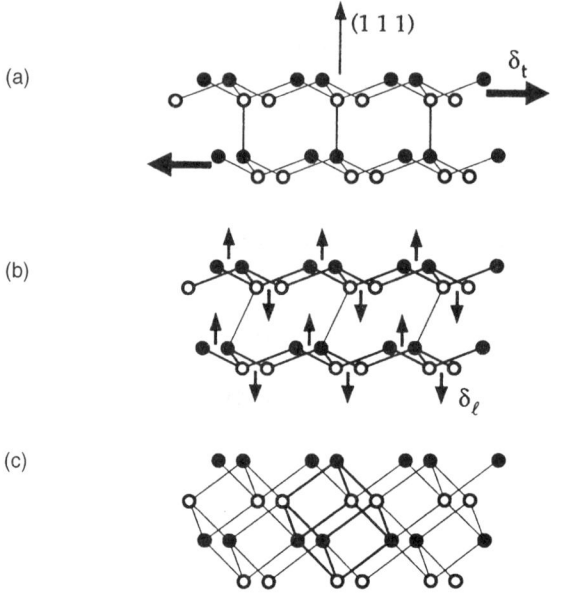

(b)

(c)

(d) $\xi_O = 0.15$

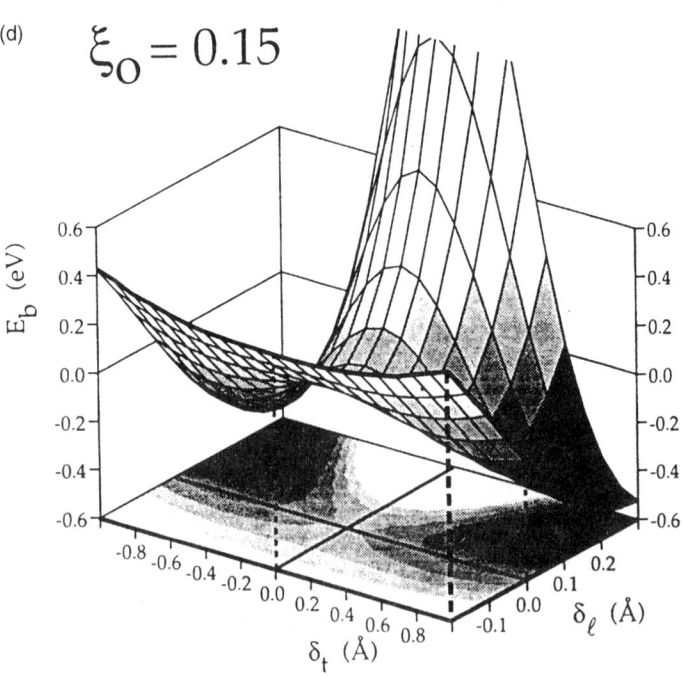

2.8 Conclusion and Outlook

While the bulk of laser-induced desorption and ablation studies treated in this review have been carried out on simple well-characterized metals, insulators, and semiconductors, there is substantial and growing interest in laser-surface interactions in more complex materials systems. Polymers, of course, have been studied in great detail; however, it is difficult to characterize the specific surface sites involved in particle ejection, so that laser-induced desorption experiments have hardly been considered in this field. On the other hand, desorption studies in complex crystals are both of fundamental interest and critical to such applications as matrix-assisted laser desorption-ionization mass spectrometry. In looking forward to the future of this field, it may be interesting to note three specific examples which illustrate the application of the general principles of this chapter to more complex materials.

In experiments which parallel those described in Section 2.5 for infrared free-electron laser irradiation, Hess et al. [199] have shown recently that CO is desorbed from cleaved natural calcite (rhombohedral $CaCO_3$) during UV irradiation. The resulting CO emissions do not appear to be thermal, for example, the CO state distributions are not equilibrated in any degree of freedom. Only a 6% population of the $v = 1$ vibrational level is observed (94% in $v = 0$). The rotational state distribution is non-Boltzmann, exhibiting a step structure every 4 or 5 rotational states and the CO velocity distribution can be characterized by a temperature ($T = 110 \pm 15$ K) significantly lower than that of the substrate ($T = 295$ K). The unusually low CO velocity distribution cannot be explained by the standard desorption models such as Menzel, Gomer, and Redhead (MGR) or Knotek and Feibelman. Further, it is argued that no equilibrium thermal desorption process can account for this phenomena and that the CO, once formed, departs the calcite surface promptly. The calcite and sodium nitrate systems display distinctly different product state distributions following similar S1 excitation. It is possible that the differences in the detailed nature of these electronic excitations produces the distinctive NO and CO state distributions

---

FIG. 49. Atomic structural transformation induced by ultrafast melting of the diamond lattice in a Si lattice. (a) Ideal diamond structure, with heavy arrows showing the transverse acoustic distortion (TA) δ_t induced by the lightwave field; (b) intermediate structure resulting from coupling between the (TA) and *longitudinal* optical (LO) phonons, with the shorter arrows marked δ_l indicating the direction of distortion; (c) the new structure resulting from the combination of longitudinal optical and transverse acoustic phonons. In (d), the cohesive energy is calculated as a function of the two parameters δ_l and δ_t. Where the stable diamond lattice structure had a deep minimum at the point $\delta_l = 0 = \delta_t$, the new melted structure has a saddle point there. From Stampfli and Bennemann, *Phys. Rev. B* **49**, 7299 (1994).

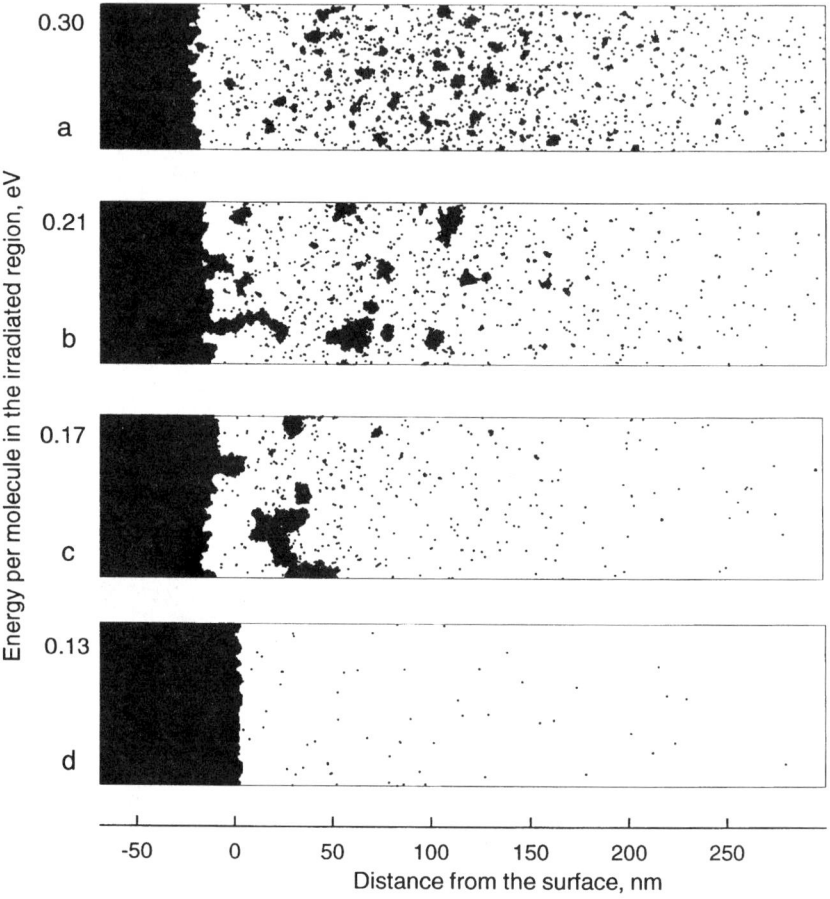

FIG. 50. Molecular dynamics simulation using the breathing sphere model, showing the two-dimensional distribution of molecules desorbed from an organic crystal as a function of distance from the surface for various values of the initial energy per molecules. From Zhigilei et al., Chem. Phys. Lett., in press (1997).

[200]. These results have been substantiated and extended by studies of photon-stimulated desorption of NO and O from $NaNO_3$, an ionic molecular crystal isoelectronic to $CaCO_3$ [201]. The yields of $O(3p)$ and $NO(^2\Pi)$ are linear in the density of electronic excitation down to fluences of $1~\mu J \cdot cm^{-2}$. As is typical of molecular desorption, the velocity distribution is bimodal and non-Boltzmann, consistent with near-surface dissociation events. This work suggests the possibility that the NO_3^- anion plays a role analogous to that played by the halogen anion in alkali halides; it may even be that an excitonic mechanism can be

TABLE VIII. Parameters for the molecular dynamics calculations of laser ablation displayed in Fig. 50. This is a two-dimensional simulation with dimensions 81 × 210 nm containing 58,800 molecules with molecular weight 100 Da. For a hypothetical surface with $5 \cdot 10^{14}$ surface sites per cm^2, a peak flux density of 10^9 photons·s^{-1}·molecule^{-1} corresponds to an intensity of approximately $3 \cdot 10^5$ W·cm^{-2}.

Case	Energy density per molecule	Penetration depth $(1/\alpha)$	Pulse duration	Peak flux density per molecule
(a)	0.30 eV	32 nm	15 ps	$5.4 \cdot 10^9$ photons·s^{-1}
(b)	0.21 eV	32 nm	15 ps	$3.8 \cdot 10^9$ photons·s^{-1}
(c)	0.17 eV	32 nm	15 ps	$3.1 \cdot 10^9$ photons·s^{-1}
(d)	0.13 eV	32 nm	15 ps	$2.3 \cdot 10^9$ photons·s^{-1}

identified which would parallel the decay of the *F-H*-center pair believed to be responsible for the laser-induced excitonic desorption of alkalis and halogens from alkali halides.

In another experiment which challenges the understanding of localization and particle ejection, the photoejection of Na monomers and dimers from solid films of polydimethylsiloxane (PDMS) has been found to vary with laser intensity and frequency [202]. The experiments are performed in PDMS-coated cells and the desorption is initiated by a continuous laser tuned to either the D_1 or D_2 resonance lines of atomic sodium. The desorption process is monitored by tracking the fluorescence induced by the 488-nm line of an Ar$^+$ laser. Even at incident power levels of 1 mW or less, strong atomic fluorescence is observed, indicating that free Na atoms and dimers are being produced in the cell. The desorption is apparently facilitated by the presence of the Si-O backgone of the siloxanes, since it does *not* occur for a number of other hydrocarbon or crown ether films. The detailed mechanism appears to involve a charge-transfer effect which, like the photoelectric effect, is dependent on the laser light frequency; the desorption of neutral atoms and dimers occurs by a Landau-Zener-type curve crossing from the charged to a neutral potential-energy surface.

Finally, we note that the kind of state-resolved molecular desorption experiments which have in the past been limited to metal surfaces are now being performed on well-characterized oxides. The technology for this involves the growth of epitaxial oxide films on Ni single-crystal surfaces in ultra-high vacuum; the surface state is characterized in detail by a variety of standard spectroscopies [203]. Desorption was initiated by KrF (248 nm) or ArF (193 nm) laser pulses; the internal state of the desorbing molecules was determined by resonant ionization mass spectrometry. On both of the surfaces probed in this experiment, a bimodal, nonthermal velocity distribution is observed. The bimodality is believed to arise from the differences in the bending vibration at the moment of excitation; the vibrational and translational temperatures are uncorrelated, indicating that the desorbing molecules are not in thermal equilibrium with the surface. Even more interesting is that there is a spin-dependence of the molecular final states which apparently arises from the differing magnetic properties of the polar and nonpolar surfaces.

These three examples suggest that, even in complex materials systems, elucidation of the desorption phenomenology at the atomic scale requires that both quantum-physical and -chemical effects be taken into account. The tools to deal with these phenomena are already at hand and should, in principle, hold the key to future studies of molecular and atomic desorption phenomena. Our understanding of the mesoscale physics of ablation, on the other hand, is much less developed at this time. As the data covering the transition from desorption to ablation on well-characterized materials become more extensive, model building based on bond-orbital theories and molecular dynamics calculations may be especially useful, since these look both inward to atomic and molecular properties and outward to the mesoscale characteristics of materials.

Acknowledgments

This work was partially supported by the Office of Naval Research through the Medical Free-Electron Laser Program, Contract Number N00014-94-C-1023. It is a pleasure to thank Drs. Hee Kuwon Park and Oguz Yavas of Vanderbilt, for their assistance with research. I owe particular debts to Professors Noriaki Itoh, Osaka Institute of Technology, and Eckhart Matthias, Free University of Berlin, for their hospitality and for many stimulating discussions.

References

1. Ready, J. F. *Effects of High Power Laser Radiation*. Academic Press, New York (1971)
2. The classic papers are Bloembergen, N. Laser-induced electric breakdown in solids. *IEEE J. Quantum Electron*. **QE-10**(3), 375–386 (1974); Bloembergen, N. Role of cracks, pores, and absorbing inclusions on laser induced damage threshold at surfaces of transparent dielectrics. *Appl. Optics* **12**(4), 661–664 (1973).
3. Matthias, E., and Green, T. A. Laser induced desorption, in *Desorption Induced by Electronic Transitions DIET IV*, G. Betz and P. Varga (eds.), pp. 112–127. Springer Verlag, Berlin, Heidelberg, 1990.
4. Kelly, R., Miotello, A., Braren, B., Gupta, A., and Casey, K. Primary and secondary mechanisms in laser-pulse sputtering. *Nucl. Instrum. Meth. B* **65**, 187–199 (1992).
5. Haglund, R. F., Jr., and Itoh, N. Electronic processes in laser ablation of semiconductors and insulators, in *Laser Ablation: Principles and Application*, John C. Miller (ed.), Springer Series in Materials Science **28**. Springer-Verlag, Berlin, 1994.

6. Kelly, P., ed. Laser modification of materials, a special issue of *Opt. Eng.* **28**, 1025–1044 (1989).
7. Itoh, N., Kanasaki, J., Okano, A. and Nakai, Y. Laser-beam interaction with defects on semiconductor surfaces: An approach to generation of defect-free surfaces. *Ann. Rev. Mater. Sci.* **25**, 97–127 (1995).
8. Miller, John C., and Haglund, R. F., Jr., eds. Laser ablation: mechanisms and applications. *Proc. of the First Conference on Laser Ablation*, Lecture Notes in Physics **389**. Springer-Verlag, Berlin, 1991.
9. Miller, John C. and Geohegan, David B., eds. Laser ablation: mechanisms and applications II. *Proc. of the Second International Conference on Laser Ablation* (COLA 1993), Knoxville, TN, April 1993. AIP Conference Proceedings **288**.
10. Fogarassy, E., Geohegan, D., and Stuke, M. Laser ablation. *Proc. of Symposium F: Third International Conference on Laser Ablation* (COLA 1995) of the 1995 E-MRS Spring Conference, Strasbourg, France, May 22–26, 1995. Elsevier, Amsterdam, 1996.
11. Chase, L. L., Laser ablation and optical surface damage, in *Laser Ablation: Principles and Applications*, John C. Miller (ed.), Springer Series in Materials Science **25**. Springer-Verlag, Berlin, 1994.
12. Wood, R. F. *Laser Damage in Optical Materials*. Adam Hilger, Bristol, 1986.
13. Battaglin, G., Boscolo-Boscoletto, A., Caccavale, F. C., Gonella, F., Mazzi, G., Mazzoldi, P., Quaranta, A., and Tramontin, L. Modification of glasses for optical purposes, in *From Galileo's Occhialino to Optoelectronics*, P. Mazzoldi (ed.), pp. 757–762. World Scientific, Singapore, 1993.
14. Taylor, A. J., Gibson, R. B., and Roberts, J. P. Two-photon absorption at 248 nm in ultraviolet window materials. *Opt. Lett.* **13**, 814 (1988).
15. Nielsen, H. B., Reif, J., Matthias, E., Westin, E., and Rosen, A. Multiphoton-induced desorption from $BaF_2(111)$, in *Desorption Induced by Electronic Transitions—DIET III*, R. H. Stulen and M. L. Knotek (eds.), p. 266. Springer-Verlag, Heidelberg, 1986.
16. Nonlinear optical properties of materials found in Table II are taken from: DeSalvo, R., Said, A. A., Hagan, D. J., Van Stryland, E. W., and Sheik-Bahae, M. Infrared to ultraviolet measurements of two-photon absorption and n_2 in wide bandgap solids. *IEEE J. Quant. Electron.* **32**, 1334–1333 (1996); Van Stryland, E. W., Vanherzeele, H., Woodall, M. A., Soileau, M. J., Smirl, A. L., Guha, S., and Boggess, T. F., Two photon absorption, nonlinear refraction, and optical limiting in semiconductors. *Opt. Eng.* **24**, 613–623 (1985); Weber, M. J., Milam, D., and Smith, W. L., Nonlinear refractive index of glasses and crystals. *Opt. Eng.* **17**, 463–469 (1978).
17. Andrews, D. L. A simple statistical treatment of multiphoton absorption. *Am. J. Phys.* **53**, 1001 (1984).
18. Lines, M. E. Bond-orbital theory of linear and nonlinear electronic response in ionic crystals: I. Linear response. *Phys. Rev. B* **41**(6), 3372–3382 (1990).
19. Lines, M. E. Bond-orbital theory of linear and nonlinear electronic response in ionic crystals: II. Nonlinear response. *Phys. Rev. B* **41**(6), 3383–3390 (1990).
20. Harrison, W. A. *Electronic Structure and the Properties of Solids*. Dover, New York, 1989. This is a reprint of the 1980 book of the same title published by W. H. Freeman, with a new introduction by the author.
21. Cox, P. A. *The Electronic Structure and Chemistry of Solids*. Oxford University Press, New York, 1987.
22. Harrison, W. A. Interatomic interactions in covalent and ionic solids. *Phys. Rev. B* **41**(9), 6008–6019 (1990).

23. Yoffa, E. J. Dynamics of dense laser-induced plasmas. *Phys. Rev. B* **21**(6), 2415–2426 (1980). For a general discussion, see Klingshirn, C. F. *Semiconductor Optics*. Springer-Verlag, Berlin, 1996.
24. Hayes, W., and Stoneham, A. M. *Defects and Defect Processes in Nonmetallic Solids*. John Wiley & Sons, New York, 1985.
25. The polaron data used in this table are taken variously from Brown, F. C. Experiments on the polaron, in *Polarons and Excitons*, C. G. Kuper and G. D. Whitfield (eds.), pp. 323ff. Oliver and Boyd, Edinburgh, 1963; Mahan, G. Polarons in heavily doped semiconductors, in *Polarons in Ionic Crystals and Polar Semiconductors*, J. T. Devreese (ed.), pp. 533–657, North-Holland, Amsterdam, 1972; Kittel, C. *Solid State Physics*, 4th ed., pp. 390–391. John Wiley & Sons, New York, 1967.
26. For detailed discussions of the plasmon polariton and its optical properties, see Kreibig, Uwe, and Vollmer, Michael. *Optical Properties of Metal Clusters*, Springer Series in Materials Science 25. Springer-Verlag, Berlin, 1995.
27. Shibata, T., Iwai, S., Tokizaki, T., Tanimura, K., Nakamura, A., and Itoh, N. Femtosecond spectroscopic studies of the lattice relaxation initiated by interacting electron-hole pairs under relaxation in alkali halides. *Phys. Rev. B* **49**, 13-255–13-258 (1994).
28. Iwai, S., Tokizaki, T., Nakamura, A., Tanimura, K., and Itoh, N. One-center small polarons as short-lived precursors in self-trapping processes of holes and electron-hole pairs in alkali iodides. *Phys. Rev. Lett.* **76**, 1691–1694 (1996).
29. Historical references and an excellent overview may be found in Knox, R. S. *Theory of Excitons*. Academic Press, New York, 1963.
30. Howard, R. E., and Smoluchowski, R. Formation of interstitials in alkali halides by ionizing radiation. *Phys. Rev.* **116**, 314–315 (1959).
31. But see Reif, J., Tepper, P., Matthias, E., Westin, E., and Rosén, A. Surface structure of cubic ionic crystals studied by optical second-harmonic generation. *Appl. Phys. B* **46**, 131–138 (1988).
32. Weber, M. J., Milam, D., and Smith, W. L. Nonlinear refractive index of glasses and crystals. *Opt. Eng.* **17**(5), 463–469 (1978).
33. See, for example, Boyd, R. W. *Nonlinear Optics*. Academic Press, San Diego, 1992.
34. For a discussion of the interpretation of experiments, see Soileau, M. J., Williams, W. E., Mansour, N., and Van Stryland, E. W. Laser-induced damage and the role of self-focusing. *Opt. Eng.* **28**(10), 1133–1144 (1989).
35. Phillips, J. C. *Bonds and Bands in Semiconductors*. John Wiley & Sons, New York, 1973.
36. Except possibly for nanometer-size metallic particles. See, for example, Bigot, J.-Y., Merle, J.-C., Cregut, O., and Daunois, A. Electron dynamics in copper metallic nanoparticles probed with femtosecond optical pulses. *Phys. Rev. Lett.* **75**(25), 4702–4705 (1995).
37. Fujimoto, J. G., Lieu, J. M., Ippen, E. P., and Bloembergen, N. Femtosecond laser interaction with metallic tungsten and nonequilibrium electron and lattice temperatures. *Phys. Rev. Lett.* **53**(19), 1837–1840 (1984).
38. Yoo, K. M., Zhao, X. M., Siddique, M., Alfano, R. R., Osterman, D. P., Radparvar, M., and Cunniff, J. Femtosecond thermal modulation measurements of electron-phonon relaxation in niobium. *Appl. Phys. Lett.* **56**(19), 1908–1910 (1990).
39. Brorson, S. D., Fujimoto, J. G., and Ippen, E. P. Femtosecond electronic heat-transport dynamics in thin gold films. *Phys. Rev. Lett.* **59**(17), 1962–1965 (1987).

40. Hohlfeld, J., Grosenick, D., Conrad, U., and Matthias, E. Femtosecond time-resolved reflection second-harmonic generation on polycrystalline copper. *Appl. Phys. A* **60**, 137–142 (1995).
41. Lannoo, M., and Bourgoin, J. *Point Defects in Semiconductors* (2 volumes). Springer-Verlag, Berlin, 1981 and 1983.
42. Nishimura, H. Luminescence and self-trapping of excitons in alkali halides, in *Defect Processes Induced by Electronic Excitation in Insulators*, N. Itoh (ed.). World Scientific, Singapore, 1989.
43. Toyazawa, Y. Theory of excitons in phonon fields, in *Excitonic Processes in Solids*, M. Ueta, H. Kanzaki, K. Kobayashi, Y. Toyozawa, and E. Hanamura (eds.). Springer-Verlag, Heidelberg, 1986.
44. Rashba, E. I., and Sturge, M. D., Eds. *Excitons*, p. 543. North Holland, Amsterdam, 1982.
45. Toyozawa, Y. Charge transfer instability with structural change. I. Two-sites two-electrons system. *J. Phys. Soc. Jpn.* **50**, 1861–1867 (1981).
46. Anderson, P. W. Model for the electronic structure of amorphous semiconductors. *Phys. Rev. Lett.* **34**, 953–955 (1975).
47. Williams, R. T., and Song, K. S. The self-trapped exciton. *J. Phys. Chem. Sol.* **51**, 679–716 (1990).
48. Williams, R. T. Photochemistry of F-center formation in halide crystals. *Semicond. and Insul.* **3**, 251–283 (1978).
49. Itoh, N., and Tanimura, K. Formation of interstitial-vacancy pairs by electronic excitation in pure ionic crystals. *J. Phys. Chem. Solids* **51**, 717–735 (1990).
50. Itoh, N., Tanimura, K., Stoneham, A. M., and Harker, A. H. The lattice relaxation energy associated with self-trapping of a positive hole and an exciton in alkali halides. *J. Phys. Condensed Matter* **1**, 3911–3918 (1989).
51. Khoo, G. S., Ong, C. K., and Itoh, N. The multi-hole localization mechanism for particle emission from semiconductor surfaces. *J. Phys.: Condens. Matter* **5**, 1187–1194 (1993).
52. Fahey, P. M., Griffin, P. B., and Plummer, J. D. Point defects and dopant diffusion in silicon. *Rev. Mod. Phys.* **61**, 289–384 (1989). The original theoretical prediction of the V center was by Baraff, G. A., Kane, E. O., and Schlüter, M., Silicon vacancy: A possible "Anderson negative-U system." *Phys. Rev. Lett.* **43**, 956–959 (1979), while the experimental demonstration came from Watkins, G. D., and Troxell, J. R. Negative-U properties for point defects in silicon. *Phys. Rev. Lett.* **44**, 593–596 (1980).
53. Theis, T. N., Mooney, P. M., and Wright, S. L. Electron localization by a metastable donor level in n-GaAs: a new mechanism limiting the free-carrier density. *Phys. Rev. Lett.* **60**, 361–364 (1988).
54. Itoh, N., and Nakayama, T. Mechanism of neutral particle emission from electron-hole plasma near solid surface. *Phys. Lett.* **92A**, 471–475 (1982).
55. Itoh, N., Nakayama, T., and Tombrello, T. Electronic-excitation mechanism in sputtering induced by high-density electronic excitation. *Phys. Lett.* **108A**, 480–484 (1985).
56. Sumi, H. *Surf. Sci.* **248**, 382–410 (1991).
57. Dexter, D. L., Klick, C. C., and Russell, G. A. Criterion for the occurrence of luminescence. *Phys. Rev.* **100**, 603–605 (1956).
58. Tanimura, K., Suzuki, T., and Itoh, N. Resonance Raman scattering of the self-trapped exciton in alkali halides. *Phys. Rev. Lett.* **68**, 635–638 (1992).

59. Henry, C. H., and Lang, D. V. Nonradiative capture and recombination by multiphonon emission in GaAs and GaP. *Phys. Rev. B* **15**, 989–1016 (1977).
60. Schou, J. Sputtering of frozen gases, *Nucl. Instrum. Methods in Phys. Res.* **B27**, 188–200 (1987).
61. Dean, P. J., and Herbert, D. C. in *Excitons*, K. Cho (ed.), Topics in Current Physics **14**, Section 3.3.3. Springer-Verlag, Berlin, 1979; Bouche-Bruevich, D. L., and Landsberg, E. G. *Phys. Stat. Solida* **29**, 9 (1968).
62. Pines, D. *Elementary Excitations in Solids*. Benjamin, New York, 1964.
63. Smith, R. A. *Semiconductors*. Cambridge University Press, Cambridge, 1978.
64. van Vechten, J. A., in *Cohesive Properties of Semiconductors Under Laser Irradiation*, L. D. Laude (ed.), p. 49. Martinus Nijhoff, Hague, 1983.
65. Chadi, D. J., and Chang, K. J. Metastability of the isolated arsenic-antisite defect in GaAs. *Phys. Rev. Lett.* **60**, 2187–2190 (1988); Dabrowski, J., and Sheffler, M. Isolated arsenic-antisite defect in GaAs and the properties of ELZ. *Phys. Rev. B* **40**, 10391–10407 (1989).
66. Shluger, A., Georgiev, M., and Itoh, N. Self-trapped excitons and interstitial-vacancy pairs in oxides. *Phil. Mag. B* **63**, 955–964 (1991).
67. Van Vechten, A., Tsu, R., and Saris, F. W. Nonthermal pulsed laser annealing of Si: plasma annealing. *Phys. Lett.* **74A**, 422–427.
68. Wautelet, M., and Laude, L. D. Cohesion of solids under laser irradiation. *Appl. Phys. Lett.* **36**, 197–199 (1980).
69. Williams, R. T., Song, K. S., Faust, W. L., and Leung, C. H. Off-center self-trapped excitons and creation of lattice defects in alkali halide crystals. *Phys. Rev. B* **33**, 7232–7240.
70. Tokizaki, T., Makimura, T., Akiyama, H., Nakamura, A., Tanimura, K., and Itoh, N. Femtosecond cascade-exciton spectroscopy for nonradiative deexcitation and lattice relaxation of the self-trapped exciton in NaCl. *Phys. Rev. Lett.* **67**, 2701–2704 (1991).
71. Manghi, F., Bertoni, C. M., Calandra, C., and Molinari, E. Theoretical study of the electronic structure of GaP(110), *Phys. Rev. B* **24**, 6029–6042 (1981).
72. Okano, A., Hattori, K., Nakai, Y., and Itoh, N. Temperature dependence of the laser-induced Ga^0 emission from a GaP(11)) surface. *Surf. Sci.* **258**, L671–L676 (1991).
73. Kanasaki, J., Okano, A., Ishikawa, K., Nakai, Y., and Itoh, N. Dynamical interaction of surface electron-hole pairs with surface defects: Surface spectroscopy monitored by particle emission. *Phys. Rev. Lett.* **70**(16), 2495–2498 (1993).
74. Avouris, Ph., and Walkup, R. E. Fundamental mechanisms of desorption and fragmentation induced by electronic transitions at surfaces. *Ann. Rev. Phys. Chem.* **40**, 172–206 (1989).
75. Stampfli, P., and Bennemann, K. H. Theory for the instability of the diamond structure of Si, Ge and C induced by a dense electron-hole plasma. *Phys. Rev. B* **42**, 7163 (1990).
76. Dickinson, J. T., Langford, S. C., Shin, J. J., and Doering, D. L. Positive ion emission from excimer laser excited MgO surfaces. *Phys. Rev. Lett.* **73**, 2630–2633 (1994).
77. Williams, R. T., and Song, K. S. Off-center self-trapped excitons and creation of lattice defects in alkali halid crystals, in *Self-Trapped Excitons*, pp. 7232–7240. Springer-Verlag, Berlin, 1994.
78. Menzel, D., and Gomer, R. Desorption from metal surfaces by low-energy electrons. *J. Chem. Phys.* **41**, 3311–3328 (1964).

79. Redhead, P. A. Interaction of slow electrons with chemisorbed oxygen. *Can. J. Phys.* **42**, 886 (1964).
80. Antoniewicz, P. R. Model for electron- and photon-stimulated desorption. *Phys. Rev. B* **21**(9), 3811–3815 (1980).
81. Knotek, M. L., and Feibelman, P. J. Ion desorption by core-hole Auger decay. *Phys. Rev. Lett.* **40**, 964–967 (1978).
82. Burgess, D., Cavanagh, R. R., and King, D. S. Laser-induced desorption: Thermal and nonthermal pathways. *J. Chem. Phys.* **88**(10), 6556–6569 (1988); Richter, L. J., Buntin, S. A., Cavanagh, R. R., and King, D. S. Non-Boltzmann rotational and inverted spin-orbit state distributions for laser-induced desorption of NO from Pt(111). *J. Chem. Phys.* **89**(8), 5344–5345 (1988); Buntin, S. A., Richter, L. J., Cavanagh, R. R., and King, D. S. Optically driven surface reactions: Evidence for the role of hot electrons. *Phys. Rev. Lett.* **61**, 1321–1325 (1988).
83. Prybyla, J. A., Heinz, T. F., Misewich, J. A., Loy, M. M. T., and Glownia, J. H. Desorption induced by femtosecond laser pulses. *Phys. Rev. Lett.* **64**, 1537–1540 (1990).
84. Budde, F., Heinz, T. F., Loy, M. M. T., Misewich, J. A., de Rougemont, F., and Zacharias, J. Femtosecond time-resolved measurement of desorption. *Phys. Rev. Lett.* **59**, 1460–1463 (1991).
85. Prybyla, J. A., Tom, H. W. K., and Aumiller, G. D. Femtosecond time-resolved surface reaction: Desorption of Co from Cu(111) in <325 fs. *Phys. Rev. Lett.* **68**, 503 (1992).
86. Misewich, J. A., Kalamarides, A., Heinz, T. F., Höfer, U., and Loy, M. M. T. Vibrationally assisted electronic desorption: Femtosecond surface chemistry of O_2/Pd(111). *J. Chem. Phys.* **100**(1), 736–739 (1994).
87. A particularly interesting desorption study is Kao, R.-J., Busch, D. G., Gnomes da Costa, D., and Ho, W. Femtosecond versus nanosecond surface photochemistry: O_2 + CO on Pt(111) at 80 K. *Phys. Rev. Lett.* **70**(26), 4098–4101 (1993).
88. Busch, D. G., Gao, S., Pelak, R. A., Booth, M. F., and Ho, W. Femtosecond desorption dynamics probed by time-resolved velocity measurements. *Phys. Rev. Lett.* **75**, 673–676 (1995).
89. Budde, F., Heinz, T. F., Kalamarides, A., Loy, M. M. T., and Misewich, J. A. Vibrational distributions in desorption induced by femtosecond laser pulses: Coupling of adsorbate vibration to substrate electronic excitation. *Surf. Sci.* **283**, 143–157 (1993). The lengthy reference list in this paper is a useful guide to other relevant literature in this field.
90. Similar effects have been observed with energetic-ion bombardment of surfaces where a high local density of electronic excitation is also attained. See Johnson, R. E., Pospieszalska, M., and Brown, W. L. Linear-to-quadratic transition in electronically stimulated sputtering of solid N_2 and O_2. *Phys. Rev. B* **44**, 7263–7272 (1991).
91. Misewich, J. A., Heinz, T. F., and Newns, D. M. Desorption induced by multiple electronic transitions. *Phys. Rev. Lett.* **68**, 3737–3740 (1992).
92. Brandbyge, M., Hedegård, Heinz, T. F., Misewich, J. A., and Newns, D. M. Electronically driven adsorbate excitation mechanism in femtosecond-pulse laser desorption. *Phys. Rev. B* **52**, 6042–6056 (1995).
93. Lee, I., Parks, J. E., Callcott, T. A., and Arakawa, E. T. Surface-plasmon-induced desorption by the attenuated-total-reflection method. *Phys. Rev. B* **39**(11), 8012–8014 (1989).
94. Lee, I., Callcott, T. A., and Arakawa, E. T. Desorption studies of metal atoms using laser-induced surface-plasmon excitation. *Phys. Rev. B* **47**, 6661–6665 (1993).

95. Hoheisel, W., Vollmer, M., and Träger, F. Desorption of metal atoms with laser light: Mechanistic studies. *Phys. Rev. B* **48**(17), 463 (1993); Vollmer, M., Weidenauer, R., Hoheisel, W., Schulte, U., and Träger, F. Size manipulation of metal particles with laser light. *Phys. Rev. B* **40**(12), 509–512 (1989). See also the earlier paper by W. Hoheisel, K. Jungmann, M. Vollmer, R. Weidenauer, and F. Träger, Desorption stimulated by laser-induced surface-plasmon excitation. *Phys. Rev. Lett.* **60**, 1649–1652 (1989).
96. The experimental details are described at length in Hoheisel, W., Schulte, U., Vollmer, M., and Träger, F. Metal particles on surfaces—Desorption, optical spectra and laser-induced size manipulation. *Appl. Phys. A* **51**, 271 (1990).
97. Monreal, R., and Apell, S. P. Electromagnetic-field-enhanced desorption of atoms. *Phys. Rev. B* **41**(11), 7852–7855 (1990).
98. Perner, M., Bost, P., Lemmer, U., von Plessen, G., Feldmann, J., Becker, U., Mennig, M., Schmitt, M., and Schmidt, H. Optically induced damping of the surface plasmon resonance in gold colloids. *Phys. Rev. Lett.* **78**, 2192–2195 (1997).
99. For a summary of studies on GaP and references to the critical early papers on this technique, see Hattori, K., Okano, A., Nakai, Y., and Itoh, N. Laser-induced electronic processes on GaP (110) surfaces: Particle emission and ablation initiated by defects. *Phys. Rev. B* **45**, 8424 (1992).
100. Hattori, K., Okano, A., Nakai, Y., Itoh, N., and Haglund, R. F., Jr. Characterization of surface defects by means of laser-induced Ga^0 emission from GaP surfaces. *J. Phys. Condens. Matter* **3**, 7001–7006 (1991).
101. Okano, A., Hattori, K., Nakai, Y., and Itoh, N. Temperature dependence of the laser-induced Ga^0 emission from a GaP(110) surface. *Surf. Sci. Lett.* **258**, L671–L675 (1991).
102. Kanasaki, J., Okano, A., Ishikawa, K., Nakai, Y., and Itoh, N. Dynamical interation of surface electron-hole pairs with surface defects: surface spectroscopy monitored by particle emissions. *Phys. Rev. Lett.* **70**, 2495 (1993).
103. Manghi, F., Bertoni, C. M., Calandra, C., and Molinari, E. Theoretical study of the electronic structure of GaP(11)). *Phys. Rev. B* **24**(10), 6029–6042 (1981).
104. Townsend, P. D., and Elliott, D. J. Defect formation in KI with ultraviolet light. *Phys. Lett.* **28A**, 587–588 (1969).
105. Kanzaki, H., and Mori, T. Photon-stimulated desorption of neutrals from silver and alkali halides. *Phys. Rev. B* **29**, 3573–3585 (1984); Stoffel, N. D., Riedel, R., Colavita, E., Margaritondo, G., Haglund, R. F., Jr., Taglauer, E., and Tolk, N. H. Photon-stimulated desorption of neutral sodium from alkali halides observed by laser-induced fluorescence. *Phys. Rev. B* **32**, 6805–6809 (1985).
106. Schmid, A., Braunlich, P., and Rol, P. K. Multiphoton-induced directional emission of halogen atoms from alkali halides. *Phys. Rev. Lett.* **35**, 1382–1385 (1975).
107. Green, T. A., Loubriel, G. M., Richards, P. M., Tolk, N. H., and Haglund, R. F., Jr. *Phys. Rev. B* **35**, 781–787 (1987); Szymonski, M., Czuba, P., Dohnalik, T., Jozefowski, L., Karawajczyk, A., Kolodziej, J., and Lesniak, R. *Nucl. Instrum. Methods* **B48**, 534–537 (1990).
108. Townsend, P. D. Defect formation by multiphoton absorption. *Phys. Rev. Lett.* **36**(14), 827–829 (1976) and references therein.
109. Puchin, V. E., Schluger, A. L., and Itoh, N. Theoretical studies of atomic emission and defect formation by electronic excitation at the (100) surface of NaCl. *Phys. Rev. B* **47**(16), 10 760–10 768 (1993).
110. Itoh, N., Nakai, Y., Hattori, A., Okano, A., and Kanasaki, J., in *Desorption Induced by Electronic Transitions V*. Springer-Verlag, Berlin, 1992, to be published.

111. Townsend, P. D., and Lama, F., in *Desorption Induced by Electronic Transitions I*, N. H. Tolk, M. M. Traum, J. C. Tully, and T. E. Madey (eds.), pp. 220–228. Springer-Verlag, Berlin, 1983.
112. Itoh, C., Tanimura, K., Itoh, N., and Itoh, M. Threshold energy for photogeneration of self-trapped excitons in SiO_2. *Phys. Rev. B* **39**, 11 183–11 186 (1989).
113. Li, X., Beck, R. D., and Whetten, R. L. Photon-stimulated ejection of atoms from alkali-halide nanocrystals. *Phys. Rev. Lett.* **68**, 3420–3423 (1992).
114. Classic references are Stoneham, A. M. *Theory of Defects in Solids: Electronic Structure Defects in Insulators and Semiconductors*. Clarendon Press, Oxford, 1975; Maradudin, A. A., Montroll, E. W., and Weiss, G. *Theory of Lattice Dynamics in the Harmonic Approximation*. Academic Press, New York, 1963. Solid state physics, Suppl. **3**.
115. Page, J. B., Jr. Defect-induced resonance modes in the asymptotic limit of low frequencies: Isotope effects and amplitude patterns. *Phys. Rev. B* **10**, 719–738 (1974).
116. Bicham, S. R., and Sievers, A. J. Intrinsic localized modes in a monatomic lattice with weakly anharmonic nearest-neighbor force constants. *Phys. Rev. B* **43**, 2339–2346 (1991).
117. Hattori, K., Okano, A., Nakai, Y., Itoh, N., and Haglund, R. F., Jr. Characterization of surface defects by means of laser-induced Ga^0 emission from GaP surfaces. *J. Phys. Condens. Matter* **3**, 7001–7006 (1991).
118. Schildbach, M. A., and Hamza, A. V. Sapphire (1120) surface: Structure and laser-induced desorption of aluminum. *Phys. Rev. B* **45(11)**, 6197–6206 (1992).
119. Schildbach, M. A., and Hamza, A. V. Clean and water-covered sapphire (1102) surfaces: Structure and laser-induced desorption. *Surf. Sci.* **282**, 306–322 (1993).
120. Kanasaki, J., Yamashita, H., Okano, A., Hattori, K., Nakai, Y., Itoh, N., and Haglund, R. F., Jr. Effects of oxygen adsorption on laser-induced sputtering from GaP(11) surfaces. *Surf. Sci. Lett.* **257**, L642–L646 (1991).
121. Kanasaki, J., Matsuura, A. Y., Nakai, Y., Itoh, N., and Haglund, R. F., Jr. Enhancement of laser-induced defect-initiated Ga^0 emission from GaAs(110) surfaces by Br adsorption. *Appl. Phys. Lett.* **62(26)**, 3493–3495 (1993).
122. Itoh, N., Hattori, K., Nakai, Y., Kanasaki, J., Okano, A., Ong, C. K., and Khoo, G. S. Theoretical studies of defect-initiated particle emission from GaP(110) surfaces: Basis for a new technique of generating perfect surfaces. *Appl. Phys. Lett.* **60**, 3271–3273 (1992).
123. Cowin, J. P., Auerbach, D. J., Becker, C., and Wharton, L. Measurement of fast desorption kinetics of D_2 from tungsten by laser induced thermal desorption. *Surf. Sci.* **78**, 54564 (1978).
124. NoorBatcha, I., Lucchese, R. R., and Zeiri, Y. Monte Carlo simulations of gas-phase collisions in rapid desorption of molecules from surfaces. *J. Chem. Phys.* **86**, 5816–5824 (1987) and Effects of gas-phase collisions in rapid desorption of molecules from surfaces in the presence of coadsorbates. *J. Chem. Phys.* **89**, 5251–5263 (1988).
125. Perry, M. D., and Mourou, G. Terawatt to petawatt subpicosecond lasers. *Science* **264**, 917–924 (1994).
126. Brau, C. A. Free-electron lasers. *Science* **239**, 1115–1121 (1988).
127. Ready, John F. *Effects of High-Power Laser Radiation*. Academic Press, New York, 1971.
128. Wang, X. Y., Riffe, D. M., Lee, Y.-S., and Downer, M. C. Time-resolved electron-temperature measurement in a highly excited gold target using femtosecond thermionic emission. *Phys. Rev. B* **50(11)**, 8016–8019 (1994).

129. Qiu, T. Q., and Tien, C. L. Short-pulse laser heating on metals. *Int. J. Heat Mass Transfer* **35**(3), 719–726 (1992); Chan, C. L., and Mazumder, J. One-dimensional steady-state model for damage by vaporization and liquid expulsion due to laser-material interaction. *J. Appl. Phys.* **63**, 4579–4586 (1987); Imen, K., Lin, J. Y., and Allen, S. D. Steady-state temperature profiles in thermally thin substrates induced by arbitrarily shaped laser beams. *J. Appl. Phys.* **66**, 488–491 (1988).
130. Chichkov, B. N., Momma, C., Nolte, S., von Alvensleben, F., and Tünnermann, A. Femtosecond, picosecond and nanosecond laser ablation of solids. *Appl. Phys. A* **63**, 109–115 (1996).
131. Corkum, P. B., Brunel, F., Sherman, N. K., and Srinivasan-Rao, T. Thermal response of metals to ultrashort-pulse laser excitation. *Phys. Rev. Lett.* **61**, 2886–2889 (1988). See also Corkum, P. B., Brunel, F., Sherman, N. K., and Srinivasan-Rao, T. Comment in *Phys. Rev. Lett.* **64**, 1847 (1990) in response to a critique by H. E. Elsayed-Ali on the preceding page.
132. Bennett, T. D., Grigoropoulos, C. P., and Krajnovich, D. J. Near-threshold laser sputtering of gold. *J. Appl. Phys.* **77**(2), 849–864 (1995).
133. Dreyfus, R. W. Cu^{o}, Cu^{+} and Cu_2 from excimer-ablated copper. *J. Appl. Phys.* **69**(3), 1721–1729 (1991).
134. Hicks, J. M., Urbach, L. E., PLummer, E. W., and Dai, H.-L. Can pulsed laser excitation of surfaces be described by a thermal model? *Phys. Rev. Lett.* **61**(22), 2588–2591 (1988).
135. Bennett, T. D., Krajnovich, D. J., and Grigoropoulos, C. P. Separating thermal, electronic, and topographic effects in pulsed laser melting and sputtering of gold. *Phys. Rev. Lett.* **76**(10), 1659–1662 (1996).
136. Shea, M. J., and Compton, R. N. Surface-plasmon ejection of Ag^+ ions from laser irradiation of a roughened silver surface. *Phys. Rev. B* **47**(15), 9967 (1993).
137. Ritchie, R. H., Manson, J. R., and Echenique, P. M. Surface-plasmon-ion interaction in laser ablation of ions from a surface. *Phys. Rev. B* **49**(4), 2963–2966 (1994).
138. Helvajian, H., and Welle, R. Threshold level laser photoablation of crystalline silver: Ejected ion translational energy distributions. *J. Chem. Phys.* **91**(4), 2616–2626 (1989).
139. Kim, H. S., and Helvajian, H. Threshold level laser photoablation of oxidized aluminum (111): photoejected ion translational energy distributions. *J. Phys. Chem.* **95**, 6623–6627 (1991).
140. Preuss, S., Matthias, E., and Stuke, M. Sub-picosecond UV-laser ablation of Ni films: strong fluence reduction and thickness-independent removal. *Appl. Phys. A* **59**, 79–82 (1994).
141. Nakayama, T., Ichikawa, H., and Itoh, N. Nonlinear photo-induced desorption of GaP. *Surf. Sci. Lett.* **123**, L693–L697 (1983).
142. Kumazaki, Y., Nakai, Y., and Itoh, N. Structural change induced by electronic excitation on GaP surfaces. *Phys. Rev. Lett.* **59**(25), 2883–2886 (1987).
143. Long, J. P., Goldenberg, S. S., and Kabler, M. N. Pulsed laser-induced photochemical decomposition of GaAs(110) studied with time-resolved photoelectron spectroscopy using synchrotron radiation. *Phys. Rev. Lett.* **68**(7), 1014–1017 (1992).
144. Brewer, P. D., Zinck, J. J., and Olson, G. L. Reversible modification of CdTe surface composition by excimer laser irradiation. *Appl. Phys. Lett.* **57**, 2526–2528 (1990).
145. Phillips, J. C. Ionicity of the chemical bond in crystals. *Rev. Mod. Phys.* **42**, 317–356 (1970).

146. Ichige, K., Matsumoto, Y., and Namiki, A. Laser-induced desorption from compound semiconductors. *Nucl. Instrum. Methods B* **33**, 820–823 (1988).
147. Biswas, R., and Ambegaokar, V. Phonon spectrum of a model of electronically excited silicon. *Phys. Rev. B* **26**(4), 1980–1988 (1982).
148. Haglund, R. F., Jr. Microscopic and mesoscopic aspects of laser-induced desorption and ablation. *Appl. Surf. Sci.* **96-98**, 1–13 (1996).
149. Kittel, C. *Introduction to Solid-State Physics*, 4th ed. Wiley Interscience, New York, 1967.
150. Preuss, S., Demchuk, A., and Stuke, M. Sub-picosecond uv laser ablation of metals. *Appl. Phys. A* **61**, 33–37 (1995).
151. Herrmann, R. F. W., Gerlach, J., and Campbell, E. E. B. Molecular dynamics simulation of laser ablation of silicon. *Nucl. Instrum. Methods in Phys. Res.* (1996).
152. Important early experiments were reported by H. W. K. Tom, G. D. Aumiller and C. H. Brito-Cruz, Time-resolved study of laser-induced disorder of Si surfaces. *Phys. Rev. Lett.* **60**, 1438 (1988); also Knox, W. H., Chemla, D. S., Livescu, G., Cunningham, J. E., and Henry, J. E. *Phys. Rev. Lett.* **61**, 1290 (1988). An historical overview and update on recent work in this field especially relevant to GaAs is by Y. Siegal, E. N. Glezer, L. Huang, and E. Mazur, Laser-induced phase transitions in semiconductors. *Ann. Rev. Mater. Sci.* **25**, 223–247 (1995).
153. Stampfli, P., and Bennemann, K. H. Dynamical theory of the laser-induced lattice instability of silicon. *Phys. Rev. B* **46**, 10, 686 (1992).
154. Stampfli, P., and Bennemann, K. H. Time dependence of the laser-induced femtosecond lattice instability of Si and GaAs: Role of longitudinal optical distortions. *Phys. Rev. B* **49**(11), 7299–7305 (1994).
155. D. Arnold and E. Cartier have shown that significant free-electron heating and impact ionization in SiO_2 requires intensities in the TW-cm^{-2} range, far above the parameters of typical laser ablation experiments. See their Theory of laser-induced free-electron heating and impact ionization in wide-band-gap solids. *Phys. Rev. B* **46**, 15102–14115 (1992).
156. Stuart, B. C., Feit, M. D., Rubenchik, A. M., Shore, B. W., and Perry, M. D. laser-induced damage in dielectrics with nanosecond to subpicosecond pulses. *Phys. Rev. Lett.* **74**(12), 2248–2251 (1995).
157. Reichling, M., Johansen, H., Gogoll, S., Stenzel, E., and Matthias, E. Laser-stimulated desorption and damage at polished CaF_2 surfaces irradiated with 532 nm laser light. *Nucl. Instrum. Methods in Phys. Res. B* **91**, 628–633 (1994).
158. Gogoll, S., Stenzel, E., Reichling, M., Johansen, H., and Matthias, E. Laser damage of $CaF_2(111)$ surfaces at 248 nm. *Appl. Surf. Sci.* **96–98**, 332–340 (1996).
159. Jones, S. C., Braunlich, P., Casper, R. T., Shen, X.-A., and Kelly, P. Recent progress on laser-induced modifications and intrinsic bulk damage of wide-gap optical materials. *Opt. Eng.* **28**, 1039–1064 (1989).
160. Arnold, D., Cartier, E., and DiMaria, D. J. Acoustic-phonon runaway and impact ionization by hot electrons in silicon dioxide. *Phys. Rev. B* **45**, 1477–1480 (1992). See also note 155.
161. Du, D., Liu, X., Korn, G., Squier, J., and Mourou, G. Laser-induced breakdown by impact ionization in SiO_2 with pulse widths from 7 ns to 150 fs. *Appl. Phys. Lett.* **64**, 3071–3073 (1994).
162. Ashkenasi, D., Varel, H., Rosenfeld, A., Noack, F., and Campbell, E. E. B. Pulse-width influence on the laser-induced structuring of CaF_2 (111). *Appl. Phys. A* **63**, 103–107 (1996).

163. Reif, J. High power laser interaction with the surface of wide bandgap materials. *Opt. Eng.* **28**, 1122–1132 (1989)
164. Stuart, B. C., Feit, M. D., Herman, S., Rubenchik, A. M., Shore, B. W., and Perry, M. D. Nanosecond-to-femtosecond laser-induced breakdown in dielectrics. *Phys. Rev. B* **53**(4), 1749–1761 (1996).
165. Hamza, A. V., Hughes, R. S., Jr., Chase, L. L., and Lee, H. W. H. Investigation of the laser-Al_2O_3 (1120) surface interaction using excitation by pairs of picosecond pulses. *J. Vac. Sci. Technol. B*. **10**, 228–230 (1993).
166. Preuss, S., Späth, M., Zhang, Y., and Stuke, M., Time resolved dynamics of subpicosecond laser ablation. *Appl. Phys. Lett.* **63**, 3049–3051 (1993).
167. Park, H. K., and Haglund, R. F., Jr. Laser-induced desorption and ablation of calcite from visible to mid-infrared wavelengths. *Appl. Phys. A* **64**, 431–438 (1997).
168. Brau, C. A., and Mendenhall, M. H., The Vanderbilt University Free-Electron Laser Center. *Nucl. Instrum. Meth. Phys. Rev. A* **331**, ABS4–ABS6 (1993). See also Brau, C. A. The Vanderbilt University Free-Electron Laser Center. *Nucl. Instrum. Meth. Phys. Res. A* **318**, 38–41 (1992).
169. Becker, K., Johnson, J. B., and Edwards, G. Broadband Pockels cell and driver for a Mark III-type free-electron laser. *Rev. Sci. Inst.* **65**, 1496 (1994).
170. Calderon, T., Aguilar, M. A., Jaque, F., and Coyll, R. Thermoluminescence from natural calcite. *J. Phys. C* **17**, 2027 (1984).
171. Dingus, R. S., and Scammon, R. J., Ablation of material by front surface spallation, in *Laser Ablation: Mechanisms and Applications*, J. C. Miller and R. F. Haglund, Jr. (eds.), pp. 180–190, Lecture Notes in Physics **389**. Springer-Verlag, Berlin, 1992.
172. Zakrevskii, V. A., and Shuldiner, A. V. Electron emission and luminescence owing to plastic deformation of ionic crystals. *Phil. Mag. B* **71**, 127 (1995).
173. de Rességuier, T., and Cottett, F., Experimental and numerical study of laser-induced spallation in glass. *J. Appl. Phys.* **77**, 3756–3761 (1995).
174. Lev, L. C., and Argon, A. S., Spallation of thin elastic coatings from elastic substrates by laser induced pressure pulses. *J. Appl. Phys.* **80**, 529–542 (1966).
175. Itzkan, I., Albagli, D., Dark, M. L., Perelman, L. T., von Rosenberg, C., and Feld, M. S. The thermoelastic basis of short pulsed laser ablation of biological tissue. *Proc. Natl. Acad. Sci. USA* **92**, 1690–1964 (1995).
176. Hurst, G. S., Payne, M. G., Kramer, S. D., and Young, P. Resonance ionization spectroscopy and one-atom detection. *Rev. Mod. Phys.* **51**, 767–819 (1979).
177. Herminghaus, S., and Leiderer, P. Nanosecond time-resolved study of pulsed laser ablation in the monolayer regime. *Appl. Phys. Lett.* **58**(4), 352–354 (1996).
178. Schilling, A., Yavas, O., Bischof, J., Boneberg, J., and Leiderer, P. Absolute pressure measurements on a nanosecond time scale using surface plasmons. *Appl. Phys. Lett.* **69**, 4159–4161 (1996).
179. Yavas, O., Maddocks, E. L., Papantonakis, M. R., and Haglund, R. F., Jr. Shockwave generation during rear and front-side ablation of calcite. *Appl. Phys. Lett.* **71**, in press (1997).
180. Chen, Julian C. *Introduction to Scanning Tunneling Microscopy*. Oxford University Press, New York, 1993.
181. Zeisel, D., Nettesheim, S., Dutoit, B., and Zenobi, R. Pulsed laser-induced desorption and optical imaging on a nanometer scale with scanning near-field microscopy using chemically etched fiber tips. *Appl. Phys. Lett.* **68**(18), 2491–2492 (1996). See also Jersch, J., Demming, F., and Dickmann, K. Nanostructuring with laser

radiation in the nearfield of a tip from a scanning force microscope. *Appl. Phys. A* **64**, 29–32 (1997).
182. Ichikawa, K., Kanasaki, J., Nakai, Y. and Itoh, N. Laser-induced bond breaking of the adatoms of the Si(111) 7 × 7 surface. *Surf. Sci. Lett.* **349**, L153–L158 (1996).
183. Hare, D. E., Franken, J., and Dlott, D. D. Coherent Raman measurements of polymer thin-film pressure and temperature during picosecond laser ablation. *J. Appl. Phys.* **77**(11), 5950–5961 (1995).
184. Lindley, R. A., Gilgenbach, R. M., and Ching, C. H. Resonant holographic interferometry of laser-ablation plumes. *Appl. Phys. Lett.* **63**(7), 888–890 (1993).
185. Geohegan, D. B., and Mashburn, D. N. Characterization of ground-state neutral and ion transport during laser abalation of $Y_1Ba_2Cu_3O_{7-x}$ using transient optical absorption spectroscopy. *Appl. Phys. Lett.* **55**, 2345 (1989).
186. Ventzek, P. L. G., Gilgenbach, R. M., Ching, C. H., and Lindley, R. A. Schlieren and dye laser resonance absorption photographic investigations of KrF excimer laser-ablated atoms and molecules from polyimide, polyethyleneterephthalate, and aluminum. *J. Appl. Phys.* **72**, 1696 (1992).
187. Okano, A., Matsuura, A. Y., Hattori, K., Itoh, N., and Singh, J. A model of laser ablation in nonmetallic inorganic solids. *J. Appl. Phys.* **73**(7), 3158–3162 (1993).
188. Westin, E., Rosén, A., and Matthias, E. Molecular cluster calculations of the electronic structure of the (111) surface of CaF_2, in *Desorption Induced by Electronic Transitions, DIET-IV*, G. Betz and P. Varga (eds.), pp. 316–321, Springer Series in Surface Science **19**. Springer-Verlag, Berlin, Heidelberg, 1990.
189. Ong, C. K., Khoo, G. S., Hattori, K., Nakai, Y., and Itoh, N. CNDO calculation of energies of Ga atom ejection from defect sites on the GaP(110) surface. *Surf. Sci. Lett.* **259**, L787–L790 (1991); Khoo, G. S., Ong, C. K., and Itoh, N. The multi-hole localization mechanism for particle emission from semiconductor surfaces. *J. Phys. Condens. Matter* **5**, 1187–1194 (1993).
190. Puchin, V. E., Shluger, A. L., and Itoh, N. The excitonic mechanism of Na-atom desorption from the (100) NaCl surface. *J. Phys.: Cond. Matter* **7**, L147–151 (1995).
191. Barker, A. S., and Sievers, A. J. Optical studies of the vibrational properties of disordered solids. *Rev. Mod. Phys.* **47**, Supplement 2, S1–S179 (1975).
192. Kiselev, S. A., Bickham, S. R., and Sievers, A. J. Anharmonic gap modes in a perfect one-dimensional diatomic lattice for standard two-body nearest-neighbor potentials. *Phys. Rev. B* **48**, 13 508–13 511.
193. Wurz, P., Sarnthein, J., Husinsky, W., Betz, G., Nordlander, P., and Wang, Y. Electron-stimulated desorption of neutral lithium atoms from LiF due to excitation of surface excitons. *Phys. Rev. B* **43**, 6729–6732 (1991).
194. Shen, X. A., Jones, S. C., Bräunlich, P., and Kelly, P. Four-photon absorption cross section in potassium bromide at 532 nm. *Phys. Rev. B* **36**, 2831–2843 (1987).
195. Shen, X. A., Bräunlich, P., Jones, S. C., and Kelly, P. Investigation of intrinsic optical damage in potassium bromide at 532 nm. *Phys. Rev. B* **38**, 3494–3504 (1988).
196. Stampfli, P., And Bennemann, K. H. Theory for the laser-induced femtosecond phase transition of silicon and GaAs. *Appl. Phys. A* **60**, 191–196 (1995).
197. Tully, J. C. Molecular dynamics with electronic transitions. *J. Chem. Phys.* **93**(2), 1061–1071 (1990).
198. Zhigilei, L. V., Kodali, P. B. S., and Garrison, B. J. Molecular dynamics model for laser ablation of organic solids. *J. Phys. Chem. B* **101**, 2028–2037 (1997).

199. Beck, K. M., Taylor, D. P., and Hess, W. P. Photostimulated desorption of CO from geologic calcite following 193 nm irradiation. *Phys. Rev. B* **55**, 13 253–13 262 (1997).
200. Beck, K. M., McCarthy, M. I., and Hess, W. P. Atomic and molecular photostimulated desorption from complex ionic crystals. *J. Electron. Materials* in press (1997).
201. Knutsen, K., and Orlando, T. M. Photon-stimulated desorption of $O(3p)$ and $NO(^2\Pi)$ from $NaNO_3$ single crystals. *Phys. Rev. B* **55**, 13 246–13 251 (1997).
202. Xu, J. H., Gozzini, A., Mango, F., Alzetta, G., and Bernheim, R. A. Photoatomic effect: Light-induced ejection of Na and Na_2 from polydimethylsiloxane surfaces. *Phys. Rev. A* **54**, 3146–3150 (1996).
203. See, for example, Menges, M., Baumeister, B., Al-Shamery, K., Freund, H.-J., Fischer, C., and Andresen, P. Dynamical studies of UV-laser-induced NO-desorption from the polar NiO(111) vs the nonpolar NiO(110) surfaces. *J. Chem. Phys.* **101**(4), 3318–3325 (1994).

3. LOW FLUENCE LASER DESORPTION AND PLUME FORMATION FROM WIDE BANDGAP CRYSTALLINE MATERIALS

J. Thomas Dickinson

Washington State University
Pullman

3.1 Introduction

In applications of high-intensity lasers to materials processing, the formation of an ablation plume is of high importance. For wide bandgap insulators (e.g., oxides, halides, and nitrides) irradiated with sub-bandgap photon energies, the route to plume formation is not well understood. For example, contrary to metals and semiconductors, inverse bremsstrahlung (IB) is not possible for a wide range of laser intensities on these materials due to insufficient photon and electron densities. Thus an alternative path to plume formation must be considered. In this chapter, we first examine the interaction of photo-emitted and thermally emitted particles from exposure to pulsed laser irradiation of surfaces of wide bandgap ionic crystals. At sub-bandgap photon energies, these emissions include photoelectrons, energetic positive ions, and neutral metal atoms. We then establish experimentally that significant portions of the distributions of these particles overlap in space and time in the near-surface region. We present a model of the collective motion of these particles and show that as laser fluence is increased we achieve sufficient densities, overlap, and kinetic energies to result in plume formation. The features examined include excitation of neutral atoms to generate plume fluorescence and eventual atomic ionization at fluences far below any inverse bremsstrahlung or catastrophic breakdown process.

Over a range of fluences below breakdown, a wide range of particle emission phenomena are observed. We begin with a description of photodesorption, which is a relatively gentle process, removing sub-monolayer quantities of material/laser pulse. The processes can fall into nonthermal (e.g., photoelectronic) and thermal categories. For semiconductors and insulators, the incident light may be sub-bandgap or may involve single-photon band-to-band transitions, depending on the laser photon energy relative to E_g.

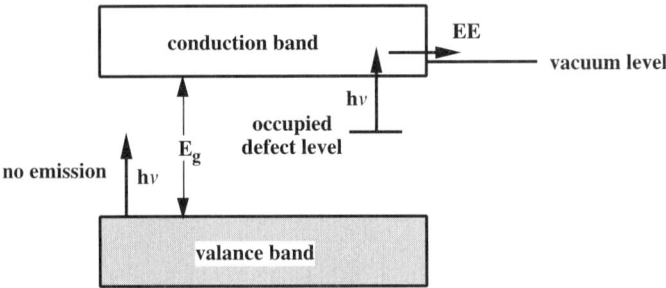

FIG. 1. Schematic diagram of energy levels for a wide bandgap insulator. EE represents photoelectron emission; E_g, the bandgap; and $h\nu$ the laser photon energy, where $h\nu < E_g$.

3.2 Photoelectron Emission

Figure 1 shows schematically the fixed band energy level diagram for a wide bandgap material. For $h\nu < E_g$, no single photoelectron emission can occur. However, with occupied surface and near-surface defect levels, excitations of these defects can couple to free electron states in the conduction band, which, if above the vacuum level, can yield electron emission (EE). Such EE has been reported for 5 eV photon irradiation of MgO (E_g = 7.8 eV) [1] and NaNO$_3$ (E_g = 10 eV) [2]. Energies of these photoelectrons determined by retarding potential analysis and by Langmuir probe techniques show values of ≤1 eV, consistent with single-photon excitation from occupied electron traps in the bandgap. Furthermore, the intensity of EE for a given laser fluence is often promoted by treatments that generate additional vacancy defects.

Multiphoton photoemission involving excitation across the gap has been reported, for example, by Petite *et al.* [3] from quartz and fused silica and from GaP and various alkali halides [4, 5] and most recently Zn chalcogenides [6] by Williams *et al.* The latter studies have benefited from the use of short pulses and energy analysis of the emitted EE. Likewise, time-resolved studies of the underlying excitations and relaxation/trapping of charge carriers in alkali halides by Tanimura *et al.* [7] and in oxides by Petite *et al.* [8] provide important mechanistic information, particularly related to multiphoton-induced effects and carrier trapping processes.

3.3 Photodesorption of Neutral Species

By far the dominant mechanism for laser-induced desorption of neutral species from surfaces involves thermal processes. Conventional laser-induced thermal desorption involves absorption of the laser by the substrate, which undergoes rapid heating. An adsorbate thus acquires thermal energy via thermally excited substrate lattice vibrations. The adsorbate may then undergo various reactions, including simple desorption from the surface. A number of surface chemistry studies on metal surfaces have employed laser-induced photodesorption to follow reaction kinetics [9, 10]. There the laser pulse provides a snapshot of the weakly bound species (which fly into a mass spectrometer) on the surface and potentially characterizes the extent of a surface reaction [11]. Under certain conditions, however, nonthermal mechanisms have been revealed to play a significant role. An example is the laser-induced decomposition of oxyanion containing ionic crystals, (e.g., nitrates and possibly carbonates). Webb et al. [12] showed that during 248 nm irradiation of $NaNO_3$ a weak coupling to the nitrate can generate NO emission due to decomposition. Typical fluences required on a cleaved crystal surface for detectable NO signals were ~ 150 mJ/cm^2. As this decomposition proceeds with repeated laser pulses, the NO emission is seen to increase (a so-called induction or incubation effect), illustrating clearly the role of defects in enhancing these emissions. Nevertheless, the original signals are seen as a consequence of direct photodecomposition via a $\pi^* \leftarrow \pi_2$ (S_1) transition that centers at 6 eV and has a long tail to lower energies at room temperature. Bradley et al. [13] have shown that NO emission accompanying irradiation of $NaNO_3$ at 213 nm (5.7 eV) is ~ 2000 times more intense (per photon) at 213 nm (5.7 eV) than at 266 nm (4.66 eV), consistent with direct dissociation following the near-resonance excitation of the $\pi^* \leftarrow \pi_2$ band. Furthermore, following ~ 6 eV excitation, the desorbed NO displays a highly excited vibrational distribution but thermally equilibrated translational and rotational distributions. These observations indicate a finite surface residence time for the newly formed NO.

Beck et al. [14] have shown recently that CO is desorbed from cleaved calcite (rhombohedral $CaCO_3$) during UV irradiation. The resulting CO emissions do not appear to be thermal; that is, the CO state distributions are not equilibrated in any degree of freedom. Only a 6% population of the $v = 1$ vibrational level is observed (94% in $v = 0$). The rotational state distribution is non-Boltzmann, exhibiting a step structure every 4 or 5 rotational states, and the CO velocity distribution can be characterized by a temperature ($T = 110 \pm 15$ K), significantly lower than that of the substrate ($T = 295$ K). The unusually low CO velocity distribution cannot be explained by the standard desorption models such as Menzel, Gomer, and Redhead (MGR) or Knotek and Feibelman [see Chapter 2 in this volume]. Further, it is argued that no equilibrium thermal desorption process can account for this phenomena and that the CO, once formed, departs the calcite surface promptly. The calcite and sodium nitrate systems display distinctly different product state distributions following similar S1 excitation. It is possible that the differences in the detailed nature of these electronic excitations lead to the differences observed in the NO and CO state distributions [15].

An important consideration in laser-induced decomposition of materials that leads to volatile products is that inside the bulk, there tends to be a strong reversability due to the constraint of the surrounding lattice; thus any increase in the number of particles greatly favors recombination due to the resulting strain energy (analogous to a gas phase reaction) [16, 17]. With sufficient internal gasification, the strain can lead to fracture of the crystal.

In contrast to these oxyanion decomposition results, Section 3.5.4 describes in more detail the laser desorption of metal atom neutrals (cations) from defect-laden insulators and show this emission exhibits principally *thermal* behavior.

3.4 Photodesorption of Positive Ions

3.4.1 Characterization of the Positive Ion Emission

An example of photodesorption of positive ions involves the emission of Mg^+ and Mg^{2+} [18] emitted from MgO during low fluence laser irradiation at 248 nm. This positive ion emission (PIE) can have surprisingly high energies— up to 20 eV for Mg^{2+}. Several lines of evidence point to an electrostatic emission mechanism: high, stable ion kinetic energies, a strongly nonlinear fluence dependence, and highly directed ion velocities (along the surface normal) [1]. The emission intensities are strongly enhanced by treatments that increase the surface defect density. We have proposed that these emissions are due to cations adsorbed on the surface at sites above occupied electron traps— we refer to these species as *adions*; emission occurs when the laser photoionizes the underlying electron trap. This process is similar to the Knotek-Feibelman mechanism for photodesorption [19], except that the Knotek-Feibelman mechanism does not involve lattice defects. Lattice defects allow for photodesorption at much lower photon energies (here 5 eV as opposed to the hundreds of eV required for core-hole production) [20, 21]. Defect-mediated mechanisms also permit the desorption of cations, whereas the Knotek-Feibleman mechanism yields anions only.

In addition to MgO, energetic ions have been observed from other ionic, wide bandgap insulators, including rhombohedral $NaNO_3$ [18, 22], $CaCO_3$ (calcite) [22], $Cs_{0.1}Na_{0.9}NO_3$ [23], and $CaHPO_4 \cdot 2H_2O$ (brushite) [23], all yielding singly charged ions with relatively high kinetic energies in the range ~2–16 eV. These materials, which all contain oxyanions, show a wide range of sensitivity to radiolysis by 248-nm photons as well as 1–3 keV electrons [24, 25]. Exposing $NaNO_3$ to keV electrons prior to laser irradiation increases the resulting ion intensities 2–3 orders of magnitude at a given fluence [26]. In contrast, MgO and CaCO$_3$ are unaffected by keV electrons. MgO and CaCO$_3$ are also resistant to photoelectronic defect production under laser irradiation at 248 nm; but high defect densities can be produced in a number of materials by mechanical treatments (abrasion, polishing) due to the accompanying plastic deformation. In each case, treatments that raise the density of surface defects also raise the ion emission intensities.

Figure 2 presents a composite of energy distributions, determined from mass-selected time-of-flight (TOF) curves, for singly charged ion emission from a variety of the nitrate and carbonate materials. We see a tendency toward two broad peaks on the order of 7 eV and 15 eV; it is not unusual to observe both of these peaks simultaneously as seen in this particular Ca$^+$ distribution from CaHPO$_4 \cdot$ 2H$_2$O.

We attribute these peaks to electrostatic repulsion of a sorbed singly charged adion located directly over an anion vacancy defect which is emptied through photoexcitation. Fig. 3(a) shows schematically this arrangement for an Mg$^+$ or Mg^{++} ion sorbed on an oxygen vacancy in MgO. If we now model the forces and kinematics of launching this ion following the removal of one or two negative charges, we can predict both the final kinetic energy and the expected trajectories for small deviations from the symmetric site (repulsion comes from the net + charge of the surrounding Mg^{++} ions in the lattice sites). We assume no relaxation and place the adion at typical lattice dimension distances from the surface plane. Strong confirmation of such sorption sites is provided by Murphy and Giamello [27] who have shown experimentally that several metals (M) sorb with significant charge transfer to surface vacancy defects on MgO. Farrari and Pacchioni [23] provide convincing theoretical support for ionic sorption states as well as unstable M$^+$ sorption on F$_s^+$ sites, where M$^+$ is a

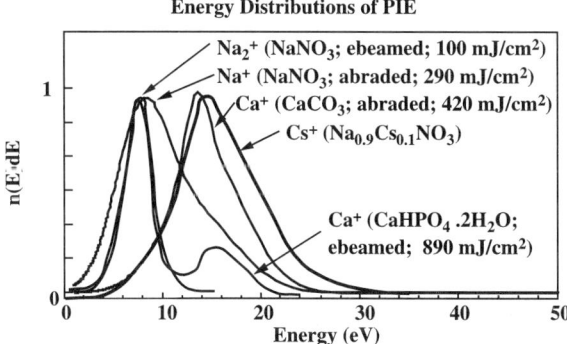

FIG. 2. Positive ion emission energy distributions for various nitrates, carbonates, and phosphates.

Model of Emission Mechanism

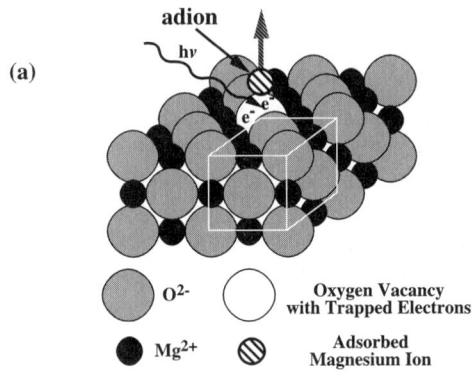

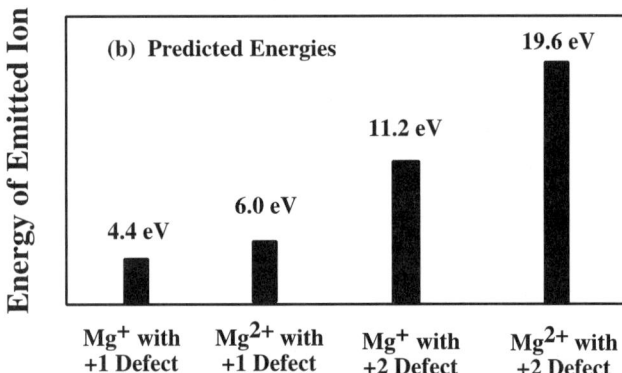

FIG. 3. (a) Model of positive ion emission mechanism involving an anion vacancy defect in MgO. (b) The predicted final kinetic energies of Mg^+ and Mg^{++} from a simple electrostatic repulsion model.

sorbed adion and F_s^+ is a surface F^+ center (a singly charged oxygen vacancy).

Fig. 3(b) shows a composite of the predicted energies for the various combinations of defect charge and adion charge from the above model. We see that two distinct energies are predicted for each adion charge; for example, for Mg^+, this simple model predicts 4.4 and 11.2 eV, where we observe a robust peak at 11 eV with a shoulder at ~6 eV [18]. A careful measurement of the angular distribution of the total ion emission (principally Mg^+) from polished surfaces strongly supports the high symmetry adion site; that is, the measured distribution is extremely narrow ($\pm 2.5°$) [1]. For Mg^{++}, we routinely obtain a two-peaked energy distribution at 7 and 18 eV, again in good agreement with the predicted values [1]. In the nitrate, carbonate, and phosphate systems where the detected cations are all singly charged, the observed (approximate) 7 and 15 eV peaks strongly suggest a singly and doubly charged defect site. We have shown that anion decomposition accompanies the formation of these defects both mechanically [22] and by radiolysis [24–26] and can result in the formation of patches of doubly charged lattice (e.g., O^{--}) and therefore doubly charged defect sites. For $NaNO_3$, we observed formation of oxide XPS signatures on surfaces of electron bombarded crystals [25]. On freshly cleaved $NaNO_3$ crystals (which lack oxide character), no evidence of a 11 eV Na^+ peak is seen, consistent with singly charged sites only [18].

To explain the observed tails at higher energies, one can envision clusters of electron traps, which also become depopulated and add additional repulsive force acting on the adion. Note that mechanically induced point defects are generated by a jog dragging mechanism, which inherently produces clusters of defects that are ideal for these collective processes. Higher kinetic energies also arise if initial binding of the adions occurs at smaller initial spacings prior to defect ionization. Finally, as positive ion densities increase, there is a further acceleration mechanism due to the expansion of the ion cloud. As we discuss below, however, we find that even at laser fluences well below ablation thresholds, this ion cloud begins to trap electrons to form a ±charge cloud which limits the repulsive forces acting on the + charges.

A serious question about the defect-moderated PIE mechanism concerns the apparent violation of energy conservation. One has a process where photons of a few eV energy are capable of releasing ions with kinetic energy that exceeds the photon energy. A clue to this dilemma is seen in the fluence dependence of the PIE intensity. Figure 4 shows this fluence dependence in a linear plot for the total PIE intensity from polished MgO at low laser fluences; note that it exhibits what appears to be a threshold effect. (This signal consists almost entirely of Mg^+ emission.) Although it appears to have a threshold behavior, in reality the rise in intensity varies with surface treatment; that is, it is not an inherent property of defect-free MgO. This apparent threshold actually arises from the ion intensity's highly nonlinear fluence dependence.

In Fig. 5(a) we present the log-log behavior of the mass-selected Mg^+ PIE intensity for two MgO crystals (polished and abraded) over a large range of fluences—clearly this emission is different for these two surfaces, both exhibit very nonlinear fluence dependencies, and the abraded surface yields higher ion intensities for any given fluence. A number of investigators have reported highly nonlinear yield curves for PIE [29, 30]. For instance, Matthias *et al.* observed nearly tenth order fluence dependence in the ion emission from BaF_2 during irradiation at 530 nm at power densities similar to those employed here [29, 30]. Although the emission mechanism was not clear, true "ten-photon" processes were discounted. Matthias *et al.* has also recently reported energetic

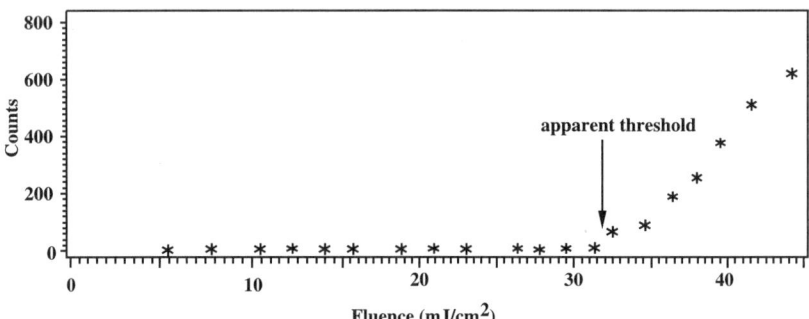

FIG. 4. Positive ion emission (Mg^+) from MgO vs. fluence; a linear plot of low fluence data.

Ca^+ and CaF^+ with high energies and a fluence dependence of order 6.8 from single crystal CaF_2 irradiated at 530 nm/8 ns pulses [31]. Kreitschitz et al. observed yield curves with slopes >6 for SrF_2 [32]. We also note that Wiedeman et al. observed laser-induced ion emission at relatively low fluences in materials associated with the high T_c superconductors [33].

The remaining curves in Fig. 5 show a wide range of nonlinear behavior of the singly charged cation emission intensities for a variety of crystals and conditions, namely, $CaCO_3$, $NaNO_3$, and $CaHPO_4 \cdot 2H_2O$. In each case, surface treatments that increase the surface defect densities (e.g., abrasion and electron irradiation) dramatically increase the emission intensities. The effect of electron irradiation has been previously demonstrated in several materials that are sensitive to keV electrons, including NaCl and SiO_2 [34]. Although MgO and $CaCO_3$ are quite resistant to defect creation by electronic processes, mechanically generated defects (abrasion) strongly enhance the ion intensities [1, 35, 22]. Recent work has demonstrated a similar mechanical effect on $NaNO_3$ [22] and $CaHPO_4 \cdot 2H_2O$ [23]. In each of these cases (carbonates, nitrates, and phosphates), the treatments lead to decomposition of the oxyanion, generating oxide-like surfaces and possibly anion vacancies. For example, from $NaNO_3$, we observe both NO and O_2 emission during electron irradiation [25] and during abrasion with a diamond [22]. From $CaHPO_4$ we observe O_2 and PO_x emission during electron irradiation and abrasion [23]. Contrary to reported electron stimulated desorption from $CaCO_3$ [36], we have observed no detectable decompositon products from during bombardment with keV electrons at current densities up to 300 $\mu A/cm^2$. In contrast, during abrasion and fracture, we see copious CO_2 emission [22, 37]. In general, these surface

Fluence Dependence of Mass-Selected Ion Emissions from Several Ionic Materials

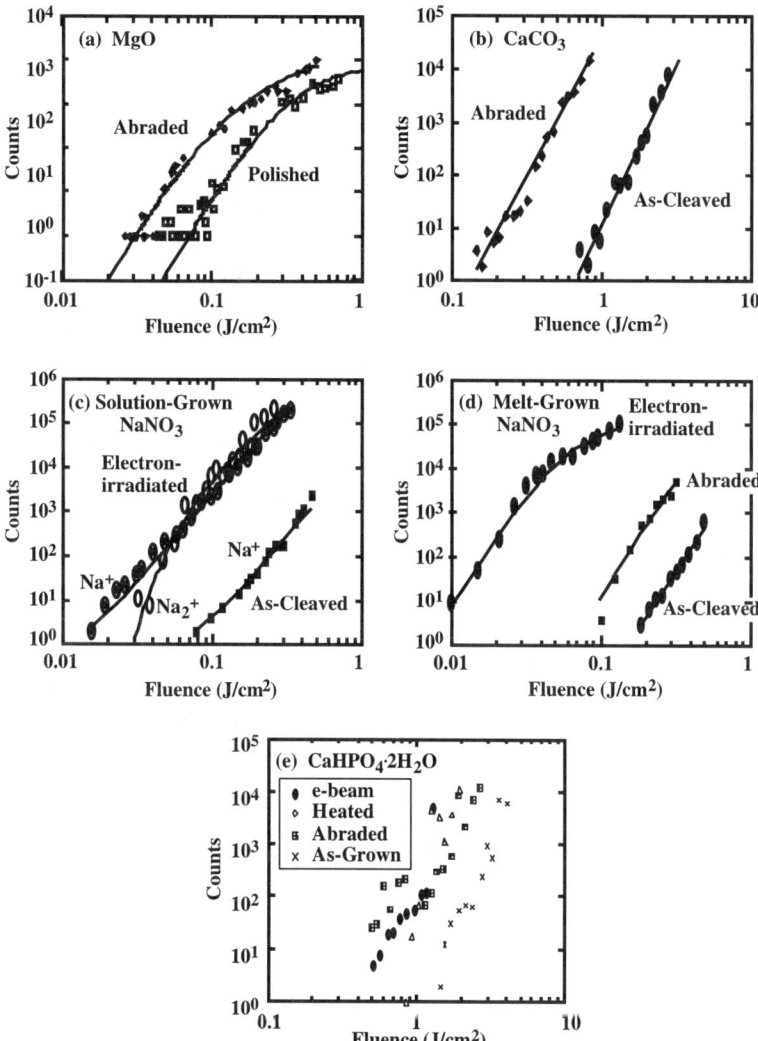

FIG. 5. Fluence dependence from a wide range of crystal surfaces.

treatments do not change the position of the ion TOF peaks; however, when more than one TOF peak is observed, the ratio of ions with different energies can be strongly affected.

A detailed analysis indicates that the low laser power densities employed in our studies are inconsistent with high-order multiphoton absorption (where simultaneous absorption of several photons are necessary to produce a single electronic excitation). Typical multiphoton absorption cross-sections are far to low to account for the observed emissions. In the case of KBr, for instance, $\sigma_4 \sim 2 \times 10^{-114}$ cm$^8 \cdot$ s^3 [38]. Even the highest power densities employed here (8×10^{25} photons $\cdot$ cm$^{-2} \cdot$ s^{-1} at 2 J/cm^2) correspond to a negligible excitation probability of 2×10^{-18} cm^{-2} per pulse. Further, excitations produced by high-order multiphoton absorption would probably not yield cation emission. Such excitations would presumably involve tightly bound, core electrons. When such excitations are produced by energetic electron or photon irradiation of ionic crystals (e.g., irradiated MgO) [20, 21], anion emission is observed (e.g., O$^+$), and not cation emission (e.g., no Mg$^+$ observed). Thus PIE due to high-order multiphoton absorption is highly improbable.

The highly ionic bonding in these materials still supports the electrostatic emission process described above. However, electrostatic cation ejection requires that sufficient negative charge be removed from a nearby anion site to produce a net repulsive force on the cation. We note that delocalized excitations, such as electrons in the conduction band or holes in the valence band, have little effect on the electrostatic forces experience by individual lattice ions. Similarly, excitons (here Frenkel excitons) should have little effect on cation binding energies. (In excitons, the decrease in binding energy when negative charge is removed from an anion is largely compensated by the increase in binding energy due to the loss of positive charge on the exciton-cation.) Thus electronic defects in intrinsic MgO (conduction electrons, valence holes, excitons) cannot sufficiently reduce cation binding energies to account for ion emission.

Lattice defects can contribute to emission processes by (1) providing weakly bound cations and (2) by providing readily ionized electron traps. In contrast with delocalized, electron/hole defects, ionized electron traps are associated with large drops in the binding energy of nearby cations.

Low energy cation binding sites would allow for emission from relatively simple defect configurations. Removing all the negative charge from a single anion site will reduce the binding energy of a nearest neighbor (NN) cation by $\sim(Ze)^2/r_0$, where Z is the charge of the lattice and r_0 is the NN distance. However, the electrostatic binding energy is on the order of $\alpha(Ze)^2/r_0$, where α is the Madelung constant for the lattice and site in question. For typical ionic crystal lattices (NaCl, calcite), $\alpha > 1$ for bulk, surface, and step sites [39, 40]. Thus cations in these sites will still be bound if all the charge is removed from a NN anion site. However, the binding energy of cations at kink sites along steps, adions along steps, and adions at terrace sites becomes negative when a NN anion site loses its charge. Calculations of the electrostatic *force* on cations in these defect sites suggest that only the last defect (a cation adsorbed along a terrace site) is actually ejected when the negative charge is removed from a NN anion [1]. (Adions associated with steps are bound in a metastable configuration by the remaining fully charged, NN anion(s).) Therefore, we restrict our attention to isolated cations adsorbed at terrace sites, presumably sitting directly above an electron trap that can be photoionized by the laser.

In the proposed scenario, adsorbed cations are emitted when an underlying trap is photo-ionized by the laser. Ionizing an electron trap adjacent to a positively charged defect, such as an adion, can be quite difficult. Assuming pointwise electrostatic interactions only, placing an adion in a NN cation site increases the binding energy of a trapped electron. Thus a surface electron trap that would normally be photo-ionized by a single 5 eV photon may require two or three photons after the addition of a positive adion. This is represented schematically in Fig. 6. If two electrons must be removed to yield ion emission, as many as six single-photon events would be required. This would account for a very strong fluence dependence.

Given sufficiently high densities of electron traps (or clusters of traps), adion emission by a sequence of single-photon absorption events could be strongly favored. When a positive adion is adsorbed onto a surface electron trap, the binding energy of other, nearby electron traps will also be enhanced by an amount proportional to $1/r$, where r is the distance between the adsorbed cation and the trap. Given the presence of electron traps situated at suitable distances r, these traps will provide a "staircase" of energy levels by which electrons can be removed from the trap beneath the adion by single-photon absorption events. This would account for high-order fluence dependence in the absence of multiphoton absorption. And it would predict that the emission kinetics are strongly dependent on defect densities and distributions, consistent with the effect of surface treatment on the emission intensities and kinetics. Because all processes proposed here are *single* photon in nature, high laser fluences are not required.

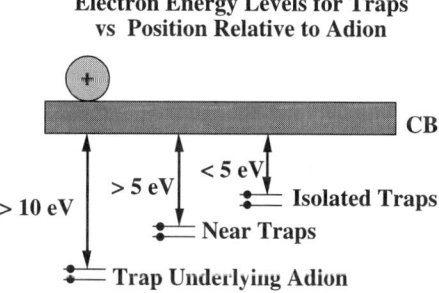

FIG. 6. Shifts in trap energies in the vicinity of a singly charged (+) adion.

3.4.2 Modeling the Fluence Dependence of the Ion Emission [18]

A numerical model of the emission process was constructed by assuming that electrons are redistributed among the traps by single-photon absorption events. For the sake of uniformity, we assume that all electron traps (denoted by the letter F) are capable of trapping two electrons and that all traps are initially full. (This poses a problem only for the singly charged $NaNO_3$ lattice, where doubly charged defects are nevertheless observed.) We also assume that electron transfer from one trap to another requires that the destination trap has lost an electron due to a previous photo-ionization event. To remove an electron from the trap underneath the adion (F_{ion}—trap under cation) requires a nearby, partially empty trap (F_N—a nearby trap). Similarly, electron transfer from an F_N trap requires a partially empty "isolated" electron trap (F_I—isolated trap). Finally, we assume that F_I centers can be directly photo-ionized by the laser. Emission occurs when both electrons are removed from F_{ion}. For MgO, we also assume that electron retrapping becomes important at high fluences. This latter process accounts for the transition from sixth-order kinetics to second-order kinetics for Mg^+ from MgO. Although retrapping could involve any or all of the above centers, for modeling purposes, it was sufficient to include only one retrapping step. The resulting sequence of events is then:

	Process	Rate Constant	
$F_I + h\nu \rightarrow F_I^+ + e^-$	emitted/conduction band e^-	A_1	(3.1a)
$F_N + F_I^+ + h\nu \rightarrow F_N^+ + F_I$	e^- transfer	A_2	(3.1b)
$F_{ion} + F_N^+ + h\nu \rightarrow F_{ion}^+ + F_N$	e^- transfer	A_3	(3.1c)
$F_{ion}^+ + F_N^+ + h\nu \rightarrow F_{ion}^{2+} + F_N$ + free ion	ion-emission	A_4	(3.1d)
$F_{ion}^+ + e^- \rightarrow F_{ion}$	reneutralization	A_5	(3.1e)

where the constants $A_i (i = 1 \cdots 5)$ refer to the rate constant for each reaction step. The essential feature of the model appears to be a series of at least six ionization events in sequence (required to explain the high-order fluence dependence at the threshold). For MgO, a retrapping step Eq. (1e) is also required to explain the transition to lower order kinetics at higher fluences.

Equation 1 can be expressed in terms of five rate equations in nine unknowns and six rate constants. Charge and mass conservation provide the three other equations required for a solution. As stated above, all electron traps are assumed to contain two electrons at the beginning of the laser pulse ($[F_I^+] = [F_N^+] = [F_{ion}^+] = [F_{ion}^{2+}] = [e^-] = 0$). The initial number of F_{ion} centers was treated as a model parameter, but the initial densities of F_N and F_I centers were arbitrarily set to $2 \times [F_{ion}]$ and $4 \times [F_{ion}]$, respectively. For the sake of simplicity, we set $A_3 = A_4$. Finally, we assume that the number of electrons available for retrapping processes at a given fluence, [e$^-$], is proportional to the detected photo-electron intensity. These assumptions reduce the total number of unknown parameters to five (four rate constants and the initial value of [F$_{ion}$]). When retrapping is neglected (NaNO$_3$, CaCO$_3$), only four parameters are required.

The resulting rate equations were integrated over a single laser pulse at a given fluence using a fourth order Runge-Kutta differential equation solver [41]. The resulting number of F$_{ion}^{2+}$ centers formed during the laser pulse was taken as the ion emission intensity. A series of solutions at different fluences was used to fit the parameters to the fluence data, using a Levenburg-Marquardt nonlinear curve-fitting procedure.

The curve fits for MgO are shown in Fig. 5(a). All five parameters were fit independently, for both polished and abraded MgO. Nevertheless, the only parameter that differed significantly in the two materials was the initial defect density (polished: 8.6×10^6 cm^{-2}; abraded: 1.8×10^7 cm^{-2}). These are high, but physically plausible, defect densities. The magnitudes of the other parameters were physically reasonable and are discussed in more detail in Reference 26.

The model is readily fit to the ~sixth-order kinetics of the Ca$^+$ intensities in Fig. 5(b). The ratio between the intensities shows that abrasion increases the density of surface adion/trap defects by at least a factor of three. The same model is applied to describe Na$^+$ emission from NaNO$_3$ both before and after electron irradiation [18]. The Na$^+$ intensities from the as-cleaved and electron-irradiated, solution-grown material in Fig. 5(c) were fit with the *same* model parameters, with the exception of the initial F$_{ion}$ densities. Similarly, the Na$^+$ intensities from the as-cleaved and electron-irradiated, melt-grown material in Fig. 5(d) again were fit with the *same* model parameters, with the exception of the initial F$_{ion}$ densities. Although the number of parameters in this model is high, the fact that only one parameter must be changed to account for the effect of electron irradiation argues for the robustness of the model. Finally, although we have not yet fit the data for the CaHPO$_4 \cdot$ 2H$_2$O (shown here in Fig. 5(e) for several different treatments), we see that the slopes of these log-log plots are ranging from ~6–8, depending on the type of "inflicted damage."

Because of the large number of fitting parameters, it clearly is inappropriate to consider these fits to be unique. However, bounded by reasonable initial values, the fact that we are able to fit the different nonlinear behavior for over ten different surfaces in a consistent manner indicates that the basic concepts proposed are reasonable.

In summary, MgO, NaNO$_3$, CaCO$_3$, and CaHPO$_4 \cdot$ 2H$_2$O surfaces exposed to low fluence 248-nm laser radiation yield energetic ions with a strongly nonlinear fluence dependence—consistent with electrostatic repulsion from surface defect sites. We attribute these emissions to the ejection of cations adsorbed atop surface electron traps when the underlying electron trap is photo-ionized. Emission requires 3−6 photons, depending on the surface. Competition between photo-ionization and retrapping accounts for the roll off to lower order at higher fluence. We now explore the possibility that as we increase the fluence, the photoelectron and positive ion emission discussed here can lead to strong coupling between these charges due to Coulomb attraction.

3.5 Low Fluence Plume Formation

3.5.1 What Is the Mechanism for Insulators at Low Fluence?

Plasma formation on insulating materials, the development of plume fluorescence, and the ionization of gas phase neutrals are important processes in the application of laser ablation to chemical analysis, thin film growth, and other technological situations. A clear understanding of the relevant mechanisms and how these mechanisms depend on material properties helps improve procedures and more accurately assess their potential in these applications. In the laser ablation of many metallic and semiconducting materials, the onset of atomic line emissions corresponds closely with the onset of laser induced "breakdown," where laser absorption in the plasma produces extensive excitation and ionization of neutrals [42]. At IR wavelengths, laser radiation is more favorably absorbed by inverse bremsstrahlung (IB) processes, wherein a three-body collision involving a photon, an electron, and a third (usually charged) particle results in photon absorption and electron acceleration. Because this three-body process depends strongly on the density of charged particles in the plume, ionization of neutrals by hot electrons can dramatically increase laser absorption, electron acceleration, and ionization, causing true "breakdown." At UV wavelengths, laser absorption by IB processes is strongly hindered. Nevertheless, laser absorption by multiphoton processes can ionize neutrals to produce similarly high densities of charged particles, given sufficient laser intensity. The production of copper ions by two-photon absorption at 193 nm and 351 nm, for instance, yields high ion and electron densities from irradiation of the metal, which eventually leads to breakdown-like behavior and strong atomic line emissions [43].

The origin of atomic line emissions during the laser ablation of wide bandgap insulators at UV laser wavelengths and nanosecond pulses is less clear [44]. At much higher fluences (and usually much shorter pulses) than we employ here, it has been shown that multiphoton absorption plus IB results in avalanche breakdown (also called impact ionization), which damages transparent dielectrics and produces visible plasmas. These intrinsic processes depend strongly on the laser fluence (often varying as the nth power of fluence, where $n > 1$), and generally show a threshold (W/cm^2) that increases as $\tau^{1/2}$, where τ is the pulse width. The IB absorption coefficient, α, involving free-free transitions among ion-electron pairs is given by [48],

$$\alpha = 3.69 \times 10^8 \frac{Z^2 n_e n_i}{T^{1/2} v^3} \left[1 - \exp\left(-\frac{h\nu}{kT}\right) \right] \text{cm}^{-1} \quad (3.2)$$

where Z is the average ion charge, n_e and n_i are the electron and ion densities in cm^{-3}, T is the electron temperature in K, and ν is the laser frequency in Hz. For the purposes of this discussion, we can take $n_e = n_i$.

From the absorption coefficient, one can estimate the average kinetic energy delivered to each electron by IB processes:

$$\Delta E_{\text{avg}} = k n_{\text{ion}} N_{\text{photon}} E_{\text{photon}} \quad (3.3)$$

where N_{photon} is the number of incident photons per cm^2, E_{photon} is the photon energy (5 eV), and $k = 3.39 \times 10^8 Z^2 T^{-1/2} \omega^{-3} [1 - \exp(h\nu/kT)]$ cm^{-5} $\approx 4 \times 10^{-39}$ cm^5 for 0.2 eV electrons. In the case of NaNO$_3$, spectral lines are first detected at a fluence of 100 mJ/cm^2, where $N_{\text{photon}} \approx 1 \times 10^{17}$ photons/cm^2. The measured ion density at this same fluence (extrapolated back to the surface) corresponds to 10^{15} cm^{-3}. Then $\Delta E \approx 10^{-6}$ eV. Electron heating via IB is truly negligible at this fluence.

What would be required to induce optical breakdown in this system? Neutral ionization becomes significant at electron temperatures of about 1 eV, so we require $\Delta E \approx 1$ eV. At a fluence of 100 mJ/cm^2 (1×10^{17} photons/cm^2 as before), this requires ion densities on the order of 10^{20} cm^{-3}. Even at a fluence of 1 J/cm^2, the required density is still 10^{19} cm^{-3}. This represents a "seed" density, which must be supplied to initiate IB. The only feasible way to attain such high ion densities is to ionize neutrals by another process, perhaps by multiphoton absorption.

Unlike metals and semiconductors, insulators (especially ionic materials) irradiated in the UV can yield intense photoelectronic ion and electron emissions at very low fluences due to defects, providing substantial numbers of charged particles. Through defect-mediated heating, temperatures near and above melting can be reached, simultaneously generating significant emissions of metal atoms. We propose that this combination of intense charged particle emissions and neutral emissions allows for the development of plume fluorescence *without* IB absorption or breakdown, thus providing an explanation for how plasma formation can arise on systems that are electron "poor" such as ionic crystals.

Here we reexamine plume formation during laser irradiation of two wide bandgap insulators, NaNO$_3$ and MgO at 248 nm [2, 49]. The plume fluorescence accompanying 248 nm irradiation of dielectrics has been previously characterized [35, 50, 51]. We will show that the particle densities at the onset of plume fluorescence are far too low for significant laser-electron interactions. Nevertheless, Langmuir probe measurements often indicate electron temperatures (T_e) of several eV. We propose that a Coulomb potential well produced by positive ions (emitted via photoelectronic processes) can trap and accelerate electrons to energies sufficient to excite metal atomic lines. Although previous modeling efforts have shown that space charge effects can accelerate a small fraction of the emitted electrons or ions to high energies [52, 53], the role of direct Coulomb interactions between photodesorbed ions and photoelectrons in the formation of a fluorescent plume has apparently not been explored. A simple two-dimensional Coulomb model that indicates electron trapping within a net positive charge cloud is presented. The accompanying electron acceleration is sufficient to account for the observed electron energies and densities. The data we show was taken in two separate vacuum systems so that distances to detectors were not the same.

3.5.2 Plume Spectra

At the level of sensitivity of a standard-image intensified Reticon detector combined with a fast, low resolution grating, the lowest fluence yielding detectable Mg° line emissions from MgO from a laser damaged surface was ~2.3 J/cm^2. A typical spectrum of the MgO plume fluorescence acquired at a fluence of 3.4 J/cm^2 appears in Fig. 7(a), where data acquisition started at the beginning of the 30-ns laser pulse and continued for 140 ns. Atomic emissions produced by the decay of excited Mg triplet states are observed at 310 nm (4d → 3p), 383 nm (3d → 3p), 518 nm (4s → 3p), and 765 nm (5p → 4s). Also observed is the Mg$^+$ line at 448 nm (4f → 3d). The Mg 285-nm (3p → 3s, singlet) resonance line was unfortunately strongly attenuated by the 285-nm cutoff filter used to suppress scattered laser light and therefore was not used in this analysis. Note that a broad photoluminescence band from the MgO crystal peaking at ~400 nm is evident [54] and has been previously assigned to anion vacancy defect clusters. Direct excitation of the observed atomic lines requires energies ranging from 5.1 eV for the Mg 518-nm line to 11.6 eV for the 448-nm Mg$^+$ line; none of these states can be produced by direct absorption of the 5.0 eV, 248-nm laser photons. Thus any model (e.g., that presented below) that is based on gas phase electron impact excitation of neutral metal atoms must explain the generation of energetic electrons.

The fluences required to generate a fluorescent plume on single crystal NaNO$_3$ is strongly dependent on the treatment prior to irradiation (e.g., cleaved vs. abraded vs. electron beam irradiated); similar to the Na$^+$ emission, plume formation is greatly aided by defect production. The acquired spectra of the fluorescent plume shown in Fig. 7(b) is all Na I in origin, principally the D-line doublet (unresolved) at 598 nm, $3s^03p^1 \rightarrow 3s^13p^0$, with an excitation energy of

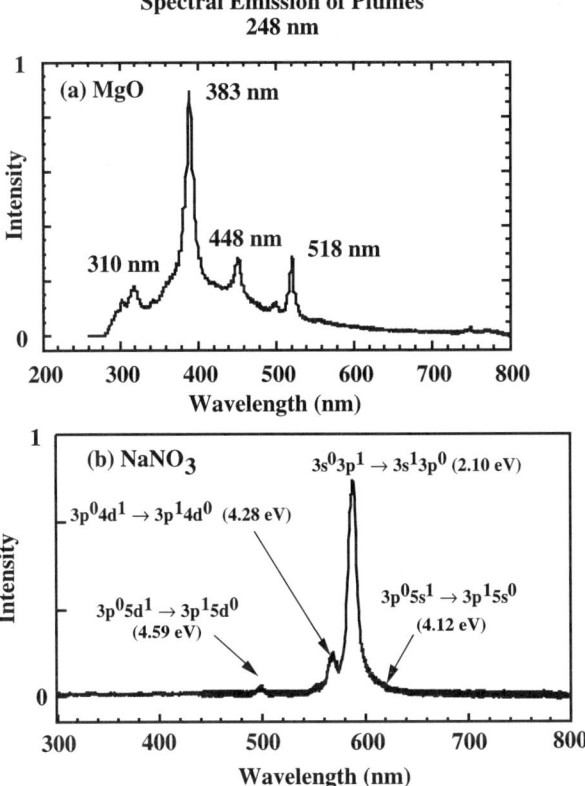

FIG. 7. The spectral emissions of the plumes generated over (a) single crystal MgO and (b) single crystal NaNO$_3$ by 248-nm incident light.

2.10 eV, which first appears at ~110 mJ/cm^2. Higher excitation levels are reached, generating the transitions $3s^04d^1 \to 3p^14d^0$ (4.28 eV), $3p^05s^1 \to 3p^15s^0$ (4.12 eV), and $3p^05d^1 \to 3p^15d^0$ (4.59 eV) as the fluence is increased. These Na I emission lines are on top of a long-lived crystal luminescence centered at 420 nm, which is not evident in Fig. 7(b).

3.5.3 Temporal Behavior of Detected Emissions

The time evolution of the Mg I 383-nm line from MgO line is shown in Fig. 8(a) for a fluence of 2.5 J/cm^2. The position of the laser pulse (30 ns) is indicated by the time intervals marked on the plot. Over the entire 1-µs time interval of data collection, the plume remained in full view of the light collection optics. Thus the detected signal reflects the spatially integrated emission over the entire volume of emitting material and is not affected by the motion of the

Time Resolved Emission from Plumes

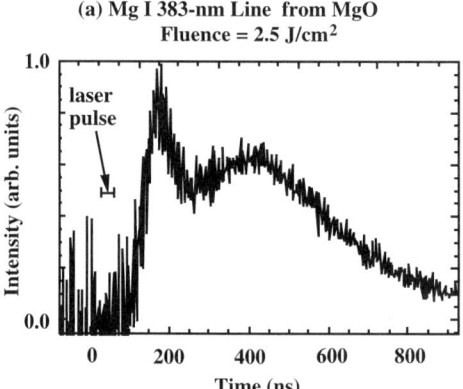

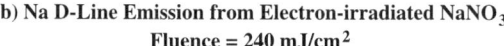

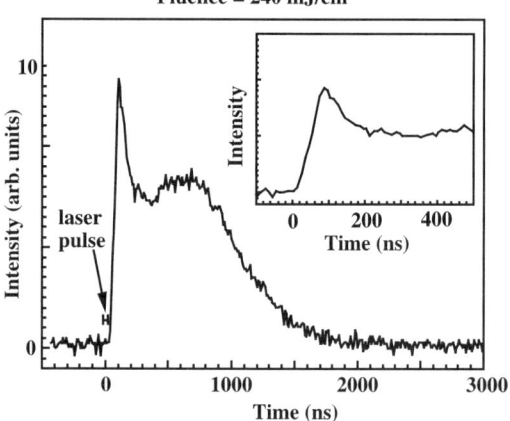

FIG. 8. The time-resolved fluorescence spectrum from the plumes generated at low fluences on (a) MgO and (b) $NaNO_3$.

plume. Surprisingly, the initial peak in emission intensity occurs *after* the laser pulse. A second, broader peak in emission intensity develops about 400 ns after the laser peak. Over the course of a microsecond (at this fluence), the light intensity gradually falls to zero. Similar measurements of the time evolution of the Mg I 285-nm and 518-nm lines showed the same double-peaked structure, with only minor variations in their time evolution.

Figure 8(b) shows the time evolution of the Na D-line from $NaNO_3$ at a fluence of 240 mJ/cm^2 where we have subtracted out the photoluminescence signal from the crystal. Again, note that the rise in Na light intensity is delayed only slightly after the laser pulse but has a double peaked structure similar to the Mg I emissions above.

In terms of mechanisms, the lack of light emission during the laser pulse is highly significant. Laser-induced breakdown produces sharply increasing ion and electron densities and electron temperatures during the laser pulse. Because breakdown requires high neutral densities and high laser irradiance, all the requirements for intense fluorescence due to breakdown are met only during the laser pulse. The lack of intense atomic fluorescence during the laser pulse is strong experimental evidence against breakdown at these fluences.

3.5.4 Neutral Metal Atom Emission

The predominant neutral species from single crystal MgO are Mg, O, and molecular MgO. At a fluence of 3.3 J/cm^2, for instance, relative intensities of 1000:10:1 are typical. Thus Mg° accounts for virtually all the neutral emissions. At the quadrupole mass spectrometer (QMS) sensitivity employed, neutral emissions were not detected at fluences below about 2.3 J/cm^2. The TOF curves for Mg° correspond to broadened Maxwell-Boltzmann distributions with some evidence of a component of emission occurring μs to ms *after* the laser pulse. The major portion of the TOF curves could be fit with a MB distribution corresponding to temperatures of ~900–3000 K, increasing with fluence. (Scanning electron microscope observations of MgO exposed to these fluences in previous work often show melted features, consistent with surface temperatures above the melting point (3125 K) [35, 50].)

The sodium atom emission from both freshly cleaved and e-beam damaged NaNO$_3$ is fit very well over a very large range of fluences with Maxwell-Boltzmann distributions [26]. If we boldly assume that the resulting fits arise from atoms in "equilibrium" with a surface at the resulting temperatures, we can push such data a little further. Fig. 9(a) shows a typical TOF fit to the Na neutrals detected with our QMS at a fluence of 149 mJ/cm^2, yielding a temperature of 857 K. (The fast peak at ~30 μs is the Na$^+$ emission that can pass through the ionizer into the QMS mass filter. For constant QMS parameters, the area under the neutral Na curve represents the yield or rate of neutral Na production. If we now take these two quantities (yield and temperature) and plot them as an Arrhenius plot (Log yield vs. 1000/T(K)) as shown in Fig. 9(b), we find a surprisingly good fit. (A similar analysis was carried out previously for the monomer emission from laser irradiation of polytetrafluoroethylene (Teflon^{TM}) [55] yielding excellent fits.) The activation energy obtained from Fig. 9(b) is 0.27 eV, which actually is rather low. In a separate experiment we looked at Na° thermal desorption from e-beamed NaNO$_3$ at heating rates of degrees/second and found an activation energy close to 1.6 eV [14]. This suggests that the

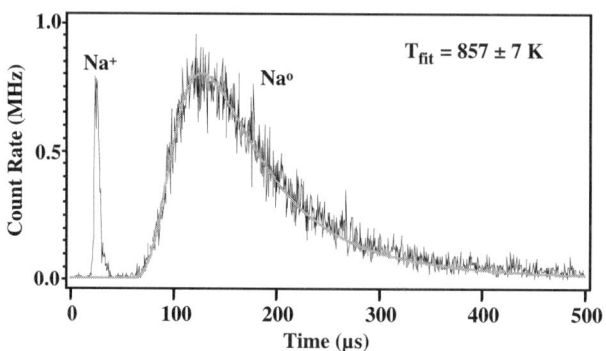

(a) Typical Na neutral emission TOF from $NaNO_3$
Sample Treatment: 36 mC/cm^2 Dose of 2500 eV Electrons

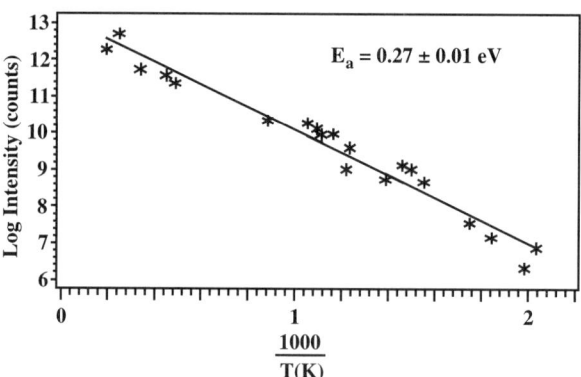

(b) Arrhenius Plot of Na Neutral Emission from $NaNO_3$
Sample Treatment: 36 mC/cm^2 Dose of 2500 eV Electrons

FIG. 9. The neutral Na emission from e-beamed $NaNO_3$. (a) The TOF curve fitted to an appropriate Maxwell-Boltzmann distribution. (b) An Arrhenius plot of the neutral Na intensity vs. $1000/T$, where T is determined from plots such as (a).

activation for laser desorbed Na° corresponds to the supply of sorbed excess Na on the surface, possibly due to diffusion of Na or F-centers to the surface. Nevertheless, a thermal emission model for neutral metal emission seems appropriate and compelling.

3.5.5 Overlap of Emissions

Typical ion and electron TOF signals detected by a biased metal collector a few cm from the sample appear in Fig. 10 for (a) MgO taken at a fluence of 2.5 J/cm^2 which is slightly above the threshold for detectable light emission, and

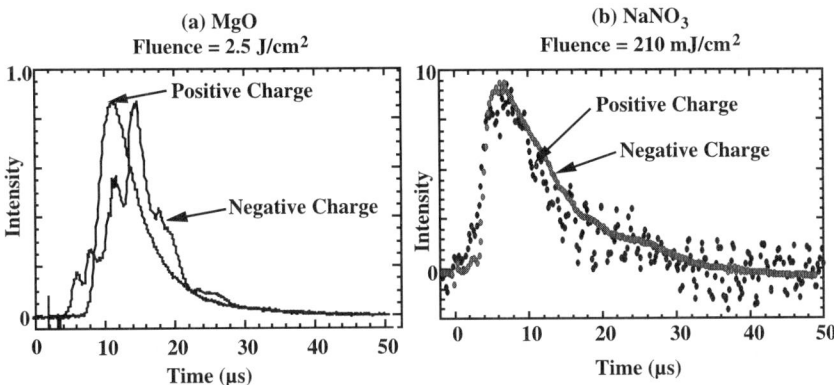

FIG. 10. Overlap in time of the positive and negative charge which has coupled together above the surfaces of (a) MgO and (b) NaNO$_3$.

(b) for NaNO$_3$ taken at 210 mJ/cm^2. The ion and electron peaks have been normalized to the same peak heights for comparison. The ion peaks are principally due to (a) Mg$^+$ and (b) Na$^+$, respectively. For the electrons, fast, sharp peaks in the electron signal are observed almost coincident with the laser on this time scale and are due to photoelectrons traveling directly from the sample to the detector at typical electron velocities. The slow electron peaks shown here are due to "slow" electrons, whose TOFs are very similar to the ion TOF. Clearly, there is a coupling of these charges, which drags electrons along with the ions. We see this coupling at fluences where the electron temperatures are much too low to produce ionization of neutrals.

We therefore attribute the slow electrons to photoelectrons that have become electrostatically coupled to the ions to form a "proto-plasma" or ±charge cloud. (At the lowest particle densities, the Debye length corresponding to the electron densities and energies is much larger than the laser spot. In this case, the charge distribution does not behave as a true plasma, thus the alternative terminology. As discussed below, these electrons are trapped in the potential well of the ion distribution and acquire considerable kinetic energy. Their actual trajectories oscillate back and forth through and/or around the central portion of the ion distribution with velocities that are much higher than the motion of the center of mass of the ±charge cloud, somewhat like a bee swarm. Since the detected ion and electron signals reflect the particle flux at the detector, a relatively large fraction of the electrons in the ±charge cloud are collected due to the higher rate of collisions with the detector because of their oscillatory motion. In contrast, the positive ions display relatively straight, line-of-sight trajectories that allow for a straightforward estimate of the local ion density. Because the plume is charged positive, the ion density provides an upper bound on the electron density. As noted below, large differences between the ion and electron densities are not expected, nor are they needed to account for electrostatic acceleration of the trapped electrons.

Many workers have observed, normally at higher charge densities, strong coupling between electrons and positive ions. Experimentally, our ±charge cloud becomes progressively more difficult to separate using electrostatic deflection, for example, repulsive grids. As expected, we find that this difficulty increases rapidly with laser fluence due to a decreasing Debye screening length. Assuming electron temperatures in the 1–10 eV range (see below), electron densities (n_e) on the order of 10^6 cm^{-3} are required to produce a Debye screening length of 200 μm. We conclude that electrostatic coupling, which at higher fluences leads to the formation of a plume, begins at fluences as low as 1 J/cm^2 for MgO and as low as 80 mJ/cm^2 for e-beam damaged NaNO$_3$.

The ion energies (assuming Mg$^+$ and Na$^+$) corresponding to the peak ion TOFs are approximately 11 eV for both MgO and NaNO$_3$, consistent with the previous mass selected measurements, which we explained by the electrostatic mechanism. We note that these energies are well above what one would expect from ionized neutrals, which should correspond to the energies of fast neutral particles. These high energies are also inconsistent with the presence of "breakdown" at fluences near the threshold for plume fluorescence. As we increase the fluence, we are seeing indications of ionization of neutrals in the angular and energy distributions of the detected ions [56].

When we compare the fluence dependencies of the plume fluorescence and the metal atom neutrals (see Fig. 11), we find for MgO that they are nearly identical, both becoming detectable at ~2.3 J/cm^2 and rising in a nonlinear fashion. Both emissions rise nonlinearly and show no tendency to saturate over this range of fluences. (The neutral intensities shown in Fig. 11(a) and (b) were acquired at reduced detector sensitivity in order to accommodate without saturation the emissions at the highest fluences.) This suggests that for MgO the fluorescence intensity is initially limited by the *availability of neutrals*. In contrast, Na° emission from NaNO$_3$ is observed at fluences well below the threshold for plume fluorescence (shown on log-log plots in Fig. 11(b)); in NaNO$_3$, the onset of plume fluorescence is not limited by the Na neutral density but rather the *electron temperature*.

3.5.6 Electron Energy Measurements

Ion densities and electron temperatures can be measured by standard Langmuir probe techniques [57]. This involves measuring the current to a shielded probe as a function of bias voltage on the probe, mounted a few mm from the sample. The analysis of the resulting I-V curves assumes a homogeneous

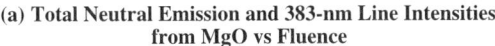
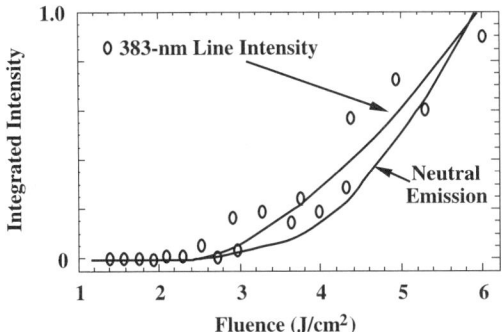
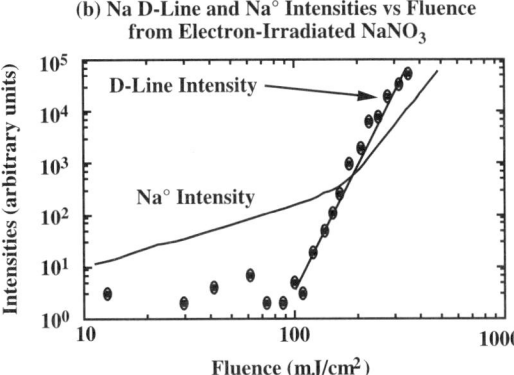

FIG. 11. The total atomic metal neutral emission and major spectral line intensity vs. fluence for (a) MgO and (b) NaNO$_3$.

±charge cloud and a Maxwell-Boltzmann electron energy distribution. The electron temperatures determined over a range of fluences near the onset of plume fluorescence are shown in Fig. 12 for (a) MgO and (b) e-beamed NaNO$_3$. For MgO, T_e increases rapidly at fluences above 1.6 J/cm^2 and exceeds 10 eV at fluences above 3 J/cm^2. Because the expansion of the +charge cloud and evaporative cooling (escape of high energy electrons) inevitably cools the electrons, these temperatures must be regarded as a lower limit on the maximum electron energies attained during the evolution of the plume. Therefore, we can conclude that the electron energies in the MgO plume are sufficient to excite Mg° lines at fluences below what is necessary for the observation of fluorescence.

In Fig. 12(b) we show T_e versus fluence for NaNO$_3$ where again we see a nonlinear increase in T_e with fluence; at the lowest reliable measurement (at

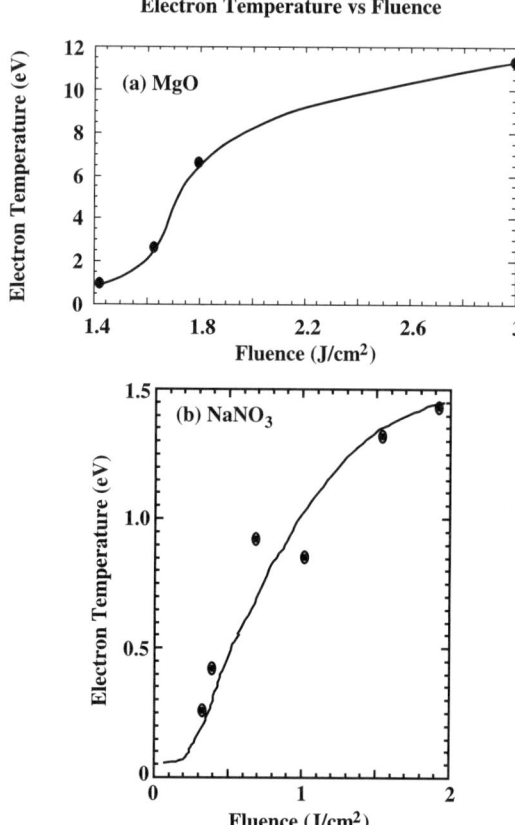

FIG. 12. Electron temperature measurements on the ±charge clouds vs. fluence for (a) MgO and (b) NaNO$_3$. These results were derived from Langmuir probe data.

300 mJ/cm^2), T_e was ~0.25 eV. As the fluence is raised from this value, the electron temperature grows gradually, reaching 1 eV at fluences between 1 and 2 J/cm^2. Due to the high energy tail in a Maxwell-Boltzmann distribution of electron energies, electron temperatures of 0.25 eV are sufficient for the excitation of emission lines in most metal atoms. Thus it not unreasonable to expect to see the Na D-line (requiring 2.1 eV to excite) at these low *electron temperatures*. The rapid increase in electron temperature with fluence at the lower fluences indicates that electron temperature is the limiting factor in the buildup of detected D-line emission intensity at similar fluences.

Although the particle densities determined in this work are far too low to support breakdown, they are nevertheless more than sufficient to account for the observed atomic emissions by electron collisions. For example, the cross-section for electron impact excitation of the Na D-line over a large range of electron energies (>2.1 eV) is about 10^{-15} cm^2 [58]. The peak electron density at 200 mJ/cm^2 is on the order of 10^{15} cm^{-2}, which corresponds to a flux of 10^{23} cm^{-2} s^{-1}. Finally, about 10^{12} neutral Na atoms are emitted per laser pulse. The D-line excitation rate is given by the product of these factors (10^{20} s^{-1}), divided by a large factor accounting for the limited overlap between ions and neutrals and the small number of electrons with energies over 2 eV. If only 1 in 10^2 neutrals and 1 in 10^4 electrons participate in excitation reactions (E > 2 eV), the excitation rate is still a very respectable 10^{13} s^{-1} (or 10^4 ns^{-1}) which could easily account for the observed emissions.

3.5.7 Proposed Electron Heating Mechanism

The Langmuir probe measurements confirm that a large fraction of the electrons have kinetic energies of eV or more at quite modest fluences. However, retarding potential analysis measurements of the directly emitted photoelectron energies (discriminated by TOF) at fluences below 1 J/cm^2 showed 95% of the electrons had energies under 1 eV. Thus electrons associated with the ±charge cloud have experienced substantial acceleration.

At the electron and ion densities observed here, electrons can be accelerated to significant velocities in the electrostatic potential well of the ±charge cloud, provided that the cloud has a net positive charge. Any excess of ions over electrons in the ±charge cloud produces a region of net positive potential, and electrons entering this region are accelerated accordingly.

Two processes ensure an excess of ions. During the early part of the laser pulse, a substantial number of electrons "outrun" the ions due to their greater velocity, leaving a region of net positive charge. Further, depending on initial electron trajectories, electrons can have angular momentum leading to orbits away from the cloud center, leaving a more positive "core." These electrons can still be accelerated to kinetic energies comparable to the depth of the potential well. They can frequently visit the tail of the ±charge cloud and will tend to provide additional overlap with the neutrals.

3.5.8 Estimate of Well Depth

For the simple case of a spheroidal ±charge cloud with a uniform charge density, the potential in the interior of the distribution, Φ, is a parabolic function of the position away from the center of the distribution [59]:

$$\Phi(r, z) = \frac{m_i \omega_i^2}{4e}(\beta r^2 + \gamma z^2) \qquad (3.4)$$

where $\omega_i^2 \equiv (n_i - n_e)e^2/m_i\varepsilon_0$, β and γ are dimensionless constants of order unity that depend on the dimensions of the ±charge cloud, and r and z are position coordinates along the major and minor axes of the spheroidal charge distribution. In the context of a plume, the z-axis lies along the surface normal through the center of the laser spot, and r is a radial coordinate, representing the radial distance from the z-axis.

The depth of this potential well can be significant at realistic charge densities. For a spheriodal charge distribution with a radius of 0.4 mm (the radius of a typical laser spot) and a length of 360 μm (the product of the velocity of a typical energetic ion and the length of the laser pulse), $\beta \approx 0.5$. The ±charge cloud tends to outrun the neutrals, so that the important electrons will lie near the edge of the cloud toward the surface, along the rim of the potential well. Thus we neglect the potential due to the z-term in Eq. (1). Then a net charge density $n_{\text{eff}} = (n_i - n_e) \sim 4.0 \times 10^{10}$ cm^{-3} (about 0.001% of the ion density, n_i, at 2.3 J/cm^2 for MgO) yields a potential well about 12 eV deep. For NaNO$_3$, a charge imbalance of $n_{\text{eff}} = n_i - n_e$ about 10^{-6} of the ion density (n_{eff} about 1×10^9 cm^{-3}) would yield a well about 1 eV deep, consistent with the smaller excitation energies needed for Na versus Mg. Therefore, a nearly neutral ±charge cloud can accelerate electrons significantly at electron/ion densities where IB heating is not significant.

3.5.9 2D Model/Electron Trajectories

Trajectory simulations show that electrons can be trapped and accelerated in regions with a net positive charge that grows during the laser pulse. We model the charge distribution as a 2-D ribbon of positive charge that grows along the surface normal, in the z-direction, during the laser pulse. The growth rate is chosen to correspond to the energy of the fastest ions observed over this range of fluences. After the laser pulse, the ribbon stops growing and moves away from the surface with the same velocity. The ribbon width in the x-direction was equal to the diameter of a typical laser spot (0.8 mm). The 2-D potential of this distribution experienced by an electron is then given by:

$$\Phi = \sigma \int_{-z_1}^{+z_1} dz' \int_{-x_1}^{+x_1} \frac{dx'}{\sqrt{(z-z') + (x-x')^2}} \quad (3.5)$$

where σ is the effective surface charge density ($\sigma \propto n_i - n_e$), z_1 is half the width of the charge sheet in the vertical dimension (parallel to the sample surface), and x_1 is half the length of the charge sheet in the horizontal dimension (perpendicular to the charge sheet). The integral can be evaluated analytically and differentiated to determine the electric field.

Single electrons were then launched from the surface with initial kinetic energies ranging from 0.1 to 5.0 eV and at various angles to the surface normal. Single-electron trajectories were calculated by solving the equations of motion using a fourth-order Runge-Kutta algorithm [60], ignoring electron-electron interactions. Particles with trajectories intersecting the surface or extending more than 20 cm from the surface were considered lost, and the calculation was terminated. A typical simulated electron trajectory for the first 60 ns after the beginning of the laser pulse is shown in Fig. 13.

In these simulations, many electrons emitted during the early part of the laser pulse escape from the region of interest. However, those that remain showed these rapid multipass oscillations through the space occupied by the ions. These bound trajectories tended to show significant acceleration, where the acquired energy depended on the initial velocity, direction, and time of "launch" during the laser pulse. Trajectories were calculated at effective densities of 10^{15} to 10^{18} cm^{-2}, corresponding to 3-D densities of roughly 10^{16}–10^{19} cm^{-3}. The associated potential well depths ranged from 0.2 to 200 eV. (The potential well depth increases with the physical size of the charge distribution, as seen in Eq. (1). Thus 3-D simulations would give deeper potential wells for a given charge density.) Except at the highest (positive) 2-D charge densities used ($\sigma_{eff} > 5 \times 10^{17}$ cm^{-2}), where high energy electrons escaped from the interaction area, trajectories were always found that yielded energies close to the depth of the potential well. Thus we have a means of "heating" or accelerating electrons without laser-plasma interactions. As long as the ±charge cloud exhibits net positive charge, electron acceleration should occur. The orbits of the rapidly moving electrons are much like bees swarming through the ion cloud, revisiting every portion of the cloud many times. The overall time dependence

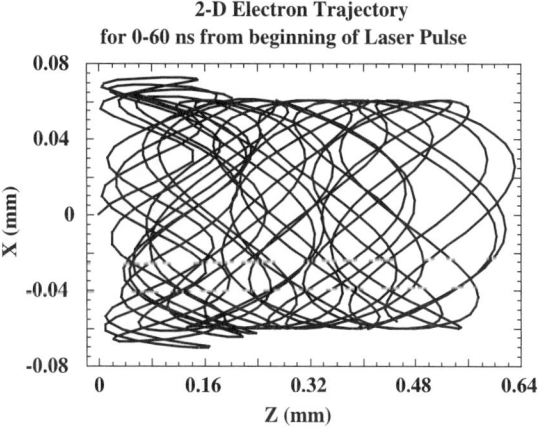

FIG. 13. Typical single-electron trajectory (simulation) for the first 60 ns after the beginning of the laser pulse (30 ns). This electron is trapped and oscillates in the ion potential well.

of the overlap of this electron "swarm" and the neutrals accounts for the unique behavior of the time evolution of the atomic line intensities.

3.5.10 Electron-Neutral Overlap

Because the accelerated electrons are electrostatically coupled to the ions, we can assume that they have the same overall spatial overlap with the neutrals as do the ions. The plume fluorescence is actually produced near the sample surface; therefore, the overlap for the first few microseconds after the laser pulse is required. At these short times, the evolution of the emission intensities during the laser pulse cannot be ignored. Unfortunately the TOF signals carry little information about the evolution of emission intensities on the submicrosecond time scale. Given some reasonable assumptions about this evolution (the "source functions") and density distributions determined from TOF data acquired with detectors at many centimeters from the surface, we can estimate the near-surface overlap. We first assume that both ion and electron emissions (which we have argued are photoelectronic in origin) peak at the maximum laser intensity. In contrast, we expect the neutral emissions to peak toward the end of the laser pulse. Because the neutral emissions are presumably emitted by thermal processes, the rate of neutral emission should depend exponentially on the surface temperature. Given the limited thermal conductivity of insulators such as MgO, the peak temperatures occur well toward the end of the laser pulse. Thus the neutral emission tends to be delayed relative to the ion emission.

For simplicity, we describe both neutral and charged particle source functions by Gaussian distributions in time, with equal amplitudes and full-width-at-half-maxima (30 ns), and we assume that the neutral emission lags the ion emission by three standard deviations ($\sim$50 ns) of the laser pulse. Thus the neutral emission peaks at the end of the laser pulse. The particle velocity distributions are then inferred from the ion and neutral density signals where we assume 1D expansion only. To portray the spatial overlap as a function of time, we integrate the product of the ion and neutral densities over all space as a function of time. The resulting "overlap integral" is displayed as a function of time in Fig. 14. (The overlap is given in units of length^{-6}—the product of two 3-D densities for (a) e^-—Mg° overlap and (b) e^-—Na° overlap where the e^- densities are assumed exactly proportional to the Mg^+, Na^+ densities, respectively.)

The computed spatial overlaps are almost identical due to similar ion and neutral kinetic energies and masses of the particles; both show a sharp initial peak *after* the peak in laser intensity, followed by a broad peak that extends some hundreds of nanoseconds after the laser pulse. These features represent surprisingly well the time dependence of all of the atomic spectral lines monitored in this work, including the 383-nm Mg line and the Na D-line in Fig. 7. Similar overlap integrals computed without the delay between the charged

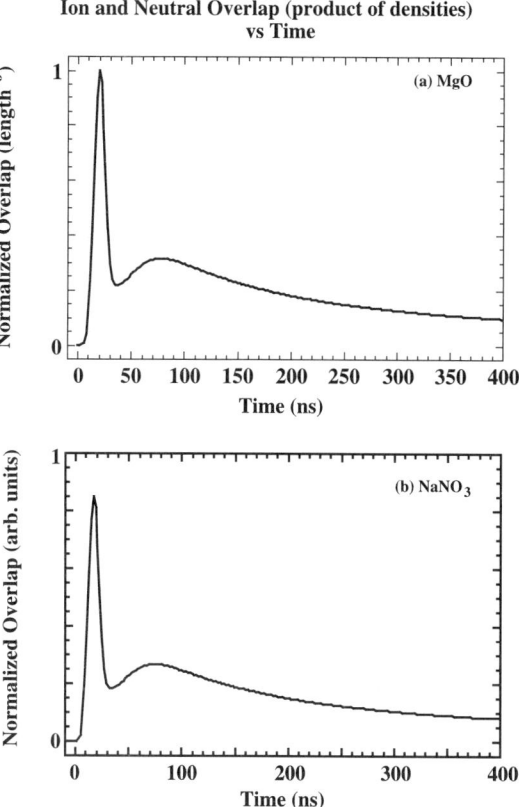

FIG. 14. Calculated overlap between the neutral and ions (and therefore electrons) as a function of time during and immediately after the laser pulse for (a) MgO and (b) NaNO$_3$.

particle and neutral source functions show only the initial peak, followed by a long decaying tail, contrary to our observations. From our particle detection measurements, we estimate the peak particle densities in the region of overlap to be $\sim 10^{14}/cm^3$ for the electrons and $\sim 10^{18}/cm^3$ for Mg°. Assuming typical cross-sections for electron impact excitation of neutral Mg ($1-5 \times 10^{-16}$ cm^{-2}) [61], these densities would correspond to a peak excitation rate of 3×10^8 excited atoms per nanosecond, which is sufficient to account for the observed fluorescence intensities.

We emphasize that plasmas formed by laser-induced ionization of neutral particles would not be expected to display the complex, two-peaked structure in the atomic line intensity versus time observed in this work. In the case of laser-induced ionization, one expects the line emission intensities to peak *during* the laser pulse [48]. The lack of line emission during the pulse provides substantial evidence against breakdown at the fluences studied here. We further note that nanosecond (long) laser pulses may well be required to establish the complex overlap between the ±charge cloud and the neutral particles observed in this work.

3.5.11 Absence of Breakdown

The absence of breakdown can be readily explained by the very high electron densities required for breakdown at 248 nm. As noted earlier, IB heating to 1 eV requires electron/ion densities of $10^{19}-10^{20}$ cm^{-3}. Under our conditions, electron emission from these insulators is defect limited and simply cannot provide these high charge densities.

Intrinsic multiphoton absorption bridging the bandgap (two-photon absorption) of the insulator would be an alternative route to forming e-h pairs. Neglecting recombination and diffusion processes, a large two-photon cross-section of $\sim 10^{-49}$ cm^2 would yield conduction band electron densities as high as 10^{19} cm^{-3} at the highest fluence employed (6 J/cm^2). However, electron hole recombination and carrier diffusion during the 30-ns laser pulse would greatly reduce this density. Using an analysis similar to Niemz [47], we estimate that at most $n_e \sim 10^{16}$ cm^{-3}. Therefore, conduction band electron densities generated by two-photon absorption are far too low to support IB.

3.5.12 Fluorescence and Ionization Mechanism

The principal mechanism for plume fluorescence from laser irradiation of MgO and NaNO$_3$ at 248 nm at relatively low fluences thus appears to be radiative decay of excited neutral metal atom states produced by electron impact. Electron-ion recombination apparently does contribute significantly to the observed emissions. Plume fluorescence is not observed until neutral emissions are detected, consistent with the excitation of neutral species, rather than the formation of excited neutrals by electron-ion recombination. Electron-ion recombination also requires low energy electrons, typically less than 0.5 eV kinetic energy [62]. Electron temperatures near the middle of the ion distribution are simply too high to permit significant electron-ion recombination.

In normal plasmas, excitation and ionization involve electrons from the high velocity tail of the velocity distributions. As noted earlier, electron temperatures of a few tenths of an eV are typically sufficient to excite atomic lines in typical metal atoms. Similarly, electron temperatures of 1 eV are generally sufficient to ionize the large majority of neutral Mg to Mg^{2+} in plasma excitation sources. Although we detect electrons temperatures of 6 eV and higher, we do not observe intense Mg^{2+}. We attribute the lack of these ions to a non-Maxwellian electron velocity distribution, first due to the acceleration mechanism, and second, due to the loss of high energy electrons from the ±charge cloud (discussed earlier). Because energies above 13.6 eV would be required to ionize Mg^+, a well depth in excess of 13.6 eV would be necessary to generate sufficiently energetic electrons, which in turn requires larger n_{eff} ($= n_i - n_e$), presumably higher than we are producing in these experiments.

3.6 Conclusions

In summary, the general requirements for neutral excitation by electron impact by the mechanism we have proposed are: (1) overlap of sufficient densities of neutral atoms and electrons, *and* (2) sufficient electron temperatures to provide the excitation. For wide bandgap materials and nanosecond laser pulses, both requirements are strongly dependent on the presence of surface and near-surface defects. Thresholds for visible fluorescence are determined by the need to establish both of these requirements simultaneously. These key factors, in turn, depend on the character (i.e., source functions, intensities, and energies) of both the photoemitted charged particles and the thermally emitted neutrals as a function of laser fluence and pulse width.

Other studies in our laboratory show that these aspects of ion, electron, and neutral emission are shared by other crystalline, ionic materials, including $CaCO_3$ and $CaHPO_4 \cdot 2H_2O$. Because similar emission mechanisms are likely to operate in other wide bandgap ionic materials, the mechanisms for plasma formation and fluorescence presented here could well apply to many other insulating materials.

Important questions and details remain for further study. A more detailed, quantitative model of the emitted particle densities and their time evolution is required to predict the time evolution of atomic line intensities relative to the laser pulse. At higher fluences, we expect the peak emissions to eventually shift toward and into the laser pulse. A better understanding of the surface and near-surface defect structures and how they evolve during the laser pulse and from pulse-to-pulse is required to understand the particle emission mechanisms at these intermediate fluences which yield ion-electron-neutral overlap. A slightly more detailed model of the ion cloud suggests that ions should be reflected back toward the surface and actually go *behind* the sample; recent low fluence measurements show significant quantities of Mg^+ from MgO at times indicating complex orbits, consistent with our modeling [63]. These types of models and measurements with refinement also can be extended to determining detailed angular distributions. Finally, these concepts should be extended in a rigorous fashion to other materials. It is conceivable that similar defect-mediated processes may control laser plume formation on covalently bonded ceramic materials, wide bandgap semiconductors, and perhaps even organic polymers.

Acknowledgments

Support by the Department of Energy under Contract DE-FG06-92ER14252 and the Pacific Northwest Laboratory under the Department of Energy's Strategic Environmental Research and Development Program is gratefully acknowledged. I also thank my students and colleagues David Ermer, J-J Shin, Mary Dawes, and Steve Langford at Washington State University for their assistance. I also thank Professor Noriaki Itoh, Osaka Institute of Technology, and Professor Eckhart Matthias, Free University of Berlin, for many helpful discussions.

References

1. Dickinson, J. T., Langford, S. C., Shin, J.-J., and Doering, D. L. *Phys. Rev. Lett.* **73**, 19, 2630 (1994).
2. Shin, J. J., Ermer, D. R., Langford, S. C., and Dickinson, J. T. *Applied Physics* **A64**, 7 (1997).
3. Petite, G., Agostini, P., Guizard, S., Martin, P., and Trainham, R., in *Laser Ablation of Electronic Materials*, E. Forgrassy and S. Lazare (eds.), pp. 21–37. Elsevier Science, Amsterdam, 1992.
4. Thoma, E. D., Yochum, H. M., Binkley, M. J., Leblans, M., and Williams, R. T., in *Defects in Insulating Materials* (*Mater. Sci. Forum* **239–242**), G. E. Williams and R. T. Williams (eds.), pp. 565–568. Trans Tech Publications, Switzerland, 1997.
5. Okano, A., Thoma, R. K. R., Williams, G. P., Jr., and Williams, R. T., *Phys. Rev. B.* **52**, 14789 (1995).
6. Leblans, M., Thoma, R. K. R., LoPresti, J. L. L., Reichlung, M., and Williams, R. T. *Materials Science Forum* **239–241**, 159 (1997).
7. Tanimura, K., and Fujiwara, H. *Materials Science Forum* **239–241**, 549 (1997).
8. Petite, G., Daguzan, P., Guizard, S., and Martin, P. *Materials Science Forum* **239–241**, 555 (1997).
9. Hall, R. B. *J. Phys. Chem.* **91**, 1007 (1987).
10. George, S. M. in *Investigations of Surfaces and Interfaces—Part A*, B. W. Rossiter and R. C. Baetzold (eds.), p. 453, Wiley, New York, 1993.
11. Hoogers, G., Papageorgopoulos, D. C., Ge, Q., and King, D. A. *Surf. Sci.* **340**, 23 (1995).
12. Webb, R. L., Langford, S. C., and Dickinson, J. T. *Nucl. Instrum. Meth. Phys. Res.* **B103**, 297 (1995).
13. Hess, W. P., German, K. A., Bradley, R. A., and McCarthy, M. I. *Appl. Surf. Sci.* **96–98**, 321 (1996).
14. Beck, M., Taylor, D. P., and Hess, W. P. *Phys. Rev. B.* (1997), in press.
15. Beck, K. M., McCarthy, M. I., and Hess, W. P. Atomic and molecular photostimulated desorption from complex ionic crystals. *J. Electron. Materials* (submitted).
16. McBride, J. M., Segmuller, B. E., Hollingworth, M. D., Mills, D. E., and Weber, B. A. *Science* **234**, 830 (1986).
17. Dickinson, J. T., Jensen, L. C., Doering, D. L., and Yee, R. *J. Appl. Phys.* **67**, 3641 (1990).
18. Ermer, D. R., Shin, J.-J., Langford, S. C., and Dickinson, J. T. *J. Appl. Phys.* **80**, 6452 (1996).

19. Knotek, M. L., and Feibelman, P. J. *Phys. Rev. Lett.* **40**, 964 (1978).
20. Kurtz, L., Sotckbauer, R., Nyholm, R., Flodström, S. A., and Senf, F. *Phys. Rev. B.* **35**, 7794 (1987).
21. Gotoh, T., Takagi, S., and Tominaga, G. In *Desorption Induced by Electronic Transitions, DIET IV*, G. Betz and P. Varga (eds.), pp. 327–332. Springer, Berlin, 1990.
22. Shin, J.-J., Kim, M.-W., and Dickinson, J. T. *J. Appl. Phys.* **80**, 7065 (1996).
23. Ermer, D., Dawes, M., Langford, S. C., and Dickinson, J. T., in preparation.
24. Webb, R. L., Langford, S. C., and Dickinson, J. T. *Nucl. Instrum. Meth. Phys. Res. B.* **103**, 297 (1995).
25. Shin, J.-J. Langford, S. C., Dickinson, J. T., and Wu, Y. *Nucl. Instrum. Meth. Phys. Res. B.* **103**, 284 (1995).
26. Dickinson, J. T., Shin, J.-J., and Langford, S. C. *Appl. Surf. Sci.* **96–98**, 326–331 (1996).
27. Murphy, D., and Giamello, E. *J. Phys. Chem.* **99**, 41, 15172–15180 (1995).
28. Ferrari, S. M., and Pacchioni, G. *J. Phys. Chem.* **100**, 21, 9032–9037 (1996).
29. Matthias, E., Nielsen, H. B., and Reif, J. *J. Vac. Sci. Technol.*, **5**, 1415 (1987).
30. Matthias, E., and Green, T. A. In *Desorption Induced by Electronic Transitions, DIET IV*, G. Betz and P. Varga (eds.), pp. 112–127. Springer, Berlin, 1990.
31. Stenzel, Bouchaala, N., Gogoll, S., Kotzbucher, T., Reichling, M., and Matthias, E. *Materials Science Forum* **239–241**, 591 (1997).
32. Kreitschitz, O., Husinsky, W., Betz, G., and Tolk, N. H. *Appl. Phys. A.* **58**, 563 (1994).
33. Wiedeman, L., Kim, H.-S., Helvajian, H. In *Laser Ablation: Mechanisms and Applications*, J. C. Miller and R. F. Haglund (eds.), pp. 350–359. Springer-Verlag, Berlin, 1991.
34. Dickinson, J. T., Langford, S. C., Jensen, L. C., Eschbach, P. A., Pederson, L. R., and Baer, D. R. *J. Appl. Phys.* **68**, 1831 (1990).
35. Webb, R. L., Jensen, L. C., Langford, S. C., and Dickinson, J. T. *J. Appl. Phys.* **74**, 2323–2337 (1993).
36. Baer, D. R., Blanchard, D. L. *Appl. Surf. Sci.* **72**, 295 (1993).
37. Dickinson, J. T., Jensen, L. C., Langford, S. C., Rosengerg, P. E., and Blanchard, D. L. *Phys. Chem. Miner.* **18**, 320 (1991).
38. Jones, C., Braunlich, P., Casper, R., Shen, X.-A., and Kelly, P. *Optical Engin.* **28**, 1039 (1989).
39. Magill, J., Bloem, J., and Ohse, R. W. *Phys.* **76**, 6227 (1982).
40. Dawes, M. The Madelung constant of the rhombohedral lattice, manuscript in preparation.
41. Press, W. H., Flannery, P., Teukolsky, S. A., and Vetterling, W. T. *Numerical Recipes in Pascal: The Art of Scientific Computing*, pp. 599–614. Cambridge University, Cambridge, 1989.
42. Radziemski, L. J., and Cremers, D. A *Laser-Induced Plasmas and Applications*. Dekker, New York, 1989.
43. Dreyfus, R. W. *J. Appl. Phys.* **69**, 1721 (1991).
44. Geohegan, D. B. In *Pulsed Laser Deposition of Thin Films*, D. B. Chrisey and G. K. Hubler (eds.), pp. 115–165. John Wiley, New York, 1994.
45. Du, D., Liu, X., Korn, G., Squier, J., and Mourou, G. *Appl. Phys. Lett.* **64**, 3071 (1994).
46. Stuart, B. C., Feit, M. D., Herman, S., Rubenchik, A. M., Shore, B. W., and Perry, M. D. *Phys. Rev. B.* **53** (4), 1749 (1996).

47. Niemz, M. H., *Appl. Phys. Lett.* **66**, 1181 (1995).
48. Leboeuf, J. N., Chen, K. R., Donato, J. M., Geohegan, D. B., Liu, C. L., Puretzky, A. A., and Wood, R. F. *Appl. Surf. Sci.* **96–98**, 14 (1996).
49. Ermer, D. R., Langford, S. C., and Dickinson, J. T. *J. Appl. Phys.* **81**, 1495 (1997).
50. Dickinson, J. T., Jensen, L. C., Webb, R. L., Dawes, M. L., and Langford, S. C. *J. Appl. Phys.* **74**(6), 3758 (1993).
51. Dirnberger, L., Dyer, P. E., Farrar, S., Key, P. H., and Monk, P. *Appl. Surf. Sci.* **69**, 216 (1993).
52. Gilton, T. L., Cowin, J. P., Kubiak, G. D., and Hamza, A. V. *J. Appl. Phys.* **68**, 4802 (1990).
53. Peurrung, A. J., Cowin, J. P., Teeter, G., Barlow, S. E., and Orlando, T. M. *J. Appl. Phys.* **78**, 481 (1995).
54. Dickinson, J. T., Jensen, L. C., and Webb, R. L. *J. Non-Cryst. Solids.* **177**, 1 (1994).
55. Dickinson, J. T., Shin, J.-J., Jiang, W., and Norton, M. G. *J. Appl. Phys.* **74**, 4729 (1993).
56. Ermer, D., Langford, S. C., and Dickinson, J. T., work in progress.
57. von Gutfeld, R. J., and Dreyfus, R. W. *Appl. Phys. Lett.* **54**, 1212 (1989).
58. Heddle, D. W. O., and Gallagher, J. W. *Rev. Mod. Phys.* **61**, 221 (1989).
59. Landau, L., and Lifshitz, E. *The Classical Theory of Fields*, 3rd rev. English ed., translated by Morton Hamermesh, p. 281. Pergamon, New York, 1971.
60. Press, W. H., Flannery, B. P., Teukolsky, S. A., and Vetterling, W. T. *Numerical Recipes in Pascal*: *The Art of Scientific Computing*, 599–614. Cambridge University, Cambridge, 1989.
61. Heddle, D. W. O., and Gallagher, J. W. *Rev. Mod. Phys.* **61**, 221 (1989).
62. Higgov, E., and Hirschberg, J. G. *Phys. Rev.* **125**, 795 (1962).
63. Ermer, D., Langford, S. C., and Dickinson, J. T., work in progress.

4. LASERS, OPTICS, AND THERMAL CONSIDERATIONS IN ABLATION EXPERIMENTS

Costas P. Grigoropoulos
Department of Mechanical Engineering
University of California, Berkeley

Lasers are used in a variety of materials processing applications, including melting and recrystallization, doping, and annealing, that do not involve mass removal from the target. Other processes, such as etching, micromachining, and pulsed laser deposition of thin films (PLD), are ablative. In all cases, design of a laser processing experiment requires consideration of the coupling of the laser energy with the target material via the optical delivery system. In this chapter, first the laser intensity and fluence variables are defined. Issues pertaining to the temperature field introduced in the target material are then recalled. An introduction to the design of the optical components used in laser processing experiments and the measurement of the laser beam characteristics is given next. The chapter concludes with a review of experimental methods available for measuring target temperature upon pulsed laser irradiation.

4.1 Definition of Laser Intensity and Fluence Variables

4.1.1 Gaussian Beams

The spatial profile of a laser beam at the laser exit aperture is determined by the geometry of the laser cavity. When the cross section of the cavity is symmetrical, as in the case of cylindrically or rectangularly shaped cavities, the spatial profiles become simple. Transverse electromagnetic modes are characterized by TEM_{mn}, where m and n indicate the number of modes in two orthogonal directions. The mode of highest symmetry is the TEM_{00} mode. Fundamental TEM_{00} operation implies that the instantaneous distribution of laser intensity across the beam is Gaussian:

$$I(r, t) = I_o(t) \exp\left[-\frac{r^2}{w^2}\right] \qquad (4.1)$$

where w is the radius of the point where the intensity drops by a factor of $1/e$ with respect to the peak intensity I_o at $r = 0$, and r is the radial coordinate. Frequently the radius where the laser beam intensity drops by a factor of $1/e^2$ is specified: $w_{1/e^2} = \sqrt{2}\, w$.

4.1.2 CW Laser Beams

Continuous wave (CW) laser beams imply that the laser intensity is constant with time, $I_o(t) = I_o$, except from transient fluctuations and the long-term drift:

$$I(r) = I_o \exp\left[-\frac{r^2}{w^2}\right] \tag{4.2}$$

The total laser power is defined by:

$$P = \int_p^\infty I(r) 2\pi r \, dr = \pi w^2 I_o \tag{4.3}$$

4.1.3 Pulsed Laser Beams

The dimensionless temporal pulse profile is characterized by:

$$p(t) = \begin{cases} \dfrac{I_o(t)}{I_{o;\max}}, & t < t_{\text{pulse}} \\ 0, & t > t_{\text{pulse}} \end{cases} \tag{4.4}$$

where $I_{o;\max}$ is the peak intensity at $t = t_{\max}$. It is common to characterize pulse lengths with the full-width-half maximum (FWHM) pulse length, t_{FWHM}. This is the temporal width of the pulse evaluated at the intensity, $I_{\text{FWHM}} = I_{o,\max}/2$.

The *local transient laser fluence* may be defined as follows:

$$F_{tr}(r, t) = \int_{-\infty}^{t} I(r, t') \, dt' \tag{4.5}$$

This quantity represents the energy per unit area incident at a specific location until the elapsed time t. The *total energy* carried by the laser pulse is given by:

$$E = \int_{-\infty}^{+\infty} \int_0^{+\infty} I(r, t') 2\pi r \, dr \, dt' \tag{4.6}$$

The *pulse fluence*, F, is defined as the pulse energy divided by an area corresponding to a circular disk of radius w.

$$F = \frac{E}{\pi w^2} \tag{4.7}$$

The intensity distribution $I(r, t)$ is:

$$I(r, t) = \frac{E}{\pi w^2} \frac{p(t)}{\int_{-\infty}^{+\infty} p(t') \, dt'} \exp\left[-\frac{r^2}{w^2}\right] \tag{4.8}$$

For a triangular temporal profile:

$$p(t) = \begin{cases} 0; & t < 0 \\ \dfrac{t}{t_{max}}; & 0 < t < t_{max} \\ \dfrac{t_{pulse} - t}{t_{pulse} - t_{max}}; & t_{max} < t < t_{pulse} \end{cases} \quad (4.9)$$

Smooth pulses whose peak intensities lie at $t = t_{max}$ can be fitted by:

$$p(t) = \left(\frac{t}{t_{max}}\right)^{\xi} \exp\left[\xi\left(1 - \frac{t}{t_{max}}\right)\right] \quad (4.10)$$

where ξ is a parameter characterizing the temporal profile.

For a sinusoidally modulated CW laser beam:

$$p(t) = 1 - \cos(\omega_{mod} t) \quad (4.11)$$

where ω_{mod} is the modulation frequency.

If a Gaussian beam of circular cross section is focused by a cylindrical lens, or if the beam is incident on the target at an oblique angle of incidence, the resulting profile is elliptical. On the other hand, the output of excimer, nitrogen, TEA CO_2, solid state lasers may have a roughly flat-top cross section. In general, the laser intensity profile is non-Gaussian and the intensity is a function of the spatial coordinates (x, y) on the irradiated target plane and time: $I(x, y, t)$. The local transient laser fluence, may be defined in a manner analogous to Eq. (4.5):

$$F_{tr}(x, y, t) = \int_{-\infty}^{t} I(x, y, t') \, dt' \quad (4.12)$$

If A is the area of the irradiated spot, the pulse fluence is defined simply as:

$$F = \frac{E}{A} \quad (4.13)$$

The laser parameters regulating laser processing are the wavelength, λ, polarization, and the intensity distribution $I(x, y, t)$ on the target surface. The temporal and spatial dependence of the intensity distribution depends on the mode structure and the external modulation through the beam delivery system.

4.2 Thermal Considerations

There are several ablative mechanisms by which material, either atomic or bulk, can be released from the surface of the target. References to "thermal" or "photothermal" ablation, generally imply a model in which laser light is converted to lattice vibrational energy before bond breaking liberates atomic material from the bulk surface. The thermal mechanism is distinct from a "photochemical" or "electronic" processes, in which laser-induced electronic excitations lead directly to bond breaking before an electronic to vibrational energy transfer has occurred. Both thermal and electronic sputtering mechanisms lead to the liberation of atomic-size material from the surface. This is different from two other ablation mechanisms, identified in the literature as "hydrodynamical" or "exfoliational," which introduce bulk material into the ablation plume. The hydrodynamical mechanism is ascribed to the liberation of micron-size droplets, following motion in the molten phase. On the other hand, exfoliation refers to an erosivelike mechanism by which material is removed from the surface in solid flakes. Separation of flakes from the surface is thought to occur along energy absorbing defects in the material. It should be pointed out that these mechanisms are not necessarily distinguishable in a specific laser ablation system, in which more than one mechanisms can occur either simultaneously, or in different phases of the ablation process and depending on the range of the laser parameters.

At a first sight, one can invoke the classical picture of thermal vaporization from a heated surface through transition to the liquid phase to describe material removal for moderate laser fluences on metallic targets at time scales that allow the establishment of local thermal equilibrium (LTE). That would be the case for nanosecond laser irradiation because relaxation times in metals are in the subpicosecond regime. Recent experiments discussed in Section 4.5 showed that laser sputtering of metals can occur at very low fluences, well below the perceived melting threshold. Sputtering experiments from the molten phase also unveiled evidence that cannot be explained through the classical thermal vaporization model. At higher laser fluences, the nascent metal vapor is photo-ionized, leading to further heating of the plasma through a cascade process as discussed, for example, in [1]. Even though laser ablation is a complex phenomenon defying a unified treatment, it is worth to recall some relevant thermal considerations.

4.2.1 Energy Absorption

The energy coupling into the target material is determined by the material optical properties, that is, the complex refractive index $n^c = n + ik$, which depends on the incident laser wavelength and the material temperature. The absorption depth into the material, $d_{abs} = 1/\gamma = \lambda/(4\pi k)$, while the volumetric energy intensity absorbed by the material at a depth z from the surface is given by

$$Q_{ab}(x, y, z, t) = (1 - R)I(x, y, t)\gamma e^{-\gamma|z|} \qquad (4.14)$$

where R is the surface normal incidence surface reflectivity, $R = (n-1)^2 + k^2/(n+1)^2 + k^2$.

More generally, for temperature dependent absorption coefficient:

$$Q_{ab}(x, y, z, t) = (1 - R(T_s))I(x, y, t)\gamma(T(z))\exp\left[-\int_0^z \gamma(T(z'))\,dz'\right] \quad (4.15)$$

where T_s is the surface temperature. Steep thermal gradients into the material alter the absorbed energy profile by introducing nonlinear effects. It is noted that for rough surfaces (RMS roughness $>\lambda/10$), the absorption of the laser energy is complicated because of surface light scattering. The light propagation in materials that are nonhomogeneous at the scale of the light wavelength is subjected to volumetric scattering. In fact, the absorption may be different than the expected contribution from the constituent components of the composite medium. It is also important to keep in mind that tabulated optical properties of materials refer to carefully prepared surfaces, usually under vacuum conditions. Surface oxides affect the material absorptivity and the coupling of the laser energy to the surface. The concept of the complex refractive index is inadequate for describing the electromagnetic energy coupling with ultra-small particles and surface features in the case of absorbing films that are thinner than $1/\gamma$. Surface modes such as the excitation of plasmons and electron scattering are dominant in this case. Finally, the simple heating picture presented does not address non-linear issues, such as thermal lensing and self-focusing in transparent media due to gradients of the real part of the complex refractive index and optical generation of free carriers in semiconductors and insulators that result in a steep increase of the light absorption with the laser energy [2].

4.2.2 Heat Transfer

For nanosecond laser pulses, the electrons and the lattice are at thermal equilibrium, characterized by a common temperature, T. The transient temperature field can then be calculated by solving the heat conduction equation:

$$(\rho C_p)(T)\frac{\partial T}{\partial t} = \nabla \cdot (K(T)\nabla T) + Q_{ab}(x, y, z, t) \quad (4.16)$$

where ρ, C_p, K, and T represent density, specific heat for constant pressure, thermal conductivity, and temperature, respectively. These properties in general are functions of temperature, but for a first estimate, constant thermal properties may be assumed to derive approximate analytical solutions. For a laser beam incident on the surface, $z = 0$, of a bulk substrate of thickness d and initial temperature T_0, the initial and boundary conditions for the heat transfer problem are:

Initial Condition:

$$T(t = 0) = T_0 \quad (4.17)$$

Boundary Conditions:

$$K\frac{\partial T}{\partial z}\bigg|_{z=0} = h_u[T(x, y, 0, t) - T_\infty] + \varepsilon_u \sigma(T(x, y, 0, t)^4 - T_\infty^4) \quad (4.18)$$

$$-K\frac{\partial T}{\partial z}\bigg|_{z=d} = h_L[T(x, y, d, t) - T_\infty] + \varepsilon_L \sigma(T(x, y, d, t)^4 - T_\infty^4) \quad (4.19)$$

In the above, h_u, h_L are the linear convective heat transfer coefficients from the top and bottom surfaces of the sample, ε_u, ε_L are the corresponding emissivities, σ is the Stefan-Boltzmann constant, and T_∞ is the ambient temperature. The thermal diffusion penetration depth into the material is given by $d_{th} = \sqrt{\alpha t_{pulse}}$, where α is the material thermal diffusivity, $\alpha = K/\rho C_p$. For $d_{th} \ll d$ (and $d_{abs} \ll d$), the target material can be considered as semi-infinite.

If $d_{abs}/d_{th} \ll 1$, the laser radiation absorption is essentially a skin surface phenomenon, which is true for metals irradiated by laser pulses of duration longer than nanoseconds. Barring plasma effects, the efficiency of energy coupling with the material is, in this case, to a considerable degree determined by the surface reflectivity. For metals, the surface reflectivity is large in the infrared (IR) range, typically over 0.90 for $\lambda > 5$ μm, but it may reduce in the near-IR and visible (VIS) ranges (e.g., [3]). Upon melting, the absorptivity $A = 1 - R$, is enhanced in a step-wise manner [4]. If the material absorption coefficient is weak and $d_{abs}/d_{th} \gg 1$, radiation penetrates deeper into the material, giving rise to shallower thermal gradients and a more uniform temperature field. A usual, yet not always accurate assumption is that of a constant absorption coefficient and surface reflectivity during the laser pulse.

4.2.3 1-D Heat Conduction

For a spatially uniform laser beam distribution, $I(x, y, t) = I(t)$, and if the radius or characteristic dimension of the laser beam is much larger than d_{abs} and d_{th}, the temperature profile in the material is one-dimensional. For $d_{abs}/d_{th} \ll 1$, negligible heat losses from the surface and assuming a laser source incident on an infinite target, the 1-D solution to Eq. (4.16) yields a surface temperature $T_{su}(t)$ as follows:

$$T_{su}(t) = \frac{2(1-R)\sqrt{\alpha/\pi}}{K} \int_0^t \frac{I(t')\,dt'}{\sqrt{t-t'}} + T_o \quad (4.20)$$

where T_o is the initial target temperature.

For a surface source and triangular temporal laser pulse profile,

$$T_{su}(t) = \frac{8\sqrt{\alpha/\pi}}{3K\tau_{max}t_{pulse}} F(1-R)\left\{t^{3/2} - \left[1 + \frac{t_{max}}{t_{pulse} - t_{max}}\right](t - t_{max})^{3/2}\right\} + T_o,$$

$$t_{max} < t < t_{pulse} \quad (4.21)$$

The peak surface temperature $T_{\text{su,pk}}$ occurs at a time, $t_{\text{pk}} = t_{\text{pulse}}^2/(2t_{\text{pulse}} - t_{\text{max}})$ and is:

$$T_{\text{su,pk}} = \frac{8F(1-R)}{3K}\sqrt{\frac{\alpha/\pi}{2t_{\text{pulse}} - t_{\text{max}}}} + T_o \quad (4.22)$$

Considering the laser beam attenuation in the medium, and for a rectangular pulse of uniform laser intensity, the surface temperature is:

$$T_{\text{su}}(t) = \frac{I(1-R)}{K}\left\{\sqrt{\frac{\alpha t}{\pi}} - \frac{1}{\gamma}[1 - e^{\gamma^2 \alpha t}\operatorname{erfc}(\gamma\sqrt{\alpha t})]\right\} + T_o, \quad t < t_{\text{pulse}} \quad (4.23)$$

In the case of a large absorption depth compared to the thermal penetration depth, $1/\gamma \gg \sqrt{\alpha t_{\text{pulse}}}$, the above equation yields peak surface temperature at the end of the laser pulse:

$$T_{\text{su,peak}} = \frac{\gamma F(1-R)}{\rho C_p} + T_o \quad (4.24)$$

It is noted that the peak surface temperature in this case is in direct proportionality with the laser pulse energy and does not depend on the pulse duration.

4.2.4 Gaussian Laser Beams

A variety of analytical expressions for the temperature rise induced by laser beams of Gaussian intensity cross sectional profiles incident on semi-infinite substrates, finite slabs, and thin films are given in [5, 6]. For negligible heat transfer losses, an approximate expression the surface temperature rise at the origin ($r = 0, z = 0$), induced by a rectangular laser pulse is:

$$T(0, 0, t) = \frac{I(1-R)\gamma w^2}{\rho C_p} \int_0^t \frac{1}{w^2 + 4\alpha t'} \operatorname{erfc}(\gamma\sqrt{\alpha t'}) \exp(\gamma^2 \alpha t') \, dt' + T_o,$$

$$t < t_{\text{pulse}} \quad (4.25)$$

If a laser beam of Gaussian cross section is considered, the heat flow is essentially in the perpendicular to the target surface direction if the beam radius is much more extensive than the thermal penetration depth, that is, $w \gg \sqrt{\alpha t_{\text{pulse}}}$. In the other extreme, if $w \ll \sqrt{\alpha t_{\text{pulse}}}$, the laser heat source may be considered as a point source and the isotherms concentric hemispherical surfaces:

$$T_s(r, 0, t) = \frac{(1-R)w^2}{\rho C_p (\pi\alpha)^{1/2}} \int_0^t \frac{I(t')}{(t-t')^2[4\alpha(t-t') + w^2]}$$

$$\times \exp\left(-\frac{r^2}{4\alpha(t-t') + w^2}\right) dt' + T_o, \quad t < t_{\text{pulse}} \quad (4.26)$$

More complicated expressions for the linear heat conduction, accounting for the laser energy absorption in the material, can be derived using Green's functions methods [7]. For the purpose of a quick estimate, it is useful to recall simple relations for the target temperature rise, such as:

$$T \approx \frac{(1-R)E}{\rho C_p V_{\text{HAZ}}} + T_o \quad (4.27)$$

where V_{HAZ} is an estimate of the heat-affected material volume:

$$V_{\text{HAZ}} = \begin{cases} \pi w^2 \sqrt{\alpha t_{\text{pulse}}}, & w \gg \sqrt{\alpha t_{\text{pulse}}} \\ \frac{4}{3}\pi (\alpha t_{\text{pulse}})^{3/2}, & w \ll \sqrt{\alpha t_{\text{pulse}}} \end{cases} \quad (4.28)$$

4.2.5 Melting Modeling

In pure element materials, the transition to the melting phase normally occurs at a specified temperature. The propagation of the solid/liquid interface is prescribed by the energy balance, which may be thought as a kinematic boundary condition. The moving interface is assumed to be isothermal at the equilibrium melting temperature, T_m, if no overheating or undercooling is assumed:

$$T_s(\mathbf{r} \in S_{\text{int}}) = T_l(\mathbf{r} \in S_{\text{int}}) = T_m \quad (4.29)$$

$$K_s \frac{\partial T_s}{\partial n}\bigg|_{\mathbf{r} \in S_{\text{int}}} - K_l \frac{\partial T_l}{\partial n}\bigg|_{\mathbf{r} \in S_{\text{int}}} = L_m V_{\text{int};n} \quad (4.30)$$

where $\partial/\partial n$ indicates the derivative of the interface along the normal direction vector $\mathbf{n}$ at any location on the interface, $\mathbf{r} \in S_{\text{int}}$, and pointing into the liquid region, while $V_{\text{int};n}$ is the velocity of the interface along $\mathbf{n}$. The above interfacial boundary conditions, together with the heat conduction equations in the solid and liquid regions, specify the "Stefan Problem." Analytical solutions are scarce and limited to the one-dimensional phase change in materials with constant properties that is driven by a surface temperature differential, the so-called "Neumann Solution" [8]. Numerical solutions implementing the exact boundary conditions are nontrivial in multiple dimensions, requiring front-fixing or front-tracking techniques.

An alternative way of modeling phase change is the enthalpy method, which circumvents the need for exact tracking of the transient interface motion. The enthalpy function is used to account for phase change:

$$H(T) = \int_{T_0}^{T} \rho_s(T) C_{p;s}(T)\, dT \qquad T < T_m \quad (4.31)$$

$$H(T) = \int_{T_0}^{T_m} \rho_s(T) C_{p;s}(T)\, dT + \int_{T_m}^{T} \rho_l(T) C_{p;l}(T)\, dT + L_m \qquad T > T_m \quad (4.32)$$

where the density ρ and specific heat C_p vary differently with temperature in the solid and liquid phases, T_m is the melting temperature and L_m the latent heat for melting. For $T = T_m$ the enthalpy function assumes values between H_s and H_l

$$H_s = \int_{T_0}^{T_m} \rho_s(T) C_{p;s}(T)\, dT \tag{4.33}$$

$$H_l = \int_{T_0}^{T_m} \rho_s(T) C_{p;s}(T)\, dT + L_m \tag{4.34}$$

The enthalpy value $H = H_s$ is assigned to solid material at the melting temperature, while $H = H_l$ corresponds to pure liquid at the same temperature. Thus there exists a region in which the melting is partial and is defined by

$$H_s < H < H_l; \quad T = T_m \tag{4.35}$$

Each point within this region can be assigned a solid fraction f_s and a liquid fraction f_l, for which

$$f_s + f_l = 1 \tag{4.36}$$

The enthalpy function during melting at $T = T_m$ is given by

$$H = H_s + f_l L_m \tag{4.37}$$

Using the enthalpy as dependent variable, along with the temperature, Eq. (4.16) is written:

$$\frac{\partial H(T)}{\partial t} = \nabla \cdot (K(T)\nabla T) + Q_{ab}(x, y, z, t) \tag{4.38}$$

The above scheme can be readily implemented numerically using either explicit or implicit schemes in multidimensional domains (e.g., [9]). Because of the relative simplicity of the numerical approach, enthalpy-based algorithms are usually preferred, unless a more accurate specification of the motion of the boundary and the driving temperature field gradients is required, as, for example, in crystal growth.

The time for the inception of melting, t_m, can be calculated using expressions for the temperature rise:

$$T_{su}(t_m) = T_m \tag{4.39}$$

For a rectangular-large area pulse incident on a bulk surface absorber, the threshold fluence, $F_{th;m}$, necessary for melting is given from Eq. (4.20):

$$F_{th;m} = \frac{1}{2(1-R)}\sqrt{\frac{\pi}{\alpha}} K(T_m - T_o)\sqrt{t_{pulse}} \tag{4.40}$$

Approximate expressions for the maximum melt depth, $\delta_{1,\max}$ are given in [5]:

(a) for fluences close to $F_{\text{th};m}$

$$\delta_{l,\max} \approx \sqrt{\frac{\alpha t_{\text{pulse}}}{\pi} \frac{F - F_{\text{th};m}}{F_{\text{th};m}}} \tag{4.41}$$

(b) for $F > F_{\text{th};m}$

$$\delta_{l,\max} \approx \sqrt{\alpha t_{\text{pulse}}} \left[\ln\left(\frac{F - F_{\text{th};m}}{F_{\text{th};m}}\right)\right]^{1/2} \approx \sqrt{\alpha t_{\text{pulse}}} \left(\frac{F - F_{\text{th};m}}{F_{\text{th};m}}\right)^{1/2} \tag{4.42}$$

(c) for $F \gg F_{\text{th};m}$, (but still below the vaporization threshold), energy balance gives:

$$\delta_{l,\max} \approx \frac{(1-R)F - q_{\text{loss}}}{\rho C_p(T_m - T_o) + L_m} \tag{4.43}$$

where the losses via conduction, convection, and radiation are lumped in q_{loss}. The sensible heat, $\rho C_p T_m$, or more accurately $\int_{T_o}^{T_m} \rho C_p T \, dT$ is in general of comparable order of magnitude with the latent heat. In the case of 1-D nanosecond laser melting of metals, the temperature in the molten layer becomes uniform rather rapidly, and most of the solidification process is driven by the thermal gradient across the solid/liquid interface into the solid material.

4.2.6 Departures from Equilibrium at the Melt Interface

The Stefan statement of the phase change problem assumes that the interface dynamics is governed by the heat flow rather than the phase transition kinetics. This assumption is true only for low melting speeds. According to the quasi-chemical formulation of crystal growth from the melt [10, 11], and for a flat interface the velocity of recrystallization ($V_{\text{int}}(T_{\text{int}}) > 0$), or melting ($V_{\text{int}}(T_{\text{int}}) < 0$) is:

$$V_{\text{int}}(T_{\text{int}}) = C \exp\left[-\frac{Q}{k_B T_{\text{int}}}\right]\left\{1 - \exp\left[-\frac{L_m \Delta T}{k_B T_{\text{int}} T_m}\right]\right\} \tag{4.44}$$

where

$$C = R_M^0 \exp\left[-\frac{L_m}{k_B T_m}\right]; \quad \Delta T = T_m - T_{\text{int}}$$

In the above, k_B is Boltzmann's constant, T_{int} is the interface temperature, Q is the activation energy for viscous or diffusive motion in the liquid and R_M^0 is the rate of melting at equilibrium. For moderate ΔT, Eq. (4.44) is linearized:

$$V_{\text{int}}(T_{\text{int}}) = \beta \Delta T \tag{4.45}$$

where β is the slope of the interface-velocity response function near T_m. On the basis of the preceding arguments, the melt front temperature is higher than T_m, while undercooling is observed in resolidification. To calculate the motion of the phase boundary, it is necessary to solve the heat conduction equation in the solid and liquid phases and apply Eqs. (4.30) and (4.44) as boundary conditions at the interface. The classical theory implies symmetry for β in the melting and recrystallization processes. Evaluation of the X-ray diffraction studies [12, 13] has challenged this argument [14] by showing asymmetry in the interface response function yielding significant undercooling in the recrystallization process. It is noted, however, that departures from equilibrium are important for determining the recrystallization process, but usually do not affect severely the overall energy balance and therefore are of relatively minor consequence to ablation.

4.2.7 Surface Vaporization

Experimentally, it is usually easier to achieve vaporization than to control melting without significant material loss to the vapor phase. For moderate laser intensities, the laser-induced peak target surface temperature is below the thermodynamic critical point and a sharp interface separates the vapor from the liquid phase. Both the sensible heat and the latent heat of melting are typically much smaller than the latent heat of vaporization, implying that evaporation is dominant in the energy balance. For a surface absorber, $1/\gamma \ll \sqrt{\alpha t_{\text{pulse}}}$, simple energy balance considerations give the following estimate for the material removal depth, δ_{ev}:

$$\delta_{\text{ev}} = \frac{(1-R)(F - F_{\text{sh}}) - q_{\text{loss}}}{\rho C_p (T_{\text{bp}} - T_o) + L_m + L_v} \quad (4.46)$$

where F_{sh} represents the fluence loss due to plasma shielding. This estimate is more appropriate for short pulses, since conduction losses become more significant for longer pulses. For low laser energy intensities, say $I < 10^8$ W/cm^2, laser energy absorption by the evaporated particles is insignificant, so that the vapor phase may be considered as transparent.

According to the thermal surface vaporization picture, the material removal rate is limited by the surface temperature. Neglecting recondensation of vapor onto the surface, the rate of evaporation from the liquid surface, j_{ev}, can be described using kinetic theory:

$$j_{\text{ev}}(t) = n_s \left(\frac{k_B T_{\text{su}}(t)}{2\pi m_a} \right)^{1/2} \exp\left(-\frac{L_v}{K_B T_{\text{su}}(t)} \right) \quad (4.47)$$

where n_s is the atom number density per unit volume at the surface, $T_{\text{su}}(t)$ the surface temperature, L_v the latent heat of vaporization. The total ablation depth,

δ_{ev}, due to surface evaporation is:

$$\delta_{ev} = \int_0^\infty j_{ev}(t) m_a/\rho \, dt \tag{4.48}$$

If it is assumed that the process thermodynamic path rides the saturation curve, calculated surface temperatures [15] may exceed the nominal atmospheric boiling temperature, as shown in Fig. 1. It is mentioned that often the concept of a surface temperature fixed at the nominal atmospheric boiling temperature is adopted, leading to the nonfounded prediction of subsurface heating and explosive material removal. The previously outlined thermal model implies that the material removal from the melt is a continuous process. However, as the liquid surface temperature increases, the ablation rate also increases dramatically. It is reasonable to assume that most of the material is removed near the peak surface temperature and that an ablation threshold, ascribed to substantial removal rates (e.g., >1 Å/pulse) can be defined.

For nanosecond laser pulses, the duration of melting is of the order of a few tens of nanoseconds. Hydrodynamic motion due to the melt instability caused by the acceleration of the molten phase following the volumetric expansion upon

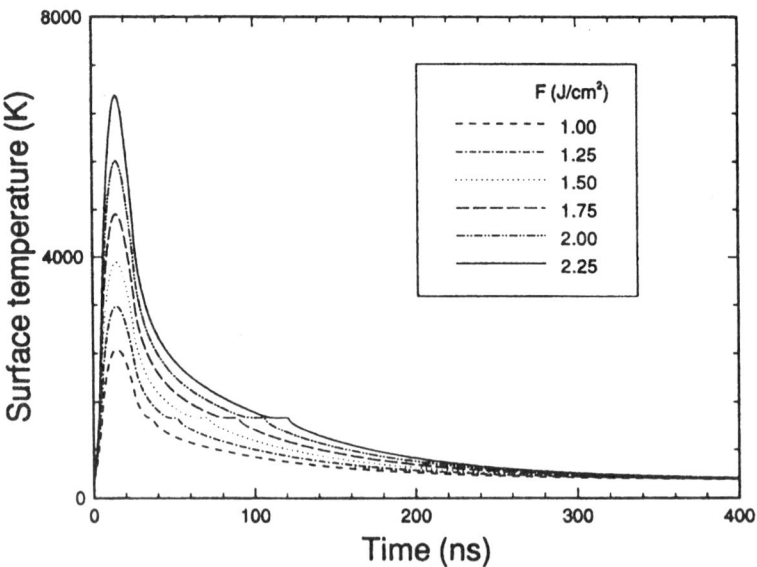

FIG. 1. Time histories of the surface temperature of a gold substrate subjected to excimer laser pulses of 26 ns duration and of different fluences. The ambient pressure is $P_\infty = 1$ atm and the laser spot radius is 1 mm [15]. Reprinted with permission from AIP.

melting may thus develop over hundreds of pulses [16, 17]. In the case of thin metal films on poorly conducting substrates, the melt duration is substantially longer, in the microsecond range, thus allowing sufficient time for the development of fluid flow and the removal of hundreds of nanometers-thick metal film material, in sharp contrast with the thermal vaporization model expectations. Figure 2 shows the average ablation depth dependence on laser fluence and background pressure for a 0.5 μm-thick gold film deposited on a quartz crystal microbalalance (QCM; [18]) and ablated by low energy excimer laser pulses at the $\lambda = 248$ nm wavelength [19].

Besides surface vaporization and material removal due to melt instabilities, boiling could be thought as providing another material removal mechanism. For ideally absorbing media that are free from impurities, voids, and structural microdefects, and for laser intensities $I < 1$ GW/cm^2, with corresponding submicron laser radiation penetration depths, it has been argued that volumetric vaporization could be significant only at temperatures exceeding tens of thousands of degrees K [20]. Such high temperatures exceed the critical point. Further thermodynamic considerations showed that volumetric vaporization is important for large radiation penetration depths.

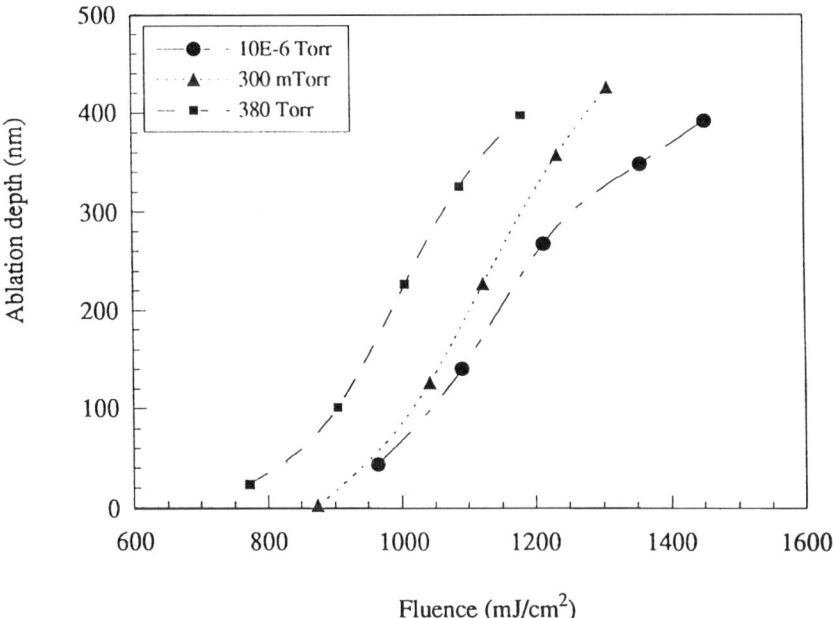

FIG. 2. Ablation rate dependence on laser fluence and argon background gas pressure for excimer laser ablation of a 0.5 μm-thick gold film on a quartz microbalance [19]. Reprinted with permission from Springer-Verlag.

4.2.8 Plasma Absorption

At high irradiances, the laser-material interaction is dominated by the formation of a plasma above the surface. The high irradiance regime is assumed to occur for laser intensities over 1 GW/cm^2. As discussed in [21], this simplification is valid for ablation of surface absorbers in a vacuum environment, by pulses that are sufficiently long, in the nanosecond range and over. Within a factor of two for metals, the intensity threshold for the initiation of plasma ignition, I_{thr}, was shown to fit the experimental results by the relation: $I_{thr}\sqrt{t_{pulse}} \geq B$, where $B \approx 4 \times 10^4$ $Ws^{1/2}$ cm^{-2}. The equilibrium electron and ion densities, n_e and n_i can be calculated using the Saha-Eggert equation [22]:

$$\frac{n_e n_i}{n_n} = \frac{Q_e Q_i}{Q_n} \frac{m_e m_i}{m_e + m_i} \exp\left(-\frac{\Psi_i}{k_B T}\right) \quad (4.49)$$

In the above, Q_e, Q_i, and Q_n represent the internal partition functions of electrons, ions and neutrals, respectively. Ψ_i is the ionization potential of the neutral atom; m_e and m_i are the electron and ion masses. Use of this equation allows calculation of possible laser-induced ionization processes. In [21], it is noted, however, that under most experimental conditions, electrons are probably present because of multiphoton ionization of atomic or molecular species and ionization from excited states in the plume, ejection of ions and electrons from the surface, and thermionic emission from the surface. For ultraviolet (UV) wavelengths, direct photoionization of the vapor atoms is expected to be a dominant ionization mechanism. Whatever the mechanism can be, the threshold for plasma initiation is significantly lower than what would be expected from equilibrium considerations. The plasma absorption coefficient, γ_{pl}, is:

$$\gamma_{pl} = \frac{2\omega}{c} Im\left[\left(1 - \frac{\omega_{pl}^2}{\omega^2[1 + (i/\tau_{ei})]}\right)^{1/2}\right] \quad (4.50)$$

where ω and τ_{ei} are the angular frequency of the laser light and the electron-ion collision time, respectively, while $\omega_{pl} = 4\pi(n_e e^2/m_e)^{1/2}$ is the electron plasma frequency. The most frequently used expression for the electron-ion collision time is taken from [22]:

$$\tau_{ei} = \frac{3}{4}\left(\frac{m_e}{2\pi}\right)^{1/2} \frac{k_B^{3/2} T^{3/2}}{Z^4 e^4 n_i \ln \Lambda} \quad (4.51)$$

In the above, Z is the average ion charge state; e, the proton charge; and $\ln \Lambda$, the Coulomb logarithm:

$$\Lambda = \frac{3}{2e^3}\left(\frac{k_B^3 T^3}{\pi n_e}\right)^{1/2} \quad (4.52)$$

The radiation absorption due to inverse bremsstrahlung is quantified by the spectral absorption coefficient, γ_{av}:

$$\gamma_{av} = \frac{4}{3}\left(\frac{2\pi}{3k_B T}\right)^{1/2}(2\pi)^3 \frac{Z^2 e^6}{hcm_e^{3/2}\omega^3} n_i n_e g_{ff}\left[1 - \exp\left(-\frac{h\omega}{2\pi k_B T}\right)\right] \quad (4.53)$$

where g_{ff} is the Gaunt factor that is introduced in a more exact quantum mechanical theory; c, the speed of light; and h, Planck's constant. As previously noted, it is important to keep in mind that close to the plasma ignition threshold, the coupling of the laser energy with the plasma is initiated via mechanisms other than inverse bremsstrahlung. However, the latter mechanism regulates the energy transfer as the laser pulse progresses. Most of the laser absorption occurs in a vapor layer confined close to the target surface. In that layer, the electron-ion density is very high, ($10^{20} - 10^{23}$ cm^{-3}) and the plasma pressure may exceed hundreds of kbar. This higher pressure region may launch a compressional wave towards the target surface [23], leading to transverse flow of the trailing ejected particles or removal of material in the form of molten droplets. The light absorption penetration depth in the hot, dense plasma region situated over the target surface may be in the micron range, producing optically thick conditions. Thus, considering one-dimensional expansion of an opaque, ideal gas plasma heated by inverse bremsstrahlung, simple, material-independent approximate relations for the ablation parameters were derived in [21]. They showed that the ablation depth is approximately equal to 40 $[I(t_{pulse})^{3/2}/\lambda]^{1/2}$ nm, the plasma temperature $3[I\lambda \sqrt{t_{pulse}}]^{1/2}$ eV, where the intensity is in W/cm^2, and the laser light wavelength, λ, in cm. The comparison of the prediction of the ablation depth with available experimental data indicated reasonable agreement over a wide range of laser wavelengths and for intensities and pulse durations spanning several orders of magnitude.

It is expected that computational fluid dynamics modeling of the coupled radiation and gas dynamics in laser ablation processes will attract significant research interest as applications become more sophisticated. The prediction of the crater formation, plasma ignition and expansion in vacuum or against a background gas is a complex and challenging task that needs the input and validation from detailed experiments.

4.3 Lasers Used in Ablation

4.3.1 Continuous Wave, Millisecond, and Microsecond Lasers

Such laser pulses are used to ablate materials either at a fixed spot (penetration material removal) or in a scanning mode where either the beam or the target are translated. Millisecond and microsecond duration pulses are produced by chopping the CW laser beam or by applying an external modulated control voltage. Fixed Q-switch solid state lasers with pulse durations from tens of microseconds to several milliseconds are often used in industrial welding and drilling applications. CW carbon dioxide lasers (wavelength $\lambda = 10.6$ μm and power in the kW range) are widely used in many industrial applications, including the cutting of bulk and thick samples of ceramics such as SiN, SiC metal matrix ceramics (e.g., [24]). CW laser radiation allows definition of continuous grooves and cuts. On the other hand, low-power CO_2 lasers in the 10 W to 150 W range are used for marking of wood, plastics, and glasses. Argon ion lasers operating in the visible range ($\lambda = 419$–514 nm) are used for thick and thin resistor trimming. In the biomedical field, various continuous-wave lasers have been used. For example, the CO_2 laser radiation is absorbed in the tissue within a layer of about 20 μm, thus achieving continuous ablation front. On the other hand, the Nd:YAG ($\lambda = 1064$ nm) and argon ion radiation penetration is of the order of millimiters, thus giving rise to explosive ablation events.

Argon laser operating in the UV ranges of 275–305 nm and 350–380 nm was used to ablate polyimide Kapton films [25]. By chopping the laser beam to produce ms and μs pulses, they showed that the ablation process scales with an intensity threshold rather than the commonly used fluence threshold. This is certainly not surprising because if the laser energy is spread over a long pulse, the beam intensity weakens and the induced temperature and structural and chemical response of the target may be of different nature. In fact, it was shown [26] that the etching of polymer films with long, ms-μs pulses leaves evidence of molten material and carbonization of the walls, but is not indicative of the ablation process that characterizes ns UV laser ablation. Microfabrication applications by direct writing can be effected by CW Ar^+ and Kr^+ lasers, using frequency doubled lines. High-power CW Nd:YAG operating on the fundamental wavelength and on frequency doubled and tripled harmonics are often used for various cutting and microprocessing applications.

4.3.2 Nanosecond Lasers

Nanosecond lasers are often used for ablation experiments. The technological development in the manufacturing of gas and solid state lasers has greatly advanced in terms of reliability and in many cases has enabled the transition from the laboratory environment to industrial applications. The workhorse of pulsed laser ablation is the excimer laser (usually KrF at $\lambda = 248$ nm, but also XeCl at $\lambda = 308$ nm and ArF at $\lambda = 193$ nm) with pulse duration in the range of 20–30 ns, maximum pulse energy in the range of 0.25 to 1 J, and pulse repetition rate typically from 5–300 Hz. Since most materials are strong absorbers to UV wavelength radiation, the excimer laser light is absorbed in a very shallow region near the irradiated material surface. On the other hand, the short duration of the laser pulse brings the peak power up to 10^{10} W/cm^2. These two features make the excimer laser a successful tool for initiating photochemical and/or photothermal ablation. Thus the excimer laser is the most efficient ablation tool operating in the UV range for precision micromachining and surface patterning [27, 28, 29], chemical or physical modification [30, 31, 32], and via hole formation in electronic circuit packaging [33]. On the other hand, the Pulsed Laser Deposition (PLD) using excimer lasers has enabled fabrication of novel thin film materials of high quality and superior properties as compared to conventional manufacturing techniques. The rapid advancement and significance of this method is comprehensively reviewed in [34].

Q-switched Nd:YAG lasers with pulse duration of about 7–10 ns, pulse energies in the near-IR wavelength of $\lambda = 1,064$ nm, typically from 10 mJ to about 1 J, and repetition rates of 10 HZ are versatile ablation tools because they can provide operation at different wavelengths. Frequency doubled pulses at $\lambda = 532$ nm, tripled at $\lambda = 355$ nm and quadrupled at $\lambda = 266$ nm, carry respectively lower energies. Pulsed laser deposition can be effected with solid state lasers; for example, production of amorphous diamond (diamondlike) films using Q-switched Nd:YAG laser radiation at the $\lambda = 1,064$ nm wavelength was reported in [35]. A relatively inexpensive ablation tool is the transversely excited atmospheric pressure (TEA) CO_2 laser, that generates low repetition rate, high energy pulses in the kJ range, while it provides low energy (~ 1 J) pulses at the 1 kHz repetition rate. The pulse has a short 100–200 ns wide high energy spike and a longer trailing component of 1–10 μs long; both parts may carry comparable energies. Another cost-effective ablation laser for applications requiring relatively low energies is the N_2 laser that operates at the $\lambda = 337$ nm wavelength, has pulse duration of 7–10 ns, pulse energies in the hundred μJ range to <10 mJ and repetition rates of about 10 Hz. Diode pumped solid state lasers such as Nd:YLF, Nd:YAG having hundreds of μJ pulse energies and operating on the fundamental or frequency-doubled wavelengths are attractive for micromachining applications because of their small size, pulse energy stability, flexibility, and high repetition rates (tens of kHz).

For ablation of biological materials, free-running solid state lasers having long pulse duration of hundreds μs and shortened by Q-switch to tens of ns, with corresponding energies in the 1 J range and tens of mJ, are often used. Typical crystals are Ho:YAG ($\lambda = 2.1$ μm) and Er:YAG ($\lambda = 2.94$ μm) modules, with respective radiation penetration depths in water of about 40 μm and 1 μm, thus achieving different absorption characteristics. For reference, it is noted that the 2.94 μm wavelength is located right at the peak absorption in water. In general, the nature of the ablation process, in terms of angular distributions and energies of the ejected particles, depends on the laser wavelength. Even though the target material structure and properties obviously affect the outcome, for comparable energies, long wavelengths usually produce thermal behavior, UV ablation exhibits nonthermal characteristics and intermediate VIS wavelengths yield results whose interpretation may be ambiguous [36].

4.3.3 Picosecond Lasers

Whereas the nanosecond time scale is much longer than the characteristic relaxation times in metallic systems, invoking the thermal picture, the picosecond regime is still longer, but comparable. It has been claimed that picosecond laser ablation of multicomponent targets offers the distinct advantage of preserving the target stoichiometry in the chemistry of the ejected plume. It is noted that collisional and chemical reaction effects in the target phase may introduce departures because conflicting evidence has been presented. Most of the ablation work with ps lasers is done with pulsed solid state lasers. For example, a 35 ps Nd:YAG laser producing 15 mJ at the $\lambda = 1,064$ nm wavelength and 10 mJ at $\lambda = 266$ nm was used to ablate Cu [37]. A mode-locked, 50 ps pulse duration Nd:YAG laser at $\lambda = 532$ nm was used to ablate freestanding metal films of 50–90 nm thickness under fluences ranging 0.6–8 J/cm^2 [38]. Fundamental studies on the picosecond laser plasma interactions were, for example, conducted by a Nd:glass laser system based on the chirped pulse amplification and compression (CPAC) technique that yielded 1.3 ps, 1.05 μm pulses with an average energy of 10 mJ [39]. For pulsed laser processing in the IR range, the free-electron laser (FEL) offers tunability and high power as demonstrated for advanced materials science [40] and medical [41] applications. For example, the Vanderbilt FEL operates in the fundamental wavelength range of 2.0–9.5 μm, yielding a long micropulse of 6 μs and 360 mJ energy, which comprises about 10^4 short pulses of 0.5 ps duration.

4.3.4 Femtosecond Lasers

In the subpicosecond or femtosecond regime, the laser pulse is shorter than the relaxation times, and the equilibrium assumption is no longer valid, necessitating treatment of the microscopic mechanisms of energy transfer via quantum mechanics. One notable characteristic of femtosecond lasers is the high radiation intensity that has the ability to create high-density plasmas. On the other hand, by beating the thermal diffusion time scale, femtosecond laser radiation can, in principle, be used for micromachining with minimal thermal damage to the surrounding area. In the UV range, KrF ($\lambda = 248$ nm) excimer lasers with typical pulse duration of 500 fs and pulse energies in the several to tens of mJ range have been demonstrated in the processing of Al and glassy C [42]; Ni, Cu, Mo, In, Au, W [43], and Ni films [44]; fused silica [45]; ceramics such as Al$_2$O$_3$, MgO, and ZrO$_2$ [46]; and for polymer ablation [47, 48]. KrF excimer lasers have also been used in studies of high-density gradient Al and Au plasmas [49] and for the production of soft X-rays from Al [50]. The latter was also accomplished from Cu and Ta targets by near-IR Ti:sapphire laser irradiation at $\lambda = 807$ nm, 120 fs pulse duration, and 60 mJ pulse energy [51]. A Ti:sapphire system with pulse duration of 150 fs and $\lambda = 770$ nm was used in studies of gold ablation [52], while pulse durations of 170–200 fs, wavelength $\lambda = 798$ nm, and energy 4 mJ ablated polymers through a multiphoton ablation mechanism. Intense, visible dye laser radiation (pulse duration 160 fs, $\lambda = 616$ nm, and energy 5 mJ) generated high-energy, X-ray emitting density Si plasmas [53]. In a biomedical application [54], a dye laser (pulse duration 300 fs, $\lambda = 615$ nm, and pulse energy >0.18 mJ) produced high-quality ablation in human corneas, characterized by less than 0.5 μm wide damage zones.

4.4 Beam Delivery

4.4.1 Focusing of Gaussian Beams

For Gaussian beams of circular cross section, the laser beam radius at a distance, z, from the focal waist is given by:

$$w(z) = w_o \left[1 + \left(\frac{\lambda z}{2\pi w_o^2} \right)^2 \right]^{1/2} \qquad (4.54)$$

or

$$w(z) = w_o \left[1 + \left(\frac{z}{z_R} \right)^2 \right]^{1/2}$$

where w_o is the laser beam radius at the focal waist given by $w_o = (2\lambda/\pi)f^{\#}$, where the f-number, $f^{\#}$, is given by $f^{\#} = f/2w_l$, where f is the focal length of the lens and w_l is the beam radius at the lens. The quantity $z_R = 2\pi w_o^2/\lambda$ indicates the distance where the diameter of the laser beam changes by a factor of $\sqrt{2}$ and is called the Rayleigh length. The depth of focus, is given by the following relation:

$$d_{\text{dof}} = \frac{\pi w_o^2}{\lambda} \sqrt{\zeta^2 - 1} \qquad (4.55)$$

By setting the acceptable focus to be within 2%, that is, $\zeta = w(z = d_{\text{dof}})/w_o = 1.02$, the depth of focus is estimated to be about 60 μm for $w_o = 10$ μm and $\lambda = 1.064$ μm. The asymptotes of the beam profile are described by $w(z) = z\theta$, where $\theta = \lambda/\pi w_o$ indicates the beam half-divergence in rads. In laser microablation systems where tight focusing is required, the beam divergence is first reduced via collimating and expanding systems that may be either converging or diverging. Focusing systems with f-numbers, $f^{\#} < 1$ require special design and high-quality optical materials. Thus, in practice, the achievable focal radius, $w_o \sim \lambda$. When the beam divergence is decreased and the spot size is decreased, the depth of focus is also decreased. This may not be advantageous in ablation applications where a relatively large focal depth is required, as, for example, in the processing of non-planar specimens. Equation (4.55) shows that when $w_o \sim \lambda$, the depth of focus, d_{dof}, becomes a fraction of λ, requiring precision positioning.

If ablation is performed in an ambient gas environment, consideration must be given to avoiding gas breakdown, the probability for which is increased by the presence of dust particles and impurities that are first removed from the target. As discussed in [2], air breakdown thresholds in the vicinity of absorbing targets are of the order of 10^7 W/cm^2 for CO_2 lasers and 10^9 W/cm^2 for Nd lasers. Direct writing is achieved by focusing the Gaussian laser beam at normal incidence on the specimen. For preserving the optical alignment in patterning operations, it is customary to translate the substrate using precision micro-positioning stages. Another practical limiting factor in using short focal length objectives is that ablation products may be deposited at the lens surfaces. Another major problem in the processing of electronic components is the redeposition of debris on the target surface. To avoid debris accumulation, nozzles are sometimes used to blow an inert gas like nitrogen on the surface.

While direct ablation of the target by irradiation from the top is the usual ablation mode, in the laser-induced forward transfer (LIFT) technique, the "source" film is transferred to the receiving wafer [55]. The irradiated film is deposited on a transparent substrate, through which the laser beam is focused. The thickness of the ablated film is on the order of 100 nm, and the gap between the receiving wafer and the film is in the tens of microns range.

4.4.2 Projection Machining

Projection machining is suitable for processing of large specimen areas, which makes it suitable for industrial-scale applications. Figure 3(a) shows the schematic diagram of an industrial-grade projection micromachining tool, a beam delivery/homogenization system, a dielectric mask with micron features, a 9-element CaF_2 transfer lens for a 2:1 image reduction (Fig. 3(b)), and a computer-controlled five-degree-freedom sample stage [56]. The overall design strategy for the excimer micromachining tool is to use the excimer laser beam to project the mask image on the wafer surface. There are two ways commonly used for masking the laser beam: absorption and reflection. Because of the strong absorption of the UV irradiation, metal masks tend to be thermally damaged by mechanisms such as sputtering, melting, and/or thermal-mechanical deformation. Conventional chromium masks cannot sustain higher excimer laser fluences in micromachining processes. Dielectric masks made from thin films on transparent substrates are far more resistant than chromium masks and usually can sustain fluences up to 1 J/cm^2. The dielectric thin film is designed in such a

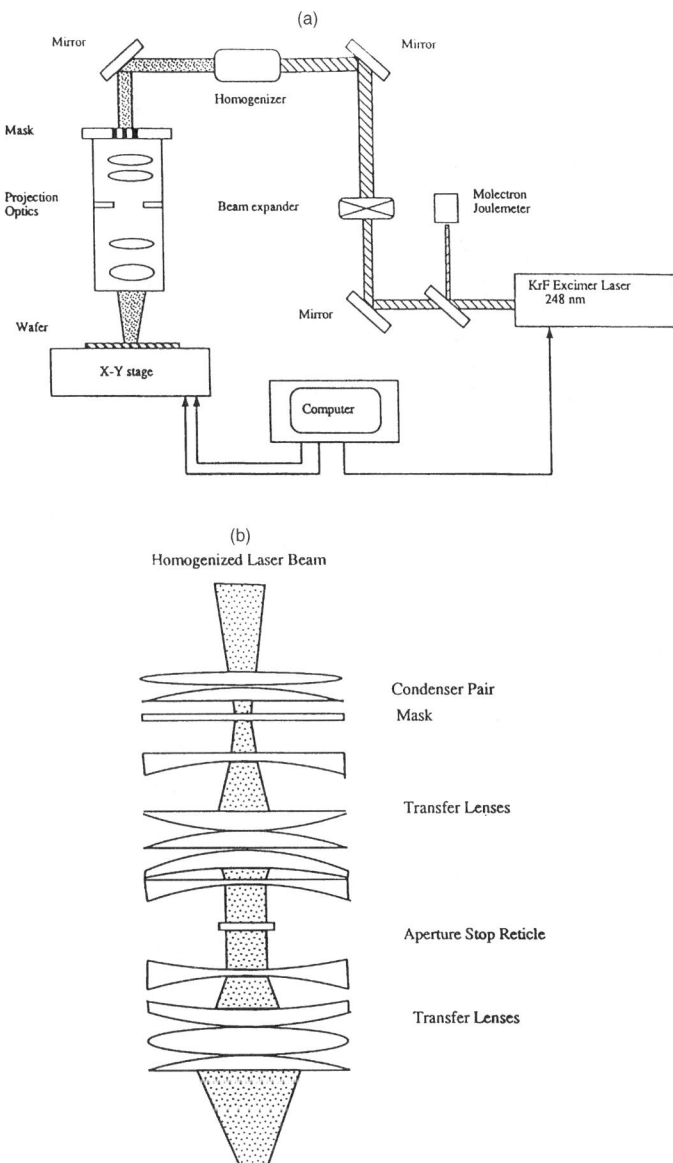

FIG. 3. (a) Schematic diagram of the excimer laser micromachining tool; (b) the 9-element CaF_2 transfer lens; (c) atomic force micrograph of a 2 μm line in width and 0.4 μm in depth micromachined by 15 pulses at laser fluence of 200 mJ/cm^2; (d) parallel lines of 5 μm in width and 1 μm in depth micromachined by 40 pulses at laser fluence of 200 mJ/cm^2. Micromachined polyimide (Probimide®) films were annealed at 400°C [56]. Reprinted with permission from IEEE.

(c)

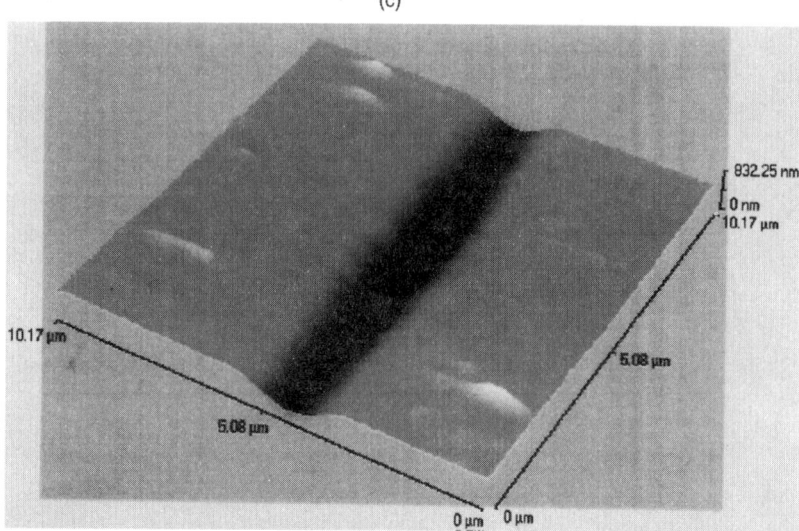

(d)

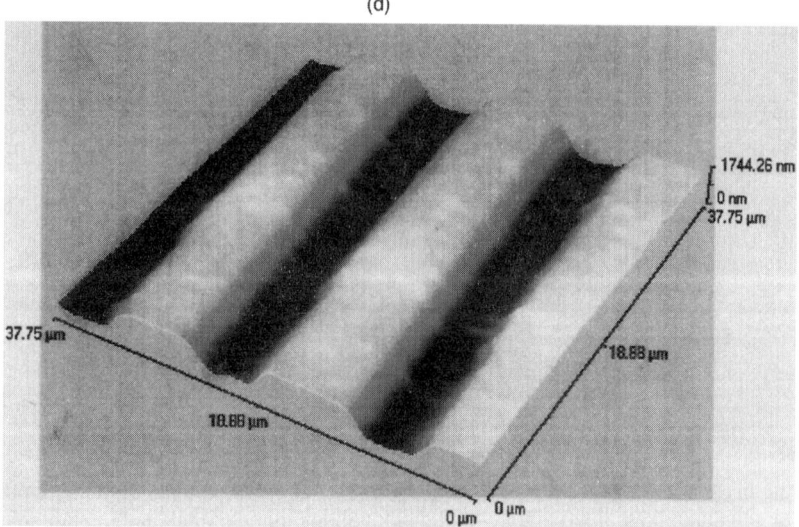

FIG. 3. *Continued*

way that the masked part reflects most of the laser energy at the specific excimer laser wavelength, so that the thermal damage induced on the mask is greatly suppressed. A three-dimensional AFM image of an enlarged 2 μm line with a 0.4 μm depth written by this excimer laser micromachining tool is shown in

Fig. 3(c). Parallel lines that are 5 μm in width and 1 μm in depth with sharp definition are also shown in Fig. 3(d).

Optical design considerations for excimer laser projection lithography systems were reviewed in [30]. In the case of imaging of equal lines and spacings with incoherent light, the absolute cutoff linewidth is $d_{co} = \lambda/4\,NA$, where λ is the wavelength and NA the numerical aperture in the image space. Typically the achievable linewidth is $d_{lw} \sim 3d_{co}$. By increasing NA, the linewidth decreases, but this is compromised by the field of depth decrease. A measure of the tolerance for the acceptable defocusing is the Rayleigh range, $z_R = \lambda/2(NA)^2$. The resolution of excimer laser projection micromachining is mainly limited by the chromatic aberration of the projection optical system. To overcome chromatic aberration, reflective optics such as Schwarzschild microscope objectives have been employed in scanning projection systems for finer patterning. Besides optical considerations, care must be exercised for the mechanical and thermal stability of the projection micromachining system.

4.4.3 Angle of Incidence and Polarization Dependence

In applications such as pulsed laser deposition, the laser beam is directed onto the target from vacuum at an angle θ_o with respect to the normal to the surface. In this case, the irradiated spot increases by a factor of $1/\cos\theta_o$ and the radiation intensity incident on the target surface is correspondingly reduced. On the other hand, the surface reflectivity and absorptivity also change, depending on the laser light polarization. For a laser beam coming from an ambient gas with unity refractive index and incident on a bulk substrate of complex refractive index, $n^c = n + ik$, the surface reflectivities for parallel (transverse electric, TE) and normal (transverse magnetic, TM) polarizations are:

$$R_{\parallel}(\theta_o) = \frac{[(n^2 - k^2)\cos\theta_o - u]^2 + [2nk\cos\theta_o - v]^2}{[(n^2 - k^2)\cos\theta_o + u]^2 + [2nk\cos\theta_o + v]^2} \tag{4.56}$$

$$R_{\perp}(\theta_o) = \frac{(n\cos\theta_o - u)^2 + v^2}{(n\cos\theta_o + u)^2 + v^2} \tag{4.57}$$

where

$$2u^2 = (n^2 - k^2 - \sin^2\theta_o) + \sqrt{(n^2 - k^2 - \sin^2\theta_o)^2 + 4n^2k^2} \tag{4.58}$$

$$2v^2 = -(n^2 - k^2 - \sin^2\theta_o) + \sqrt{(n^2 - k^2 - \sin^2\theta_o)^2 + 4n^2k^2} \tag{4.59}$$

For a bulk metal having components of the complex refractive index such that $n^2 + k^2 \gg 1$, the surface reflectivities for parallel and normal polarization are:

$$R_{\parallel}(\theta_o) = \frac{(n^2 + k^2)\cos^2\theta_o - 2n\cos\theta_o + 1}{(n^2 + k^2)\cos^2\theta_o + 2n\cos\theta_o + 1} \tag{4.60}$$

$$R_\perp(\theta_o) = \frac{(n^2 + k^2) - 2n\cos\theta_o + \cos^2\theta_o}{(n^2 + k^2) + 2n\cos\theta_o + \cos^2\theta_o} \qquad (4.61)$$

Consider a laser beam incident on a multilayer stack of films, $z_o \leq \cdots \leq z_{j-1} \leq z \leq z_j \leq \cdots \leq z_N$. The plane of incidence is (y, z), with z being the direction of stratification. In the classical case of exponential decay of the electromagnetic field with depth, the radiation properties of thin films can be calculated [57] if the material complex refractive index and the film thickness are known. The case of arbitrary polarization can be treated as a superposition of TE (transverse electric) and TM (transverse magnetic) polarized waves. For TE-polarized light, the electrical and magnetic field components are:

$$E_y = U(z) \exp\left[i\left(\frac{2\pi}{\lambda} y \sin\theta - \omega t\right)\right] \qquad (4.62)$$

$$H_y = V(z) \exp\left[i\left(\frac{2\pi}{\lambda} y \sin\theta - \omega t\right)\right] \qquad (4.63)$$

$$H_z = W(z) \exp\left[i\left(\frac{2\pi}{\lambda} y \sin\theta - \omega t\right)\right] \qquad (4.64)$$

Using Maxwell's equations, it is found that W is linearly dependent on U, and that the solution can be expressed in the form of a characteristic transmission matrix, defined by:

$$\begin{Bmatrix} U_o \\ V_o \end{Bmatrix} = M^c(z) \begin{Bmatrix} U(z) \\ V(z) \end{Bmatrix} \qquad (4.65)$$

where

$$M^c(z) = \begin{pmatrix} \cos\left(\frac{2\pi}{\lambda} n^c z \cos\theta^c\right) & -\frac{i}{p^c}\sin\left(\frac{2\pi}{\lambda} n^c z \cos\theta^c\right) \\ -ip^c \sin\left(\frac{2\pi}{\lambda} n^c z \cos\theta^c\right) & \cos\left(\frac{2\pi}{\lambda} n^c z \cos\theta^c\right) \end{pmatrix} \qquad (4.66)$$

In the above expression, $p^c = n^c \cos\theta^c$ for TE wave, and by $p^c = \cos\theta^c/n^c$ for a TM wave. The angle θ^c is complex for absorbing films and is defined by the generalized Snell's law of refraction:

$$n^c \sin\theta^c = \sin\theta_o \qquad (4.67)$$

The multilayer system transmission matrix is:

$$M^c = \prod_{j=1}^{N} M^c(z_j - z_{j-1}) \qquad (4.68)$$

The Fresnel coefficients for reflection and transmission are:

$$r = \frac{(M^c(1,1) + M^c(1,2)p_b)p_o - (M^c(2,1) + M^c(2,2)p_b)}{(M^c(1,1) + M^c(1,2)p_b)p_o + (M^c(2,1) + M^c(2,2)p_b)} \quad (4.69)$$

$$t = \frac{2p_o}{(M^c(1,1) + M^c(1,2)p_b)p_o + (M^c(2,1) + M^c(2,2)p_b)} \quad (4.70)$$

where for the backing medium $z > z_N$, $p_b = n_b^c \cos\theta_b^c$ or $p_b = \cos\theta_b^c/n_b^c$ for TE or TM wave polarization and $p_o = \cos\theta_o$. The structure reflectivity is:

$$R = |r|^2 \quad (4.71)$$

The time-averaged power flow per unit area that crosses the plane perpendicular to the z-axis is given by the magnitude of the Poynting vector.

$$\mathbf{S}(z) = \tfrac{1}{2}Re[\vec{\mathbf{E}}(z) \times \mathbf{H}^*(z)] \quad (4.72)$$

The local laser energy absorption per unit volume is:

$$Q_{ab}(x, y, z, t, T_j) = \frac{dS(z)}{dz} \quad (4.73)$$

The local variation of the energy absorption due to interference effects in thin films was demonstrated [58].

4.4.4 Lenses, Windows, Mirrors and Beamsplitters

Spherical lenses are most commonly used in ablation systems. In case that beam expansion along one direction is needed, cylindrical lenses can be used. For example, the raw excimer laser beam usually is of rectangular-elliptical cross section, with an aspect ratio of ~3 to 5; a cylindrical lens can be used for expanding the shorter dimension to give a square cross section. The lens performance, with regard to the theoretical prediction, depends on lens aberrations: spherical aberration, coma, astigmatism, field curvature, and distortion. To reduce spherical aberration, apertures can be used to attenuate the beam rays diverging from the optical axis. Alternatively a condensing planoconvex lens can be used, with the convex surface facing the incoming laser beam. Coma that results from imaging light rays from off-axis points as ring-like structures can be eliminated by control over the lens shape, as well as by using apertures, again at the expense of some power loss. For demanding applications, custom-made lenses may be necessary, providing the necessary corrections. In any case, the choice of the materials for optical components depends on the laser wavelength, energy level, and pulse repetition rate.

Operation in the wavelength range from 350 to 1000 nm can be handled with quartz, pyrex, or other glasses. In the UV range, from 190 to 350 nm, UV-grade fused silica is adequate for relatively low repetition rates, but under prolonged operation at high repetition rates, color formation and a significant transmission loss have been observed [59]. In this case, single crystalline quartz, CaF_2 or MgF_2 lenses must be used. For the near-IR Nd:YAG, Nd:YLF, and Nd:Glass lasers and the like, anti-reflection, coated quartz and glass lenses are normally adequate. However, these materials are not transparent farther in the infrared. Crystalline alkali halides NaCl and KCl, and various semiconductor materials such as ZnSe, CdTe, GaAs, and Ge are highly transmitting in the far-IR, in the 10 μm spectral range of CO_2 laser radiation. Although transmission is a major concern, high thermal conductivity, hardness, smoothness and chemical resistance are also desirable, when coatings need to be applied. Thin film anti-reflective (AR) coatings typically reduce the surface reflectance to about 0.01.

The selection of mirrors must also be done carefully to avoid damage by the incident laser radiation. Multilayer dielectric coatings are designed to enhance the reflectivity to 0.99 for the particular wavelength and laser beam incidence angle. It should be cautioned that use of these mirrors at other wavelengths and incidence directions may have detrimental effects, as the reflectance is decreased. Beamsplitters are also used either to sample part of the beam for temporal profile measurement or to divert a portion of the beam for pump-probe schemes, or to share the laser beam in different experiments.

For controlled ambient pressure and composition, but also to provide the appropriate chemical reaction environment for example in PLD systems, experiments are being conducted in vacuum chambers. The laser windows through which the laser enters the deposition chamber have to be made from the above-mentioned high-quality optical materials. To avoid losses by scattering, the window surfaces must be optically smooth ($\lambda/10$ flatness). Deposition of ablated particles on the windows must also be avoided because it leads to gradual transmission loss and potentially to local damage.

4.4.5 Beam Shaping and Homogenization

In several applications, it is necessary to produce uniform irradiation at the target surface through homogenization. In the case of coherent laser beams that have Gaussian profiles, this is accomplished by diffractive optics using diffraction gratings [60], phase plates, or holographic techniques.

The raw beam emerging from the excimer laser is incoherent and spatially nonuniform. To avoid hot spots for quantitative experiments, it is necessary to provide some means of homogenizing. The simplest homogenizer is a tunnel type, whose internal surfaces are polished to about 0.2 μm roughness. The tunnel material of preference is aluminum because it can be polished easily, yielding a reflectance in the UV range of about 90%. The excimer laser beam is focused slightly outside the tunnel opening. Very tight focusing may lead to material damage at the tunnel entrance. The size of the opening is determined by the desired laser fluence range and shape from the cross sectional profile of the laser beam. On the other hand, the length of the homogenizer is designed to provide an adequate number of bounces and beam mixing at the tunnel exit. Use of diffuser plates upstream of the homogenizer entrance may improve beam quality. If the beam fluence needs to be characterized accurately, one may try to eliminate the falling crests of the laser beam by placing an aperture after the exit of the homogenizer. The beam homogenizer provides a simple and flexible means for improving the beam quality at the expense of a power reduction that can reach 40% in the case of tight focusing and multiple bounces on the tunnel walls. Figure 4 compares laser beam intensity profiles before and after the tunnel type homogenizer [61].

Another alternative is to use a fly's-eye type beam homogenizer, which is often used in mask projection lithography. Shown in Figure 5, is the homogenizer employed in the micromachining tool, described previously [56], which consists of two arrays of cylindrical lenses that are parallel to each other. The spatially nonuniform incoming laser beam is first divided into many bundles,

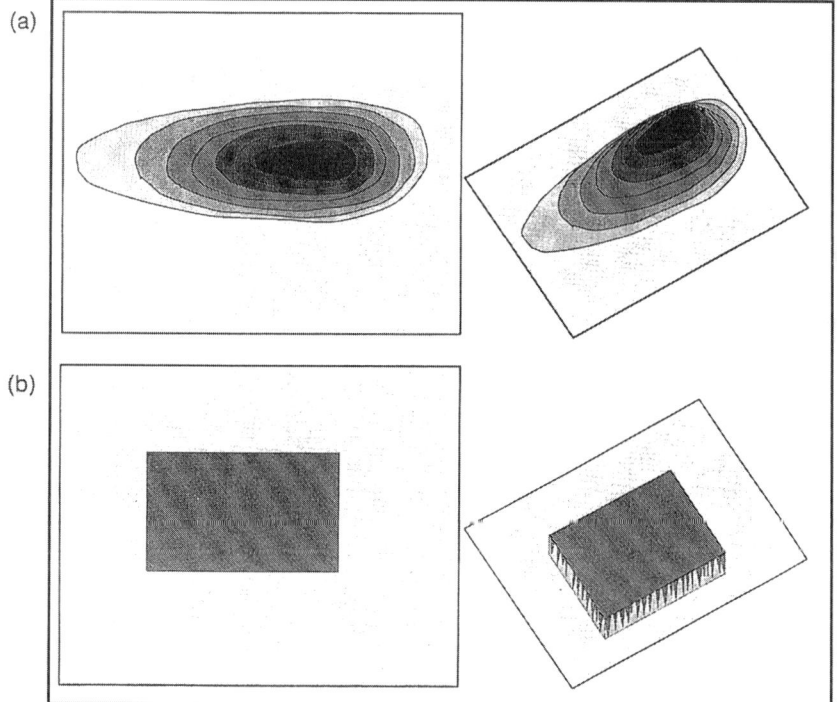

FIG. 4. Spatial laser intensity maps (a) at the laser exit aperture, and (b) at the tunnel-type beam homogenizer exit [61]. Reprinted with permission from IEEE.

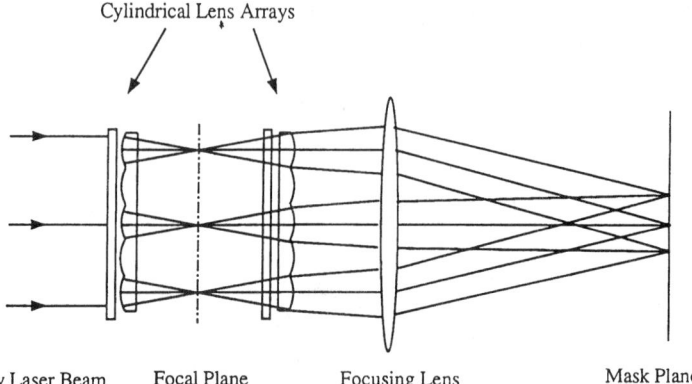

FIG. 5. Schematic diagram of a fly's-eye type homogenizer consisting of two cylindrical lens arrays and showing the optical path forming a uniform spot on the mask plane, and the system of transfer lenses to image the spot on the target [56]. Reprinted with permission from IEEE.

which form arrays of point images when focused on a plane. The laser light is further diverged by a second array of cylindrical lenses and then refocused by a spherical lens. To keep the power losses low, it is important for all optical elements to be coated by AR thin films and for the fabrication finish of the various optical components to be precise. For example, gaps between the cylindrical lens elements of the homogenizer due to poor machining and surface matching may lead to significant reduction of the available power. With careful design, the fly's-eye beam homogenizer allows large optical elements that facilitate the high-throughput processing of large specimen areas.

4.4.6 Beam Profile and Power

Commercial beam analyzers are available based on charge coupled device (CCD) cameras in the UV-visible range; pyroelectric matrix cameras are used in the infrared range. Linear or two-dimensional arrays of photodiodes can also be sensing elements. These analyzers are equipped with frame grabbers and image processing systems to provide precise measurements of the laser beam profile for CW as well as pulsed lasers for laser beam diameters exceeding tens of microns. Such information may be important for evaluating of the stability and performance of the laser system and to ensure process quality, for example, in micromachining applications. In the case of CW laser beams, the beam profile can be estimated by chopping the laser beam using a mechanical chopper, or by mechanically scanning a sharp knife edge [62], pinhole or slit through the laser beam. For focused laser beams with spot size in the micron range, the variation of the laser power, $P_o(x)$, measured by a single element detector yields the laser beam profile as a function of the distance of the knife edge from the spot center:

$$P_d(x) = \int_x^{+\infty} \int_{-\infty}^{+\infty} I_o e^{-((x^2 + y^2)/w^2)} \, dx \, dy = \frac{P}{2} \text{erfc}\left(\frac{x}{w}\right) \quad (4.74)$$

where P is the laser beam power. The beam radius, w, is obtained by fitting $P_d(x)$.

The laser power can be measured by calibrated commercial power meters and Joulemeters that accept beams extracted from the main ablation beam via beamsplitters. Solid state lasers in general offer pulse to pulse stability, whereas gas lasers may be subject to pulse-to-pulse variations that can be within ±10%. If ablation is performed at low fluences, it is perhaps a better strategy to run the laser at higher laser energies and then attenuate the beam. For polarized solid state lasers, this can be done by rotating polarizer beamsplitter cubes or liquid crystal devices. For excimer lasers, one may try combinations of quartz beamsplitter plates. A usual practice in the IR regime is to use acousto-optical (AO) modulators for temporal and amplitude adjustment of laser pulses. The beam power for CW lasers and pulse profile can be monitored by reading the signal deflected from beamsplitters or simply the energy scattered from optical components to sufficiently fast detectors. For short pulse lasers, several techniques based on phase-sensing have been presented.

4.5 Temperature of the Target

Knowledge of the temperature field development in laser interactions with materials is essential, not only for better control of ablation and desorption experiments, but also for delineating fundamental issues on the evolution of the material ejection process from the molten phase as the incident laser energy is increased. Several techniques have been developed to probe the transient temperature field during pulsed laser processing. However, as will become apparent in this brief review, these techniques have inherent limitations that render them suitable only for particular materials and temperature ranges, while subject to spatial constraints.

4.5.1 Surface Reflectivity and Transmissivity Probes

Surface reflectivity measurements are often used to probe the surface conditions in pulsed laser processing. Among the notable applications of time-resolved surface reflectivity measurement is the detection of surface melting in pulsed laser irradiation of semiconductors, which is facilitated by the large

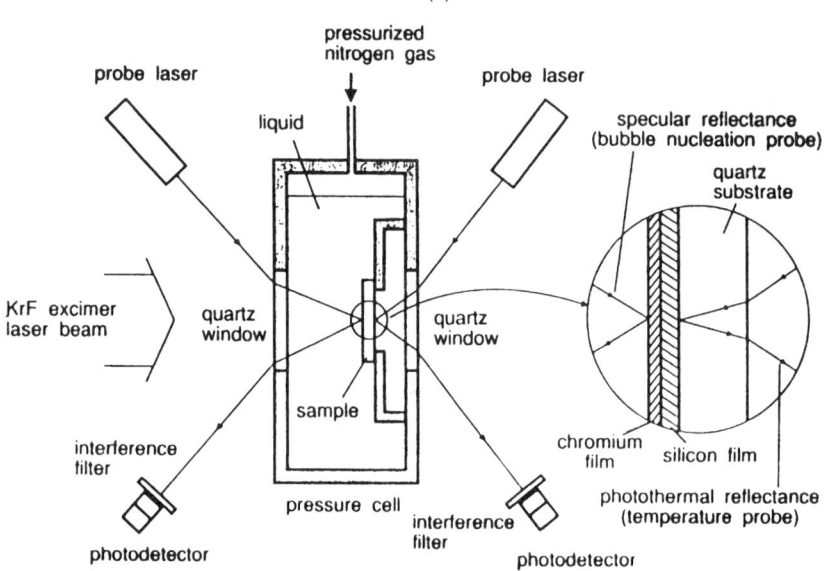

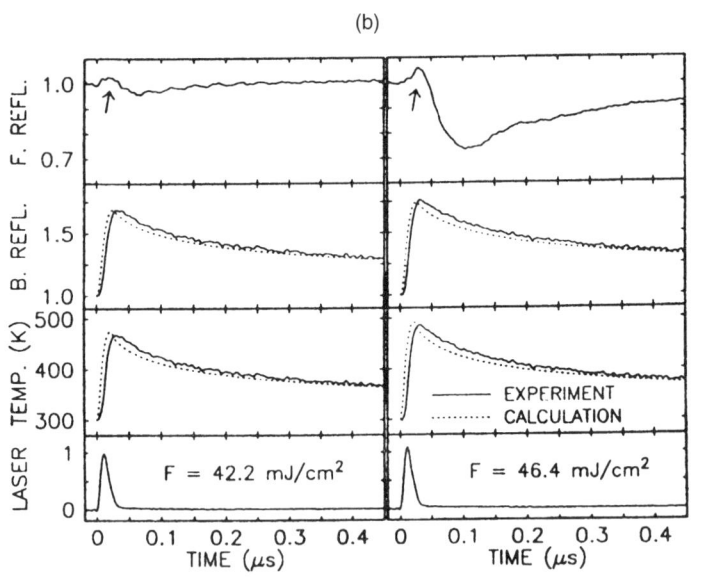

increase of the reflectivity upon melting [63]. In principle, because the complex refractive index depends on temperature, the temporal variation of the reflectivity can provide the surface temperature. A picosecond pump and probe optical reflectivity technique was used to measure the thermal diffusivity of thin metal films [64]. For metals, the variation of the normal incidence reflectivity with temperature, $\Delta R/\Delta T$ is of the order of 10^{-5}; this results in a reflectivity variation from room temperature to the melting point of only a few percent. Thus the measurement requires the use of sophisticated electronics and signal averaging. If the irradiated solid is a thin film of appropriate thickness, optical interference effects may enhance the surface reflectivity.

If the irradiated sample is semi-transparent, a transmissivity probe could give temperature information. Optical transmissivity and reflectivity measurements on thin silicon samples irradiated by laser pulses of picosecond duration were conducted in [65]. Optical properties of thin films may be different than their bulk counterparts and have to be measured, together with an accurate determination of the film thickness. It is noted that, in the case of surface heating of a bulk semi-transparent material, the refractive index depends on the temperature distribution in the heat affected zone. This variation can be handled by assuming a multilayer system of continuously varying refractive index [66] as described in the previous section.

Other variances of optical reflection techniques have been developed, including an interferometric technique to capture temperature-induced optical path reflectance changes from the back-side of a sample film [67]. Figure 6 shows the experimental setup for back-surface photoreflectance monitoring and transient temperature measurement of the superheat in the boiling of liquids on a pulsed laser-heated surface [68]. Although optically smooth surfaces are certainly required, surface topography may develop due to melt instabilities or due to gratings induced by electromagnetic interference and nonlinear effects, introducing a roughening and decrease of the reflectivity through scattering. Surface reflectivity techniques offer the advantage of high spatial resolution by point-by-point mapping of the surface. The main shortcoming of optical reflection and

---

FIG. 6. (a) Schematic of the experimental setup. The probe lasers are a HeNe laser ($\lambda = 632.8$ nm) for photothermal reflectance at the backside and an Ar^+ laser ($\lambda = 488$ nm) for specular reflectance of phase change at the front side. (b) The experimental reflectance curves (solid lines) for water are shown in both top panels for the front-side reflectance and both second panels for the back-side photothermal reflectance. The dotted lines in the second and third panels are calculated transient reflectivity response and surface temperature, respectively. Shown in solid lines in the third panels are resultant surface temperature traces from the measured reflectances. The bottom panels show the excimer laser pulses, $F = 42.2$ mJ/cm^2 and 46.4 mJ/cm^2 [68]. Reprinted with permission from AIP.

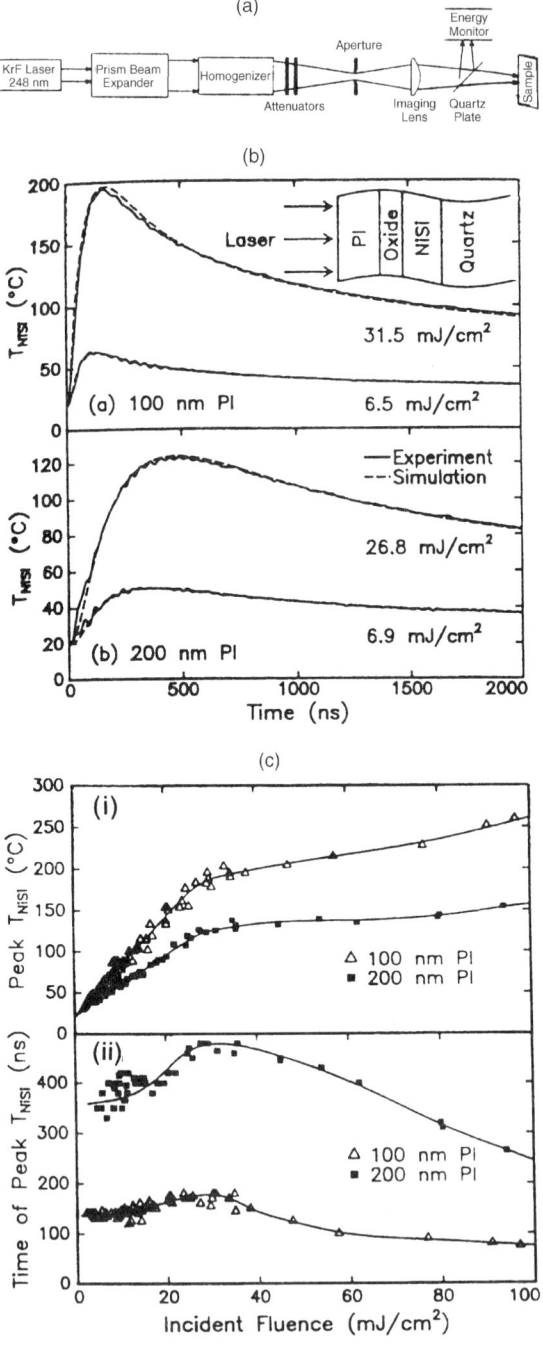

transmission techniques stems from the fact that the surface must be optically smooth and have temperature-dependent optical properties.

4.5.2 Thermistor Techniques

Temperature measurement by thermistors is effected by embedding thin film sensors under the irradiated area to construct miniature thermistor or thermocouple junctions. The sensitivity and accuracy of the temperature measurement in these systems have to be evaluated considering the thermal properties of the structure. Thermal properties of thin films may be unknown and interfacial thermal contact resistance effects may impede the heat flow. Parametric studies, considering for example lower thermal conductivities for the thin films, may provide a reasonable indication of these effects on the temperature measurement. Obviously direct heating by the irradiating source has to be avoided. Also, as electrical isolation is required, the thermistor has to be placed underneath a poor thermal conductor and well below the radiation penetration depth. Additionally the construction of a small thermocouple junction requires lithographic and film fabrication techniques. Spatial resolution usually restricts the application of the method to planar laser beam spots because interpretation of the experimental data becomes more complex if two- and three-dimensional effects need to be considered. Transient temperature during nanosecond Nd:glass laser irradiation was measured using an iron/constantan thin film thermocouple [69]. A similar thermistor technique [70] was applied to study the interface temperature of Si-As alloys during planar-interface solidification induced by a pulsed XeCl excimer laser. This approach was applied to obtain the temperature in excimer laser-irradiated polymer films [71]. The thermistor was a 140 nm thick serpentine NiSi film, fabricated by e-beam evaporation on a quartz wafer and annealing. The thermistor was equipped with Al contacts and covered by a 50 nm SiO_x protective layer; polyimide films of 100 to 200 nm thickness were subsequently sputtered on top of this structure. Analysis of the experimental temperature profiles (Fig. 7) indicated the existence of a threshold temperature of about 1660 K for ablation. Measurement beyond the ablation threshold becomes

---

FIG. 7. (a) Schematic diagram of the experimental system. (b) Plots of simulated and measured thermistor profiles. Except for fluence and polyimide thickness, simulation parameters are identical at all times. Solid lines correspond to experiment and dashed lines are simulated profiles. The oscillations in the experimental profiles arise from laser-induced rf noise. (c) Peak NiSi temperature (i) and time of maximum NiSi temeprature observed as function of incident fluence for two poyimide thicknsses. Solid lines are guides to the eye only. Note that the data in (i) for the two polyimide thicknesses lie on different curves, but the break in linear behavior occurs at the same fluence [71]. Reprinted with permission from AIP and M. O. Thompson, Cornell University.

difficult to interpret because the progressing ablation front implies a change in the film thickness altering the thermal path.

4.5.3 Pyrometry

Pyrometry is based on the measurement of the thermal emission for the deduction of the temperature. Application of pyrometry for static temperature distributions is quite well established. However, in applying this method to transient temperature fields, it is necessary to consider the variation of the surface emissivity due to the temperature dependence of the optical properties, but also because of possible surface compositional alterations and oxidation. A concern may arise as to what extent the surface temperature is governed by a Planckian distribution. In the case of a metallic target, and for molten semiconductors that also exhibit a metallic behavior, it is likely that the thermal emission corresponds to a blackbody temperature under conditions of local thermal equilibrium (LTE) that are expected in the nanosecond time scale. Time-resolved emission spectrometry of a silicon surface during irradiation by a XeCl ($\lambda = 308$ nm) excimer laser in a variable density Cl_2 environment was presented in [72]. The experimental results showed that for various applied pressures, the surface temperature decreased with fluence. On the other hand, as the gas pressure increased, the surface temperature increased, being solely a function of the Cl_2 coverage of the surface. It is observed that this experiment was conducted in the spectral range between 490 and 710 nm.

Rapid phase change phenomena in bulk silicon and undoped p-Si thin films were probed in the experiment described in [73]. Surface temperature information is obtained by multicolor near-IR pyrometry (Fig. 8). The desired temperature range is between 1500 and 3500 K. In this temperature range, thermal emission is strongest at wavelengths between 1 and 2 μm. To obtain temperature information from the measured thermal emission requires the temperature dependence of the spectral emissivity. The emissivity is generally a function of temperature and wavelength. The surface conditions may alter the emissivity drastically. The emissivity of a real sample surface can be much different from the reported literature data for "ideal" surfaces. In this work, the emissivity of the sample was independently measured by direct spectral reflectivity measurements. The detector collects thermal radiation through a solid angle (θ_1 to θ_2, ϕ_1 to ϕ_2) and over a wavelength bandwidth (λ_1 to λ_2). In fact, the voltage signal recorded on the oscilloscope, V, represents an integration of the thermal emission in this solid angle and wavelength band, modified by the emissivity of the material, the transmission of the optical components, and the detector spectral response:

$$V \sim \int_{\lambda_1}^{\lambda_2} \int_{\theta_1}^{\theta_2} \int_{\phi_1}^{\phi_2} \varepsilon_\lambda(\lambda, \theta, \phi, T) \tau(\lambda) D(\lambda) e_{\lambda b}(\lambda, T) \, d\phi \, d\theta \, d\lambda \, dA \quad (4.75)$$

where $\tau(\lambda)$ is the spectral transmittance of the lenses and filters in the optical

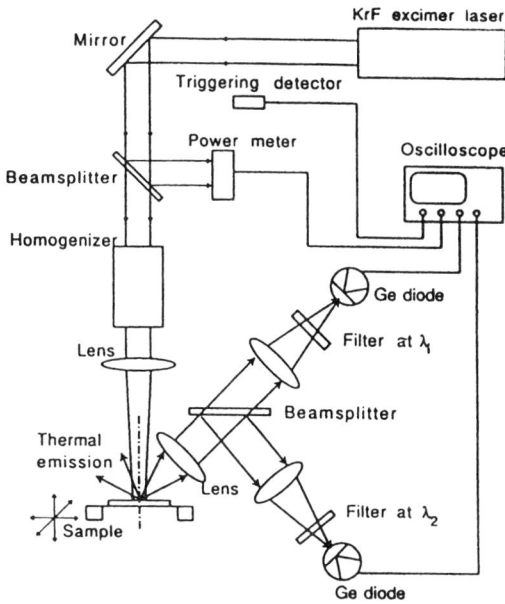

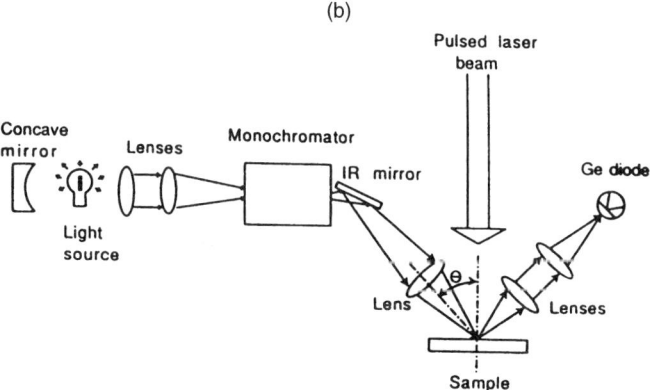

FIG. 8. (a) Experimental set-up for transient thermal emission measurement and the *in situ* spectral reflectivity measurement. (b) Measured and calulated surface temperature histories during melting of a bulk, crystalline Si sample with an excimer laser pulse of fluence, $F = 0.9$ J/cm^2 [73]. Reprinted with permission from Springer-Verlag.

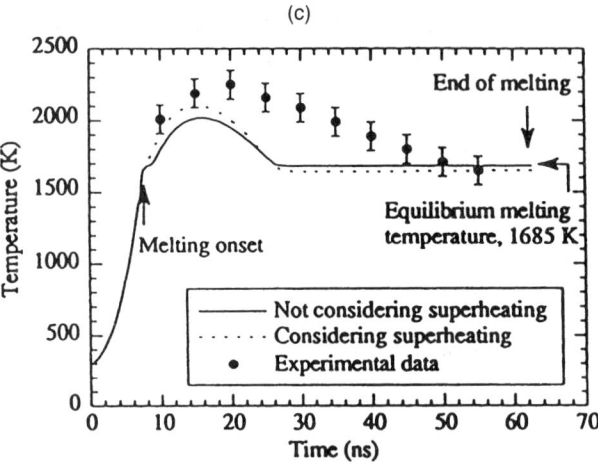

FIG. 8. *Continued*

path, $D(\lambda)$ is the responsivity of the diode, $\varepsilon_\lambda(\lambda, \theta, \phi, T)$ is the directional spectral emissivity, and dA is the area on the sample. The blackbody emissive power, $e_{\lambda b}$:

$$e_{\lambda b} = \frac{2\pi C_1}{\lambda^5 \exp(C_2/\lambda T) - 1} \quad (4.76)$$

where C_1 and C_2 are blackbody radiation constants.

The emissivity is derived using the measured transient reflectivity data. Once the emissivity is determined and the thermal radiation emission of the target material is measured, the temperature can be obtained. The effect of a temperature gradient into the surface is considered in deducing the surface temperature. The measured thermal radiation is emitted from a skin liquid layer whose thickness is only a few optical absorption depths. Because there exists a falling temperature gradient into the liquid layer, the temperature assigned to the measured thermal emission may be lower than the actual surface temperature by a few tens of degrees K. The issue of the internal thermal gradient contributions to the thermal emission warrants further attention, particularly when steep temperature profiles are encountered in low thermal conductivity materials.

4.5.4 Time-of-Flight Measurements

Time-of-flight (TOF) measurements characterize the translational kinetic energies of the ejected plume. In the absence of collisional and plasma effects or chemical reactions in the plume, it would appear that the TOF kinetic energy distributions can be referred to the surface thermal condition. At this point, it is worthy to recall the basic principles of such measurements.

The expectation of a Boltzmann energy distribution in TOF measurements can be obtained from classical gas dynamics. The number density of vapor particles having velocities between

$$\mathbf{u} = \langle u_x, u_y, u_z \rangle \quad \text{and} \quad \mathbf{u} + d\mathbf{u} = \langle u_x + du_x, u_y + du_y, u_z + du_z \rangle$$

released from the surface is:

$$dn_s(\mathbf{u}) = dn_s(\mathbf{u})/u_x = n_s f(\mathbf{u}) \, du_x \, du_y \, du_z \quad (4.77)$$

where n_s is the total vapor number density at the surface and the Maxwellian velocity distribution function is given by:

$$f(\mathbf{u}) = \left(\frac{m}{2\pi k_B T}\right)^{3/2} e^{-m(u_x^2 + u_y^2 + u_z^2)/2k_B T} \quad (4.78)$$

When the surface flux temporal behavior is approximated by a delta function, the number density velocity distribution becomes spatially resolved as higher velocity particles move farther from the surface than lower velocity particles in a given period of time. If the surface flux is approximated by a point source, and the particle stream is collisionless, and $u_x^2 \gg u_y^2$, u_z^2 and $u_x = x/t$, the detector signal is:

$$N_d(t) \sim \frac{1}{t^4} \exp\left(-\frac{m(x/t)^2}{2k_B T}\right) \quad (4.79)$$

The measured density signal, $N_d(t)$, yields the energy probability function, $P(E)$:

$$P(E) \sim E \exp\left(-\frac{E}{k_B T}\right) \quad (4.80)$$

In the limit of a small surface vapor flux, the vapor expansion into vacuum is collisionless and the mean translational energy should be indicative of the surface phase temperature, $\bar{E} = 2k_B T$.

The above relations were derived under the assumption that the nascent distribution is preserved, i.e., in the limit of collisionless sputtering. If, for a hard-sphere model, the scattering cross section is considered independent of the absolute and relative velocities, the number of near-surface collisions per atom within the laser pulse of duration t_{pulse} is derived with the aid of elementary gas kinetics:

$$n_c \sim t_{\text{pulse}} P_v \Sigma \frac{2}{3} \left(\frac{\pi}{m k_B T}\right)^{1/2} \quad (4.81)$$

where P_v is the vapor pressure at the condensed phase temperature. For temperatures below the atmospheric boiling temperature, tens of nanoseconds long laser pulses, and for metals with typical atomic cross sections of the order of 10^{-16} cm^2, n_c is less than 10. With the development of a stream velocity $\tilde{u}$, the flight distance relative to coordinates moving at $\tilde{u}$ becomes $\tilde{x} = x - \tilde{u}t$, so that Eq. (4.78) becomes:

$$N_d(t) = \frac{1}{t^4} \exp\{-\tilde{G}^2(x - \tilde{u}t)^2/t^2\} \qquad (4.82)$$

where $\tilde{G}^2 = m/(2k_B \tilde{T})$. Using Eq. (4.81) the temporal behavior of the density signal can be converted into a flux sensitive energy distribution:

$$\tilde{P}(E) \sim E \exp\{-E/k_B \tilde{T} + 2\tilde{G}\tilde{u}(E/k_B \tilde{T})^{1/2} - \tilde{G}^2 \tilde{u}^2\} \qquad (4.83)$$

Kelly and Dreyfus (1988b) have argued that $\tilde{u}$ is at least the sonic velocity at the outer edge of the Knudsen layer. If this is assumed, then for a monatomic perfect gas, $\tilde{u} = (5/6)^{1/2}\tilde{G}$, and Eq. (4.82) becomes:

$$\tilde{P}(E) \sim E \exp\{-E/k_B \tilde{T} + 2(5E/6k_B \tilde{T})^{1/2} - 5/6\} \qquad (4.84)$$

The mean kinetic energy of the above distribution is $\bar{E} = 3.67 k_B \tilde{T}$ [17].

A quadrupole mass spectrometer (QMS) in a time-of-flight (TOF) arrangement was used to determine the kinetic energies of evaporated Si and GaAs targets by 20 ns long pulsed ruby laser irradiation [74, 75]. Based on these energy distributions, they extracted the temperature of the evaporated atoms by fitting Maxwellian distributions from gas kinetic theory. Furthermore, by assuming that this temperature represented the lattice temperature, they concluded that the process was thermal. Highly sensitive resonance ionization mass spectrometry (RIMS) was applied to investigate the emission of neutral and ionized Fe atoms induced by N_2 laser irradiation ($\lambda = 337$ nm) in ultrahigh vacuum (10^{-10} torr) [76]. The experiment was conducted at low laser fluences (<275 mJ/cm^2), well below plasma formation regime. The kinetic energies of the neutral atoms derived from the time-of-flight distributions at different heights from the irradiated surface yielded most probable velocities and temperatures in broad agreement with the thermal expectations (Fig. 9). The thresholds for detection of 100 atoms or ions were 25 mJ/cm^2 and about 100 mJ/cm^2, correspondingly. On the other hand, the number of Fe$^+$ ions released per pulse from the surface at the highest laser fluence of 275 mJ/cm^2 was about three orders of magnitude less than the number of neutrals (Fig. 10). The same method of investigation was used to probe soft laser sputtering of InP(100) surfaces [77]. For low fluence values (<190 mJ/cm^2), it was found that sputtering results mainly from absorption and excitation of defect sites. At higher laser fluences, the process assumes a thermal-like behavior. Measurements of the kinetic energies of the neutral ablated In and P atoms, indicated that

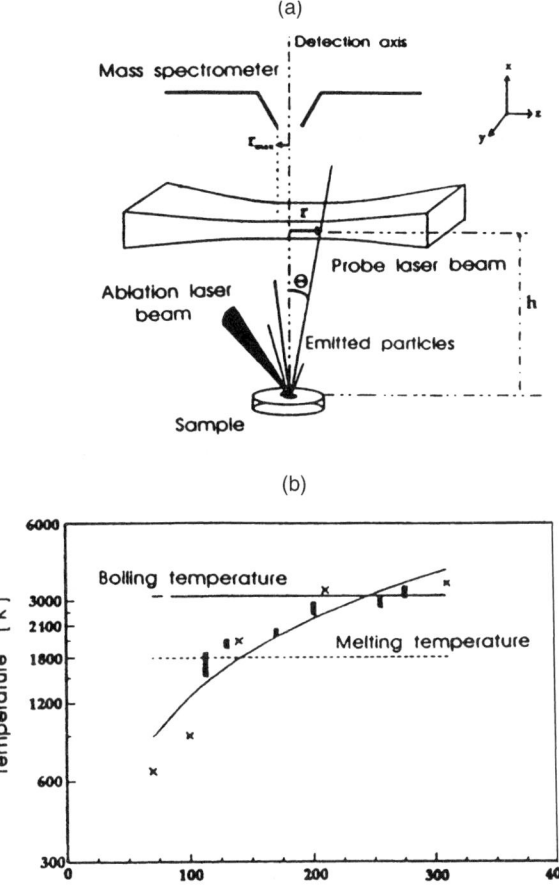

FIG. 9. (a) Laser sputtering experiment and RIMS detection geometry. (b) Time-of-flight distribution of sputtered Fe atoms for different heights above the target surface and two laser fluences (upper figure: 275 mJ/cm^2; lower figure: 170 mJ/cm^2). The relative Fe atom density values are deduced from the RIMS signals. the curves are the best fits by half-range Maxwellian functions with respective temperatures 3130 K and 2210 K (lower figure). (c) Temperature of the sputtered Fe atoms as a function of the laser fluence. -^- Kinetic temperature deduced from the time-of-flight distributions. -×- Excitation temperature deduced from the sublevel population distributions. -——- Fit by the solution of the 1-D heat-flow equation laser sputtering yields. (d) Number of neutral and ionized Fe atoms sputtered per laser shot as a function of the laser fuence. The neutral atom emission yield repersents the yield for the a^3D_4 ground state only and does not take into account the contribution of higher excited states when the temperature increases. The dotted line gives an indication of the total atom yield. Note that at the threshold, about 100 atoms are detected by RIMS [76]. Reprinted with permission from AIP. and T. Gibert, CNRS-Université d'Orléans.

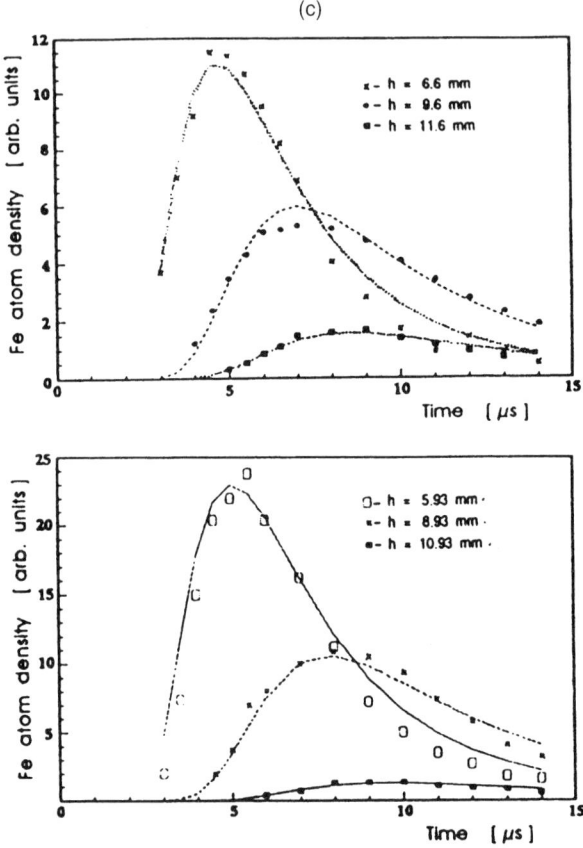

FIG. 9. *Continued*

the transition between these regimes occurred at a temperature of ~1400 K, which is close to the 1340 K melting temperature of InP.

Recent experiments have provided evidence that electronic excitations may be involved in desorption processes from metals at low photon energies. Metal atoms can be desorbed from small metal clusters by low-power CW laser radiation, as shown in [78]. The dependence on wavelength, power, and cluster size indicated that collective plasmon excitations play a role in stimulating desorption. Bimodal energy distributions of gold atoms desorbed from a continuous film with pulsed laser irradiation at the $\lambda = 532$ nm wavelength with $t_{\text{pulse}} = 7$ ns duration were observed and attributed the 0.3 eV high-energy peak to an electronic de-excitation process involving surface plasmons [79]. It is noted that the experimental geometry used in this work was suitable for

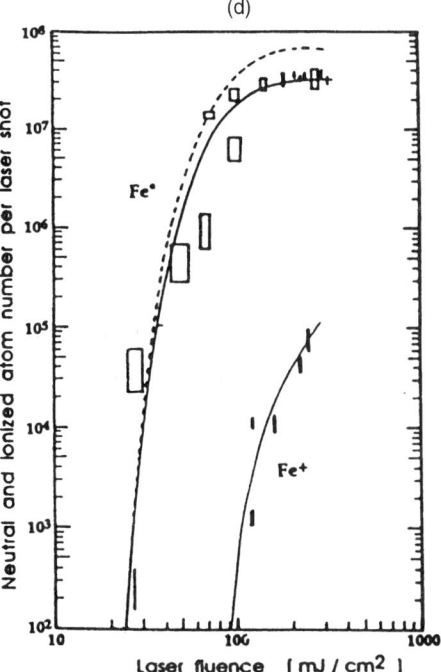

Fig. 9. Continued

plasmon excitation, as the gold film was coated on a glass prism and total internal reflection (TIR) was used to enhance plasmon excitation via the evanescent wave. Energetic desorption of ~0.7 eV energetic Ag^+ ions from continuous films using both a TIR geometry and a direct surface irradiation scheme was observed in [80].

It was previously remarked that the temperature field development in the heating of a bulk metal target with an excimer laser beam of mm size is normally understood as a one-dimensional phenomenon. However, surface topography development has been observed in various systems. In metals under atmospheric conditions, surface topography develops under multiple pulse irradiation [16]. Pulsed laser heating and melting of gold in a 10^{-6} torr vacuum produces steady state surface topography growing from the molten phase in several hundreds of pulses [17]. The heat transfer modeling of the near-surface thermal conditions and a hydrodynamic stability analysis traced the origin of the observed characteristic periodicity to the acceleration that the molten layer experiences due to the volumetric expansion upon melting. The characteristics of the translation energy distribution in the ablation plume released from the

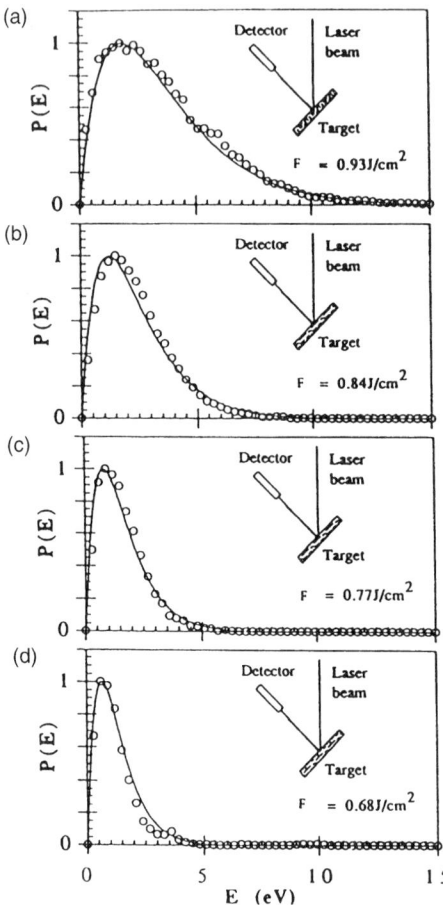

FIG. 10. (a) Effect of laser fluence on the translational energy distribution of neutral Au atoms ($\theta_o = 45°$, $\theta_d = 0°$). The dots correspond to the inverted TOF signal, while the solid line represents a Boltzmann distribution having the same mean energy as the inverted TOF signal [17]. Reprinted with permission from AIP.

surface showed that both the mean translational energy and the ablation yield achieve a steady state, together with the surface topography. It is noted that in that study the material removal rates were kept small (≤ 10 Å/pulse), and it was verified that the laser intensity was not sufficient to ignite plasma effects in the vapor plume. Figures 10(a–d) demonstrate the effect of laser fluence on the steady state kinetic energy distribution in the plume. The laser fluence has been varied from a near-threshold fluence (~ 0.68 J/cm^2) to about 1.0 J/cm^2 over the course of this experiment. For these quadrupole mass spectrometry (QMS) measurements, the detector was located normal to the surface ($\theta_d = 0°$) and centered over the ablation area, and the laser beam was incident at $\theta_o = 45°$ from the surface normal. Details of the experimental apparatus are given in [81]. Depicted in the figures are inverted TOF measurements in comparison with the theoretically predicted Boltzmann distribution. The first observation to be made from these figures is the apparent success of the Boltzmann distribution in describing the energy distribution in the plume. The second finding is that the measured mean kinetic energies (up to several eV) correspond to temperatures far exceeding the thermal expectations (1 eV $\sim$ 11,620 K). It is noted that the temperature in Eq. (4.80), implicitly refers to the vapor temperature rather than the surface temperature. Nevertheless, it was shown that the Boltzmann distribution fits better the experimental results than the "stream velocity-corrected" distribution derived in Eq. (4.84). The implication is that an order of 10 or 20 collisions per particle that is obtained by a simple estimate for the material removal rates considered is not sufficient to impart a significant stream velocity.

The size of the surface features developed in the preceding study was comparable to the depth of the molten zone and the thermal diffusion depth in the material, thereby destroying the assumption of one-dimensional field development and homogeneous surface temperature. For this purpose, an experiment was devised [82] where, after the cleaning and chemical etching procedure, the target was preconditioned by surface melting effected by $N = 30$ pulses delivered on the gold target which was preheated statically at the $T_o = 1,100$ K temperature. This treatment produced a reasonably flat surface, thus ensuring accuracy of peak surface temperature prediction within a 200 K confidence window. The average yields and mean translational energies show little dependence on the partitioning ratio between the energy delivered by the laser and the steady target preheating as the calculated peak surface temperature was kept constant. Similar behavior was observed for Ag, Cu, and a CuAu alloy, while Ni exhibited a thermal-like behavior.

Following the initial development of the thermal expectation for the mass efflux from the target in Eq. (4.47), it can be argued that the integrated TOF signal, Y, can be related to the vapor pressure, P_v, and temperature:

$$Y \sim P_v \sim \exp\left(-\frac{L_v}{k_B T}\right) \tag{4.85}$$

Hence,

$$\ln(Y) \sim -\frac{L_v}{k_B T} + \text{const.} \tag{4.86}$$

As shown in Fig. 11(a) the measured mean kinetic energies for gold samples have different initial temperatures. They greatly exceed the anticipated energies obtained on the basis of calculated peak surface temperatures. Figure 11(b)

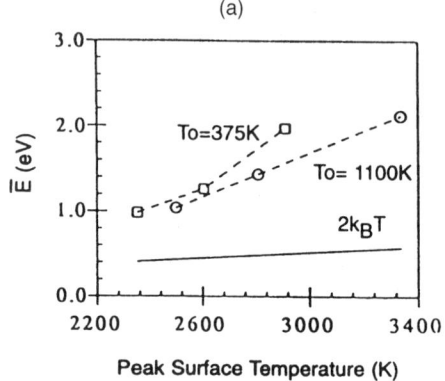

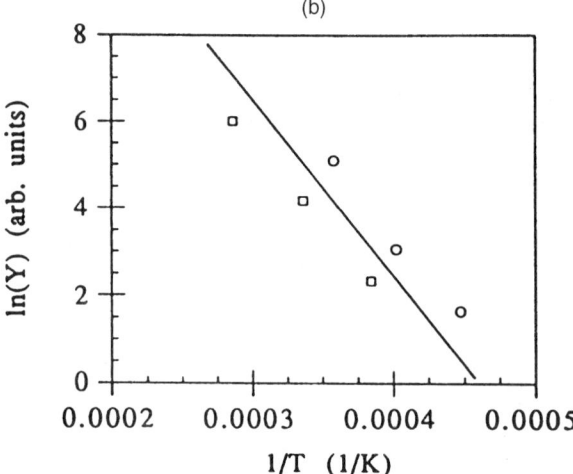

FIG. 11. (a) Measured and thermally anticipated ($2k_B T$) translational energies as a function of calculated peak surface temperature. The data pertains to the first 10 laser pulses on the annealed target surface. (b) Natural log of TOF integrated signal from the polycrystalline Au target versus the reciprocal of the calculated peak surface temperature. The solid line corresponds to the anticipated slope $-L_v/k_B T$ for a thermal process [82]. Reprinted with permission from APS.

shows that this relation fits the experimental results for polycrystalline Au. With the exception of Ni, it was indeed noted that the experimental data appear to support the view that the surface vaporization is thermally mediated, in accordance to the observation on excimer laser ablation of Cu [1]. From the results of this study, it can be stated that the use of the mean translational energy to extract the peak surface temperature is questionable. On the other hand, the ablation yield obeys the behavior anticipated by thermal expectations, but then again it is difficult to obtain precise measurement of the ablation yield in absolute terms.

4.5.5 Beam Deflection Probes

The beam deflection schemes are based on the detection of changes in the refractive index of the medium due to thermal, pressure, or concentration gradients. These probes have been demonstrated for applications, including absorption spectroscopy [83, 84, 85] and thermal diffusivity measurements (a survey is given in [86]). The deflection angle transients are described by:

$$\phi(t) = \left| \int_{\mathbf{1} \in \text{path}} \frac{1}{n} \nabla n(\mathbf{1}, t) \times d\mathbf{1} \right| \quad (4.87)$$

where n is the refractive index of the medium and $d\mathbf{1}$ is an incremental distance along the probe beam path. In general, the defection angle has one component parallel and one component prependicular to the sample surface. The deflection angle is detected by a position sensitive detector such as a bi-cell or a quadrant detector and a knife edge. In the case of a bi-cell photodetector,

$$\phi \propto \frac{V_1 - V_2}{V_1 + V_2} \quad (4.88)$$

where V_1 and V_2 denote the signal amplitudes from cells 1 and 2. The perpendicular and parallel components of the deflection signal can be resolved by a quadrant detector.

When a time-modulated or pulsed laser beam irradiates the target, the temperature field diffuses in the ambient backing medium and produces photothermal deflection (PTD) of the probing beam in a time scale representative of the evolution of thermal transport in the system. Even in the simple heating case, below the melting and ablation thresholds, the PTD response is sensitive to the boundary conditions on the temperature field; the material properties; the deflecting medium; the shape, temporal duration, or modulation frequency of the irradiating beam; and the location and size of the probing laser beam. If the profile of the laser beam is Gaussian with dependence on the radial coordinate, r, the temperature field in the backing medium depends on (r, z, t) and the calculation of the probe beam deflection has to take into account the corresponding refractive index changes, along with the finite size and profile of the probing laser beam. Because the deflection signal carries information developed along the path of the beam, a simple ray tracing approach may not be sufficient for a precise calculation of the transient deflection angle, particularly as the probe beam cross section dimensions approach the extent of the heat affected zone. Temporal and spatial effects were examined in [87], where it was shown that, in addition to the magnitude and phase of the deflection signal, the temporal shape of the deflection signal provides fundamental information about the transient heat flow in the target and the backing medium. In a subsequent study [88], it was shown that the PTD method could indicate the transition to the molten phase, which was manifested by a kink in the signal temporal profile.

The thermoelastic expansion launches in the backing medium a pressure pulse that disturbs the probing laser beam, producing photoacoustic deflection (PAD). The compressional PAD pulse scales with the transient heating, whereas the cooling time is generally slower, producing a much weaker rarefaction wave. If the pressure dependence of the refractive index of the backing medium is known, the deduction of absolute values of the pressure field can be accomplished, provided the spatial shape of the acoustic wave is accounted for [89]. If the laser beam energy exceeds the ablation threshold, the ablation products interfere with the probing laser beam. If ablation is performed in a backing atmosphere higher than 1 torr, the ablation products push the background gas, creating a shock wave that travels with supersonic speed. The strength of the shock wave and the traveling speed decay as the distance from the target is increased. By varying the probe beam axis separation from the target surface, deflection schemes can yield the shock speed [90]. The ablation products trail the leading shock front, but the deflection scheme is not species specific to provide clear information. Probe deflection to analyze density gradients in laser ablation in vacuum [91, 92]. The probe beam deflection due to the change of the refractive index in plasmas is negative, and proportional to the electron density, n_e [91]:

$$\delta n = -\left(\frac{e^2}{2\pi m v^2}\right) n_e \qquad (4.89)$$

where e and m are the electron mass and v the probing laser frequency. On the other hand, changes in concentrations of neutral species are positive. In general, probe deflection diagnostics provide an appealing means for analyzing thermal conditions of the target and the generated ablation plume, but an interpretation of the results requires isolation of the physical phenomenon.

Acknowledgments

Support provided by the National Science Foundation, under Grant CTS-9402911 is acknowledged. Research described in this chapter was done with Hee K. Park, Ted Bennett, Jeng-Rong Ho, Dongsik Kim, Xiang Zhang, and Xianfan Xu of the University of California at Berkeley. The author is especially thankful to Andrew C. Tam and Douglas J. Krajnovich of the IBM Almaden Research Center for their contributions to collaborative work.

References

1. Dreyfus, R. W. Cu^0, Cu^+, and Cu_2 from excimer-ablated copper. *J. Appl. Phys.* **69**, 1721–1729 (1991).
2. Von Allmen, M. *Laser-Beam Interactions with Materials.* Springer-Verlag, Heidelberg, 1987.
3. Palik, E. D. *Handbook of Optical Constants of Solids, Vol. I and II.* Academic, London 1985, 1991.
4. Miller, J. C. Optical properties of liquid metals at high temperatures. *Phil. Magaz.* 1115–1132 (1969).
5. Baüerle, D. *Laser Processing and Chemistry.* Springer-Verlag, Heidelberg, 1996.
6. Prokhorov, A. M., Konov, V. I., Ursu, I., and Mihailescu, I. N. *Laser Heating of Metals.* Adam Hilger, New York, 1990.
7. Özisik, M. N., *Heat Conduction.* 2nd ed. John Wiley & Sons, New York, 1993.
8. Carslaw, H. S., and Jaeger, J. C. *Conduction of Heat in Solids*, 2nd ed. Oxford University Press, Oxford, Gr. Britain, 1959.
9. Grigoropoulos, C. P., Rostami, A. A., Xu, X., Taylor, S. L., and Park, H. K. Localized transient surface reflectivity measurements and comparison to heat transfer modeling in thin film laser annealing. *Int. J. Heat Mass Transfer* **36**, 1219–1229 (1993).
10. Jackson, K. A., and Chalmers, B. Kinetics of solidification. *Can. J. Phys.* **34**, 473–490 (1956).
11. Jackson, K. A. Theory of melt growth, in *Crystal Growth and Characterization*, R. Ueda, and J. B. Mullin (eds.). North-Holland, Amsterdam, 1975.
12. Larson, B. C., White, C. W., Noggle, T. S., and Mills, D. Synchrotron X-ray diffraction study of silicon during pulsed laser-annealing. *Phys. Rev. Lett.* **48**, 337 (1982).
13. Larson, B. C., Tischler, J. Z., and Mills, D. M. Nanosecond resolution time-resolved X-ray study of silicon during pulsed-laser irradiation. *J. Mater. Res.* **1**, 144–154 (1986).
14. Peercy, P. S., Thompson, M. O., and Tsao, J. Y. Dynamics of rapid solidification in silicon, in *Proceedings, Materials Research Society*, M. O. Thompson, S. T. Picraux, and J. S. Williams (eds.). MRS, Pittsburgh, **75**, 15–30, 1987.
15. Ho, J.-R., Grigoropoulos, C. P., and Humphrey, J. A. C., Computational study of heat transfer and gas dynamics in the pulsed laser evaporation of metals,. *J. Appl. Phys.* **78**, 4696–4709 (1995).
16. Kelly, R., and Rothenberg, J. E. Laser sputtering: Part III. The mechanism of the sputtering of metals at low energy densities. *Nuc. Instrum. Meth. Phys. Res.* **B7/8**, 755–763 (1985).
17. Bennett, T. D., Grigoropoulos, C. P., and Krajnovich, D. J. Near-threshold laser sputtering of gold. *J. Appl. Phys.* **77**, 849–864 (1995).
18. Denes, E. Improved quartz microbalance technique. *J. Appl. Phys.* **56**, 608–626 (1984).
19. Zhang, X., Chu, S. S., Ho, J. R., and Grigoropoulos, C. P. Excimer laser ablation of thin gold films on a quartz crystal microbalance at various argon background pressures. *Appl. Phys. A* **64**, 545–552 (1997).
20. Rykalin, N., Uglov, A., Zuev, I., and Kokora, A. *Laser and Electron Beam Materials Processing Handbook*, MIR Publishers, Moscow, pp. 194–232, 1988.
21. Phipps, C. R., and Dreyfus, R. W. The high irradiance regime, in *Laser Ionization Mass Analysis*, A. Vertes, R. Gijbels, and F. Adams (eds.). John Wiley & Sons, New York, 1993.

22. Spitzer, L., Jr. *Physics of Fully Ionized Gases.* John Wiley Interscience, New York, 1962.
23. Ho, J.-R., Grigoropoulos, C. P., and Humphrey, J. A. C. Gas dynamics and radiation heat transfer in excimer laser ablation of aluminum. *J. Appl. Phys.* **79**, 7205–7215 (1996).
24. Duley, W. W. *Laser Processing and Analysis of Materials.* Plenum, New York, 1983.
25. Srinivasan, R., Hall, R. R., Loehle, W. D., Wilson, W. D., and Albee, D. C. Chemical transformations of the polyimide kapton brought about by ultraviolet laser radiation. *J. Appl. Phys.* **78**, 4881–4887 (1995).
26. Srinivasan, R. Ablation of polymer films with ultraviolet laser pulses of long (10–400 μs) duration. *J. Appl. Phys.* **72**, 1651–1653 (1992).
27. Brannon, J. H. Micropatterning of surfaces by excimer laser projection. *J. Vac. Sci. Techn. B* **7**, 1064–1071 (1989).
28. Horiike, Y., Hayasaka, N., Sekine, M., Arikado, T., Nakase, M., and Okano, H. Excimer-laser etching on silicon. *Appl. Phys. A* **44**, 313–322 (1987).
29. Patzel, R., and Endert, H. Excimer laser-a reliable tool for microprocessing and surface treatment. *Proceedings AIP* **288**, pp. 613–618, 1993.
30. Rothschild, M., and Ehlrich, D. J. A review of excimer laser projection lithography. *J. Vac. Sci. Techn. B* **6**, 1–17 (1988).
31. Phillips, H. M., Wahl, S., and Sauerbrey, R. Submicron electrically conducting wires produced in polyimide by ultraviolet laser irradiation. *Appl. Phys. Lett.* **62**, 2572–2574 (1993).
32. Srinivasan, R., and Braren, B. Ultraviolet laser ablation of organic polymers. *Chem. Rev.* **89**, 1303–1306 (1989).
33. Lankard, J. R., Sr., and Wolbold, G. Excimer laser ablation of polyimide in a manufacturing facility. *Appl. Phys. A* **54**, 355–359 (1992).
34. Chrisey, D. B., and Hubler, G. K., eds. *Pulsed Laser Deposition of Thin Films.* Wiley-Interscience, 1994.
35. Davanloo, F., Juengerman, E. M., Jander, D. R., Lee, T. J., Collins, C. B., and Matthias, E. Mass flow in laser-plasma deposition of carbon under oblique angles of incidence. *Appl. Phys. A* **54**, 369–372 (1992).
36. Sappey, A. D., and Nogar, N. S. Diagnostic studies of laser ablation for chemical analysis, in *Laser Ablation*, J. C. Miller (ed.). Springer-Verlag, Heidelberg, pp. 157–183, 1994.
37. Mao, X. L., Chan, W. T., Shannon, M. A., and Russo, R. E. Plasma shielding during picosecond laser sampling of solid materials by ablation in He versus Ar atmosphere. *J. Appl. Phys.* **74**, 4915–4922 (1993).
38. Bostangoglo, O., Niedrig, R., and Wedel, B. Ablation of metal films by picosecond laser pulses imaged with high-speed electron microscopy. *J. Appl. Phys.* **76**, 3045–3048 (1994).
39. Chen, H., Soom, B., Yaakobi, B., Uchida, S., and Meyerhofer, D. D. Hot-electron characterization from $K\alpha$ measurements in high-contrast, p-polarized picosecond laser-plasma interactions. *Phys. Rev. Lett.* **70**, 3431–3434 (1993).
40. Brau, C. A. Free-eElectron lasers. *Science* **239**, 1115–1121 (1988).
41. Danly, B. G., Temkin, R. J., and Bekefi, G. Free-electron lasers for medical applications. *IEEE J. Quant. Electron.* **QE-23**, 1739–1750 (1987).
42. Sauerbrey, R., Fure, J., Le Blanc, S. P., van Wonterghem, B., Teubner, U., and Schäfer, F. P., Reflectivity of laser-produced plasmas generated by a high intensity ultrashort pulse. *Phys. Plasmas* **1**, 1635–1642 (1994).

43. Preuss, S., Demchuk, A., and Stuke, M. Sub-picosecond UV-laser ablation of metals. *Appl. Phys. A* **61**, 33–37 (1995).
44. Preuss, S., Matthias, E., and Stuke, M. Sub-picosecond UV-laser ablation of Ni films. *Appl. Phys. A* **59**, 79–82 (1994).
45. Ihlemann, J., Schmidt, H., and Wolff-Rottke, B. Excimer laser micromachining. *Adv. Mater. Opt. Electron.* **2**, 87–92 (1993).
46. Ihlemann, J., Scholl, A., Schmidt, H., and Wolff-Rottke, B. Nanosecond and femtosecond excimer-laser ablation. *Appl. Phys. A* **60**, 411–417 (1995).
47. Bor, Z., Racz, B., Szabo, G., Xenakis, D., Kalpouzos, C., and Fotakis, C. Femtosecond transient reflection from polymer surfaces during femtosecond UV photoablation. *Appl. Phys. A* **60**, 365–368 (1995).
48. Wolff-Rottke, B., Ihleman, J., Schmidt, H., and Scholl, A. Influence of the laser-spot diameter on photo-ablation rates. *Appl. Phys. A* **60**, 13–17 (1995).
49. Fedosejevs, R., Ottmann, R., Sigel, R., Kühnle, G., Szatmari, S., and Schäfer, F. P. Absorption of femtosecond laser pulses in high-density plasma. *Phys. Rev. Lett.* **64**, 1250–1253 (1990).
50. Teubner, U., Bergmann, J., van Wonterghem, B., Schäfer, F. P., and Sauerbrey, R. Angle-dependent X-ray emission and resonance absorption in a laser-produced plasma generated by a high intensity ultrashort pulse. *Phys. Rev. Lett.* **70**, 794–797 (1993).
51. Kmetec, J. D., Gordon, III, C. L., Macklin, J. J., Lemoff, B. E., Brown, G. S., and Harris, S. E. MeV X-ray generation with a femtosecond laser. *Phys. Rev. Lett.* **68**, 1527–1530 (1992).
52. Pronko, P. P., Dutta, S. K., and Du, D. Thermophysical effects in laser processing of materials with picosecond and femtosecond pulses. *J. Appl. Phys.* **78**, 6233–6240 (1995).
53. Murnane, M. M., Kapteyn, H. C., and Falcone, R. W. High density plasmas produced by ultrafast laser pulses. *Phys. Rev. Lett.* **62**, 155–158 (1989).
54. Kautek, W., Mitterer, S., Kruger, J., Husinsky, W., and Grabner, G. Femtosecond-pulse laser ablation of human corneas. *Appl. Phys. A* **58**, 513–518 (1994).
55. Kántor, Z., Tóth, Z., and Szörenyi, T. Metal pattern deposition by laser-induced forward transfer. *Appl. Surf. Sci.* **86**, 196–201 (1995).
56. Zhang, X., Grigoropoulos, C. P., Krajnovich, D. J., and Tam, A. C. Excimer laser projection micromachining of polyimide thin films annealed at different temperatures. *IEEE Trans. Comp., Packag. Manuf. Techn. C* **19**(3), 201–213 (1996).
57. Born, M., and Wolf, E. *Principles of Optics* 6th ed., Pergamon, Exeter, United Kingdom, 1980.
58. Grigoropoulos, C. P., Park, H. K., and Xu, X. Modeling of pulsed laser heating of thin silicon films. *Int. J. Heat Mass Transfer* **36**, 919–924 (1993).
59. Krajnovich, D. J., Pour, I. K., Tam, A. C., Leung, W. P., and Kulkarni, M. K. 248 nm lens materials: Performance and durability issues in an industrial environment. *Proceedings SPIE* **1848**, 544–560 (1992).
60. Veldkamp, W. B. Laser beam profile shaping with interlaced binary diffraction gratings, *Appl. Opt.* **21**, 3209–3212 (1982).
61. Park, H. K., Grigoropoulos, C. P., Leung, W. P., and Tam, A. C. A practical excimer laser-assisted cleaning tool for removal of surface contaminants. *IEEE Trans. Comp. Packag. Manuf. Techn. A* **17**, 631–643 (1994).
62. Suzaki, Y., and Tachibana, A. Measurement of the µm sized radius of Gaussian laser beam using the scanning knife-edge. *Appl. Opt.* **14**, 2809–2810 (1975).
63. Jellison, G. E., Jr., Lowndes, D. H., Mashburn, D. N., and Wood, R. F. Time-resolved reflectivity measurements of silicon and germanium using a pulsed excimer KrF laser heating beam. *Phys. Rev. B* **34**, 2407–2415 (1986).

64. Paddock, C. A., and Eesley, G. L., Transient thermoreflectance from thin metal films. *Opt. Lett.* **11**, 273–275 (1986).
65. Lompré, L. A., Liu, J. M., Kurz, H., and Bloembergen, N. Time-resolved temperature measurement of picosecond laser irradiated silicon. *Appl. Phys. Lett.* **43**, 168–170 (1983).
66. Park, H. K., Xu, X., Grigoropoulos, C. P., Do, N., Klees, L., Leung, P. T., and Tam, A. C. Temporal profile of optical transmission probe for pulsed laser irradiation of amorphous silicon films, *Appl. Phys. Lett.* **61**, 749–751 (1992).
67. Saenger, K. L. An Interferometric calorimeter for thin-film thermal diffusivity measurements. *J. Appl. Phys.* **65**, 1447–1452 (1989).
68. Park, H. K., Grigoropoulos, C. P., Poon, C. C., and Tam, A. C. Optical probing of the temperature transients during pulsed-laser induced vaporization of liquids. *Appl. Phys. Lett.* **68**, 596–598 (1996).
69. Baeri, P., Campisano, S. U., Rimini, E., and Zhang, J. P. Time-resolved temperature measurement of pulsed laser irradiated germanium by thin film thermocouple. *Appl. Phys. Lett.* **45**, 398–400 (1984).
70. Kittl, J. A., Reitano, R., Aziz, M. J., Brunco, D. P., and Thompson, M. O. Time-resolved temperature measurements during rapid solidification of Si-As alloys induced by pulsed-laser melting. *J. Appl. Phys.* **73**, 3725–3733 (1993).
71. Brunco, D. P., Thompson, M. O., Otis, C. E., and Goodwin, P. M. Temperature measurements of polyimide during KrF excimer laser ablation. *J. Appl. Phys.* **72**, 4344–4350 (1992).
72. Baller, T. S., Kools, J. C. S., and Dieleman, J. Surface temperature measurements using pyrometry during excimer laser etching of silicon in a Cl_2 environment. *Appl. Surf. Sci.* **46**, 292–298 (1990).
73. Xu, X., Grigoropoulos, C. P., and Russo, R. E. Nanosecond time resolution thermal emission measurement during pulsed excimer laser interaction with materials. *Appl. Phys. A* **62**, 51–59 (1996).
74. Stritzker, B., Pospieszcyck, A., and Tagle, J. A. Measurement of lattice temperature of silicon during pulsed laser annealing. *Phys. Rev. Lett.* **47**, 356–358 (1981).
75. Pospieszczyk, A., Harith, M. A., and Stritzker, B. Pulsed laser annealing of GaAs and Si: combined reflectivity and time-of-flight measurements. *J. Appl. Phys.* **54**, 3176–3182 (1983).
76. Gibert, T., Dubreil, B., Barthe, M. F., and Debrun, J. L. Investigation of laser sputtering of iron at low fluence using resonance ionization mass spectrometry. *J. Appl. Phys.* **74**, 3506–3513 (1993).
77. Dubreuil, B., and Gibert, T. Soft laser sputtering of InP(100) surface. *J. Appl. Phys.* **76**, 7545–7551 (1994).
78. Hoheisel, W., Vollmer, M., and Träger, F. Desorption of metal atoms with laser light: Mechanistic studies. *Phys. Rev. B* **48**, 17463–17476 (1993).
79. Lee, I., Calcott, T. A., Arakawa, E. T. Desorption studies of metal atoms using laser-induced surface-plasmon excitation. *Phys. Rev. B* **47**, 6661–6666 (1993).
80. Kim, H.-S., and Helvajian, H. Laser-induced ion species ejection from thin silver films. *AIP Proceedings*, J. C. Miller and D. B. Geohegan (eds.), pp. 38–43, 1994.
81. Krajnovich, D. J. Laser sputtering of highly oriented pyrolytic graphite at 248 nm. *J. Chem. Phys.* **102**, 726–743 (1995).
82. Bennett, T. D., Krajnovich, D. J., and Grigoropoulos, C. P. Separating thermal, electronic, and topography effects in pulsed laser melting and sputtering of gold. *Phys. Rev. Lett.* **76**, 1659–1662 (1996).

83. Boccara, A. C., Fournier, D., and Badoz, J., Thermo-optical spectroscopy: Detection by the mirage effect. *Appl. Phys. Lett.* **36**, 130–132 (1980).
84. Sigrist, M. W. Laser generation of acoustic waves in liquids and gases. *J. Appl. Phys.* **60**(7), R83–R122 (1986).
85. Tam, A. C. Applications of photoacoustic sensing techniques. *Rev. Mod. Phys.* **58**, 381–431 (1986).
86. Park, H. K., Grigoropoulos, C. P., and Tam, A. C. Optical measurements of thermal diffusivity of a material. *Int. J. Thermophys.* **16**, 973–995 (1995).
87. Shannon, M. A., Rostami, A. A., and Russo, R. E. Photothermal deflection measurements for monitoring heat transfer during modulated laser heating of solids. *J. Appl. Phys.* **71**, 53–63 (1992).
88. Shannon, M. A., Rubinsky, B., and Russo, R. E. Detecting laser-induced phase change at the surface of solids via latent heat of melting with a photothermal defelction technique. *J. Appl. Phys.* **75**, 1473–1485 (1994).
89. Diaci, J. Response functions of the laser beam deflection probe for detection of spherical acoustic waves. *Rev. Sci. Instrum.* **63**(11), 5306–5310, (1992).
90. Sell, J. A., Heffelfinger, D. M., Ventzek, P. L. G., and Gilgenbach, R. M. Photoacoustic and photothermal beam deflection as a probe of laser ablation of materials. *J. Appl. Phys.* **69**, 1330–1335 (1991).
91. Enloe, C. L., Gilgenbach, R. M., and Meachum, J. S. A fast, sensitive laser deflection system suitable for transient plasma analysis. *Rev. Sci. Instrum.* **58**, 1597–1600 (1987).
92. Chen, G., and Yeung, E. S. A Spatial and temporal probe for laser-generated plumes based on density gradients. *Anal. Chem.* **60**, 864–868 (1988).

5. PLUME FORMATION AND CHARACTERIZATION IN LASER-SURFACE INTERACTIONS

ROGER KELLY and ANTONIO MIOTELLO
Dipartimento di Fisica
Università di Trento, Italy

ALDO MELE and ANNA GIARDINI GUIDONI
Dipartimento di Chimica
Università di Roma, La Sapienza, Italy

5.1 Introduction

5.1.1 General Comments

An important development in the history of sputtering (or ablation or desorption or emission) due to laser pulses concerns measurements of the motion of the emitted particles above the target surface where the particles are said to form a plume. Examples go back at least to 1963, when plume fluorescence was photographed [1]. With polymers, as well as $YBa_2Cu_3O_{7-\delta}$ (YBCO), the plume was imaged by firing a second laser pulse through it towards photographic film and forming an image by loss of light intensity [2, 3]. This loss of intensity is due to such effects as scattering or absorption, and is also termed "shadowgraphy" [4]. With metals, it has been more usual to image the plume by measuring either the ground-state neutral density [4, 5], the intensity of excited states by spectroscopy of particular transitions [6], or the overall light intensity (fluorescence) due to excited-state decay [7–11].

Plumes show well-defined gas-dynamic effects, the existence of which requires that a basic distinction be made in the matter of the mechanisms of laser sputtering. The primary action of the laser on the solid surface leads to the release of particles by what have been termed the *primary* mechanisms [12, 13]. These include electronic processes, normal vaporization, normal boiling, phase explosion (or explosive boiling or vapor explosion) [14–16], and subsurface heating [17, 18].

5.1.1.1 Electronic Processes.
Electronic processes can, in general, be expected with all insulators. Nevertheless we will discuss in Section 5.3.1.2 experimental evidence that with AlN electronic processes appear to initiate the laser interaction, but, provided that either incubation pulses are supplied or the fluence is sufficiently high, thermal processes such as normal vaporization then dominate.

5.1.1.2 Normal Vaporization.
We use the term "normal vaporization" to indicate vaporization in the sense of particle emission from the extreme outer surface, with a particle flux given by the Hertz-Knudsen equation [15, 19]:

$$flux = \alpha(p_{sv} - p_v)(2\pi m k_B T)^{-1/2} \quad \text{particles/cm}^2 \text{ s} \quad (5.1)$$

Here α is the condensation (or vaporization) coefficient [20], p_{sv} is the equilibrium (or saturated) vapor pressure, p_v is the partial pressure of *vapor* permanently present in the ambient, and m is the particle mass. Normal vaporization can be expected to occur transiently with both insulators and metals whenever the laser pulse is sufficiently energetic [15, 19]. A special case arises when an initially electronic process evolves to a thermal process (Section 5.3.1.2).

5.1.1.3 Normal Boiling and Phase Explosion.
There is some doubt as to whether normal boiling, characterized by *heterogeneous* nucleation of vapor bubbles, will occur on a short time scale [15, 21, 23]. In addition, it is unclear if the density of heterogeneous nuclei is sufficient, given the atomically small value of the laser penetration depth. However, as will be seen in Section 5.2.1.3, the subject is still in flux. Phase explosion, pioneered in work by Martynyuk [14, 21, 22], by Seydel *et al.* [15, 23, 24], and by Avedisian [16], resembles normal boiling except that it occurs near the *thermodynamic critical temperature* (T_{tc}) and involves *homogeneous* vapor nucleation. Somewhat similar to phase explosion, at least in the final results, is the subsurface heating model [17, 18, 25]. It will be shown in Section 5.2.2, however, that there is some doubt about the correctness of this model [19, 26].

Laser sputtering processes sometimes involve a sufficiently low yield that the emitted particles enter immediately into free flight and therefore retain information on the primary mechanism. There are many examples, however, where the yields are sufficiently high that the particles go through the typical sequence of forming a *Knudsen layer* (*KL*) [27, 28] and only then entering into free flight (Fig. 1(a)). This is apparently the situation with CH_3Br which was bombarded by 15 ns laser pulses in work by Bourdon *et al.* [29] (Fig. 2). If adsorbed on LiF, the CH_3Br showed a distribution similar to $\cos \theta$, as for free flight, but if in the form of an ice the distribution was similar to $\cos^5 \theta$, as for a *KL* [27]. Or else both a *KL* and an *unsteady adiabatic expansion* (*UAE*) are formed (Fig. 1(b)). One recognizes such a sequence for high coverages of CH_3Br on LiF, the forward peaking being similar to $\cos^{30}\theta$ [29], indicative of a combined *KL* and *UAE* [30]. Cowin *et al.* [31] studied the desorption of D_2 from W by 30 ns laser pulses and found that the emitted D_2 was significantly forward peaked, which is a typical indication of gas-dynamic effects, for the release of as little as ~ 1 monolayer per pulse. The work of Cowin *et al.* was subsequently analyzed in molecular dynamics simulations [32, 33], and the correctness of the idea that desorption of ~ 1 monolayer per pulse would trigger gas-dynamic effects was confirmed (Fig. 3).

INTRODUCTION

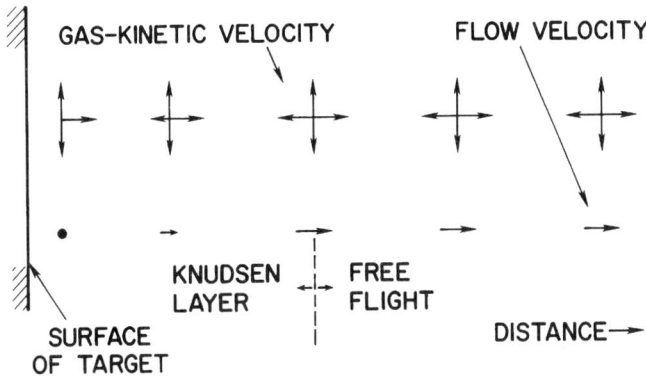

Fig. 1(a). Schematic representation of particles emitted from a laser-bombarded target with sufficient gas density that they form first a Knudsen layer (KL) and only then enter into free flight. A continuous, one-dimensional column of gas is assumed to be released from the target surface. The gas nearest the surface is characterized by a gas-kinetic velocity (v) which, normal to the surface, is given by $v_x > 0$ and by a flow velocity (u) which, again normal to the surface, is given by $u = 0$. At the KL boundary, on the other hand, the gas shows $-\infty < v_x < +\infty$ and, normal to the surface, $u = u_{KL}$. The KL is the near-surface region where the particles come to equilibrium with each other [27, 38, 80, 81].) A surprisingly small number of collisions (~ 3) is sufficient to establish the KL [32, 33]. The particles finally go into free flight and the velocities then persist unchanged. Due to Kelly et al. [12].

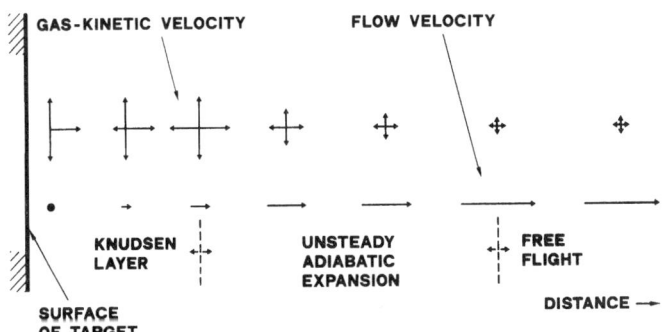

Fig. 1(b). Schematic representation of KL formation followed by an unsteady adiabatic expansion (UAE) and, finally, free flight as with Fig. 1(a). A continuous, one-dimensional column of gas is again assumed to arise. As always the gas nearest the surface is characterized by $v_x > 0$ and $u = 0$, while at the KL boundary the gas shows $-\infty < v_x < +\infty$ and $u = u_{KL}$. What is new here is that if the number of collisions is more than is needed for KL formation, a UAE occurs so that, with increasing distance, v decreases and at the same time u increases beyond u_{KL}. A passage to free flight finally occurs, and v and u then persist unchanged. Due to Kelly et al. [39].

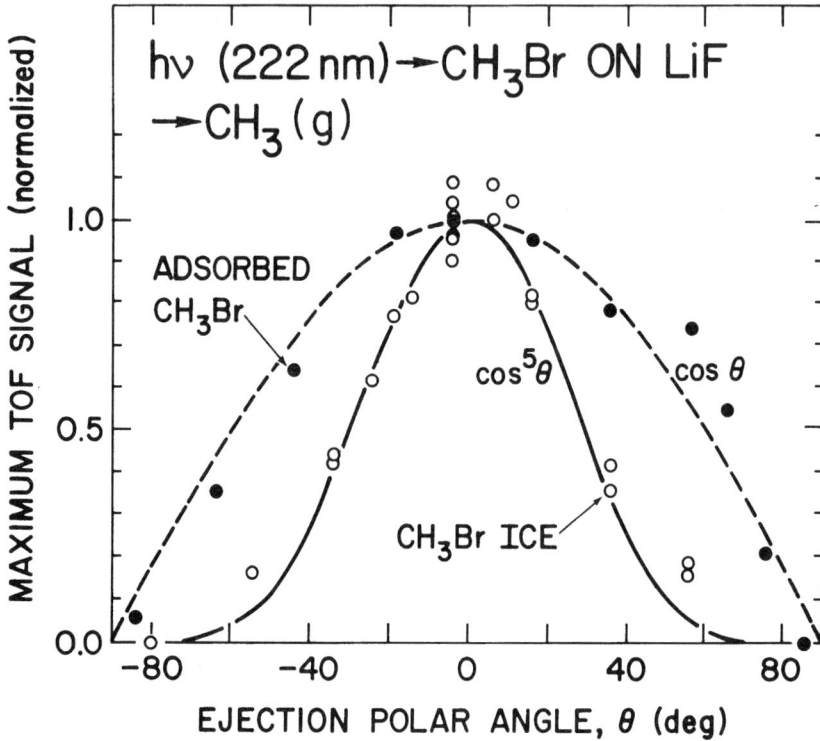

FIG. 2. Experimental time-of-flight (*TOF*) yields versus the ejection polar angle for laser-pulse desorption of CH_3 from CH_3Br adsorbed on LiF at 110 K and then maintained at 110 K. A KrCl excimer laser was used (222 nm, 15 ns, ≪1 J/cm^2, 85° incidence). The quantity removed per pulse in the case of the points labeled "adsorbed CH_3Br" can be inferred (because of the similarity to $\cos \theta$) to be substantially less than a monolayer (ML), while that removed for the points labeled "CH_3Br ice" was (because of the similarity to $\cos^5\theta$) probably of order 1 ML. The significant aspect is the pronounced forward peaking which sets in when a quantity similar to 1 ML per pulse is removed. Adapted from a figure due to Bourdon et al. [29].

We note that a *KL* is the region in which the vaporized particles, initially having only positive velocities normal to the surface (v_x), develop negative velocities, but, mainly in order that momentum be conserved, the particles also develop a positive center-of-mass (or flow) velocity (u_{KL}). The distribution function then changes from the usual Maxwell-Boltzmann form [27, 28] to a form in which the normal velocity v_x is replaced by $v_x - u_{KL}$. A *KL* could also be said to be the region where the particles come to equilibrium with each other.

Whenever a *KL* or both a *KL* and *UAE* occur, the particles will tend to lose memory of the primary process, and can then be said to show a *secondary*

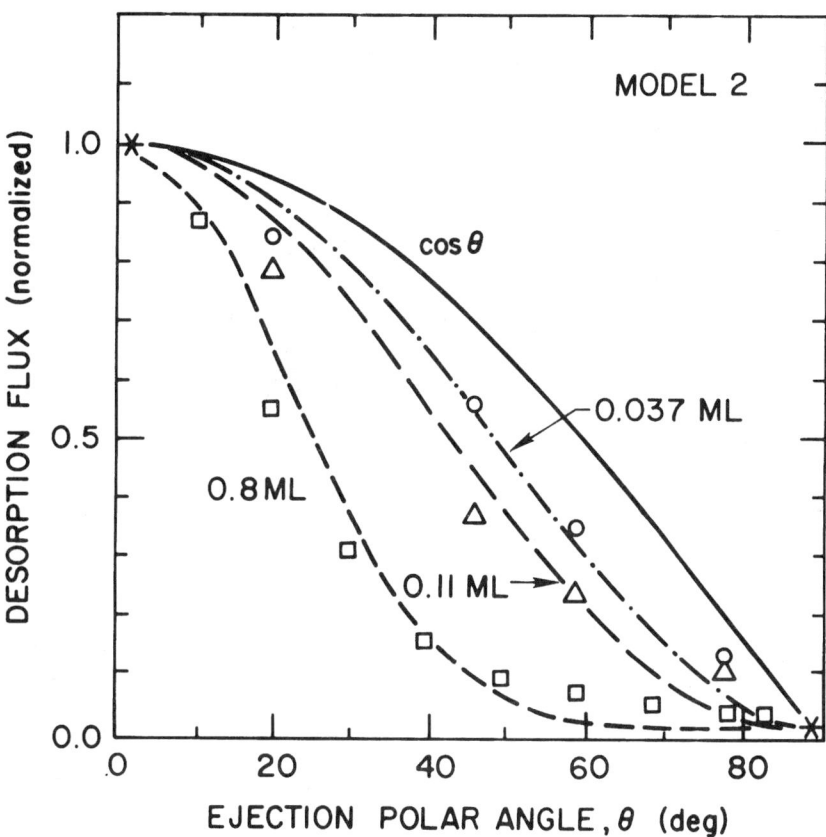

FIG. 3. Calculated angular distribution of the desorption flux obtained using "model 2," i.e. second-order desorption kinetics combined with an intrinsic angular effect [32]. The points represent the experimental results of Cowin et al. [31] for surface coverages in units of ML as indicated. A Nd-glass laser was used (1060 nm, 30 ns, 0.15–1.5 J/cm^2, 40° incidence). For reference, a $\cos\theta$ distribution is shown, as for collisionless vaporization. The experimental distributions are approximated as $\cos^n\theta$, with $n = 1.04 \pm 0.04$ for 0.037 ML, 1.28 ± 0.15 for 0.11 ML, and 5.41 ± 0.80 for 0.8 ML. Adapted from a figure due to NoorBatcha et al. [32].

mechanism. The latter is normally analyzed in terms of the flow equations of gas-dynamics [34] and at least three situations must be taken into account:

1. *Effusion-like process.* If the surface of the laser-bombarded target is sufficiently hot, *normal vaporization* will take place. (We note once again that it is unclear whether *normal boiling* is possible on a short time scale [15, 21] or when the heating depth is atomically small (Section 5.2.1.3).) The particles

that are released are said to originate from an effusion-like process [34]. They then form a KL, a terminating UAE, and finally undergo free flight (Fig. 1(b)).

2. *Normal outflow.* A second possibility is that a terminating UAE (without a KL) initiates by normal outflow from a reservoir (also termed *wall removal*), which is equivalent to the experimentally popular techniques of breaking a diaphragm [35] or setting off an explosion [36] in order to initiate an expansion.

3. *Phase explosion.* A third possibility is that phase explosion occurs. We are not aware of a precedent in the literature for a discussion of the gas dynamics appropriate to phase explosion. We will, however, argue in Section 5.3.1.3 that it is sometimes correct to think in terms of a variant of normal outflow that occurs from an *isolated spherical reservoir*.

5.1.2 Abbreviations and Terminology

We will use throughout the sections that follow certain abbreviations and terminology. Those that require comment include:

1. CF = contact front. This is a well-defined plane in an idealized one-dimensional UAE that separates an expanding gas from an ambient gas but, in real systems, will be more or less diffuse [37]. u_{CF} is the flow velocity of the CF.
2. *explosive boiling.* See *phase explosion.*
3. gamma = $\gamma = C_p/C_V$. The heat-capacity ratio.
4. $ICCD$ = intensified charge-coupled device.
5. KL = Knudsen layer. "KL" will be used also as a subscript, e.g., u_{KL} to mean the flow velocity at the KL "boundary." We use quotation marks on "boundary" because, in a rigorous sense, a KL is infinitely thick [38].
6. LOC = line of contact. This is a well-defined plane in an idealized one-dimensional UAE where there is an abrupt change of slope.
7. *normal boiling* is boiling that occurs at or somewhat above T_b and involves heterogeneous nucleation of vapor bubbles. The nucleation is commonly thought to occur beneath the surface [15], but there is some indication that the surface itself may be a very efficient source of nuclei [39–41]. It is to be contrasted with phase explosion, where the nucleation is homogeneous.
8. *normal outflow.* See *outflow.*
9. *normal vaporization* is thermal particle emission (atoms or molecules) from the extreme outer surface of a condensed phase. Vapor nucleation does not enter, although formation of surface defects such as ledges and kinks do enter [42]. We do not distinguish between normal vaporization, vaporization, evaporation, or sublimation, although we do suggest that an

unambiguous terminology (namely, normal vaporization) be used. There is currently a strong abuse of terminology, for example, with vaporization and boiling being confused, or else with vaporization and evaporation being distinguished.

10. *outflow*. *Normal* outflow has been defined as an expansion that originates from the removal of a wall, often (but not necessarily) a planar wall [34]. Outflow from an *isolated spherical reservoir* will be discussed in Section 5.3.1.3.
11. PBSCCO = $Pb_{0.2}Bi_{1.8}Sr_2Ca_1Cu_2O_x$. This substance becomes a superconductor at low temperatures.
12. *phase explosion* or *explosive boiling* or *vapor explosion* is very rapid boiling that occurs at $\sim 0.90 T_{tc}$ and involves homogeneous vapor nucleation beneath the surface. It is to be contrasted with normal boiling, where the nucleation is heterogeneous.
13. SW = shock wave. u_{SW} is the flow velocity of the *SW*.
14. T_b = the boiling temperature appropriate to p_b. p_b is the "boiling" gas pressure, i.e., the partial pressure of *nonvapor* permanently present in the ambient and which is often (but not necessarily) similar to 0.1 MPa. We note that T_b *is not* a rigorously fixed quantity because it depends not only on p_b but, for short enough times, also on the time scale [15, 16].
15. T_m = the melting temperature. Unlike T_b, the quantity T_m is more or less fixed unless the ambient pressure departs greatly from 0.1 MPa.
16. T_s = the surface temperature. For example, if a target undergoes normal boiling then T_s is given by $T_s \approx T_b$. There is currently a tendency (completely wrong!) to assume that T_s is fixed also for normal vaporization.
17. T_{tc} = the thermodynamic critical temperature. This is identical with what is commonly termed simply the "critical temperature" but we add "thermodynamic" to avoid confusion with the critical temperature of superconductivity or the various critical temperatures that appear on phase diagrams.
18. T_v = the "vaporization temperature". This is a misleading concept. We note that for normal vaporization a quantity T_v does not exist in any sense, while for normal boiling the terminology T_b is strongly preferable. The use of T_v should, therefore, be avoided.
17. UAE = unsteady adiabatic expansion.
18. *vapor explosion*. See *phase explosion*.
19. x = the spatial coordinate normal to the target. y and z are the lateral coordinates, i.e., the spatial coordinates parallel to the target. Note that an alternative notation is also common, in which z corresponds to the present x. v_x, v_y, and v_z are the corresponding gas kinetic velocities, and u_x, u_y, and u_z are the corresponding flow (or center-of-mass) velocities.
20. YBCO = $YBa_2Cu_3O_{7-\delta}$. Like PBSCCO, this substance becomes a superconductor at low temperatures.

5.2 Emission Mechanisms and Plume Formation

5.2.1 The Primary Mechanisms

5.2.1.1 Electronic Sputtering. Electronic sputtering is not a unique process but rather a group of processes having the common feature of involving some form of excitation or ionization [43]. It arises, for example, due to events as varied as ion explosion [44], the "hole-pair" mechanism [45], defect formation [46, 47], or surface plasmon excitation [48, 49]. It will normally fail to reveal itself in an explicit way in scanning electron microscopy (*SEM*) but will, by contrast, often be *suggested* by temperature information, especially by the inequality, $T \gg T_{tc}$. We write "suggested" because evidence based on temperature is not as good as it might seem, given the fact that the inequality, $T \gg T_{tc}$ could also indicate laser-plume interaction or phase explosion [50]. It is normally confined to dielectrics and wide bandgap semiconductors, except when it occurs with metals due to surface plasmons.

Nevertheless there is experimental evidence that with many substances (e.g., AlN), electronic processes serve mainly to initiate the laser interaction. Provided either incubation pulses are supplied or the fluence is sufficiently high, thermal-spike processes such as normal vaporization or phase explosion then dominate (Sections 5.2.1.2 and 5.2.1.4).

Let us now consider the specific example of Al_2O_3 [51]. For high enough laser-pulse energy fluences, dense electron excitation can be expected, perhaps as high as $N_e \approx 10^{22}$ cm^{-3}, where N_e is the excited electron number density. These electrons will increase the total energy of each atom by an amount similar to $N_e E_{gap}/N$, where E_{gap} is the energy gap and N is the number density of target atoms [52]. For a system like Al_2O_3, where E_{gap} is ~9 eV, N is 4.7×10^{22} Al/cm^3, and the depth of the potential well is 5.7 ± 0.1 eV, it is clear that a value of N_e such as 10^{22} cm^{-3} would raise the energy of each atom by about 2 eV and thus increase the vapor pressure by orders of magnitude or even render the lattice unbound.

We regard this model as related to the *rapid energy deposition* model [53, 54], which was devised to explain the unusual response of solids to incident particles ranging from laser pulses, to fission fragments, to electron pulses, to small accelerated dust particles. A possible description is that a rapid (i.e., nonadiabatic) transition takes individual ions directly into anti-bonding states. The result is that the system makes a transition from a tightly bound solid to a densely packed, repulsive gas, and that particles are expelled energetically.

Rapid energy deposition is not the only point of view, however, especially for low laser-pulse energies. As emphasized in recent work [46, 47], defects can form at and beneath the target surface, including self-trapped excitons, the decay products of excitons, as well as damage revealed by low-energy electron diffraction (*LEED*). These defects, if they are formed at the surface or migrate to

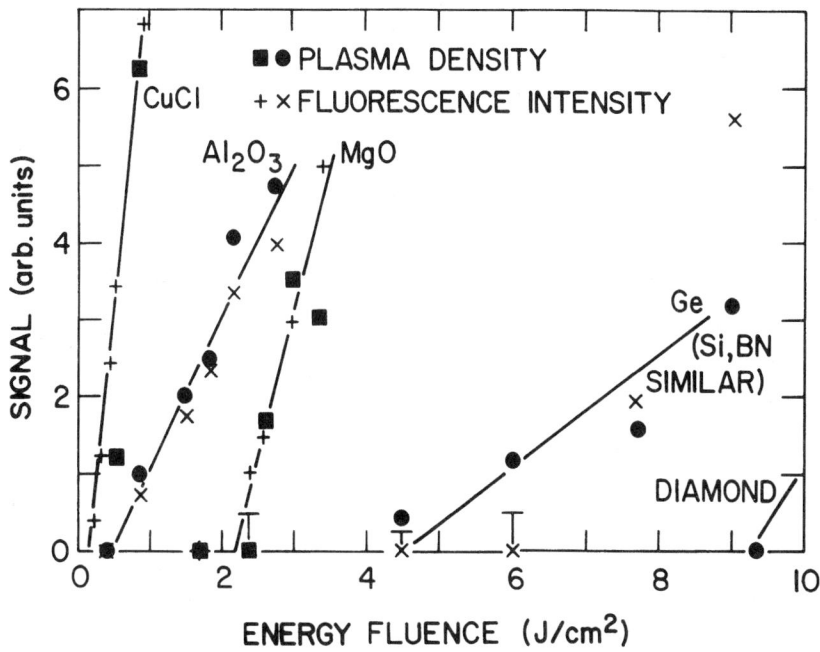

FIG. 4. Comparison of the experimental threshold fluences for particle emission from a number of solids that were bombarded *in vacuum* with a KrF excimer laser (248 nm, 12 ns, normal incidence). In the case of plasma density, the threshold is for the measured free-electron density accompanying the particles; in the case of fluorescence, the threshold is for the intense optical emission. Due to Dreyfus et al. [51].

the surface, lead to the energetic expulsion of individual atoms. Sometimes the expulsion is directional [55].

The details of just what constitutes an electronic process are thus not straightforward nor is there a clear "finger-print." Thus, although the particles that are emitted will be characterized by a number of unusual features, there is a remarkable level of ambiguity:

1. Existence of a threshold energy fluence for particle emission (Fig. 4), a detail which, however, may denote almost any other mechanism [47].
2. Kinetic energies of order of a few eV. As noted above, high energies can arise also from laser-plume interaction and from phase explosion [50].
3. Low internal energies (rotation, vibration) owing to a lack of equilibration during the energetic transition referred to in point number 2 [51].
4. High degree of directionality due to the crystallographic orientation of the relevant defects [55], a comportment which is, however, easily confused with forward peaking due to gas dynamics (Figs. 2 and 3).

5. An apparently excessive (i.e., apparently non-thermodynamic) yield of molecules and ions, AlO and Al$^+$ in the case of Al$_2$O$_3$. Molecules can, however, also form directly by vaporization and ions by laser-plume interaction. For example, the vapor composition of Al$_2$O$_3$ includes AlO, Al$_2$O, AlO$_2$, and Al$_2$O$_2$ [56, 57].

5.2.1.2 Normal Vaporization.
Normal vaporization, which belongs to a group of processes having in common a thermal-spike rather than electronic origin, can operate at essentially any laser fluence and pulse length. (We use the term "thermal spike", rather than simply "thermal", because the high temperature is *transient* [39].) The target emits atoms and molecules from the extreme outer surface, and nucleation of vapor bubbles at [40, 41] or beneath [15] the surface does not enter. The flux (particles/cm^{2}s) is governed by the Hertz-Knudsen equation (Eq. (5.1)), which if multiplied by m/ρ (equivalent to λ^3, ρ being the target mass density and λ being the mean atomic spacing of the target) gives the velocity of surface recession in a one-dimensional situation:

$$(\partial x/\partial t)|_{x=0} = \alpha(p_{sv} - p_v)(2\pi m k_B T)^{-1/2}\lambda^3 \quad \text{cm/s} \tag{5.2}$$

$$= \alpha(p^{sv}_{atm} - p^v_{atm})\left(\frac{1000}{T} \times \frac{100}{M}\right)^{1/2} \times 5.28 \times 10^7 \quad \text{monolayer/s} \tag{5.3}$$

$$\approx \alpha\left[p_b \exp\left\{\frac{\Delta H_v m}{k_B}\left(\frac{1}{T_b} - \frac{1}{T}\right)\right\} - p_v\right](2\pi m k_B T)^{-1/2}\lambda^3 \tag{5.4}$$

Here p^{sv}_{atm} is p_{sv} in units of atmosphere, p^v_{atm} is p_v in units of atmosphere, p_b is the "boiling" gas pressure (i.e., the partial pressure of *nonvapor* permanently present in the ambient), ΔH_v is the heat of vaporization (units of J/g), and it is assumed that there is no recondensation [15]. Equation (5.4) expresses p_{sv} in terms of the Clausius-Clapeyron equation and is shown as being approximate because it assumes a precisely exponential relation between p_{sv} and $1/T$. (A better approximation is given by Fucke and Seydel [15].) Because the vapor pressure is nonzero at all temperatures exceeding 0 K, it follows that for normal vaporization the surface temperature (T_s) *is not* fixed. Claims to the contrary, that is, that a "vaporization temperature" (T_v) exists, are therefore wrong, except in a limited sense. For a given time scale and a given sensitivity of measurement, vaporization *will not be sensed* until a particular temperature is reached. However, additional heat input will always cause T_s to exceed this temperature and a true T_v will therefore not exist.

We note that the appearance of T_b in Eq. (5.4) does not mean that there has been confusion between vaporization and boiling. It is simply a mathematical device to express $p_{sv}(T)$ as a function of $p_{sv}(T_b)$.

5.2.1.3 Normal Boiling.
A second type of thermal-spike process requires that the pulse length be sufficiently long for *heterogeneous* nucleation of vapor

bubbles to occur. Provided p_{sv} exceeds p_b, the target undergoes what we will term *normal boiling* (a term occasionally used also elsewhere [21, 24]) in a zone extending from the surface to a depth related either to the absorption length ($1/\mu$, μ being the absorption coefficient [58]) or to the thermal diffusion length ($[\kappa t]^{1/2}$, κ being the thermal diffusivity). In this case, T_s is fixed at (or somewhat above [15, 16]) T_b, and the temperature gradient at and beneath the surface is $\partial T/\partial x \approx 0$. We note, however, that even if the necessary heterogeneous nucleation sites exist, their density ($\sim 10^6$ kg^{-1} [15]) is far too low for boiling to be possible.

Unfortunately the argument is incomplete. Nuclei are also believed to form at the target *surface*, beginning at 13 ns in one particular case [40, 41], and could in principle show a much higher density than 10^6 kg^{-1}. This possibility must be studied further, and we here note only that a negative answer, implying that normal boiling is in general impossible in laser-surface interaction, would constitute an enormous simplification of the problem of mechanisms.

5.2.1.4 Phase Explosion. A third type of thermal-spike process requires that the laser fluence be sufficiently high and the pulse length sufficiently short that the target reaches $\sim 0.90 T_{tc}$ at and beneath the surface. *Homogeneous* nucleation of vapor bubbles therefore occurs, and the target makes a rapid transition from superheated liquid to a mixture of vapor and equilibrium liquid droplets (Fig. 5). The vapor and droplets are, however, well separated in time, vapor being detectable for short times (<500 ns) and droplets for long times (>25 μs) [59, 60]. As with normal boiling, one can expect that at and beneath the surface the condition $\partial T/\partial x \approx 0$ will apply. Much of the historical work was done by Martynyuk [14, 21, 22], by Seydel *et al.* [15, 23, 24], and by Avedisian [16], and the terms *phase explosion, explosive boiling,* and *vapor explosion* were all introduced. Because the rate of homogeneous vapor nucleation rises catastrophically near T_{tc}, the necessity that nuclei form does not constitute a kinetic obstacle. Thus, as shown by Martynyuk [61], the density of critical nuclei passes from $\sim 10^{-25}$ to $\sim 10^{25}$ cm^{-3} s^{-1} in the temperature range 0.88 to 0.92 T/T_{tc}.

It will be shown in Section 5.3.1.3 that one can expect phase explosion (or an equivalent process leading to high particle energies), together with a lesser role for normal vaporization, to be important with AlN for short, energetic laser pulses.

5.2.1.5 Subsurface Heating Model. Whether or not a fourth thermal-spike process, the *subsurface heating model*, exists is less clear. There have been repeated claims for a mechanism based on heating the target surface to a sufficiently high temperature that rapid vaporization occurs, the surface is subject to cooling, and the subsurface region therefore retains a much higher temperature with maximum $\hat{T}$ (Fig. 6). We will analyze this process in Section 5.2.2, where it will be shown that the analysis given in [17, 18, 25, 62, 63] is probably wrong.

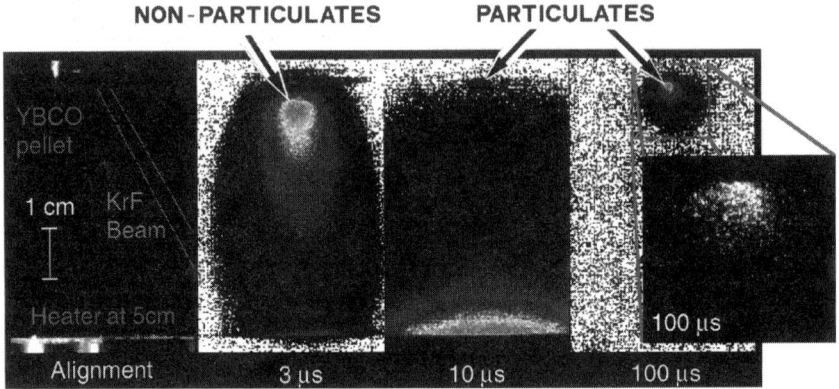

FIG. 5. *ICCD* photographs showing the plasma plume which arises when YBCO is bombarded *in vacuum* ($\sim 10^{-4}$ Pa) with a KrF excimer laser (248 nm, 30 ns, 6.6 J/cm^2, 30° incidence). Each photograph involved a single pulse. The first image shows the orientation of the target and *cold* heater, the target being at the top. With a time delay of 3 μs, one sees only the nonparticulate part of the plume (marked with an arrow), the expansion front of which moved at 2.0×10^6 cm/s. With a time delay of 10 μs, one recognizes that the nonparticulates have reached (and rebounded from) the cold heater. But glowing particulates, marked with an arrow, are also evident near the target, and move about 100 times slower than the nonparticulates (0.03×10^6 cm/s). The glowing particulates (again marked with an arrow) are well resolved at 100 μs. Due to Geohegan [59].

5.2.1.6 Other Mechanisms.

Exfoliational sputtering, as when flakes detach from a target owing to repeated thermal shocks caused by individual laser pulses, shows an obvious and characteristic topography easily seen by *SEM* [13, 64, 65]. It can be expected to occur whenever the system has a high linear thermal expansion ($\Delta L/L_o$), where L_o is the thickness that is heated and ΔL is the change in thickness, a high Young's modulus (E), a high T_m, and the laser-induced temperature excursions approach but do not exceed T_m. The thermal shocks would occur repeatedly and, because they were not relieved by melting, would finally lead to cracking.

To describe exfoliational sputtering it is convenient to identify the thermal shock with the thermal stress,

$$stress = E\Delta L/L_o \quad (5.5)$$

One finds, for example, that exfoliational sputtering will occur with high melting metals such as W (therefore, not Au) and with oxides having large $E\Delta L/L_o$ such as Al_2O_3 (therefore, not SiO_2).

A type of exfoliation is also well known in heavy-ion bombardments [66]. The mechanism is, however, unrelated to the thermal effects described by Eq. (5.5), being due instead to stresses caused by precipitation of the incident ions.

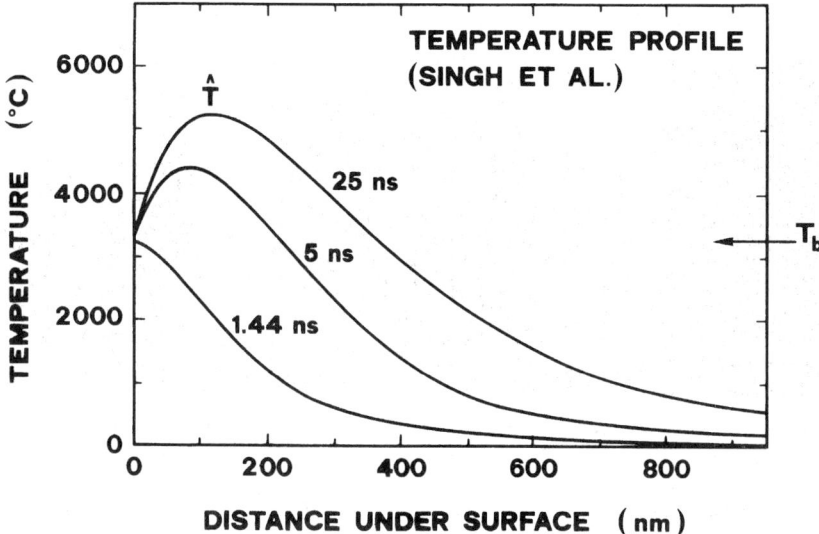

FIG. 6. Calculated near-surface temperature profiles appropriate to the model of Singh *et al.* [25] for three different time delays: 1.44 ns ("evaporation onset," a concept that is understandable only if "evaporation" means "normal boiling"), 5 ns, and 25 ns as developed in Si *after* irradiation with a 25 ns laser pulse having a fluence of 10 J/cm^2 and incident *probably* normally. An absorption coefficient of 1×10^5 cm^{-1} was assumed. The condition $T = T_b$ was imposed at the surface as for normal boiling with $p_b = 0.1$ MPa. Furthermore the velocity of surface recession was identified with Eq. (5.7) (which is incorrect!) instead of Eqs. (5.1)–(5.4). The values of $\hat{T} - T_s$, which are a measure of the subsurface temperature maxima, will be noted to be enormous and it was concluded that explosive expulsion of material (both vapor and liquid droplets approximately as in Fig. 5) would occur. Adapted from a figure due to Singh *et al.* [25].

Hydrodynamic sputtering refers to the process in which droplets of material are formed and expelled from a target as a consequence of the transient melting [13, 64, 67]. Such a process is distinct from phase explosion (which also leads to droplet expulsion, Section 5.2.1.4) and, in addition, has no analog in ion-surface interactions. Evidence for hydrodynamic effects is thus far limited mainly to metals. It is true that droplets form also on polymer surfaces [68, 69], but it is not known whether they are expelled from the target in the same way as happens with metals.

To describe hydrodynamic sputtering, we recognize that asperities (i.e., bumps) tend to develop on a laser-bombarded surface, especially if melting occurs [67, 70]. The asperities will be accelerated away from the melted substrate during each pulse, owing to the combination of the volume change on melting followed by thermal expansion of the liquid. The asperities will finally separate from the surface as droplets when the product *mass × acceleration*

($\propto r^3$) exceeds the restraining force,

$$f = -\partial(4\pi r^2\sigma)/\partial r = -8\pi r\sigma \tag{5.6}$$

where r is the droplet radius and σ is the liquid surface energy (units of J/cm^2). The droplets are found experimentally to have radii lying in the interval 0.5 to 5 µm.

We would finally note that thermal effects resembling phase explosion have been advocated for ion-surface interactions. These include *thermal-spike sputtering*, as well as the conceptually similar *thermal-spike mixing* [39]. With sputtering, the claims ceased about 15 years ago owing to a number of studies that either failed to find a predicted *temperature dependence* of the yield [71, 72] or else showed explicitly that a supposedly observed temperature dependence was due to normal vaporization. For example, a recent reevaluation [39] (Fig. 3) suggests that the increasing sputtering yield of SiO$_2$ in the interval 100–900°C can be explained in terms of *nonstoichiometric* vaporization: Si(ℓ) + SiO$_2$(ℓ) = 2SiO(g), where ℓ means liquid and g means gas. The alternative, thermal-spike sputtering, can therefore be dismissed. With mixing, on the other hand, the claims persist to this day [73, 74]. Not all authors (e.g., [39, 75]) accept the importance of thermal-spike mixing, however, and it is interesting that one of the arguments against such mixing, namely, that the temperature dependence is too marked [76], is the precise opposite to that sought in order to support thermal-spike sputtering. In other words, the *absence* of temperature dependence was claimed to disprove thermal-spike sputtering, whereas the *presence* of temperature dependence was claimed to disprove the corresponding mixing!

5.2.2 Critique of the Subsurface Heating Model

Starting with work by Dabby and Paek [17] and Gagliano and Paek [18] there have been repeated claims for a laser sputtering mechanism based on heating the target surface either to T_m [62], or to T_b (assigned the value appropriate to $p_b = 0.1$ MPa even for experiments carried out in vacuum!) [25, 77], or to an ill-defined T_v [17, 18, 63]. Rapid vaporization then occurs, the surface is subject to cooling with its temperature fixed at T_s (i.e., one of T_m, T_b, or T_v), and the subsurface region therefore retains a much higher temperature, with maximum $\hat{T}$. As a result, the pressure is much greater beneath the surface, and a type of explosion, leading to similar results as with phase explosion, takes place. For example, both vapor and liquid droplets are claimed to be expelled (Fig. 5).

The problem is discussed elsewhere [19, 26, 39], where it is shown that the argument is faulty. First, a surface boundary condition $T_s = T_m$ [62] makes no sense. Second, a boundary condition $T_s = T_b$ implies that normal boiling is occurring, which in turn requires that the time scale be long enough for heterogeneous nucleation. Supposing that this condition is met (which is not at all self-evident for the shortest pulse lengths), the temperature gradient at and beneath the surface will be given by $\partial T/\partial x \approx 0$, whereas it was unrestricted in [17, 18, 25, 62, 63, 77]. Third, a boundary condition $T_s = T_v$, provided T_v refers to *normal vaporization*, is incomprehensible because for normal vaporization T_s is in no sense fixed. If T_v means T_b problems arise with $\partial T/\partial x$ as just discussed. Finally, if T_v means T_{tc} (or a temperature near T_{tc}) phase explosion occurs. Subsurface heating is then impossible, if only because the time scale of phase explosion is so short [61]. One can, however, also expect the condition $\partial T/\partial x \approx 0$ to hold as in Fig. 7(b), where T_s was similar to T_{tc}.

A further problem with the subsurface heating model is that surface recession was defined in terms of heat conduction:

$$\rho \Delta H_v \frac{\partial x}{\partial t}\bigg|_{x=0} = K \frac{\partial T}{\partial x}\bigg|_{x=0} \qquad (5.7)$$

where K is the thermal conductivity (units of J/s·cm·K). This is wrong because, for normal vaporization, surface recession must always be defined in terms of the Hertz-Knudsen equation (Eqs. (5.1)–(5.4)). A further problem with Eq. (5.7) is that it is evidently mathematically resolvable only if T_s is fixed [78, 79]. But T_s is *not* fixed for normal vaporization, which is the only thermal-spike process that is relevant to subsurface heating.

We finally note that the equation of heat diffusion was recently solved for laser-pulse impact on a target surface having thermal and optical parameters appropriate to Al [19, 26]. The erosion of the target was assumed to occur by normal vaporization, the recession velocity being taken as in Eqs. (5.2)–(5.4) with $p_v = 0$, and incorrect boundary conditions of the type $T_s = T_b$ were avoided. The final result for a 30 ns pulse having an energy 4.0 J/cm^2 was that there was no subsurface maximum at all (Fig. 7(a)), while for a 6.0 ns pulse having an energy 4.0 J/cm^2, $\hat{T} - T_s$ had the negligible value $\sim$2 K (Fig. 7(b)). For a 3.0 ns pulse, the value of $\hat{T} - T_s$ was $\sim$20 K, but the result was unacceptable because of the inequality $T > T_{tc}$.

5.2.3 The Secondary Mechanisms

5.2.3.1 The Knudsen-layer *(KL)* Phase.
We suppose that photons strike a solid surface and that particles are, therefore, emitted in a pulse. When the density of emitted particles is small enough, the particles show a gas-kinetic velocity (v) normal to the target surface (v_x), which is given by $v_x > 0$, and escape without interaction. In effect, they go into *free flight* and are described by whatever velocity distribution is appropriate to the *primary* sputtering mechanism. For example, a Maxwellian would be appropriate for normal vaporization or normal boiling but not otherwise.

When the quantity of particles per unit time is larger, of order 0.5 monolayer in nanoseconds [32, 33], the particles collide sufficiently that they come to

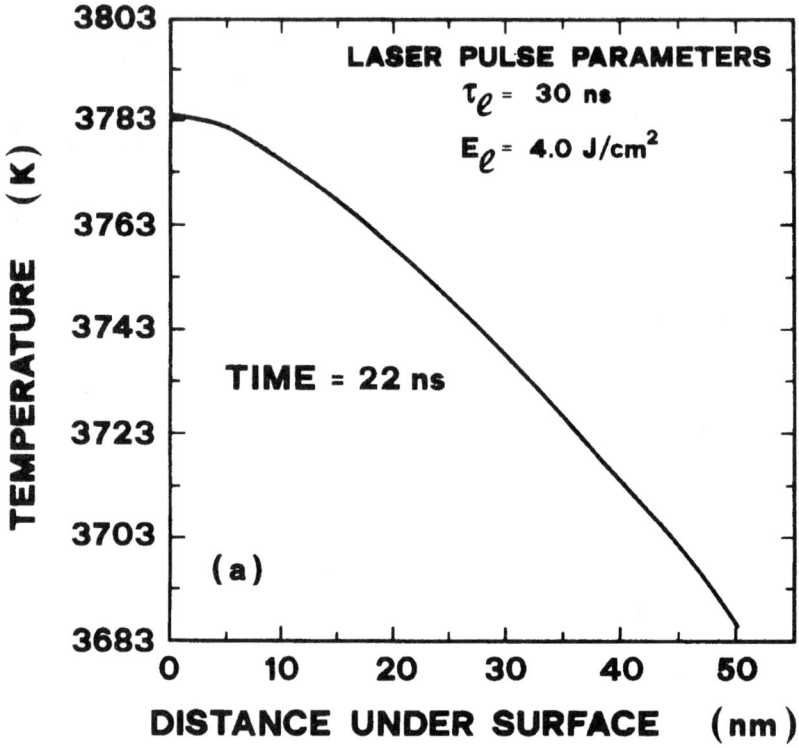

FIG. 7(a). 30 ns laser-pulse length (τ_ℓ). The pulse shape is irrelevant because the calculation was one-dimensional.

Calculated near-surface temperature profiles appropriate to the model of Kelly and Miotello [19] for three different laser-pulse lengths, the times being chosen which in each case maximized $\hat{T} - T_s$. The target parameters were taken as those for Al and the fluence was 4.0 J/cm^2. The temperatures were derived by numerical solution of the heat-diffusion equation [19, Eq. (7)]. The recession velocity $(\partial x/\partial t)|_{x=0}$ was evaluated with Eq. (5.2) and not (as in [25]) with Eq. (5.7). The source term was evaluated with [19, Eq. (6a)]. A surface boundary condition of the type $T = T_b$ (as in [25]) was avoided. The spatial step in the calculation was 2.0 nm and the time step was 1×10^{-13} s. Due to Kelly and Miotello [19].

equilibrium [27, 38, 80, 81], the region in which this happens being termed the KL (Fig. 1(a)). They then have all v_x, i.e., $-\infty < v_x < +\infty$, show a flow velocity (u) normal to the target surface (u_{KL}) which is given by

$$u_{KL} = a_{KL} = (\gamma k_B T_{KL}/m)^{1/2} \qquad (5.8)$$

and show forward peaking with an angular distribution of the flux similar to $\cos^4\theta$ (Fig. 2). Here a_{KL} is the sound speed at the KL boundary. As is apparent from Eq. (5.8), the KL has an outer boundary characterized by a Mach number

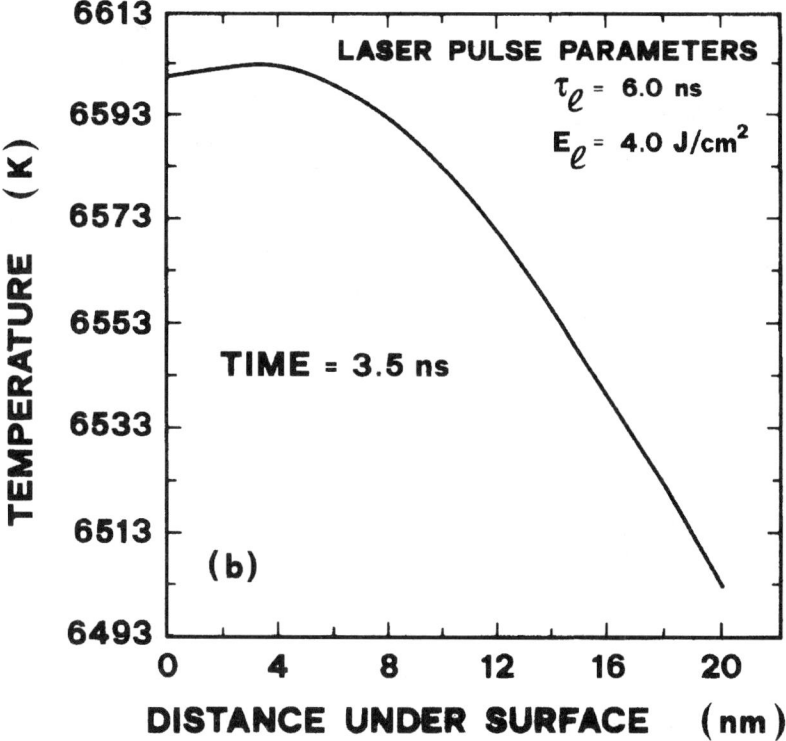

FIG. 7(b). 6.0 ns laser-pulse length. A similar calculation for a 3.0 ns laser-pulse length is given in [19].

$M = u_{KL}/a_{KL} = 1$. Rather independently of the primary velocity distribution, a Maxwellian form with v_x replaced by $v_x - u_{KL}$ can be expected whenever a KL forms. This is also termed a "shifted Maxwellian," the corresponding velocity distribution having the form:

$$f_{KL}^{\pm}(v_x, v_y, v_z) = \left(\frac{m}{2\pi k_B T_{KL}}\right)^{3/2} \frac{E_I^{(j/2-1)}}{\Gamma(j/2)(k_B T_{KL})^{j/2}}$$
$$\times \exp\left[-\frac{2E_I + m\{(v_x - u_{KL})^2 + v_y^2 + v_z^2\}}{2k_B T_{KL}}\right] dv_x\, dv_y\, dv_z \quad (5.9)$$

Here E_I is the total internal energy and j is the number of internal degrees of freedom of the emitted particles accessible at T_{KL}. An important consequence of Eq. (5.9) is that, under conditions of collision, a Maxwellian form (with or without shifting) does not necessarily denote a thermal primary mechanism. This distinction is not always respected.

Without going into detail, we note that KL theory defines the ratios T_{KL}/T_s and ρ_{KL}/ρ_s for each value of the heat-capacity ratio, γ. It also shows that a fraction, F^-, of the emitted particles is scattered back to the target surface and is conventionally assumed to recondense. Under conditions of KL formation, a time-of-flight spectrum with maximum velocity $\hat{v}$ might be expected to yield a temperature

$$k_B T_s = (m\hat{v}^2/2)(T_s/T_{KL}) \qquad (5.10)$$

but a more careful treatment in which the detailed form of the spectrum is considered [27, 81] yields

$$k_B T_s = (m\hat{v}^2/2)(T_s/\eta T_{KL}) \qquad (5.11)$$

where η is a parameter depending on γ and θ, and the product $\eta T_{KL}/T_s$ has been tabulated [13, 30, 81].

5.2.3.2 Effusion-like Release. When the quantity of particles released by a laser pulse exceeds that necessary for a KL alone, a combined KL and terminating UAE are formed (Fig. 1(b)). When, in addition, the target surface is sufficiently hot that the release occurs mainly by *normal vaporization*, the particles are said to originate from an *effusion-like process* [34]. (We neglect normal *boiling* in this context on the grounds that it will be impossible on a short time scale [15, 21] unless nuclei can form at the target surface [40, 41].) This UAE will tend to be one-dimensional provided there is sufficient forward peaking. The details will, however, differ according to the surface boundary condition that operates when the release process terminates. If particles scattered back to the target surface *reflect*, we have the situation of Fig. 8; whereas if they *recondense*, we have the situation in [39, Fig. 18(a)]. It should be noted that in both cases the particle density is initially monotonically decreasing but, when the release terminates, it takes on a form having a maximum that moves away from the surface.

The velocity of the *expansion front* $\hat{u}$, applicable when there is no ambient gas, represents the maximum possible flow velocity of particles emitted in an effusion-like process (Fig. 8). For motion normal to the target, the expansion front moves with a velocity given by the sum of that for a KL (a_{KL}) and that for outflow:

$$\hat{u}_x = a_{KL} + \frac{2}{\gamma - 1} a_{KL} = \frac{\gamma + 1}{\gamma - 1} a_{KL} \qquad (5.12)$$

a result valid even for $\gamma = 1$ [82]. Equation (5.12) corresponds to energies given by

$$\tfrac{1}{2} m \hat{u}_x^2 = \frac{m(\gamma+1)^2}{2(\gamma-1)^2} a_{KL}^2 = \frac{\gamma(\gamma+1)^2}{3(\gamma-1)^2} E_{KL} \qquad (5.13)$$

where $E_{KL} = 3k_B T_{KL}/2$ is the mean kinetic energy per particle at the KL boundary.

5.2.3.3 Normal Outflow. A terminating UAE can also initiate by *normal outflow* from a reservoir (also termed *wall removal*), which is equivalent to the experimentally popular techniques of breaking a diaphragm [35] or setting off an explosion [36] in order to initiate an expansion. An important detail is that, as seen in Fig. 9, a UAE due to normal outflow *does not* show a density maximum like that of Fig. 8.

The velocity of the *expansion front* $\hat{u}$, applicable when there is no ambient gas, again represents the maximum possible flow velocity of the emitted particles. For both normal and lateral motion, thence for outflow with any

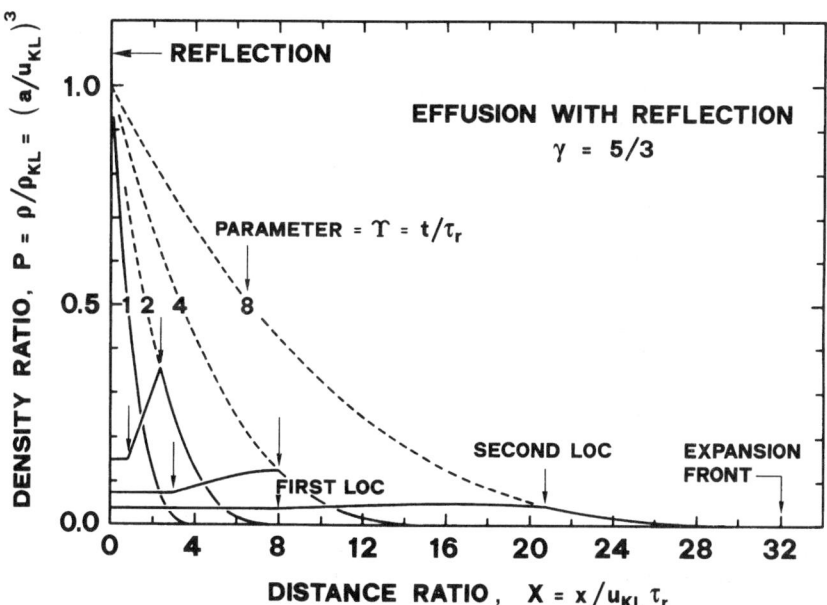

FIG. 8. Calculated values of the density ratio, $P = \rho/\rho_{KL} = (a/u_{KL})^3$, versus the distance ratio, $X = x/u_{KL}\tau_r$, for *effusion-like* release of atoms ($\gamma = 5/3$) when particles scattered back to the target surface at $x = 0$ *reflect*. The KL, characterized for $0 \leq t \leq \tau_r$ by density ρ_{KL} and flow velocity u_{KL}, is assumed to have zero thickness. The problem is equivalent to that of a semi-infinite reservoir of gas that is able to effuse into vacuum for $0 \leq t \leq \tau_r$, τ_r being the particle release time, the effusion then terminating abruptly. The indicated expansion front is that of the curve for time ratio $Y = t/\tau_r = 8$. The two lines of contact (*LOC*) are marked explicitly for $Y = 8$ and implicitly for $Y = 2$ and 4. The method of calculating Fig. 8 is discussed in [34, Sect. 1.3]. The figure itself is due to Kelly and Miotello [39].

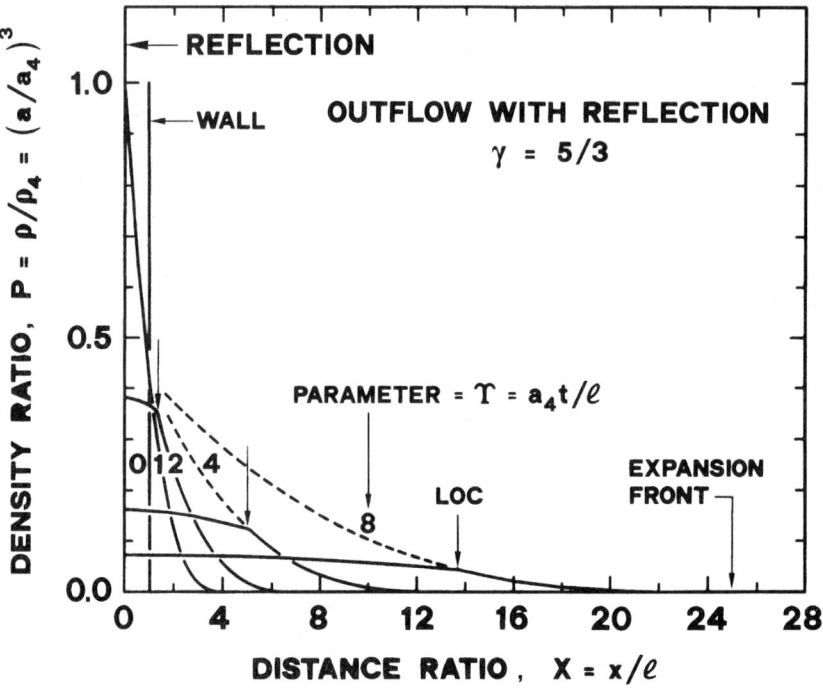

FIG. 9. Calculated values of the density ratio, $P = \rho/\rho_4 = (a/a_4)^3$, versus the distance ratio, $X = x/\ell$, for the release of atoms ($\gamma = 5/3$) due to *outflow* from a finite reservoir ("4") into vacuum when particles scattered to the back of the reservoir with length ℓ *reflect*. There is no *KL*. The problem is equivalent to that of a finite reservoir of gas that flows into vacuum when a wall is removed at $t = 0$. The indicated expansion front is that of the curve for time ratio $Y = a_4 t/\ell = 8$. The single *LOC* is marked explicitly for $Y = 8$ and implicitly for $Y = 2$ and 4. The method of calculating Fig. 9 is discussed in [129, Sect. 2.1]. The figure itself is due to Kelly and Miotello [39].

dimensionality [83], the expansion fronts move with a velocity given by

$$\hat{u}_i = \hat{u}_x = \hat{u}_y = \hat{u}_z = \frac{2}{\gamma - 1} a_4 \qquad (5.14)$$

Here $a_4 = (\gamma k_B T_4/m)^{1/2}$ is the sound speed initially applicable in the reservoir. Equation (5.14) corresponds to energies given by

$$\tfrac{1}{2} m \hat{u}_i^2 = \frac{2m}{(\gamma - 1)^2} a_4^2 = \frac{4\gamma}{3(\gamma - 1)^2} E_4 \qquad (5.15)$$

where $E_4 = 3k_B T_4/2$ is the mean kinetic energy per particle initially applicable in the reservoir.

5.2.3.4 Phase Explosion. Except for work by Kools *et al.* [84, 85], there is little precedent for a discussion of the gas dynamics appropriate to phase explosion. We will, however, argue in Section 5.3.1.3 that a variant of normal outflow that occurs from an *isolated spherical reservoir* has a limited experimental support with AlN. At least, this is true for short enough times (<500 ns) before liquid droplets appear (>25 μs) (Fig. 5). Provided the reservoir is motionless (or moves sufficiently slowly), the relevant velocities will have a similar analytical form as for normal outflow, where it will be recalled that $\hat{u}_i$ is the same for any dimensionality (Eq. (5.14)).

Whether an isolated spherical reservoir is relevant to Cu and PBSCCO, on the other hand, cannot be decided on the basis of what is known (Section 5.3.2.4).

5.3 Plume Characterization

5.3.1 The AlN Plume

5.3.1.1 General Comments on the Motion of the Expansion Front.

Normal Outflow into Vacuum. We have already noted (Section 5.2.3.3) that for one-dimensional outflow into vacuum, the velocity of the *expansion front*, $\hat{u}_x$, represents the maximum possible flow velocity of emitted particles (Figs. 9 and 10(a)). Also, for both normal and lateral motion, thence for outflow with any dimensionality, the expansion fronts move in accordance with Eq. (5.14).

Normal Outflow into an Ambient Gas. Outflow into an ambient gas is considerably more complicated and is discussed elsewhere, and even then only for one-dimensional motion [86]. As seen by comparing Figs. 10(a) and 10(b), it is necessary to distinguish the reservoir ("4"), the region between the terminal *line of contact (LOC)* and the *contact front (CF)* ("3"), the region of disturbed ambient gas between the *CF* and the *shock wave (SW)* ("2"), and the region of undisturbed ambient gas beyond the *SW* ("1"). The velocity of the *CF*, u_{cf}, rather than $\hat{u}_x$ now represents the maximum possible flow velocity of the emitted particles. Furthermore, a monotonically decreasing density profile (Fig. 10(b)) will be noted to *not be* the only possibility (Fig. 10(c)).

The quantities a_3 and u_3 follow, respectively, from the general relations [30] $p \propto a^{2\gamma/(\gamma-1)}$ and $u = [2/(\gamma - 1)][a_4 - a]$, the results being

$$a_3 = a_4(p_3/p_4)^{(\gamma-1)/2\gamma} \tag{5.16}$$

$$u_3 = \frac{2}{\gamma - 1}(a_4 - a_3) \tag{5.17}$$

The *CF* between the two gases is described by $u_{cf} = u_3$. The point that we want to make with Eq. (5.17) is that it demonstrates how the ambient gas introduces a limited deceleration of the particles with respect to what happens in vacuum.

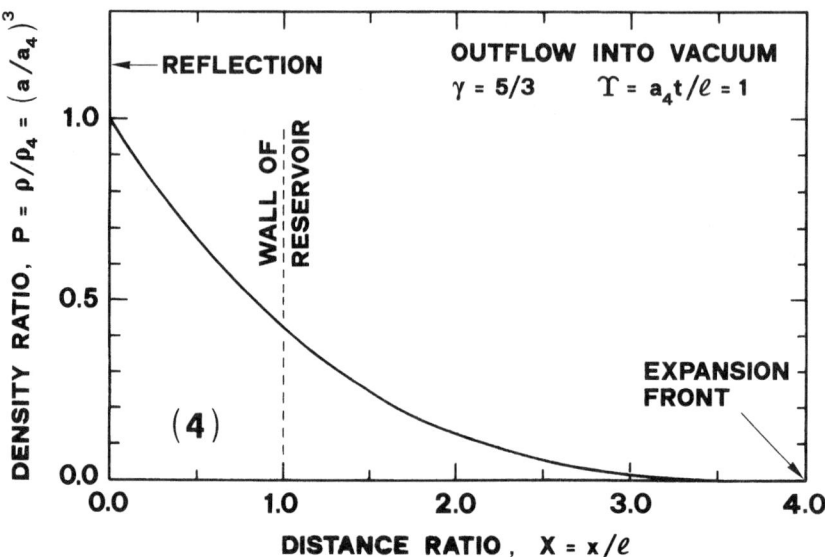

FIG. 10(a). Calculated values of the density ratio, $P = \rho/\rho_4 = (a/a_4)^3$, versus the distance ratio, $X = x/\ell$, for the release of atoms ($\gamma = 5/3$) due to *outflow* from a finite reservoir ("4") into vacuum. The time ratio has been taken as $\Upsilon = a_4 t/\ell = 1$, and the curve is therefore identical with the corresponding curve of Fig. 9. The expansion front, at $X = x/\ell = 3a_4 t/\ell = 3$ as measured with respect *to the wall*, is described by Eq. (5.14) with $\gamma = 5/3$, as for ground-state atoms. (Thus the expansion front *is not* measured with respect to the *back* of the reservoir. Moreover, the real value of γ is probably less than 5/3 for atoms that are both ionized and excited [101].) The method of calculating Fig. 10(a) is discussed elsewhere [34, Eq. (1)].

(Compare the expansion front of Fig. 10(a) with the *CF* of Figs. 10(b) and 10(c).) The effect *would not*, however, lead to the extreme deceleration observed experimentally with AlN and to be described in Section 5.3.1.3. If the imageable limit of the plume is the *CF*, E_4 can be deduced from a relation closely similar to Eq. (5.17):

$$\tfrac{1}{2}mu_{cf}^2 = \tfrac{1}{2}mu_3^2 = \frac{2m}{(\gamma-1)^2} a_4^2 (1 - a_3/a_4)^2 = \frac{4\gamma}{3(\gamma-1)^2} E_4 (1 - a_3/a_4)^2 \quad (5.18)$$

Allowance for Lateral Expansion. For large enough p_2/p_1, the velocities u_2, u_3, and u_{sw} all scale as $(p_2/p_1)^{1/2}$ [35]. Furthermore, p_2/p_1 will decrease in response to the overall expansion of the plume. By assuming that the plume is spherical (approximately true in most cases) and uniform (less acceptable), we can write for any of u_2, u_3, or u_{sw}

$$u_i \propto (p_2/p_1)^{1/2} \propto (\text{volume})^{-1/2} \propto (u_i t + r_o)^{-3/2} \quad (5.19)$$

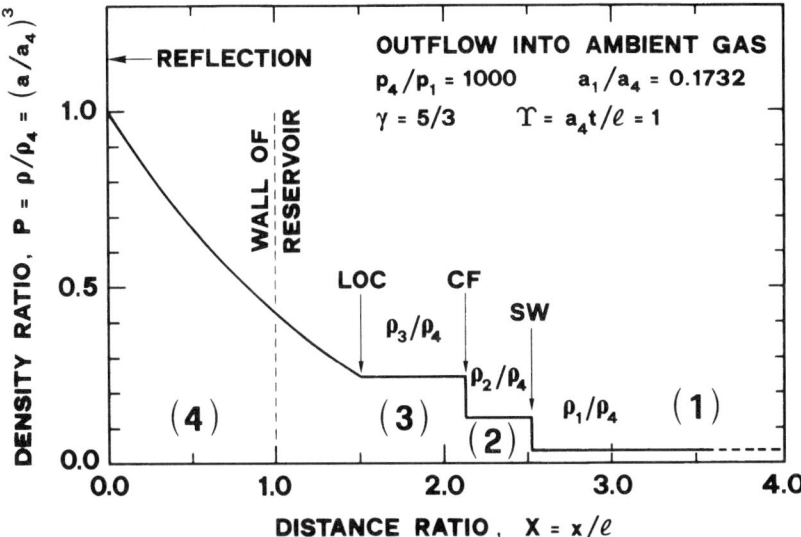

FIG. 10(b). Like Fig. 10(a) but for *outflow* from a finite reservoir ("4") into an ambient gas ("1"). The time ratio was again $\Upsilon = a_4 t/\ell = 1$, and the reservoir pressure and temperature have been taken as leading to values of p_4/p_1 and a_1/a_4 as indicated. A significant deceleration with respect to Fig. 10(a), as manifested in the position of the contact front (*CF*), will be noted to have occurred. The notation *SW* = shock wave has been used. The method of calculating Fig. 10(b) is discussed elsewhere [86].

where r_o is the initial radius, from which follows the important scaling law,

$$u_i \propto (t + r_o/u_i)^{-3/5} \sim \propto t^{-3/5} \quad (5.20)$$

We conclude that the assumption of one-dimensionality made above with, for example, Eqs. (5.16) and (5.17) significantly *underestimates* the deceleration caused by the ambient gas. When the lateral expansion of the emitted particles is taken into account, it follows from Eq. (5.20) that the deceleration is much more severe.

5.3.1.2 Optical Properties of AlN.

A fundamental problem regarding optical absorption might appear to enter with AlN. In fact, the band gap of AlN is of order 5.9–6.2 eV [87–89] or somewhat lower, 5.5–5.8 eV, for AlN which is either amorphous [90] or prepared by deposition techniques [91]. It is true that there is a prominent tail at the photon energy that was used in the experiments to be discussed (5.0 eV) due to effects that include disorder, strain, and oxygen impurity [87–89]. One-photon absorption was therefore possible to a limited extent, but in fact the absorption coefficient is 10–100 times smaller than that of a metal. The absorption should therefore be much extended beneath the surface, and a high temperature would not be expected.

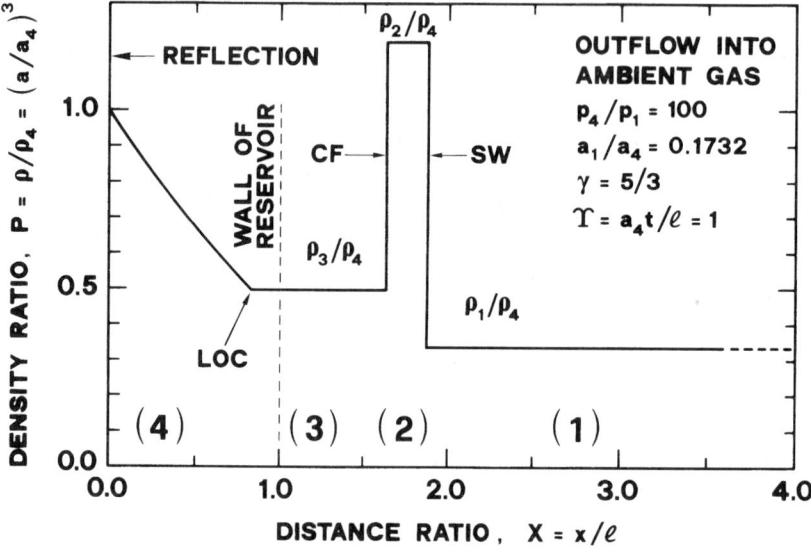

FIG. 10(c). Like Fig. 10(b) but for *outflow* from a finite reservoir into an ambient gas with a value of p_4/p_1 a factor of 10 lower than that of Fig. 10(b). Because there is a lesser pressure difference between the reservoir and ambient gas, the deceleration is even more marked. The important points to note are that a monotonically decreasing density profile as in Fig. 10(b) *is not* the only possibility with outflow and, moreover, that an even more significant deceleration with respect to Figs. 10(a) and 10(b) is evident.

It is therefore important to realize that, as has been well-established in recent experiments, the threshold laser fluence for AlN to lose N_2 at the surface and therefore be metallized is only $\sim$1.5 J/cm^2 (40 ns), the necessary temperature being $\sim$2600 K [92, 93]. Such metallization is not observed in furnace heatings [94], but only when the heating time scale is so short that normal vaporization is limited in accordance with the Hertz-Knudsen equation (Eqs. (5.1)–(5.4)). (A limitation of this sort is not always appreciated, leading to impossible claims for vaporization with for example, Si and SiO$_2$ [39, (Table 7), 95, 96]. Likewise, the threshold fluence for AlN to melt is very similar, $\sim$1.5 J/cm^2 (40 ns) [92, 93], the necessary temperature being $\sim$3050 K [97, 98]. There was no incubation effect under the conditions of the experiments to be discussed ($\sim$20 J/cm^2, 30 ns), but incubation periods of $\sim$10 pulses occur at lower fluences (5 J/cm^2) [92]. The absorption behavior of AlN, provided the fluence is sufficient and incubation pulses are supplied, is thus similar to that of a metal, where $\sim$2.3 J/cm^2 gives a temperature of 2500 ± 400 K [67].

There are a number of reasons why the metallization and melting thresholds should be so low, for example, bulk point defects, surface point defects, surface mechanical defects, dislocations, and microcrystallinity. What is important is that there is little doubt that, even if (as is likely) laser-surface interaction with AlN begins as an electronic process, it rapidly evolves to a thermal-spike process. In fact, the fluence that was used, ~ 20 J/cm^2, would lead to temperatures greatly exceeding those both for N_2 loss and for melting, i.e., the temperatures would exceed ~ 3050 K. The AlN surface can be reasonably assumed to have approached T_{tc} of both Al (5400 K [22]) and AlN (~ 8000 K [99]).

We now discuss recent experiments of Mele *et al.* [100] in terms of the gas-dynamic ideas summarized in Section 5.3.1.1.

5.3.1.3 The Experiments.
The experiments of Mele *et al.* [100] involved bombardment of AlN with a KrF excimer laser (248 nm, ~ 30 ns, ~ 20 J/cm^2, 45° incidence). The spot size was 0.2×0.8 mm^2. The fluence (which will be noted to be quite high) was typical of work intended for film deposition, and the system was filled with a controlled pressure of N_2 ranging from 0.13 to 27 Pa. The AlN target was in the form of a sintered disc and was located 25 mm from the substrate. The excited states in the plume emerging from the AlN surface were recorded with an *ICCD* camera having a temporal gate of ~ 5 ns. A monochromator was *not* used, and the measurements therefore related to overall fluorescence in the interval 400–800 nm rather than to particular spectral transitions. This was basically a defect in the method because it hid information on possible spatial separation of states with different excitation or charge.

Fig. 11(a) shows the fluorescence contours for six different time delays with 0.13 Pa of N_2 constituting the ambient gas. In each case there was a well-defined evolution of the fluorescence from a low value at the target surface to a maximum, and then to a low value beyond the maximum. It is clear that the expansion was not of the type termed "normal outflow" because the latter does not give a maximum (Fig. 9). It can also be argued that the expansion was not effusion-like as in Fig. 8, even though a fluorescence maximum exists. There are two reasons to assert that this is so. First, effusion-like behavior implies normal vaporization, but it was shown that E_4 was too high (even given the uncertainty in γ and other uncertainties) for normal vaporization *by itself* to be relevant. (It will, of course, occur transiently at the beginning and end of each pulse no matter what is the dominant mechanism. See below; see also the discussion of Fig. 13 in Section 5.3.1.4.) Second, effusion-like behavior requires that the position of the maximum move with time as in Fig. 8, whereas experimentally it was nearly stationary at 0.13 Pa ambient gas pressure starting at ~ 400 ns (Fig. 11(a)). The imageable limit of the plume, which should be similar to the *CF*, is easily located in Fig. 11(a) and it is found that not only does it move (u_{cf}) but that there is only a slight tendency for u_{cf} to decrease with time (Fig. 12 and Table I). Considering the existence of a nearly stationary fluorescence maximum yet a freely moving *CF*, one concludes that the expansion is like that for outflow from an *isolated spherical reservoir*.

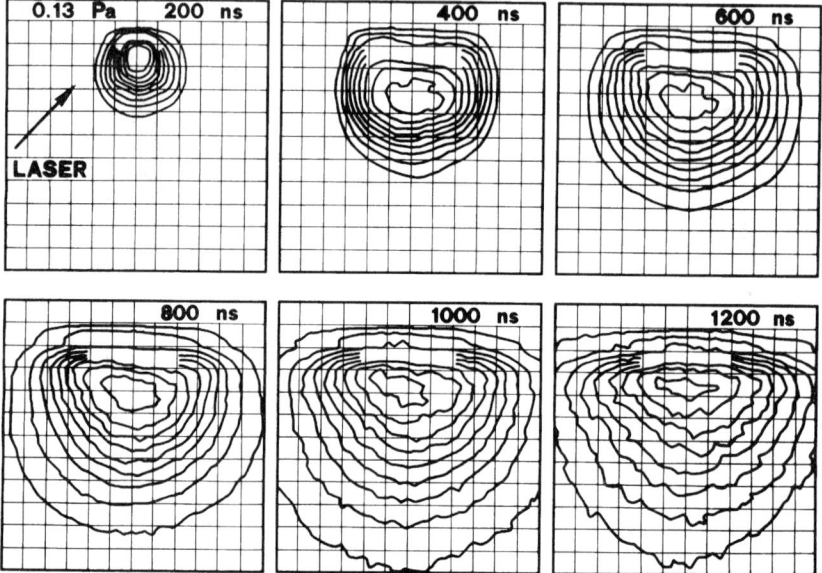

FIG. 11(a). Bombardment of AlN in the form of a sintered disc in the presence of N_2 with a KrF excimer laser (248 nm, ~30 ns, ~20 J/cm^2, 45° incidence). The overall light intensity (fluorescence) of the plume from 400 to 800 nm was measured with an *ICCD* camera having a temporal gate of ~5 ns. The laser spot had dimensions of 0.2 × 0.8 mm^2 on the y and z axes, respectively. The light intensity is shown in terms of two-dimensional density contours, which have an interval of 10% and are superimposed on a grid with a spacing of 2 mm. The time delay with respect to the laser pulse is given in ns and the ambient N_2 pressure was 0.13 Pa (1 mTorr). Note that the fluorescence maximum present is effectively stationary, and that there is only a weak tendency for the *CF* to decelerate. Due to Mele *et al.* [100].

The mean kinetic energies in the reservoir, E_4, were deduced in two ways: from the *CF* by evaluating $\frac{1}{2}m\langle u_{cf}\rangle^2$ as in Eq. (5.18) and from the initial expansion at 0–200 ns by evaluating $\frac{1}{2}m\hat{u}^2$ as in Eq. (5.15). The values of E_4 were thus found to lie in the interval 1.5–2 eV (Table 2). One concludes that the reservoir energy E_4 exceeds $3kT_{tc}/2$ (~1.0 eV for $T_{tc} \approx 8000$ K) to some extent and that, as already noted, a primary release mechanism involving normal vaporization *by itself* can therefore be excluded. The obvious alternatives are an *electronic sputtering process* (Section 5.2.1.1) or *phase explosion* (Section 5.2.1.4) or *subsurface heating* (Section 5.2.1.5) or (independently of the mechanism) laser-plume interaction [50]. The argument attempting to justify subsurface heating has, however, been shown to be faulty [19, 26] (Section 5.2.2). It is not possible to exclude an electronic process or laser-plume interaction, but, given the fact that the fluence was ~20 J/cm^2 and only 1.5 J/cm^2 is sufficient to raise AlN to

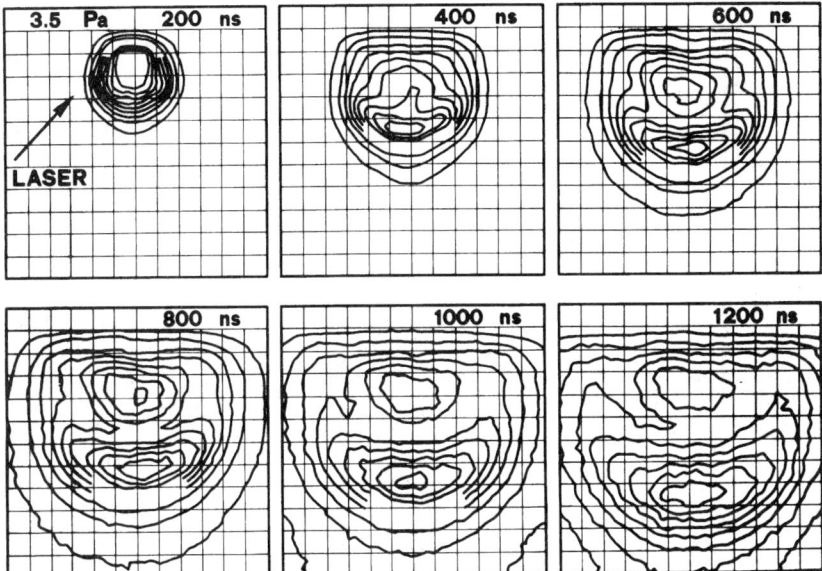

FIG. 11(b). Like Fig. 11(a), but the ambient N_2 pressure was 3.5 Pa (26 mTorr). Both stationary and moving fluorescence maxima are present starting at 400 ns. Again there is only a weak tendency for the CF to decelerate.

~3050 K [92, 93], it is reasonable to assume that phase explosion occurred. In this connection, it should be remembered that the fact that AlN was heated to ~3050 K is based on experimental observation and not on the vagaries of a calculation in which laser-plume interaction exists but cannot be quantified.

The fluorescence contours at 1.9 Pa were similar to those at 0.13 Pa, both as far as the kinetic energies were concerned (~2 eV, Table II) as well as for the low extent of deceleration of the CF (Fig. 12 and Table I). However, an important difference was apparent starting at 1200 ns, namely, the appearance of a second, weak fluorescence maximum that moved.

Fig. 11(b) shows the fluorescence contours for 6 different time delays with 3.5 Pa of N_2 constituting the ambient gas. As in the preceding cases, there was still a clear tendency for a fluorescence maximum that did not move although it was no longer dominant. Also E_4, as deduced both from $\langle u_{cf} \rangle$ and from the initial expansion at 0–200 ns (2–3 eV, Table II), still indicated a primary sputtering mechanism that did not involve normal vaporization *by itself*. But the moving fluorescence maximum was now dominant and, unlike the CF (Fig. 12), showed a moderate deceleration (Table I).

The main novelties at 13 Pa (Fig. 11(c)) were that a stationary fluorescence maximum could no longer be resolved, that strong (but not total) deceleration

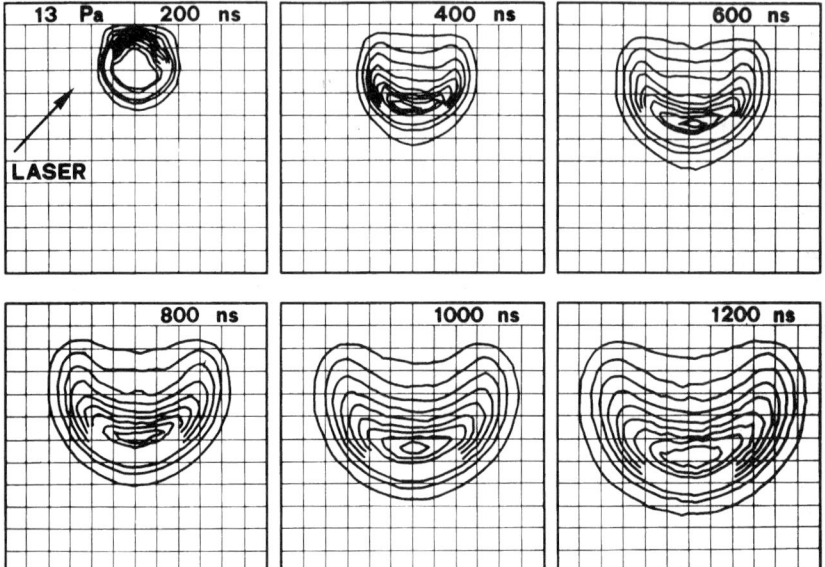

FIG. 11(c). Like Fig. 11(a), but the ambient N_2 pressure was 13 Pa (100 mTorr). Only a moving fluorescence maximum is now resolved, but a stationary maximum is probably also present. Deceleration of the *CF* is strong but not total.

was evident both with the *CF* (Fig. 12 and Table I) and with the moving fluorescence maximum (Table I), and that only E_4 as derived from the initial expansion appeared to be reliable (1.5–2 eV, Table II).

The behavior at 6.7 and 27 Pa in general resembled that at, respectively, 3.5 and 13 Pa.

5.3.1.4 Discussion.

General Comments. A surprising amount of mechanistic information on the high-fluence laser sputtering of AlN is contained in fluorescence contours such as those seen in Figs. 11(a)–11(c) and in the corresponding values of the reservoir particle energies, E_4 as in Table 2. Note, for example, that the initial values of E_4 (i.e., the values before deceleration was important) were moderately high, 1.5–3 eV. By contrast, the target temperature was by necessity limited to T_{tc}, corresponding to a kinetic energy of ~1 eV in the case of AlN [99].

The interpretation was as follows [100]. If a fluence of ~1.5 J/cm^2 (40 ns) can bring the surface of AlN to a temperature sufficient for metallization and melting (~3050 K, Section 5.3.1.2), it follows that the impact of far more energetic (~20 J/cm^2, 30 ns) pulses can heat AlN even more strongly. This suggests that phase explosion was relevant as the *primary* sputtering mechanism and that both electronic sputtering and laser-plume interaction can be excluded as reasons for

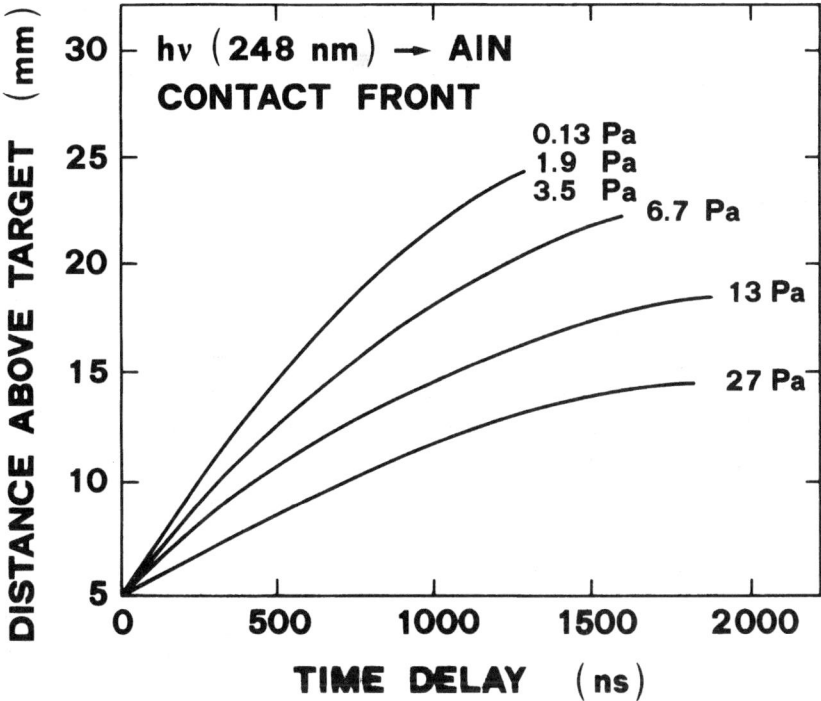

FIG. 12. Experimental time dependence of the dimension $x(t)$, i.e. the distance above the target, of the plume generated from an AlN target under the conditions of Fig. 11. The term "contact front" is reasonable but not quite correct for two reasons: (i) the plume position was measured from the 10% density contour and (ii) the particle density was approximated in terms of the overall fluorescence. Due to Mele et al. [100].

TABLE I. Summary of the Experimental Trends for the Plume of Laser-Sputtered AlN [100].

Ambient N_2 pressure		Stationary fluorescence maximum	Time of appearance (ns)	Moving fluorescence maximum	Time of appearance (ns)	Deceleration of contact front (CF)	Deceleration of moving fluorescence maximum
(Pa)	(mTorr)						
0.13	1	dominant	400	absent	...	slight	...
1.9	14	dominant	400	weak	1200	slight	...
3.5	26	moderate	400	dominant	400	slight	moderate
6.7	50	weak	~400	dominant	~200	moderate	moderate
13	100	not resolved	(never)	dominant	~200	strong	strong
27	200	not resolved	(never)	dominant	~200	strong	strong

TABLE II. Experimental mean velocities and mean kinetic energies for the plume of laser-sputtered AlN [100] for all pressures of N_2. The velocities were derived with $M = 27$ u, as for Al, and $\gamma \approx 1.25$, as for a monatomic gas which is both ionized and excited [101]. The uncertainty in the value of γ must be regarded as a major obstacle in quantifying E_4.

Ambient N_2 pressure (Pa)	$\langle u_{cf} \rangle$ (10^6 cm/s)	E_4 from $\langle u_{cf} \rangle^{(a)}$ (eV)	$\hat{u}$ (10^6 cm/s)	E_4 from expansion at 0–200 ns$^{(b)}$ (eV)
0.13	1.7	>1.5	2.0	2.1
1.9	1.9	>1.9	1.9	1.8
3.5	1.8	>1.7	2.3	2.8
6.7	1.20	(>0.8)	1.8	1.7
13	0.75	(>0.3)	1.9	1.9
27	0.50	(>0.13)	1.7	1.5

$^{(a)}$ E_4 was obtained here from Eq. (5.18). It is shown as a lower limit since the factor $(1 - a_3/a_4)^2$ could not be evaluated. The three values in brackets are artificially low because of deceleration.
$^{(b)}$ E_4 was obtained here from the initial expansion by evaluating $\hat{u}$ from the plume radius at 200 ns and then using Eq. (5.15).

the high particle energies. It was therefore proposed that particles were expelled first by normal vaporization, then (when the surface approached T_{tc}) by *phase explosion*, and finally (as the target cooled) again by normal vaporization. Obviously the assumption is here contained that phase explosion leads to high particle energies, a detail that lacks a formal proof. A possible representation of the mechanism is a sequence of particles near the target surface ("A," "B," "C") as in Fig. 13, although the fluorescence contours apparently did not distinguish the normally vaporized ("A" and "C") from the phase-exploded ("B") particles.

The expelled particles tend to decelerate due to the presence of ambient gas, provided the relevant pressure is high enough. This deceleration was, however, well defined only for the *CF*, being minimal at 0.13–3.5 Pa and important at 6.7–27 Pa (Fig. 12 and Table I). These pressures must be regarded as being surprisingly low though not unique. Thus values reported in other work that confirmed deceleration include 10^{-3} and 66 Pa [102], 0.03 and 133 Pa [4], 4–400 Pa [103], 13 Pa [7,8], 650 and 2000 Pa [104], and 1300 Pa and 1 atm [3]. See, for example, Fig. 14, where the ambient pressure was described as "vacuum" [4].

Important gas-dynamic information is contained in the fact that, at 0.13–3.5 Pa, the near-surface fluorescence maximum ("B" in Fig. 13) was nearly stationary, yet the *CF* moved freely. Evidently extreme deceleration has occurred in the interior of the plume and the overall expansion, thence the

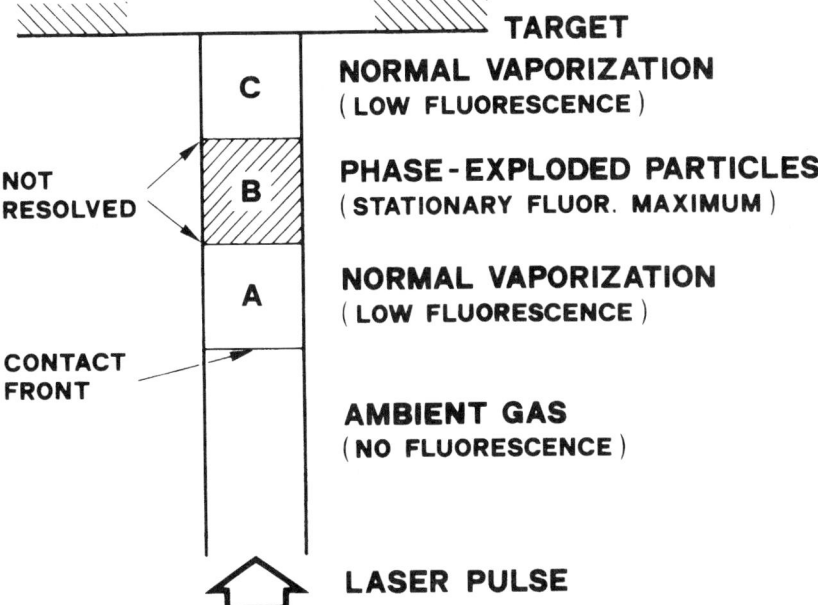

FIG. 13. Sketch in one dimension of the sequence of particles that is probably present when a target is laser-sputtered at high fluence (e.g., 20 J/cm^2) in the presence of an ambient gas. The temperature of the emitted particles is first well below T_{tc}, such that normal vaporization occurs ("A"), then approaches (but does not reach) T_{tc}, such that phase explosion occurs ("B"), and finally decreases again ("C"). The normally vaporized particles ("A"), which initially can have a high pressure [105], in principle impede the motion of the phase-exploded particles ("B") leading finally to a stationary fluorescence maximum. Much weaker braking is caused by the ambient gas at all pressures so the CF always moves (as observed experimentally, Figs. 11(a)–11(c) and 12). Both types of braking are supplemented by lateral expansion (Section 5.3.1.1). It has been assumed that the measured fluorescence profiles as in Fig. 11 do not resolve the boundary between the normally vaporized and phase-exploded particles. It has also been assumed (on the basis of the high fluence) that neither electronic sputtering nor laser-plume interaction plays a role. Remember, in this regard, that ~3050 K is reached at ~1.5 J/cm^2 [92, 93].

secondary sputtering mechanism, is like that for outflow from an *isolated spherical reservoir* [39, Fig. 20].

The Stationary Fluorescence Maximum. How can it happen that the CF moved yet the fluorescence maximum was nearly stationary? There are at least three possibilities [100] but we will here discuss only the *collisional* model.

According to the collisional model, a role is played both by the ambient gas and by particles emitted by normal vaporization ("A" in Fig. 13) before the

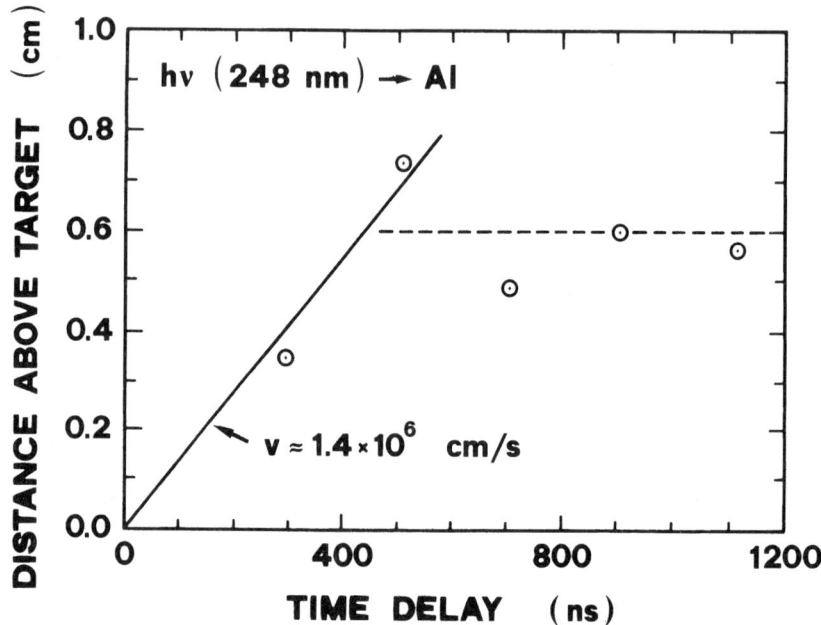

FIG. 14. Experimental time dependence of the dimension $x(t)$, i.e. the distance above the target, of the plume generated from an Al target irradiated *in vacuum* (0.03 Pa) with a KrF excimer laser (248 nm, 40 ns, 2.4 J/cm^2, normal incidence). The measurements, which were made with the 1×10^{14} cm^{-2} equicontour line for ground-state neutrals, at first showed a velocity of 1.4×10^6 cm/s (very similar to the YBCO of Fig. 5 and the AlN of Fig. 12), but then a remarkable deceleration occurred. Each point involved a single pulse. Adapted from a figure due to Lindley et al. [4].

phase explosion ("B"). The ambient gas was assumed to affect mainly the *CF* (cf. Eq. (5.17)) and evidently permitted an ongoing expansion with at most moderate deceleration. The particles emitted by normal vaporization ("A"), on the other hand, were assumed to affect explicitly the particles contained in the phase explosion ("B"), causing the latter to rapidly lose their velocity. Why this might be is clear from the nature of normally vaporized particles. They can have *initial* particle densities similar to 100 μg/cm^3, equivalent to $\sim 10^5$ Pa [105] and leading to unusually strong braking for the first 50–100 ns. (Remember also that, when lateral expansion is taken into account, braking effects are much greater (Section 5.3.1.1).) The particles of the phase explosion, which are deposited rather far away from the surface (~ 3 mm), thenceforth expand as for an *isolated spherical reservoir*; that is, the particle density (thence fluorescence) maximum does not move perceptibly in the experimental time scale (200 ns). For example, if the vaporized particles are at 3000 K, their velocity (assuming M = 27 u) will be $\sim 2 \times 10^5$ cm/s and the motion in 200 ns will be only 0.4 mm.

Good precedent for the collisional model is found with the laser sputtering of polyimide [106, 107]. Here there is a unique situation in which 96% of the sputtered mass is contained in light particles [108], and the latter therefore easily impede the motion of the 4% of the particles having high mass. As a result, the latter particles tend to expand laterally (Section 5.3.3).

Finally, it was pointed out [100] that the evidence for deceleration of the phase-exploded particles might be faulty, due to observing excited states. That is, a fluorescence maximum could either represent a true particle-density maximum or a region with a high rate of re-excitation. It is therefore important to note that, in related work, in which an Al target was laser sputtered at 2.1–2.4 J/cm^2, Lindley et al. [4] confirmed that a stationary particle-density maximum was present by observing *ground states* (Fig. 14). Likewise Gilgenbach et al. [109] confirmed a stationary maximum in work where AlN was laser sputtered at 2.5–3.1 J/cm^2 and Al$^+$ *ions* were observed. Re-excitation is clearly irrelevant both to ground states and to ions.

The Moving Fluorescence Maximum. At this point, a major problem enters: Why, for ambient N$_2$ pressures of 3.5 Pa and greater, was there a moving *fluorescence* maximum? Does this constitute a formal transition from a stationary to a moving *particle-density* maximum? Again there is precedent for this effect in other work, for example, that of Geohegan [7, 8] and again several models are possible. We will here discuss only the *electron-scattering* model.

Horwitz [110], Ohkoshi et al. [102], and Mehlman et al. [111] have proposed that electrons emitted from the surface of a laser-bombarded target will, in the presence of sufficient ambient gas (e.g., 40 Pa [111]), scatter back to the plume and cause a subsidiary fluorescence maximum to appear. Moreover, this maximum would move because it is near the *CF* and, as we have already noted (Fig. 12 and Table 1), the latter shows an ongoing expansion at all pressures such that the deceleration is never total as in Fig. 14. Such a maximum must, however, be regarded as *false* in the sense that it does not relate to a particle-density maximum.

Are the electrons sufficiently abundant? In fact, it has been proposed that a high yield of electrons is emitted from laser-bombarded surfaces, either if the temperature approaches T_{tc} and massive thermionic emission takes place [112] or by nonthermal (probably electronic) processes in the case of insulators [109]. For example, for AlN bombarded at 4 J/cm^2 the maximum electron density at 100 ns was found to be 1.2×10^{18} cm^{-3}, corresponding to a total number of order 10^{16}. The electron density remained significant up to 2.5 mm, thence within the region where the phase-exploded particles had their fluorescence maximum.

5.3.1.5 Conclusions. At first sight, AlN, with a bandgap near 6 eV, should absorb 5 eV photons only weakly. In fact, it is well established [92, 93] that a fluence near 1.5 J/cm^2 (40 ns) is sufficient, provided incubation pulses are supplied [94], to induce in AlN both metallization and melting. A temperature near 3050 K has therefore been reached, and it is clear that at the fluence used in the work being discussed ($\sim$20 J/cm^2, 30 ns), the temperature of the AlN will have approached T_{tc} ($\sim$8000 K [99]). This scaling (i.e., 20 *versus* 1.5 J/cm^2) was taken as the prime indication for the relevance of phase explosion and for the neglect both of electronic sputtering and of laser-plume interaction. Also, incubation pulses were not necessary because of the high fluence.

The sequence of events was thus tentatively as follows:

1. Particles were released by normal vaporization ("A" in Fig. 13).
2. Further particles were released by phase explosion ("B").
3. A final group of particles was released by normal vaporization ("C").
4. A strong deceleration of the phase-exploded particles occurred due to collisions so that the system resembled an *isolated spherical reservoir*.
5. A moving fluorescence maximum appeared, which was probably due to electron scattering by the ambient gas and which was, therefore, not related to a particle-density maximum.

5.3.2 The $Pb_{0.2}Bi_{1.8}Sr_2Ca_1Cu_2O_x$ (PBSCCO) and Cu Plumes

5.3.2.1 General Comments. As further examples of plume characterization, the superconductor PBSCCO [9, 10] and metallic Cu [10, 11] are now discussed. The analyses of these plumes were more detailed than with AlN in that they were done in 3 dimensions instead of 2. As a result information was obtained not only on the primary and secondary emission mechanisms, but also on what will be termed the "rotation" effect. It was again concluded that phase explosion was possibly the *primary* sputtering mechanism. As with AlN, this conclusion was supported most explicitly by the rather high fluences (8–12 J/cm^2), with both an electronic process and laser-plume interaction being tentatively excluded. In the spirit of Section 5.2.2, the subsurface heating model was also excluded.

The *secondary* mechanisms relate to the gas-dynamic aspects of the plume, especially (with PBSCCO and Cu) to an understanding of the spatial distribution. This depends to some extent on the laser parameters (wavelength, pulse duration, and energy fluence), but, as will be seen in what follows, it depends also on the size and shape of the laser spot. The methods typically used to measure the plume shape include probing the plume by means of small detectors [60], analyzing the *deposited film* thickness on the substrate [113], and analyzing the *recondensed film* thickness on the target in those cases (such as polyimide) where recondensation occurs (Section 5.3.3). It may be assumed that the plume shape will affect many properties of the film, including the uniformity of thickness, the symmetry, and even the composition.

5.3.2.2 The Experiments. The experiments of [9–11] involved bombardment of PBSCCO and Cu with a KrF excimer laser (248 nm, ~30 ns, 8–12 J/cm^2, normal incidence). Different spot sizes and shapes were used, variously rectangular, circular, or square. The fluences (which were quite high) were again, as with AlN, typical of work intended for film deposition, but the system was this time evacuated to 10^{-3} Pa. The PBSCCO was in the form of a sintered disc whereas the Cu target was in the form of a metal foil. Both were located far enough from the substrate that (as distinct from AlN) plume–substrate interactions were avoided.

An *ICCD* camera was used to analyze laterally in two orthogonal directions the profile of the excited states in the plume as it moved from the target to the substrate. The spatial and temporal resolution of the *ICCD* camera were 150 μm and 5 ns. Because a monochromator was *not* used, it follows that (as with AlN) the measurements related to overall fluorescence in the interval 400–800 nm rather than to particular spectral transitions. The experiments were carried out both as a function of time delay (200–2200 ns) and as a function of the size and shape of the laser spot

Results for Large Rectangular Laser Spots (1.2 × 3.0 mm^2 or 1.0 × 2.4 mm^2). Figure 15 serves both to define the axes of the laser pulses as well as to indicate the approximate lateral symmetry of the expanding plume. Consider Figs. 16(a) and 16(b), which show orthogonal views of the PBSCCO (1.0 × 2.4 mm^2 spot) plume at 600 ns time delay [9]. In Fig. 16(a), the plume is viewed laterally in the *xy* plane, whereas in Fig. 16(b) it is viewed in the *xz* plane. It is clear that at 600 ns the plume has an ellipsoidal shape with *three* distinct axes, the expansions being $x = 22$ mm, $y/2 = 9$ mm, and $z/2 = 6$ mm. The corresponding dimensions at 1000 ns were $x = 34$ mm, $y/2 = 12$ mm, and $z/2 = 8$ mm, whereas Figs. 16(c) and 16(d) show that at 2200 ns all dimensions are considerably enlarged without, however, a loss of the characteristic ellipsoidal shape. A correlation between the laser *spot shape* and the lateral spread of the luminous material as viewed by the *ICCD* camera is indicated. As far as the fluorescence maximum is concerned, it does not show the major deceleration seen with AlN (Table I) but is uniformly slow, moving at about 25% of the velocity of the expansion front. The latter value applies also to the moving fluorescence maximum of AlN. We are not, however, prepared to extract mechanistic information from the motion of the fluorescence maximum in view of the uncertainty as to why there is such a maximum (see Section 5.3.1.4).

It will be noted that the lateral expansion was defined in terms of $y/2$ and $z/2$ rather than y and z. This choice was made on the assumption that the motion occurred with respect to the midpoint of the plume. The argument is not fully rigorous, however, because the lateral expansion is subject to a time delay.

Results for Cu (1.0 × 2.4 mm^2 spot) were in general equivalent [11]. For example, for a 600 ns time delay the expansions were $x = 12$ mm,

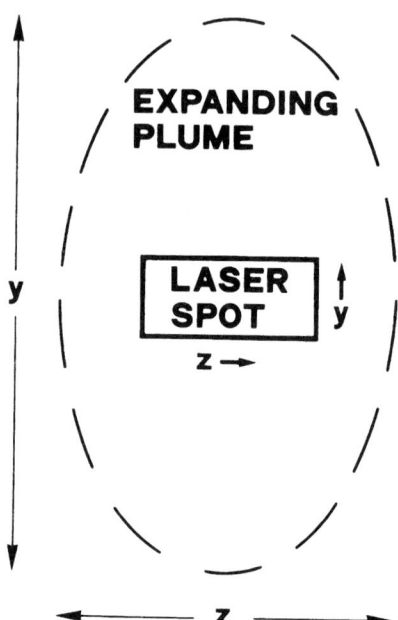

FIG. 15. *Solid*: Sketch showing the lateral (yz) axes of a rectangular laser spot appropriate to a pulse directed at normal incidence towards a target. *Dashed*: Approximate yz symmetry of the expanding plume, showing that the larger side of the spot becomes the smaller side of the plume. Due to Mele et al. [9, 10].

$y/2 = 6.0$ mm, and $z/2 = 3.2$ mm. Also, the fluorescence maximum moved and, as with PBSCCO, did so uniformly and at about 20% of the velocity of the expansion front.

The evolution of the PBSCCO plume with time is shown in Fig. 17 [9]. A linear expansion with different velocities on the three axes is evident. These data allowed the plume shape to be measured at various time delays after the laser pulse and thus at various distances from the target. The anisotropic shape is seen to be kept for all time delays exceeding ~250 ns. Similar information was obtained also for Cu [10].

Results for Small Laser Spots. In other experiments, it was found that irradiation of a PBSCCO target with either a small circular spot (0.7 mm^2, 532 nm, 6.3 J/cm^2) or a small rectangular spot (0.5 × 1.0 mm^2, 248 nm, ~8 J/cm^2) yielded a plume with ellipsoidal shape but having only *two* distinct axes [9]. That is, the profiles in the xy and xz planes were almost the same. Results for Cu, when irradiated with a circular spot having a diameter of either 3.0 or 0.2 mm, are shown in Figs. 18(a) and 18(b). In both cases, there were again only *two distinct axes*.

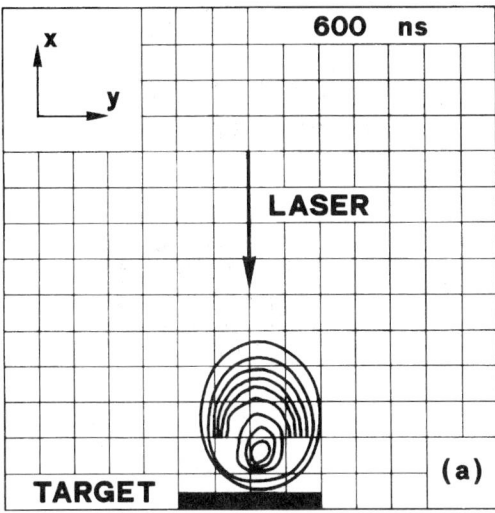

FIG. 16(a). View in xy plane, where the y coordinate contains the smaller dimension or the laser spot but the larger dimension of the plume. Time delay 600 ns. The target is not drawn to scale.

Bombardment of PBSCCO *in vacuum* ($\sim$0.001 Pa) with a KrF excimer laser (248 nm, 30 ns, $\sim$8 J/cm^2, normal incidence). The overall fluorescence of the plume from 400 to 800 nm was measured with an *ICCD* camera having a temporal gate of $\sim$5 ns. The laser spot had dimensions of 1.0×2.4 mm^2 on the y and z axes, respectively. The light intensity is shown in terms of two-dimensional density contours, which have an interval of 10% and are superimposed on a grid with a spacing of 5 mm. Note that the maximum intensity decreases with time delay, with that for 2200 ns being only 25% of that for 600 ns. Due to Mele *et al.* [9].

The evolution of the Cu plume with time under the conditions of Figs. 18(a) and 18(b) is shown in Figs. 19(a) and 19(b). These data show an interesting aspect of the rotation effect, such that the *normal* expansion is essentially independent of the spot size but the *lateral* expansion is distinctly more rapid for the *smaller* spot. Whether one considers PBSCCO or Cu, a correlation of the lateral spread of the luminous plume material not only with the laser *spot shape*, but also with the *spot size*, is indicated.

5.3.2.3 Numerical Simulation.
The present results confirm what has already been well established in other work (e.g. [114–116]), namely, that the plume symmetry correlates closely with the symmetry of the laser spot. Thus, after a time delay of order 250 ns, the plume from a large rectangular laser spot has the appearance of being "rotated" at 90° with respect to the spot. The larger side of the spot becomes the smaller side of the lateral cross section of the plume and vice versa. This correlation is shown in Fig. 15 schematically, in Figs. 16(a)–16(d) and 17 by contrasting the y and z dimensions, and in Figs. 18 and 19

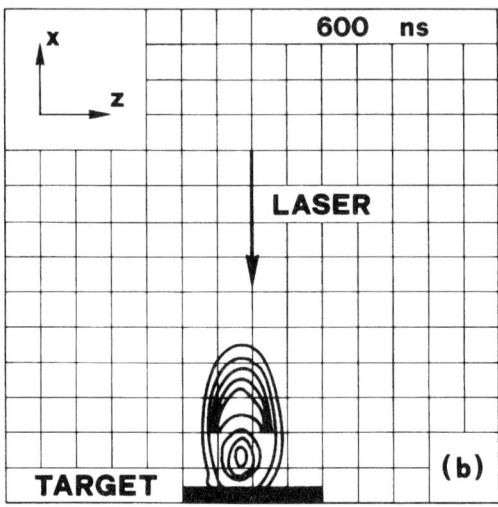

FIG. 16(b). View in xz plane, where the z coordinate contains the larger dimension of the spot but the smaller dimension of the plume. Time delay 600 ns. Again the target is not drawn to scale.

by contrasting large and small circular spots. What is new in these figures is that the rotation effect is being studied as a function of time delay, thence with respect to distance from the target. It can be asserted that the effect is already present at ~250 ns and that it persists to at least 2200 ns. It is therefore not a surprise that the films deposited on substrates normally retain the rotation [85, 114, 115].

It is interesting to ask why the rotation effect should occur. For example, is flow normal to the central portion of the y and z axes less divergent than at the corners, i.e., does the flow velocity tend to be normal to the surface? To this end, the 2-dimensional gas-dynamic (flow) equations were considered [9, 116]:

$$\frac{\partial \rho}{\partial t} + \frac{\partial \rho u_y}{\partial y} + \frac{\partial \rho u_z}{\partial z} = 0 \qquad (5.21)$$

$$\frac{\partial \rho u_y}{\partial t} + \frac{\partial (p + \rho u_y^2)}{\partial y} + \frac{\partial \rho u_y u_z}{\partial z} = 0 \qquad (5.22)$$

$$\frac{\partial \rho u_z}{\partial t} + \frac{\partial \rho u_y u_z}{\partial y} + \frac{\partial (p + \rho u_z^2)}{\partial z} = 0 \qquad (5.23)$$

Here $\rho \propto a^{2/(\gamma-1)}$ is the particle number density, a (which does not appear explicitly in Eqs. (5.21)–(5.23)) is the sound speed, u_y is the y-direction flow velocity, u_z is the z-direction flow velocity, and p is the pressure. For purposes of making numerical calculations, Eqs. (5.21)–(5.23) were used in normalized

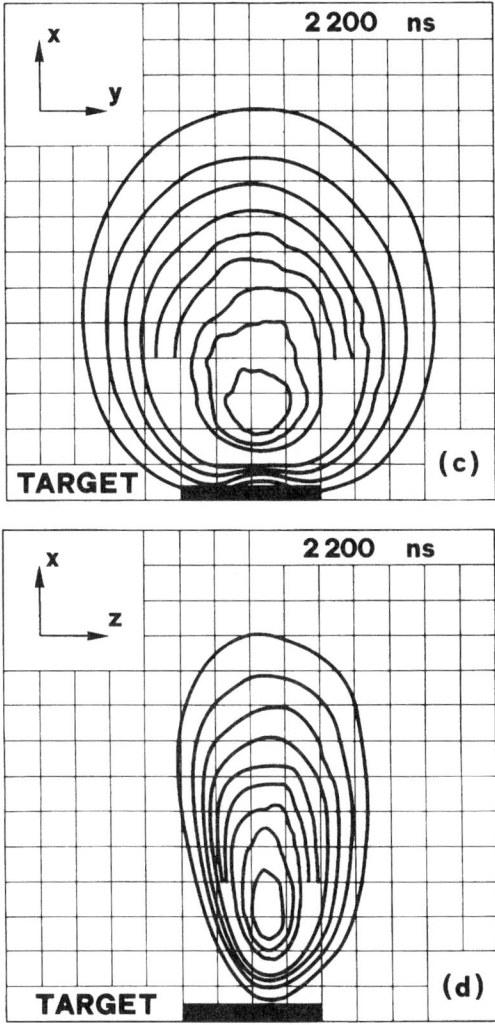

FIG. 16(c), 16(d). Like (a) and (b) but with a 2200 ns time delay.

(thence unitless) form. For example, ρ was replaced with ρ/ρ_n and t with t/t_n, where ρ_n and t_n serve to normalize. The details of deducing the *normalizers* is discussed elsewhere [9, 116].

Numerical solutions were obtained mainly for a square (rather than rectangular) reservoir of particles, as would develop from a square laser spot, because the calculations were more stable for a square geometry. The original configuration was an 8 × 8 array of points with $\rho/\rho_n = 1$. The spatial steps were

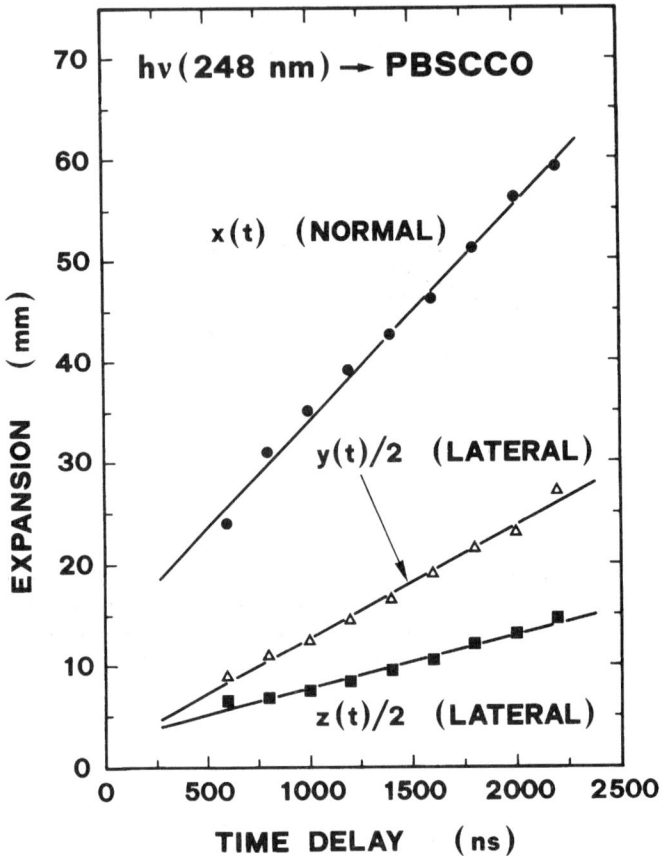

FIG. 17. Experimental time dependence of the dimensions $x(t)$, $y(t)/2$, and $z(t)/2$ of the plume generated from a PBSCCO target under the conditions of Fig. 16 (1.0 × 2.4 mm^2 spot). $y(t)$ and $z(t)$ refer to the largest lateral dimensions of the plume but have been reduced by a factor of 2 on the assumption that the relevant motion was with respect to the midpoint of the plume. Due to Mele et al. [9].

$\Delta(x/x_n) = \Delta(y/y_n) = 0.05$ and the time step was $\Delta(t/t_n) = 0.008$. At time $t/t_n = 0$ the four confining walls of the reservoir were removed as is appropriate for *normal outflow* (Section 5.2.3.3) [34, 100, 116] and expansions as in Figs. 20(a)–20(e) occurred. One sees in Fig. 20(a) that, for the shortest time delay used ($t/t_n = 0.048$), there is a marked tendency for flow to be normal to the central portions of the y and z axes. That is, the symmetry is largely "unrotated" though there is already a loss of density at the corners. Fig. 20(b) shows, however, that for a time delay only slightly greater ($t/t_n = 0.072$), the particles are more extended in space near the central portions of the axes. That is,

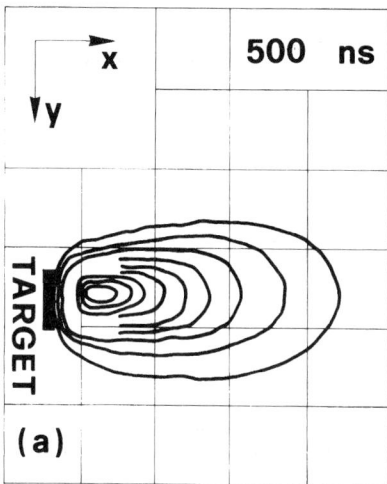

FIG. 18(a). Circular laser spot with a diameter of 3.0 mm.
Bombardment of Cu *in vacuum* (~0.001 Pa) with a Nd-YAG laser (532 nm, 6 ns, 2.7 J/cm^2, normal incidence). The overall fluorescence of the plume from 400 to 800 nm was measured with an *ICCD* camera having a temporal gate of ~5 ns. The light intensity is shown in terms of two-dimensional density contours which have an interval of 10% and are superimposed on a grid with a spacing of 4 mm. Due to Mele *et al.* [11].

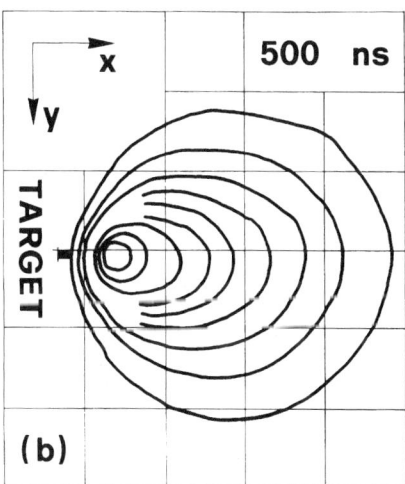

FIG. 18(b). Circular laser spot with a diameter of 0.2 mm. Because the smaller spot was obtained with a diaphragm, there was no change of fluence.

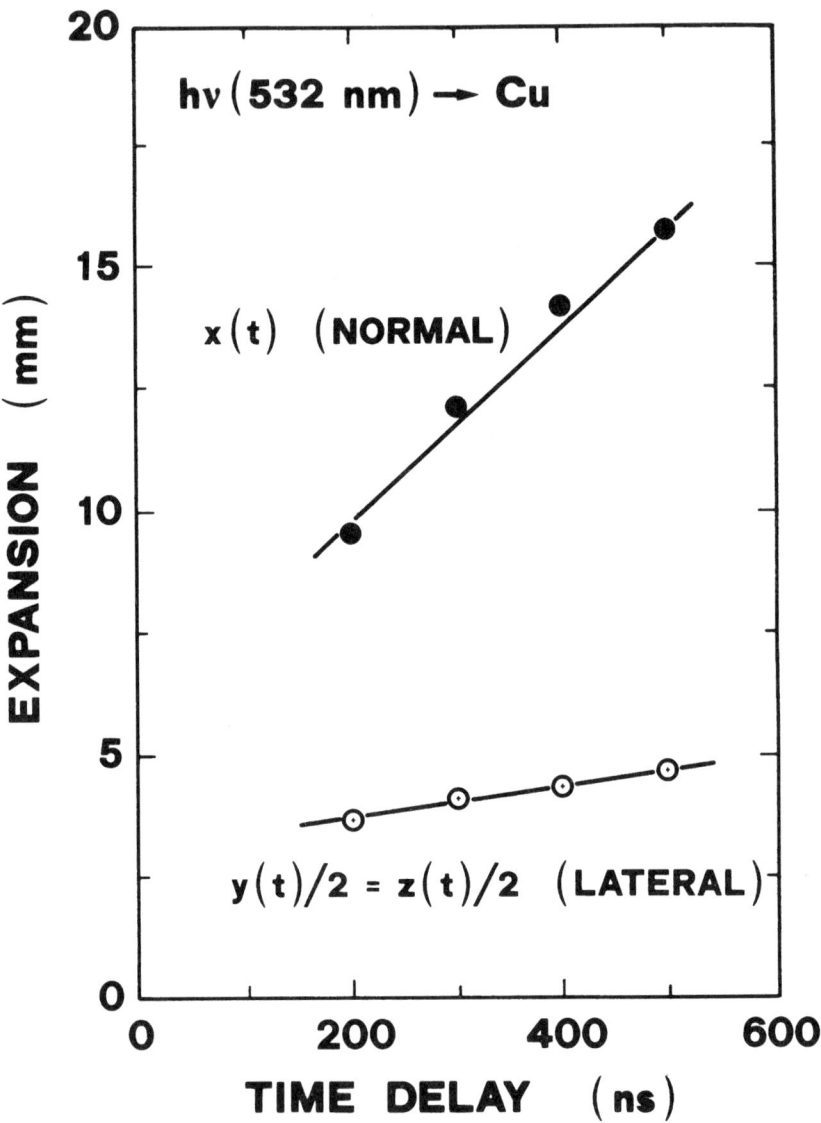

FIG. 19(a). Circular laser spot with a diameter of 3.0 mm.

Experimental time dependence of the dimensions $x(t)$ and $y(t)/2 = z(t)/2$ of the plume generated from a Cu target under the conditions of Fig. 18. As in Fig. 17, $y(t) = z(t)$ refers to the largest lateral dimensions of the plume but have been reduced by a factor of 2 on the assumption that the relevant motion was with respect to the midpoint of the plume. The x expansion is largely independent of the laser spot size but the y ($=z$) expansion is distinctly larger for the smaller spot. Due to Mele et al. [11].

"rotation" by 45° has begun to occur. Figs. 20(c) and 20(d) relate to still longer time delays ($t/t_n = 0.16$ and 0.32). In both cases, pronounced "rotations" were apparent. Unlike a flow governed by the diffusion equation, there is thus no strong tendency for the particles, at moderately long delay times, to lose

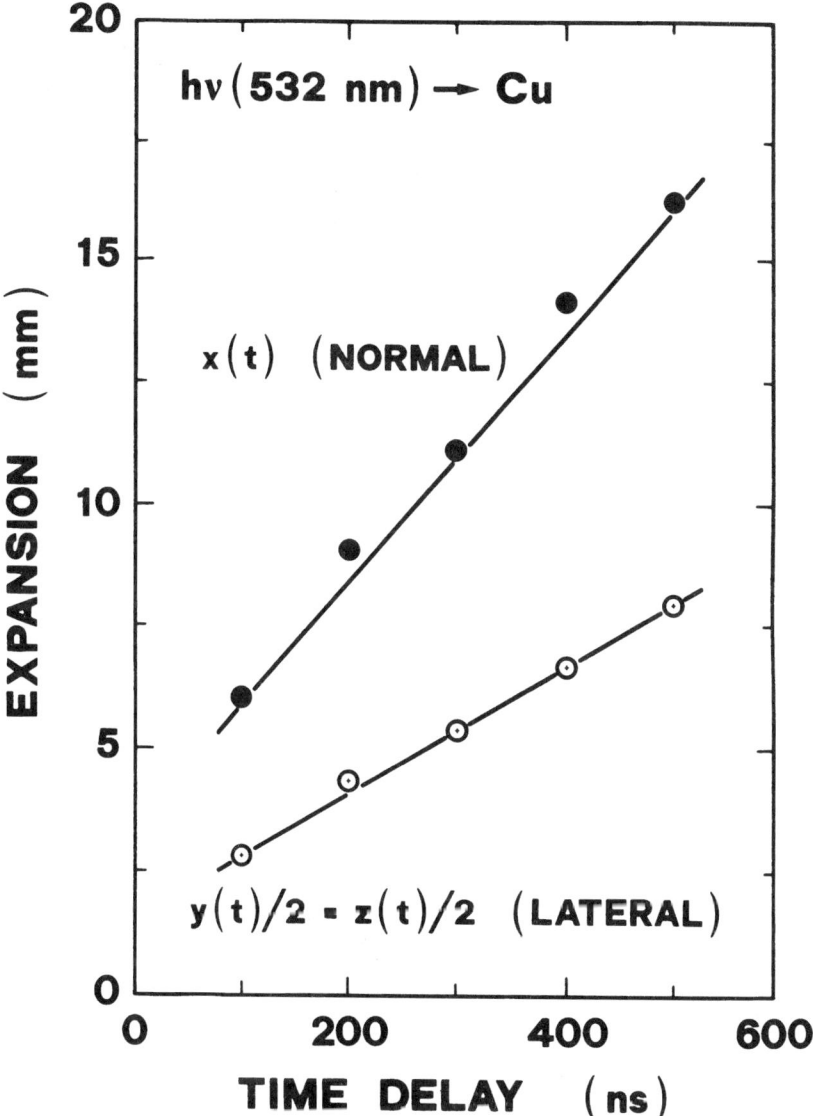

FIG. 19(b). Circular laser spot with a diameter of 0.2 mm.

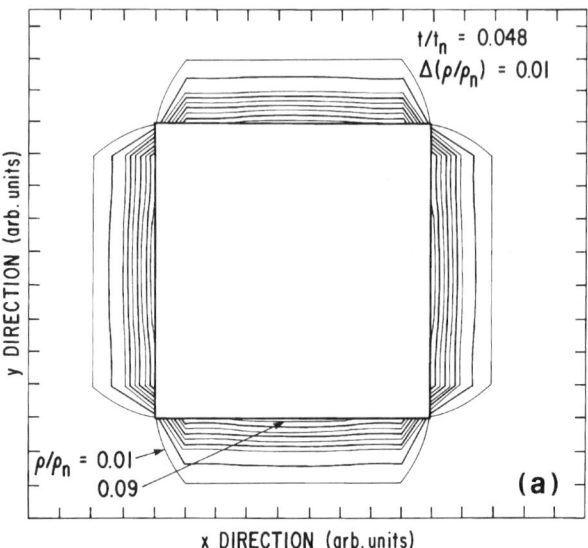

FIG. 20(a). $t/t_n = 0.048$. This constitutes a very short time delay, of order 200–700 ns after the removal of the wall. Such times are inferred in the experiments with PBSCCO (Fig. 17) to give unrotated plumes, and indeed "rotation" is minimal.

Particle-density contours calculated numerically with Eqs. (5.21)–(5.23) for an initially square reservoir of gas, which begins to expand in 2 dimensions when the four confining walls are removed at $t = 0$. The density contour intervals are indicated by $\Delta(\rho/\rho_n)$. The quantities ρ_n and t_n serve to normalize ρ and t so that the results will be unitless. Due to various sources [9, 107, 116].

memory of the reservoir shape. Figure 20 shows furthermore that the "rotations" were not a formal twisting of the plume but rather an accumulated shape change. (This is why quotation marks were put around the word "rotation.") Information is also available on the time scale. It was shown [9] that, to within a factor-of-two uncertainty in the reservoir temperature, the value $t/t_n = 0.048$ corresponds to the real time interval of 200–700 ns, whereas $t/t_n = 0.16$ corresponds to 600–2000 ns. The times for the calculations were thus very similar to those used in the experiments.

An important conclusion is that the rotation effect is a purely gas-dynamic phenomenon connected with the occurrence in Eqs. (5.21)–(5.23) of flow velocities u_y and u_z.

Fig. 20(e) relates to $t/t_n = 0.64$. An interesting and important effect is *tentatively* observed. The particles have not yet gone into free flight, and there is an incipient tendency for them to lose memory of the reservoir shape. The real times now lie in the interval 3000–9000 ns [9]. Similar effects occur also for heat diffusion, such that a square heat source cools more rapidly at the corners.

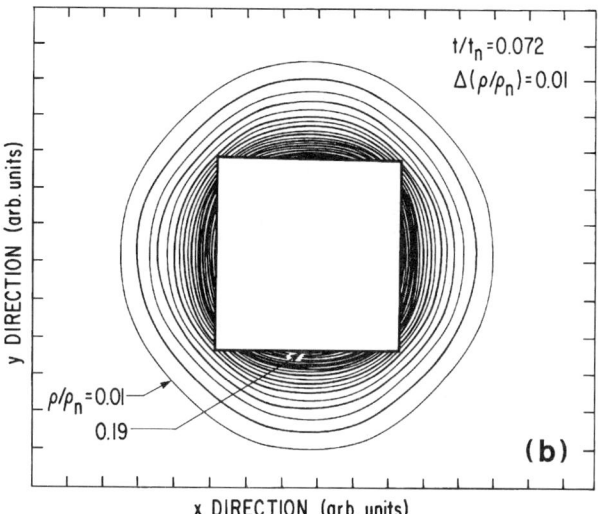

FIG. 20(b). $t/t_n = 0.072$. Also this is a short time delay.

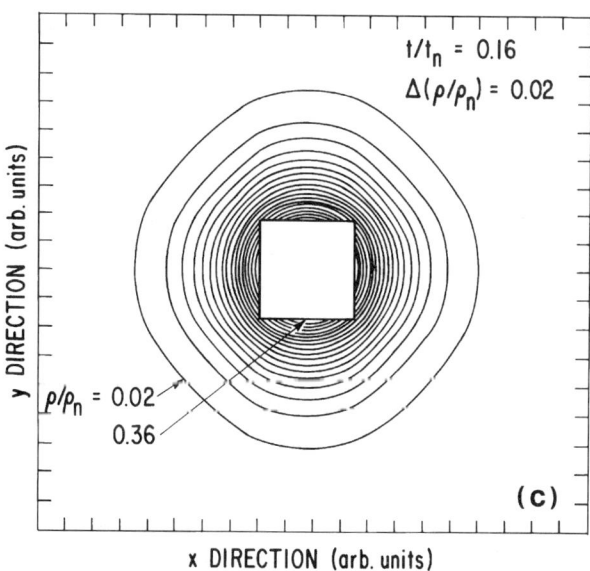

FIG. 20(c). $t/t_n = 0.16$. The time delay, of order 600–2000 ns, is here similar to the range of values used experimentally (Fig. 17). The "rotation" effect is fully developed.

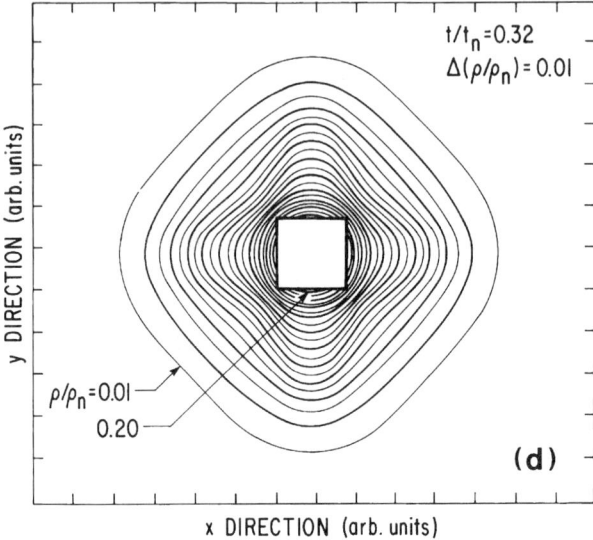

FIG. 20(d). $t/t_n = 0.32$. Similar to (c).

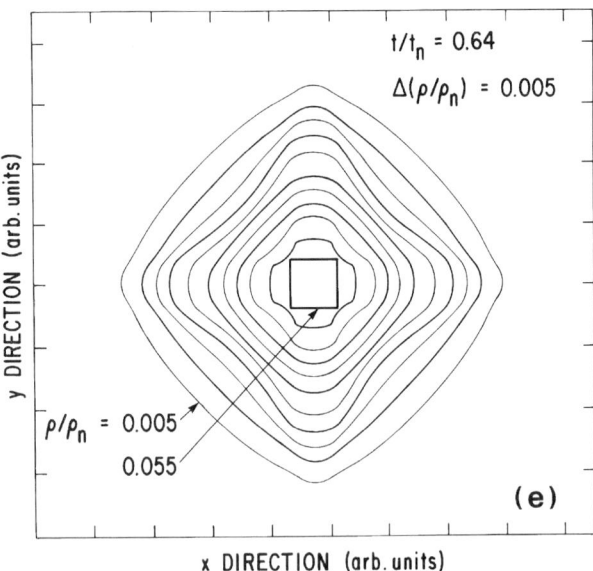

FIG. 20(e). $t/t_n = 0.64$. The time delay, of order 3000–9000 ns, is here significantly larger than that used experimentally. Interestingly, the plume particles are starting to lose memory of the reservoir shape. Thus even a gas-dynamic expansion finally behaves as would be expected for motion governed by the diffusion equation.

The change of symmetry in diffusion is much less dramatic, however, and in fact as time increases, a circular rather than rotated pattern develops. We regard the difference as being due to the fact that in a flow process there is a well-defined flow direction, for example, normal to the midpoint of each side in the case of a square, triangle, or rectangle.

5.3.2.4 Discussion.

The Primary Mechanism. Limited information on the primary mechanism relevant to PBSCCO and Cu follows from the motion in the x-direction. The plume is distinctly more extended in this direction by a factor of order 2.3 or 4.2 for PBSCCO and 3.4 or 4.1 for Cu. We first note that particles moving in the x-, y-, or z-directions will, provided that they are emitted by *normal outflow*, have a maximum flow velocity ($\hat{u}_i$) located at the expansion fronts ($\hat{x}$, $\hat{y}$, and $\hat{z}$) and given by Eq. (5.14):

$$\hat{u}_i = \hat{u}_x = \hat{u}_y = \hat{u}_z = \frac{2}{\gamma - 1} a_4 \qquad (5.14)$$

Eq. (5.14) describes the velocity of the expansion front when there is no ambient gas. Because it is valid for any dimensionality, there is no difficulty in applying it to plumes, even to those of complicated shape. The ratio of the x extension to those for $y/2$ or $z/2$ therefore follows as:

$$\frac{x \text{ extension}}{y/2 \text{ or } z/2 \text{ extension}} \approx 1.0 \qquad (5.24)$$

where we have taken into account that the motion in the y- or z-directions is (approximately) with respect to the midpoint of the plume.

For particles moving in the x-direction and originating by *effusion-like release* (equivalent to normal vaporization or boiling), a KL forms at the surface and the corresponding maximum flow velocity is slightly higher, being given by Eq. (5.12). Although it is unclear how to describe the sound speed applicable to the lateral motion, i.e., the particles moving in the y- or z-direction, it is likely that it would be only minimally lower than that in the x-direction. Eq. (5.24) therefore remains approximately valid.

For release by *phase explosion*, as well as for release due to an electronic sputtering process or (independently of the mechanism) when laser-plume interaction occurs [50], $\hat{u}_x$ will possibly be much higher than $\hat{u}_y$ and $\hat{u}_z$. Unfortunately it is not clear how to predict $\hat{u}_x$ because the results depend on the details of the particle-release mechanism and on how important the deceleration process is (Section 5.3.1.4). It is therefore best simply to assert that there is a tentative indication (on the basis of the observed expansion ratios) that the expansion normal to the target is significantly larger than the lateral expansion, as if the emission process were unusually energetic.

Additional mechanistic information is, however, contained in the absolute value of the expansion-front velocity normal to the target, which will be here identified with $3a_4$. In the case of PBSCCO (Fig. 17), we take the mean cation mass of the target to be 109 u. Then the relation $a_4 = (\gamma k_B T_4/m)^{1/2}$ gives $T \approx 600{,}000$ K, and we note simply that such temperatures cannot be real because real temperatures for a condensed phase are limited by T_{tc}. The highest known value of T_{tc} is that for W, $\sim 14{,}000$ K [22]. Similar reasoning applied to Cu (Fig. 19) gives $T \approx 500{,}000$ K, and similar conclusions are reached.

We conclude from all this that the sputtering mechanism is neither *normal* vaporization nor *normal* boiling. Likewise, the short length of the pulses (30 ns) suggests in a very general sense that *normal* boiling can be excluded, while we have already emphasized in Section 5.2.2 that the subsurface heating model is unlikely. It follows that what is involved is either phase explosion or an electronic sputtering process or (independently of the mechanism) laser-plume interaction. A similar situation was apparent also with AlN, except that here the unusually high fluence (~ 20 J/cm^2) favored phase explosion (Section 5.3.1.3).

The Secondary Mechanism. The work of [9–11] does not give explicit information on the secondary mechanism relevant to PBSCCO and Cu. In effect, it is not possible to choose between effusion-like release or normal outflow or outflow from an isolated spherical reservoir (Section 5.2.3). The reason is that information of the type seen in Figs. 11 and 12 is not available for PBSCCO and Cu.

5.3.2.5 Conclusions. We would finally emphasize that the work discussed here was apparently the first instance in which the rotation effect was directly studied by using three-dimensional fast photography. It was also the first attempt in which the effect was studied as a function of time. It was possible to conclude that the "rotation" was already present at ~ 250 ns and continued to evolve up to at least 2200 ns, when the x-direction expansion front was ~ 6 cm away from the target. (This distance applies explicitly to PBSCCO, Fig. 17, and by extrapolation of Fig. 19 also to Cu.) In addition, limited information on the primary emission mechanism for the particular targets, PBSCCO and Cu, and for the particular wave length, 248 nm, was obtained. It was argued that the ratios of the x to the $y/2$ and $z/2$ extensions conveyed little information except for a hint that the emission process was unusually energetic.

Interestingly, however, the x expansion-front velocities inferred impossible reservoir temperatures, $\sim 600{,}000$ K for PBSCCO and $\sim 500{,}000$ K for Cu. It is therefore possible to exclude both *normal* vaporization and *normal* boiling as the mechanisms of the laser sputtering. This leaves phase explosion or an electronic sputtering process or (independently of the mechanism) laser-plume interaction.

We would also emphasize that gas-dynamic solutions to the problem of the expansion of a square reservoir were examined. For short enough times, there was no "rotation" (as must be true); whereas for intermediate times, "rotation" was well developed (also reasonable). For the longest time used, however, there was a hint that the plume particles were losing memory of the reservoir shape.

5.3.3 The Polyimide (PI) Plume

5.3.3.1 General Comments.
Singleton et al. [108, 117] have studied the laser-pulse sputtering of the five polymers, polyimide (PI), polyethylene terephthalate (PET), polymethyl methacrylate (PMMA), polystyrene (PS), and polyethylene (PE). In each case, two groups of particles were identified, a first group with small enough molecular weight to pass through 0.2 or 1 μm sieves and a second group ("ejecta") that remained on the sieves. With PI and PET the light particles dominated, whereas with PMMA, PS, and PE the heavy particles dominated. In related work, it was shown that with PMMA the ejecta underwent a largely one-dimensional expansion away from the target surface [2, 12] whereas with PI [106–108, 116, 118–120] and PET [118] they were in part recondensed around the bombarded spot. It was assumed that the behavior of PI and PET was associated with the one-dimensional *outwards* expansion as in Fig. 11 being coupled to an unusually strong *sideways* expansion, which was rationalized in terms of the force exerted on the heavy particles by the light ones [106]. More precisely, it was proposed that the parameter (*force*) ∝ (*pressure*)/(*distance*) ∝ $m_1^{2/3} p_1$ was relevant, where m_1 is the particle mass of the ambient gas (air), p_1 is the ambient gas (air) pressure and "pressure" and "distance" relate to the light particles. This unusual situation was subsequently confirmed in images of the plume (Fig. 21), where one recognizes that there were, effectively, two *CF*. Because the velocities were not so extremely different as those relevant to Fig. 5, the two groups of particles were in contact and the ejecta were constrained to expand laterally.

Particles recondensed around the bombarded spot will be subsequently referred to as "debris." An important aspect of the debris was that, for bombarded spots with other than circular shape, the symmetry of the debris differed from that of the spot in such a way as to imply that the flow pattern involved a "rotation". (As in Section 5.3.2, we put quotation marks around the word "rotation" because a formal twisting does not occur.) Such "rotation" is closely related to that of PBSCCO and Cu (Section 5.3.2), except that here applied to recondensed particles, whereas with PBSCCO and Cu it applied to the plume at up to 6 cm above the target. The "rotated" pattern was relatively simple for low to medium fluences, up to perhaps 1 J/cm^2, but at high fluences became complicated. "Rotation" was not evident in the work of Taylor et al. [118] because the spot was nearly circular but was well developed in other work for square or triangular spots (Figs. 22(a) and 22(b)). For a rectangular spot (Fig. 22(c)) the symmetry was again "rotated" (see the superimposed ellipse) but

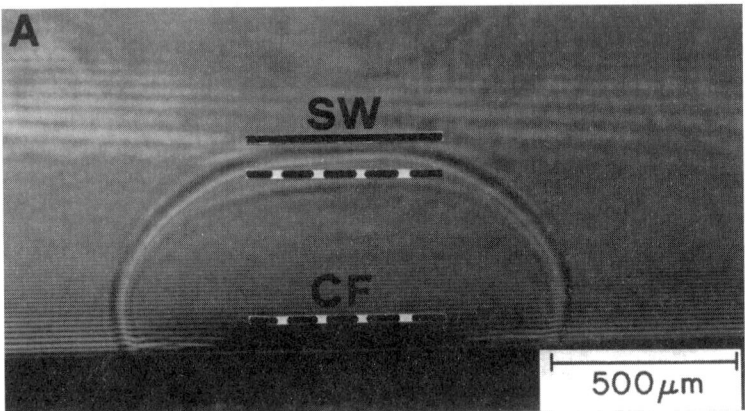

FIG. 21(a). Time delay of 190 ns.

Bombardment of polyimide (PI) *in air* with 1 pulse from a XeCl excimer laser (308 nm, ~20 ns, 2.3 J/cm^2, pulse diameter 800 μm, normal incidence). The PI had a thickness of 125 μm. The ejecta (dark region near surface), *CF* of the ejecta (first dashed line), *CF* of the light particles (second dashed line or lines), and the shock wave (*SW*) caused by the light particles (solid line) were photographed by firing parallel to the target surface a second ("probe") laser (596 nm, ~1 ns). Due to Kelly et al. [106, 107].

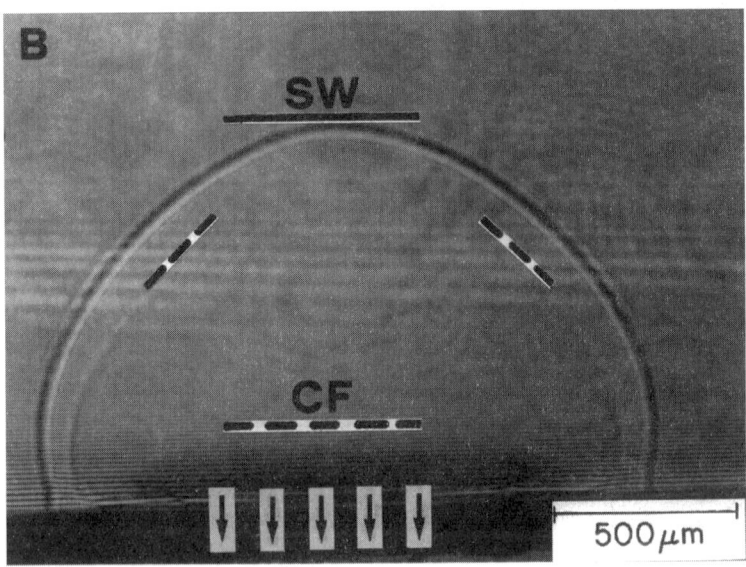

FIG. 21(b). Time delay of 410 ns, the arrows here representing the recondensation process.

showed the complicated form normally associated with high fluences. Closely related to the debris pattern found on the target is the symmetry of the plume (Section 5.3.2) and of a film deposited on a substrate. For example, an elliptical spot gave a film with a 90° "rotation" resembling Fig. 22(c) but without the complications evident in the latter [85, 114, 115].

Historically, the debris seen in Fig. 22 posed a significant problem for the fabrication of microcircuits, and it was asked whether extraneous effects such as charging [121] or temperature gradients were involved. It was also wondered to what extent debris formation correlated with the complexity of the particles involved. And we will again answer these questions by examining numerical solutions of the flow equations and will find that "rotation" of debris is, like "rotation" of a plume, a purely gas-dynamic effect. It follows that it *cannot* be avoided.

5.3.3.2 Pure Sideways Expansion. The real problem of describing debris formation is three-dimensional, involving a simultaneous outwards and sideways expansion into an ambient gas coupled with recondensation on the target surface (Fig. 23(a)). Because the real problem is prohibitively complex, it

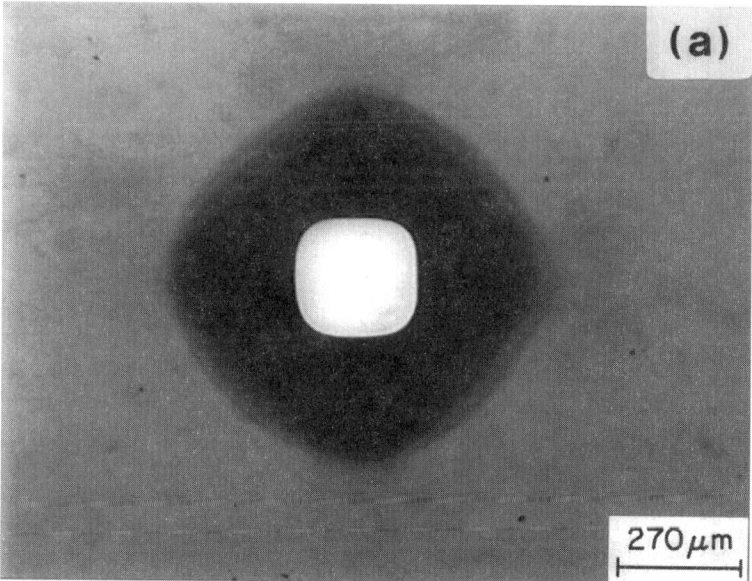

FIG. 22(a). Square mask with hole having a width of 270 μm.
Debris (i.e. recondensed particles) on the surface of a PI target with a thickness of 8 μm that has been bombarded *in air* with roughly 50 pulses from a XeCl excimer laser (308 nm, ~20 ns, ~0.25 J/cm^2, normal incidence). The light region near the center is a hole sputtered through the entire target. The dark region beyond the hole is debris. Due to various sources [106, 107, 116], including unpublished work by Braren [120].

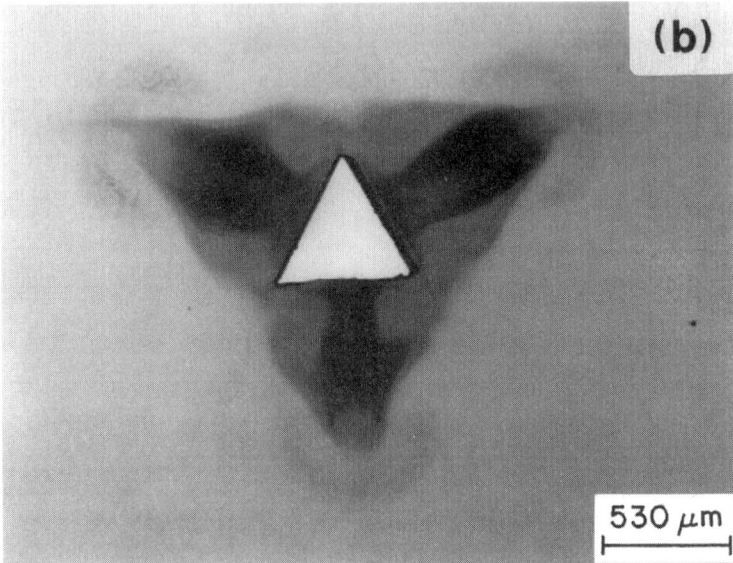

FIG. 22(b). Triangular mask with hole having a width of 530 μm.

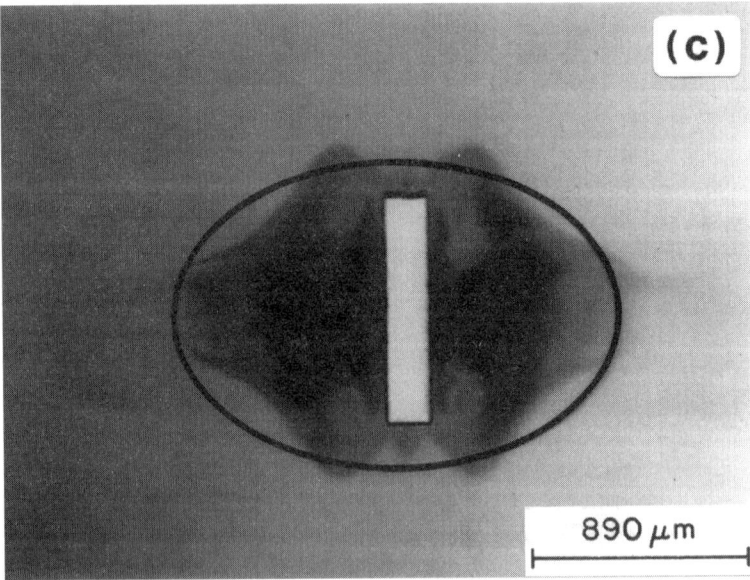

FIG. 22(c). Rectangular mask with hole dimensions of 165 × 890 μm. The superimposed ellipse serves to emphasize the "rotated" nature of the debris.

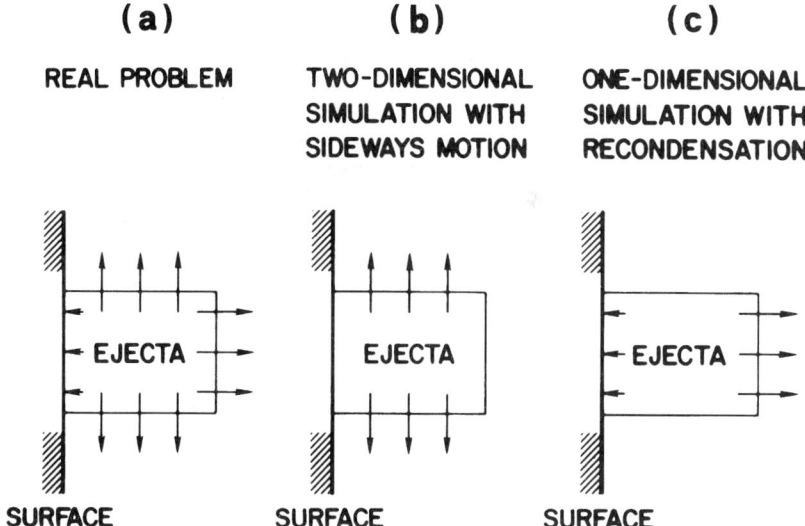

FIG. 23. The geometries relevant to the flow problems to be considered. (a) The real geometry applicable to all possible situations of laser sputtering, namely, a combined outwards and sideways expansion. (b) The simplified geometry used to analyze the expansion (including the *inferred* recondensation) for laser sputtering of materials for which the plume shows an unusually strong sideways expansion. (c) The simplified geometry used to analyze the expansion (including the *explicit* recondensation) for laser sputtering when the plume is reasonably one-dimensional. Due to Kelly and Miotello [107].

was divided into two parts: sideways (Fig. 23(b)) and outwards (Fig. 23(c)) expansion [116]. The sideways expansion was proposed as a means of simulating an ambient gas that impeded motion normal to the target (x-direction). Likewise, the outwards expansion, because it was combined with recondensation, served to quantify the recondensation process.

Numerical solutions of Eqs. (5.21)–(5.23) appropriate to sideways expansion of a square bombarded spot were already considered in Section 5.3.2.4 (Fig. 20). It was concluded that the "rotation" effect was a purely gas-dynamic phenomenon connected with the occurrence in Eqs. (5.21)–(5.23) of flow velocities u_y and u_z, thence of particle motion that tends to be normal to the central portions of the y and z axes. Table 3 summarizes the extent to which a square reservoir was exhausted for various values of t/t_n. It follows that the "rotations" were already apparent when the reservoir was roughly half empty.

Figs. 24(a) and 24(b) show numerical calculations for the sideways expansion of a rectangular reservoir of gas. The original configuration was a 4 × 16 array of points with $\rho/\rho_n = 1$, and the same spatial and time steps were used as for the

TABLE III. Numerical calculations of the extent to which a plume simulated by a square reservoir (8 × 8 array of points with $p/p_n = 1$) or by a rectangular reservoir (4 × 16 array of points with $p/p_n = 1$) and which expands sideways (y- and z-directions) into vacuum is exhausted for various values of t/t_n. t_n is the normalizer for time, being equivalent to the time constant for the reservoir to empty. It is assumed that there is no motion normal to the target (x-direction).

t/t_n	Exhaustion of a square (8 × 8) reservoir	Exhaustion of a rectangular (4 × 16) reservoir
0	1	1
0.024	0.980	0.974
0.048	0.933	0.916
0.072	0.868	0.833
0.16	0.601	0.498
0.32	0.302	0.214
0.64	0.073	0.050

square reservoir (Section 5.3.2.4). Rectangular symmetry was found to be preserved only for the shortest times (Fig. 24(a)), while for longer times the model showed a well-defined 90° "rotation" (Fig. 24(b)). Such a symmetry change is well known for deposits formed when using an elliptical bombarded spot [85, 114, 115] and, if the somewhat complicated form is overlooked, also for recondensed debris (Fig. 22(c)). Included in Table III is also a summary of the extent to which a rectangular reservoir was exhausted for each value of t/t_n.

The apparent "rotation" of the flow pattern seen in Fig. 22 (like that seen in Fig. 20) is, therefore, a purely gas-dynamic effect, i.e. it obeys the laws of flow as seen in Eqs. (5.21)–(5.23). Particles released from the corners can be regarded as moving into a two-dimensional region, so that they rapidly lose their density. That is, the flow is strongly divergent. Those emitted from the sides move more nearly in one dimension, so that the flow is less divergent. There was some indication, however, that for the longest times the particles began to lose memory of the reservoir (Fig. 20(e)).

5.3.3.3 Outwards Expansion Combined with Recondensation.

To obtain information on the extent of recondensation, a purely outwards expansion was considered (Fig. 23(c)) [116]. That is, the real three-dimensional expansion was simplified by restricting the motion to one-dimensional flow into vacuum normal to the target surface (x-direction). The flow equations were, accordingly, like Eqs. (5.21)–(5.23) but with only one flow velocity:

$$\frac{\partial \rho}{\partial t} + \frac{\partial \rho u_x}{\partial t} = 0 \qquad (5.25)$$

$$\frac{\partial \rho u_x}{\partial t} + \frac{\partial (p + \rho u_x^2)}{\partial x} = 0 \qquad (5.26)$$

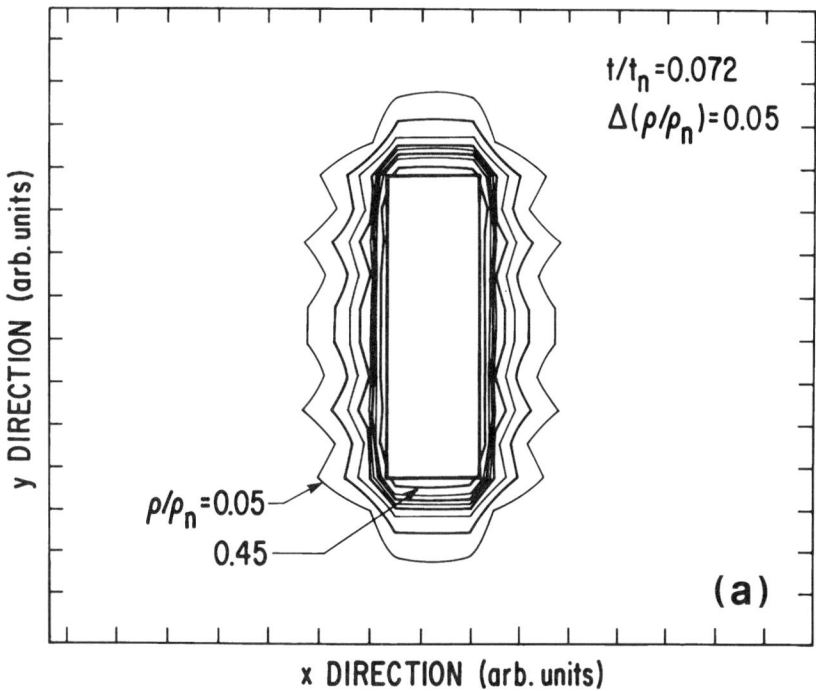

FIG. 24(a). $t/t_n = 0.072$.
Particle-density contours calculated numerically with Eqs. (5.21)–(5.23) for an initially rectangular reservoir of gas, which begins to expand in 2 dimensions when the four confining walls are removed at $t = 0$. Due to Kelly and Miotello [107].

Five different values of γ were used to simulate the behavior of particles ranging from atoms ($\gamma = 5/3$) to large molecules ($\gamma = 1$). Instead of confining walls being removed at $t = 0$, effusion-like expansion (Section 5.2.3.2) was permitted for $0 \leq t \leq \tau_r$. The surface was then abruptly sealed and, making the postulate that debris formation was equivalent to recondensation [106, 116] rather than an inward moving shock wave [122], recondensation was simulated by imposing the boundary condition $\rho = 0$ for the first spatial step *beneath* the surface. This model overestimates the extent of debris formation by a factor of about two, i.e., by the ratio (*debris area* + *spot area*)/(*spot area*). The results, summarized in Table IV, showed that recondensation increased significantly with the complexity of the particles involved. Also, the extent of recondensation approached a limit for t/τ_r exceeding about 10. To anticipate possible confusion, it should be emphasized that t_n and τ_r are different and unrelated. t_n is the normalizer for time, being effectively the *time constant* for a reservoir to empty, whereas τ_r is

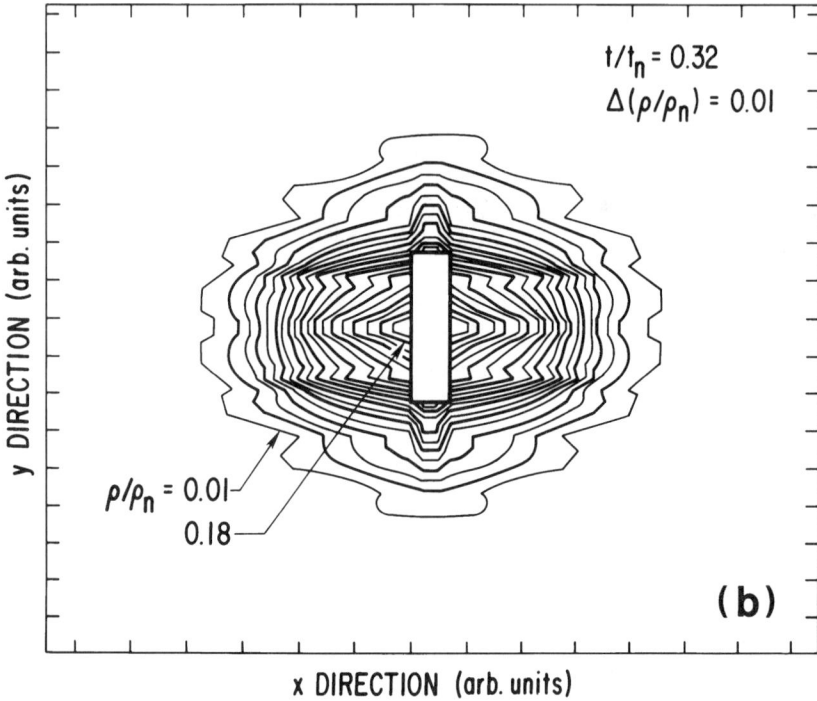

FIG. 24(b). $t/t_n = 0.32$.

TABLE IV. Numerical calculations of the extent of recondensation as would occur if the gas of a plume expanded outwards (x-direction) into vacuum away from the target surface instead of sideways (y- and z-directions) as in Table III.

	Extent of recondensation as a fraction of all emitted particles				
t/τ_r	$\gamma = 5/3$	$\gamma = 7/5$	$\gamma = 9/7$	$\gamma = 11/9$	$\gamma = 1$
1	0	0	0	0	0
2	0.024	0.040	0.048	0.048	0.070
4	0.052	0.080	0.094	0.102	0.14
8	0.072	0.110	0.13	0.14	0.19
14	0.083	0.13	0.15	0.17	0.23
20	0.089	0.14	0.17	0.18	0.26
30	0.095	0.15	0.18	0.21	0.29
$\infty^{(a)}$	0.104	0.174	0.215	~0.27	0.43

[a] Analytical values from [107].

the *release time*, governed by such considerations as the length of the laser pulse as well as heat loss from the target.

5.3.3.4 Conclusions. In summary, when the polymers PI (and probably also PET) were laser sputtered, debris was found around the bombarded spot. If the spot was square or triangular, and if the fluence was not too large, the debris showed remarkable symmetry changes as if the flow patterns involved a "rotation" (Figs. 22(a) and 22(b)). The debris for a rectangular spot was more complicated but still showed "rotation" (Fig. 22(c)). Debris formation was identified with recondensation, and it was shown by numerical solution of the flow equations that the changed symmetry was a purely gas-dynamic effect. Thus it was not necessary to consider extraneous effects such as might be related to an electric field [121] or temperature gradients. It was also shown that the numerical extent of recondensation (thence of debris formation) increased with the complexity of the particles involved, and it can be concluded that debris would respond by varying, for example, the laser fluence [123] or the chemical nature of the ambient gas [117].

5.4 General Conclusions

Laser-pulse sputtering differs from ion sputtering in that thermal-spike processes can be very important. We here use the term "thermal spike," rather than simply "thermal," because the high temperature is *transient* [39]. The thermal-spike processes include the following (Section 5.2.1):

1. Normal vaporization based on the Hertz-Knudsen equation (Eqs. (5.1)–(5.4)) and which would lead to temperature profiles as in Figs. 7(a) and 7(b).
2. Normal boiling at $T = T_b$, which is caused by heterogeneous bubble nucleation and which would lead to a temperature profile with the form $\partial T/\partial x \approx 0$ at and beneath the target surface. Normal boiling may, however, possibly be bypassed for two reasons: because the time scale is too short for heterogeneous nucleation [15, 21, 23] and because the density of heterogeneous nucleation sites is too small [15]. A major qualification must be made, however: if nuclei can form at the target *surface* [40, 41] then the lack of *heterogeneous* nuclei would be irrelevant.
3. Phase explosion, also termed explosive boiling or vapor explosion, of the type analyzed by Martynyuk [14, 21, 22, 61], by Fucke and Seydel [15, 23, 24], and by Avedisian [16]. This occurs at $T \approx 0.90 T_{tc}$, is caused by homogeneous nucleation, and gives a temperature profile with the form $\partial T/\partial x \approx 0$ at and beneath the target surface (Fig. 7(b), where T_s was similar to T_{tc}). As a result, there is a more or less violent expulsion of both vapor and equilibrium liquid droplets (Fig. 5). The rate of homogeneous nucleation rises catastrophically near T_{tc} [21] and therefore does not constitute a kinetic

obstacle. (On the other hand, the concept of a "kinetic obstacle" is very important for ion-beam mixing as caused by thermal spikes [39].)
4. Explosive release due to a subsurface temperature maximum ($\hat{T}$). This was analyzed incorrectly in early work [17, 18], although it has been repeatedly invoked in later work [25, 62, 63, 124, 125]. A more nearly correct analysis was given here in Section 5.2.2, where it was shown that the subsurface temperature maximum was a very minor effect provided it was calculated correctly (Fig. 7). This remained true even at temperatures above T_{tc} [19], temperatures that are not permitted *at the surface* of a condensed phase.

In addition to thermal-spike processes, electronic and collisional processes also exist [39]. That there should be electronic processes is not a surprise as the initial photon-target interaction is totally electronic for visible and ultraviolet photons. In fact, thermal-spike processes are important only because a major part of the electronic excitation is degraded to thermal energy on a very short time scale. Collisional (ballistic) interactions in the sense of momentum transfer in direct beam-surface interactions obviously cannot occur with laser pulses. To show this, it is sufficient to consider the maximum energy transfer, $\hat{E}_2$, in a binary collision between an incident particle with energy E_1 and rest mass m_1, and a target atom with mass m_2. If relativistic effects are *included*, the expression in units of eV is

$$\hat{E}_2 = 2E_1^2/m_2c^2 + 4M_1M_2E_1/(M_1+M_2)^2$$
$$= (E_1^2/M_2) \times 2.147 \times 10^{-9} + 4M_1M_2E_1/(M_1+M_2)^2 \quad \text{eV} \quad (5.27)$$

Values of $\hat{E}_2$ for photons, electrons, and He$^+$ colliding with Al atoms are compared in Table V. We note that photons (unless having a very high energy) transfer negligible energy, that the electrons of a high-voltage electron microscope can displace a small number of atoms, and that He$^+$, depending on the

TABLE V. Maximum energy transfer for binary collisions between photons, electrons, and He$^+$ with Al atoms. Calculated with Eq. (5.27).

Particle	Incident energy (eV)	Relativisitic part of maximum energy transfer, $\hat{E}_2$ (eV)	Total maximum energy transfer, $\hat{E}_2$ (eV)
Photon	10	8.0×10^{-9}	8.0×10^{-9}
	500,000	20	20
Electron	10	8.0×10^{-9}	8.1×10^{-4}
	500,000	20	61
He$^+$	10	8.0×10^{-9}	4.5
	500,000	20	225,000

energy, either does or does not cause significant displacements. In reaching these conclusions we assume a displacement energy threshold of $E_d = 25$ eV.

Indirect collisional effects do, however, exist with photons. It has been shown that if a plasma forms for any reason during laser-surface interaction either in vacuum or in an ambient gas, an explicit laser-plasma interaction begins. Neutrals in the plasma are found to have energies of ~5–50 eV, whereas the ions have energies in the range 100–1000 eV [126]. The ions have probably undergone explicit acceleration, while the energies of the neutrals can be understood in terms of conversion of ions into neutrals by charge exchange. The final result is that nearby surfaces are ion-bombarded [127, 128] (Fig. 25).

In pulsed-laser sputtering it is, however, necessary to distinguish between *primary* and *secondary* mechanisms (Sections 5.2.1 and 5.2.3). The former include the various processes discussed above, together with exfoliational and

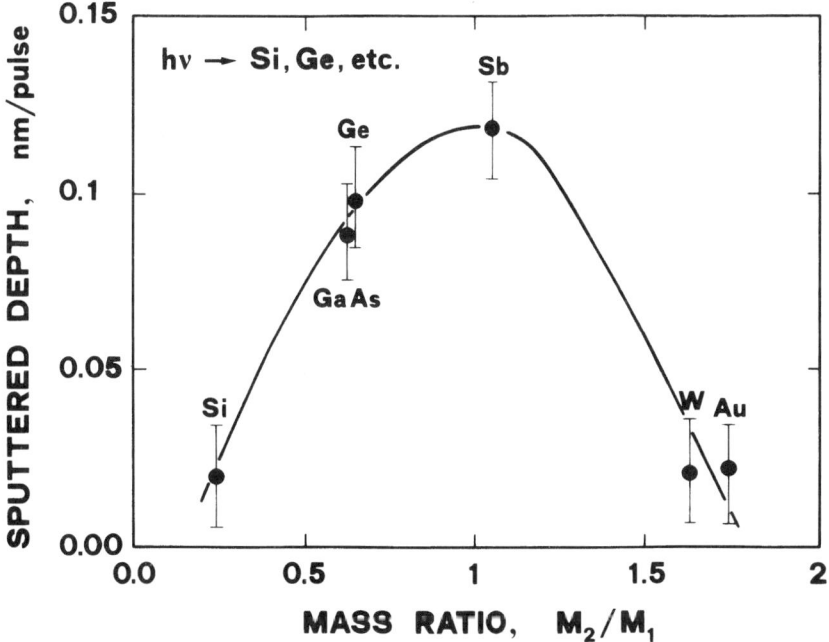

FIG. 25. Bombardment of Cd *in vacuum* with a large number of pulses from a Nd laser (1060 nm, 30 ns, 12–30 J/cm^2, incidence not specified). The plume particles interacted with various substrates. The important detail is that the substrates were held at 400°, the impacting Cd particles vaporized between pulses, and a cumulative sputtering process (rather than a deposition process) resulted. This enabled the thickness sputtered per pulse to be deduced, and it was found, as seen in the figure, that the sputtering scaled with the ratio $M_2/M_1 = M_2/112$. The sputtering in question was, by necessity, collisional. Adapted from a figure due to Akhsakhalyan *et al.* [127].

hydrodynamic sputtering. Interestingly, of these processes, only the hydrodynamic, equivalent to droplet emission [13, Fig. 3.9] does not have a close analog in ion sputtering [13, Table 3.1].

The *secondary* mechanisms include various types of pulsed flow (i.e., plume formation) which differ both depending on whether the release is from the surface ("effusion-like") or from a reservoir ("outflow"), and also depending on whether particles which are scattered back to the target surface reflect or recondense [13, Table 3.2]. To a limited extent similar processes are found with ion sputtering, but only (as with condensed-gas or organic targets) when the yield is unusually high. The flow processes often cause the system to lose memory of the primary mechanism. For example, rather independently of the primary velocity distribution, a shifted Maxwellian form as in Eq. (5.9) will tend to apply. Conversely, observing a shifted Maxwellian distribution does not necessarily indicate a thermal primary mechanism!

It is an attempt to describe the pulsed flow (i.e., the plume), which was the main object of the present work. (By contrast, mechanisms are discussed in [13, 39].) In the past, the plume was most often imaged photographically [1–4], and one then had a limited possibility of distinguishing different secondary processes. For example, the particles flowing from laser-bombarded YBCO showed a pattern like that for *outflow* [13, Fig. 3.14], while those from laser-bombarded polymethylmethacrylate (PMMA) showed a pattern more nearly like that for *effusion* [13, Fig. 3.16]. Such photographs do not distinguish between reflection and recondensation. However, the relevance of recondensation is demonstrated explicitly in those instances where there is sideways expansion and debris therefore extends beyond the bombarded spot (Fig. 22).

Very recently a more subtle approach to plume analysis was developed, based variously on measuring the ground-state neutral density [4, 5], light emission from particular spectral transitions [6], or the overall light intensity (fluorescence) [7–11]. This can be done from several points of view (Section 5.3):

1. With AlN (Section 5.3.1), the overall light intensity was measured in two dimensions (Fig. 11) and various features were identified. The limit of the light in the x-direction was identified with the contact front (CF). The CF was subject to deceleration (Fig. 12), which is easily understood in terms of the presence of ambient N_2 in the particular experiments. Somewhat surprisingly, however, there was also a *stationary* fluorescence maximum near the target. It can be tentatively understood as due to braking (i.e., collisions) as schematized in Fig. 13. We recognize a sequence of normally vaporized particles ("A"), phase-exploded particles ("B"), and again particles originating by vaporization ("C"). Braking is also well known in other work, the example of Lindley et al. [4] which is seen in Fig. 14 being particularly

relevant because there was no ambient gas at all. This *fluorescence* maximum is probably also a *particle-density* maximum. Finally, it was noted that a *moving* fluorescence maximum occurred near the *CF* with AlN at higher ambient N_2 pressures. This maximum can be tentatively understood in terms of electron scattering by the ambient gas in the region of the *CF* and, as such, *would not* represent a particle-density maximum.
2. With PBSCCO and Cu (Section 5.3.2), the overall light intensity was measured in three dimensions (Figs. 16 and 18). In both cases, there was a well-defined *CF* but, as is appropriate when there is no ambient gas, the *CF* was not subject to deceleration (Figs. 17 and 19). The only fluorescence maximum moved slowly at about 25% of u_{cf} and was located near the target. Significantly, there was no moving fluorescence maximum near the *CF*. This result strongly supports the idea that the feature observed with AlN was a false maximum related to the presence of ambient gas. As indicated above in connection with AlN, the ambient gas causes electrons to be scattered back to the plume.

Because the measurements were in three dimensions rather than two, important additional information was obtained with PBSCCO and Cu. When rectangular laser spots were used the *lateral* expansion of the plume was more marked normal to the smaller spot dimension (Figs. 15–17). When circular laser spots of different size were used, the *lateral* expansion was more marked for the smaller spot (Figs. 18 and 19). It was shown by solving the flow equations numerically in two dimensions that the origin of the lateral expansion was a purely gas-dynamic effect and, furthermore, that there *was no* formal rotation (Figs. 20 and 24).
3. Finally, polyimide (PI) (Section 5.3.3) was considered. It was already established in other work [106, 107, 116, 120] that the PI plume was rather unusual in that it consisted of slow-moving heavy particles ("ejecta") and fast-moving light particles (Fig. 21). The velocities were, however, not so extremely different as those relevant to Fig. 5, where the nonparticulates moved about 100 times faster than the particulates. As a result, the two groups were in contact (Fig. 21), the ejecta were constrained to expand laterally (Fig. 23(b)), and (because of the recondensation) the ejecta finally appeared as a deposit on the target (Fig. 22) But the same effect as was observed for the plumes of PBSCCO and Cu and for deposits on substrates [85, 114, 115] again took place, such that the lateral expansion was not uniform. Another way of describing this nonuniformity is that a "rotation" effect sets in, as in Fig. 15. For example, the deposit from a square spot shows an apparent "rotation" of 45°, while triangular and rectangular spots show "rotations" to other extents (Figs. 20, 22, and 24). As already pointed out above in connection with PBSCCO and Cu this is a purely gas-dynamic effect and, furthermore, does not involve a formal rotation.

References

1. Harris, T. J. *IBM J. Res. Develop.* **7**, 342 (1963).
2. Braren, B., Casey, K. G., and Kelly, R. *Nucl. Instr. Meth.* **B58**, 463 (1991).
3. Gupta, A., Braren, B., Casey, K. G., Hussey, B. W., and Kelly, R. *Appl. Phys. Lett.* **59**, 1302 (1991).
4. Lindley, R. A., Gilgenbach, R. M., Ching, C. H., Lash, J. S., and Doll, G. L. *J. Appl. Phys.* **76**, 5457 (1994).
5. Gilgenbach, R. M., and Ventzek, P. L. G. *Appl. Phys. Lett.* **58**, 1597 (1991).
6. Pietsch, W., Dubreuil, B., and Briand, A. *Appl. Phys.* **B61**, 267 (1995).
7. Geohegan, D. B. *Appl. Phys. Lett.* **60**, 2732 (1992).
8. Geohegan, D. B. *Thin Sol. Films* **220**, 138 (1992).
9. Mele, A., Giardini Guidoni, A., Kelly, R., Miotello, A., Orlando, S., and Teghil, R. *Appl. Surf. Sci.* **96–98**, 102 (1996).
10. Mele, A., Giardini Guidoni, A., Kelly, R., Miotello, A., Orlando, S., Teghil, R., and Flamini, C. *Nucl. Instr. Meth.* **B116**, 257 (1996).
11. Mele, A., Giardini Guidoni, A., Kelly, R., Miotello, A., and Orlando, S. *Mater. Res. Soc. Symp.* **397**, 87 (1996).
12. Kelly, R., Miotello, A., Braren, B., Gupta, A., and Casey, K. *Nucl. Instr. Meth.* **B65**, 187 (1992).
13. Kelly, R., and Miotello, A. Chapter 3 of *Pulsed Laser Deposition of Thin Films*, D. B. Chrisey and G. K. Hubler (eds.), p. 56. Wiley, New York, 1994.
14. Martynyuk, M. M. *Sov. Phys. Tech. Phys.* **21**, 430 (1976).
15. Fucke, W., and Seydel, U. *High Temp.-High Pressures* **12**, 419 (1980).
16. Avedisian, C. T. *J. Phys. Chem. Ref. Data* **14**, 695 (1985).
17. Dabby, F. W., and Paek, U.-C. *IEEE J. Quantum Optics* **QE-8**, 106 (1972).
18. Gagliano, F. P., and Paek, U.-C. *Appl. Optics* **13**, 274 (1974).
19. Kelly, R., and Miotello, A. *Appl. Surf. Sci.* **96–98**, 205 (1996).
20. Pound, G. M. *J. Phys. Chem. Ref. Data* **1**, 135 (1972).
21. Martynyuk, M. M. *Sov. Phys. Tech. Phys.* **19**, 793 (1974).
22. Martynyuk, M. M. *Russ. J. Phys. Chem.* **57**, 494 (1983).
23. Seydel, U., and Fucke, W. *J. Phys. F: Metal Phys.* **8**, L157 (1978).
24. Seydel, U., Bauhof, H., Fucke, W., and Wadle, H. *High Temp.-High Pressures* **11**, 635 (1979).
25. Singh, R. K., Bhattacharya, D., and Narayan, J. *Appl. Phys. Lett.* **57**, 2022 (1990).
26. Miotello, A., and Kelly, R. *Appl. Phys. Lett.* **67**, 3535 (1995).
27. Kelly, R., and Dreyfus, R. W. *Surf. Sci.* **198**, 263 (1988).
28. Miotello, A., Peterlongo, A., and Kelly, R. *Nucl. Instr. Meth.* **B101**, 148 (1995).
29. Bourdon, E. B. D., Das, P., Harrison, I., Polanyi, J. C., Segner, J., Stanners, C. D., Williams, R. J., and Young, P. A. *Faraday Disc. Chem. Soc.* **82**, 343 (1986).
30. Kelly, R. *J. Chem. Phys.* **92**, 5047 (1990).
31. Cowin, J. P., Auerbach, D. J., Becker, C., and Wharton, L. *Surf. Sci.* **78**, 545 (1978).
32. NoorBatcha, I., Lucchese, R. R., and Zeiri, Y. *J. Chem. Phys.* **86**, 5816 (1987).
33. NoorBatcha, I., Lucchese, R. R., and Zeiri, Y. *J. Chem. Phys.* **89**, 5251 (1988).
34. Kelly, R., and Miotello, A. *Appl. Phys.* **B57**, 145 (1993).
35. Liepmann, H. W., and Roshko, A. *Elements of Gasdynamics*, p. 79. Wiley, New York, 1957.
36. Freiwald, D. A. *J. Appl. Phys.* **43**, 2224 (1972).

37. Siano, S., Pini, R., Salimbeni, R., and Vannini, M. In *Laser Material Processing*, p. 585. SPIE Vol. 2207, 1994.
38. Ytrehus, T. In *Rarefied Gas Dynamics*, J. L. Potter (ed.), Vol. II, p. 1197. AIAA, New York, 1977.
39. Kelly, R., and Miotello, A. *Nucl. Instr. Meth. B* **122**, 374 (1997).
40. Park, H. K., Grigoropoulos, C. P., Poon, C. C., and Tam, A. C. *Appl. Phys. Lett.* **68**, 596 (1996).
41. Tam, A. C. IBM Almaden Research Center, San Jose, U.S.A. Personal communication, 1996.
42. Hirth, J. P., and Pound, G. M. *J. Chem. Phys.* **26**, 1216 (1957).
43. Haglund, R. F., and Dickinson, J. T. Chapter 2 of this Volume.
44. Fleischer, R. L., Price, P. B., and Walker, R. M. *J. Appl. Phys.* **36**, 3645 (1965).
45. Itoh, N., and Nakayama, T. *Phys. Lett.* **92A**, 471 (1982).
46. Nakai, Y., Hattori, K., Okano, A., Itoh, N., and Haglund, R. F. *Nucl. Instr. Meth.* **B58**, 452 (1991).
47. Haglund, R. F., and Kelly, R. *Kgl. Danske Vid. Selsk., Mat. Fys. Medd.* **43**, 527 (1993).
48. Helvajian, H., and Welle, R. *J. Chem. Phys.* **91**, 2616 (1989).
49. Hoheisel, W., Vollmer, M., and Träger, F. In *Laser Ablation: Mechanisms and Applications*, J. C. Miller and R. F. Haglund (eds.), p. 77. Springer-Verlag, Berlin, 1991.
50. Mele, A., Giardini Guidoni, A., Kelly, R., Flamini, C., and Orlando, S. *Appl. Surf. Sci.* **109/110**, 584 (1997).
51. Dreyfus, R. W., Walkup, R. E., and Kelly, R. *Rad. Effects* **99**, 199 (1986).
52. Wautelet, M., and Laude, L. D. *Appl. Phys. Lett.* **36**, 197 (1980).
53. Jöst, B., Schueler, B., and Krueger, F. R. *Z. Naturforsch.* **37a**, 18 (1982).
54. Kissel, J., and Krueger, F. R. *Appl. Phys.* **A42**, 69 (1987).
55. Townsend, P. D., Browning, R., Garlant, D. J., Kelly, J. C., Mahjoobi, A., Michael, A. J., and Saidoh, M. *Rad. Effects* **30**, 55 (1976).
56. Drowart, J., DeMaria, G., Burns, R. P., and Inghram, M. G. *J. Chem. Phys.* **32**, 1366 (1960).
57. Farber, M., Srivastava, R. D., and Uy, O. M. *J. Chem. Soc., Faraday I* **68**, 249 (1972).
58. Palik, E. D., ed. *Handbook of Optical Constants of Solids*, pp. 280–406. Academic, Orlando, 1985.
59. Geohegan, D. B. *Appl. Phys. Lett.* **62**, 1463 (1993).
60. Geohegan, D. B. Chapter 5 of *Pulsed Laser Deposition of Thin Films*, D. B. Chrisey and G. K. Hubler (eds.), p. 156. Wiley, New York, 1994.
61. Martynyuk, M. M. *Phys. Combustion and Explosion* **13**, 178 (1977).
62. Otsubo, S., Minamikawa, T., Yonezawa, Y., Morimoto, A., and Shimizu, T. *Jap. J. Appl. Phys.* **29**, L73 (1990).
63. Bhattacharya, D., Singh, R. K., and Holloway, P. H. *J. Appl. Phys.* **70**, 5433 (1991).
64. Kelly, R., Cuomo, J. J., Leary, P. A., Rothenberg, J. E., Braren, B. E., and Aliotta, C. F. *Nucl. Instr. Meth.* **B9**, 329 (1985).
65. Rothenberg, J. E., and Kelly, R. *Nucl. Instr. Meth.* **B1**, 291 (1984).
66. Braun, M., Whitton, J. L., and Emmoth, B. *J. Nucl. Mat.* **85/86**, 1091 (1979).
67. Kelly, R., and Rothenberg, J. E. *Nucl. Instr. Meth.* **B7/8**, 755 (1985).
68. Novis, Y., Pireaux, J. J., Brezini, A., Petit, E., Caudano, R., Lutgen, P., Feyder, G., and Lazare, S. *J. Appl. Phys.* **64**, 365 (1988).

69. Nakata, T., Kannari, F., and Obara, M. *Optoelectron., Devices Technol.* **8**, 179 (1993).
70. Anisimov, S. I., Gol'berg, S. M., Malomed, B. A., and Tribel'ski, M. I. *Sov. Phys. Doklady* **27**, 130 (1982).
71. Besocke, K., Berger, S., Hofer, W. O., and Littmark, U. *Rad. Effects* **66**, 35 (1982).
72. Hofer, W. O., Besocke, K., and Stritzker, B. *Appl. Phys.* **A30**, 83 (1983).
73. Cheng, Y.-T., van Rossum, M., Nicolet, M.-A., and Johnson, W. L. *Appl. Phys. Lett.* **45**, 185 (1984).
74. Lieb, K.-P., Bolse, W., and Uhrmacher, M. *Nucl. Instr. Meth.* **B89**, 277 (1994).
75. Miotello, A., and Kelly, R. *Surf. Sci.* **314**, 275 (1994).
76. Knystautas, E. J., Lo Russo, S., Kelly, R., and Miotello, A. *J. Appl. Phys.* **80**, 2702 (1996).
77. Guillot-Noël, O., Gomez-San Roman, R., Perrière, J., Hermann, J., Craciun, V., Boulmer-Leborgne, C., and Parboux, P. *J. Appl. Phys.* **80**, 1803 (1996).
78. Flamini, C. Dipart. di Chimica, Univ. di Roma, Roma, Italy. Personal communication, 1996.
79. Luk'yanchuk, B. S. General Physics Institute, Moscow, Russia. Personal communication, 1996.
80. Cercignani, C. In *Rarefied Gas Dynamics*, S. S. Fisher (ed.), Vol. I, p. 305. AIAA, New York, 1981.
81. Kelly, R., and Dreyfus, R. W. *Nucl. Instr. Meth.* **B32**, 341 (1988).
82. Sibold, D., and Urbassek, H. M. *Phys. Fluids* **A4**, 165 (1992).
83. Greenspan, H. P., and Butler, D. S. *J. Fluid Mech.* **13**, 101 (1962).
84. Kools, J. C. S., and Dieleman, J. *J. Appl. Phys.* **74**, 4163 (1993).
85. Kools, J. C. S., Baller, T. S., De Zwart, S. T., and Dieleman, J. *J. Appl. Phys.* **71**, 4547 (1992).
86. Kelly, R., Mele, A., Miotello, A., Giardini Guidoni, A., Hastie, J. W., Schenck, P. K., and Okabe, H. (in preparation).
87. Pastrňák, J., and Roskovcová, L. *Phys. Stat. Sol.* **26**, 591 (1968).
88. Yim, W. M., Stofko, E. J., Zanucchi, P. J., Pankove, J. I., Ettenberg, M., and Gilbert, S. L. *J. Appl. Phys.* **44**, 292 (1973).
89. Aita, C. R., Kubiak, C. J. G., and Shih, F. Y. H. *J. Appl. Phys.* **66**, 4360 (1989).
90. Radhakrishnan, G. *J. Appl. Phys.* **78**, 6000 (1995).
91. Aita, C. R., and Tait, W. S. *Nanostruct. Mater.* **1**, 269 (1992).
92. DeSilva, M. J., Pedraza, A. J., and Lowndes, D. H. *J. Mater. Res.* **9**, 1019 (1994).
93. Cao, S., Pedraza, A. J., and Allard, L. F. *J. Mater. Res.* **10**, 54 (1995).
94. Dreger, L. H., Dadape, V. V., and Margrave, J. L. *J. Phys. Chem.* **66**, 1556 (1962).
95. Liu, J. M., Yen, R., Kurz, H., and Bloembergen, N. *Appl. Phys. Lett.* **39**, 755 (1981).
96. Liu, J. M., Lompre, L. A., Kurz, H., and Bloembergen, N. *Appl. Phys.* **A34**, 25 (1984).
97. *Engineering Property Data on Selected Ceramics. Volume 1, Nitrides*. Metals and Ceramics Inform. Center, Columbus, 1976.
98. Vinogradov, V. L., Kostanovskii, A. V., and Kirillin, A. V. *High. Temp.-High Pressures* **23**, 685 (1991).
99. Kelly, R., and Miotello, A. (unpublished, 1997).
100. Kelly, R., Miotello, A., Mele, A., Giardini Guidoni, A., Hastie, J. W., Schenk, P. K., and Okabe, H. *Appl. Surf. Sci.* (submitted, 1997).

101. Dyer, P. E., Greenough, R. D., Issa, A., and Key, P. H. *Appl. Phys. Lett.* **53**, 534 (1988).
102. Ohkoshi, M., Yoshitake, T., and Tsushima, K. *Appl. Phys. Lett.* **64**, 3340 (1994).
103. Scott, K., Huntley, J. M., Phillips, W. A., Clarke, J., and Field, J. E. *Appl. Phys. Lett.* **57**, 922 (1990).
104. Dyer, P. E., Issa, A., and Key, P. H. *Appl. Phys. Lett.* **57**, 186 (1990).
105. Peterlongo, A., Miotello, A., and Kelly, R. *Phys. Rev.* **E50**, 4716 (1994).
106. Kelly, R., Miotello, A., Braren, B., and Otis, C. E. *Appl. Phys. Lett.* **60**, 2980 (1992).
107. Kelly, R., and Miotello, A. *Nucl. Instr. Meth.* **B91**, 682 (1994).
108. Singleton, D. L. National Research Council of Canada, Ottawa. Personal communication, 1992.
109. Gilgenbach, R. M., Ching, C. H., Lash, J. S., and Lindley, R. A. *Phys. Plasmas* **1**, 1619 (1994).
110. Horwitz, J. S. Naval Research Lab., Washington, U.S.A. Personal communication, 1994.
111. Mehlman, G., Chrisey, D. B., Horwitz, J. S., Burkhalter, P. G., Auyeung, R. C. Y., and Newman, D. A. *Appl. Phys. Lett.* **63**, 2490 (1993).
112. Martynyuk, M. M. *Radio Eng. and Electronic Phys.* **25**, 100 (1980).
113. Pérez Casero, R., Kerhervé, F., Enard, J. P., Perrière, J., and Regnier, P. *Appl. Surf. Sci.* **54**, 147 (1992).
114. Singh, R. K., and Narayan, J. *Phys. Rev.* **B41**, 8843 (1990).
115. Afonso, C. N., Serna, R., Catalina, F., and Bermejo, D. *Appl. Surf. Sci.* **46**, 249 (1990).
116. Miotello, A., Kelly, R., Braren, B., and Otis, C. E. *Appl. Phys. Lett.* **61**, 2784 (1992).
117. Singleton, D. L., Paraskevopoulos, G., and Irwin, R. S. *J. Appl. Phys.* **66**, 3324 (1989).
118. Taylor, R. S., Leopold, K. E., Singleton, D. L., Paraskevopoulos, G., and Irwin, R. S. *J. Appl. Phys.* **64**, 2815 (1988).
119. Küper, S., and Brannon, J. *Appl. Phys. Lett.* **60**, 1633 (1992).
120. Braren, B. IBM Research Center, Yorktown Heights, NY. Personal communication, 1990.
121. von Gutfeld, R. J., and Srinivasan, R. *Appl. Phys. Lett.* **51**, 15 (1987).
122. Koren, G., and Oppenheim, U. P. *Appl. Phys.* **B42**, 41 (1987).
123. Becker, C. H., and Pallix, J. B. *J. Appl. Phys.* **64**, 5152 (1988).
124. Dyer, P. E., Farrar, S. R., and Key, P. H. *Appl. Surf. Sci.* **54**, 255 (1992).
125. Dyer, P. E., Farrar, S. R., and Key, P. H. *Appl. Phys. Lett.* **60**, 1890 (1992).
126. Akhsakhalyan, A. D., Bityurin, Yu. A., Gaponov, S. V., Gudkov, A. A., and Luchin, V. I. *Sov. Phys. Tech. Phys.* **27**, 969 (1982).
127. Akhsakhalyan, A. D., Bityurin, Yu. A., Gaponov, S. V., Gudkov, A. A., and Luchin, V. I. *Sov. Phys. Tech. Phys.* **27**, 973 (1982).
128. Gaponov, S. V., Luskin, B. M., Nesterov, B. A., and Salashchenko, N. N. *Sov. Phys. Sol. State* **19**, 1736 (1977).
129. Kelly, R. *Nucl. Instr. Meth.* **B46**, 441 (1990).

6. SURFACE CHARACTERIZATION

Marika Schleberger, Sylvia Speller, and Werner Heiland
FB Physik
Universität Osnabrück, Germany

6.1 Introduction

Of major interest for the analysis of thin films are, beside the thickness and uniformity of the films, their chemical composition and their structure. The analytical methods are essentially surface analytical tools. In the context of laser ablation and the use of laser ablation for the making of thin films, it is useful to distinguish between in situ and ex situ methods. For example the "ultimate" surface structural methods, e.g., the atomic force microscopy (AFM) and scanning tunneling microscopy (STM), are not in situ methods. A typical in situ structural probe is RHEED (Reflection High Energy Electron Diffraction), also used for MBE (molecular beam epitaxy) film growth. A second aspect is the compatibility of the analytical methods with the laser ablation experiment or technical setup. A method that intereferes with the film-making process is rather useless.

Table I lists surface analytical methods by the basic phenomena used, e.g., elastic diffraction, charge exchange, scattering, bond breaking, and ionization. Most of the elastic scattering techniques have the advantage of being "remote" probes, e.g., the ion-scattering methods. They can be set up without interfering with the film making. The disadvantages, for instance, of MEIS (medium energy ion scattering) or RBS (Rutherford backscattering), are the installation and use of an accelerator—a rather expensive utility.

Electron energy loss spectroscopy is a very fine tool, but due to the need for short electron transport lengths from the source to the sample and from there to the analyzer, it cannot be used in situ. Scanning electron microscopy is also a well-established, however, ex situ, technique that can in combination with either X-ray detection or Auger electron spectroscopy (AES) afford laterally resolved chemical information. This discussion leads to other important aspects, e.g., the combination of different tools. In most cases, the techniques listed have excellent properties for the analysis of one surface property but comparably poor characteristics for others. The STM is certainly the ultimate tool for the analysis of the topography of surfaces, but with respect to chemical information, it has severe restrictions and is almost unable to provide in-depth information.

This chapter is organized according to the entries in Table I. A selection of the techniques will be introduced and described. EELS, HREELS, SEM, and

Table I. Surface Analysis Methods for UHV Thin Films

Physical Phenomenon	Technique	Acronyms	Probe	Def. Particle
Diffraction	Low Energy Electron Diffraction	LEED	electron	electron
	Reflection High Energy Electron D	RHEED	electron	electron
Electron tunneling	Scanning Tunneling Microscopy	STM	tip	electron
Mech. Forces		AFM		
Elastic scattering	Ion Scattering Spectroscopy	ISS, MEIS, RBS	ion	ion
	Electron Energy Loss Spectroscopy	EELS, HREELS	electron	electron
	Scanning Transmission Electron Microscopy	SEM, TEM	electron	electron
Sputtering	Secondary Ion Mass Spectrometry	SIMS, SNMS	ion	ion
Ionization	Photoelectron Spectroscopy	XPS, UPS	photon	electron
	Auger Electron Spectroscopy	AES	electron	electron
	Atomic Force Microscopy	AFM		

TEM will not be discussed in detail. To a certain extent, the in situ vs. ex situ question is discussed, as well as the compatibility problem. Possible merits with respect to quantitative/qualitative analysis with respect to composition and/or structure are discussed. A "tentative" comparison is also given.

6.2 LEED and RHEED

Low energy electron diffraction (LEED) is the oldest surface analysis tool [1]. The first experiment proved the wave nature of the electron. It was about 30 years before LEED was used for surface structure studies. Since then, LEED has become the standard technique for structural analysis [2, 3]. For use in systems for thin film growth, LEED can be used to follow epitaxial growth, to study the possible phase transitions on a surface during annealing.

LEED is not a proper in situ method because the large screen needed close to the sample (Fig. 1) prevents direct evaporation. Possible solutions to the problem are multiport systems or simply a 180° rotation of the target combined with a shutter for the LEED system to prevent contamination. Thus a shutter protects the LEED screen from being covered with the film material. In many systems, rear view LEED systems are used combined with a video camera read out setup. Microchannel plate systems combined with a fluorescent screen are also in use. Both with a video system and microchannel plate system, the primary electron beam current can be reduced to avoid the build up of contamination or damage. With high-resolution systems, transfer lengths of 2000 Å

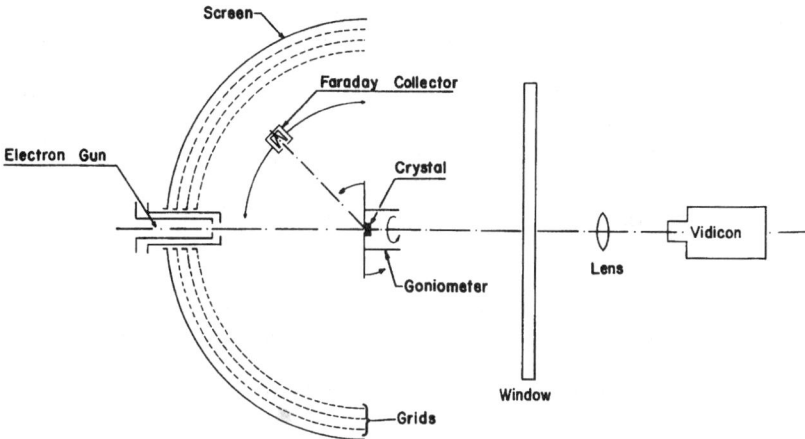

FIG. 1. Scheme of a LEED experiment equipped with a fluorescent screen, a three-grid system, and a Faraday cup. The diffraction spots are observed visually or registered with a vidicon system. With the Faraday cup, the intensity of the diffracted beams are measured directly [5]. (Reprinted with permission of Academic Press.)

are achieved [4, 5]. Spot profile analysis affords information about defect structures on surfaces. An early example is the study of the growth of Ge on GaAs [6]. For a quantitative analysis of the crystallographic structure, rather elaborate experiments and calculations are necessary. These are based on the so-called "I-V curves," i.e., the diffraction spot intensity vs. primary electron energy (voltage) dependence.

The basic principle of LEED is the diffraction of the electrons at the 2-D surface mesh following Bragg's equation with the electron wavelength estimated from de Broglie's equation. Usual electron energies are between 20 and 200 eV.

When considering the diffraction spot intensity, we observe that multiple electron scattering occurs and that not only the top surface layer contributes to the scattering. This apparent complication has the advantage that the crystallography of the top three layers (in general) can be evaluated [2, 3]. Figure 2 shows LEED patterns of clean Pt(110) in the (1 × 2) and (1 × 4) modification [7]. These structures are easy to recognize by simply counting the "superstructure" spots that appear in addition to the spots of the regular, unreconstructed Pt(110) surface.

Like LEED, RHEED is based on the basic phenomenon of the diffraction of electrons. Higher electron energies are used, 10–20 keV, and therefore, to achieve surface sensitivity, grazing incidence of the primary beam is necessary [5, 8]. The technique is widely used for the study of epitaxial film growth because it is a true in situ method owing to the scattering geometry (Fig. 3) [5].

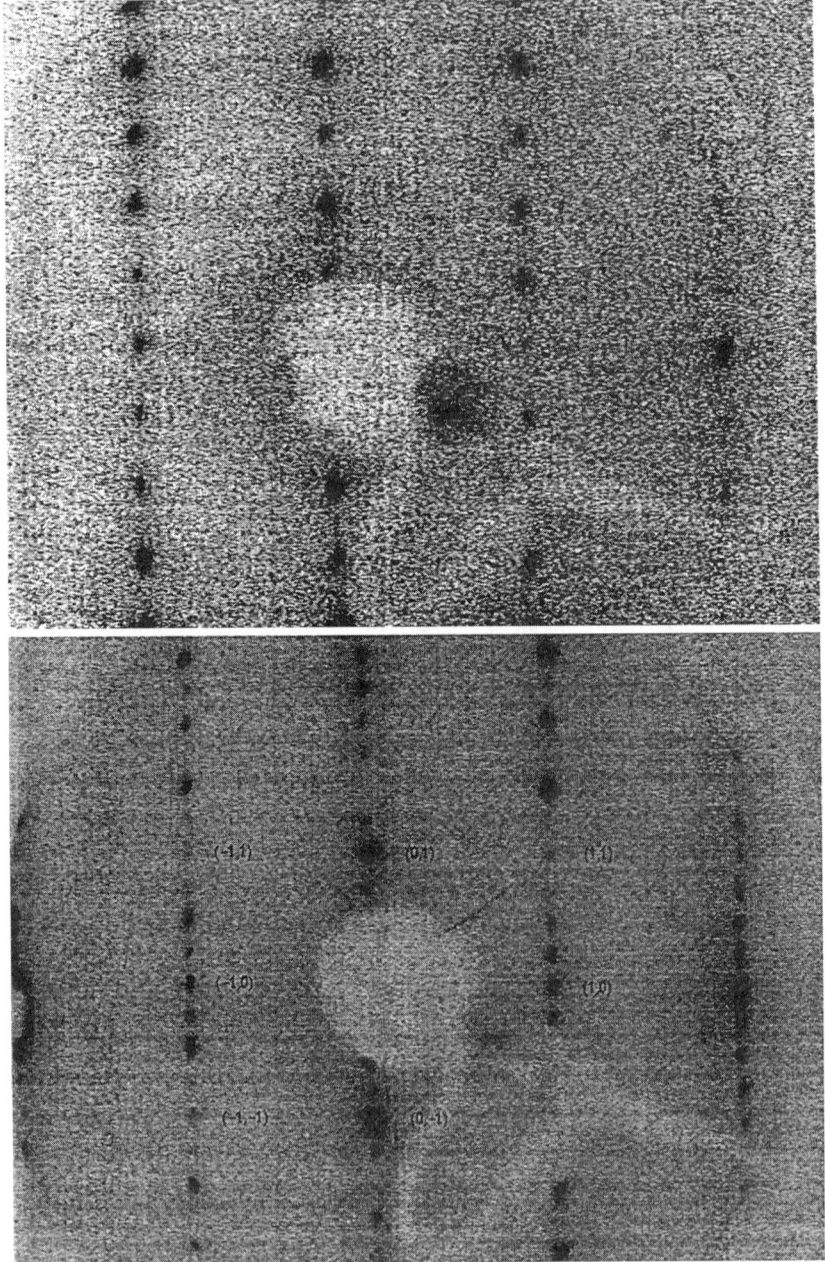

FIG. 2. LEED patterns of Pt(110) (1 × 2) and Pt(110) (1 × 4) [7]. (Reprinted with permission of Elsevier Science.)

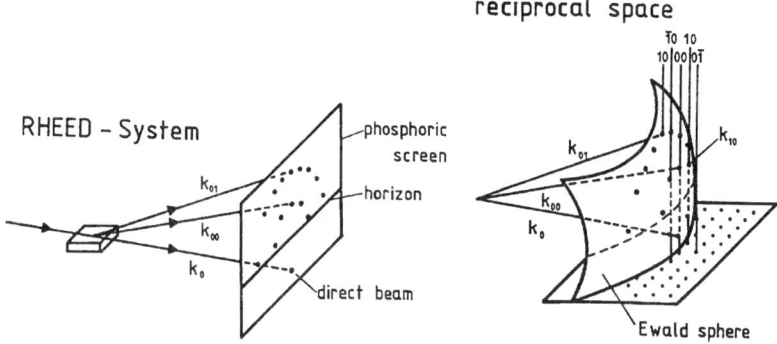

FIG. 3(a). Scheme of a RHEED system in real space (left) and in k-space (right) with the Ewald sphere concentration and the Miller indices for the different beams, which correspond to intersections of diffraction rools with the Ewald sphere [5]. (Reprinted with permission of Academic Press.)

Furthermore, due to the higher electron energy, the electron source and the observation screen can be placed far enough from the sample surface that the evaporation procedure is not disturbed. Figure 4 shows the temperature dependence of the RHEED pattern of Pt(110) (1×2) when passing through the

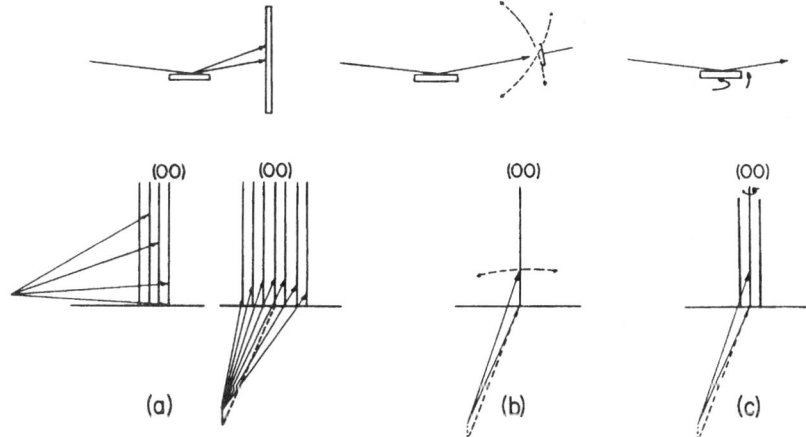

FIG. 3(b). Different experimental schemes realized in RHEED in real and reciprocal space [5]. (a) The registration of patterns as in Fig. 3(a); (b) is the measurements of angular profiles; (c) are rotation plots or rocking curves. The dashed line in (b) shows the movement of the detector or the path of the diffracted beam across a fixed detector (screen). Rocking curves are obtained in (c) for a set of fixed crystallographic directions, usually low index directions, and varying the glancing angle between 0° to 5° equivalent to 0 mrad to 100 mrad. (Reprinted with permission of Academic Press.)

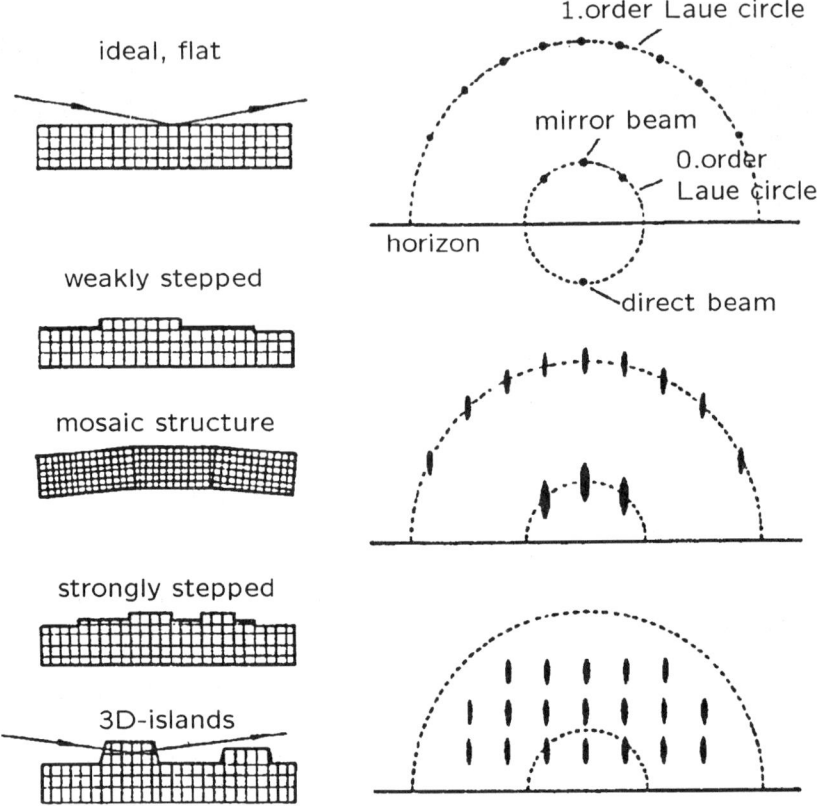

FIG. 3(c). Schematic RHEED patterns for different types of defects on a single crystal surface [5]. (Reprinted with permission of Academic Press.)

$(1 \times 2) \rightleftharpoons (1 \times 1)$ phase transition at 950 ± 30 K [9]. As in LEED, the "half order" or superstructure spots indicate the (1×2) surface reconstruction. The half order spots disappear with the disappearance of the reconstruction, i.e., phase transitions can be observed directly. It is worth noting that RHEED is possible at elevated temperatures. Figure 5 is an example of the application of RHEED in a film growth experiment [10]. Here the intensity of the [0, 0] or specular beam is followed as the film grows. The model for the explanation of the intensity oscillations is rather simple if there is layer-by-layer growth and the proper glancing angle is chosen. Whenever a layer is filled, the beam intensity reaches a maximal value. For an incomplete layer, the interference between the beams from the flat, full layer and the growing incomplete layer causes the decrease in intensity. The example in Fig. 5 is the growth of Si on Si [10]. If the growth is not layer-by-layer or van der Merwe growth, this is also evident from

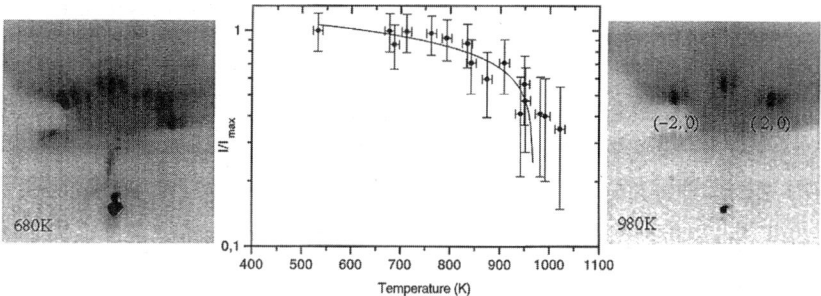

FIG. 4. RHEED intensity as a function of T in the range of the Pt(110) (1×2($\rightleftharpoons$ Pt(10) (1×1) order-disorder phase transition. The insets are RHEED patterns below and above $T_c = 960$ K [9]. (Reprinted with permission of Elsevier Science.)

RHEED patterns as has been discussed in detail previously [5, 8]. As indicated in Fig. 3(c), for example, islands on a surface cause diffraction patterns as in a forward Laue experiment. More quantitative information can be obtained from the rocking curves (Fig. 3(b)). By comparing experimental rocking curves, i.e., intensity vs. angle of incidence, with model calculations, surface crystallographic data are obtained. Nonperiodic structure can be analyzed using the diffuse intensity around the diffraction spots from polar profiles, i.e., intensity vs. polar exit angle [11]. The example studied is again Pt(110) and the $(1 \times 2) \rightleftharpoons (1 \times 1)$ order-disorder transition (Fig. 6).

More recently [12], the imaging with LEED beams was developed and labeled LEEM (low-energy electron microscopy). This type of imaging technique belongs to the imaging modes in cathode lens microscopy [13]. With a lens close to the sample, electrons excited or scattered from the surface layers are imaged. Depending on the primary excitation, e.g., by electrons, photons or ions, a different type of information is obtained. The contrast observed is due to work function differences, field distortion by the surface topography, and "macroscopic" shadowing caused by the oblique illumination with the primary beam. LEEM contrast is caused by two- and three-dimensional structural differences, thickness variations of thin films, and surface topography, i.e., steps. The strength of these imaging techniques is the rather large field of view of several μm with a lateral resolution below 100 nm. Furthermore, the imaging is fast, such that time-dependent measurements are possible. Because the imaging is possible essentially at all temperatures, annealing processes, surface chemical reactions, magnetic surface effects, and so on can be followed in real time. The film growth proper cannot be followed due to the instrumental conditions, e.g., the cathode lens electron microscopies are ex situ techniques for film growth. Chemical information is naturally obtained when using photoelectrons (PEEM) or Auger electrons (AEEM) for the imaging.

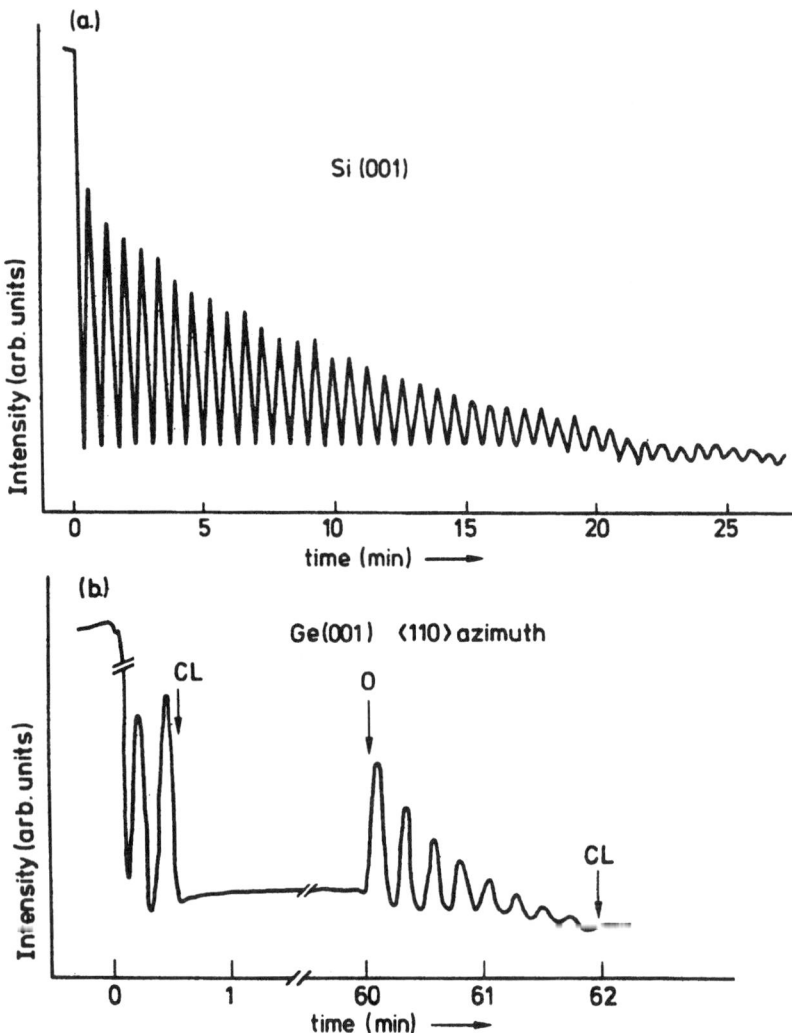

FIG. 5. Oscillations of the intensity of the specular beam in a RHEED experiment in homodeposition growth experiments of (a) Si on Si(001) and (b) Ge on Ge(001). At CL in b the shutter is closed and at 0 reopened [10]. (Reprinted with permission of Plenum Press.)

In summary, RHEED and LEED are very convenient tools to observe the structure and structural changes of surfaces. RHEED is a proper in situ method. Quantitative data can be obtained with rather large computational efforts. Chemical information is practically not available. It is worth mentioning that

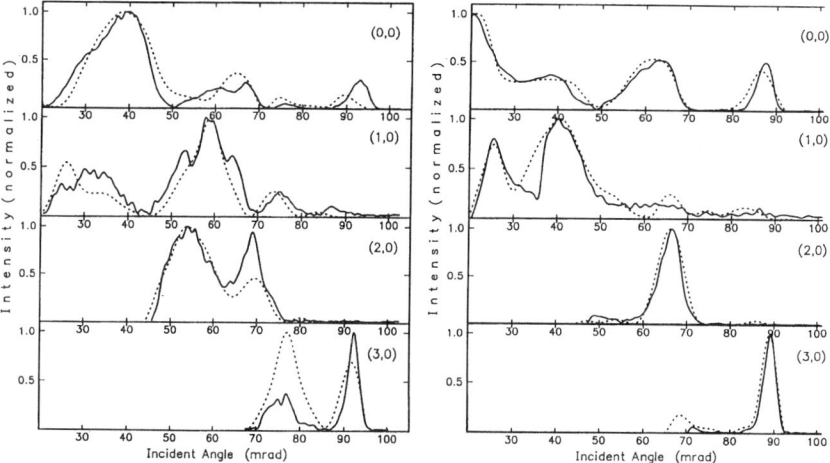

FIG. 6. Comparison of experimental and calculated intensities of rocking curves of Pt(110) below (373 K) and above (1048 K) of (110) (1 × 2) ⇌ Pt(110) (1 × 1) order-disorder phase transition [11]. (Reprinted with permission of Elsevier Science.)

having a several keV electron gun for the RHEED experiment, Auger electron spectroscopy (AES) can be added to the system using an electrostatic electron energy analyser.

6.3 Scanning Tunneling Microscopy (STM) and Atomic Force Microscopy (AFM)

Like LEED and RHEED, STM and AFM are mainly structural probes, i.e., chemical information is not directly measured. Whereas the diffraction methods are k-space probes, STM and AFM operate in real space. STM is a pure UHV technique; AFM can be operated in UHV, air, and liquids. Both instruments are difficult to be arranged as in situ methods. Usually some means for target transfer from the evaporation system to the STM or AFM has to be provided.

STM is a rather "new" instrument, first publications are from 1982/3 [14–16]. Many more review papers are available [17–21]. The physical principal of STM is the tunneling effect between a metal tip and the surface (Fig. 7(a)). The basic experimental scheme is shown in Fig. 7(b) [14]. Depending on the bias between tip and surface, the electrons tunnel from the tip to empty states of the surface or vice versa.

Figure 7(a) shows a scheme of the situation for a metal surface. The tunnel current I_t is approximately given by the work function ϕ, the tip-surface distance

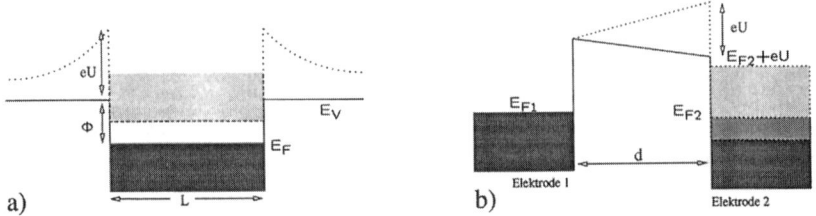

FIG. 7(a). Scheme of the tunnel gap between two metallic electrodes, i.e., tip and solid. $E_{F1,2}$ are the Fermi energies, U is the variable external bias-voltage, and d is the tip-solid distance.

d, and the applied voltage V_t

$$I_t \simeq \frac{e^2}{4\pi^2} \frac{\sqrt{\phi}}{\hbar} \frac{V_t}{d} k \exp(-2k\sqrt{\phi}\,d)$$

with $k = \sqrt{2m}/\hbar$. Typical values for metals are $\phi = 4$ eV and $k \cdot \sqrt{\phi} \simeq 1$ Å^{-1}. Hence small changes of the distance d cause large variations of the tunnel

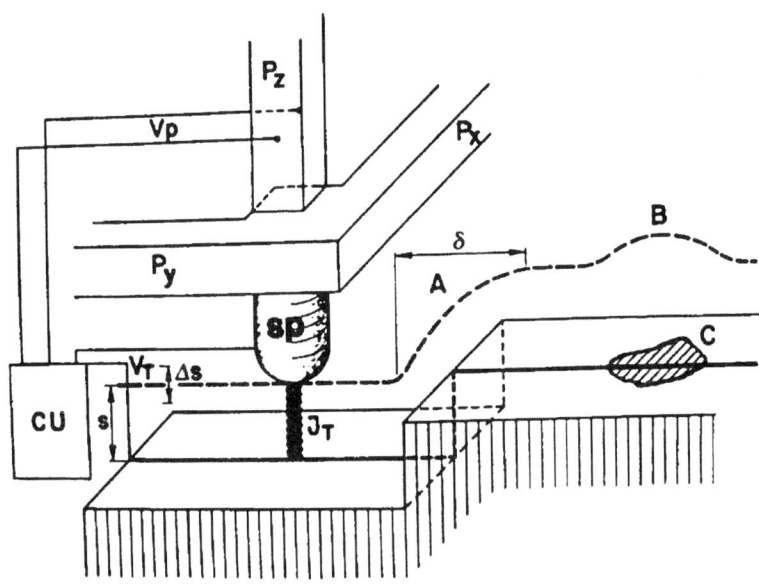

FIG. 7(b). Scheme of an STM experiment with the tip, sp, the piezo-electric tripod, P_x, P_y, P_z, and the control unit, CU. The tip-sample distance is s, which is controlled via the piezos, the current I_T, and the bias voltage V_T. The dashed line marks the pass of the tip over a step (A) and an impurity C(B). The bump at C may be due to topography or electron density effects [15]. (Reprinted with permission of the American Physical Society.)

current. So, in the constant current mode (CCM) of STM, the tip current is kept constant and the change of the control $\delta I/\delta x$ of the z-piezo, if x is the scan direction, gives the measure of the topography. This topographic map corresponds to an equipotential surface. In constant height mode (CHM), the length of the z-piezo is kept constant, and the tip current is the measure for the surface topography. The CHM is faster than the CCM, but with the tip-surface distance being constant, the tip may "crash" into some obstacle on the surface. Figure 8 shows the Pt(110) (1 × 2) and Pt(110) (1 × 4) surface [7] as shown above (Figs. 2 and 4) using LEED and RHEED. The real-space image gives immediately the "mesoscopic" information, i.e., in the cases here, the step structure of the surface. The (1 × 2) surface forms the so-called "fish scale" pattern [22, 23], whereas the (1 × 4) forms large flat terraces separated by "cliffs" with heights of about 30 Å to 40 Å. The step heights on the (1 × 2) surface are monoatomic, equal to the interplanar spacing. The structure of the (1 × 4) surface is recognized from height scans. It is, in fact, a (1 × 2) structure with a pairing of alternating pairs of [1$\bar{1}$0] rows. This is an example of the power of STM for solving structural problems. Even though atomic resolution is not easily achieved on metals, in most cases a clear correlation exists between the topography and the atomic positions in the surface.

On semiconductor surfaces, Fig. 9 shows the famous Si(111) (7 × 7), which was *the* historical breakthrough for the method [14], the electron density distribution on the surface may be shifted relative to the atom core positions [24]. The literature has many examples of the use of STM in thin film production, metal on metal, oxide films on metal, semiconductors on semiconductors, and metals on semiconductors. For epitaxially grown crystalline films, STM is certainly the ultimate structural probe. A review of all possible modifications of the use of STMs, spectroscopy, magnetic applications, light emission, and so on goes beyond the scope of the present paper. To control epitaxial growth, STM as a topographic instrument in the CCM and/or the CHM is sufficient. The one setback of STM is the insulators. Only very thin films are accessible with STM.

AFM has no such limitations. It was first described in 1986 [25]. Its operational principle is mechanics. A tip is steered over a surface by the identical means as an STM tip, i.e. a proper choice of piezo controls. The tip is mounted on a cantilever. With a laser beam and an optical detector, the movement of the cantilever is monitored (Fig. 10(a)) [26]. The signals are processed into images, again, as with an STM. Atomic resolution is not achieved routinely. The strengths of AFM are the possibilities of scanning in air, scanning larger areas, and being applicable to all surfaces including insulators [27].

Another technical realization of AFM uses the "needle sensor" (Fig. 10(b)) [26]. The needle sensor is a micromechanical device making use of the

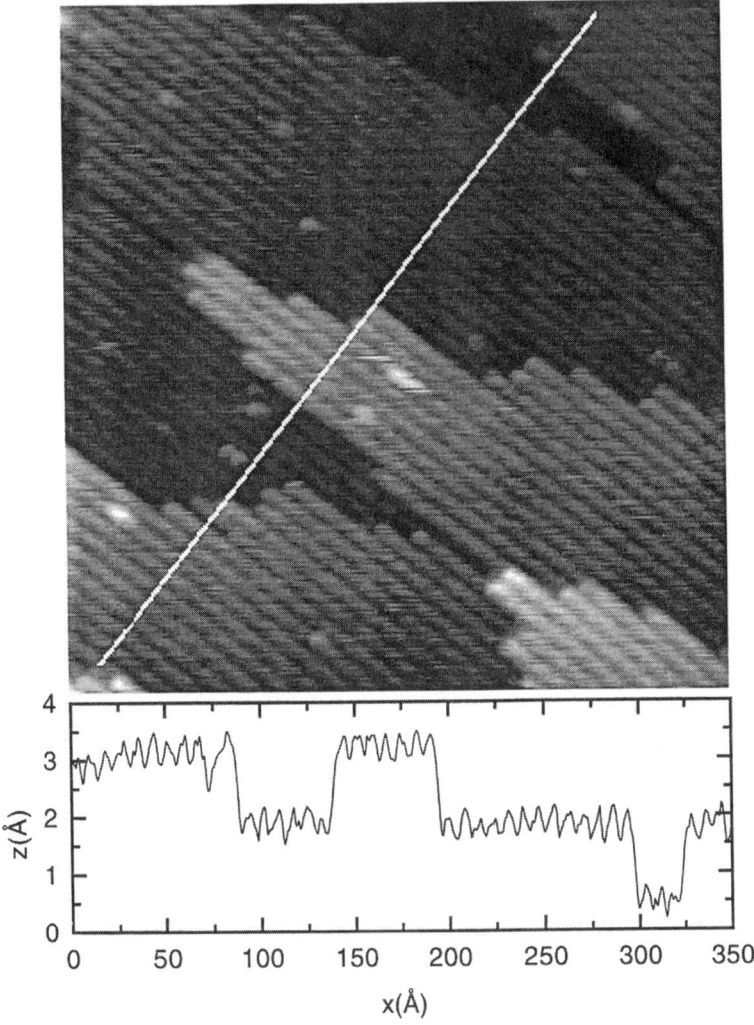

FIG. 8. STM pattern of a Pt(110) (1 × 2) (a) and a Pt(110) (1 × 4) (b) surface. The LEED patterns in Fig. 2 are from the same surfaces [7]. (Reprinted with permission of Elsevier Science.)

piezoelectric properties of quartz. The mechanical forces between the needle and the surface cause shifts of the resonance of the oscillating quartz, as in the case of a quartz microbalance where the deposited material shifts the frequency of the balance. A scheme of the forces acting between a tip and a surface is shown in Fig. 11 [26]. With the piezoelectric tube scanners combined with the cantilever

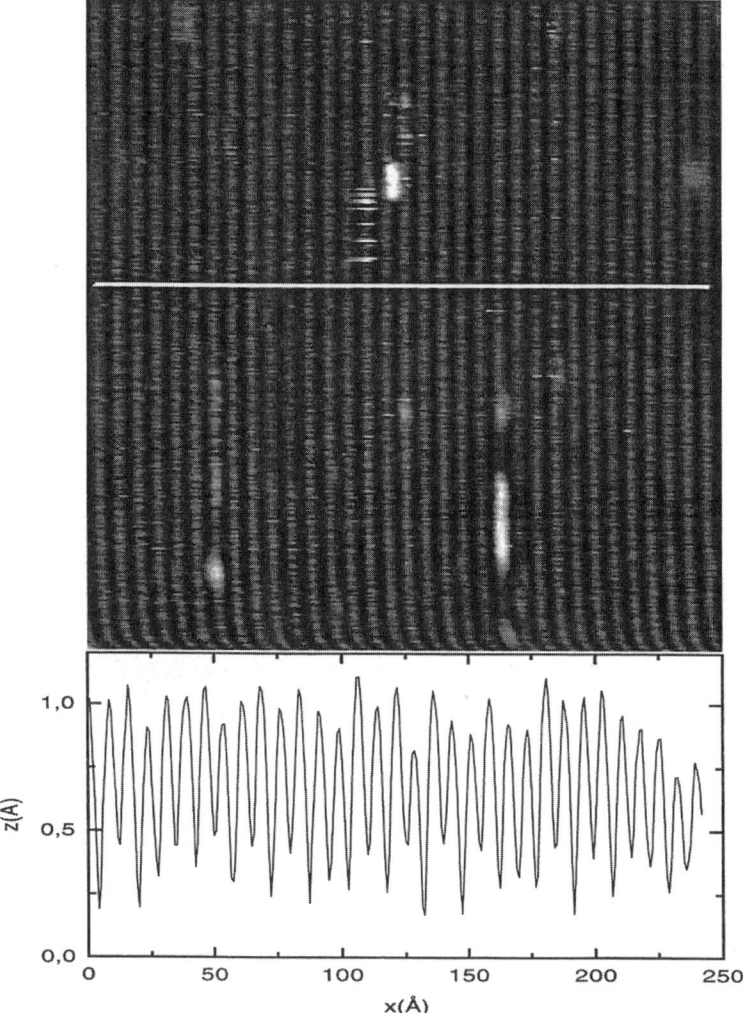

FIG. 8. *Continued*

system, areas of the order of 10 μm × 10 μm can be scanned. The depth resolution reaches 10^{-2} nm and the lateral resolution 0.01 nm [26]. Although AFM is a typical ex situ technique, it bridges the gap between the atomic resolution of STM and the resolution reached in scanning Electron Microscopy (SEM), which is very important for the imaging of device structures. For further AFM reviews and other extensions of STM, see [27, 28].

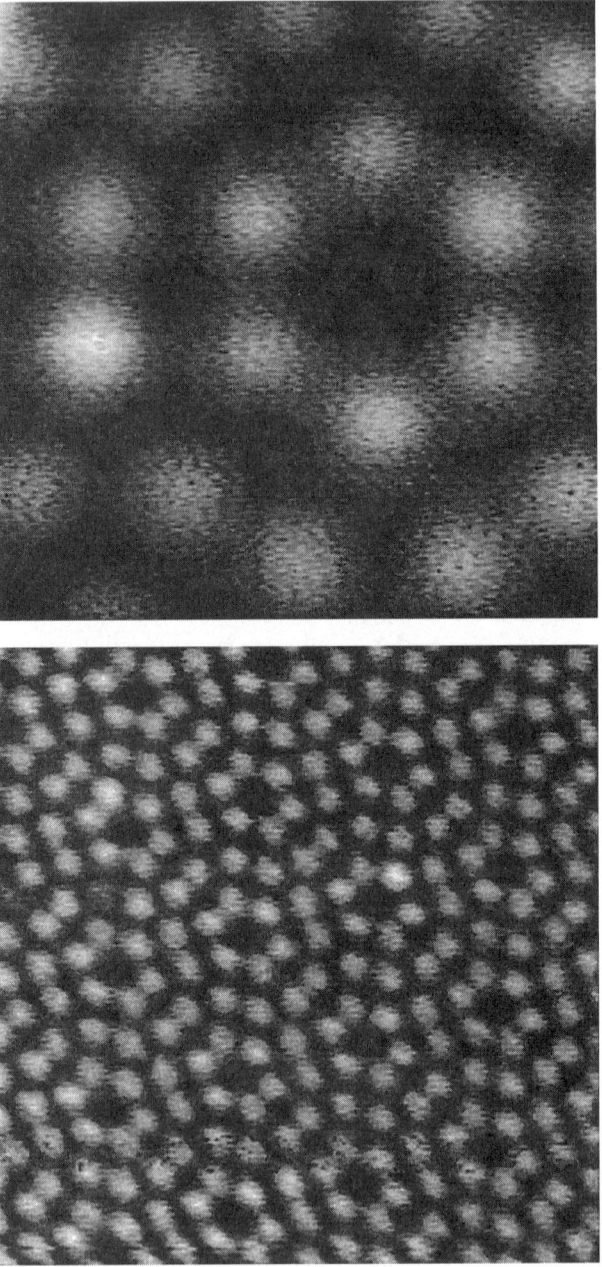

FIG. 9. STM pattern of a Si(111) (7 × 7) surface [Ch. Röthig, Thesis Osnabrück 1995, unpublished].

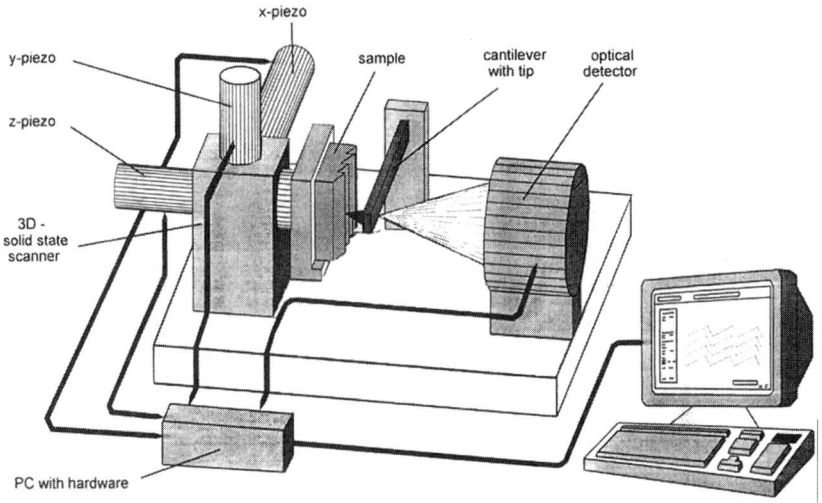

FIG. 10(a). AFM with a conventional cantilever system [26]. (Reprinted with permission of Taylor & Francis.)

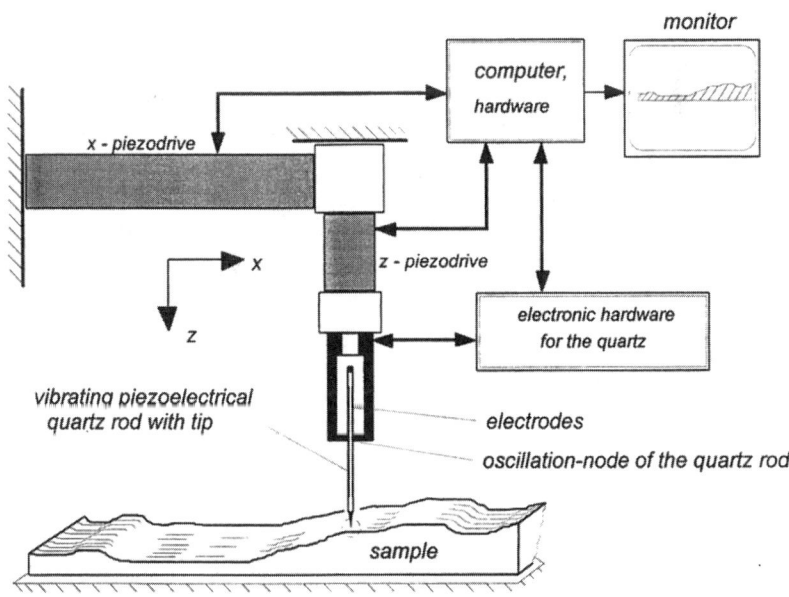

FIG. 10(b). AFM with a "needle sensor," i.e., a micromechanic quartz rod as the tip drive [26]. (Reprinted with permission of Taylor & Francis.)

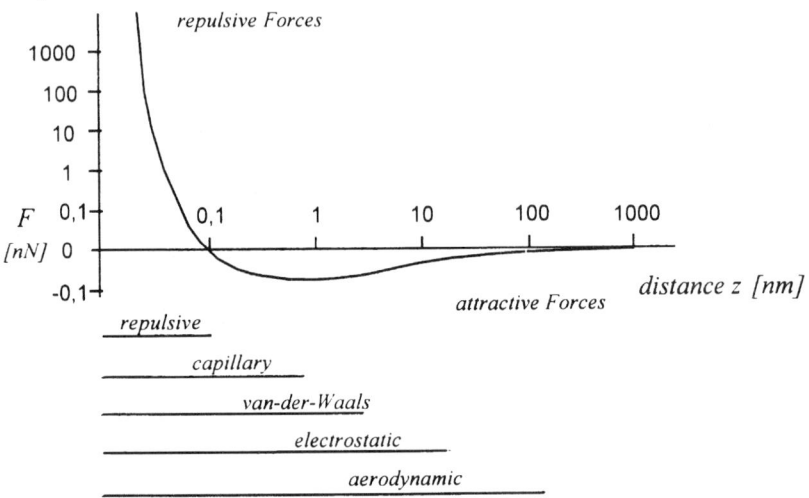

FIG. 11. Schematic representation of the mechanical forces between a tip and a solid [26]. (Reprinted with permission of Taylor & Francis.)

6.4 Ion Scattering Spectrometry

The application of ions beams for surface analysis is based on the "elastic" and "inelastic" interaction. The elastic interaction is the scattering of the projectile ions from the atoms of the target. It was first observed in experiments with α-particles interacting with a Au target by Rutherford [29]. Much later, the effect was introduced as analytical technique [30]. The technique is now called Rutherford backscattering (RBS) when MeV ions are used. The elastic scattering cross-section is determined with high accuracy by the Coulomb potential acting between the nuclei. At energies of around 100 keV, the acronym MEIS (medium energy ion scattering) is used. At low energies, LEIS (low EIS) or simply ISS (ion scattering spectrometry) is in use. Typical ion energies are between 500 eV and 3 keV. The applicability of low-energy ion scattering for surface analysis was first shown in 1967 [31]. In all energy regimes, both chemical analysis and structural analysis is afforded. With decreasing energy, the depth of information decreases and the surface sensitivity increases. The chemical analysis is due to the mass dependence of the elastic energy loss. The energy E_1 of a projectile with mass M_1 and primary energy E_0 scattered from a target atom with mass M_2 is given by the kinematic factor

$$E_1/E_0 = \left[\frac{(M_2^2 + M_1^2 \sin^2\Theta)^{1/2} + M_1 \cos \Theta}{M_2 + M_1}\right] \quad (6.1)$$

where Θ is the laboratory scattering angle. At high energies, the spectra of backscattered ions show steps that appear at the energies estimated from Eq. (6.1). See Fig. 12 [32]. The structure of the spectra is due to the fact that ions are backscattered not only from surface atoms, but also from the bulk of the target. At low energies, the penetration of primary ions is reduced and ions scattered from the bulk are neutralized with high probability. Hence the spectra show peaks and not steps at the energies corresponding to the masses of the target atoms (Fig. 13). The different behavior with respect to penetration and backscattering depth, respectively, leads to different types of information when thin films are analyzed [33, 34]. When using high-energy ions (MEIS and RBS), the film is penetrated and particles are scattered from the film and the substrate. In the case of a high Z material on a low Z substrate, the film is seen as a separate

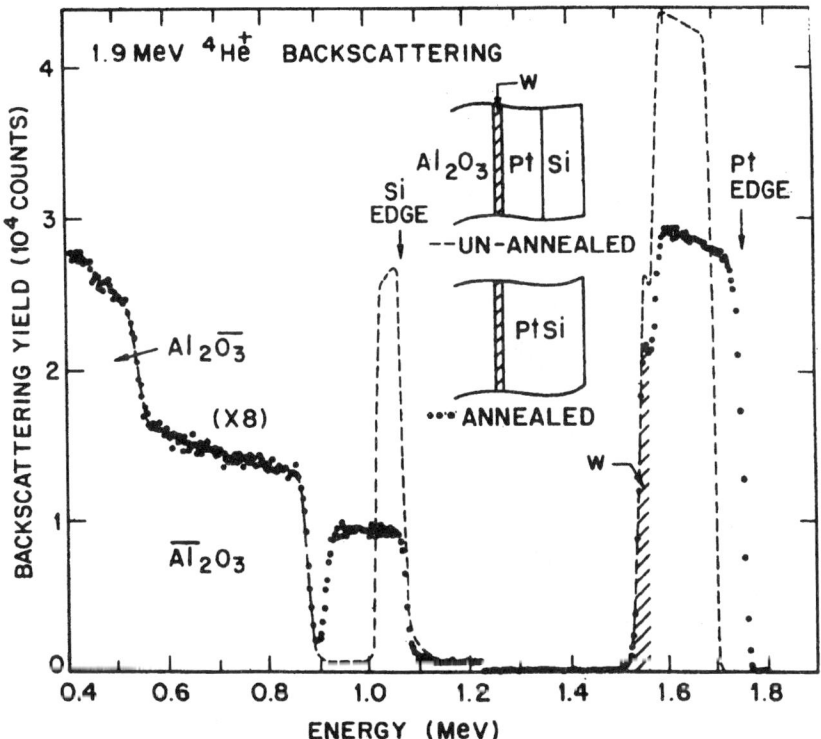

FIG. 12. RBS spectrum of a Pt/Si layer system on a Al_2O_3 substrate. The dashed line is the spectra as deposited; the dotted line is after annealing and the formation of a PtSi compound. W is used as a thin layer between the Al_2O_3 and the original Pt layer. The A_2O_3 backing is seen as a step of $\bar{Al}$ (higher Z) and $\bar{O}$ (lower Z). Note that the Si and Pt areas, i.e., the number density of atoms/cm^2, is not changed in the annealing process [32]. (Reprinted with permission of the American Institute of Physics.)

peak at high energy in the spectrum (Fig. 12). At low ion energies, the result of such an experiment depends on the type of film growth and is limited to a few monolayers. If layer-by-layer growth (van der Merwe) prevails, the substrate peak decreases as the film peak increases. When the first layer is "closed," the substrate peak is zero and the film peak saturates (Fig. 13). In the case of Volmer-Weber growth, i.e., an island formed on the first, closed monolayer, there is no difference. In the case of island or Stranski-Krastanov growth, the substrate signal decreases much more slowly until the film is closed. In the case of RBS, such details can be analyzed to a certain extent from the details of the spectra [35]. The reason for this behavior is the fact that, with RBS, the mass density of the target is measured essentially independently of the structure when considering the elastic scattering only. The coverage can be evaluated quantitatively based on the Rutherford cross-section high energies. Film thickness are estimated from the inelastic loss of the scattered particles.

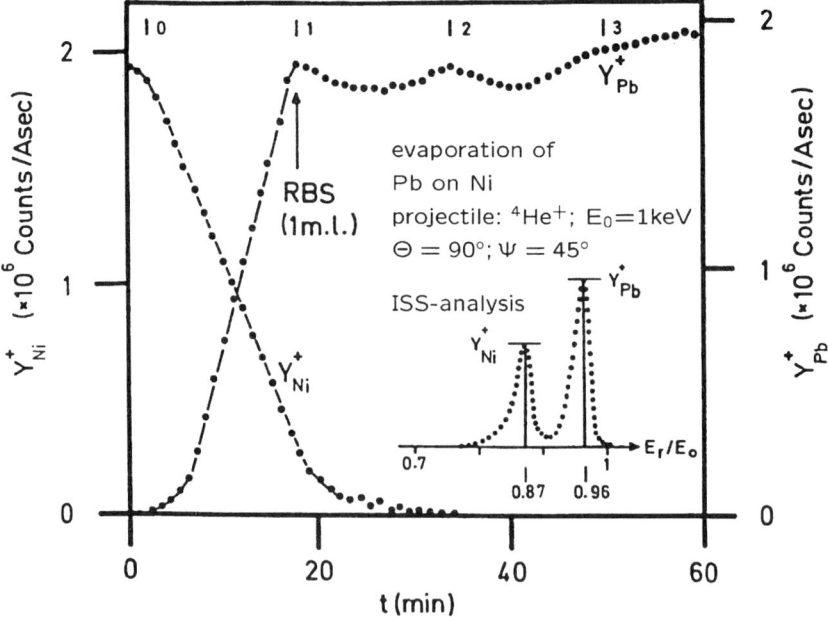

FIG. 13. Surface coverage of Pb on Ni during Pb evaporation measured by 1 keV He^+ ISS at scattering angle of 90°. The first monolayer calibrated by RBS (1 ml) covers the substrate completely. The inset shows the IS-spectra at 0.5 ml Y_{Ni}^+ and Y_{Pb}^+ are the yield of He^+ ions scattered from Ni and Pb, respectively [A. Zartner, Thesis, Tech. Univ. Munich 1978, unpublished].

6.4.1 Quantitative Thin Film Analysis with RBS

In Fig. 14, a typical trajectory of a MeV light ion through matter is shown schematically [33]. The particle with a primary energy E_0 loses the energy ΔE_{in} on the ingoing path and comes to the elastic collision with $E_t = E_0 - \Delta E_{in}$, where the energy loss is $\Delta E_{el} = (1 - K)E_t$. The particle then loses the energy ΔE_{out} on the outgoing path. The inelastic losses ΔE_{in} and ΔE_{out} are usually estimated using tabulated energy loss data dE/dx (eV/Å) [36]. Note that for the outgoing path, the pathlength is $t/\cos \Theta$, where t is the depth of the target atom considered and Θ is the laboratory scattering angle.

The outgoing energy E_1 of a particle scattered at the depth t is

$$E_1(t) = -t\left(K\frac{dE}{dx}\bigg|_{in} + \frac{1}{|\cos \Theta|}\frac{dE}{dx}\bigg|_{out}\right) + KE_0 \tag{6.2}$$

The energy loss ΔE_{in} is here replaced by

$$\Delta E_{in} = \int_0^t \frac{dE}{dx}dx \simeq \frac{dE}{dx}\bigg|_{in} \cdot t \tag{6.3}$$

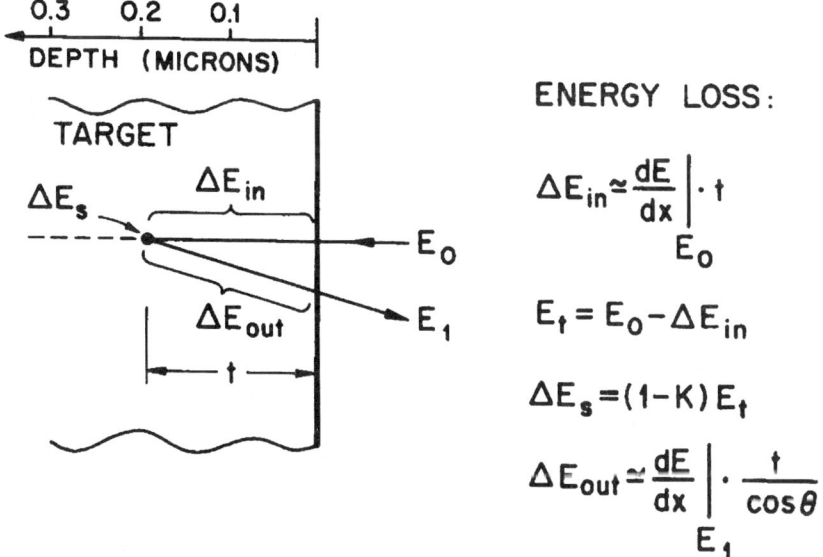

FIG. 14. Scheme of a high-energy ion trajectory in a solid and the energy loss components encountered by the ion. On the way in, the particle loses energy, ΔE_i, via electron stopping, then ΔE_s in the large-angle elastic collision and ΔE_{out} via electronic stopping again. The energy measured by an external detector is $E_1(\Theta, t) = E_0 - \Delta E_{in} - \Delta E_s - \Delta E_{out}$, where Θ is the laboratory scattering angle and t is the depth where the elastic collision occurs [33]. (Reprinted with permission of Elsevier Science.)

taking an average energy between the incident energy E_0 and the energy at depth t, which is $E(t) = E_0 - t \cdot dE/dx|_{in}$.

The energy width ΔE of a thin film of thickness Δt is related via

$$\Delta E = \Delta t \left(K \frac{dE}{dx}\bigg|_{in} + \frac{1}{|\cos \Theta|} \frac{dE}{dx}\bigg|_{out} \right) \quad (6.4)$$

For thin films and/or short path length, dE/dx is constant in good approximation and the "surface energy approximation" can be used:

$$\Delta E_0 = \Delta t[s_0] = \Delta t \left[k \frac{dE}{dx}\bigg|_{E_0} + \frac{1}{|\cos \Theta|} \frac{dE}{dx}\bigg|_{KE_0} \right] \quad (6.5)$$

Computer programs are available for calculating backscattering spectra and depth profiles [37]. The subscripts denote the "surface energy approximation." $[s_0]$ is the backscattering energy loss factor.

The yield of the scattered particles I_s from a thin layer Δt is given, using the Rutherford scattering cross-section $\sigma(\Theta)$, the measured number of incident particles I_0 and the solid angle $\Delta \Omega$ subtended by the detector, by

$$I_s = T_0 \sigma(\Theta) N \Delta t \cdot \Delta \Omega \quad (6.6)$$

where $N \Delta t$ is the number of target atoms/cm^2 in the layer.

Because the Rutherford scattering cross-section is a good approximation, the number of target atoms can be evaluated quantitatively.

For medium energy ions (MEIS), thin film analysis is based on equal formulas [38]. Corrections of the Rutherford cross-section due to the screening by the target atom electrons are necessary. The accessible depth is smaller, as in the case of RBS, due to the shorter range at lower energies. A higher depth resolution is possible at lower energies due to the lower energy loss straggling at lower energies. Corrections for neutralization are also necessary for quantitative analysis.

6.4.2 Low-Energy Ion Scattering

The elastic energy loss of low-energy ions is also calculated from Eq. (6.1). Due to the high degree of neutralization, ions survive as ions essentially only if scattered by atoms on the topmost layer [34]. The yield of backscattered ions is estimated from Eq. (6.6). The cross-section $\sigma(\vartheta)$ is calculated using a screened Coulomb potential. Figure 15 [34] shows $\sigma(\vartheta)$ values for some projectile-target combinations and scattering angles of $\vartheta = 90°$ and $\vartheta = 142°$. For 90°, Eq. (6.1) reduces to

$$W_1/E_0 = \frac{M_2 - M_1}{M_2 + M_1} \quad (6.7)$$

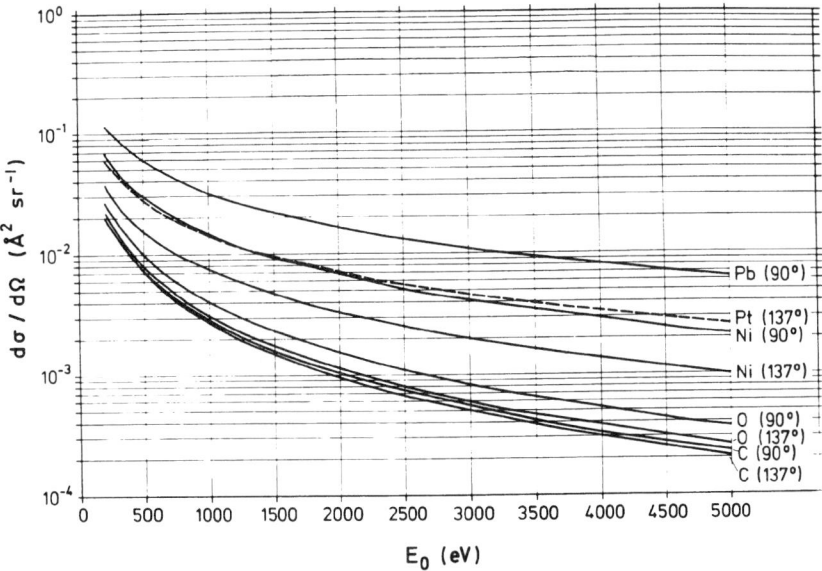

FIG. 15. Elastic scattering cross-sections $d\sigma$ calculated for a screened Coulomb potential (Thomas-Fermi-Moliere potential) for He scattering of C, O, Ni, Pt, and Pb for $\Theta = 90°$ and $\Theta = 137°$. The energy range and the scattering angles are typical for ISS [34]. (Reprinted with permission of Academic Press.)

for $M_1 < M_2$. Therefore, $\vartheta = 90°$ is very often used in ISS experiments. The angle $\vartheta = 142°$ is the fixed scattering angle of the cylindrical mirror analyzers (CMA), as used for Auger electron spectroscopy (AES). By inverting the voltage on a CMA, it can be used alternatively for ISS and AES. The ISS ion yield is given by $Y = N_t \sigma \cdot P$, where N_t is the target atom density, σ is the scattering cross-section, and P is the ion survival probability.

Normalized experimental ion yields are collected in Fig. 16 [39], which allow estimates of the sensitivity of ISS. For a quantitative analysis, a calibration from "clean" surfaces is the practical approach. The ion yield depends linearly on the coverage in many experiments (Fig. 13). This is an empirical observation that has to be checked, of course, for "new" substrate-adsorbate combinations.

An interesting aspect of ISS is the immediate, qualitative recognition of the possible growth at the submonolayer level. Furthermore, the cases of deposition of material on a substrate, the incorporation in the first layer, or the coverage of the deposited material by the substrate atoms can be distinguished (Fig. 13). To make sure that the deposition is on top, the angle of incidence is changed. The backscattered intensity from the substrate decreases rapidly in such a case due to the shadowing by the deposited atoms. If the substrate atoms are replaced by deposited atoms and come to rest at the same height, no strong angular effects

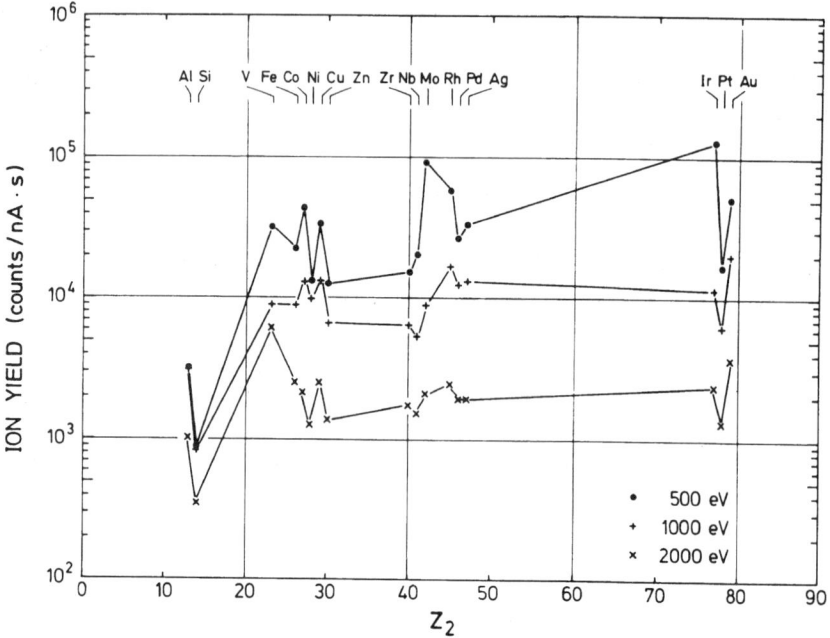

FIG. 16. Experimental ion yields for He$^+$ scattering at three energies for different targets. The deviations from the monotonic behavior of $d\sigma$ (Fig. 15) are due to neutralization effects, i.e., the ion yield $Y \alpha N\, d\sigma P$, where N is the target atom density, $d\sigma$ is the scattering cross-section, and P is the ion survival probability [39]. (Reprinted with permission of Plenum Press.)

are observed. The covering of the deposited atoms by substrate atoms has the surprising effect that the deposited atoms are not "seen" by ISS. (For recent cases, see [40–42].) It is obvious that in the case of coverage by the substrate, additional analytical tools (AES and XPS) are useful to measure the deposited material.

6.5 Structural Analysis Using Ion Beams

The basis for structural analyses with ion beams is essentially due to the "shadow-cone" effect or "Rutherfords classical shadow" [43]. Behind an atom that is hit by a parallel beam of ions exists a conical region which is free of projectiles. The envelope of the cone has a parabolic shape. For a Coulomb potential, the shadow-cone radius is given by

$$R_c = 2(Z_1 Z_2 e^2 d/E_0)^{1/2} \qquad (6.8)$$

where d is the distance behind the atom. If the first atom is followed by a chain of atoms, these atoms cannot be hit by ions as long as they are inside the shadow cone of the preceding atom. From these considerations follows the channeling effect, first found in a computer simulation [44]. For channeling to occur, the angle of incidence has to be lower than

$$\psi = (2Z_1 Z_2 e^2 / E_0 a_{hkl})^{1/2} \tag{6.9}$$

where a_{hkl} is the interatomic distance in the $[hkl]$ row. For non-coulombic potentials, i.e., at lower energies, the shadow-cone radius and the critical angle are modified [40]. Important to note is the dependence on the primary energy E_0, i.e., the critical angles are smaller at high energies as are the shadow-cone radii.

In general, high-energy channeling and blocking, the "inverse" effect, are used for structural analysis. At low energies, it is mostly the shadow-cone radius that is directly used for surface crystallography. A typical channeling application is the study of epitaxial growth (Fig. 17) [33]. Au on Ag is an example of perfect epitaxy for several layers. Hence the Ag surface peak decreases as estimated by theory. In the case of Au on Pd, where the lattice mismatch is 4.7% compared to <0.2% in the case of Au on Ag, the layer-to-layer growth stops after the first layer. The accuracy of the channeling experiments affords very detailed measurements of lattice distortions, which are caused by the lattice mismatch. Well-known cases involve Si-Ge layers [33]. Another obvious application is the determination of lattice sites of impurities in thin films.

As can be seen from Fig. 17, an impurity placed substitutionally is not "seen" under the given angular conditions. The incident beam can be directed into other channels where the impurity is blocking the channel. Again the accuracy of channeling is sufficient to obtain very detailed information. Applications are found for semiconductors, optical materials, and high T_c superconductors.

At low energies, surface channeling can be visualized in forward and backward geometry (Fig. 18) [45]. For the forward case, a position sensitive detector is placed in the downstream ion beam. The intensity distribution obtained is an image of the elastic potential energy surface perpendicular to the beam direction for one surface half channel averaged over all channels within the beam diameter. In the case of the backward geometry, the angle of incidence is kept and the azimuthal angle is varied. In that case, minima are found for all low index surface half channels. The peaks are caused by scattering from the ridges between the channels. In this geometry, structural changes are easily recognized. In this case, it is the surface melting of Pb(110), previously analyzed by MEIS [46].

The third scattering geometry makes direct use of the shadow cone. In this so-called ICISS (Impact Collision ISS) [47, 48], the impact angle varies from grazing (channeling) to larger values. The intensity increases when the ions at the edge of the shadow cone strike the neighboring atoms of each atom in a surface chain. The shadow-cone radius can be calculated or calibrated using a known surface structure. Surface lattice constants, adatom positions, and relaxations between first and second layer can be measured in this geometry. Examples of structural analyses using ICISS are found in a recent review [40].

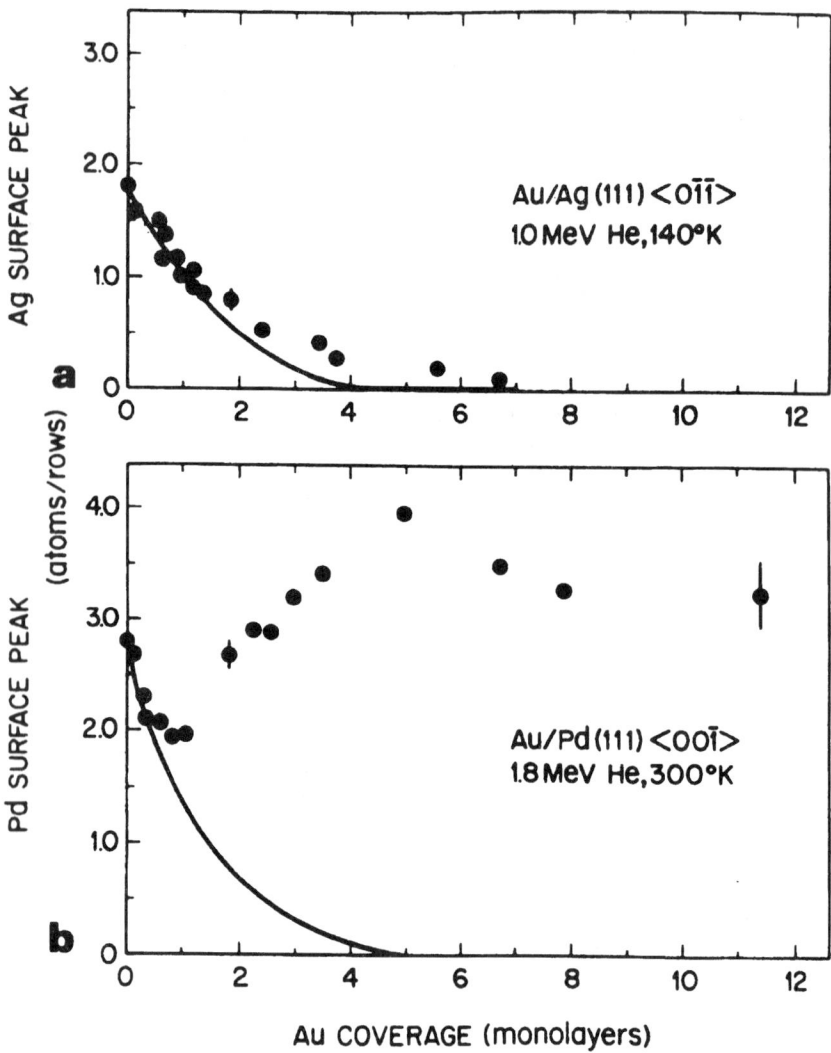

FIG. 17. RBS film growth measurements of Au on Ag(111) (a) and Au on Pd(111). On Ag, the Au grows epitaxially; on Pd, the epitaxial growth stops at 1 ml due to the lattice constant mismatch [33]. (Reprinted with permission of Elsevier Science.)

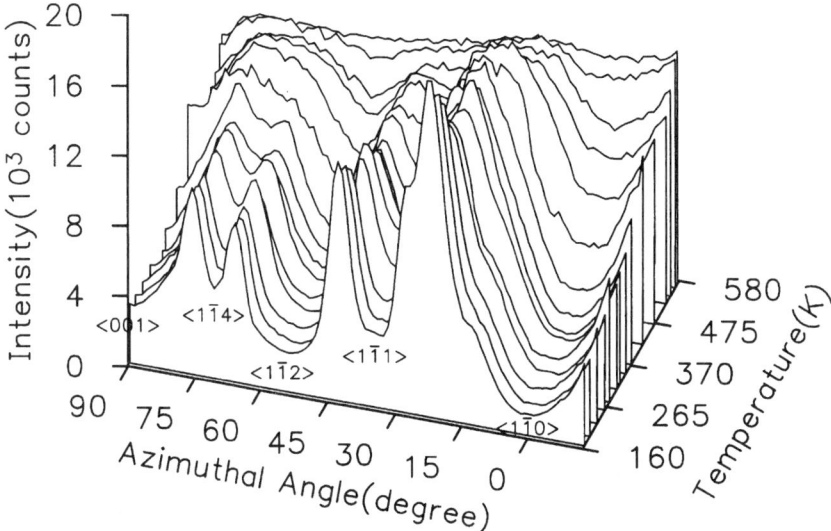

FIG. 18. Low-energy ion backscattering experiment (2 keV Ne$^+$) of Pb(110) from 160 K to 580 K, i.e., the surface melting temperature. The scattering geometry is using a small angle of incidence $\Psi = 11°$ and a large backscattering angle $\Theta = 165°$. The minima at low index surface crystallographic directions are due to the (forward) surface channeling [45]. (Reprinted with permission of the American Institute of Physics.)

In summary, ion scattering methods are quantitative tools for both structural and chemical analyses. They are real-space methods with respect to structure analysis. In principle, the methods can be applied in situ. A 100 keV or 1 MeV accelerator is, however, a rather large capital investment compared, e.g., to an RHEED setup or even an STM. Low-energy ion guns are very often used as sputter guns in experimental surface science experiments. Combined with an electrostatic analyzer, the ISS experiment is complete.

6.6 Secondary Ion Mass Spectrometry (SIMS)

6.6.1 Introduction

SIMS has been reviewed previously in this series [34]. It is in many cases very sensitive, more sensitive than any of the methods discussed previously. The method is destructive in principle and hence is a bulk analysis method. The physical effect used is sputtering, which is directly related to the nuclear stopping power [49]. When the incident ions hit a target, atomic energy is transferred and the recoil energy E_2 is given by

$$E_2/E_0 = \frac{4M_1 M_2}{(M_1 + M_2)^2} \cos^2\vartheta_2 \tag{6.10}$$

where ϑ_2 is the angle of the recoil motion with respect to the incident beam direction. The recoils can be analyzed directly [50]. For SIMS, the further distribution of the recoil energy and momentum is important. If other target atoms are hit, a "collision cascade" develops. If the collision cascade is intersected by the surface, target atoms are released into the vacuum. From these simple considerations it follows that the sputter yield, i.e., number of released atoms per incident ion, has a maximum—at low energies, the primary beam energy is not sufficient to break surface bonds, i.e., $E_0 < U_B$, the surface binding energy; and at high energies, the ions penetrate deeply into the solid such that the collision cascades do not reach the surface. For the most common sputter gas Ar, the sputter yield maxima are in the keV range [51, 52]. Beside the high sensitivity, SIMS provides straightforward depth analysis and the possibility of imaging.

An inherent problem of SIMS is quantification. The yield of singly charged ions of a species depends strongly on the chemical environment. Metal ion yields from oxides may reach values close to 1.0, whereas the same metal sputtered from a clean metal sample may have an ion yield as low as 10^{-6}. Alkali metal impurities have high ion yields due to their low ionization energies. Without going into a detailed discussion of the "matrix-effect," we instead mention the experimental possibilities to overcome this setback. One general approach is the post-ionization by photons, electrons, or a plasma. Using photons, i.e., laser beams, sputtered particles can be analyzed very specifically [53]. In the case of post-ionization by electrons or plasmas, the acronym SNMS (Secondary Neutral Mass Spectrometry) is used [54]. The plasma can be used as a source for the primary ions as well. With an HF plasma, the sputtering on insulating samples is achieved. With SNMS, a uniform sensitivity for all elements within a concentration range from 1.0 to 10^{-3} can be achieved. With laser ionization schemes, lower concentrations are reached due to the higher ionization efficiencies of up to 1.0, compared to 10^{-1} to 10^{-3} in plasmas. For Fe in Si, a ^{56}Fe detection limit of 2 ppb has been estimated using resonant ionization by two lasers [53].

Another more recently developed technique uses Cs^+ as primary ions [55]. The experimental findings are that especially semiconductor materials, e.g., Si-H, Si-C, Si-N, GaAs, and InP, show practically no "matrix" effects when looking at MCs^+ ions, where M denotes the target atom in question. Because Cs^+ ion beams can be focused very well when produced with a field ion source, this scheme is also advantageous for imaging.

For good depth profiling, "raster gating" is necessary. With this technique, the primary beam is scanned over a certain area and the detector is gated to an inner region of that area. The technique avoids artefacts caused by signals from the walls of the crater. Crater effects are also important when sputtering is combined with ISS, AES, or XPS for depth profiling, i.e., the probe measuring the surface composition has to be limited to the inner part of the sputter crater.

6.6.2 Chemical Analysis with SIMS and SNMS

SIMS properly measures the bulk concentration as a function of time or depth and not the surface concentrations. Under steady state conditions, the sputtered flux must represent the bulk concentrations for reasons of conservation of mass and particle numbers. Therefore, the yield of components A and B with concentrations C_A and C_B are given by

$$\frac{Y_A}{Y_B} = \frac{C_A}{C_B} \tag{6.11}$$

at steady state. Initially, starting at $t = 0$, there are "preferential" sputter effects, i.e., lighter atoms are sputtered more easily due to the energy transfer (Eq. (6.10)). The depletion effects of oxygen from oxides under ion bombardment are well known [34]. This effect has to be taken into account when measuring depth profiles by sputtering in combination with a surface sensitive probe like ISS, AES, or XPS. Figure 19 [54] shows as an example the SNMS depth profile of a W-Si multilayer system with a W-Si double layer thickness of 3.6 nm. In the

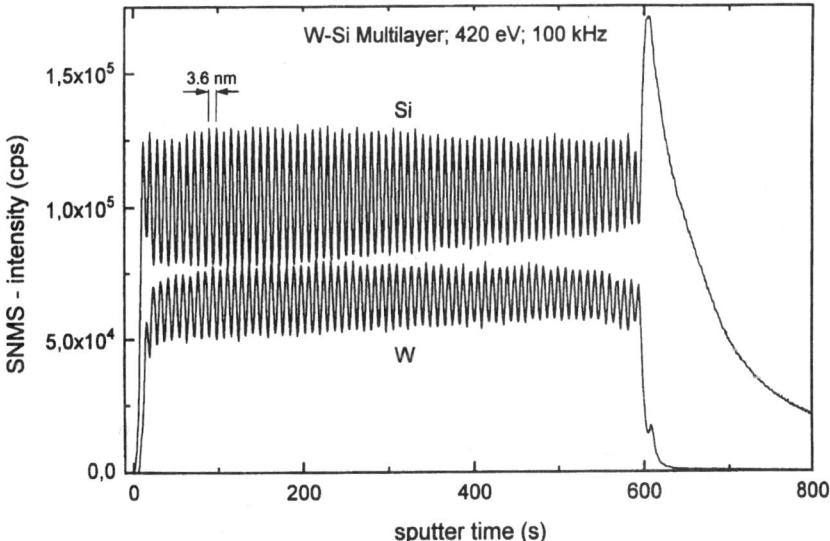

FIG. 19. SNMS depth profile of Si-W layers [54]. (Reprinted with permission of Elsevier Science.)

case of a dielectric sample like W-Si, the sputtering is performed using a 100 kHz Ar hf plasma. The plasma is used for post-ionization, too. The mass spectra are measured with a quadrupole mass filter and an electrostatic energy analyzer. With the plasma sputtering, very flat craters are produced due to the low ion energy distribution of the plasma.

For ion imaging, i.e., ion microprobed, beams of higher energy are necessary to obtain good focusing. Typical parameters are 20 keV Ga^+ in the nA range with a lateral resolution of about 100 nm. It is worth noting that SIMS, SNMS, and recoil spectrometry are the only techniques that can detect hydrogen. Another important aspect of SIMS is the absolute mass identification at high mass resolution (see Section 6.7 and [79]).

6.7 XPS and AES

The techniques described in this section are core level spectroscopies. That is, due to the photoelectric effect, inner shell electrons are excited by a source—X-rays, UV light, an electron beam, or an ion beam. This leads directly or via the Auger effect to the ejection of electrons with a certain kinetic energy, related to their binding energy. From the kinetic energies measured with an energy-resolving detector, the binding energies of the emitted electrons are derived, which are characteristic of the atomic species. Thus the elemental composition of the sample can be determined.

Although other methods are related to the same principle, we will discuss in detail here only X-ray photoelectron spectroscopy (XPS) and Auger electron spectroscopy (AES) because they are by far the most widely used techniques. Both techniques take advantage of the short mean free path of electrons and the fact that the core hole binding energies are element specific. Nevertheless, they are somewhat complementary because XPS is more sensitive to chemical states and AES is faster and has a much higher spatial resolution.

6.7.1 XPS

XPS as it is used today was invented by the nobel prize winner K. Siegbahn and his co-workers in Uppsala almost 40 years ago. It took the group about 10 years to develop the method from the first spectra taken in 1955 to the powerful nondestructive analytical tool it is today [56, 57]. Siegbahn named the method Electron Spectroscopy for Chemical Analysis (ESCA), but we shall refer to it using the acronym XPS.

The fact that XPS machines are commonly used in industrial laboratories as well as in university research facilities is due to the ready availability of the method once the instrument is calibrated and the straightforward interpretation of the data as long as only qualitative chemical information is needed. The basic theory for XPS is very well established and hundreds of reference spectra are available [58]. However, using XPS for precise quantitative surface analysis turns out to be much more difficult, as will be discussed later.

Like many other surface analytical tools, XPS became applicable only with the routine achievement of UHV conditions. XPS is sensitive to the outermost surface layers and thus cannot be operated in an environment where the surface cannot be kept clean during the measurement. Additionally, the excited electrons can be detected only in vacua better than 10^{-5} torr, due to their finite elastic mean free path.

The method can be used in situ, although it does not give the results directly because a scan needs some time and in general more than one scan is necessary to obtain a satisfying signal-to-noise ratio. The source and the detector can be installed in a wide range of angles with respect to the sample, so that no constructional limitations are imposed on the setup for in situ measurements. The distance between the detector (source) and the sample, however, must not be too large and is variable only to some extent. Furthermore, the X-ray source cannot be operated at too high pressures, which might occur during laser ablation or evaporation. In most setups, XPS is not used in situ.

Apart from the UHV conditions that should be met as discussed above, XPS requires basically two things, an X-ray source and an energy-resolving detector. The most common X-ray source is a twin anode design, offering the advantage of two different X-ray lines. As for the anode material, mostly Mg and Al are used, which gives rise to the characteristic K_α lines at $MgK_\alpha = 1253.6$ eV and $AlK_\alpha = 1486.6$ eV, respectively. The linewidth of both is small enough (≤ 0.85 eV) that they do not limit the energy resolution of the system. The use of a monochromator to remove the bremsstrahlung continuum that comes with the characteristic lines is possible, but not common. Another possible source is of course synchrotron radiation, but it is usually not available in a normal laboratory.

There are various types of analyzers but the most common ones are the cylindrical mirror analyzer (CMA), the concentric hemispherical analyzer (CHA), and the retarding field analyzer (RFA), all being of the electrostatic type. Energy resolution better than 1 eV is possible with all three designs.

The best obtainable spatial resolution with XPS is about 5–10 μm, but it makes a specially designed system necessary [59, 60]. Any ordinary system will usually have a spatial resolution in the order of mm^2.

The basic principle of photoelectron spectroscopy is expressed in Einstein's formula describing the photoelectric effect:

$$E_{kin} = E_{h\nu} - E_{bind} - \phi \tag{6.12}$$

where E_{kin} is the kinetic energy of the electron that is measured, $E_{h\nu}$ is the energy of the exciting light, E_{bind} is the binding energy of the excited electron, and ϕ is the workfunction of the material. According to this formula, the binding energy can be determined if the kinetic energy of the electron leaving the sample is measured. Because this energy can be detected very accurately once the sample is Fermi edge coupled to the probe, chemical shifts in the binding energy can be resolved. These shifts are due to the fact that the core levels of the atoms change according to their chemical surroundings and they are in the order of 1–5 eV. From the chemical shifts, information about the bonding can be derived, and thus it is possible to yield elemental as well as chemical information with XPS.

Compositional analysis is normally done measuring the peak area and comparing it to either references or to areas from other elemental peaks in the same spectrum. The detection limit is 1–10% of a monolayer, depending on the material. To calculate the peak areas, reliable procedures for background subtraction are needed because the elastic peak (electrons with zero energy loss) is accompanied by a rather strong background due to electrons that suffered from either inelastic or elastic energy losses on their way through the solid.

The most widely used procedure for background subtraction is the Shirley method [61], where it is assumed that each electron contributing to the elastic peak gives rise to a flat background of losses at lower energies. This easy-to-use procedure is based on mathematical principles only. Furthermore, the results of the procedure may vary from user to user, so that no reproducible background correction can be performed. With this method, typical errors of 10%–50% in the quantitative result are possible [62].

To obtain a more accurate method for the background subtraction, Tougaard *et al.* have developed a method which is based on a physical model for the energy loss of the excited electrons [63]. Figure 20, shows as an example of this method applied to the XPS (Mg K_α) spectrum of Au. Shown are the Au 4f and 4d peaks (solid line) and the calculated inelastic background (dotted line). After subtraction of the latter, the intrinsic Au peaks are obtained (dashed line), which can then be used for quantitative analysis.

Although some may consider XPS to be more of a bulk technique, it is certainly not. XPS's information depth can be as low as a few Å and is seldom higher than 100 Å. XPS owes its surface sensitivity to the fact that the excited electrons suffer from energy losses on their way through the solid. Though the X-rays may excite electrons in depths up to a few 100 Å, the excited electrons will not be able to penetrate the solid if they have been excited more than a few 10 Å below the surface. The energy losses that limit the escape depth of the excited electrons are mainly inelastic in nature and give rise to the inelastic background in XPS spectra. As pointed out earlier, this background has to be and can be removed if a qualitative compositional analysis of the sample is required.

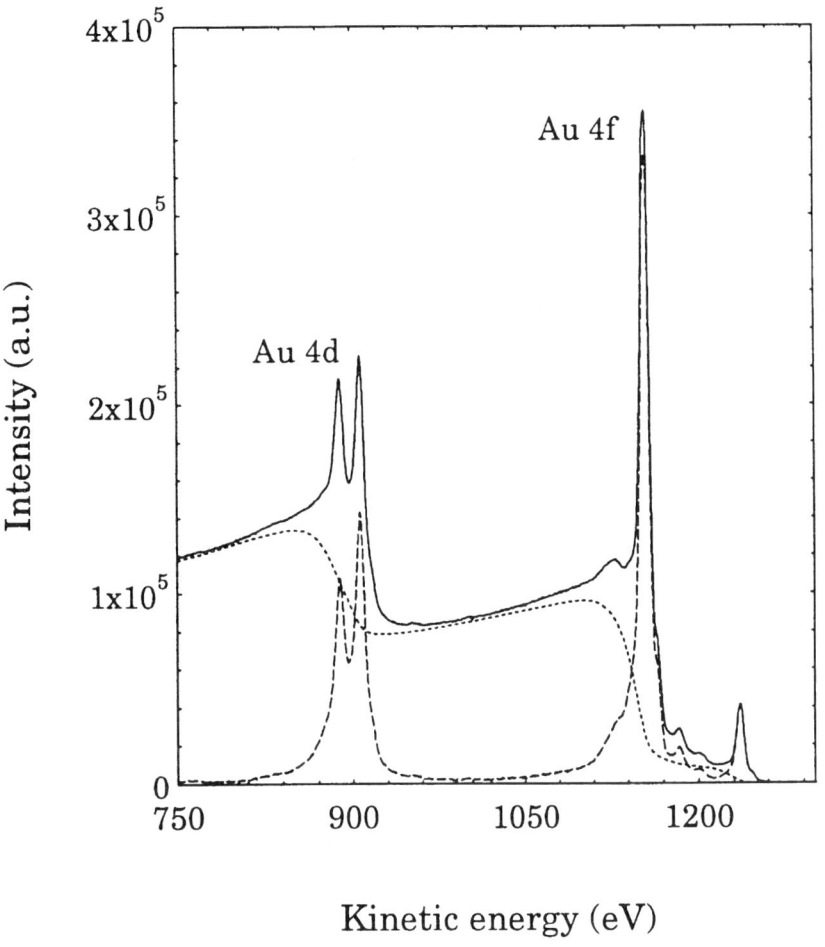

FIG. 20. X-ray photoelectron spectrum from gold (Au 4d and 4f). The solid line represents the recorded spectrum. The dashed line is the background due to inelastic scattering of the electrons calculated with the Tougaard algorithm. After subtraction of the latter, the intrinsic spectrum as excited in the solid is obtained (dotted line).

Recently, Tougaard *et al.* could show that the inelastic background itself can be used to obtain structural information about the sample surface [64]. The idea behind this is that the exact shape and intensity of the inelastic background depends strongly on the in-depth concentration profile of the electron emitters. By comparing a reference spectrum from a pure sample to the background-corrected measured spectrum of the surface to be characterized, the in-depth concentration profile and thus the nanostructure of the sample can be determined. The method is, therefore, extremely well suited to study thin films [65, 66], without the need for conventional destructive sputter depth profiling. An example of this technique is given in Fig. 21 [67]. Shown is a spectrum taken from an Si sample onto which a thin film of 6 Å Pt was evaporated. By applying this method, the amount of Pt and its nanostructure could be determined, in this case, a single island growth mode.

To a certain degree, all materials can be analyzed by XPS; they do not even have to be in the solid phase. Some problems arise, however, if the sample under investigation is an insulator. Even if the charging problem can be solved by, e.g., an electron flood gun, the difficulty of defining the electronic energy scale with respect to that of the spectrometer will remain. Because Fermi edge coupling of the sample and the detector is not possible in the case of a nonconducting sample, it may be impossible to determine the binding energies according to Eq. (6.12). The number of samples, however, that cannot be Fermi edge coupled is small. For thin films deposited onto conducting substrates, the problem does not occur anyway.

In conclusion, XPS is a very helpful nondestructive tool for surface analysis in connection with thin film growth and one of the very few methods that can obtain elemental, chemical, and structural information.

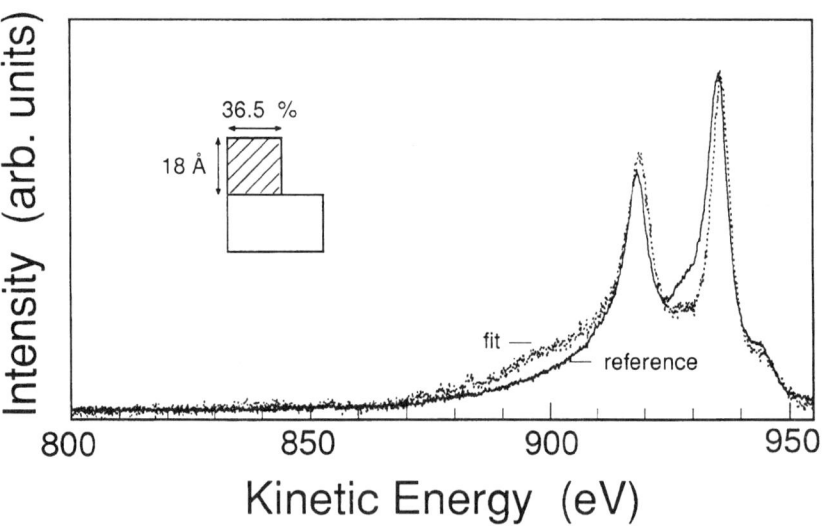

FIG. 21. Structural analysis with XPS: The recorded spectrum of Pt deposited on Si(111) is corrected for an inelastic background caused by an assumed nanostructure of a single island being 18 Å high and covering 36.5% of the surface (dashed line). The solid line represents the intrinsic spectrum for a pure homogeneous platinum sample. The good agreement between the two curves shows that the assumed nanostructure (see inset) is correct. (Adapted from M. Schleberger *et al.* [67], by permission of Elsevier Science.)

6.7.2 AES

AES is an offspring of one of the oldest analytical tools, namely, LEED [1]. LEED setups, which had been successfully used since the 1930s, had all that was needed to do AES—an electron gun and a retarding field analyzer. Therefore, many of the LEED systems could easily be modified to operate as AES systems, after Harris had realized the potential of Auger electrons to perform surface analysis [68]. Due to poor energy resolution at that time, AES was used for elemental analysis only.

Today AES is one of the most widely used routine tools in surface analysis. The low experimental requirements are one reason why AES is a standard technique in almost any surface science laboratory. As with XPS, the theoretical background is well understood and support data is easily available [69, 70]. Although AES can be done using a number of excitation sources, we will concentrate on electron-excited Auger spectroscopy.

AES requires UHV conditions due to the surface sensitivity and the relatively short elastic mean free path of the electrons. The technique can be used in situ. In many systems however, the LEED setup is used for AES and, therefore, the same restrictions that applied for LEED apply here, too. Because it is relatively easy to produce a high-intensity electron beam, the time for data acquisition is considerably shorter than in XPS. For the same reason though, AES is not as nondestructive as XPS. To study thin films, the intensity must be kept below the threshold for surface damage. The depth of information is comparable to that in XPS, that is, in the order of a few 10 Å.

As mentioned above, only an electron gun and an energy-resolving detector are necessary to do AES. The detector is often a retarding field type detector because it can also be used in LEED measurements, but other types like CMA and CHA are also common. AES spectra were usually recorded as differentiated spectra because the Auger peaks are superimposed on a very strong background of secondary electrons. Today the detection technique is advanced enough to allow for the recording of direct spectra.

The choice of excitation source depends on what is already available in the system, but it will most likely be an electron gun capable of producing typically 1–5 eV electron beams. X-ray sources can also be used for excitation and in principle even ion guns.

The major virtue of AES is its superior spatial resolution due to the easy focusing and scanning of the primary electron beam. This makes it possible to employ AES in a scanning mode, sometimes called chemical imaging, with a spatial resolution below 20 nm. A resolution as high as this, however, makes rather high beam currents necessary and sample damage may occur.

With the photoelectric effect, the energy of the excited electrons depends on the energy of the exciting light. Auger electrons are from a de-excitation process and are therefore independent of the energy of the exciting beam. The Auger effect, named after Pierre Auger, is a two-electron process, where an ionized atom is de-excited by means of a nonradiative process.

As in XPS, inner shell electrons are ejected. The core hole is filled up by an electron from a higher shell. The excess energy in this transition is taken by an electron from another level. The last shallowly bound electron is ejected from the atom and may be able to leave the solid. The energy balance for the Auger process is thus

$$E_{kin} = E_X - E_Y - E_Z \qquad (6.13)$$

where E_{kin} is the kinetic energy of the emitted electron, E_X and E_Y are the binding energy of the singly ionized state, and E_Z is the binding energy for the doubly ionized state.

The energy-resolved spectrum exhibits peaks unique to each element. Only hydrogen and helium cannot be detected. The technique is surface senitive because only electrons from the topmost layer contribute to the spectrum. The sensitivity is better than 1 atom percent of a monolayer.

From the beginning, AES was used as a tool for elemental analysis only and it was thought that the Auger signal yielded only little chemical information. Unlike in XPS, where this information is easily obtained by measuring the chemical shift, the chemical information in AES peaks cannot be obtained so straightforward. In principle, both peak shape and energy shifts can be used to differentiate between different chemical states [71–73], but only little effort has been made to exploit the chemical information in Auger spectra. The energy resolution necessary to study shifts in Auger peaks is generally higher and more detailed studies may make an X-ray source for excitation necessary.

Qualitative elemental analysis is fast and easy with AES. As an example, the Auger spectrum of Pt is shown in Fig. 22 [7]. Regarding precise quantitative analysis, we face similar problems as with XPS. Quantification is mainly done by comparing peak areas, applying the idea that the intensity of the signal is proportional to the molar fractional content.

Because the background in AES is significantly larger than in XPS and has its origin in different sources [74, 75], it is much more difficult to remove. However, often only relative changes in elemental peaks are used. Additionally, a rough estimate of the peak area is often sufficient to obtain the required information. Using calibrated standards, quantifications to better than 10% can be achieved in cases where the material is homogenously distributed. The thickness of thin films can be estimated by following the attenuation of the AES signal as a function of depositing time (assuming a constant rate of deposition and layer-by-layer growth) [74, 75].

The major advantage of AES in elemental analysis is certainly the fact that it can be used as a scanning microprobe [76, 77]. Spatially resolved two-

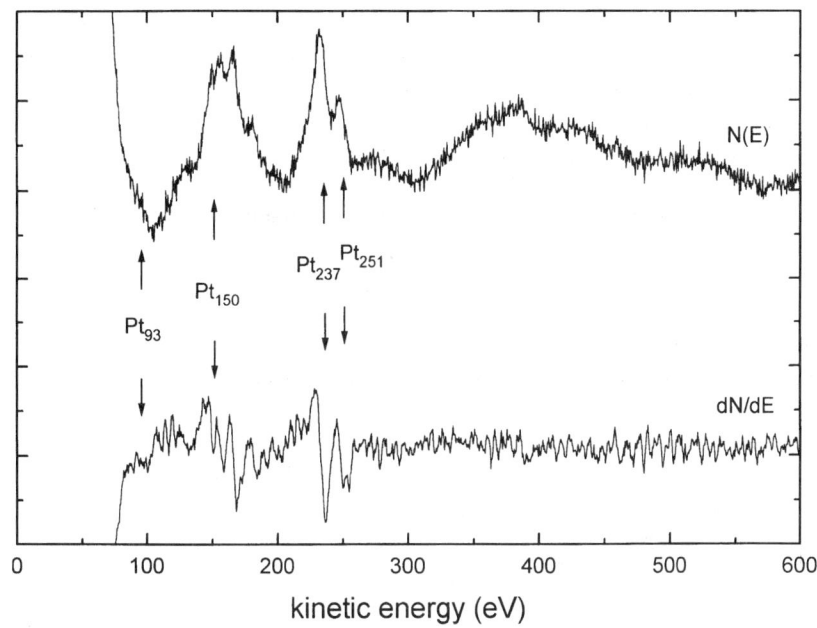

FIG. 22. AES spectrum of a clean Pt(110) (1 × 2) surface as in Figs. 2, 4, and 8 [7]. (Reprinted with permission of Elsevier Science.)

dimensional maps of the elemental composition of the surface can be obtained in this way [78] (see Fig. 23). Compared to other scanning probes (e.g., SIMS), however, AES is rather time consuming.

Analysis of sample structures by conventional sputter depth profiling is possible with AES and in fact widely used [79], but it is, of course, destructive. The AES signal alone yields little structural information. The angular distribution of the emitted electrons could be useful for structural analysis [80], but the interpretation of those measurements is quite complicated.

The Tougaard method could in principle be applied to Auger spectra, too, but additional difficulties arise. First, the large background of secondary and backscattered primary electrons has to be removed before the spectrum can be analyzed. There is no reliable procedure to do that. Second, the depth of excitation for electrons is much lower than for X-rays. So, unless X-ray excited AES is performed, information about the depth excitation function would be necessary [81].

The biggest problems encountered with AES are sample damage due to high currents and charging of the sample if it is an insulator. These problems can partly be solved by adjusting the experimental conditions accordingly.

326   SURFACE CHARACTERIZATION

In summary, AES and especially the scanning Auger microprobe are very useful for qualitative and quantitative analysis with respect to the elemental composition of surfaces, although precise quantification is difficult to achieve. As for structural and chemical analysis, AES can in principle be used, but XPS is better suited for this purpose.

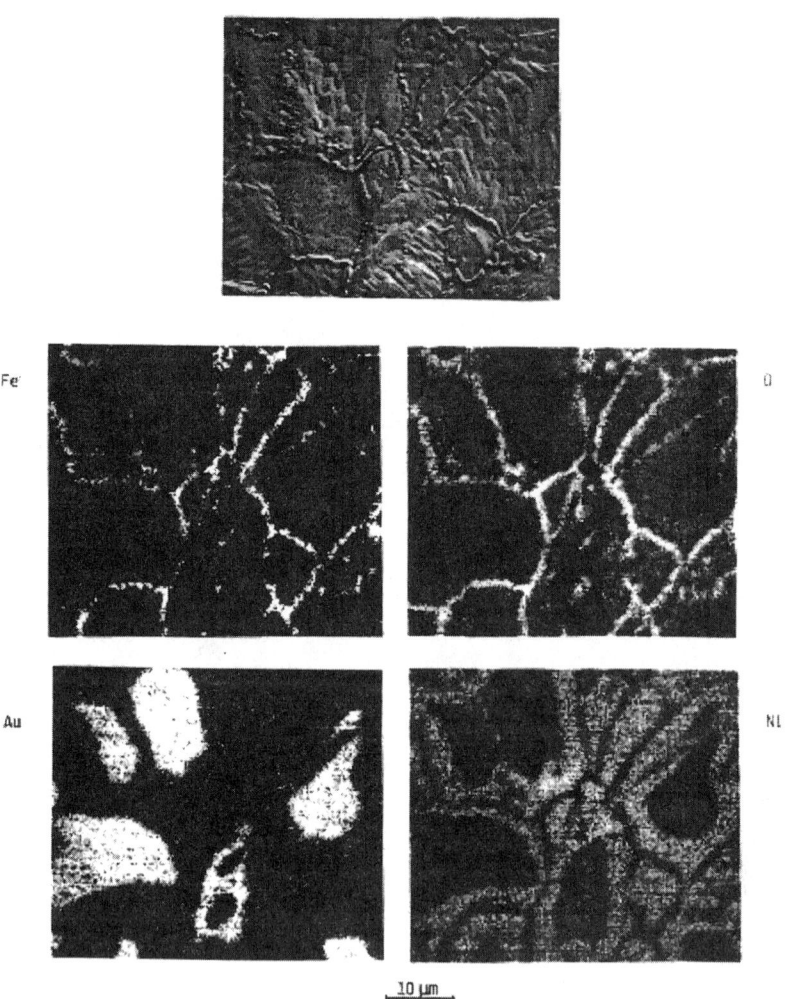

FIG. 23. Scanning electron microscopy picture (upper) and elemental maps (lower) of gold coated Alloy 42. Scanning auger microscopy allows for a detailed study of the substrate diffusion through grain boundaries. Shown are the micrographs of iron, oxygen, gold, and nickel. (Reproduced from D. W. Harris et al. [78], by permission of John Wiley & Sons.)

TABLE II. Characteristic of Surface Analytical Methods

Method	Excitation/ Detection	Information		Depth Range	Depth Analysis	In Situ	Real Time	Lateral Resolution
		Chemical	Structural					
LEED	e → e	−	+	1 nm	−	−	+	1 nm^2
RHEED	e → e	−	+	1 nm	−	+	+	1 nm^2
LEEM	e → e	−	+	1 nm	−	−	+	10 nm^2
STM	tip → current	−	+	<1 nm	−	−	−	0.1 nm^2
AES	tip → forces	−	+	<1 nm	−	−	−	1 nm^2
RBS	ion → ion	+	+	3 µm	+	+	+	1 µm^2
ISS	ion → ion	+	+	<1 nm	−	+	−	1 nm^2
SIMS	ion → ion	+	−	<1 nm	+	−	−	100 nm^2
XPS	hv → e	+	0	10 nm	+	+	0	10 µm^2
AES	e → e	+	0	1 nm	−	−	−	100 nm^2

6.8 Summary

The properties of the analytical methods are summarized in Table II. The data shown have a qualitative and average weight. In special cases, "better" values are quite possible. With the label "in situ," direct compatibility with a film growth system is indicated. A minus sign then means that the substrate or the fabricated film has to be transported at least within the same vacuum chamber over some distance. It is worth noting, for example, that not all STM and AFM systems "accept" large area samples, and most LEED systems operate with UHV manipulators that accept samples of the order of 1 cm^2. Details of these calibers are neglected here.

For the techniques not discussed at all, i.e., EELS and SEM, see [82–85].

References

1. Davisson, C. J., and Germer, L. H. Diffraction of electrons from a crystal of nickel. *Phys. Rev.* **30**, 705–740 (1927).
2. Duke, C. B. Interaction of electrons and positions with solids: From bulk to surface in thirty years. *Surf. Sci.* **299/300**, 24–33 (1994); Tsong, S. Y. Electron diffraction for surface studies—The first 30 years. *Surf. Sci.* **299/300**, 358–374 (1994); Pendry, J. B. Multiple scattering theory of electron diffraction. *Surf. Sci.* **299/300**, 375–390 (1994).
3. MacLaren, J. M., Pendry, J. B., Rous, P. J., Sldin, D. K., Somorjai, G. A., van Hove, M. A., and Vredensky, P. P. *Surface Crystallography Information Service.* Reidel Publ., Dordrecht (1987).
4. Scheithauer, G., Meyer, G., and Henzler, M. A new LEED instrument for quantitative spot profile analysis. *Surf. Sci.* **178**, 441–451 (1986).
5. Lagally, M. Diffraction techniques. *Meth. in Exp. Phys.* **22**, 237–298 (1985).

6. Clearfield, H. M., Welkie, D. G., Lu, T.-M., and Lagally, M. G. LEED investigation of extended defects at the surface of the films grown epitaxially on GaAs (110). *J. Vac. Sci. Tech.* **19**, 323–330 (1981).
7. Speller, S., Kuntze, K., Rauch, T., Bömermann, J., Huck, M., Aschoff, M., and Heiland, W. The (1 × 2) and (1 × 4) structure on clean Pt(110) studied by STM, AES, and LEED. *Surf. Sci.* **366**, 251–259 (1996).
8. Larson, P. K., and Dobson, P. J. *Reflection High Energy Electron Diffraction and Reflection Imaging of Surfaces*, NATO ASI Series, Vol. 188. Plenum Press, New York, 1988.
9. Kuntze, J., Rauch, T., Bömermann, H., Speller, S., and Heiland, W. The Pt(110) surface studied by STM and RHEED. *Surf. Sci. Lett.* **355**, L300–L304 (1996).
10. Aarts, J., and Larsen, P. K. RHEED studies of growing G and Si surfaces. In ref. [8] 449–461 (1988).
11. Korte, U., and Meyer-Ehmsen, G. The structure of the Pt(110) 1 × 2 surface and its (1 × 2) ⇌ (1 × 1) structural transition. I. RHEED from the periodic part of the structure. *Surf. Sci.* **271**, 616–640 (1992); Korte, U., and Meyer-Ehmsen, G. The structure of the Pt(110)1 × 2 surface and its (1 × 2) ⇌ (1 × 1) structural transition. II. RHEED from the non-periodic part of the structure. *Surf. Sci.* **277**, 109–122 (1992).
12. Telieps, W., and Bauer, E. The (7 × 7) ⇌ (1 × 1) phase transition on Si(111). *Surf. Sci.* **162**, 163–169 (1985).
13. Bauer, E. Low energy electron microscopy and nanostructures. *Phys. Stat. Sol.* **192**, 375–387 (1995).
14. Binnig, G., Rohrer, H., Gerber, Ch., and Weibel, E. Tunneling through a controllable vacuum gap. *Appl. Phys. Lett.* **40**, 178–180 (1982).
15. Binnig, G., Rohrer, H., Gerber, Ch., and Weibel, E. Surface studies by scanning tunneling microscopy. *Phys. Rev. Lett.* **49**, 57–61 (1982).
16. Binnig, G., Rohrer, H., Gerber, Ch., and Weibel, E. (111) Facets as the origin of reconstructed Au(110) surfaces. *Surf. Sci.* **131**, L379–L384 (1983).
17. Tersoff, J., and Lang, N. D. Theory of scanning tunneling microscopy. *Meth. Exp. Phys.* **27**, 1–29 (1993).
18. Song-Il Park, and Barrett, R. C. Design considerations for an STM system. *Meth. Exp. Phys.* **27**, 31–76 (1993).
19. Stroscio, J. A., and Feenstra, R. M. Methods of tunneling spectroscopy. *Meth. Exp. Phys.* **27**, 96–147 (1993).
20. Feenstra, R. M. Scanning tunneling spectroscopy. *Surf. Sci.* **299/300**, 965–979 (1994).
21. Rohrer, H. Scanning tunneling microscopy: A surface tool and beyond. *Surf. Sci.* **299/300**, 956–964 (1994).
22. Gritsch, T., Coulman, D., Behm, R. J., and Ertl, G. A scanning tunneling microscopy investigation of the structure of the Pt(110) and Au(110) surfaces. *Surf. Sci.* **257**, 297–306 (1991).
23. Gimzewski, J., Berndt, R., and Schittler, R. Scanning-tunneling-microscopy study of antiphase domain bombardies, dislocations and local mass transport on Au(110) surfaces. *Phys. Rev. B* **45**, 6844–6857 (1992).
24. Becker, R., Wolkow, R., Feenstra, R. M., and Stroscio, J. A. Semiconductor surfaces. *Meth. Exp. Phys.* **27**, 149–276 (1993).
25. Binnig, G., Quate, C. F., and Gerber, Ch. Atomic force microscope. *Phys. Rev. Lett.* **56**, 930–933 (1986).

26. Bartzke, K., Antrack, T., Schmidt, K.-H., Daumann, E., and Schatterny, Ch. The needle sensor—A micromecharical detector for atomic force microscopy. *Int. J. Optoelectronics* **8**, 669–676 (1993).
27. Wickramasinghe, H. K. Extensions of the STM. *Meth. Exp. Phys.* **27**, 7794 (1993).
28. Quate, C. F. The AFM as a tool for surface imaging. *Surf. Sci.* **299/300**, 980–995 (1994).
29. Rutherford, E. The scattering of α and β particles by matter and the structure of the atom. *Phil. Mag.* **21**, 669–688 (1911).
30. Rubin, S. Surface analysis by charged particle spectroscopy. *Nucl. Instr. Meth.* **5**, 177–183 (1959).
31. Smith, D. P. Scattering of low-energy noble gas ions from metal surfaces. *J. Appl. Phys.* **38**, 340–347 (1967).
32. Liau, Z. L., Mayer, J. W., Brown, W. L., and Poate, J. M. *J. Appl. Phys.* **49**, 5295–5305 (1978).
33. Feldman, L. C., and Mayer, J. W. *Fundamentals of Surface and Thin Film Analysis.* North Holland, Amsterdam, 1986.
34. Heiland, W., and Taglauer, E. Ion scattering and secondary-ion mass spectrometry. *Meth. Exp. Physics* **22**, 299–348 (1985).
35. Zinke-Almang, M., and Feldman, L. C. Clustering in surfaces. *Surf. Sci. Rep.* **16**, 377–463 (1992).
36. Andersen, H. H. *The Stopping and Ranges of Ions in Matter*, pp. 1–207. Pergamon Press, New York, 1977.
37. Doolittle, L. R. Algorithm for the rapid simulation of Rutherford backscattering spectra. *Nucl. Instr. Meth.* **B9**, 344–351 (1985).
38. Mashkova, E. S., and Molchanov, V. A. Medium energy ion scattering from solids. *Mod. Probl. Cond. Matt. Phys.* **11**, 1–444 (1985).
39. Taglauer, E. Ion scattering spectroscopy, in *Ion Spectroscopies for Surface Analysis*, A. W. Czanderna and D. M. Hercules (eds.). Plenum Publ. Co., New York, 1991.
40. Niehus, H., Heiland, W., and Taglauer, E. Low-energy ion scattering at surfaces. *Surf. Sci. Rep.* **17**, 213–304 (1993).
41. Overbury, S. H., Mullins, D. R., and Wendelken, J. F. Surface structure of stepped NiAl(111) by low energy Li^+ ion scattering. *Surf. Sci.* **236**, 122–134 (1990).
42. Weiland, P., Jelinek, B., Hofer, W., and Varga, P. $Pt_{25}Ni_{75}$ (100) and (110) single crystals preferential sputtering and segregation reversal. *Surf. Sci.* **307–309**, 416–421 (1994).
43. Lindhard, J. Influence of crystal lattice on motion of charged particles. Kongelige Danske Vindensk Selsk. *Matematisk-Fysiske Medd.* **34**, 1–65 (1965).
44. Robinson, M. T., and Oen, D. S. The channeling of energetic atoms in crystal lattices. *Appl. Phys. Lett.* **2**, 30–33 (1963).
45. Speller, S., Schleberger, M., Niehof, A., and Heiland, W. Structural effects on the Pb(110) surface between 160 and 580 K observed low-energy ion scattering. *Phys. Rev. Lett.* **68**, 3452–3455 (1992).
46. Frenken, J. W. M., and Van der Veen, J. F. Observation of surface melting. *Phys. Rev. Lett.* **54**, 134–137 (1985).
47. Aono, M. Quantitative surface structure analysis by low-energy ion scattering. *Nucl. Instr. Meth.* **B2**, 375–383 (1984).
48. Niehus, H. Characterization of metal alloy systems by scanning tunneling microscopy and low-energy ion backscattering. *Phys. Stat. Sol.* **B92**, 357–374 (1995).
49. Sigmund, P. Theory of sputtering. I. Sputtering yield of amorphous and polycrystalline targets. *Phys. Rev.* **184**, 383–416 (1969).

50. Sung, M. M., Bykoi, V., Al-Bayahi, A., Kim, C., Todorov, S. S., and Rabalais, J. W. From scattering and recoiling spectrometry to scattering and recoiling imaging. *Scanning Microscopy* **9**, 321–330 (1995).
51. Behrisch, R., ed. Sputtering by particle bombardment, I and II. *Topics Appl. Phys.* (Springer, Berlin) **47** (1981) and **52** (1983).
52. Behrisch, R., and Wittmaack, K. Characteristics of sputtered particles technical applications. *Topics Appl. Phys.* (Springer, Berlin) **64** (1991).
53. Young, C. E., Pellin, M. J., Calaway, W. F., Jørgensen, B., Schreiber, E. L., and Gruen, D. M. Laser-based secondary neutral mass spectroscopy: Useful yields and sensitivity. *Nucl. Instru. Meth.* **B27**, 119–129 (1987).
54. Oechsner, H. Secondary neutral mass spectrometry (SNMS)—Recent methodical progress and applications to fundamental studies in particle/surface interaction. *Int. J. Mass. Spectr. and Ion Proc.* **143**, 271–282 (1995).
55. Gnaser, H. Improved quantification in secondary-ion mass spectrometry detecting MCs^+ molecular ions. *J. Vac. Sci. Tech.* **A12**, 452–456 (1994).
56. Siegbahn, K., Nordling, C. N., Fahlmann, A., Nordberg, R., Hamrin, K., Hedmann, J., Johansson, G., Bermark, T., Karlsson, S. E., Lindgren, I., and Lindberg, B. *ESCA Atomic Molecular And Solid State Structure Studied by Means of Electron Spectroscopy*. Almqvist and Wiksell, Uppsala, 1967.
57. Siegbahn, K. From X-ray to electron spectroscopy and new trends. *J. Electron. Spectrosc. Relat. Phenom.* **51**, 11–36 (1990).
58. Wagner, C. D., Riggs, W. M., Davis, L. E., Moulder, J. F., and Muilenberg, G. E. *Handbook of X-ray Photoelectron Spectroscopy*. Perkin Elmer Corp., Minnesota, 1979.
59. Coxon, P., Krizek, J., Humperson, M., and Wardell, I. R. M. The ESCASCOPE—A new imaging photoelectron spectrometer. *J. Electron. Spectrosc. Relat. Phenom.* **52**, 821–836 (1990).
60. Forsyth, N. M., and Coxon, P. Use of parallel imaging XPS to perform rapid analysis of polymer surfaces with spatial resolution <5 μm. *Surf. Interface Anal.* **21**(6–7), 430–434 (1994).
61. Shirley, D. A. High-resolution X-ray photoemission spectrum of the valence band of gold. *Phys. Rev.* **B5**(12), 4709–4714 (1972).
62. Jansson, C., Tougaard, S., Beamson, G., Briggs, D., Davis, S. T., Rossi, A., Hauert, R., Hobi, G., Brown, N. M. D., Merenan, B. J., Anderson, C. A., Repoux, M., Malitesta, C., and Sabbatini, L. Intercomparison of algorithms for background correction in XPS. *Surf. Interface Anal.* **23**, 484–494 (1995).
63. Tougaard, S., and Sigmund, P. Influence of elastic and inelastic scattering on energy spectra of electrons emitted from solids. *Phys. Rev.* **B25**(7), 4452–4466 (1982).
64. Tougaard, S. Quantitative analysis of the inelastic background in surface electron spectroscopy. *Surf. Interface Anal.* **11**, 453–472 (1988).
65. Tougaard, S., and Hansen, H. S. Non-destructive depth profiling through quantitative analysis of surface electron spectra. *Surf. Interface Anal.* **14**, 730–738 (1989).
66. Schleberger, M., Fujita, D., Scharfschwerdt, C., and Tougaard, S. Nanostructure of thin metal films on Si(111) investigated by X-ray photoelectron spectroscopy: Inelastic peak shape analysis. *J. Vac. Sci. Technol.* **B13**(3), 949–953 (1995).
67. Schleberger, M., Fujita, D., Scharfschwerdt, C., and Tougaard, S. Growth and in-depth distribution of thin metal films on silicon (111) studied by XPS: Inelastic peak shape analysis. *Surf. Sci.* **331–333**, 942–947 (1995).
68. Harris, L. A. Analysis of materials by electron-excited Auger electrons. *J. Appl. Phys.* **39**, 1419–1427 (1968).

69. Davies, L. E., Macdonald, N. C., Palmberg, P. W., Riach, G. E., and Weber, R. E. *Handbook of Auger Electron Spectroscopy*. Physical Electronics Division of Perkin Elmer Corp., Minnesota, 2nd ed., 1976.
70. Mogami, A., and Hirata, K. *Handbook of Auger Electron Spectroscopy*. JEOL, Tokyo, 1982.
71. Haas, T. W., Grant, J. T., and Dooley, G. J. Chemical effects in Auger electron spectroscopy. *J. Appl. Phys.* **43**, 1853–1860 (1972).
72. Rye, R. R., Houston, J. E., Jennison, D. R., Madey, T. E., and Holloway, P. E. Chemical state effects in Auger electron spectroscopy. *J. Chem. Phys.* **69**(4), 1504–1512 (1978).
73. Rye, R. R., Houston, J. E., Jennison, D. R., Madey, T. E., and Holloway, P. E. Chemical information in Auger electron spectroscopy. *Ind. & Eng. Chem., Prod. Res. Dev.* **18**(1), 2–7 (1979).
74. Sickafus, E. N. Linearized secondary-electron cascades from the surfaces of metals. I. Clean surfaces of homogeneous specimens. *Phys. Rev. B* **16**(4), 1436–1447 (1977).
75. Sickafus, E. N. Linearized secondary-electron cascades from the surfaces of metals. II. Surface and subsurface sources. *Phys. Rev. B* **16**(4), 1448–1458 (1977).
76. Powell, B. D., and Woodruff, D. P. Anisotropy in grain boundary segregation in copperbismuth alloys. *Phil. Mag.* **34**, 169–176 (1976).
77. Powell, B. D., Woodruff, D. P., and Griffiths, B. W. A. A scanning Auger electron microscope for surface studies. *J. Phys. E* **8**(7), 1548–1551 (1975).
78. Harris, D. W., and Nowicki, R. S. Applications of AES in microelectronics, in *Practical Surface Analysis*, 2nd ed., Vol. 1, pp. 257–310. John Wiley & Sons Ltd., 1990.
79. Oechsner, H., Ed. Thin films and depth profile analysis. *Topics in Current Physics*, Vol. 37. Springer, Berlin, 1984.
80. Woodruff, D. P. Photoelectron and Auger electron diffraction. *Surf. Sci.* **299/300**, 183–198 (1994).
81. Fujita, D., Schleberger, M., and Tougaard, S. Extraction of depth distributions of electron-excited Auger electrons in Fe, Ni, and Si using inelastic peak-shape analysis. *Surf. Sci.* **357–358**, 180–185 (1996).
82. Ibach, H., and Mills, D. L. *Electron Energy Loss Spectroscopy and Surface Vibrations*. Academic Press, N.Y., 1982.
83. Johari, O., and Samudra, A. V. Scanning electron microscopy, in *Characterization of Solid Surfaces*, P. F. Kane and G. B. Larrabel (eds.), pp. 107–131. Plenum Press, N.Y., 1974.
84. Reimer, L. *Transmission Electron Microscopy*, 2nd ed. Springer, Heidelberg, 1989.
85. Lee-Deacon, O., Le Gressus, C., and Massignon, D. Analytical scanning electron microscopy for surface science, in *Electron Beam Interactions with Solids*, D. F. Kyser, H. Niedrig, D. F. Newburg, and R. Schimizu (eds.), pp. 271–279. SEM Inc., 1984.

7. SURFACE MODIFICATION WITH LASERS

Zane Ball and Roland Sauerbrey

Institut für Optik und Quantenelektronik
Friedrich-Schiller-Universität Jena, Germany

7.1 Introduction

In recent years, surface modification with lasers has taken on increased sophistication and now offers tremendous promise for applications. Laser processing of materials can produce many changes to a surface, including novel microstructures and alloys, micromachined morphology, chemical reactions, and spontaneous periodic structures. To meet stringent engineering requirements, it is desirable for surface and bulk to have contrasting properties where either it would be impossible to use a single material or doing so would incur unreasonable costs. Traditionally simple coating through electroplating and other means has allowed inexpensive materials to acquire the properties of rare and expensive materials for corrosion resistance or catalysis. Using lasers, a whole new level of ability to modify the surface is becoming available, particularly in microelectronics where tight thermal budgets prevent or limit the use of traditional modification techniques. As laser pulses allow for high local temperatures with low energy and operation at wavelengths that allow control over chemistry as well, it is easy to see that laser processing offers many advantages. The many combinations of materials and laser sources in various intensity regimes produce a truly vast range of novel effects that can be exploited for industrial or scientific application.

We first examine the laser sources commonly used in surface modification techniques, the absorption of an intense laser pulse, and explore the possible pathways through which the surface returns to equilibrium. In the second section, we divide the effects of the laser pulse into two categories, physical and chemical. Physical effects include surface wave optical phenomena, shock waves, phase transitions, and thermal diffusion, and chemical reactions include thermally activated processes, photochemical bondbreaking, and charge transfer reactions. Specific processes based on these physical and chemical effects are then discussed: ablation, annealing, lithography, and laser-induced electrical conductivity. The list of processes discussed is certainly not exhaustive but is rather intended to be exemplary while providing information on the major results of recent investigations. Finally, we examine the role of surface modification in lithography. As excimer laser lithography is already in use for 0.25 µm processes [1], a large effort toward understanding and improving control over the interaction with photoresists continues, and the main results and considerations are presented here.

7.1.1 Laser Sources for Surface Modification

Although much research has been performed with continuous wave (cw) lasers such as Ar^+ lasers and CO_2 lasers [2], we will constrict ourselves here to pulsed laser systems, which form the core of the most active research areas today.

The traditional pulsed laser source for materials applications is the Nd:YAG laser because of its reliability and high output energy. At 1.06 µm these lasers couple to a material through electronic and vibrational excitation in free molecules and inter- and intraband transitions in solids. Through harmonic generation, wavelengths as short as 213 nm (fifth harmonic) can be produced with substantial energies. The laser can be operated CW, Q-switched for nanosecond pulses, and mode-locked for sub-ns pulses. For processes where the 1.06 µm fundamental wavelength is acceptable, the Nd:YAG laser is almost always the laser of choice, and it has therefore been widely used in industrial environments [3]. Microsecond pulsed CO_2 lasers (10.6 µm) have also traditionally been used in high-power materials applications for laser welding and other industrial processes, though most newer effects that will be discussed here benefit from shorter wavelengths and shorter pulse durations. The primary advantages of CO_2 lasers are their high average power and commonplace use in industrial environments.

Great progress has been made in recent years with excimer laser technology such that excimers can now operate profitably in an industrial environment. The most common excimer lasers (ArF, 193 nm; KrF, 248 nm; and XeCl, 308 nm) provide pulses with energies of up to 1 J in pulses of 10 to 30 ns in length [4]. Modern excimer lasers can operate at repetition rates as high as 1 kHz and for up to a billion pulses on a single gas fill. These short wavelengths are absorbed by direct electronic transitions and therefore allow for manipulation of the target through photochemistry as well as thermal effects with very high spatial resolution. It is the opportunity for increased spatial resolution in photolithography that has provided a prime motivation for major improvements in excimer technology. Other applications have also appeared, such as in opthamology [5] and via creation in electronics packaging [6]. Specialized excimer systems even allow for the production of sub-picosecond pulses with peak intensities as high as 10^{17} W/cm^2.

Newly available Ti:Sapphire lasers may find great application in surface modification as well. These lasers exhibit tunability over a wide wavelength range (720 nm to 1010 nm) and can generate femtosecond pulses at a number of wavelengths through a combination harmonic generation and various cavity configurations. Pulse duration can also be controlled over many orders of magnitude using pulse compression techniques that allow for investigation of intensity dependences of laser materials interactions [7]. Like the Nd:YAG laser and other solid state systems, Ti:Sapphire lasers have obvious reliability advantages.

7.1.2 Absorption of Laser Pulses at Moderate Intensities

Most laser-based materials processing techniques are a result of the energy density absorbed in a surface layer of the target material. Explosive thermal decomposition, shock waves, and phase transitions are all directly related to the deposited energy density as are many thermally induced chemical reactions. The energy balance is governed by two considerations: the amount of energy deposited and the volume in which it is deposited. Although the former may at first appear to be a simple laboratory measurement, changes in optical properties during irradiation can significantly alter the amount of energy that is actually absorbed. The depth in which the energy is absorbed is the quantity that is most dependent on material properties, however. The determination of the absorption depth is certainly nontrivial and is essential in understanding any light matter interaction. The small signal absorption coefficient, which is readily measured, serves as only a starting point to understanding the interaction.

Different intensity regimes are dominated by different processes. At low intensities, the small signal absorption coefficient and Beer's Law provide an accurate measure. At moderate intensities such as those used in nanosecond excimer or solid state laser ablation, saturation of the ground state and excited state absorption may play a role [8]. In the case of chemically complex materials such as polymers, chemical reactions may occur during irradiation that alter absorption properties. Even in simpler species such as metals or semiconductors, phase transitions can change materials properties remarkably, particularly reflectivity [9]. At high intensities, plasma physics must be employed as entirely different absorption mechanisms become important [10]. At the highest intensities with ultrashort pulses, effects such as resonance absorption of p-polarized light have been observed [11]. Here we focus on the absorption of moderate intensity pulses commonly associated with nanosecond laser ablation. As pulsewidths of 10 to 30 ns are almost universal in practical commercial laser systems, effects are often discussed in terms of the laser fluence. Typical fluences used in laser ablation of a wide variety of materials vary from 10 mJ/cm^2 to 100 J/cm^2, corresponding to intensities ranging from 1 MW/cm^2 to 10 GW/cm^2.

7.1.2.1 Absorption in Polymers. Taking the most conservative end of this intensity range as a numerical example, we consider a 30 ns excimer pulse at a fluence of 50 mJ/cm^2 incident on polyimide (an important thermopolymer that is a popular insulating material in the microelectronics industry), where the small signal absorption coefficient is 10^5 cm^{-1} [8] producing a substantial energy density of 5000 J/cm^3. This situation is depicted in Fig. 1. Such an

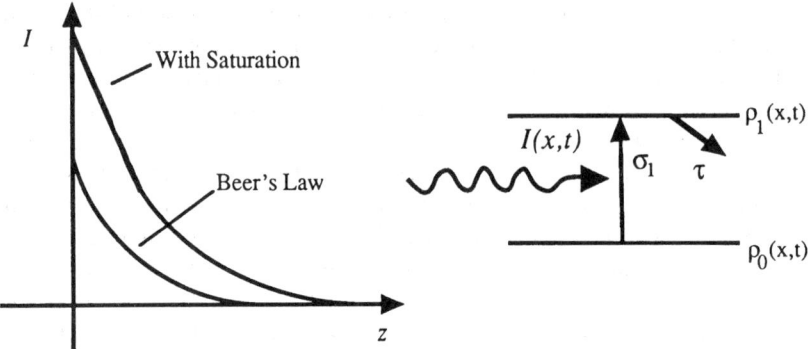

FIG. 1. Schematic of two-level model for absorption in polymers. For sufficiently high intensities, the profile deviates from an exponential, falling initially only linearly with depth.

energy density is large enough to cause some material removal and certainly thermal decomposition, though many effects such as plasma formation are reserved for much higher intensities. If we consider the KrF excimer laser (a common source for polyimide ablation) which emits 5 eV photons (248 nm), this energy density corresponds to 6×10^{21} photons/cm^3 delivered in a 30 ns timeframe (an intensity of 1 MW/cm^2). The possibility exists, then, that the number of ground state electrons promoted to an excited state has become large enough to limit absorption from the depleted ground state. In such an instance, the absorption depth would no longer be governed simply by the small signal absorption coefficient and the laser energy will then be absorbed to a considerably greater depth. As the fluence is increased, the importance of the effect grows and an effective fluence dependent absorption depth must be defined in order to arrive at the correct deposited energy density. If the excited state has a large absorption cross-section for the irradiating wavelength excited state, absorption must also be considered. This can lead to a decreasing penetration depth of the laser radiation, even below the depth given by the small signal absorption coefficient.

A rate equation model is commonly used to approximate the situation. Following the method of Petitt et al. [8], we consider the simple two-level system depicted in Fig. 1 where the cross-section for stimulated emission and absorption is defined in the usual way, and rate equations for the photon density and ground state population are written as:

$$\frac{\partial \rho_1}{\partial t} = -\frac{\partial \rho_0}{\partial t} = \sigma_1 [\rho_0(x,t) - \rho_1(x,t)] I(x,t)$$

$$\frac{\partial I}{\partial x} = -\sigma_1 [\rho_0(x,t) - \rho_1(x,t)] I(x,t).$$

(7.1)

Where $I(x, t)$ is the photon flux from the laser pulse as a function of depth into the material x and time t, ρ_0 is the ground state population density, ρ_1 is the upper state population density, and σ_1 is the stimulated emission or absorption cross-section. Equation (7.1) is similar to a two-level laser, the important difference being that relaxation is through nonradiative processes rather than emission. The model assumes that the absorption cross-section remains constant through irradiation even though chemical and structural changes are occurring at the fluences where saturation is important. The model also neglects absorption in the excited state and assumes that the lifetime of the upper state is long compared to the laser pulse. These latter two can be easily incorporated into Eq. (7.1) with additional terms, though their addition necessitates numerical solutions. To specify the solution of Eq. (7.1), we impose the conditions that $\rho_0(x, 0) = \rho$ and $\rho_1(x, t) = 0$, which is simply to say that initially all of the population is in the ground state. $I(0, t)$ is just the incident photon flux. The first equation of Eq. (7.1) can then be integrated directly:

$$\rho_0(x, t) - \rho_1(x, t) = \rho \cdot \exp\left[-2\int_0^t \sigma_1 I(x, t')\, dt'\right] \quad (7.2)$$

Substituting Eq. (7.2) into the second of Eq. (7.1), we obtain an expression for $I(x, t)$.

$$\frac{\partial I}{\partial x} = -\sigma_1 \rho I(x, t) \exp\left\{-2\sigma_1 \int_0^t I(x, t')\, dt'\right\} \quad (7.3)$$

Now if we consider the total photon flux, $S(x)$, which is just the integral of $I(x, t)$ over time, we find:

$$\frac{\partial S}{\partial x} = \int_0^\infty \frac{\partial I(x, t')}{\partial x}\, \partial t' \quad (7.4)$$

Employing Eq. (7.4) with Eq. (7.3), we can then write the dependence of the total photon slux on depth:

$$\frac{\partial S}{\partial x} = -\frac{\rho}{2}[1 - \exp\{-2\sigma_1 S(x)\}] \quad (7.5)$$

which reduces to the familiar form of Beer's law when $S(x) \ll 1/\sigma_1$, and the exponential can be developed:

$$\frac{\partial S}{\partial x} = -\sigma_1 \rho S(x) \quad (7.6)$$

Clearly such a result cannot be relied on in all situations, but with suitable modifications to Eq. (7.1) based on detailed knowledge of the target material, numerical solutions yield a reliable picture of light absorption at these intensities. For many systems, including 248 nm KrF irradiation of polyimide,

Eq. (7.5) can be used directly. This yields for the etch depth as a function of fluence [8]:

$$d_1 = \frac{2}{\rho}(S_0 - S_{th}) + \frac{1}{\rho\sigma_1} \ln\left[\frac{1 - \exp(-2\sigma_1 S_0)}{1 - \exp(-2\sigma_1 S_{th})}\right] \quad (7.7)$$

where $S_0 = S(0)$ is the photon flux at the surface and $S_{th} = F_{th}/(hv)$ is the threshold photon flux.

Complex molecules such as polyimide monomers may have more than one absorption site or chromophore each with a different absorption cross-section. A straightforward averaging process can be applied in such a case to give an effective cross-section for use in a simple two- or three-level model. Such a model likewise ignores the absorptive properties of the upper state. These can be easily implemented into the same framework, but only if some information about the cross-section and lifetime of the upper state is known.

For ultrashort laser pulses, absorption may be dominated by multiphoton absorption. A similar treatment as that outlined in Eqs. (7.1)–(7.6) leads to the following expression for the etch depth per pulse [8]:

$$d_n = \frac{2}{n\rho} \int_{S_{th}}^{S_0} \frac{dS}{1 - \exp(-2\sigma_n K_n S^n)} \quad (7.8)$$

where n is the order of the absorption process (i.e., $n = 2$ for two-photon absorption), σ_n is the multiphoton absorption cross-section and $K_n = A_n/\tau^{n-1}$, where A_n is a pulse shape dependent numerical factor of the order of 1 and where τ is the laser pulse duration ($K_1 = 1$). Good agreement was found between experiments on two-photon absorption of fs KrF laser pulses in Teflon [8] or multiphoton absorption of Ti-Sapphire pulses [12] in polymers and Eq. (7.8). Although the cross-sections that are used in Eqs. (7.7) and (7.8) are usually not known, these relations can be used to parametrize experimental results and predict etch depths as a function of laser parameters.

In Fig. 2, a schematic of the energy pathways present in ultraviolet processing of polymers is shown, and it is clear from the figure that as polymer ablation and modification involves actual chemical transformation of the material, absorption can change with laser processing. Some polymers initially only lightly absorbing become absorbing after the first few laser pulses, and ablation does not begin until this "incubation" is accomplished.

7.1.2.2 Metals and Semiconductors. In metals, an effectively infinite supply of electrons is available for absorption such that the penetration depth is given by the skin depth until the intensity is high enough that plasma is formed [13]. Absorption is then dominated by the physics of the laser-produced plasma. Absorption in semiconductors comes from electron-hole pair creation for excitation above the bandgap, though below the bandgap, a whole host of effects are

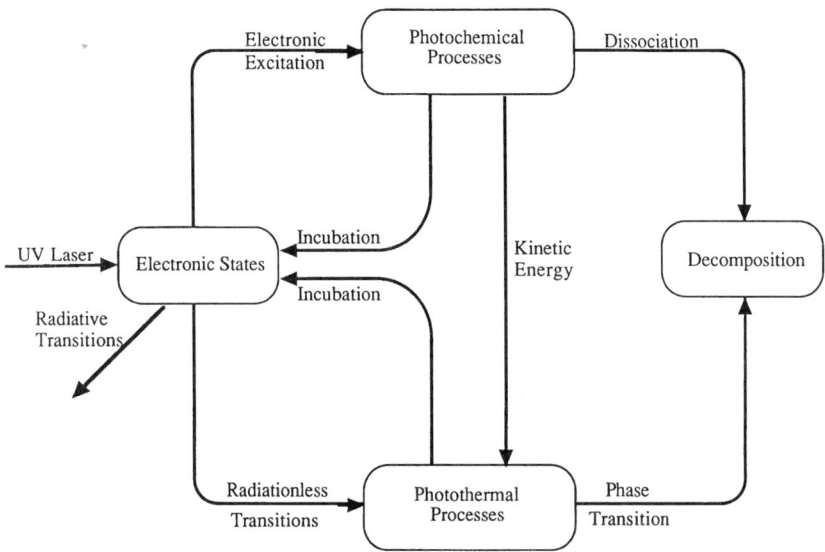

FIG. 2. Diagram showing various energy pathways in ultraviolet processing of polymers. Photochemical and photothermal mechanisms can both lead to decomposition as well as changes in absorption properties.

dominant, depending on the doping and defects [13, 14]. Cross-sections for two-photon absorption in semiconductors can be quite large, and two-photon absorption is observed in some cases. The large temperature dependence of electron and hole densities in semiconductors necessitates detailed thermal modeling to understand the exact absorption properties [14].

7.1.2.3 Other Absorption Mechanisms. In many systems, the nature of absorption may be completely different than the simple situation described by Eq. (7.1). In particular, defects may become the major source of absorption and thus give a lateral spatial distribution to absorption. Such effects are most notably observed in systems with a wide bandgap such as MgO [15]. In such a case, a cleaved surface always has much higher ablation threshold than a cut surface due to the greater defect density of the cut surface. In PMMA, for example, absorption at 248 nm is much smaller than in a polymer such as polyimide. After a small number of preliminary or "incubation" pulses, the absorption of PMMA increases dramatically and ablation begins [16]. Another interesting system is diamond for which the bandgap is almost equal to the energy of the 248 nm photons (5 eV). In this case, the small signal absorption at 248 nm is strongly temperature dependent. Ablation, however, is largely temperature *independent* [17] because surface states absorb initially and cause a phase transformation into graphite of a thin surface layer, which serves to couple the laser light into the diamond on subsequent laser pulses, making the small signal absorption properties completely irrelevant for laser ablation.

7.1.2.4 Conclusion. An understanding of radiation transport in a system alone can help us explain many experimental observations. For example, the rather large variation in ablation thresholds for different materials and wavelengths can be understood when the different absorptive properties are considered. Smaller absorption depths lead to higher energy densities and thus lower ablation thresholds, although the depth of material ablated per pulse will then be small. For situations with large absorption depths, a higher fluence is necessary to produce the high-energy density necessary for ablation. However, once the ablation threshold is reached, the etch rates will be higher. If surface defects are determined to dominate absorption, irregular ablation in patches and a strong dependence on surface preparation are explained immediately. To understand the effects of laser irradiation in any material, the details of the absorption process must be known. From this point, temperature modeling and other considerations can be used to understand the wide variety of effects that can be induced by laser processing.

7.2 Physical and Chemical Effects of Laser Irradiation

Once the absorptive properties of a material are understood for the appropriate wavelength and intensity, the energy density produced can be calculated. In the case of excimer ablation of insulators where the source photons are in the ultraviolet, this energy is absorbed into electronic states. The excited electronic state can result in bond cleavage or chemical reaction, or the energy is converted to heat, usually on timescales much shorter than the nanosecond pump pulses (Fig. 2). Infrared pulses (e.g., from a CO_2 laser) interact with the target through vibrational states either localized or delocalized. Thus we normally view the ablating pulse as essentially a process of instantaneous heating that follows the intensity profile of the incident pulse. However, the purely thermal picture is not always appropriate, particularly for deep uv pulses such as with ArF (193 nm) or F_2 (157 nm) lasers where photochemical reactions can occur directly. In the case of polymer ablation, there has been a great deal of discussion over the role of direct photochemical bond breaking in the ablation process. In the case of metals or semiconductors, excitation of localized lattice modes can produce anomalous effects, namely, desorbed atoms with energies well beyond those accounted for by uniform temperature distributions [13]. Also, many surface modification effects with semiconductors involve laser-stimulated interactions with a chemically reactive ambient. We have, therefore, divided the interaction phenomena into two groups, physical and chemical, meaning roughly those that involve chemical reactions and those that do not.

7.2.1 Physical Effects in Surface Modification

Physical effects fall into two categories, those concerned with optical phenomena and those concerned with thermal phenomena. Turning first to optical effects, two primary phenomena are of great importance and are also related: laser speckle (of crucial concern for lithography) and laser-induced, periodic surface structures or LIPSS. Both are wave optical in nature and create an inhomogeneous spatial intensity distribution across the sample, however, speckle occurs with a single pulse, whereas LIPSS build up over a number of pulses. Changes in optical properties of the surface (i.e., the complex refractive index) also effect energy absorption through reflectivity changes.

7.2.1.1 Laser Speckle and LIPSS. Lasers typically emit light with a high degree of spatial coherence because of the properties of modes of an optical cavity. In continuous wave, single mode operation, spatial coherence is virtually perfect, such that a Young's two-slit experiment will yield clear fringes even as the slit separation is made larger than the beam waist [18]. Such a high degree of spatial coherence allows scattered light from random irregularities in a surface to interfere even when those irregularities are far from one another. The resulting pattern is known as speckle. In laser ablation, the speckle pattern can be faithfully reproduced in the etched surface and thus constitutes a serious problem when any sort of patterning is desired [19]. In lithography applications, speckle is a major concern in the move to laser-based techniques. Excimer lasers, fortunately, have poor spatial coherence compared to most other lasers, though still considerably more than incoherent sources. This is due to the high gain and consequently small number of passes in the resonator. In excimer lasers, the gain lifetime is only about 20 ns, which in turn requires a large output coupling for optimum energy extraction [20].

Similar to speckle, LIPSS occur from the interference of scattered light from a rough surface with the incident laser beam. The exact solutions to Maxwell's equations from which the LIPPS phenomenon are derived have been termed "radiation remnants" and have proved an interesting physical problem [21, 22]. If the intensity of the light is high enough, the interference from the "radiation remnants" and the incident beam is etched into the target surface. On the first pulse, this is a random roughening of the surface. However, subsequent pulses then create intensity distributions that are in response to the modified surface. This positive feedback then results in periodic ripples in the surface after a sufficient number of pulses. The structures come in three types and have been given the designation S_+, S_-, and c type fringes, which correspond to types of remnants found from the solution of Maxwell's equations. As they are an interference phenomenon, the ripple structures have a definite period related to the wavelength of the incident light and have a direction related to the polarization of the light. The S_+ and S_- type fringes appear perpendicular to the polarization and have a period given by the following relation:

$$\Lambda_{\pm} = \frac{\lambda}{1 \pm \sin(\theta)} \quad (7.9)$$

Where Λ is the fringe period; θ, the angle of incidence; and λ, the laser wavelength. The S_- is usually dominant in experiments. A rarer form of LIPSS are the c type fringes, which run parallel to the polarization and have a period given by:

$$\Lambda_c = \frac{\lambda}{\cos(\theta)} \quad (7.10)$$

The phenomenon of LIPSS is really quite general and has been observed in all types of materials and with many laser sources. An example of LIPSS is shown in Fig. 3, where strong c type fringes were produced with modification of a Ge surface with an Nd:YAG laser at an angle of 60°.

The key difference between laser speckle and LIPSS is that speckle is a function of the spatial coherence of the laser and comes from interference effects on the surface regardless of whether the light intensity is strong enough to modify the surface. LIPSS, in contrast, only form when the light can modify the surface and when successive pulses continue to modify the surface with positive feedback. The surface modification need not be direct ablation, and it can come from photochemical etching or any other process that modifies the surface in response to the spatial intensity distribution present at the surface [23]. If the

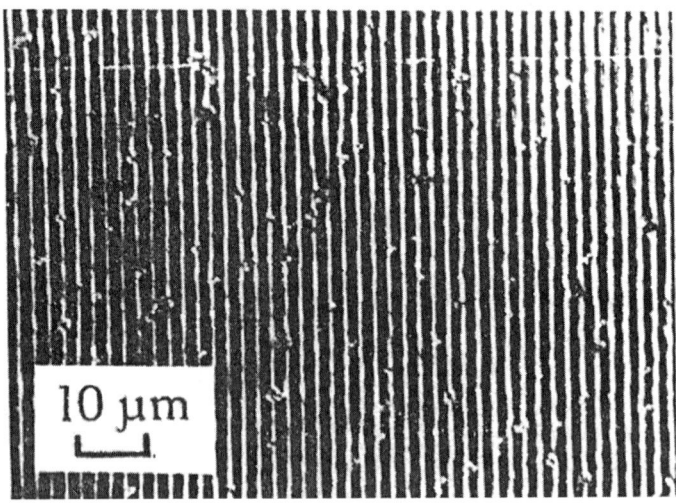

FIG. 3. Example LIPPS formation on Nd:YAG laser irradiated Ge.

ultimate surface morphology is dominated by effects that do not exactly follow the light intensity distribution at the surface, LIPSS will not be observed. Holographic irradiation has also been shown to suppress LIPSS formation by imposing an external periodicity on the surface, which breaks down the necessary positive feedback [24].

7.2.1.2 Reflectivity. The extreme conditions produced by laser irradiation can affect the refractive index and roughness of the surface. The melting transition in metals and semiconductors causes a drastic change in reflectivity. In most metals reflectivity is sharply reduced, and in silicon the reflectivity sharply increases [9, 25, 26]. In excimer laser ablation of polyimide, a transient change in refractive index results in strong reduction in reflectivity [27], while plasma formation can produce a "mirror" effect where reflectivity is greatly increased [28].

7.2.1.3 Thermal Effects. The primary effect of an intense laser pulse absorbed into a target materials is to heat the surface layer to relatively high temperatures in only a few nanoseconds. The heating rates are typically on the order of 10^{11} Kelvins/s. These are extreme thermodynamic conditions and produce extreme results. One of the principle problems with a theoretical understanding of laser material interaction is that the system is very far from thermodynamic equilibrium when all interesting effects are occurring. Phase transitions, decomposition temperatures, and physical constants that are well understood close to equilibrium should be viewed only as estimates and be correlated with experimental results from a variety of characterization techniques.

The most important parameter in understanding thermal effects is, of course, the temperature distribution in the material and its development in time. Reasonable estimates of the subsurface temperature distribution can be made from solutions to the thermal diffusion equation. Such solutions suffer from their reliance on physical parameters that are determined from equilibrium experiments. Nevertheless, good agreement between experiment and calculation is obtained with such models. As a starting point, we consider the thermal diffusion length $l(t)$, which gives an estimate of the length that heat will diffuse in a time t,

$$l(t) - \sqrt{\frac{\kappa t}{\rho c}} \quad (7.11)$$

where κ is the thermal conductivity, ρ is the density, and c is the specific heat. Equation (7.11) allows us to estimate in a very simple way how far a significant amount of heat should diffuse during a given amount of time. Of great importance is the diffusion length over the time of the laser pulse as this can affect the energy density if the length is greater than the absorption depth. In thermally conductive targets such as metals and semiconductors, thermal diffusion is of demonstrated importance. For example, the ablation threshold of many metals can be decreased by two orders of magnitude in fluence if an ultra-short pulse (<1 ps) laser source is used where thermal diffusion during the pulse is negligible. During the course of a nanosecond laser pulse, heat can diffuse to many times the absorption depth in these materials and is thus the dominant factor in determining the absorbed energy density and thus the maximum subsurface temperature. Table I lists thermal diffusion lengths of a variety of materials calculated from Eq. (7.11) together with their room temperature physical constants. Noting the large differences in values for the thermally insulating polymers compared to the conductive metals and semiconductors, we see that thermal diffusion during the pulse is less important in polymer ablation but crucial in metals and semiconductors. This is due to both the shorter diffusion length *and* the larger absorption depths. If tightly focused beams are used for processing of metals, lateral diffusion of heat will require the problem to be treated as a three-dimensional one.

Detailed temperature modeling begins with the complete thermal diffusion equation [2],

$$\rho c \frac{\partial T}{\partial t} - \nabla[\kappa(T)\nabla T] = Q \qquad (7.12)$$

where $\kappa(T)$ is a temperature-dependent thermal conductivity, T is the subsurface temperature distribution, and Q is the deposited heat (per unit volume and time). The specific heat also depends on tempature, although this is usually less significant than the temperature dependence of the thermal conductivity. From Eq. (7.12), we can begin to make approximations and arrive at solutions for particular situations. In some cases, it is still not possible to have an analytic or

TABLE I. Thermal Diffusion Lengths in Various Materials

Material	ρ (g/cm^3)	c (J/gK)	κ (W/mK)	$1/\alpha$ (nm)*	l (t = 15 ns) nm
Aluminum[a]	2.702	0.897	2.37	10	121
Nickel[29]	8.90	0.444	0.907	8	59
Copper[29]	8.92	0.385	4.01	14	132
Tungsten[29]	19.35	0.132	1.74	7	101
Silicon[29]	2.32	0.705	1.48	†	117
polyimide[b30]	1.42	1.09	0.12	46[31]	34
PMMA[32]	1.19	1.40	0.21	200 μm[31,**]	42
PTFE[32]	2.18	1.045	0.25	200 μm[31]	57
PET	1.41[33]	1.425[32]	0.15[32]	33[33]	47

* For irradiation at 248 nm.
** Dominated by incubation effects.
† Strongly temperature, wavelength, and dopant dependent.

simple numeric integration of the equation, and numerical techniques such as finite difference and element analysis can be used. Implicit in Eq. (7.12) is that convection and blackbody radiation from the surface do not play a significant role, which is almost always a good approximation for pulsed laser irradiation. The first and most important approximation we make to Eq. (7.12) is to consider only one-dimensional heat flow into the bulk substrate. Such a picture is valid so long as the thermal diffusion length $l(t)$ for the timescale of interest is much smaller than the irradiated feature size. A one-dimensional partial differential equation of even a high degree of complexity can be solved efficiently with an ordinary microcomputer. The form of Q and the boundary and initial conditions specify the solution to Eq. (7.12). A typical form for Q that assumes that the laser energy is deposited following an exponential decay according to Beer's law and circular Gaussian beam is given by [2]:

$$Q = \alpha F(1 - R) \exp\left\{\frac{-2r^2}{w_0^2} - \alpha z\right\} P(t) \tag{7.13}$$

where F is the laser fluence (the energy divided by the area of the beam waist), R is the reflectivity, r is the beam radius coordinate, z is the direction into the material, $P(t)$ is the temporal pulse shape ($\int_0^\infty P(t)\,dt = 1$), α is the absorption coefficient, and w_0 is the beam waist at the sample. With the one-dimensional approximation, the Gaussian beam portion ($\exp(-2r^2/w_0^2)$) is simply left out. Q is often put at a delta function in z for highly absorbing materials ($\alpha > 10^4$ cm^{-1}), which simplifies solutions, although analytic solutions using the exponential decay are still possible in many cases.

As an example, we take the thermal conductivity and diffusivity to be constants and integrate Eq. (7.12) directly via Green's function for the case of a semi-infinite medium. In such a case, we can solve the spatial part of the problem analytically, leaving only a simple temporal integral to be solved numerically. The appropriate Green's function, is then,

$$G(z, t; \xi, \tau) = \frac{1}{2\sqrt{\pi}\, l(t - \tau)} \left[\exp\left\{-\frac{(\xi - z)^2}{4l^2(t - \tau)}\right\} - \exp\left\{-\frac{(\xi + z)^2}{4l^2(t - \tau)}\right\}\right] \tag{7.14}$$

which, using the boundary conditions that $dT/dz = 0$ at the surface and that the temperature should go to zero at infinite depth, is a solution to:

$$\frac{\partial G}{\partial t} - \frac{\kappa}{\rho c}\frac{\partial^2 G}{\partial z^2} = \delta(z - \xi)\delta(t - \tau) \tag{7.15}$$

The validity of the semi-infinite medium approximation is established when the diffusion length $l(t)$ is much smaller than the thickness of the medium. Thus even thin films can sometimes be "semi-infinite" if the timescale under consideration is small. To find the subsurface temperature distribution, we simply multiply Green's function by the source term and integrate over ξ and τ.

$$T(z, t) = \iint d\xi \, d\tau \, G(z, t; \xi, \tau) Q(\xi, \tau) \tag{7.16}$$

which for the spatial integration yields,

$$T(z, t) = \frac{\alpha F}{4\rho c} \int_0^t d\tau P(\tau) \left[\left(erfc\left(\alpha l(t - \tau) + \frac{z}{2l(t - \tau)} \right) \right) (\exp(\alpha^2 l^2 (t - \tau) + \alpha z)) \right.$$
$$\left. + \left(erfc\left(\alpha l(t - \tau) - \frac{z}{2l(t - \tau)} \right) \right) (\exp(\alpha^2 l^2 (t - \tau) - \alpha z)) \right] \tag{7.17}$$

where $erfc(x)$ denotes the complimentary error function, and $l(t - \tau)$ is as defined in Eq. (7.11). Such a solution can be easily implemented on a computer. Fig. 4 shows a set of solutions for different times where for each point in space the time integral in Eq. (7.17) is performed numerically. These equations yield

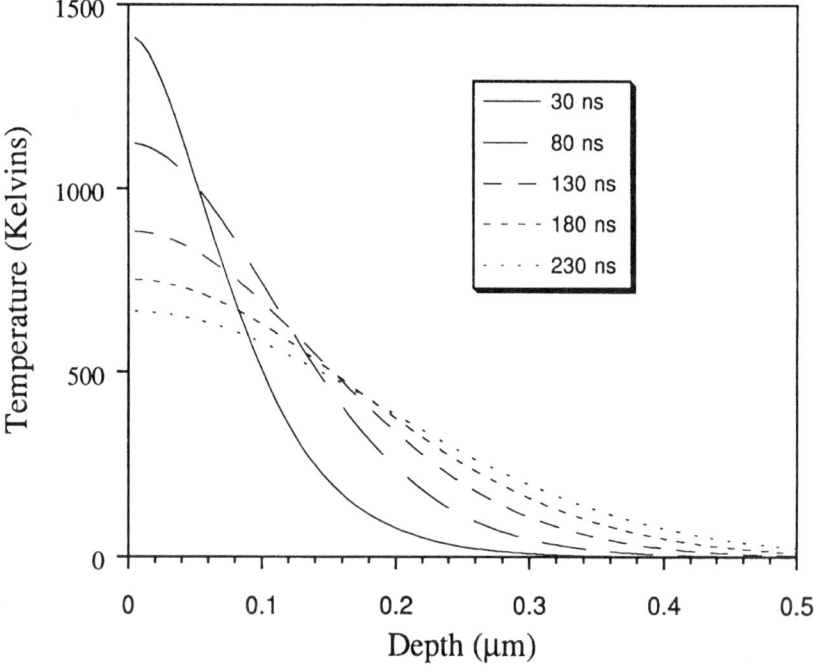

FIG. 4. Temperature shown as a function of depth for several different values of time given by a Green's function analysis for a semi-infinite medium.

good results for polymers and metals where the thermal diffusivity is only slightly dependent on the temperature. Larger errors are found for semiconductors because the diffusivity is strongly temperature dependent. Silicon, in particular, has been extensively studied, and the literature provides the results of detailed numerical analysis.

Three basic effects, excluding chemical reactions, are related to the large temperature increase produced by laser irradiation: phase transitions, the formation of a powerful shock wave, and material removal. All three of these are interrelated, complicating a rigorous analysis.

Phase Transitions. Phase transitions provide some of the most interesting opportunities to use surface modification by lasers to produce novel properties. The primary advantage of laser techniques is the ability to create high temperatures in a surface layer in a short time without depositing a large amount of heat into the target as a whole. To model a phase transition, we use the thermal diffusion Eq. (7.12) for both phases separately, although the introduction of the change of state significantly complicates the solution. The desired phase transition in metals and semiconductors is generally melting for the introduction of a dopant, a planarized surface, or crystallization. Because the diffusion coefficients for mass transport are several orders of magnitude higher for the liquid state compared to the solid state, it is most necessary to know to what depth the melt is present and its duration.

Numerous theories of pulsed laser melting have been put forward [34–40], both analytical and numerical. All are based on the same general framework of Neumann's solution [34] where a moving reference frame corresponding to the melting front is chosen for the heat diffusion equation with appropriate boundary conditions. The desired "solution" is not so much a temperature distribution as the trajectory of the melt front. Analytical solutions for pulsed laser absorption can be obtained, although speed and convenience of modern microcomputers and analysis software may make numerical analysis a more flexible and even more intuitive approach. Different approaches focus on special difficulties associated with a given target material. Silicon has been the most studied and presents the additional problem of a strongly temperature-dependent absorption coefficient. (How such laser-induced phase transitions are exploited to produce novel properties such as dopant diffusion and crystallization are discussed in Section 7.5 on surface modification of metals and semiconductors.)

As an example, we can consider the simplest case of a solid where the surface is held at a constant temperature T_s greater than the melting point T_m. A schematic of the problem is shown in Fig. 5. Following Carlslaw and Jaeger [34], we identify the parameters D_1, c_1, κ_1, ρ, and $T_1(x, t)$ with the solid and D_2, c_2, κ_2, ρ, and $T_2(x, t)$ with the liquid, where these quantities are the thermal diffusivity, specific heat, thermal conductivity, mass density, and temperature, respectively. An additional assumption here is a constant density for both

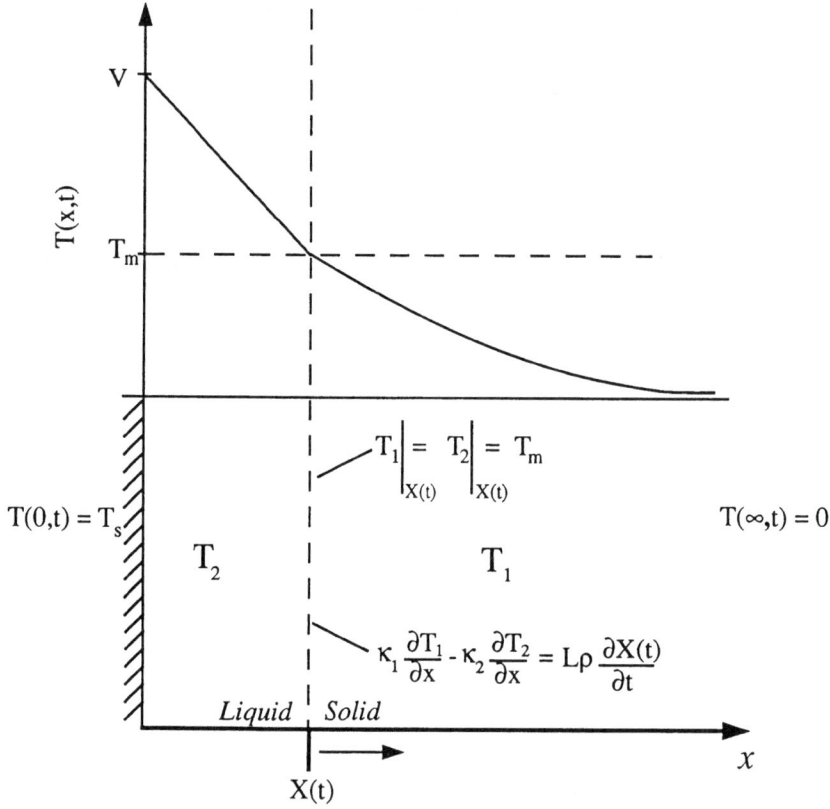

Fig. 5. Illustration of Neumann's solution to the melting problem.

phases. We define L as the latent heat of fusion (the amount of energy per unit mass consumed by melting). Confining ourselves to one dimension, we define $X(t)$ as the position in time of the solid liquid interface. At this interface, both the solid and liquid temperature distributions must be at the melting temperature, yielding the boundary condition:

$$T_1(X(t), t) = T_2(X(t), t) = T_m \qquad (7.18)$$

As the solid begins to melt, a quantity of heat $L\rho\, dX$ per unit area is absorbed, yielding a second boundary condition at the interface,

$$\kappa_1 \frac{\partial T_1}{\partial x} - \kappa_2 \frac{\partial T_2}{\partial x} = L\rho \frac{dX}{dt} \qquad (7.19)$$

We then write the homogeneous temperature equation for both temperature distributions,

$$\frac{\partial^2 T_1}{\partial x^2} - \frac{1}{D_1}\frac{\partial T_1}{\partial t} = 0$$

$$\frac{\partial^2 T_2}{\partial x^2} - \frac{1}{D_2}\frac{\partial T_2}{\partial t} = 0 \quad (7.20)$$

Where the additional boundary conditions necessary to specify the problem are

$$T_1(\infty, t) = 0$$
$$T_2(0, t) = T_s \quad (7.21)$$

Which give the well-known solutions,

$$T_2(x, t) = T_s - Aerf\left(\frac{x}{2\sqrt{D_2 t}}\right)$$

$$T_1(x, t) = Berfc\left(\frac{x}{2\sqrt{D_1 t}}\right) \quad (7.22)$$

We then apply the boundary condition Eq. (7.18) to Eq. (7.22)

$$T_s - Aerf\left(\frac{X(t)}{2\sqrt{D_2 t}}\right) = Berfc\left(\frac{X(t)}{2\sqrt{D_1 t}}\right) = T_m \quad (7.23)$$

which is clearly only true if $X(t)$ is proportional to $t^{1/2}$, so the trajectory of the melt front is then given by,

$$X(t) = 2\lambda\sqrt{D_2 t} \quad (7.24)$$

where λ is a parameter given by application of the boundary condition Eq. (7.19). A and B are also then found from Eq. (7.23) in terms of λ to define the temperature profile as well. Upon application of Eq. (7.19), λ is then the root of the following equation:

$$\frac{\exp(-\lambda^2)}{erf(\lambda)} - \frac{\kappa_1 \sqrt{D_2}\, T_m \exp\left(-\lambda^2 \frac{D_2}{D_1}\right)}{\kappa_2 \sqrt{D_1}\,(T_s - T_m) erfc\left(\lambda\sqrt{\frac{D_2}{D_1}}\right)} = \frac{\lambda L \sqrt{\pi}}{c_2(T_s - T_m)} \quad (7.25)$$

Naturally the addition of an exponential heat source, a temporal pulse shape, and temperature-dependent thermal constants severely complicates the solution. Using this basic framework, however, intuitive numerical approaches can be devised to deal with many situations. An important tie in to experimental measurements is that the melted surface's decreased reflectivity can be detected with a temporally delayed probe beam and the results—fluence dependence and time dependence—compared to a model calculation.

Shock Waves and Acoustic Phenomena. An acoustic wave can be produced at low laser fluences through thermoelastic phenomena, although once fluences above the ablation threshold are reached, the acoustic signal is dominated by the recoil of ablating material and at high fluences by stress from plasma formation and expansion [40]. In the latter two cases, the acoustic signal is in reality a supersonic shock wave.

Thermoelastic expansion of the surface is a significant phenomenon and is exploited in laser cleaning. Following Tam *et al.* [41], we can estimate this effect by considering the normal expansion of the surface ξ, which is given by $\xi = \mu l(\tau) \Delta T$, where μ is the coefficient of thermal expansion, and $l(\tau)$ is the thermal diffusion length for the pulse width τ of the laser pulse. Taking typical numbers of $\Delta T = 1000$ K, $l(\tau) = 100$ nm, and $\mu = 2 \times 10^{-5}$ K^{-1}, we find $\xi = 2$ nm. This distance may seem insignificant, but because this expansion takes place in only 20 ns, it corresponds to a velocity of $\xi/\tau = 10$ cm/s and an acceleration of $\xi/\tau^2 = 5 \times 10^8$ cm/s^2, about 500,000 times the acceleration due to gravity.

If ablation is occurring, the recoil from ejected material creates stress transients of much larger proportions. This is of enormous concern is medical applications where transients in these shock waves can be large enough to damage the target material away from the ablated spot. Studies in water indicate that the range of such shocks is about 100–200 μm at which point the shock wave has become an ordinary acoustic wave [42]. Damage from such phenomenon away from the treatment area is a focus of attention in medical applications as damage to surrounding tissue is unacceptable, although in some cases such as kidney stone fragmentation, the shock is actually exploited for treatment [43]. In strongly absorbing organic media, peak stresses in acoustic waves away from the interaction region are observed in the range 10^4 to 10^9 Pa [40] where the larger end of this range can easily damage a fragile or otherwise brittle material. In medical applications of laser ablation, using low fluences and short pulses can minimize the effect of intense stress transients. Mode-locked Nd:YAG systems are reliable sources of picosecond pulses at high repetition rates and may prove the most favorable system for medical application where collateral damage from shock waves is an issue.

Material Removal. Material removal in the case of metals and semiconductors is generally the result of further heating beyond the melting transition to vaporization. Such a phase transition, accompanied by a large change in density in short times, is generally explosive in nature. A purely thermal picture of material removal does not describe all results, and excitation of localized lattice vibrations that can break bonds is also present [13]. In the case of polymers, chemical processes dominate as shown in Fig. 2, whether photothermal degradation or direct photochemical bond breaking [44].

7.2.2 Chemical Effects of Laser Irradiation

Potentially, some of the most useful applications of laser surface modification may come from the ability of the laser to induce chemical reactions at the surface such as in top-surface, imaged photoresists in deep uv lithography. The laser can induce a chemical reaction in two ways, when the temperature rise of the substrate (and/or reactants) exceeds the activation energy of a reaction, or direct electronic excitation from the photons results in bond breaking. These are termed pyrolytic and photolytic processes, respectively. There are four types of reactants: adsorbates, gases, liquids, and solids, each of which defines a set of experimental techniques. Much work has been done with cw lasers (usually Ar^+ lasers) using a focused spot for direct writing. This work is well reviewed in the litature [2], and we will look primarily at pulsed laser applications.

7.2.2.1 Pyrolytic Reactions.
The most important feature of any chemical reaction is the rate at which the reaction occurs. In laser processing, two possibilities can limit the reaction rate: reaction kinetics or mass transport [2]. The mass transport is strongly affected by the phase and density of the reactants and products while the kinetics are given by the chemistry of the reaction itself and the laser energy. At lower laser energies and temperatures, kinetics are usually more important; at higher energies and temperatures, mass transport tends to become more important.

In the kinetic limited case, the reaction rate is given by an Arrhenius type equation:

$$R(t) = kn(t) \exp\left\{-\frac{E_a}{k_B T(t)}\right\} \tag{7.26}$$

where k is a constant, $n(t)$ is the density of reactants at the surface, E_a is the activation energy of the slowest step in the given reaction, and k_B is the Boltzman constant. The temperature $T(t)$ is the surface temperature. What has been omitted in Eq. (7.26) is that there is, of course, a spatial dependence to the surface temperature and the reactant density given by the irradiated pattern. This spatial dependence of great importance in determining the spatial resolution of a given process and determining what edge effects are present at the boundaries of the irradiation.

In the mass-transport limited case, the diffusion of reactants into the reaction zone is considered and is particularly important for interactions with gas and liquid phases. In this case, we assume the reaction rate is directly proportional to the density of reactants in some region about the surface where laser irradiation is present and then we solve for this density by computing the flux of reactants into the reaction region with the diffusion equation. In such a case, sharp differences can exist between pulsed and cw processing. In cw processing the steady state considerations are all that is necessary, but in pulsed irradiation the time dependence of the processes must be investigated, although in some cases, a steady state picture may also be appropriate for nanosecond pulses. If broad area irradiation is considered, only the one-dimensional diffusion equation need be considered; but if small feature sizes or focused spots are used, a fully three-dimensional picture may be necessary. Three-dimensional diffusion of reactants results in far greater rate of diffusion of reactants into the reaction zone such that the mass transport limited case occurs at much higher temperatures for gas and liquid phase reactions. Mass diffusion is of primary consideration in self-terminating reactions like thermal oxidation where oxide growth on the surface, the desired effect, is itself the diffusion barrier that prevents the oxygen from reaching the bulk material.

7.2.2.2 Photolytic Reactions.
Direct photochemical bond breaking in an ambient, absorbate, or substrate occurs when a photon of sufficient energy or multiple photons whose sum is sufficient energy are absorbed by a molecule. Bond energies, typically several eV in magnitude, then require uv photons or multiphoton absorption of longer wavelengths. The mechanism of photodissociation is an electronic transition from a bound ground state to an excited dissociative state, although this electronic transition is normally accompanied by vibrational excitation as well. Because the timescale of the dissociation is in the tens of femtoseconds, relaxation processes and collisions can all be safely ignored [45].

Most photolytic processes that have been studied involve etching or deposition [2]. Deposition occurs as a complex molecule is dissociated by the laser in the vicinity of the substrate. One of the dissociation products attaches to the surface, and the other products leave the reaction area in the gaseous phase. Etching is similar, except that the laser photodissociation creates a highly reactive radical such as Cl or F, which combines with and removes molecules on the surface.

Photolytic processes often termed "surface modifications" are laser-enhanced oxidization and nitridation [2, 14]. Oxide and nitride growth on semiconductors for passivation and use as an electrical insulator are of crucial importance for semiconductor device manufacturing, and traditional furnace-based oxide growth techniques detract significantly from the thermal budget of a given process. It has been observed that under laser processing, SiO_2 growth rates can be substantially increased. Though there is likely a strong thermal component to the enhancement, it is believed that electron-hole pair creation catalyzes the growth process. laser processing is usually less economical than rapid thermal annealing (RTA) and CVD techniques and has thus not been significant in manufacturing.

More recent research has shown photolytic reactions with polymers, particular fluoropolymers, can alter reactive, mechanical, and adhesive properties [46, 47]. Such results are of tremendous interest in the microelectronic industry because the low dielectric constant of such polymers together with their chemical and thermal stability would be ideal as interlayer dielectrics for packaging and metallization applications [48]. We discuss this research in detail in Section 7.4.

7.3 Surface Morphology Modification

One of the primary applications of laser processing of materials is to manipulate their surface morphology. The laser offers the ability to control surface morphology with submicron lateral spatial resolution and with even finer depth resolution. Lasers are equally effective at removing morphological structures as creating them, and many important applications related to polishing and cleaning of surfaces have arisen in recent years. This section focuses on specific examples from recent work and also on the limitations of laser techniques.

7.3.1 Direct Laser Ablation

Virtually all materials can be patterned and etched by direct laser ablation. The primary advantage of the technique is its simplicity. Essentially all that is necessary is to direct a pulse of sufficiently high fluence at the target and a layer of material is removed. Pulsed lasers allow for a small but significant removal of material for each pulse, starting from a fraction of a monolayer to about 10 µm, and are thus ideal for projection patterning with a high degree of depth resolution. Continuous wave lasers are generally used in direct writing schemes where a focused beam is scanned to produce a desired pattern.

Morphology modification by direct ablation with excimer lasers has been most successful with polymers. The highly conjugated structure of organic polymers absorbs uv radiation strongly for low ablation thresholds and highly controllable etching. Excimer ablation of polyimide has been implemented in semiconductor packaging manufacturing as a method to create vias between dielectric layers [6]. The excimer laser can provide high aspect ratios at low cost without the need for wet chemical etching and photolithography.

As an example of direct ablation with polymers, we consider the experiment of Phillips *et al.* [49] where a Talbot interferometer was used to produce structures of nanometer scale dimensions on polyimide. The interferometer is shown in Fig. 6, where the $+1$ and -1 orders of a diffraction grating are combined using an interferometrically polished quatz block as the mirrors of the interferometer. Such a technique allows for a high-visibility interference pattern despite the poor spatial coherence of the excimer laser used. The result of

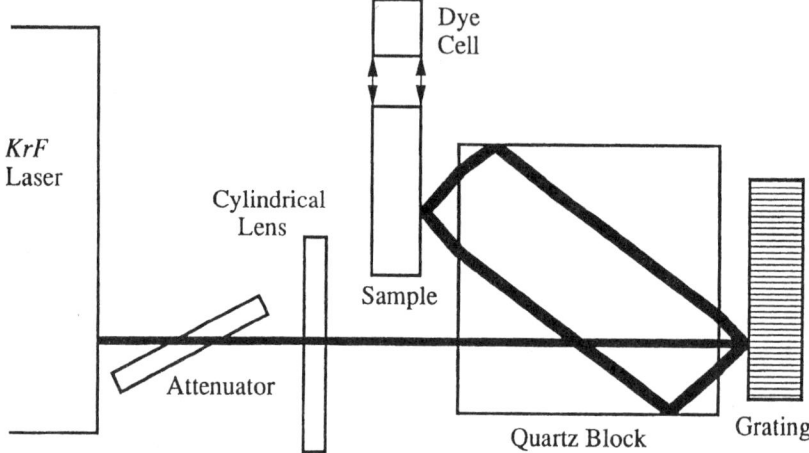

FIG. 6. A novel interferometer for the production of nanostructures in polymers. The +1 and −1 orders of the diffraction grating are combined via total internal reflection in a polished quartz block.

irradiation is shown in Fig. 7, where the 166 nm period of the interference pattern has been directly ablated into the polymer. By adjusting the fluence, the linewidth could be controlled and structures as small as 35 nm were produced, considerably smaller than the excimer wavelength of 248 nm. Such small linewidths were attainable because of the sharp threshold in polymer ablation. At a lower fluence therefore only the tops of the interference maxima are above the threshold. Such structures are approaching the thermal diffusion length in polyimide, and smaller patterns are likely not realizable. Clearer patterns were produced with a similar technique but using a specialized laser system with sub-picosecond pulses where thermal diffusion is negligible. The spatial resolution obtained in these experiments is a good approximation to the ultimate resolution limit obtainable with direct laser ablation.

7.3.2 Laser-Assisted Chemical Etching

Many of the problems with direct laser ablation can be overcome with laser-assisted chemical processing, although the techniques are significantly more complex. The basic technique is illustrated in Fig. 8. Here an ambient gas, such as Cl_2, is dissociated with a uv laser pulse. The highly reactive Cl radicals attach to atoms of the suface and remove them. Assuming a large absorption cross-section, the laser power is kept relatively low and the surface is etched without damage or significant heating.

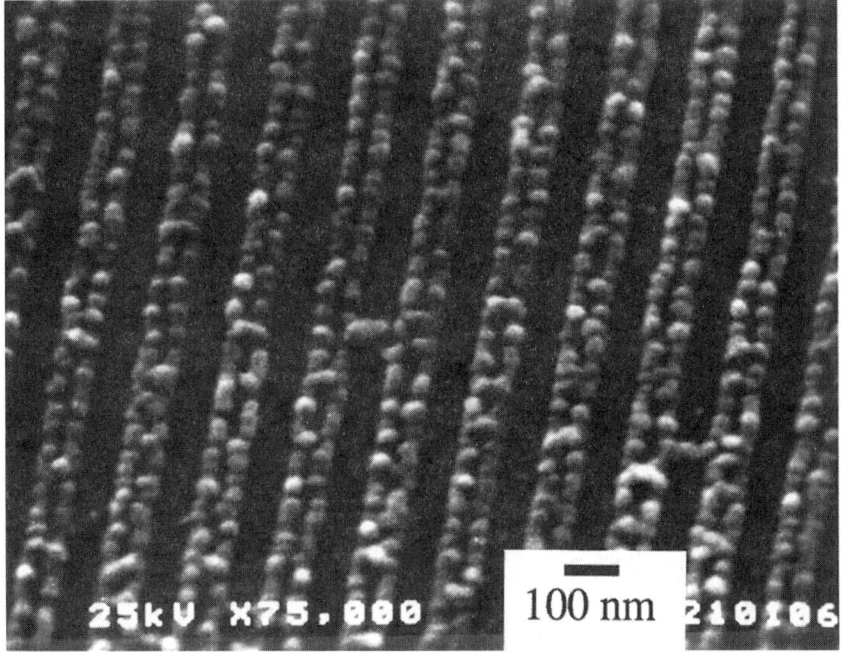

FIG. 7. SEM micrograph of the result of laser ablation with the interferometer shown in Fig. 5. Line structures with sub-100 nm linewidths and a period of 166 nm were produced.

Good spatial resolution can be obtained as in the experiment by Toyoda et al. [23]. Similar to the direct ablation experiment of Phillips et al., an interference pattern was produced with the fourth harmonic of an Nd:YAG laser (266 nm) in a CH_3Br ambient. High-quality etching of GaAs substrates was achieved. In the same work, LIPSS were etched into the substrate with the chemical etching procedure. Important here is that uniform etching of small features over large areas was obtained. As dissociation-based etching requires only a low-intensity pulse, the 0.1 W output of the laser was sufficient. Much work has been done with laser-assisted chemical etching and deposition by direct writing with cw lasers, but such techniques have a 10-µm resolution limit and a low throughput compared to projection methods [2]. In addition, the ability to create grating structures in optically active semiconductors may find applications as an inexpensive processing technique to produce distributed feedback semiconductor lasers.

Laser-induced chemical etching has been successfully demonstrated by Foulon and Green [50] using projection techniques, where up to 2 µm resolution was readily achieved. The result of their experiment is shown in Fig. 9, where a

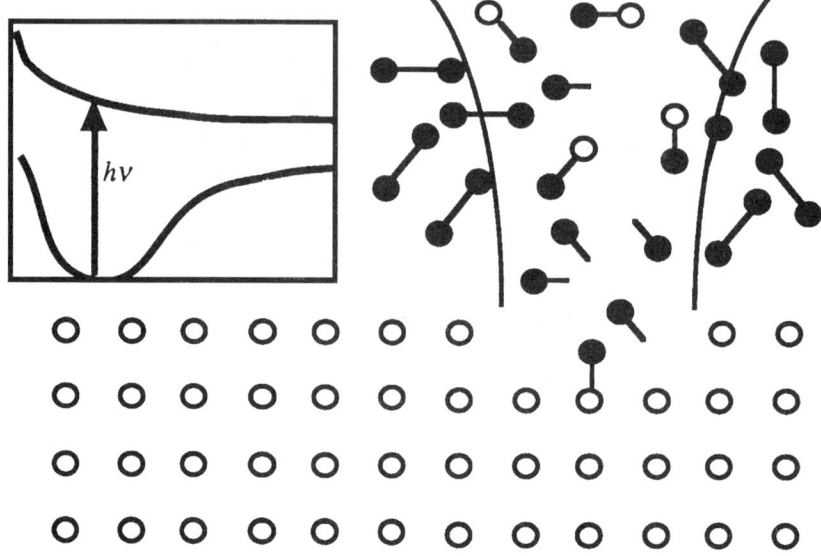

FIG. 8. Schematic of laser-induced chemical etching.

test mask pattern was etched into silicon. Thermal diffusion in these photothermally activated reactions were the primary limitation on resolution. One important observation was a strong dependence of feature size on pulse energy, although feature size did stabilize at higher energies. Such an effect is similar to

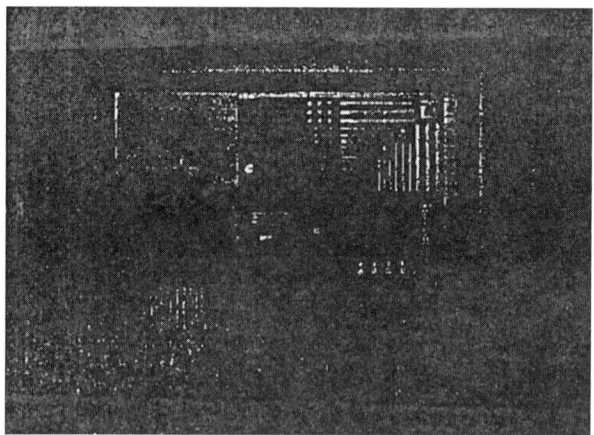

FIG. 9. Test mask pattern etched intoa silicon substrate via laser-induced chemical etching.

the linewidths observed with direct laser ablation by Phillips *et al.* [42], where etching occurs only when the peaks of the temperature distribution exceed a threshold value. In the case of direct ablation, this is the ablation threshold where, for photothermal chemical etching, the threshold is a surface temperature corresponding to the activation energy of the chemical reaction.

7.3.3 Laser Cleaning

Cleaning processes are crucial in semiconductor manufacturing and other industries. Conventional wet cleaning processes generally involve organic solvents that are environmentally problematic, particularly ultrasonic cleaning with chlorofluorocarbons, CFCs [51]. As a large portion of CFCs in the environment come from their commonplace use as industrial solvents, it is very desirable to have an effective environmentally friendly technique. In addition, these traditional techniques begin to lose effectiveness for submicron-sized particles because the forces adhering the particle to the surface (van der Waals attraction, capillary action with adsorbed moisture, and electrostatic forces) scale with the diameter, d, of the particle, whereas the particle mass scales with d^3 [41]. Consequently, increasingly larger accelerations are necessary to remove smaller particles from surfaces. Laser-based processing is a good candidate for such applications and has also been implemented for cleaning of optical surfaces [52].

A typical setup is shown in Fig. 10 and consists of an optical system, a pulsed vapor source to coat the sample with a thin layer of liquid (mostly water), and a means of monitoring and controlling the process. The simplicity of the process is evident, but the exact mechanisms for particle removal can be manipulated by choice of liquid film and laser wavelength and fluence. Typically the laser wavelength is chosen such that either particles or substrate absorb strongly, although one may more easily optimize the system to the substrate's properties than to the particle's properties. If no liquid is used, the only removal mechanism is the rapid thermal expansion of the substrate or particles in response to the laser-induced temperature rise, which ejects particles from the surface. The use of the liquid film provides additional mechanisms for particle ejection from the vaporization process. It is best if the vaporization takes place at the substrate-liquid interface, so strong substrate absorption with low film absorption is desirable.

To enhance the acceleration of the surface from thermal expansion, it is also clearly advantageous to use nanosecond pulsewidths. Ultraviolet wavelengths have yielded the best results due to the strong ultraviolet absorption of metal and semiconductor substrates [41, 45], so excimer lasers offer the best combination of wavelength and pulsewidth. IR cleaning has also been performed with good results as well, but absorption is only at the surface of the liquid film and some types of particles (e.g., gold) would melt before ejection thresholds were reached

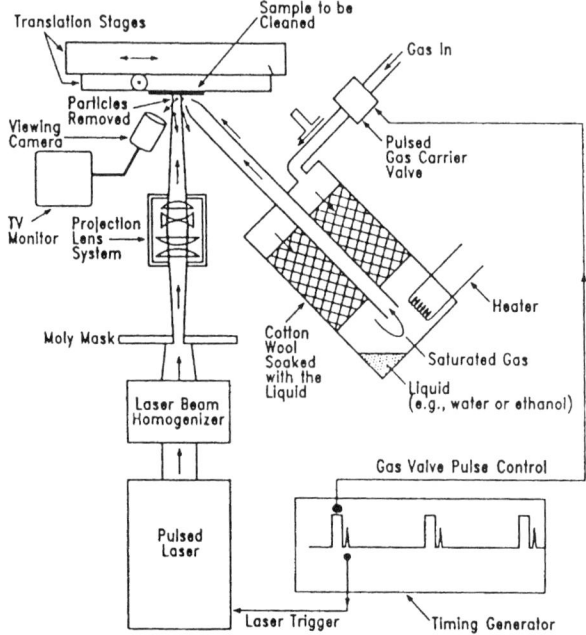

FIG. 10. Experimental arrangement for excimer laser cleaning using liquid thin film evaporation.

[53, 54]. Laser cleaning is also advantageous as direct mechanical contact with the surface is not necessary. If the target must be in a vacuum or other special environment, laser cleaning may be the only available option.

Laser cleaning may find application in a number of areas, and one particularly novel application has been in the restoration of centuries old paintings where a variety of corrosive films and particles can be removed with excimer laser irradiation [55].

7.3.4 Laser Polishing

Laser processing for producing flat polished surfaces is important as an inexpensive way to machine optical surfaces to the required high tolerances as well as to machine thin films of very hard materials such as CVD diamond.

Laser polishing of diamond [56] offers many advantages including reduced surface roughness when compared to mechanical or chemical-based techniques, less formation of nondiamond carbon forms, and a two order of magnitude decrease in processing times. Diamond films are limited in their potential for tribological applications because of the large grain size produced when growth is on a nondiamond substrate. The ability to polish these films adds low friction coefficients to the already demonstrated properties of extreme hardness and high thermal conductivity and should open the way for industrial applications.

7.4 Surface Modification of Polymers

The ease with which lasers can manipulate polymers was noted in Section 7.3 with regard to surface morphology, but recent work has shown that laser processing can go way beyond the etching structures toward changing the surface's chemical and electrical properties. Here, we discuss the laser processing of two particularly important polymers, polyimide and teflon. There is currently great interest in these two materials for semiconductor applications where they can serve as interlayer dielectrics in interconnection schemes. The primary advantage of using a polymer is its low dielectric constant, which can dramatically reduce RC propagation delay. It is expected that RC delay will be the primary limitation on device performance for design rules of 0.35 μm and smaller if low dielectric constant materials are not used [57]. Polyimide is already in use for such applications on a larger scale in multichip modules. Polymers also have a number of disadvantages related to water absorption and adhesion with metal layers for which laser modified surfaces may prove practical solutions. Teflon has many ideal properties for these applications, but its near total chemical inertness is a hindrance to conventional processing techniques.

7.4.1 Excimer Laser-Induced Electrical Conductivity in Polyimide

As excimer laser ablation is already used as a manufacturing process for the formation of vias in polyimide for packaging, it is natural to consider how ultraviolet laser processing of polyimide can be extended beyond morphology modification. In the last few years, research has shown that the electrical conductivity of polyimide can be permanently increased by as much as 17 orders of magnitude through ultraviolet laser processing [58, 59], and this process has been demonstrated with submicron spatial resolution [60]. Similar results have also been obtained with proton beams [61] and UV cw argon ion irradiation [62], although the nature of the change in conductivity with cw radiation is significantly different, involving the formation of glassy carbon. Also, as cw irradiation is accomplished through direct writing with a focused beam, line-widths are limited to about 15 μm.

The conductivity change with excimer irradiation occurs when the polymer absorbs a critical number of laser pulses just above the ablation threshold of the polyimide. The energy of the laser effects a temperature rise, which decomposes

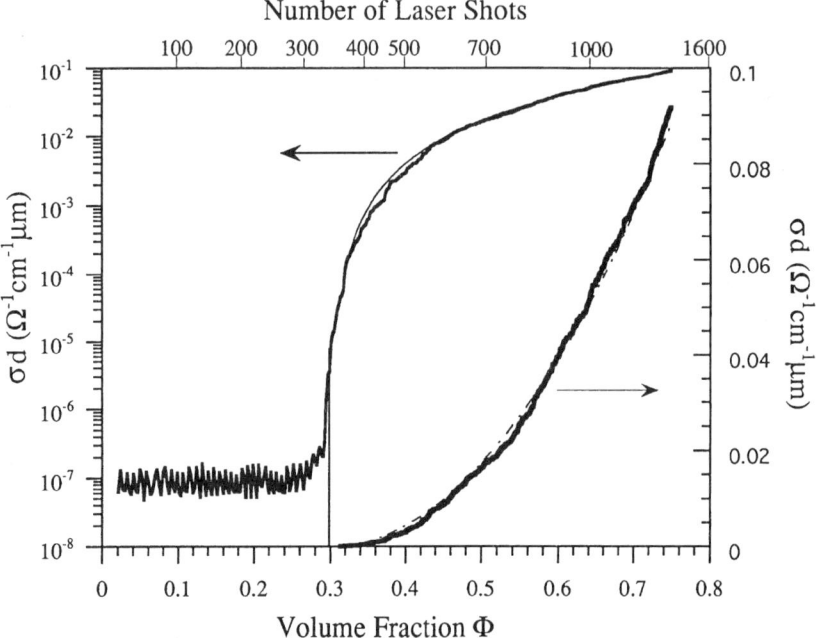

FIG. 11. Conductivity of an excimer laser-treated polyimide surface as a function of laser pulses. A critical number of laser shots must be used before any conduction can be observed, and the conductivity subsequently increases as a power law in agreement with a percolation model.

a thin surface layer of the polymer (0.5 μm t 2 μm) into carbon clusters of about 50 to 100 nm in diameter [63]. The density of these clusters increases with continued irradiation until a conducting path is formed across the sample surface. In Fig. 11, the conductivity of a treated foil is shown as a function of the number of laser pulses where the conductivity was monitored during laser processing. The noise level in the figure represents the limits of the measurement technique where separate measurements have shown the below-threshold conductivities to be comparable to the untreated foil. The transition from nonconducting to conducting material is extremely sharp, reflecting the geometrical nature of the transition. This transition has been modeled successfully with percolation theory and has been used to extract the critical exponents with great accuracy [63]. Surfaces prepared near the critical point exhibit unique properties that can be understood through percolation theory such as nonlinear I-V characteristics. To fully transform the surface into a conducting carbon layer requires approximately 1000 pulses, which given commercial systems with repetition rates of 1 kHz is not excessive [4].

The laser-treated surface has been analyzed with Fourier transform IR spectroscopy [59], transmission electron microscopy [64], Raman spectroscopy [65], parallel electron energy loss spectroscopy [66], and other techniques, all of which indicate that the process is essentially one of thermal decomposition where concentrations of noncarbon elements in the surface layer decrease with increasing irradiation. Thermal coupling between pulses has been shown to be important for the effect to occur. This is attributed to the need for an increased background temperature to promote the formation of a conducting arrangement of carbon atoms within the clusters, i.e., polymerized six-member carbon rings [67].

The surface roughness after conductivity transformation can be controlled by adjusting the laser fluence, and it is the increase in roughness combined with the carbon clusters that limits the ultimate depth that can be ablated in polyimide at lower fluence ranges. One application of laser-induced electrical conductivity may be the plating of copper onto a conducting pattern for interconnection at a chip or packaging level and it is expected that the ability to control surface roughness and conductivity together would allow for greatly enhanced adhesive properties. This control, combined with the high spatial resolution possible with excimer lasers, may be advantageous for advanced metallization applications [67].

7.4.2 Surface Modification of Teflon

Two primary types of teflon are of technological interest and laser modification of both has been achieved. These are Teflon AF, which is an amorphous polymer, and the more common PTFE (polytetrafluoroethylene). Teflon, widely known for its chemical inertness, has found innumerable applications in various industries and consumer products. Not surprisingly, adhesion to teflon of anything is a considerable problem for which ultraviolet laser treatment is an important solution with the principal advantage of high spatial resolution.

First we consider the work of Okoshi *et al.* [68], who irradiated a PTFE surface with an ArF excimer laser is various solutions and atmospheres where the 193 nm photons could break the F-C bonds and promote a reaction with the ambient. The surfaces could then be studies with X-ray photoelectron spectroscopy to determine the nature and extent of surface modification. A solid piece of teflon irradiated in this way could then be epoxied directly onto metallic material (in this case, steel) with a degree of adhesion completely uncharacteristic of teflon.

Several groups [68, 69, 70] have reported modification with gas phase reactants, including Hydrazine and $B(CH_3)_3$. In the former case, activation of the surface for selective electroless plating of Ni was achieved, although intermediate steps were required. All such modifications resulted in hydrophilic behavior with a measured decrease in contact angle with increasing fluence and number of laser shots. Changes in contact angle (water or benzene) were on the order of 100° or more in these experiments. It should be emphasized that these modifications were performed, at least with the lower fluences studied, without change to the surface morphology of the material and without visible damage. Similar results were obtained by Matienzo *et al.* [71] using VUV radiation from helium microwave plasma. Multiphoton processing of Teflon using ultrashort pulse lasers is a possible alternative to UV/VUV laser processing and has been demonstrated by Kumagai *et al.* [12] using a high-power Ti : Sapphire laser.

7.5 Surface Modification of Metals and Semiconductors

Surface modification of semiconductor materials has been the most active research area in the laser processing of materials and has been reviewed previously in the literature [2, 14]. Laser processing offers many unique abilities to manipulate a surface without depositing a large amount of heat into the system. The most important of these is laser crystallization of amorphous silicon (a-Si), which can be used for thin film transistor (TFT) fabrication for active matrix videographic displays. Laser-induced doping and alloying in metals have also been important for industrial applications.

7.5.1 Surface Modification of Metals

It is often desirable for a material's surface to have contrasting properties with the bulk such that we can take advantage of an inexpensive material's mechanical properties and avoid what may be poor surface qualities such as lack of hardness or poor resistance to corrosion. Simple coatings often are employed through electroplating or other means, but in particularly high performance applications, a surface layer may not be adequate due to poor adhesion at high temperatures or cracks that can form in the coating. In these cases, a true intermixture of surface and bulk materials is required, and because high temperature processing can lead to deterioration of bulk properties, low temperature processes are very desirable. Two primary low-temperature processes that have been explored for metal processing are ion implantation and laser-induced diffusion.

A good example is surface modification of stainless steel, which has been investigated recently by Jyumonji *et al.* [72]. In this work, Si was simultaneously deposited from a silane ambient and diffused into the steel substrates with an excimer laser. The 248 nm excimer laser pulse melts a thin surface layer for about 100 ns and allows Si atoms to diffuse rapidly into the melted surface because the diffusion constant of the melted surface is several orders of magnitude higher than the solid. RBS spectra confirmed an intermixing layer of about 100 nm in thickness, and a linear variation of the thickness with the number of laser pulses. The concentration profile of the Si could be predicted by a simple mass diffusion model and followed the resulting error function distribution to good accuracy. At a fluence of 400 mJ/cm^2, a dosage of $2 \cdot 10^{17}$ atoms/cm^2 was achieved, and the profile could be controlled with gas pressure, fluence, and number of pulses. One particular advantage is the agreement between relatively simple models of dopant profiles and experimental results allows parameters to be adjusted in a straightforward way to produce a desired doping profile.

Generally speaking, when thin submicron layers are desired, short wavelenths and short pulses are necessary; whereas for deeper intermixture layers, longer pulses in the infrared region can be used. CO_2 lasers or the fundamental of an Nd:YAG laser are typically used in this case. One recent example investigated by Haferkamp et al. [73] is alloying of pure copper without various metals to improve surface hardness. Pure copper is a desirable material for its electrical conductivity and is often used in electric motors and dynamos. Wear on copper components from brush contacts and thermal loading is a serious concern for soft copper surfaces. The alloying element was either deposited as a mixture with turpentine oil before laser processing or in a one-step process using thermal spraying. Laser alloying of copper with various dopants was achieved at 1.06 µm with 5 ms pulses where interdiffusion depths as large as 1 mm were possible, yielding significant increases in hardness for the dopants Sn, Al, Cr, and B, although only a marginal increase in hardness was found for Ni.

One potentially very useful application of laser-induced melting is reflowing and planarization of metal interconnection layers in the semiconductor industry [74]. A short laser pulse melts the deposited metal layer, typically aluminum, and in the short duration of the melt, the metal completely fills vias and leaves a planarized surface. A cross-section of a sample before and after laser planarization is shown in Fig. 12 where excellent results were obtained. Such highly planar surfaces are possible due to the high surface tension of the melted Al. The primary obstacle to implementing the technique in manufacturing lines is that careful control over the laser energy is required to avoid ablation while supply-

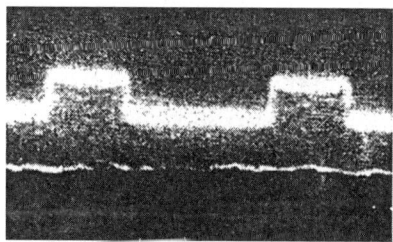
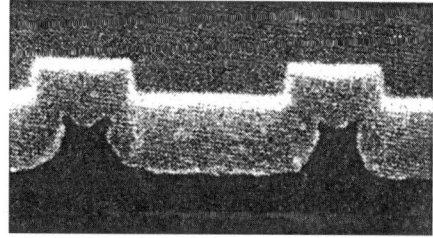

FIG. 12. Cross-sections of Al interconnections before and after laser planarization.

ing enough energy to melt the entire layer. Best results were obtained with the short wavelengths and nanosecond pulses of excimer lasers, although dye lasers have also been used.

7.5.2 Laser Annealing of Implanted Seminconductors

One of the principle technologies of modern semiconductor manufacturing is ion implantation of dopants, although lattice damage and amorphization of the surface are serious problems that become worse with higher doping densities. Annealing is, therefore, necessary to recrystallize the lattice, but bulk annealing can cause redistribution of dopants. In laser annealing, only the surface layer is melted and for only a short time, and subsequent epitaxial regrowth has been demonstrated with a number of laser sources.

Epitaxial regrowth after laser melting was conclusively demonstrated by Auston et al. [9], where 50 ns ND:YAG laser pulses were used to anneal amorphous, ion-implanted Si. Full annealing was accomplished for fluences greater than 3 J/cm^2. Transient reflectivity measurements were used to verify surface melting and determine the melt duration, which was about 300 ns. The work of Venkatesan et al. [75] has shown that absorption of the 1.06 μm pulses and annealing characteristics have a strong dependence on the implant dose, complicating the process.

As thermal growth of oxide is the highest temperature component of semiconductor processing, laser annealed, ion-implanted oxide layers are a possibility. A particular advantage of the laser annealed method over more common CVD techniques is the ability to create buried oxide layers hundreds of nanometers into the surface [14].

7.5.3 Laser-Induced Diffusion in Semiconductors

Laser-induced melting of semiconductors can result in epitaxial regrowth of the lattice and presents an opportunity for laser-controlled introduction of dopants. Laser-induced diffusion has some advantages over conventional ion implantation particularly with regard to the lack of lattice damage and the very sharp concentration gradients that are possible. Thin doped regions occur essentially because pulsed laser irradiation only allows a thin melted layer where diffusion rates are significant. Although for most applications ion implantation is clearly the preferred technique, laser-induced diffusions may find practical application in areas such as ohmic contact formation, passivation, and metallization of GaAs and other compound semiconductors for optical communications applications.

Excimer laser doping of semiconductors was demonstrated more than a decade ago by Deutsch et al. [76–78] where ohmic contacts were made in InP and GaAs while *p-n* junctions fabricated in Si and GaAs could be made into efficient solar cells. Similar to results above for doping of metals, depths were thin (<1 μm) and had a sharp profile. Dopant material was simultaneously deposited from an ambient gas by photodissociation of organometallic molecules in the case of Si and by both photodissociation and thermal activation of H_2S for GaAs. In both materials, quality photocells could be constructed. The technique was of particular interest for GaAs where traditional techniques of ion implantation and laser annealing suffer significant problems due to material loss (owing to the low vapor pressure of arsenic), requiring a capping layers, which can itself diffuse into the substrate.

More recently these techniques have been used for dope passivation of GaAs surfaces to reduce interface state density for field effect transistor fabrication. Zhang et al. [79] used a two-step method where the surface was slowly etched with an $(NH_4)S_x$ solution. Samples prepared in this fashion retained a few monolayers of sulphur on the passivated surface and were then placed in vacuum chamber for irradiation. Secondary ion mass spectrometry (SIMS) measurements and RBS characterization of lattice damage were performed, and good quality surfaces were obtained.

Laser-induced diffusion may take on much greater importance in the future as device sizes shrink to levels where the inability of ion implantation to create shallow highly doped regions becomes significant.

7.5.4 Laser Crystallization of a-Si

Investigations of laser annealing of implanted silicon wafers in the late 1970s revealed two fluence thresholds for laser processing, a lower one where polycrystalline material was formed and a higher one where true epitaxial growth was observed, which, for annealing of ion implantation damage, is the desired effect. In the last few years, processes for the production of high-quality polycrystalline Si have taken on great importance for the production of thin film transistors (TFTs) on glass substrates for flat panel display technology. It is in this area that laser processing of semiconductors is currently seen as a likely manufacturing process.

TFT technology generally consists of CVD deposited a-Si on quartz substrates, which is then crystallized via furnace or laser techniques, and, by controlling the CVD gas mixture, doping is accomplished simultaneously with deposition. Key problems with furnace-based techniques are that low substrate temperatures must be maintained to avoid deformation of the glass and that low-temperature, solid state annealing is prohibitively slow. The high defect densities associated with polycrystalline techniques.

In excimer laser crystallization of a-Si, the excimer pulse melts the amorphous layer and a polycrystalline phase appears during solidification. The process is also reversible if higher laser energies (dependent on thin film thickness) are used; this is termed "excimer laser amorphization." Comparisons of excimer laser and furnace-based recrystallization techniques have demonstrated lower in-grain defect densities in laser crystallized films and higher carrier mobilities [80]. Further improvements have been found in combinations of laser- and furnace-based techniques [81]. Furnace annealing is not particularly economical, requiring ~8 hours [82], and a laser crystallization and rapid thermal annealing (RTA) are seen as the most likely technologies where excimer laser methods yield better device properties but with poorer reproducibility than RTA methods.

In the work of Carluccio *et al.* [81], comparisons for different irradiating conditions were made where initial furnace annealing was followed by laser annealing. Grain sizes of up to 1.5 μm could be produced with low densities of in-grain defects and low surface roughness (5 nm), although such slow processes do little to increase throughput, one of the key advantages of excimer laser recrystallization.

Laser annealing of a-Si is not only limited to TFT technology. El-Kader *et al.* [83] have recently demonstrated the formation of luminescent "porous" silicon from hydrogenated amorphous thin films. In this work, a XeCl excimer laser was used to irradiate a-Si thin films under vacuum at fluences from 0.3 J/cm^2 to 0.8 J/cm^2, where only a few laser pulses were required to produce a stable layer with voids and craters about 2 μm in diameter. Photoluminescence spectra peak around 700 nm and are comparable to those prepared through HF processing. Such work illustrates the interesting behavior that can be observed due to the nonequilibrium nature of the rapid heating and cooling induced by nanosecond laser pulses. Because these experiments correspond to the boundary between crystallization and amorphization, the final (optically active) phase is likely a combination of both phases.

7.6 Excimer Laser Lithography

Excimer laser lithography [1, 84, 85] is the most advanced optical lithography method in use today, the first to use a coherent source, and the designated successor to mercury lamp–based methods. As KrF (248 nm) lithography is currently being brought into production for 0.25 μm processes such as for 256 Mbit DRAMs, it is clear that most difficulties have been overcome. However, many problems remain research considerations, particularly with regard to the laser-polymer interaction. Further reduction in design rule to 0.18 μm is still a research issue with the possibility of improved KrF technology or a move to ArF excimers at 193 nm [1]. The only alternative to excimer laser-based methods would be a major paradigm shift to proximity X-ray lithography, a technology that becomes more attractive as lithography stepper costs soar.

7.6.1 Laser System Requirements

Initially the principal obstacles to excimer-based lithography were poor laser reliability, low beam quality, high cost, and a lack of adequate achromatic optics. Each of these problems can now be considered solved and no longer an issue for applications. First, new laser systems have been developed with high reliability and low maintenance costs. These lasers can operate for up to 10^8 pulses on a single gas fill, and components are robust for orders of magnitude more pulses.

Beam quality [84] has been addressed through beam homogenization techniques, primarily the "fly's eye" type beam homogenizer that uses an array of small lenses to produce a "top hat" beam profile. This has the highly desirable side effect of destroying the spatial coherence of the beam across the mask, eliminating speckle.

Because fluorite-based optical materials are expensive and have difficulty meeting standards of reliability necessary for mass production, achromatic optical systems have largely been rejected in favor of line narrowing of the laser. Thus conventional high-quality fused silica optics are used in most systems. To eliminate chromatic aberration, the laser linewidth must be reduced to 3 pm or less [84], far less than the natural linewidth of the excimer. This is accomplished by making one mirror of the resonator highly frequency selective. Initial designs used etalons to provide frequency selection, but poor reliability of the etalons has motivated designs to use hybrid grating-prism-etalon configurations to provide high-frequency selection and long component lifetime. Another option is achromatic reflective optics, although this approach has not been used in commercial systems.

Thus all primary problems associated with the integration of excimer sources into a manufacturing line have been overcome with considerable success. However, the development of resists has become the primary limitation on excimer laser lithography.

7.6.2 Resists for Excimer Laser Lithography

The essential problem to the move to excimer sources for resists is the high absorption of organic molecules at excimer wavelengths. Process requirements dictate that 1 µm resist layers be patterned. The absorption length in most polymers at 248 nm is more than 10 times smaller than this. The laser must be able to expose homgeneously a full 1 µm layer of resist. The key to full penetration of the laser light is bleachable absorption characteristics. Recalling the laser absorption discussion in Section 7.1, the material must have saturable absorption properties at low fluences. Nonsaturable absorption must be minimal and excited state absorption cross-sections should be insignificant.

The class of resists now used in excimer laser lithography are called "chemical amplification resists" [84, 85], where deprotection of the resist comes from a uv-absorbing photoacid. These resists have been successful, but they feature a number of problems such as the tendency for a soluble layer to form at the substrate resist boundary, resulting in undercutting or insoluble layer formation, resulting in sloped walls.

Perhaps the most serious problem with resist technology results from the temporal coherence of the laser, which is an unavoidable consequence of linewidth narrowing [85]. Full exposure of the resist implies that significant amounts of laser light reach the resist-substrate interface. A portion of this light is then reflected, resulting in multiple reflection interference. These beams are reflected at considerable angles and thus expose resist and limit resolution. Depending on the substrate material, multiple reflection interference can erode resolution by as much as 0.05 μm. One solution has been to spin on a soluble anti-reflection coating before resist application. Variations in resist thickness cause similar problems.

7.6.3 193 nm Lithography

193 nm lithography is considered a prime candidate for the move to 0.18 nm design rules. Most of the problems associated with KrF lithography exist for ArF lithography as well but with greater severity [1]. Absorption properties at 193 nm are significantly different and will require fundamentally different resist materials. ArF laser systems are less powerful and less reliable than KrF lasers, but the same technology used in KrF steppers to extend gas and component lifetimes are probably adequate for ArF systems as well.

The traditional resist for ArF and other extremely short wavelengths is PMMA (Poly(methylmethacrylate)) [1]. Processing of PMMA with excimer lasers has been studied for a long time, primarily in terms of laser ablation. In lithography, however, photochemical modification does not have to lead to material removal, just significant degradation for chemically based development. For an effective resist, a methacrylate type resin is used with a photoacid generator to produce a relatively transparent amplification resist for 193 nm. Such resists have the added advantage that they should work equally well at 248 nm. The limitations of methacrylate-based resists are two-fold, poor resistance to plasma-based etching and the lack of a suitable AR coating.

Top-surface imaged (TSI) resists [1] have also been developed for ArF lasers. These processes do not require light penetration through the whole volume of resist. In a TSI resist, a silyl amine is selectively in-diffused into a phenolic polymer. This diffusion process creates a silyl ether, and development takes place in the form of an oxygen plasma etch, sometimes termed "dry developing," Depth of focus limitations are thus avoided because exposure is necessary only at the surface of the resist layer, and the resolution of the etching process determines the final resist profile.

The future of photolithography in the longer term may depend on the ability to use innovative approaches such as TSI resists. If devices are to scale below 0.1 μm, entirely new techniques may need to be relied on such as X-ray and XUV lithography or atomic probe–based patterining [86].

7.7 Conclusion

Laser processing of materials encompasses many fields and applications. Common to all of these are the physical and chemical processes outlined in Section 7.2. In particular, the ability of a laser to deposit heat in a shallow surface layer in short time creates a unique thermodynamic situation that can be exploited for applications. Of these, phase transitions such as melting, crystallization, and decomposition are the most important and have found their way into industrial processes. The ability of laser-induced diffusion to produce shallow regions of high doping may also provide important solutions to the limitations of ion implantation in future generations of deep submicron semiconductor processing. The use of the laser to manipulate surface chemistry such as with TSI resists in photolithography, laser-induced oxidation, and laser-induced chemical etching are active areas of research and may develop into important technological advances as well.

References

1. Rothschild, M., Forte, A. R., Horn, M. W., Kunz, R. R., Palmateer, S. C., and Sedlacek, J. H. C. 193 nm lighography. *IEEE J. of Selected Topics in Quantum Electrons* **1**, 916 (1995).
2. Bäuerle, D. *Laser Processing and Chemistry*, second edition. Springer, Berlin, 1996.
3. Chester, R. B., and Geusie, J. E. Solid-state ionic lasers, in *Laser Handbook*, F. T. Arechi and E. O. Schulz-Dubois (eds.), Vol. 1, p. 325. North Holland, Amsterdam, 1972.
4. Lacour, B., Brunet, H., Desaucele, H., Gagnol, C., and Vincent, B. 500 Watts industrial excimer laser at high repetition rate. *SPIE Proc.* **2206**, 41 (1994).
5. Gartry, D. S. *Br. Med. J.* **31**, 979 (1995).
6. Lankard, J. R., Sr., and Wolbold, G. Excimer laser abalation in a manufacturing facility. *Appl. Phys.* **A54**, 355 (1992).
7. Pronko, P. P., Dutta, S. K., Du, D., and Singh, R. K. Thermophysical effects in laser processing of materials with picosecond and femtosecond pulses. *J. Appl. Phys.* **78**, 6233 (1995).
8. Pettit, G. H., Ediger, M. N., Hahn, D. W., Brinson, B. E., and Sauerbrey, R. Transmission of polyimide during pulsed ultraviolet laser irradiation. *Appl. Phys.*

A58, 573 (1994); Sauerbrey, R., and Pettit, G. H. Theory for the etching of organic materials by ultraviolet laser pulses. *Appl. Phys. Lett.* **55**, 421 (1989).
9. Auston, D. H., Surko, C. M., Venkatesan, T. N. C., Slusher, R. E., and Golovchenko, J. A. Time resolved reflectivity of ion-implanted silicon during laser annealing. *Appl. Phys. Lett.* **33**, 437 (1978).
10. Ginsburg, V. L. *The Propagation of Electromagnetic Waves in Plasmas.* Pergamon, new York, 1970; Kruer, W. L. *The Physics of Laser Plasma Interaction.* Addison-Wesley, new York, 1988.
11. Sauerbrey, R., Fure, J., LeBlanc, S. P., van Wonterghem, B., Teubner, U., and Schäfer, F. P. Reflectivity of laser-produced plasmas generated by a high intensity ultrashort pulse. *Phys. Plasmas* **1**, 1635 (1994).
12. Kumagi, H., Midorikawa, K., Toyoda, K., Nakamura, S., Okamoto, T., and Obara, M. Ablation of polymer films by a femtosecond high-peak-power Ti : sapphire laser at 798 nm. *Appl. Phys. Lett.* **65**, 1850 (1994).
13. Haglund, R. F., Jr., and Kelly, R. *Matmatisk-fysiske Meddelelser* **43**, 527 (1993).
14. Boyd, I. W. *Laser Processing of Thin Films and Microstructures.* Springer, Berlin, 1987.
15. Dickinson, J. T., Jensen, L. C., Webb, R. L., Dawes, M. L., and Langford, S. C. Mechanisms of excimer laser ablation of wide band gap materials: the role of defects in single crystal MgO. *MRS Proceedings* **285**, 131 (1993).
16. Pettit, G. H., and Sauerbrey, R. Pulsed ultraviolet laser ablation. *Appl. Phys.* **A56**, 51 (1993).
17. Patterson, M. J., Margrave, J. L., Hauge, R. H., Ball, Z., and Sauerbrey, R. *Proceedings of the Electrochemical Society: Diamond Materials IV* **95-5** (1995).
18. Milonni, P. W., and Eberley, J. H. *Lasers*, pp. 56–58. Wiley, New York, 1988.
19. Jain, K. *Excimer Laser Lithography*. SPIE Bellingham, 1990.
20. Sauerbrey, R., and Ball, Z. Threshold condition and optimum output coupling for a pulsed laser. *Optics Communications* **95**, 153 (1993).
21. Sipe, J., Preston, J. F., and van Driel, H. M. Laser induced periodic surface structures I: theory. *Phys. Rev.* **B27**, 1141 (1983).
22. Van Driel, H. M., Sipe, J. E., and Young, J. F. Laser-induced periodic surface structure on solids: a universal phenomenon. *Phys. Rev. Lett.* **49**, 1955 (1982).
23. Kumagai, H., Toyoda, K., Machida, H., and Tanaka, S. Exponential growth of periodic surface ripples generated in laser-induced etching of GaAs. *Appl. Phys. Lett.* **39**, 2974 (1991).
24. Heitz, J., Arenholz, E., Bäuerle, D., Sauerbrey, R., and Phillips, H. M. Femtosecond excimer-laser-induced structure formation on polymers. *Appl. Phys.* **A59**, 289 (1994).
25. Walters, C. T., and Clauer, A. H. Transient reflectivity behavior of pure aluminum at 10.6 μm. *Appl. Phys. Lett.* **33**, 713 (1978).
26. Chun, M. K., and Rose, K. Interaction of high-intensity laser beams with metals. *J. Appl. Phys.* **41**, 614 (1970).
27. Ball, Z., Hopp, B., Csete, M., Ignácz, F., Rácz, B., Sauerbrey, R., and Szabo, G. Transient optical properties of excimer-laser-irradiated polyimide. *Appl. Phys.* **A61**, 547 (1995).
28. Bor, Z., Rácz, B., Szabo, G., Xenakis, D., Kalpouzos, C., and Fotakis, C. Femtosecond transient reflection from polymer surfaces during femtosecond UV photoablation. *Appl. Phys.* **A60**, 365 (1995).

29. Weast, R. C., Lide, D. R., Astle, M. J., and Beyer, W. H. eds. *Handbook of Chemistry and Physics* 70th Edition. CRC Press, Boca Raton, Florida, 1989.
30. Kapton, T. M. *Summary of Properties*. Dupont Corporation.
31. Philip, H. R., Cole, H. S., Liu, Y. S., and Sitnik, T. A. Absorption of some polymers in the region 170 nm–250 nm. *Appl. Phys. Lett.* **48**, 192 (1986).
32. Marki, H. F., Bikales, N. M., Overberger, C., and Menges, G., eds. *Encyclopedia of Polymer Science and Engineering*, Vol. 16, p. 585.
33. Furzikov, N. P. Approximate theory of highly absorbing polymer ablation by nanosecond laser pulses. *Appl. Phys. Lett.* **56**, 1638 (1990).
34. Carslaw, F. H., and Jaegar, J. R. *Heat Conduction in Solids*, pp. 282–288. Oxford University Press, Oxford, 1959.
35. Baerii, P., Campisano, S. U., Foti, G., and Rimini, E. A melting model for pulsing laser annealing of implanted semiconductors. *J. Appl. Phys.* **50**, 788 (1979).
36. Meyer, J. R., Kruer, M. R., and Baroli, F. J. Optical heating in semiconductors: Laser damage in Ge, Si, InSb, and GaAs. *J. Appl. Phys.* **51**, 5513 (1980).
37. Bertolotti, M., and Sibilia, C. Depth and velocity of the laser-melted front from an analytical solution of the heat conductione quation. *IEEE J.* **QE-17**, 1980 (1981).
38. Wood, R. F. Macroscopic theory of pulsed-laser annealing. III. Nonequilibrium segregation effects. *Phys. Rev.* **B25**, 2786 (1982).
39. Kwong, D. L., and Kim, D. M. Pulsed laser heating of silicon: the coupling of optical absorption and thermal conduction during irradiation. *J. Appl. Phys.* **54**, 366 (1983).
40. Zweig, A. D., Venugopalan, V., and Deutsch, T. F. Stress generated in polyimide by excimer-laser irradiation. *J. Appl. Phys.* **74**, 4181 (1993).
41. Tam, A. C., Leung, W. P., Zapka, W., and Ziemlich, W. Laser cleaning techniques for removal of surface particulates. *J. Appl. Phys.* **71**, 3515 (1992).
42. Zysset, B., Fujimoto, J. G., and Deutsch, T. F. *Appl. Phys.* **B48**, 139 (1989).
43. Doukas, A. G., Zweig, A. D., Frisoli, J. K., Birngruber, R., and Deutsch, T. F. *Appl. Phys.* **B53**, 237 (1991).
45. Levine, I. N. *Molecular Spectroscopy*, pp. 296–314. Wiley, New York, 1975.
46. Niino, H., and Yabe, A. Surface modification and metallization of fluorocarbon polymers by excimer laser processing. *Appl. Phys. Lett.* **63**, 3527 (1993).
47. Okoshi, M., Murahara, M., and Toyoda, K. Photochemical modification of polytetrafluroethylene in oleopholic property using an ArF excimer laser. *J. Mater. Res.* **7**, 1912 (1992).
48. Cho, C.-C., Smith, D. M., and Anderson, J. Low dielectric-constant insulators for electronics applications. *Materials Chemistry and Physics* **42**, 91 (1995).
49. Phillips, H. M., Callahan, D. L., Sauerbrey, R., Szabó, G., and Bor, Z. Sub-100 nm lines produced by direct laser ablation in polyimide. *Appl. Phys. Lett.* **58**, 2761 (1991).
50. Foulon, F., and Green, M. Projection-patterned etching of silicon in chlorine atmosphere with a KrF excimer laser. *Appl. Phys.* **A61**, 655 (1995).
51. Lu, Y. F., Takai, M., Komru, S., Shikowa, T., and Aoyagi, Y. Surface cleaning of metals by pulsed-laser irradiation in air. *Appl. Phys.* **A59**, 281 (1994).
52. Lu, Y. F., Komuru, K., and Aoyagi, Y. *Jpn. J. Appl. Phys.* **33**, 4691 (1994).
53. Lee, S. J., Imen, K., and Allen, S. D. CO_2 laser assisted particle removal threshold measurements. *Appl. Phys. Lett.* **61**, 2314 (1992).
54. Zapka, W., Ziemlich, W., and Tam, A. C. Efficient pulsed laser removal of 0.2 μm sized particles from a solid surface. *Appl. Phys. lett.* **58**, 2217 (1991).

55. Hontzopoulos, H., Fotakis, C., and Doulgeridis, M. *SPIE*, C. Fotakis (ed.), vol. 1810, p. 749, 1992.
56. Bhushan, B., Subramanian, V. V., Malshe, A., Gupta, B. K., and Ruan, J. Tribological properties of polished diamond films. *J. Appl. Phys.* **74**, 4174 (1993).
57. Cho, C.-C., Smith, D. M., and Anderson, J. Low dielectric-constant insulators for electronics applications. *Materials Chemistry and Physics* **42**, 91 (1995).
58. Schumann, M., Sauerbrey, R., and Smayling, M. C. Permanent increase of the electrical conductivity of polymers induced by ultraviolet laser radiation. *Appl. Phys. Lett.* **58**, 428 (1991).
59. Feurer, T., Sauerbrey, R., Smayling, M. C., and Storey, B. J. Ultraviolet laser induced permanent electrical conductivity in polyimide. *Appl. Phys.* **A56**, 275 (1993).
60. Phillips, H. M., Wahl, S., and Sauerbrey, R. Submicron electrically conducting wires produced in polyimide by ultraviolet laser irradiation. *Appl. Phys. Lett.* **62**, 2572 (1995).
61. Feurer, T., Wahl, S., and Langhoff, H. Modification of polyimide surfaces using intense proton pulses. *J. Appl. Phys.* **74**, 3523 (1993).
62. Srinivasan, R., Hall, R. R., Wilson, W. D., Loehle, W. D., and Albee, D. C. *Synthetic Metals* **66**, 301 (1993).
63. Ball, Z., Phillips, H. M., Callahan, D. L., and Sauerbrey, R. Percolative metal-insulator transition in excimer laser irradiated polyimide. *Phys. Rev. lett.* **73**, 2099 (1994).
64. Phillips, H. M., Feurer, T., Callahan, D. L., and Sauerbrey, R. Excimer laser induced electrical conductivity in polymers. *MRS Proc. No.* **285**, 175 (1993).
65. Gu, X. J. Raman spectroscopy and the effects of ultraviolet irradiation on polyimide film. *Appl. Phys. Lett.* **62**, 1568 (1993).
66. Bentley, J., Phillips, H. M., Callahan, D. L., and Sauerbrey, R. Electron energy-loss spectrometry of laser-irradiated kapton polyimide. *Proc. of the 51st Annual Meeting of the Microscopy Soc. of American* (1993).
67. Ball, Z., Feurer, T., Callahan, D. L., and Sauerbrey, R. Thermal and mechanical coupling between successive pulses in KrF excimer laser ablation of polyimide. *Appl. Phys.* **A62**, 203 (1996).
68. Okoshi, M., Murahara, M., and Toyoda, K. Photochemical modification of polytetrafluoroethylene in oleopholic property using an ArF excimer laser. *J. Mater. Res.* **7**, 1912 (1992).
69. Niino, H., and Yabe, A. Surface modification and metallization of fluorocarbon polymers by excimer laser processing. *Appl. Phys. Lett.* **63**, 3527 (1993).
70. Bor, Z. Private communication.
71. Matienzo, L. J., Zimmernmann, J. A., and Egitto, F. D. Surface modification of fluoropolymers with vacuum ultraviolet irradiation. *J. vac. Sci. Technol.* **A12**, 2662 (1994).
72. Jyunmonji, M., Sugioka, K., Takai, H., Tshiro, H., and Toyoda, K. Mechanism of silicon implant-deposition for surface modification of stainless steel 304 using KrF-excimer laser. *Appl. Phys.* **A60**, 41 (1995).
73. Haferkampf, K., Marquering, M., and Ebsen, H. Alloying of copper surface with a pulsed Nd:YAG laser. *SPIE Prov. Vol.* **2207**, 690 (1994).
74. Wolf, S. *Silicon Processing for the VLSI Era:Vol. 2 Process Integration*, pp. 255–256 and references cited therein. Lattice Press, Sunset Beach, CA, 1992.
75. Venkatesan, T. N. C., Golovchenko, J. A., Poate, J. M., Cowan, P., and Celler, G. K. Dose dependence in the laser annealing of arsenic-implanted silicon. *Appl. Phys. Lett.* **33**, 429 (1978).

76. Deutsch, T. F., Ehrlich, D. J., Osgood, R. M., Jr., and Liau, Z. L. Ohmic contact formation on InP by pulsed laser photochemical doping. *Appl. Phys. Lett.* **36**, 847 (1980).
77. Deutsch, T. F., Fan, J. C. C., Turner, G. W., Chapman, R. L., Ehrlich, D. J., and Osgood, R. M., Jr. Efficient Si solar cells by laser photochemical doping. *Appl. Phys. Lett.* **38**, 144 (1981).
78. Deutsch, T. F., Fan, J. C. C., Turner, G. W., Chapman, R. L., and Gale, R. P. Efficient GaAs solar cells formed by UV laser chemical doping. *Appl. Phys. Lett.* **40**, 722 (1982).
79. Zhang, S. K., Sugioka, K., Fan, J., Toyoda, K., and Zou, S. C. Studies on excimer laser doping of GaAs using sulphur adsorbate as dopant source. *Appl. Phys.* **A58**, 191 (1994).
80. Sameshima, T. *Solid State Phenomena* **37–38**, 269 (1994).
81. Carluccio, R., Stoemenos, J., Fortunato, G., Meakin, D. B., and Bianconi, M. Microstructure of polycrystalline silicon films obtained by combined furnace and laser annealing. *Appl. Phys. Let.* **66**, 1394 (1995).
82. Duhamel, N., and Loisel, B. *Solid State Phenomena* **37–38**, 535 (1994).
83. El-Kader, K. M. A., Oswald, J., Kocka, J., and Cháb, V. Formation of luminescent silicon by laser annealing of a Si : H. *Appl. Phys. Lett.* **64**, 2555 (1994).
84. Nakase, M. Recent progress in KrF excimer laser lithography. *ICICE Trans. Electron.* **E76-C**, 26 (1993).
85. Sasago, M., Endo, M., Tabi, Y., Kobayashi, S., Koizumi, T., Matsuo, T., Yamashita, K., and Nomura, N. Quarter micron KrF excimer laser lithography. *IEICE Trans. Electron.* **E76-C**, 582 (1993).
86. Bokor, J., Neureuther, A. R., and Oldham, W. G. Advanced lithography for ULSI. *Semiconductor International*, January 1996, p. 11.

8. CHEMICAL ANALYSIS BY LASER ABLATION

Richard E. Russo and Xianglei Mao

Lawrence Berkeley National Laboratory
Berkeley, California

Introduction

Chemical analysis is an important decree for our world. In forensics, health, environmental, nonproliferation, and many other disciplines, the chemical composition of a sample can be consequential. For the semiconductor industry, a trace amount of impurity may cost millions of dollars in lost processing. The billion-dollar effort to remediate contaminated environmental sites requires venerable chemical analysis. For nonproliferation, chemical analysis may foretell covert development of nuclear or chemical weapons. In health, chemical analysis can save lives. This chapter discusses the use of laser ablation for chemical analysis. Emphasis is placed on direct solid-phase analysis; when a pulsed laser beam is focused onto any solid sample, a portion of the material is instantaneously transformed into vapor phase constituents. Chemical analysis can be performed directly within the luminous laser-induced plasma at the sample surface, or by transporting the ablated mass to a secondary excitation source for analysis. The goal is to break (ablate) the sample into a fine vapor with a stoichiometric representation of the solid phase.

8.1 Direct Solid-Phase Chemical Analysis

Direct solid-phase chemical analysis is a challenge [1, 2]. Most analytical techniques are based on solution samples; a portion of the solid must be physically removed and dissolved, generally in acid. The risk of introducing contaminants or losing volatile components exists, personnel may be exposed to hazardous samples, and hazardous wastes may be generated. Some of the technologies investigated for direct solid-phase chemical analysis include: slurry nebulization [3]; arc and spark sampling [4]; electro-thermal vaporization [5-8]; direct insertion into an analytical source using graphite cups [9-11]; exploding films [12]; and reduced pressure rf discharges [13-16].

Laser ablation provides myriad capabilities for direct solid-phase chemical analysis. Photon-initiated processes govern sample removal; therefore, any sample (conducting, nonconducting, organic, inorganic, biological, radioactive, opaque, transparent, and so on) can be placed in the path of the laser beam, which converts a portion of the solid sample into the vapor phase. There is no sample size requirement and no sample preparation. Chemical analysis using laser ablation requires a smaller amount of sample (<micrograms) than that required for solution nebulization (milligrams). Depending on the analytical detection system, picogram to femtogram quantities may be sufficient for analysis, providing a pseudo "non-destructive" analytical technology. Finally, a focused laser beam permits spatial characterization of heterogeneities in solid samples. Because of these capabilities, laser ablation has received a great deal of attention; numerous analytical studies are described and reviewed in the literature [17–30]. Specific applications are not emphasized in this chapter. Instead, this chapter addresses laser ablation processes and their influence on chemical analysis. Ablation processes include: evaporation; ejection of atoms, ions, molecular species, and fragments; hydrodynamic expulsion; shock waves; plasma initiation and expansion; plasma-solid interactions; and a hybrid of these and other unknown processes [22, 31–36]. These processes influence the amount and composition of ablated mass and must be understood for accurate and sensitive chemical analysis.

8.2 Laser-Induced Plasmas

The direct approach to chemical analysis using laser ablation involves monitoring the ablated mass within the laser-induced plasma (LIP). Optical emission, absorption, fluorescence, and resonance ionization spectroscopy have been used with laser-induced plasmas [19, 22, 32, 37–46]. These spectroscopic technologies have been used since the discovery of laser plasmas, not only for chemical analysis, but also for diagnostics of laser ablation processes [47–59]. By understanding the characteristics of the plasma and the influence of experimental conditions, analytical capabilities can be optimized while minimizing limitations.

8.2.1 Laser-Induced Plasma Emission

Laser-induced plasma emission consists of atomic and ionic spectral lines superimposed on a broad-band continuum. The continuum emission results from electron-ion recombination, and it can be enhanced by laser beam absorption during collisions among atoms, ions, electrons, and the gas species [22, 31, 34, 35, 60, 61]. For multicomponent samples, atomic and ionic emission lines exist simultaneously for the constituent species. Identification of the spectral lines and their intensities provides qualitative and quantitative characterization of the sample, respectively. Ablation of a pure metal sample is presented here to demonstrate LIP characteristics. A repetitive pulsed laser beam is focused onto a solid sample in a quartz "ablation" chamber and emission from the plasma is imaged onto a monochromator with a CCD (charged-coupled device) detector (Fig. 1). For homogeneous samples, reproducibility in the amount of ablated mass can be poor with single pulsed ablation due to variations of the surface chemical and physical state, instability in the power and spatial mode qualities of the laser, and instabilities in the nonlinear laser material interaction. Precision is improved significantly with repetitively pulsed laser ablation because the ablation depends on a laser-conditioned surface, and analysis is averaged from repetitive ablation events. Improved precision provides enhanced analytical sensitivity and allows fundamental studies of ablation processes.

Laser-induced plasma emission from a copper sample ablated using a power density of 4.93 GW/cm^2 is shown in Fig. 2 [62]. Emission intensity was integrated for 5 seconds during repetitive laser ablation at a fixed location on the

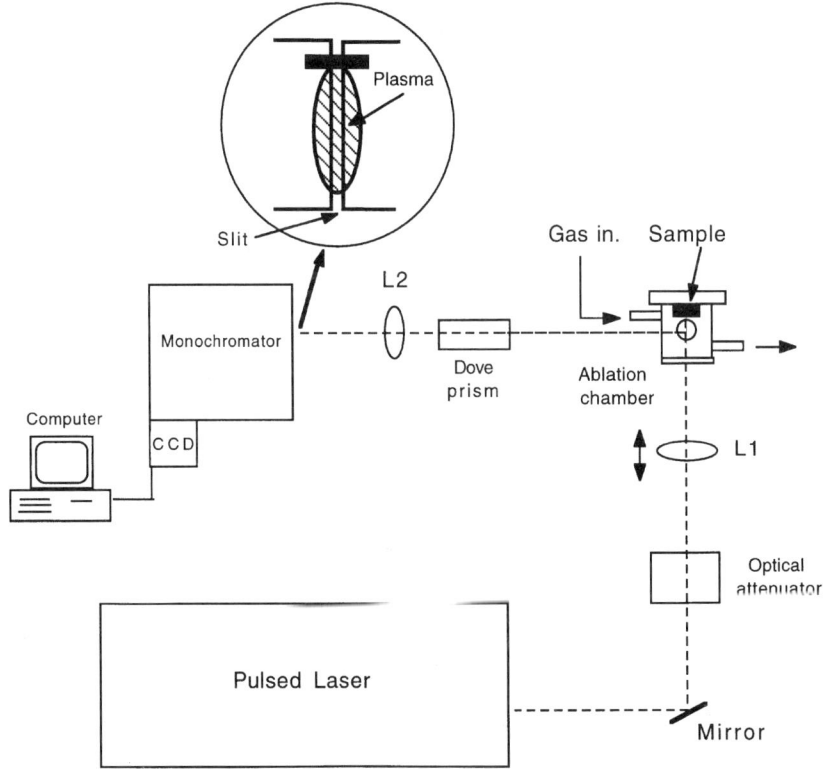

FIG. 1. Experimental diagram for initiating a laser-induced plasma and measuring optical emission spectroscopy. Inset shows the region of the plasma that is imaged onto the monochromator.

Cu sample surface. Cu atomic emission lines at 510.55, 515.32, 521.82, and 529.25 nm are observed in this spectral region. Atomic emission intensity is strong compared to ionic emission intensity because of its short lifetimes in the plasma and the time-integrated measurement technique. Cu ionic emission has been reported to exist for approximately 200 ns after the laser pulse, whereas Cu atomic emission can last for several hundred μs [43, 63]. Both integrated and time-resolved measurements have been used for chemical analysis in laser-induced plasmas [19, 22, 32, 37–46]. For multicomponent samples, the plasma's emission is characteristic of the atomic and ionic spectral lines of the sample constituents and the time resolution of the measurement technique. Identification of these lines allows qualitative chemical analysis of unknown samples, even at remote locations from the laser [32, 64–66].

Quantitative analysis is somewhat more difficult; spectral emission intensity in the LIP is not only influenced by the elemental concentration in the sample, but also by the laser properties (energy, power density, wavelength). The quantity of mass ablated does not scale linearly with power density (discussed in detail in the ICP section). The geometrical shape of the plasma and spatial emission intensity profile are strongly dependent on the laser power density. For

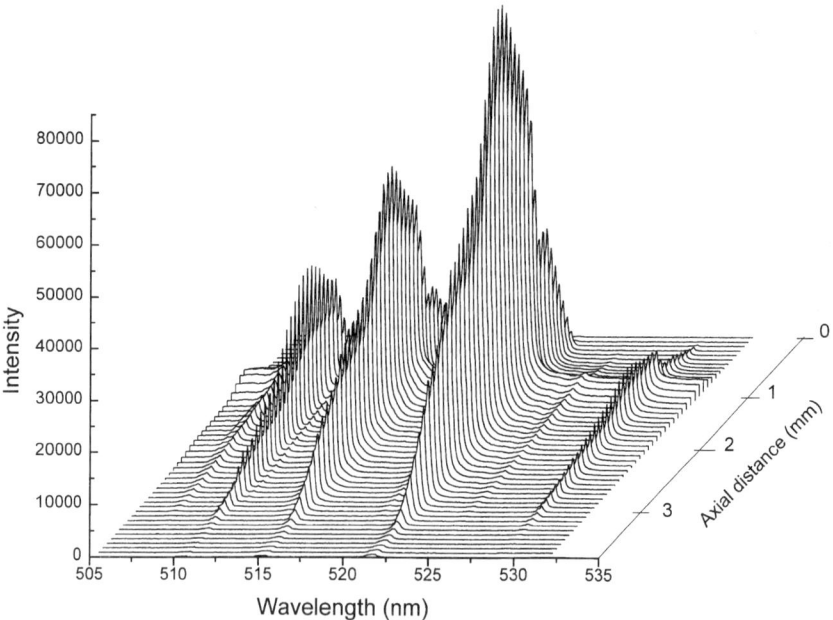

FIG. 2. Spectral emission measured from the laser-induced plasma during ablation of a copper sample at a power density of 4.93 GW/cm^2. Emission intensity was integrated for 5 seconds during repetitive laser ablation at a fixed location on the Cu sample surface.

the 4.93 GW/cm^2 laser power density used in Fig. 2, copper emission extends approximately 3 mm above the sample surface, with the peak emission intensity at approximately 1.5 mm. The dependence of spatial extent can be seen by measuring the 521.82 nm Cu emission intensity at laser power densities from 0.4 to 5 GW/cm^2 (Fig. 3). The maximum emission intensity and its spatial position change with power density [62]. If the analytical imaging system is based on fixed spatial coordinates, changes in the plasma emission extent will be manifested as an erroneous change in concentration of the element, unless the observation region encompasses and spatially integrates the entire geometrical extent of the plasma for all power densities. For experimental conditions in which these condition are satisfied, chemical analysis has been successful in providing qualitative and semi-quantitative analysis on a routine basis.

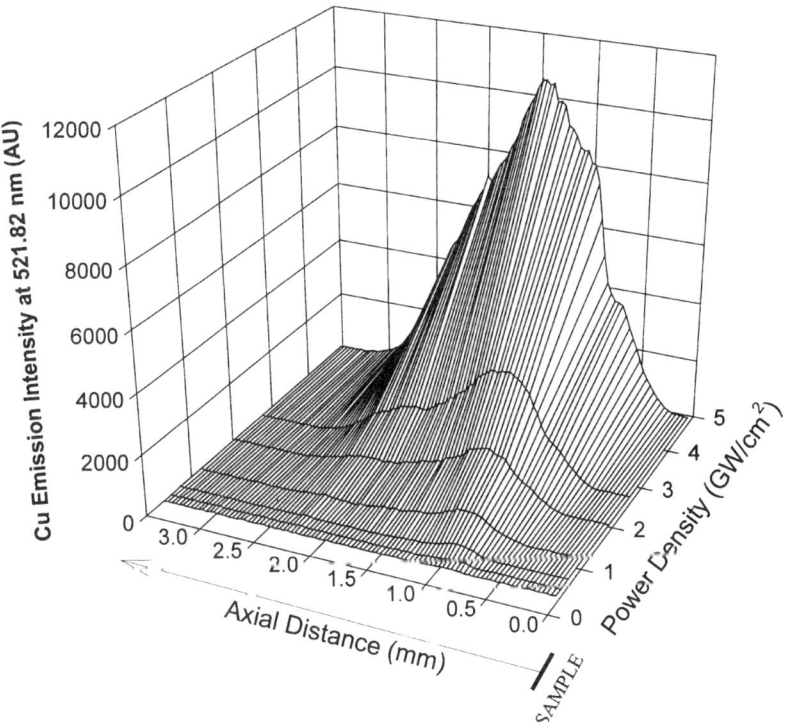

FIG. 3. Spatial Cu emission intensity at 521.82 nm versus laser power densities from 0.4 to 5.0 GW/cm^2. Emission intensity was integrated for 5 seconds during repetitive laser ablation at a fixed location on the Cu sample surface.

8.2.2 Laser-Induced Plasma Temperature

The dissociation and excitation of the ablated mass and, therefore, spectral emission intensity are dependent on the plasma temperature. Assuming Boltzmann conditions in the plasma, the ratio of spectral emission intensities can be used as a measurement of the excitation temperature. Excitation temperatures calculated from the measured 521.82 nm and 510.55 nm atomic Cu emission lines at a power density of 0.5 GW/cm^2 and 4.93 GW/cm^2 are shown in Figs. 4 and 5, respectively [62]. The temperature is approximately 10,000K or less, depending on the power density and spatial position in the plasma. These temperature values are consistent with previous studies, although the actual temperature is significantly influenced by laser power density, as can be seen in Figs. 4 and 5. Higher power density may be beneficial for increased analytical sensitivity as long as self-absorption and temperature broadening do not interfere with the spectral analysis.

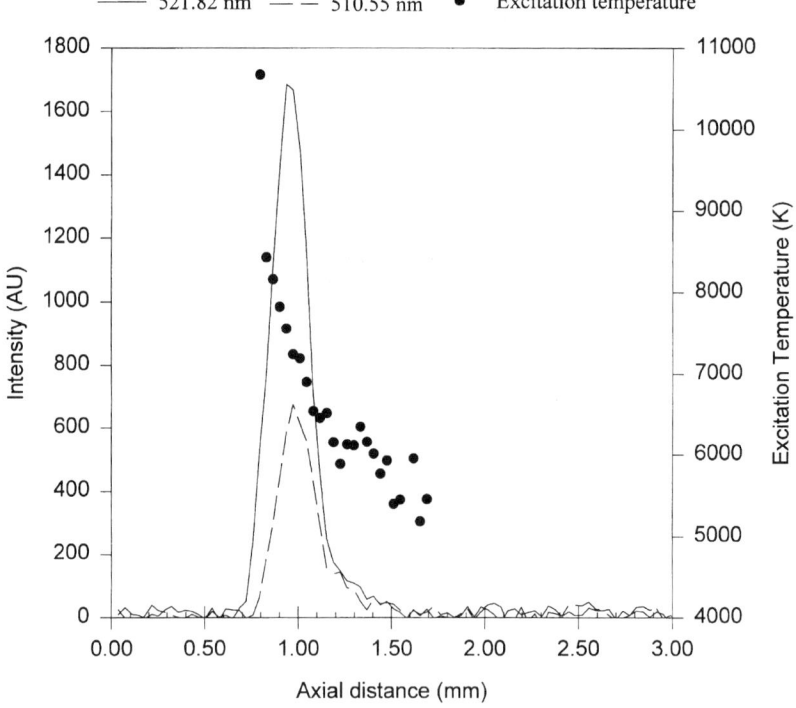

FIG. 4. Measure 521.82 nm and 510.55 nm atomic Cu emission line intensities and calculated (Boltzman) excitation temperature in the laser-induced plasma at a power density of 0.5 GW/cm^2.

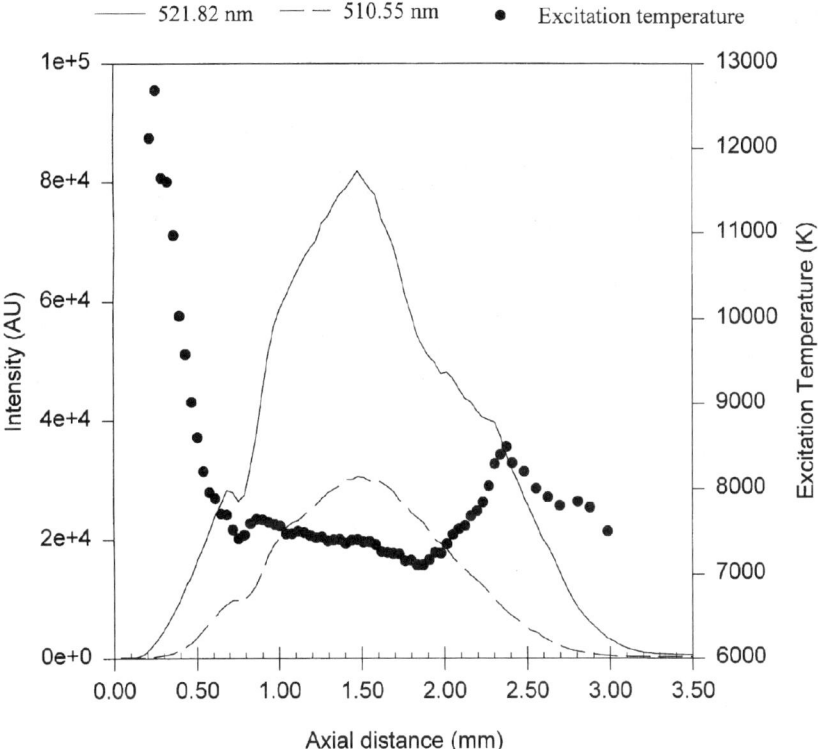

FIG. 5. Measured 521.82 nm and 510.55 nm atomic Cu emission line intensities and calculated (Boltzman) excitation temperature in the laser-induced plasma at a power density of 4.93 GW/cm^2.

Unique spatial behavior is observed in these figures; the maximum temperature does not coincide with the maximum spectral emission intensity and the offset increases with laser power density. Plasma excitation and recombination mechanisms influence these spatial emission intensity measurements. However, these mechanisms alone cannot accurately explain the offset nor the temperature increase at the distal end of the plasma, especially for the high power density case (cf. Fig. 5). These data could be a manifestation of plasma expansion, stagnation, and reflection processes occurring during time-integrated measurements. A shock-wave model for plasma expansion is presented here to demonstrate how spatial emission intensity can be affected by time-integrated measurements, and how these integrated emission line intensities can affect excitation temperature calculations.

8.2.3 Plasma Emission During Expansion

Plasma expansion along its central axis is considered to be one-dimensional for this illustration of time integration on measured emission intensities and calculated temperature profiles. The integrated measured intensity I of emission line i at position x is:

$$I_i(x) = \int_{t_0}^{t_1} e_i(x, t)\, dt \tag{8.1}$$

where $e_i(x, t)$ is emissivity of line i, and t_0 to t_1 is the integration period. The emissivity is giving by [68]:

$$e_i(x, t) = \left(\frac{hc}{4\pi}\right) \frac{A_i g_i}{\lambda_i Q} \eta(x, t) e^{-E_i/(k_B T(x,t))} \tag{8.2}$$

where A_i is the Einstein transition probability for spontaneous emission, E_i is the energy of the transition, g_i is the statistical weight of level E_i, T is the excitation temperature, λ_i is the wavelength, η is the emitting atom number density, h is Planck's constant, k_B is Boltzmann's constant, c is the speed of light, and Q is the internal partition function. For a one-dimensional expansion, the width of the plasma is assumed to be 2d as it moves along the x-axis during time-integrated measurements. At time t, the temperature and number density inside the plasma are assumed to be homogeneous. To calculate $I_i(x, t)$, relationships between x and t must be known or assumed. When the plume expands as a function of time (t) following shock-wave behavior, the position (x) can be expressed as [49, 54, 68–71]:

$$x = k\left(\frac{E_l}{\rho_0}\right)^{1/5} t^m \tag{8.3}$$

where E_l is the laser energy coupled to the expanding material during the ablation process, and ρ_0 is the ambient gas density. The coefficient k is an experimentally derived constant that has units of $\{s^{0.4-m}\}$. The exponent m theoretically equals 0.4, but it varies from 0.4 to 0.6 in experiments [54]; m is chosen to be 0.5 for these calculations. The times when the leading and trailing edges of plasma reach position are t_0 and t_1, respectively, and can be found as a function of x from Eq. (8.3). The intensity of an atomic emission line at position x will be:

$$I(x) = \left(\frac{hc}{4\pi}\right)\left(\frac{A_i g_i}{\lambda_i Q}\right) \int_{t_0}^{t_1} \eta(x, t) e^{-E_i/k_B T(x,t)}\, dt \tag{8.4}$$

with

$$t_0 = \frac{(x-d)^2}{k^2(E_i/\rho_0)^{2/5}} \quad \text{and} \quad t_1 = \frac{(x+d)^2}{k^2(E_i/\rho_0)^{2/5}} \tag{8.5}$$

To integrate Eq. (8.4), functional relationships for T and η with x must be assumed. For demonstration, a monotonic profile is chosen to represent the decreasing temperature such that:

$$T = \frac{T_o}{(1 + (x/L)^2)} \tag{8.6}$$

where T_o is the initial plasma excitation temperature at the sample surface and L represents an effective thermal diffusion length ($L \propto [(t_1 - t_0)\alpha_{\text{diffusivity}}]^{1/2}$). Atomic emission intensity is an exponential function of temperature Eq. (8.2); by assuming temperature decreases and by neglecting heating due to a shock front, any increase in emission intensity with x must be due primarily to time integration of the atomic number density. The atomic number density is largest initially because the target atoms are constrained from expanding beyond the shock front. Therefore, the upper limit of η in the 2d thick plume would be the initial emitting atom density, η_0, and η would be constant in space. For this case, the time integrated intensity versus x from Eqs. (8.4)–(8.6).

$$I_i(x) = \left(\frac{hc}{2\pi}\right)\left(\frac{A_i g_i}{\lambda_i Q}\right)\left(\frac{k_B T_o}{E_i}\right) \frac{L^2 \eta_o}{k^2(E_i/\rho_0)^{2/5}} e^{-(E_i/k_B T_o)(1 + ((x^2 + d^2)/L^2))} \sinh\left(\frac{2 E_i x d}{k_B T_o L^2}\right) \tag{8.7}$$

Emission intensities for the 521.82 and 510.55 nm Cu lines were calculated as a function of position over the thermal diffusion parameter, x/L. The degeneracies and Einstein coefficients were taken from Reader and Corliss [72]. The calculated emission and temperature profiles are shown in Fig. 6 for an initial temperature of 7800 K (average temperature in Fig. 5). By using this upper bound of constant atom density, the measured peak emission intensity will always occur away from the surface, even though the temperature is highest at the surface and decreases with distance x. Cu intensity in the plasma is proportional to the density of atoms and exponentially proportional to temperature. Initially the plasma is moving fast, so the effective density of Cu atoms during the measurement time is small, even though the temperature is highest at this time. At later times, when the plasma velocity is slower, the integrated intensity can be greater, even though the temperature has decreased.

A lower limit on η was evaluated by assuming that the atomic density decreases monotonically during plasma expansion [62], such that

$$\eta = \frac{\eta_o}{1 + x/x_o} \tag{8.9}$$

where x_o is the total distance that the plume expands during the integration. Although the detailed mathematical expressions are not presented here, the lower bound solution to the shock-wave model shows that the peak emission intensity from time-integrated measurements occurs later than the peak temperature of the plasma, in all cases. The position of the maximum intensity

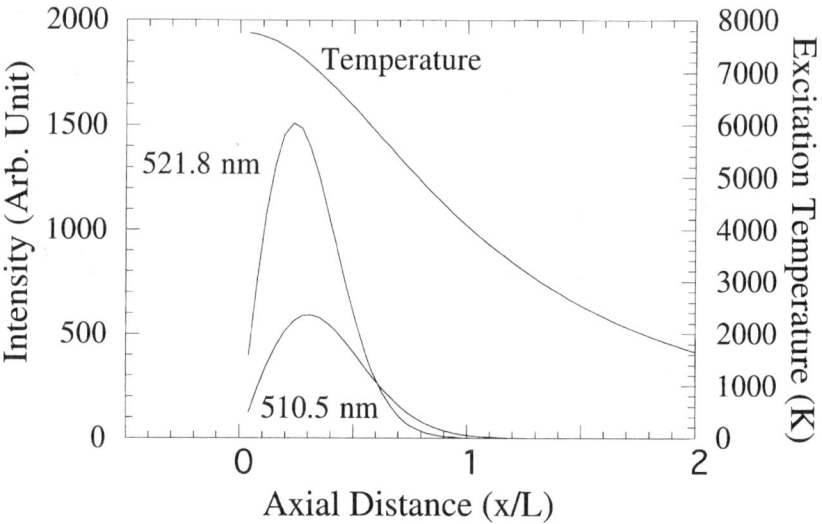

FIG. 6. Emission intensities for the 521.82 and 510.55 nm Cu lines and temperature calculated as a function of position over the thermal diffusion parameter, x/L. The calculated emission and temperature profiles are shown for an initial temperature of 7800 K (average temperature in Fig. 5).

occurs closer to the target surface for higher versus lower transition energies, as observed experimentally for the 521.8 nm and 510.5 nm Cu lines. Also, as the laser power increases, raising the initial temperature of the plasma, the location of the peak emission intensity moves away from the target surface. The effect of integration time on the measured spectral emission line intensity depends on the expansion velocity of the plasma. Even for fast time-resolved measurements, finite integration during plasma expansion can exist. By better understanding plasma behavior in time, suitable measurement technologies can be developed for sensitive and accurate chemical analysis in laser-induced plasmas.

8.3 Laser Ablation—Inductively Coupled Plasma

An alternate method for chemical analysis using laser ablation involves the transport of the ablated mass to a secondary excitation source. High-voltage sparks [22, 73], microwave plasmas [74–76], direct current plasmas [24, 77], graphite furnaces [25], and the inductively coupled plasma (ICP) have been used as excitation sources with laser ablation. The ICP is the most prevalent excitation source for chemical analysis using laser ablation at this time [17–23, 26–30, 78–102]. Argon gas is introduced into a three-concentric tube torch. A copper coil around the torch delivers rf energy to the gas, and by seeding the gas with electrons/ions from a spark, the plasma is sustained by the continuous rf power. The center tube is used to carry gas-entrained sample vapor into the ICP where it is heated and excited to optical emission. The ICP is the preferred excitation source for laser ablation chemical analysis because of its high temperature, relative freedom from matrix effects, and ability to dissociate and excite refractory samples, a potential limitation with cooler excitation sources or during analysis in the laser-induced plasma. Chemical analysis in the ICP is achieved by using atomic emission spectroscopy (ICP-AES) or by directing this inductively coupled plasma into a mass spectrometer (ICP-MS). ICP-MS provides enhanced sensitivity and isotopic analysis compared to ICP-AES. However, laser ablation processes are germane to both measurement technologies.

8.3.1 ICP Spectral Response to Laser Ablation

For chemical analysis, the elemental composition and concentration are the important parameters. Similar to the LIP, elemental composition is easily determined by identification of the spectral emission lines within the ICP. The intensity of the emission lines is an indication of the quantity of mass ablated and the concentration of constituents in that mass. For quantitative measurements, chemical analysis based on spectral emission relies on standards (samples with known compositions and similar physical properties, but slightly different concentrations) to calibrate intensity measurements. A series of standards is analyzed, and a graph of intensity versus concentration is plotted. Ideally, spectral emission intensity will respond linearly with changes in concentration and mass. When the unknown sample is analyzed, its emission intensity is related to the standard plot to determine concentration. For laser ablation chemical analysis, standards need to be close replicas of the unknown because sample properties can significantly influence the amount of ablated mass. In some cases, component and concentration differences in the sample may be small, and mass ablation will be similar to the standards. For some applications, calibration curves can be established for routine quantitative chemical analysis of "expected" samples. However, for unknown samples, the mass ablation processes need to be understood because standards are not available or impractical to fabricate (e.g., environmental mixed-waste or radioactive samples).

8.3.2 ICP Emission Versus Laser Power Density

The laser power density is the most critical parameter in defining the quantity of mass ablated. For accurate and sensitive chemical analysis, the ablated amount should be known and reproducible from pulse to pulse. However, mass ablation is not linearly related to laser power density. Zn emission intensity in

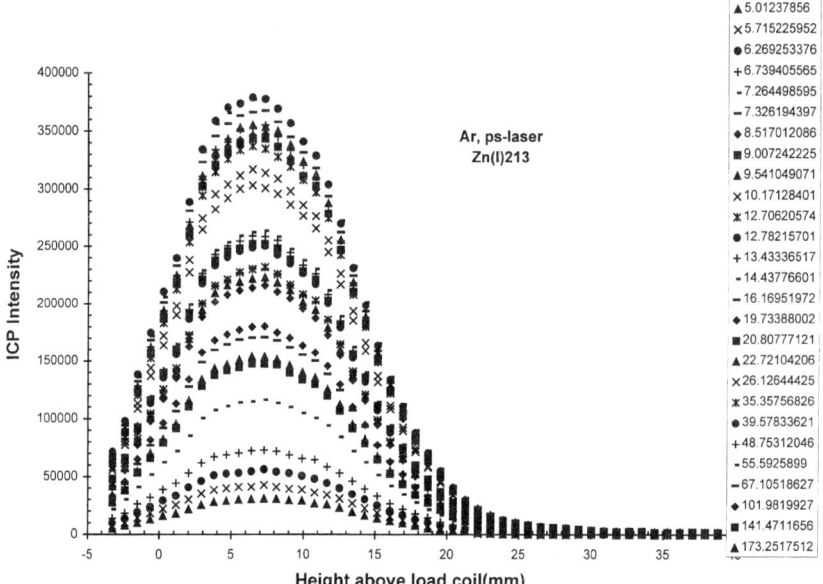

FIG. 7. Zinc emission intensity in the ICP as a function of laser power density during ablation of a brass sample. Each profile was measured using a 5-second integration time during repetitive laser ablation using 35-picosecond laser pulses from a Nd:YAG laser ($\lambda = 1064$ nm).

the ICP was monitored during laser ablation of a brass sample to demonstrate the influence of laser power density on these measurements (Fig. 7) [78]. The charged-coupled device detector with a monochromator was used to observe emission along the central channel of the ICP (height above the rf coil) (Fig. 8). Each profile was measured using a 5-second integration time during repetitive laser ablation using 35-picosecond laser pulses from a Nd:YAG laser ($\lambda = 1064$ nm). The spectral emission intensity profile is governed by the ICP temperature and excitation characteristics, the transport velocity of the ablated mass through the plasma, and time integration of the detection system. As the laser power density increases, the intensity increases due to a larger quantity of ablated mass from this homogeneous sample. However, a plot of peak emission intensity versus laser power density shows the nonlinear mass ablation behavior; mass ablation plateaus and decreases at higher power densities (Fig. 9(a)). Chemical analysis is influenced by this nonlinear behavior. Specifically, the absolute mass ablated per laser pulse can vary significantly because power density changes as a crater develops or for analysis over an irregular or rough sample surface. Understanding this nonlinear mass ablation behavior is important to achieving good accuracy and sensitivity for chemical analysis.

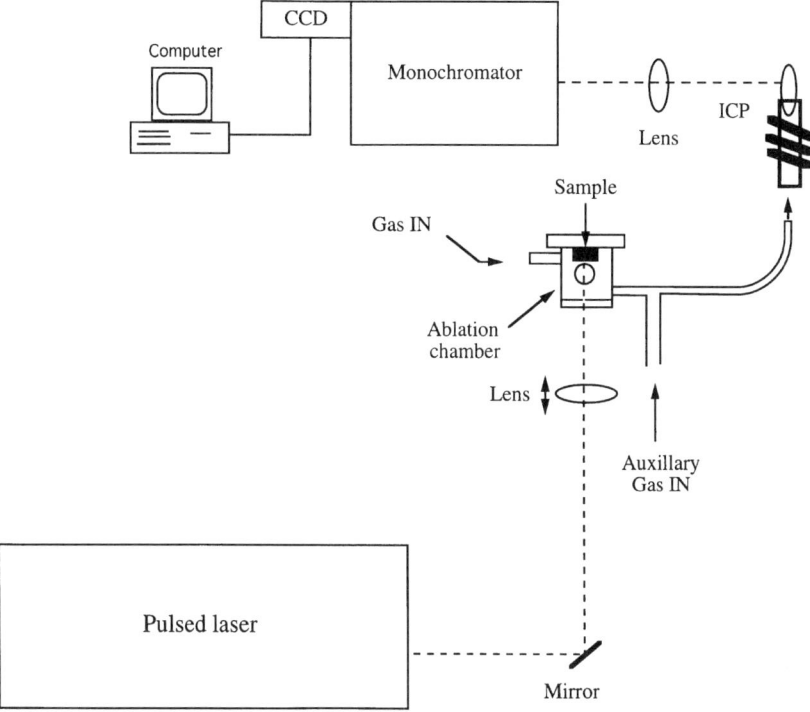

FIG. 8. Experimental system used to measure spectral emission in the inductively coupled plasma (ICP). The charged-coupled device (CCD) detector with a monochromator was used to observe emission along the central channel of the ICP (height above rf coil). Gases can be introduced into the ICP via the ablation chamber or an auxiliary inlet.

8.3.3 Mass Ablation Rate

The mass ablation behavior can be explained partially by the reduced laser beam area on the sample as the power density was increased in the previous studies; the mass ablation rate can be found by normalizing these data to laser beam area (Fig. 9(b)). Emission intensity I normalized to incident laser beam area A is related to the mass flux from the sample and exhibits exponential dependence with laser power density ϕ: $I/A \approx \phi^n$. This exponential dependence has been measured from diverse samples, when using nanosecond and picosecond lasers with ultraviolet (UV), visible (Vis), and infrared (IR) wavelengths, in different gas atmospheres, and for two cases: when the laser energy is increased with fixed spot sizes and when the laser energy is constant and the spot size is reduced [17, 62, 79, 80, 89, 93]. In all these experiments, two distinct mass ablation rates were measured versus ϕ. For UV nanosecond pulsed laser ablation, when ϕ is less

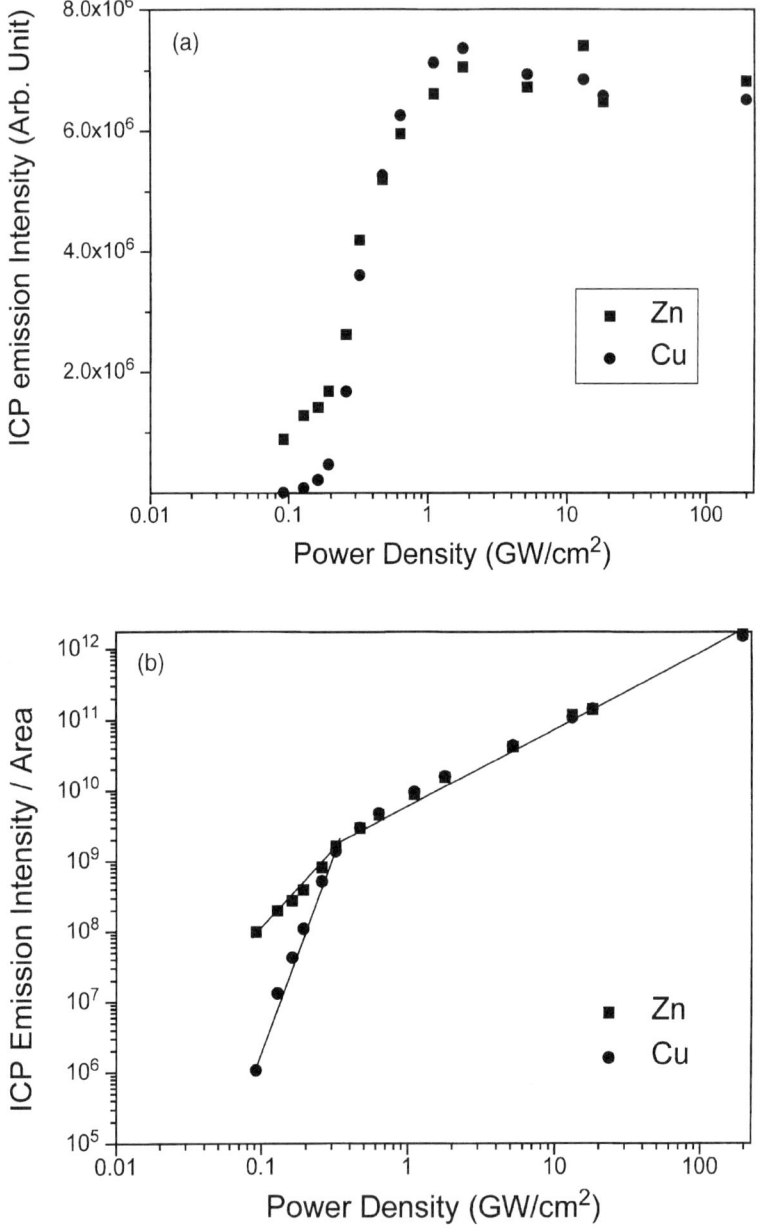

FIG. 9. Spectral emission intensity in the ICP for Zn and Cu during repetitive ablation of brass versus laser power density (a), and the mass ablation rate behavior when these data are normalized to laser beam area (b).

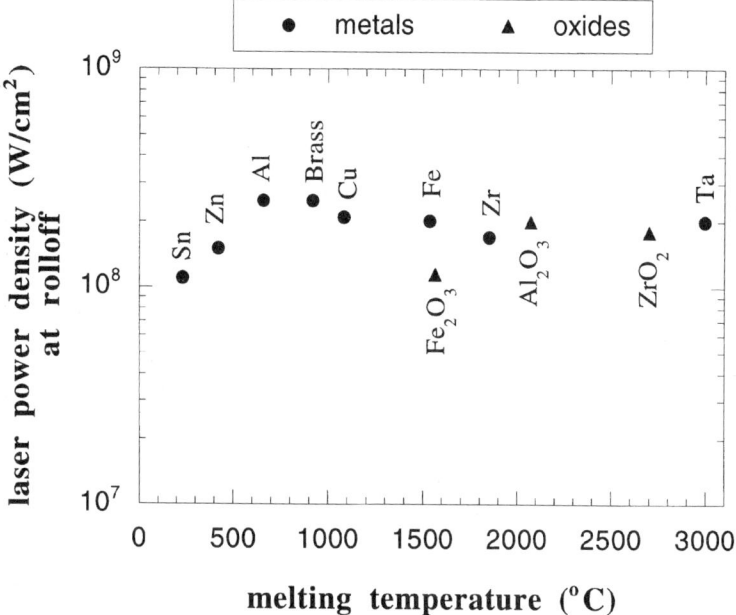

FIG. 10. The power density at which the mass ablation rate changes (roll-off) versus melting temperatures for various samples when using UV nanosecond laser ablation.

than approximately 3×10^8 W/cm^2, the coefficient n ≳ 2. At laser intensities above 3×10^8 W/cm^2, the mass ablation rate rolls off to $n \lesssim 1$. Similar behavior was measured for nanosecond laser ablation from brass (Zn and Cu), Al, Al$_2$O$_3$, Fe, Fe$_2$O$_3$, Ta, Zn, Zr, ZrO$_2$, Sn, glass, and high temperature superconducting samples [62, 99–102]. For these samples, the power density at which the rate changes (roll-off) versus melting temperatures is shown in Fig. 10. Regardless of the sample's electronic, chemical, physical, or thermal properties, the laser power density at roll-off occurred between 0.1 to 0.3 GW/cm^2.

The fundamental processes describing this exponential dependence and roll-off are not well understood. The slopes (rates) represent the average mass ablated from nonlinear processes occurring during and after the laser-pulse interaction with the sample, including multiphoton absorption, photo-ionization, thermionic emission, shock-wave and stress power, spallation, hydrodynamic expulsion, plasma formation and expansion, and so on. For pure vaporization without a pressure change, the slopes would be unity. At this time, there is no obvious correlation of the slopes to a direct photon process, especially because they are similar for diverse elements in diverse sample matrices. Nonlinear slopes are observed using nanosecond and picosecond pulses, UV, Vis, and IR, wavelengths, and for alloys as well as ceramic samples. Elucidation of these nonlinear mechanisms is an ongoing investigation and is essential to the success of laser ablation for chemical analysis and many other applications.

8.3.4 Ablation Rate Roll-off

The change in slope or rate represents a decrease in the efficiency of laser energy used for mass ablation. Several mechanisms have been proposed to describe the decrease in mass ablation rate vs. ϕ, with primary emphasis on plasma shielding via inverse bremsstrahlung [17, 35, 61, 103–105]. The weak dependence of roll-off on melting temperature (cf. Fig. 10) is consistent with plasma shielding of a nanosecond pulsed laser beam through inverse bremsstrahlung: absorption of laser photons during collisions among ablated sample atoms and ions. As the amount of ablated mass increases, inverse bremsstrahlung acts to effectively truncate the laser pulse duration. The threshold for nanosecond inverse bremsstrahlung above solid targets in noble gases is approximately 10^8 W/cm^2, which is consistent with the measured values in this work [103]. A preliminary thermal-based model with inverse bremsstrahlung plasma absorption predicts a roll-off in the vaporization of Zn and Cu from brass at approximately 3×10^8 W/cm^2 power density [106].

The model also predicts that laser transmission to the sample surface begins to attenuate once the power density exceeds approximately 0.3 GW/cm^2. Recent experiments monitored the laser beam through a micron-sized hole in the sample and found that the transmitted pulse duration decreased as a function of power density, indicating laser energy shielding by the plasma [107, 108]. Roll-off in mass ablation observed for picosecond laser pulses also can be described by inverse bremsstrahlung processes, but due primarily to collisions of fast electrons with ambient gas species [109]. It is necessary to verify that this mass ablation rate behavior is due to a change in the laser sample interaction and not an artifact of the ICP-AES sampling process. Once verified, procedures are necessary to enhance the efficiency of mass ablation for improved chemical analysis accuracy and sensitivity.

8.4 Correlation of ICP Emission to Ablated Mass

A known quantity of mass must be ablated from the sample so that measured changes in emission intensity accurately represent concentration changes. In addition, a reproducible quantity of mass must be ablated to ensure good sensitivity. A normalization or real-time monitor of the ablated mass is, therefore, required to standardize chemical analysis based on laser ablation. To compensate for the nonlinear and nonreproducible mass ablation behavior, several techniques have been demonstrated to independently monitor the ablated mass, thereby improving accuracy and sensitivity. Such techniques include light scattering [110] a combination of solution nebulization and laser ablation [19, 111, 112] using a mass monitor to collect a portion of the vapor [112] and to measure spectral emission intensity in the ICP and LIP simultaneously [113]. The absolute method to monitor mass loss is to weigh the sample before and after ablation. A sensitive microbalance was used in-situ to measure the mass loss from gold films during laser ablation [114], and these data showed a roll-off of mass ablated at approximately 0.3 GW/cm^2.

The volume of the crater formed after repetitive laser ablation also provides a direct measure of the mass removed, and crater volume measurements show the same mass ablation rate behavior and roll-off as observed for the ICP emission intensities [79]. These direct mass measurements confirm that the nonlinear rate behavior and roll-off are due to laser ablation processes and not an artifact of the ICP-AES technology. However, these measurement technique are tedious and may not be accurate because of the small quantity (<μg) of mass ablated for each laser pulse; direct mass measurements are not recommended for routine mass compensation.

8.4.1 Acoustic Emissions

Acoustic emissions in the sample [79, 115–117] or in the ambient medium [115–118] can be measured with stress transducers and used to compensate for changes in ablated mass. The shock wave formed in the ambient medium and the momentum recoil or stress power in the sample from this high velocity shock are related to the quantity of mass ablated. An experimental setup for monitoring shock pressure and stress power is shown in Fig. 11. A piezoelectric transducer is attached to the back of the sample to detect bulk longitudinal and shear waves. A microphone transducer detects the shock wave propagating through the ambient medium. Figure 12 shows the ratio of the shock pressure to stress power in an aluminum sample as a function of power density. At approximately 0.2 GW/cm^2, the amount of laser energy coupled to the sample plateaus indicating a change in the efficiency of laser energy delivered to the sample. Acoustic emissions from the laser ablation processes can be used to calibrate mass ablation on a routine, real-time basis.

8.4.2 Laser-Induced Plasma and ICP Emissions

An independent external method to monitor the ablated mass involves simultaneously measuring the spectral emission intensity in the LIP and ICP [113]. Spatially and temporally integrated emission intensity from the LIP was measured to exhibit exponential dependence with power density, similar to that observed in the ICP. Figure 13 shows the mass ablation behavior simultaneously measured from Cu in the ICP and LIP as a function of power density; the inset

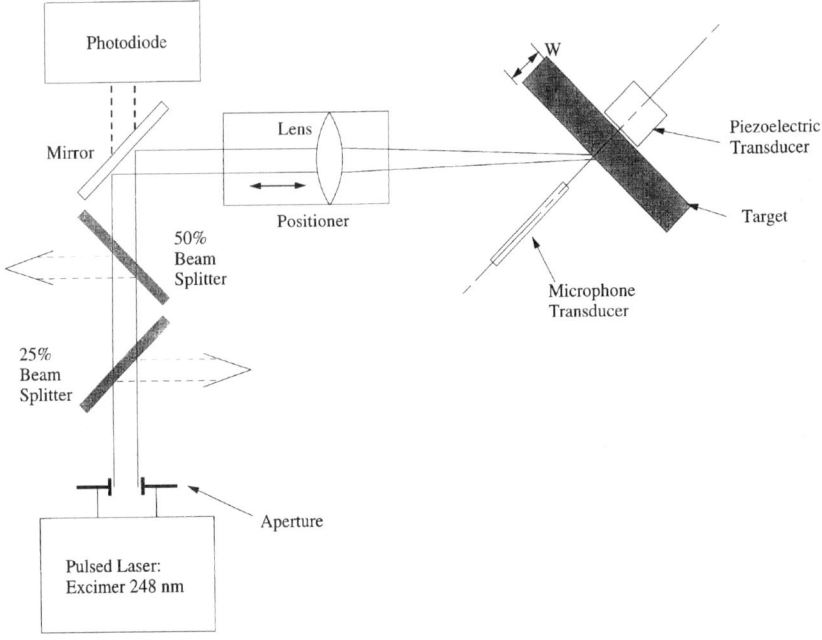

FIG. 11. Experimental system for monitoring shock pressure in the ambient medium above the sample surface and stress power in the sample. A piezoelectric transducer is attached to the back of the samples to detect bulk longitudinal and shear waves. A microphone transducer detects the shock wave propagating through the ambient medium.

shows the excellent correlation for these two sources. Although the absolute values of the exponents are not the same for the ICP and LIP because of temperature and excitation characteristics, both sources depend on the amount of ablated mass. Simultaneous spectral emission intensity measurements were performed using copper, brass, aluminum, zinc, and aluminum oxide samples. Ablation from these samples exhibited the two distinct mass ablation regions versus power density in the LIP and ICP, and a good correlation was observed when either the central channel or the entire LIP was imaged onto the monochromator. This dual emission monitoring approach can be used to compensate for mass ablation changes and improve precision in real-time [113].

8.4.3 Internal Standards

A relative measure of mass ablated can be achieved by using an internal standard, simultaneously measuring emission from the analyte and a common matrix element [18, 19, 80]. The spatial distribution of the internal standard

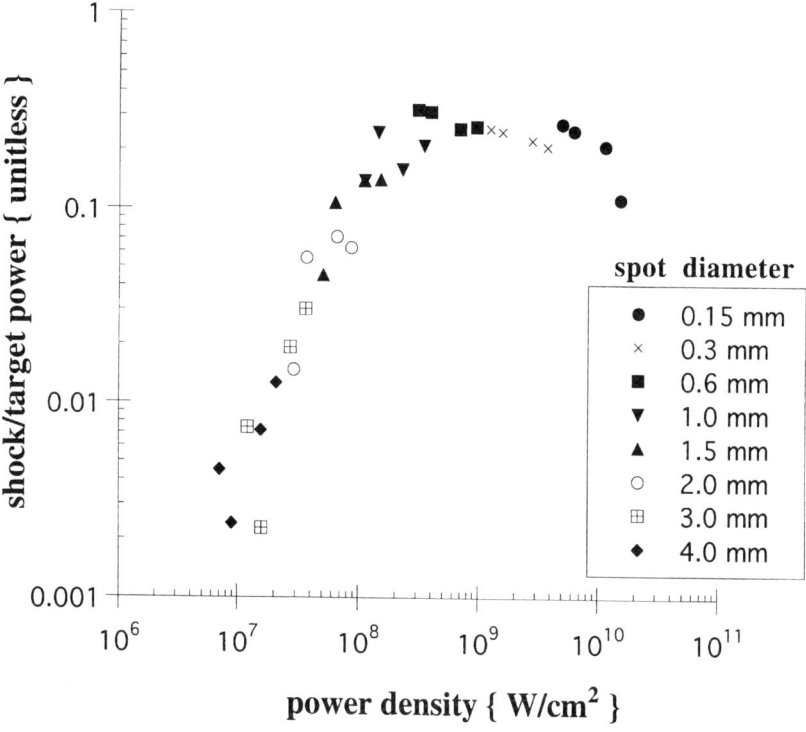

FIG. 12. Measured ratio of the shock pressure to stress power in an aluminum sample as a function of laser power density. At approximately 0.2 GW/cm^2, the amount of laser energy coupled to the sample plateaus indicating a change in the efficiency of laser energy delivered to the sample.

must be homogeneous, and the internal standard and analyte must be equally affected by the ablation. In general, it is not definite that the internal standard and the desired element will exhibit the same mass ablation rate behavior. For trace elements in some homogeneous samples, this approach has been used to improve measurement precision to better than 1% [80].

8.5 Ablation in Noble Gases

The preceding studies confirm that roll-off in the mass ablation rate is indigenous to the laser sample interaction. If the primary mechanism responsible for this change in efficiency is plasma shielding, it may be possible to minimize the plasma influence by varying the gas environment, i.e., changing the ionization potential of the ambient medium [17, 31, 35, 109, 113, 119–121]. To study

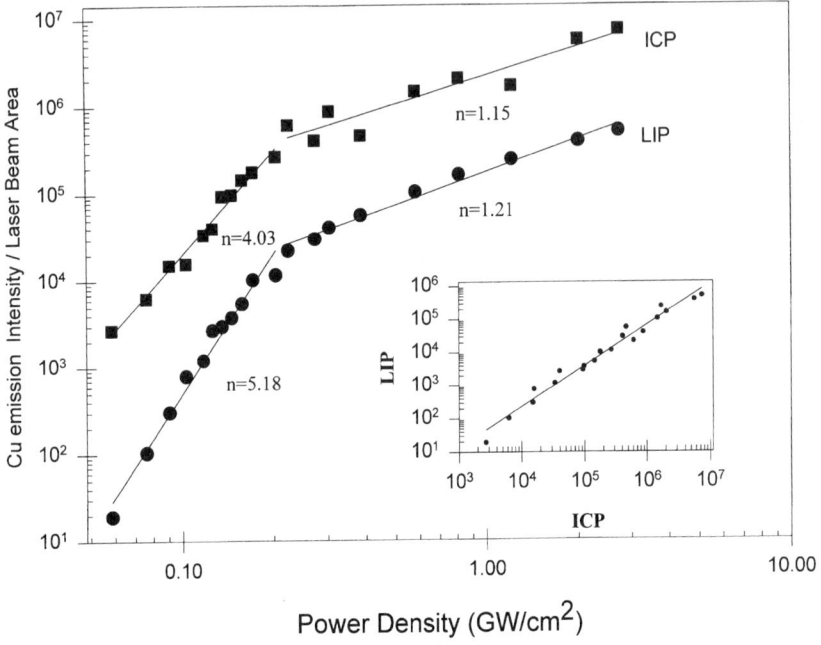

FIG. 13. Mass ablation rate behavior simultaneously measured from Cu in the inductively coupled plasma and laser-induced plasma as a function of power density; the inset shows the correlation for these two sources.

laser ablation in different gases, the total ICP gas composition must be equal; the ablation gas environment can be varied as long as another gas is introduced downstream from the chamber using an auxiliary gas inlet port. Therefore, the gas mixture from the T-connector to the ICP (cf. Fig. 8) was kept constant and the effect of gases on laser ablation processes is reported relative to argon. Spatial emission intensity profiles were measured and verified that the ICP remained constant when switching gases in the ablation chamber with the T-connector.

Laser ablation in Ar versus He, Ne, Kr, and Xe was studied by alternating these gases through the ablation chamber and into the T-connector [121]. The brass sample was ablated using UV nanosecond and picosecond laser pulses, and the effect of gas was found to be a function of power density, as shown in Figs. 14 and 15, respectively. For the nanosecond laser pulses (Fig. 14), Xe and Kr provided a lower mass ablation efficiency compared to an increase with Ne and He, relative to Ar. The enhancement or depression is power density dependent, with the greatest effect occurring in He at 0.25 GW/cm^2; a factor of approximately 2.5 enhancement in emission intensity. In support of the enhanced ICP-AES data, crater volumes were measured to be greater in the He

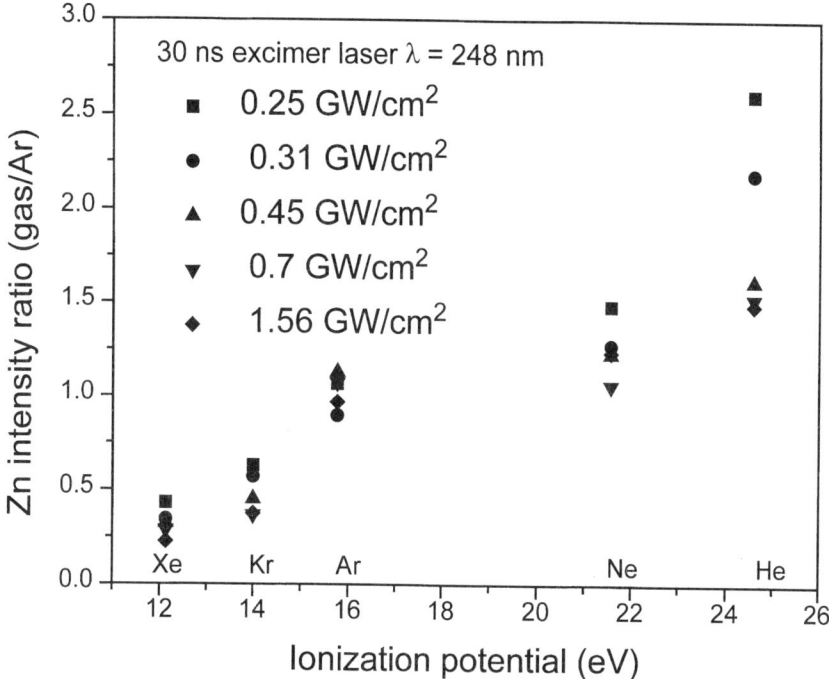

FIG. 14. Laser ablation in He, Ne, Kr, and Xe versus Ar. The brass sample was ablated using UV nanosecond laser pulses at several laser power densities.

atmosphere. The largest enhancement is measured using picosecond laser ablation in He, a factor of 6 increase in emission intensity. However, there is very little influence on the mass ablation behavior for the other gases.

The gas influence versus power density is different for the nanosecond and picosecond cases (see the data in Figs. 14 and 15). In the nanosecond case, the enhancement is reduced as the power density increases past the roll-off, whereas the enhancement is improved with increasing power density for the picosecond case. This different influence of gas atmosphere is due to the laser pulse duration because the photon energy is essentially the same. This difference in plasma shielding for picosecond and nanosecond laser ablation can be partially explained based on collisions among atoms, ions, and electrons in the atmosphere above the target surface. During the picosecond laser pulse, emitted atoms/ions travel only a few hundred Angstroms from the surface, assuming velocities on the order of 10^6 cm/s [122]. In contrast, high-energy (>100 eV) electrons are generated during picosecond interactions, and these electrons can acquire velocities on the order of 10^9 cm/s. On the picosecond timescale, these fast electrons travel several hundred microns during the laser pulse and undergo

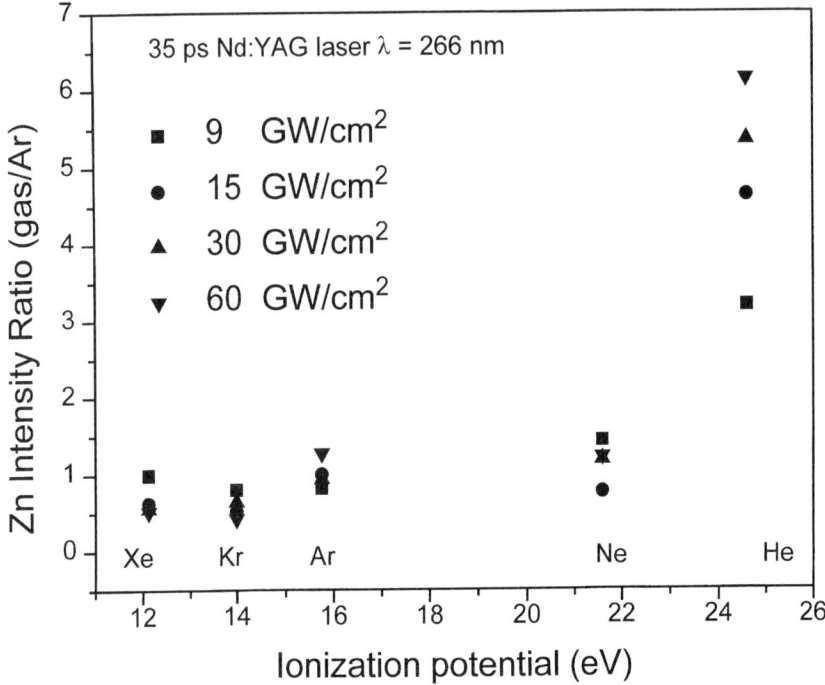

FIG. 15. Laser ablation in He, Ne, Kr, and Xe versus Ar. The brass sample was ablated using UV picosecond laser pulses at several laser power densities.

many more collisions with the *gas* atoms than the ejected atoms or ions [123]. The fast electrons absorb laser photons during collisions with the ambient gas atoms. Ar and He at pressures ranging from 10^{-5} torr to 1 atmosphere were used to demonstrate plasma shielding based on an inverse bremsstrahlung model involving fast electrons and the ambient gas atoms [17, 109]. The model predictions exhibited good agreement with measured changes in the ablated mass.

During the nanosecond laser pulse, the distance that atoms and ions travel from the sample surface is several hundred microns. Fast photoelectrons have not been observed for nanosecond laser ablation. The sample atoms/ions collide with each other as they expand into the gas, absorbing photons from the laser beam. Collisions of gas species with ejected atoms and ions are negligible during the nanosecond laser pulse. For fundamental elucidation of laser ablation processes, a complete understanding of gas effects requires in-depth studies of thermal properties, collisional cross-sections, as well as ionization potentials. Understanding these interactions and reducing plasma shielding will be

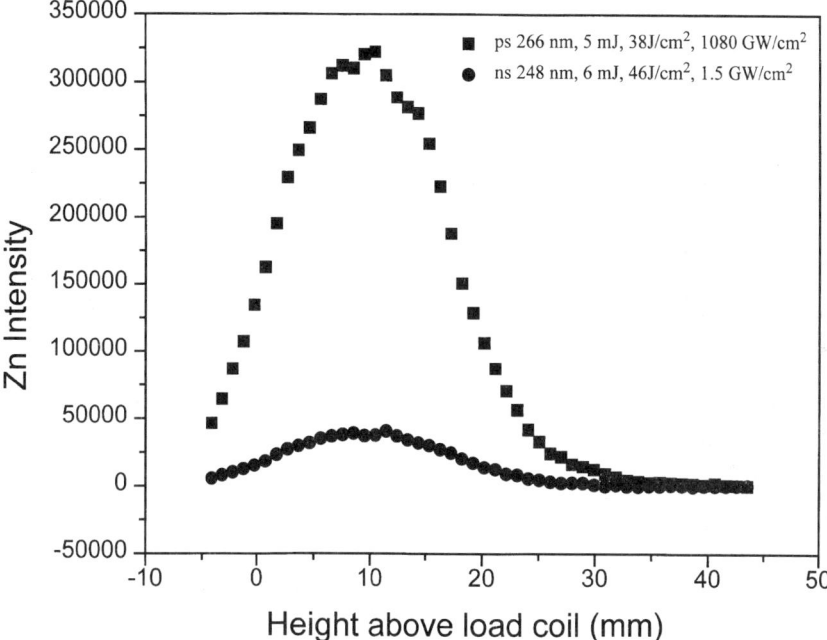

FIG. 16. Zinc emission spatial profiles in the inductively coupled plasma during ablation of brass. Mass ablation enhancement measured for picosecond versus nanosecond lasers with essentially the same energy, wavelength, and spot area.

beneficial to laser ablation for chemical analysis, by improving sensitivity (increased ablated mass) and minimizing preferential ablation from plasma heating.

An additional enhancement of the picosecond versus nanosecond measurements is seen in Fig. 16. Zn emission intensity is about a factor of ten greater than that obtained from the nanosecond laser. Both lasers have essentially the same energy, wavelength, and spot area. Crater volume measurements confirmed the increased mass removal for the picosecond case. Therefore, enhanced chemical analysis sensitivity is achieved with the picosecond laser by the more efficient coupling of laser energy to the sample, and used for mass ablation.

8.6 Quantitative Analysis

Accuracy is essential in chemical analysis; the ablated mass composition must represent the sample. Ideally, the entire laser-irradiated volume would be ablated by a spallation process, with no melting of adjacent sample regions; the ablated mass would be stoichiometric. Preferential ablation (mass removal based on thermal properties) can occur [17, 79, 89, 93, 106, 124, 125]. Direct laser heating and laser-induced plasma radiation can contribute to mass removal and preferential ablation. The brass sample is used to demonstrate preferential ablation. The composition of the brass sample is 65% copper and 35% zinc, whose boiling points are 2567 K and 907 K, respectively. The ratio of Zn to Cu emission in the ICP versus power density for the nanosecond pulsed laser is shown in Fig. 17. As power density increases, the Zn/Cu ratio reduces until it approaches that characteristic of the actual brass composition (calibrated ICP). Chemical analysis accuracy is best at approximately 0.2 GW/cm^2, the same power density at which the mass ablation rate rolls off for both Zn and Cu (cf. Fig. 9(b)). Preferential ablation occurs over the remainder of the power density range (see the inset in Fig. 17). This behavior may represent a balance between the direct thermal component from laser heating and that due to plasma

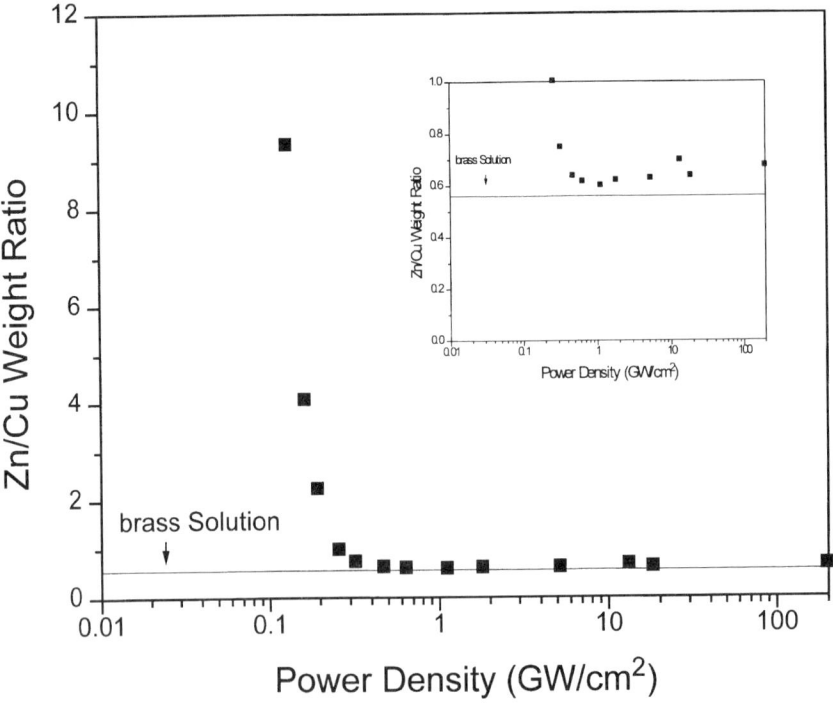

FIG. 17. The ratio of Zn to Cu emission in the inductively coupled plasma versus power density during nanosecond pulsed laser ablation of the brass sample. Chemical analysis accuracy is best at approximately 0.2 GW/cm^2, the same power density at which the mass ablation rate rolls-off for both Zn and Cu (cf. Fig. 9(b)).

radiation. With the picosecond laser pulses, the analysis is essentially accurate over the entire power density range, although slightly off in the lower power density region (Fig. 18). The picosecond laser provides better accuracy than the nanosecond laser without exhibiting a significant influence on power density over a wide range, throughout the two distinct mass ablation rate regions and roll-off.

A new technology based on laser ablation is pulsed laser deposition (PLD). A bulk sample is ablated and the ablated mass is collected onto a substrate as a thin film [126–129]. Chemical analysis can be used to characterize the ablated mass composition, but it cannot assure that the film composition will be similar to the bulk because of preferential "sticking" of the vapor species on the heated substrate. ICP-AES emission intensity was measured simultaneously from the elements in a BiSrCaCuO sample, during UV nanosecond and picosecond laser ablation [89]. The power density at the sample was 1.6×10^8 W/cm^2 (similar to

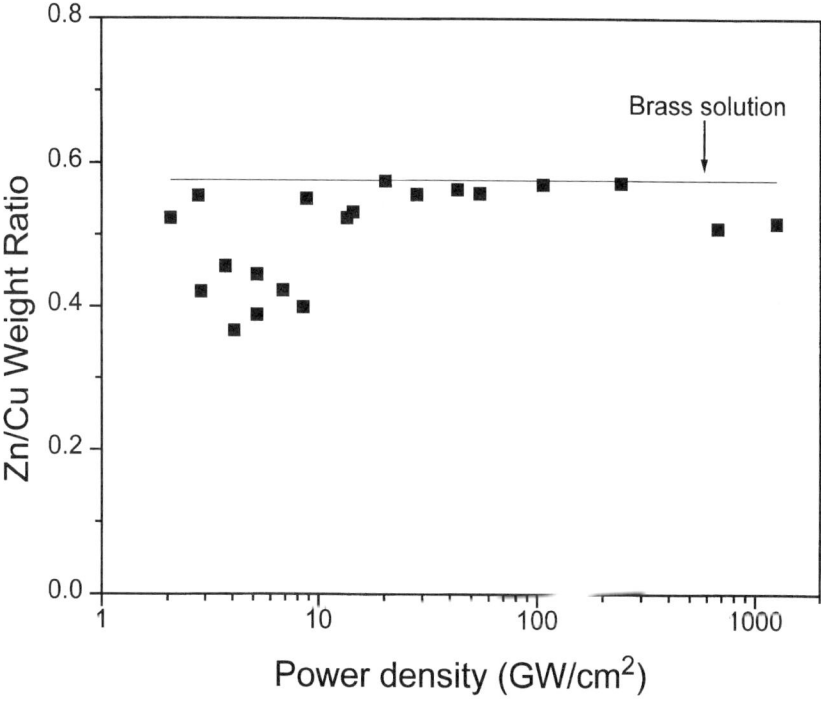

FIG. 18. The ratio of Zn to Cu emission in the inductively coupled plasma versus power density during picosecond pulsed laser ablation of the brass sample. The picosecond laser provides better accuracy than the nanosecond laser without exhibiting a significant influence on power density over a wide range; throughout the two distinct mass ablation rate regions and roll-off. Scale is equivalent to inset in Fig. 17.

TABLE I. Composition of Bi-Sr-Ca-Cu-O Sample and the Ablated Mass

	Bi, %	Ca, %	Cu, %	Sr, %
Nominal	28.6	23.8	31.8	15.8
Solution nebulization	29.3	20.7	33.9	16.2
Excimer laser	39.3	8.2	44.7	7.8
ps Nd:YAG laser	38.9	10.5	40.8	9.8
Collected vapor	41.7	14.4	37.0	6.9

that used in PLD) and 2.1×10^{10} W/cm^2 for the nanosecond and picosecond lasers, respectively. There is considerable difference in the measured composition of the ablated mass versus the bulk sample using both lasers; a reduction of Ca and Sr and an enhancement of Bi and Cu was measured from the ablated mass compared to the sample (Table I). Bi_2O_3 and CuO have much lower melting points than CaO and SrO, which indicates a significant thermal component to the ablation process. Even though the picosecond pulsed laser provided accurate chemical analysis of the brass alloy, there is still considerable preferential ablation for the low melting point components (Bi and Cu) in the ceramic oxide sample. A better understanding of laser ablation processes is necessary so that laser power density can be tailored to provide stoichiometric chemical analysis.

To ensure that the preferential ablation is not due to differential transport efficiency to the ICP, ablation of the BSCCO sample was performed in a sealed quartz chamber. After repetitive pulsing for 10 minutes, the chamber was coated with a layer of powder, which was washed off with nitric acid and chemically analyzed. The ablated mass showed the same enhancement of Bi and Cu and reduction of Ca and Sr as measured in the ICP (Table I). It is intriguing that preferential ablation can occur when a crater several hundred microns in depth is developed. A phenomenological process can explain these data. Droplet residues near the crater, analyzed using EDAX, are found to be rich in Ca and Sr compared to the bulk (Table II). As the laser irradiated volume is heated, vaporization begins and the volatile elements evaporate earlier during the laser pulse. Once vaporization begins and establishes the shock wave, recoil pressure operating on the sample can act to hydrodynamically flush out the remaining

TABLE II. Composition of Bi-Sr-Ca-Cu-O Bulk Sample and Ablated Crater

	Bi, %	Ca, %	Cu, %	Sr, %
Bulk	29.36	18.12	36.39	16.13
Ablated crater	22.13	27.23	29.05	21.60

molten pool of material. The splattered droplets enriched in Ca and Sr observed adjacent to the crater supports such a process.

Superconducting thin films were fabricated by laser ablation (PLD) of $Bi_{1.6}Sr_2Ca_3Cu_4$ on MgO at 750°C. The composition of the film was $Bi_{1.2}Sr_{1.6}Ca_{3.2}Cu_4$ as determined by EDAX. These films had similar Ca and Cu content as the sample, but lower Bi and Sr concentration. Although Bi and Cu are enriched in the ablated vapor, they can be vaporized at the heated substrate surface. A unique laser power density exists for all materials in which the composition of the sample and ablated mass provides the desired composition in the films, depending on the substrate temperature. BiSrCaCuO was chosen here to demonstrate a worst-case scenario. For YBaCuO and many other materials, stoichiometric films are routinely fabricated by PLD [126–129]. To date, the best quality (highest critical current density) YBaCuO film has been prepared by PLD. The laser power density is critical for stoichiometric films with such high critical current densities. The fortuitous overlap of preferential ablation with selective sticking coefficients occurs for YBaCuO at a fluence of approximately 3 J/cm^2 using nanosecond UV pulsed laser ablation.

8.7 Particles

In addition to atomic and molecular vapor, laser ablation produces particles in the μm-size range. The number of particles and their size distribution depends on the laser pulse duration, wavelength, fluence, ambient gas and pressure, and physical properties of the sample [31, 34, 35]. Excitation and emission in the LIP is affected by particles, complicating chemical analysis. The entrainment of ablated mass into the gas and transport to the ICP depends on particle size. Large particles may not be entrained, and those that are may not completely decompose in the ICP. Particle sizes should be less than about 2 μm for efficient transport and excitation in the ICP [111, 130–132]. For liquid samples, extensive studies have been carried out on the design of nebulization chambers for accurate and sensitive chemical analysis. Laser ablation chambers and dry-vapor transport studies are only preliminary and not conclusive in defining general parameters for chemical analysis. New studies are needed to understand particle generation and transport during ablation processes. Such studies will be beneficial not only to chemical analysis but also to PLD, in which case particles can destroy film properties.

A concern with chemical analysis measurements using ICP-AES is that as the particle size distribution changes because of the laser and/or sample properties, the peak emission intensity in the ICP may shift. AES detection systems have a fixed imaging region of the ICP; spatial changes in peak emission would be manifested as a change in sample concentration. To address this concern, spatial

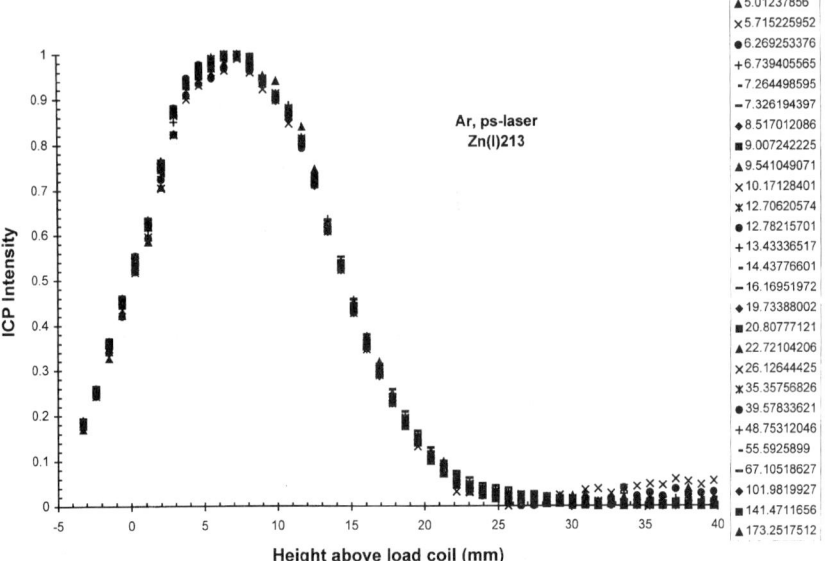

FIG. 19. Normalized Zn emission intensity versus height in the ICP during picosecond-pulsed ablation of brass with power density from approximately 5 to 62 GW/cm^2 (from data in Fig. 7).

emission intensity profiles were measured for diverse laser and sample conditions [133]. Zn emission intensity versus height in the ICP during ablation of brass with power density from approximately 5 to 62 GW/cm^2 was shown in Fig. 7. The emission intensity changed as the laser power density increased, but the profile remained constant, as shown when these data are normalized to unit intensity (Fig. 19).

Nanosecond-pulsed ablation produces a different particle size distribution than that using picosecond-pulsed ablation [31, 34, 35]. The nanosecond excimer laser was used to ablate the same brass sample using power densities from approximately 0.1 to 6.5 GW/cm^2. The spatial emission intensity profile remained constant throughout the diverse mass ablation region and resembled the profile from picosecond laser ablation. The spatial emission intensity profile also remained constant for both UV and IR laser ablation, even though the particle size distribution is shifted to smaller values at lower wavelengths [134].

Ablation from metals and alloys generally produces smooth molten spherical particles, whereas ablation from ceramics and glasses can produce rough, broken, irregularly shaped particles [31, 34, 80, 111, 131, 132]. By using the wide range of power densities from the nanosecond and picosecond lasers, the spatial emission intensity profile in the ICP remained constant for similar elements from diverse sample materials. An example is shown in Fig. 20 for aluminum emission in the inductively coupled plasma during ablation of aluminum metal (Fig. 20(a)) and aluminum-oxide ceramic (Fig. 20(b)), using the nanosecond excimer laser. From these diverse samples, the change in particle size distribution did not manifest a change in the spatial emission profile within the ICP. Either transport selectively filters that portion capable of entrainment, the ICP properties (temperature, excitation, ionization) govern the spatial profile for the particle size range reaching the ICP, or time integration washes out any particle influence on the spatial profile.

Spatial profiles were not influenced by the CCD integration time from 5 milliseconds to 5 seconds; the shorter integration time being limited by the sensitivity of detection. By using a fast photomultiplier tube and fast data acquisition, spectral emission intensity spikes can be observed in the ICP, indicative of larger particles undergoing excitation and emission as they vaporize through the high-temperature plasma [93]. However, these emission spikes only represent a small portion of the overall integrated intensity. It may be possible to correlate these emission spikes to particle size distribution, although these data will be biased by transport characteristics. These emission spikes may also be used as a sensitive means of chemical analysis [135]. Overall, the constant spatial emission intensity profiles from diverse laser and sample conditions are beneficial for chemical analysis using laser ablation, in that emission intensity measurements will represent concentration or mass changes and not changes due to variation in the particle size distribution. A fixed ICP-AES imaging system should be suitable for all laser and sample conditions. However, further studies of particle generation are necessary in order to improve mass transport and to achieve enhanced chemical analysis sensitivity.

8.8 Conclusion

Laser ablation provides significant benefits and capabilities for chemical analysis, as evidenced by the rapidly increasing number of applications and commercially available systems. Diverse solid samples can be directly ablated and analyzed by emission spectroscopy from the laser-induced plasma or by transporting the ablated mass to an inductively coupled plasma. To ensure accurate and sensitive analysis for these applications, laser ablation mechanisms must be better understood. Elucidation of laser ablation mechanisms is necessary in order to efficiently couple the laser beam into the sample, ablate a reproducible quantity of mass, control the amount of mass ablated, minimize preferential ablation and plasma shielding, produce stoichiometric ablation, and control particle size distribution. Through these studies, laser ablation can

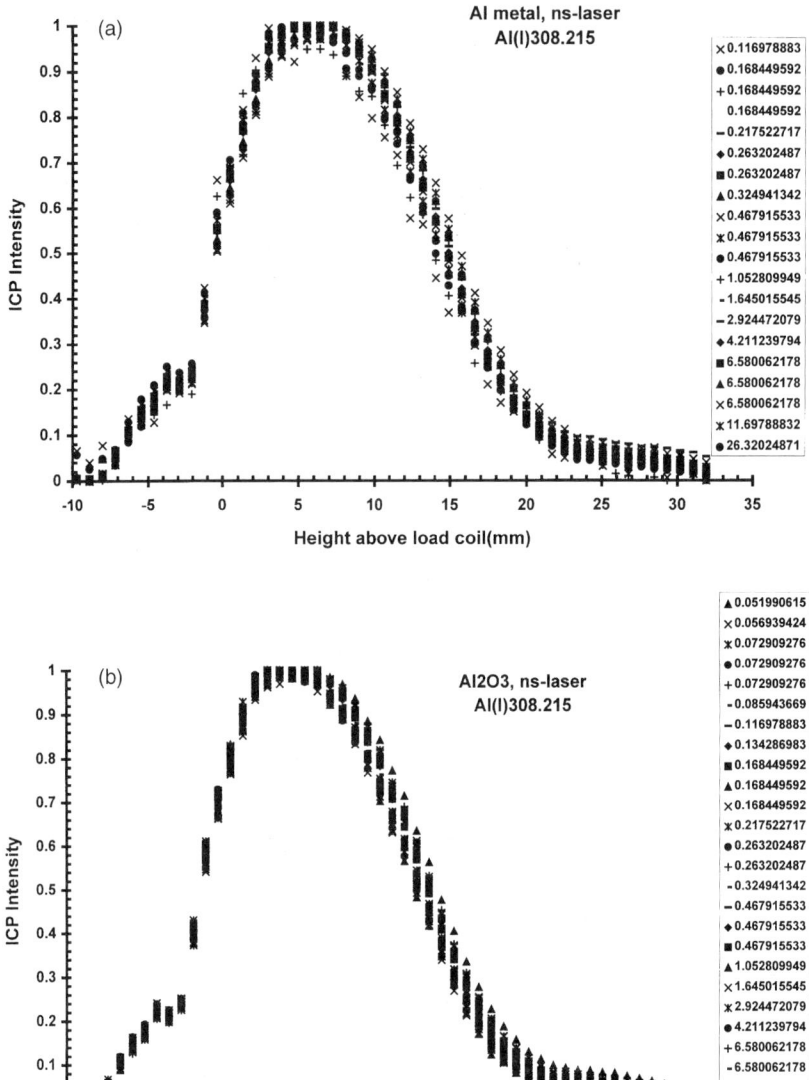

FIG. 20. Spatial emission intensity profiles in the inductively coupled plasma for aluminum emission during ablation of aluminum metal (a) and aluminum-oxide ceramic (b), using the nanosecond excimer laser.

provide routine chemical analysis for environmental, health, forensics, nonproliferation, and other applications of critical importance.

Acknowledgments

This work would not be possible without the efforts due to W.-T. Chan, M. Shannon, A. Fernandez, and M. Caetano for contributions related to references within this chapter. Supported by the U.S. Department of Energy, Office of Basic Energy Sciences, Chemical Sciences Division, Processes & Techniques Branch, under Contract No. DE-AC03-76SF00098.

References

1. Browner, R. F., and Boorn, A. W. Sample introduction techniques for atomic spectroscopy. *Analyt. Chem.* **56**, 875A–888A (1984).
2. Kantor, T. New approaches to the separation of evaporation and atomization-excitation in atomic spectrometry. *Spectrochim. Acta* **38B**, 1483–95 (1983).
3. Darke, S. A., Long S. E., Pickford, C. J., and Tyson, J. F. A study of laser ablation and slurry nebulization sample introduction for the analysis of geochemical materials by inductively coupled plasma spectrometry. *Fresenius J. Analyt. Chem.* **337**, 284–289 (1990).
4. Pak, Y. N., and Koirtyohan, S. R. Direct solid sample analysis in a moderate-power argon MIP with spark generation. *J. Analyt. Atom. Spectrom.* **9**, 1305–1310 (1994).
5. Ren, J. M., Rattray, R., Salin, E. D., and Gregoire, D. C. Assessment of direct solid sample analysis by graphite pellet electrothermal vaporization ICP-MS. *J. Analyt. Atom. Spectrom.* **10**, 1027–1029 (1995).
6. Richner, P., Evans, D., Wahrenberger C., and Dietrich V. Applications of laser ablation and electrothermal vaporization as a sample introduction technique for ICP-MS. *Fresenius J. Analyt. Chem.* **350**, 235–241 (1994).
7. Holcombe, J. A., and Wang, P. X. Direct solid sample analysis using pressure regulated electrothermal atomization with AAS. *Fresenius J. Analyt. Chem.* **346**, 1047–1053 (1993).
8. Rossi, G., Omenetto, N., Pigozzi, G., and Vivian, R. Analysis of radioactive waste solutions by atomic absorption spectrometry with electrothermal atomization. *At. Spectrosc.* **4**, 113–117 (1983).
9. Rattray, R., and Salin, E. D. Aerosol deposition direct sample insertion for ultratrace elemental analysis by ICP-MS. *J. Analyt. Atom. Spectrom.* **10**, 829–836 (1995).
10. Liu, X. R. and Horlick, G. Direct analysis of materials using sample insertion devices and mixed gas ICP-AES. *J. Analyt. Atom. Spectrom.* **9**, 833–840 (1994).
11. Chan, W. T., and Horlick, G. Some signal characteristics in direct sample insertion ICP-AES. *Appl. Spectrosc.* **44**, 525–530 (1990).
12. Suh, S. Y., and Sacks, R. D. Excitation temperature, degrees of ionization of added iron species, and electron density in an exploding thin film plasma. *Spectrochim. Acta* **36B**, 1081–1096 (1981).
13. Ratliff, P. H., and Harrison, W. W. Time-resolved studies of the effects of water vapor in glow discharge mass spectrometry. *Appl. Spectrosc.* **49**, 863–871 (1995).

14. Markus, R. K., Harville, T. R., Mei, Y., and Shick, C. R. RF-powered glow discharges—elemental analysis across the solid spectrum. *Analyt. Chem.* **66**, A902–A911 (1994).
15. Klingler, J. A., Savickas, P. J., and Harrison, W. W. The pulsed glow discharge as an elemental ion source. *J. Amer. Soc. Mass Spectrom.* **1**, 138–143 (1990).
16. Marcus, K. R., and Harrison, W. W. The hollow cathode plume-a plasma emission source for solids. *Spectrochimica Acta* **40B**, 933–941 (1985).
17. Russo, R. E. Laser ablation. *Appl. Spectrosc.* **49**, A14–A28 (1995).
18. Moenke-Blankenburg, L. Laser-ICP-spectrometry. *Spectrochim. Acta Rev.* **15**, 1–37 (1993).
19. Moenke-Blankenburg, L. Laser micro analysis, in *A Series of Monographs on Analytical Chemistry and its Applications*, J. D. Winefordner and I. M. Kolthoff (eds.), Vol. 105. John Wiley & Sons, New York, 1989.
20. Darke, S. A., and Tyson, J. F. Interaction of laser radiation with solid materials and its significance to analytical spectrometry. *J. Analyt. Atom. Spectrom.* **8**, 145–209 (1993).
21. Dittrich, K., and Wennrich, R. Laser vaporization in atomic spectrometry. *Progress in Analyt. Atom. Spectrosc.* **7**, 193–198 (1984).
22. Laqua, K. Chapter 2 Analytical spectrsocopy using laser atomizers, in *Analytical Laser Spectroscopy*, N. Omenetto (ed.). John Wiley and Sons, New York, 1979.
23. Carr, J. W., and Horlick, G. Laser vaporization of solid samples into an inductively coupled plasma. *Spectrochim. Acta* **37B**, 1–15 (1982).
24. Mitchell, P. G., Sneddon J., and Radziemski, L. J. Direct determination of copper in solids by direct current argon plasma emission spectrometry with sample introduction using laser ablation. *Appl. Spectrosc.* **41**, 141–148 (1987).
25. Wennrich, R. and Dittrich, K. Simultaneous determination of traces in solid samples with laser-AAS. *Spectrochim. Acta* **37B**, 913–919 (1982).
26. Raith, A., Hutton, R. C., Ablee, I. D., and Crighton, J. Non-destructive sampling method of metals and alloys for laser ablation ICP-MS. *J. Analyt. Atom. Spectrom.* **10**, 591–594 (1995).
27. Cromwell, E. F., and Arrowsmith, P. Semiquantitative analysis with laser ablation ICP-MS. *Analyt. Chem.* **67**, 131–138 (1995).
28. Moenke-Blankenburg, L., Schumann, T., and Nolte, J. Direct solid analysis by laser ablation ICP-AES. *J. Analyt. Atom. Spectrom.* **9**, 1059–1062 (1994).
29. Jarvis, K. E., and Williams, J. G. Laser ablation ICP-MS—A rapid technique for the direct quantitative determination of major, trace, and rare-earth elements in geological samples. *Chem. Geology* **106**, 251–262 (1993).
30. Lee, Y. I., and Sneddon, J. Direct and rapid determination of potassium in solid glasses by excimer laser ablation atomic emission spectrometry, *Analyst* **119**, 1441–1443 (1994).
31. Bloembergen, N. in *Laser Solid Interaction and Laser Processing*, S. D. Ferris, H. J. Leamy, and J. M. Poate (eds.). American Institute of Physics, New York, 1979.
32. Cremers, D. A., and Radziemski, L. J. Chapter 5, Laser plasmas for chemical analysis, in *Laser Spectroscopy and its Applications*, L. J. Radziemski, R. W. Solarz, and J. A. Paisner (eds.). Marcel Dekker, Inc., New York, 1987.
33. Miller, J. C., ed. Laser ablation: Principles and applications. *Springer Series in Materials Science* **28**. Springer-Verlag, Berlin, Heidelberg, 1994.
34. von Allmen, M. Chapter 5, Evaporation and plasma formation, in *Laser-Beam Interactions with Materials—Physical Principles and Applications*. Springer-Verlag, New York, 1987.

35. Ready, J. F. Chapter 4, Laser-induced particle emission, in *Effect of High-Power Laser Radiation*. Academic Press, New York, 1971.
36. Duley, W. W. *Laser Processing and Analysis of Materials*. Plenum Press, NY, 1983.
37. Pietsch, W., Dubreuil, B., and Briand, A. A study of laser-produced copper plasma at reduced pressure for spectroscopic applications. *Appl. Phys. B*, **61**, 267–275 (1995).
38. Oki, Y., Tani, T., Kidera, N., and Maeda, M. Trace element analysis by laser ablation atomic fluorescence spectroscopy. *Opt. Comm.* **110**, 298–302 (1994).
39. Andre, N., Geertsen, C., Lacour, J. L., Mauchien, P., and Sjostrom, S. UV laser ablation optical emission spectrometry on aluminum alloys in air at atmospheric pressure, *Spectrochim. Acta* **49B**, 1363–1372 (1994).
40. Simeonsson, J. B., and Miziolek, A. W. Spectroscopic studies of laser-produced plasmas formed in CO and CO_2 using 193, 266, 355, 532 and 1064 nm laser radiation. *Appl. Phys. B* **59**, 1–9 (1994).
41. Joseph, M. R., Ning, Xu, and Majidi, V. Time-resolved emission characteristics and temperature profiles of laser-induced plasma in helium. *Spectrochim. Acta* **49B**, 89–103 (1994).
42. Sjostrom, S., and Mauchien, P. Laser atomic spectroscopic techniques—The analytical performance for trace elemental analysis of solid and liquid samples. *Spectrochim. Acta Rev.* **15**, 153–180 (1993).
43. Autin, M., Briand, A., Mauchien, P., and Mermet, J. M. Characterization by emission spectrometry of a laser-produced plasma from a copper target in air at atmospheric pressure. *Spectrochimica Acta*, **48B**, 851–862 (1993).
44. Niemax, K., and Sdorra, W. Optical emission spectrometry and laser-induced fluorescence of laser produced sample plumes. *Appl. Opt.* **29**, 5000–5006 (1990).
45. Quentmeier, A., Sdorra, W., and Niemax, K. Internal standardization in laser induced fluorescence spectrometry of microplasmas produced by laser ablation of solid samples. *Spectrochim. Acta.* **45B**, 537–546 (1990).
46. Lewis, A. L., and Piepmeier, E. H. Chemical and physical influences of the atmosphere upon the spatial and temporal characteristics of atomic fluorescence in a laser microprobe plume. *Appl. Spectrosc.* **37**, 523–530 (1983).
47. Xu, X., Grigoropoulos, C. P., and Russo, R. E. Nanosecond time resolved thermal emission measurements during pulsed excimer laser interaction with materials. *Appl. Phys. A—Materials Science & Processing* **62**, 51–59 (1996).
48. Nakata, Y., Kumuduni, W. K. A., Okada, T., and Maeda, M. Two-dimensional laser-induced fluorescence imaging of non-emissive species in pulsed-laser deposition of $YBa_2Cu_3O_{7-x}$. *Appl Phys Lett.* **66**, 3206–3208 (1995).
49. Kurniawan, H., Tjia, M. O., Barmawi, M., and Yokoi, S. A time-resolved spectroscopic study on the shock wave plasma induced by the bombardment of a TEA CO_2 laser. *J. Phys.* **D28**, 879–883 (1995).
50. Parigger, C., Lewis, J. W. L., and Plemmons D. Electron number density and temperature measurements in a laser induced hydrogen plasma. *J. Quant. Spectrosc. & Rad. Trans.* **53** 249–255 (1995).
51. Parigger, C., Plemmons, D. H., and Lewis, J. W. L. Spatially and temporally resolved electron number density measurements in a decaying laser induced plasma using hydrogen-alpha line profiles. *Appl. Opt.* **34**, 3325–3330 (1995).
52. Sergienko, G. V., Stotsky, G. I., and Zykova, N. M. Measurement of the target surface temperature in the presence of a laser induced plasma. *Meas. Sci. Tech.* **5**, 1448–1452 (1994).

53. Lindley, R. A., Gilgenbach, R. M., Ching, C. H., and Lash, J. S. Resonant holographic interferometry measurements of laser ablation plumes in vacuum, gas, and plasma environments. *J. Appl. Phys.* **76**, 5457–5472 (1994).
54. Puretzky, A. A., Geohegan, D. B., Haufler, R. E., Hettich, R. L., Zheng, X. Y., and Compton, R. N. Laser ablation of graphite in different buffer gases, in *Laser Ablation: Mechanisms and Applications-II*, J. C. Miller and D. B. Geohegan (eds.). AIP Press, New York, 1993.
55. Otis, C. E., and Goodwin, P. M. Internal energy distributions of laser ablated species from $YBa_2Cu_3O_{7-x}$. *J. Appl. Phys.* **73**, 1957–1964 (1993).
56. Sappey, A. D., and Gamble, T. K. Planar laser-induced fluorescence imaging of Cu(2) in a condensing laser ablated copper plume. *J. Appl. Phys.* **72**, 5095–5107 (1992).
57. Izumi, H., Ohata, K., Sawada, T., Morishita, T., and Tanaka, S. Direct observation of ions in laser plume onto the substrate. *Appl. Phys. Lett.* **59**, 597–599 (1991).
58. Grant, K. J., and Paul, G. L. Electron temperature and density profiles of excimer laser induced plasmas. *Appl. Spectrosc.* **44**, 1349–1354 (1990).
59. Dyer, P. E., Issa, A., and Key, P. H. Dynamics of excimer laser ablation of superconductors in an oxygen environment. *Appl. Phys. Lett.* **57**, 186–188 (1990).
60. Fabbro, R., Fabre, E., Amiranoff, F., and Garban-Labaune, C. Laser-wavelength dependence of mass-ablation rate and heat-flux inhibition in laser-produced plasmas. *Phys. Rev.* **A26**, 2289–2292 (1982).
61. Richter, A. Characteristic features of laser-produced plasmas for thin film deposition. *Thin Solid Films* **188**, 275–292 (1990).
62. Mao, X. L., Shannon, M. A., Fernandez, A. J., and Russo, R. E. Temperature and emission spatial profiles of laser-induced plasmas during ablation using time-integrated emission spectroscopy. *Appl. Spectrosc.* **49**, 1054–1062 (1995).
63. Iida, Y. Atomic emission characteristics of laser-induced plasmas in an argon atmosphere at reduced pressure. *Appl. Spectrosc.* **43**, 229–234 (1989).
64. Yamamoto, K. Y., Cremers, D. A., Ferris, M. J., and Foster, L. E. Detection of metals in the environment using a portable laser induced breakdown spectroscopy instrument. *Appl. Spectrosc.* **50**, 222–233 (1996).
65. Davies, C. M., Telle, H. H., Montogomery, D. J., and Corbett, R. E. Quantitative analysis using remote laser-induced breakdown spectrometry. *Spectrochim. Acta.* **50B**, 1059–1075 (1995).
66. Cremers, D. A., Barefield, J. E. and Koskelo, A. C. Remote analysis by laser-induced breakdown spectroscopy using a fiber-optic cable. *Appl. Spectrosc.* **49**, 857–860 (1995).
67. Mermet, J. M. Chapter 10, Spectroscopic diagnostics: basic concepts, in *Inductively Coupled Plasma Emission Spectroscopy, Part 2, Application and Fundamentals*, P. W. J. M. Boumans (ed.). John Wiley & Sons, New York, 1987.
68. Chen, K. R., Leboeuf, J. N., Wood, R. F., and Geohegan, D. B. Accelerated expansion of laser-ablated materials near a solid surface. *Phys. Rev. Lett.* **75**, 4706–4709 (1995).
69. Geohegan, D. B. Imaging and blackbody emission spectra of particulates generated in the KrF-laser ablation of BN and $YBa_2Cu_3O_{7-x}$. *Appl. Phys. Lett.* **62**, 1463–1465 (1993).
70. Zel'dovich, Y. B., and Raizer, Y. P. *Physics of Shock Waves and High Temperature Hydrodynamic Phenomena*. Academic Press, New York, 1966.
71. Liberman, M. A., and Velikovich, A. L. *Physics of Shock Waves in Gases and Plasmas*. Springer, Berlin, 1986.

72. Reader, J., and Corliss, C. H. Wavelengths and transition probabilities for atoms and atomic ions, U.S. Government Printing Office, Washington, DC, 1980.
73. Wennrich, R., Dittrich, K., and Bonitz, U. Matrix interference in laser atomic absorption spectrometry. *Spectrochim. Acta* **39B**, 657–666 (1984).
74. Hiddemann, L., Uebbing, J., Ciocan, A., Dessenne, O., and Niemax, K. Simultaneous multielement analysis of solid samples by laser ablation microwave-induced plasma optical emission spectrometry. *Analyt. Chim. Acta.* **283**, 152–159 (1993).
75. Ciocan, A. Hiddemann, L., Uebbing, J., and Niemax, K. Measurement of trace elements in ceramic and quartz by laser ablation microwave-induced plasma atomic emission spectrometry. *J. Analyt. Atom. Spectrom.* **8**, 273–278 (1993).
76. Ciocan, A., Uebbing, J., and Niemax, K. Analytical applications of the microwave induced plasma used with laser ablation of solid samples. *Spectrochim. Acta.* **47B**, 611–617 (1992).
77. Mitchell, P. G., Sneddon J., and Radziemski, L. J. A sample chamber for solids analysis by laser ablation/DCP Spectrometry. *Appl. Spectrosc.* **40**, 274–279 (1986).
78. Caetano, M., Mao, X. L., and Russo, R. E. Spatial emission intensity profiles in the ICP for laser ablation sampling versus power density. *Spectrochimica Acta B* **51**, 1473–1485 (1996).
79. Shannon, M. A., Mao, X. L., Fernandez, A., Chan, W.-T., and Russo, R. E. Laser ablation mass removal versus incident power density during solid sampling for ICP-AES. *Analyt. Chem.* **67**, 4522–4529 (1995).
80. Russo, R. E., Chan, W.-T., Bryant, M. F., and Kinard, W. F. Laser ablation sampling with ICP-AES for the analysis of prototypic glasses. *J. Analyt. Atom. Spectrom.* **10**, 295–301 (1995).
81. Goodall, P. S., and Johnson, S. G. Isotopic uranium determination by ICP-AES using laser ablation sample introduction. *J. Analyt. Atom. Spectrom.* **11**, 57–60 (1996).
82. Allen, L. A., Pang, H. M., Warren, A. R., and Houk, R. S. Simultaneous measurement of isotope ratios in solids by laser ablation with a twin quadrupole ICP-MS. *J. Analyt. Atom. Spectrom.* **10**, 267–271 (1995).
83. Cousin, H., Weber, A., Magyar, B., Abell, I., and Gunther, D. An Auto-Focus system for reproducible focusing in laser ablation inductively coupled plasma mass spectrometry. *Spectrochim. Acta* **50B**, 63–66 (1995).
84. Gunther, D., Longerich, H. P., Forsythe, L., and Jackson, S. E. Laser ablation microprobe—ICP-MS. *American Lab.* **27**, 24–29 (1995).
85. Evans, R. D., Outridge, P. M., and Richner, P. Applications of laser ablation ICP-MS to the determination of environmental contaminants in calcified biological structures. *J. Analyt. Atom. Spectrom.* **9**, 985–989 (1994).
86. Walder, A. J., Abell, I. D., Platzner, I., and Freedman, P. A. Lead isotope ratio measurements of NIST 610 glass by laser ablation ICP-MS. *Spectrochim. Acta* **48B**, 397–402 (1993).
87. Durrant, S. F. Multielement analysis of environmental matrices by laser ablation ICP-MS. *Analyst* **117**, 1585–1592 (1992).
88. Vandeweijer, P., Baeten, W. L, Bekkers, M. H., and Vullings, P. J. Semiquantitative survey analysis of solids by laser ablation ICP-MS. *J. Analyt. Atom. Spectrom.* **7**, 599–603 (1992).
89. Chan, W.-T. Mao, X. L., and Russo R. E. Differential vaporization during laser ablation/deposition of Bi-Sr-Ca-Cu-O superconducting materials. *Appl. Spectrosc.* **46**, 1025–1031 (1992).

90. Yasuhara, H., Okano, T., and Matsumura, Y. Determination of trace elements in steel by laser ablation ICP-MS. *Analyst* **117**, 395–399 (1992).
91. Moenke-Blankenburg, L., Schumann, T., Gunther, D., Kuss, H.-M., and Paul, M. Quantitative analysis of glass using ICP-AES: Laser micro-analysis ICP-AES and laser ablation ICP-MS. *J. Analyt. Atom. Spectrom.* **7**, 251–254 (1992).
92. Imai, N. Microprobe analysis of geological materials by laser ablation inductively coupled plasma mass spectrometry. *Analytica Chim. Acta* **269**, 263–268 (1992).
93. Chan W.-T., and Russo, R. E. Study of laser material interactions using ICP-AES. *Spectrochim. Acta* **46B**, 1471–1478 (1991).
94. Furuta, N. Fundamental studies of laser ablation for the introduction of powdered solid samples into an ICP. *Appl. Spectrosc.* **45**, 1372–1376 (1991).
95. Hwang, Z.-W., Teng, Y.-Y., and Sneddon, J. A modified torch system for direct solid analysis by laser ablation ICP-AES. *Microchem. J.* **43**, 42–45 (1991).
96. Marshall, J., Franks, J., Abell, I., and Tye, C. Determination of trace elements in solid plastic materials by laser ablation ICP-MS. *J. Analyt. Atom. Spectrom.* **6**, 145–150 (1991).
97. Denoyer, E. R., Fredeen, K. J., and Hager, J. W. Laser solid sampling for ICP-MS. *Analyt. Chem.* **63**, 445A–457A (1991).
98. Darke, S. A., Long, S. E., Pickford, C. J., and Tyson, J. F. Laser ablation system for solid sample analysis by ICP-AES. *J. Analyt. Atom. Spectrom.* **4**, 715–719 (1989).
99. Arrowsmith, P. Laser ablation of solids for elemental analysis by inductively coupled plasma mass spectrometry. *Analyt. Chem.* **59**, 1437–1444 (1987).
100. Gray, A. L. Solid sample introduction by laser ablation for inductively coupled plasma source mass spectrometry. *Analyst* **110**, 551–556 (1985).
101. Ishizuka T., and Uwamino, Y. Inductively coupled plasma emission spectrometry of solid samples by laser ablation. *Spectrochim. Acta* **38B**, 519–527 (1983).
102. Thompson, J., Goulter J. E., and Sieper, F. Laser ablation for the introduction of solid samples into an ICP for atomic emission spectrometry. *Analyst* **106**, 32–39 (1981).
103. Phipps C. R., and Dreyfus, R. W. *Laser Ionization Mass Analysis*, A. Vertes, R. Gijbels, and F. Adams (eds.). Wiley and Sons, New York, 1993.
104. Hora, H. *Plasmas at High Temperature and Density, Applications and Implications of Laser-Plasma Interactions.* Springer-Verlag, Berlin, Heidelberg, 1991.
105. Weyl, G. M. Physics of laser-induced breakdown: An update, in *Laser-Induced Plasmas and Applications*, L. J. Radziemski and D. A. Cremers (eds.). Dekker, New York, 1989.
106. Mao, X. L., Chan, W. T., Caetano, M., Shannon, M., and Russo R. E. Preferential vaporization and plasma shielding during nanosecond and picosecond laser ablation. *Appl. Surface Science* **96–98**, 126–130 (1996).
107. Ihlemann, J., Scholl, A., Schmidt, H., and Wolff-Rottke, B. Nanosecond and femtosecond excimer laser ablation of oxide ceramics. *Appl. Phys. A* **60**, 411–417 (1995).
108. Wolff-Rottke, B., Ihlemann, J., Schmidt, H., and Scholl, A. Influence of the laser spot diameter on photo-ablation rates, *Appl. Phys.* **A60**, 13–17 (1995).
109. Mao, X. L., Chan, W. T., Shannon, M. A., and Russo R. E. Plasma shielding during picosecond laser sampling of solid materials by ablation in He versus Ar atmosphere. *J. Appl. Phys.* **74**, 4915–4922 (1993).
110. Richner, P., Borer, M. W., Brushwyler, K. R., and Hieftje, G. M. Comparison of different excitation sources and normalization techniques in laser ablation AES using a photodiode based spectrometer. *Appl. Spectrosc.* **44**, 1290–1296 (1990).

111. Thompson, M., Chernery, S., and Brett, L. J. Nature of particulate matter produced by laser ablation—implications for tandem analytical systems. *J. Analyt. Atom. Spectrom.* **5**, 49–55 (1990).
112. Baldwin, D. P., Zamzow, D. S., and D'Silva, A. P. Aerosol mass measurements and solution standard additions for quantitation in laser ICP-AES. *Analyt. Chem.* **66**, 1911–1917 (1994).
113. Fernandez, A. Mao, X. L. Chan, W. T. Shannon, M. A., and Russo R. E. Correlation of spectral emission intensity in the ICP and laser induced plasma during laser ablation of solid samples. *Analyt. Chem.* **67**, 2444–2450 (1995).
114. Perez, J., and Weiner, B. R. The laser ablation of gold films at the electrode surface of a quartz crystal microbalance. *Appl. Sur. Sci.* **62**, 281–285 (1992).
115. Shannon M. A., and Russo R. E. Laser-induced stresses versus mechanical stress power measurements during laser ablation of solids. *Appl. Phys. Lett.* **67**, 3227–3229 (1995).
116. Shannon M. A., and Russo R. E. Monitoring stress power during high power laser material interactions. *Appl. Surface Sci.* **96–98**, 149–153 (1996).
117. Shannon, M. A., Rubinsky, B., and Russo R. E. Far-field mechanical stresspower measurements during high-power laser ablation. *J. Appl. Phys.* **80**, 4665–4672 (1996).
118. Pang, H. M., Wiederin, D. R., Houk, R. S., and Yeung, E. S. High-repetition rate laser ablation for elemental analysis in an ICP with acoustic wave normalization. *Analyt. Chem.* **63**, 390–394 (1991).
119. Iida, Y. Effects of atmosphere on laser vaporization and excitation processes of solid samples. *Spectrochim. Acta* **45B**, 1353–1367 (1990).
120. Durrant, S. F. Feasibility of improvement in analytical performance in laser ablation ICP-MS by addition of nitrogen to the argon plasma. *Fresenius J. Analyt. Chem.* **349**, 768–771 (1994).
121. Russo, R. E., Mao, X. L., Shannon, M. A., and Caetano M. Fundamental characteristics of laser-material interactions (ablation) in inert gases at atmospheric pressure using ICP-AES. *Appl. Surface Sci.* **96–98**, 144–148 (1996).
122. Landen, O. L., and Alley, W. E. Dynamics of picosecond-laser-pulse plasmas determined from the spectral shifts of reflected probe pulses. *Phys. Rev. A* **46**, 5089–5100 (1992).
123. Farkas, G., and Toth, C. Energy spectrum of photoelectrons produced by picosecond laser-induced surface multiphoton photoeffect. *Phys. Rev.* **A41**, 4123–4126 (1990).
124. Cromwell E. F., and Arrowsmith, P. Fractionation effects in laser ablation ICP-MS. *Appl. Spectrosc.* **49**, 1652–1660. (1995).
125. Mochizuki, T., Sakashita, A., Tsuji, T., Iwata, H., Ishibshi, Y., and Gunji, N. Selective vaporization in laser ablation solid sampling for the ICP-AES and MS of steels. *Analyt. Sci.* **7**, 479–481 (1991).
126. Venkatesan T. V. Chapter 4, Pulsed laser deposition of high-temperature superconducting thin films, in *Laser Ablation: Principles and Applications*, J. C. Miller (ed.). Springer Series in Material Science **28** (1994).
127. Chrisey, D. B., and Hubler, G. K., eds. *Pulsed Laser Deposition of Thin Films.* John Wiley and Sons, New York, 1994.
128. Reade, R. P., Berdahl, P., Schaper, L. W., and Russo R. E. YBaCuO multilayer structures with amorphous dielectric layers for multichip modules using ion-assisted pulsed laser deposition. *Appl. Phys. Lett.* **66**, 2001–2003 (1995).
129. Reade, R. P., Church, S., and Russo R. E. Ion-assisted pulsed laser deposition. *Rev. Sci. Instrum.* **66**, 3610–3614 (1995).

130. Arrowsmith, P. and Hughes, S. K. Entrainment and transport of laser ablated plumes for subsequent element analysis. *Appl. Spectrosc.* **42**, 1231–1239 (1988).
131. van Heuzen, A. A. Analysis of solids by laser ablation ICP-MS I. Matching with a glass matrix. *Spectrochim. Acta* **46B**, 1803–1817 (1991).
132. van Heuzen, A. A., and Morsink, J. B. W. Analysis of solids by laser ablation ICP-MS II. Matching with a pressed pellet. *Spectrochim. Acta* **46B**, 1819–1828 (1991)
133. Caetano, M., Mao, X. L., and Russo R. E. Spatial emission intensity profiles in the ICP for laser ablation sampling versus power density. *Spectrochim. Acta B* **51B**, 1473–1485 (1996).
134. Geertsen, C., Briand, A., Chartier, F., Lacour, J. L., Mauchien, P. Sjostrom, S., and Mermet, J. M. Comparison between IR and UV laser ablation at atmospheric pressure—Implications for solid sampling ICP spectrometry. *J. Analyt. Atom. Spectrosc.* **9**, 17–22 (1994).
135. Liu X. R., and Horlick, G. In situ laser ablation sampling for ICP-AES. *Spectrochim. Acta* **50B**, 537–548 (1994).

9. MATRIX-ASSISTED LASER DESORPTION AND IONIZATION

James A. Carroll
Monsanto Company

Ronald C. Beavis
Department of Pharmacology
New York University Medical Center

9.1 Introduction

Matrix-assisted laser desorption/ionization (MALDI) is the most successful solid-state biopolymer ion source. It is the culmination of a twenty-year evolution in solid state ion sources, with each new generation having greater sensitivity and easier sample preparation methods. MALDI has the capability of selectively producing ions representative of a mixture of biopolymer molecules present at a concentration of one micromolar, in the presence of other organic and inorganic ions and molecules at one million–fold higher concentrations. The ability to selectively ionize biopolymers (particularly polypeptide-based polymers) out of complicated mixtures is the single feature of MALDI that sets it apart from all other solid-state and liquid ion sources. This introduction outlines the development of solid-state ion sources for biopolymer analysis.

Solid-state ion sources have developed in parallel with liquid phase ion sources over the course of the last twenty years. The two types of ion sources are quite distinct in philosophy as well as practice. In a liquid phase ion source, a solution containing the sample of interest is introduced as droplets into a vacuum system through some series of baffled vacuum pumping chambers. The droplets dry as they traverse the ion source, resulting in bare analyte ions at the ion source outlet. Direct liquid introduction, thermospray, and electrospray ion sources all use this strategy to produce ions. The ion production mechanism in these sources is a near-equilibrium, reversible process—the relatively slow drying of small droplets results in vacuum-isolated ions.[1] The ions produced have roughly the same charge state that they had in the solution during evaporation [1], so no additional charge transfer mechanism is necessary to explain the observed charge on these ions.

In a solid-state ion source, the analyte is deposited by some special preparation method, which frequently involves the introduction of a "matrix" material

---

[1] See following page for footnote.

to improve ion production. This matrix can be either a viscous fluid (glycerol) or a solid. The sample preparation method effectively removes all of the volatile components from the original material, resulting in an involatile deposit. This deposit is then irradiated by energetic particles or photons that effect the production of vacuum-isolated ions near the surface of the deposit that can be extracted with an electric field. The ion production mechanism in this case is nonreversible and nonequilibrium: relatively large amounts of energy must be deposited in a short-range interaction with the solid deposit.

The ion formation process has two distinct components. First, there must be some mechanism for ejecting large molecules from the deposit with sufficient kinetic energy to escape the surface without introducing large amounts of vibrational or electronic excitation to the molecules themselves. The second step involves some mechanism for transferring charge to the molecules during their ejection. This first portion of the mechanism is inherently nonequilibrium. If the translational energy involved in ejecting the molecule from the surface was in equilibrium with the internal degrees of freedom of the molecule, the molecule would disintegrate before it could be extracted from the ion source. The fact that two discrete mechanisms are involved with the formation of observable products, i.e., emission and ionization, has made the study of the fundamental processes involved in ion production difficult.

The evolutionary process leading to MALDI began in the 1970s, with the introduction of the plasma desorption solid-state ion source by Macfarlane and Torgesson [2]. In this ion source, a peptide sample was deposited on a thin metal film by drying a few microliters of a peptide solution. The metallic film was then placed in a vacuum chamber and irradiated by high-energy ions produced by ^{252}Cf spontaneous fission. Surprisingly the interaction of 100 MeV fission fragments with the deposited peptide molecules resulted in the production of intact peptide ions from the surface. These ions could then be accelerated away from the surface and analyzed in a mass spectrometer. If the number of fission fragments passing through the sample was kept relatively low (10–1000/second), the results of individual particle interactions could be followed using time-of-flight mass measurement. Other sources for high-energy, highly charged primary ions

---

[1] Throughout this text the phrase "vacuum-isolated" will be used for ions produced by various sources. These ions are frequently referred to as "gas phase" ions in the literature, but this usage suggests that the ions are present in an equilibrium phase with the properties of conventional van der Waals gases. This association with conventional gases is untrue and very deceptive, leading to spurious questions such as "What is the temperature of the ions produced by a MALDI source?" The ions produced by the sources considered here do not interact with each other and are never present as an equilibrium phase, even though the ion production mechanism may be near-equilibrium (such as electrospray). At best, the vacuum-isolated ions can be considered a highly metastable phase, which disappears after any meaningful interaction with the ion's environment.

were used, such as the output of linear accelerators, but none of these sources could compete with the simplicity and low cost of ^{252}Cf. Plasma desorption produced the largest intact ions from proteins and peptides, giving it a special place in biopolymer analysis during the 1980s [3].

At about the same time as plasma desorption was being developed, secondary ion sources were being developed to produce high mass ions. In these sources, a beam of low energy (10 keV) primary ions irradiates a solid sample, producing vacuum-isolated ions from the sample [4]. In the first secondary ion sources, high primary beam currents were used, resulting in the erosion of the sample. These high beam currents heated the surface, resulting in damage to any fragile molecules on the surface, making them unsuitable for biopolymer analysis. Benninghoven et al. [4] used a very low current primary ion beam to examine only the molecules on the surface of the deposit, resulting in spectra that were remarkably similar to those obtained by plasma desorption ion sources. Ion sources with low current primary ion beams were called "static" sources because they did not damage the surface of the sample quickly, as opposed to "dynamic" sources that produced surface erosion.

Static ion sources suffered from very low output currents: they could not produce a large enough flux of ions to use with magnetic sector mass spectrometers, with their inherently low current transmission. This problem was solved in 1981 by Barber et al. [5] with the use of a fluid sample rather than a completely solid one. The analyte of interest was mixed with a very viscous, relatively involatile fluid and the fluid deposited on a surface. This fluid was then irradiated with a beam current that was high enough to damage a solid surface.[2] The fluid did not have a fixed surface—convection and diffusion continuously removed the damaged layer from the surface and created a fresh layer for producing ions. This scheme for producing a sample that would stand up to dynamic conditions but produce ions like a static source was given the peculiar name "fast atom bombardment" (or FAB). This name was the result of the mistaken impression that this new sample preparation method required a neutral atomic primary beam to produce secondary ions and that the most important part of the new source was the neutral atom source. Subsequent studies proved that the neutral beam was not required; the self-healing property of viscous fluid surfaces was the most significant feature of this new ion source. The fast atom bombardment source could produce much higher ion currents than static sources, and this led to the great popularity of the ion source on magnetic sector and quadrupole mass spectrometers during the 1980s and early 1990s [6]. These sources were never as sensitive as static methods, but the availability of an installed base of mass spectrometers compatible with the new source led to its

---

[2] The fluid used was subsequently identified as glycerol, although no mention of glycerol occurs in the original paper.

dominance in biological applications. Also, effects caused by the surface activity of certain analytes dissolved in glycerol led to a minor specialty in analytical science—the development of recipes for producing good spectra from particular classes of compounds.

Laser desorption ion sources were physically very similar to secondary ion sources, except that the surface of the deposit was illuminated by photons rather than irradiated by primary ions. A wide variety of lasers have been used as the photon source, with very similar results. Organic materials deposited onto surfaces can be examined by laser desorption, producing intact ions with masses up to approximately 1000 Da. Above this mass, the ions produced are almost always dissociation products of the original molecule [7]. Therefore, laser desorption sources were considered to be inferior to secondary ion sources for high mass biopolymers.

Two separate groups attempted to solve this problem. Tanaka *et al.* [8] developed a method based on a viscous fluid sample. Simply using the fast atom bombardment method of dissolving the analyte in glycerol was not effective; glycerol was transparent to the ultraviolet lasers used to illuminate the sample. The first good results were obtained when finely divided metal powder was added to the glycerol. The metal powder was fine enough to absorb power from the laser light, providing a method of coupling the light with the glycerol. Surprisingly the combination of this crude coupling method with a pulsed UV laser to provide illumination produced significant ion currents of intact molecules with masses in excess of 30,000 Da, breaking the high mass record set by plasma desorption sources. It was also clear from the first spectra that this type of ion source was going to be much more sensitive and produce better signals than could be obtained from existing ion sources.

At roughly the same time, Franz Hillenkamp's group was developing a modified laser desorption ion source based on their observations of previous laser desorption experiments [9]. They had observed that for most analyte deposits, very little if any laser light was being absorbed by the sample deposit. The samples were almost completely transparent: almost all of the absorption was occurring in the substrate. The energy deposited in the substrate was subsequently producing ions from the surface, presumably by either an induced shock wave or a quasi-equilibrium emission of ions from a transiently heated surface. Hillenkamp's group thought that the ion emission process could be improved if the transparent analyte and absorbing substrate could be mixed somehow, forming a composite material. The absorbing material in this composite was conceived as a "matrix" holding the analyte molecules in place and protecting them from the harsh effects of the laser.

The problem then became the formation of the composite material. The first set of experiments were not encouraging, and the project may have been shelved, except for the success of Tanaka's method. Subsequently a number of modifications were made to the original instrument used in Hillenkamp's lab to improve its sensitivity for high mass ions and a screening program instituted to find the right small organic molecule to use as a matrix for 266 nm laser light (the fourth harmonic of a neodymium:yttrium aluminium garnet laser). Relatively quickly, nicotinic acid was discovered to have the desired properties of their postulated matrix material [10]. With nicotinic acid, they were able to produce ions from proteins with molecular masses in excess of 100,000 Da, requiring only a few picomoles of protein to produce the samples. The sensitivity of the solid matrix method was 500–1000 times greater than anything demonstrated by Tanaka, and it produced the ion signals of higher quality. For these reasons, the solid matrix method demonstrated by Hillenkamp and Karas eclipsed the fluid method of Tanaka, and it is the solid matrix method that is commonly referred to as "matrix-assisted laser desorption/ionization."

9.2 Protein-Doped Matrix Crystals

9.2.1 General Principles

The production of ions using matrix-assisted laser desorption/ionization depends on the production of a suitable composite material, consisting of the matrix and analyte biopolymer. The true discovery of Hillenkamp and Karas was that there is a very simple way to dope crystals of small organic molecules with much larger polypeptide-based molecules or proteins.[3] An aqueous solution of the matrix compound was prepared, and the analyte protein was added to the solution. A droplet of this solution was then dried, resulting in a solid deposit of analyte-doped matrix crystals [10].

This recipe is as simple as it sounds. The only difficulty that faced Hillenkamp and Karas was to find a matrix molecule that would actually dry out of solution with analyte (protein) molecules in the resulting matrix crystals, rather than elsewhere in the deposit. The molecule that they chose would have to be soluble in solvents appropriate for proteins and would have to absorb light more strongly than a protein at 266 nm. The solid deposit formed on drying would also have to be vacuum-stable for long enough to obtain ions; materials

---

[3] The word "protein" refers to the function of a biopolymer, rather than its chemical structure. Proteins are based on highly ordered linear polypeptide molecules that are produced by ribosomes and then modified by a wide variety of enzymes to produce an active protein. The active protein may incorporate several identical or different polypeptides, with a variety of additional groups attached, such as carbohydrates, phosphates, and long chain fatty acids. The mass spectra of "proteins" obtained by MALDI are actually mass spectra of the individual polypeptide chains that make up a protein, commonly referred to as "subunits."

with very rapid sublimation rates are unsuitable matrix choices. They solved this problem by screening a number of candidate compounds until they discovered nicotinic acid (3-pyridine carboxylic acid). It is soluble in aqueous solutions and forms large, protein-doped crystals that absorb 266 nm light strongly. The crystals eventually sublime in vacuum, but the sublimation rate is low enough that good mass spectra can be obtained from the ion source.

Nicotinic acid had several drawbacks as a matrix. Its major failing was the fact that it reacted with proteins when excited by the desorption laser. Nicotinic acid easily loses COOH when photochemically excited, leaving a very reactive pyridyl group [11]. This group can attach to nearby protein molecules during the emission process, resulting in ions containing one or more pyridyl groups (called adducts) attached to the protein of interest. Analytically this is a major drawback because the resulting modified protein molecules appear at a variety of molecular masses, depending on the extent of pyridyl attachment. The average number of adduct groups attached to the protein increases with molecular mass, resulting in the broad, poorly defined signals shown in Karas and Hillenkamp's first paper on MALDI [10].

Several matrix compounds have been discovered that have much better properties than nicotinic acid for producing protein ion source crystals (see Fig. 1). Several cinnamic acid derivatives, particularly 3,5-dimethoxy-4-hydroxycinnamic acid (sinapic or sinapinic acid), have been found to produce much lower intensity adducts that are the result of the loss of -OH from the intact matrix, rather than -COOH [11]. One benzoic acid derivative, 2,5-dihydroxybenzoic acid (gentisic acid), has also been found to have a low matrix adduct formation rate, and it also forms adducts via the loss of -OH from the matrix molecule [12]. One currently popular matrix, α-cyano-4-hydroxycinnamic acid, produces almost no observable adduct peaks, presumably because of the replacement of the hydrogen on the α-carbon with a -CN group [13]. This derivatization should block the formation of the reactive ketene intermediate thought to be responsible for adduct formation in cinnamic acid matrices.

With the practical issue of adduct formation aside, the most remarkable property of successful matrix compounds is their ability to selectively incorporate polypeptide molecules into their crystals under very benign conditions. A number of studies have been performed to investigate this phenomenon. Polypeptide-doped sinapic acid or gentisic acid crystals can be grown under a wide variety of conditions. The doping can be blocked by the presence of strong ionic detergents (e.g., sodium dodecylsulfate), high concentrations of polyvalent anions (e.g., phosphate or sulphate ions), or anionic polymers (e.g., heparin). The level of doping is affected by temperature, with higher doping levels occurring at lower temperatures. The rate of crystal growth may also alter doping levels (see the following discussion). Note that systematic studies on the

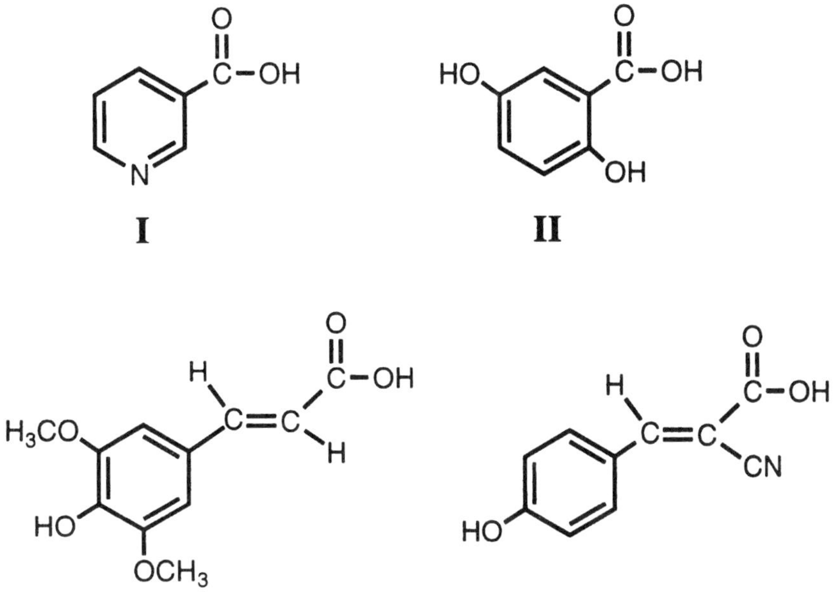

FIG. 1. Structures of some common MALDI matrix compounds. I. Nicotinic Acid II. 2,5-Dihydroxy Benzoic Acid III. Sinapinic (or Sinapic) Acid IV. α-Cyano-4-hydroxy Cinnamic Acid.

effect on doping levels of temperature and crystal growth rates have not been done; the results to date have been for very specialized crystal growth conditions and only for a very limited range of protein dopants.

To understand the features displayed by a particular protein doped matrix crystal and to manipulate the doping levels in these crystals, it is necessary to think about the crystal growth and doping process. The important parameters in any matrix crystal doping experiment are: the concentrations of the solutes and the solvent composition of the mother liquor, the relative diffusion rates of the matrix and the proteins present, and the affinity of the proteins to adsorb to the matrix crystals in that solvent composition. The doping levels in crystal growth that is slow with respect to the diffusion of protein molecules in the mother liquor will be dominated by the rate of adsorption of proteins to the growing crystal. When the crystal growth is fast with respect to protein diffusion, the doping levels will only depend on the bulk concentration of the proteins present.

In the case of very slow matrix crystal growth in a nearly equilibrium solution, as the crystal grows, protein molecules that adsorb to a growing face will become incorporated into the resulting volume of the crystal as inclusions. Assume that the crystals are grown in a mother liquor containing the matrix compound and a mixture of m different protein species, with the ith type of protein molecule represented by P_i. Also, assume that the rate of protein absorption is too low to strongly affect the growth of the crystal. Then the number of a particular type of protein molecules adsorbed to the surface of a growing crystal face per unit area, $\sigma_p(P_i)$, in a near equilibrium solution is given by

$$\sigma_p(P_i) = k(P_i)[P_i] \qquad (9.1)$$

where $k(P_i)$ is the rate constant for the adsorption, per unit area. This surface area density will be maintained in the doped crystal. The equation assumes near equilibrium conditions, in which the rate of protein diffusion in the solution is large compared to the local rate at which proteins are being removed from the solution into the growing matrix crystal. The number of protein molecules per unit area illuminated in an ablation event $(n_a(P_i))$ can then be calculated as

$$n_a(P_i) = \sigma_p(P_i) \cdot (d/l) \quad \text{or} \quad n_a(P_i) = k(P_i)[P_i] \cdot (d/l) \qquad (9.2)$$

where d is the depth of the crystal ablated and l is the thickness of a layer of matrix molecules in the crystal. Note that a protein molecule must remain adsorbed to the surface of the crystal for long enough to become trapped in the growing crystal. If the protein molecules exchange back and forth between the adsorbed phase and the solution phase more rapidly than the crystal can overgrow them, they will not become included into the crystal.

In the situation where the crystal growth is very rapid compared to the rate of protein diffusion, proteins will become trapped in the growing crystals at approximately the same volume concentration as in the original solution. This situation is not as unlikely as it might sound. The ratio of diffusion constants between the matrix material and hen's egg lysozyme (a small globular protein) is approximately 20 : 1. Therefore, if the matrix crystal growth was limited only by the rate of matrix crystal diffusion onto the growing face, even a small protein would be unable to escape from the advancing crystal face. The number of protein molecules of a particular protein species (with diffusion constant $D(P_i)$) available in an ablation event for crystals grown under these conditions is simpler than in the previous case, namely,

$$n_a(P_i) = D(P_i) \cdot [P_i] \qquad (9.3)$$

In most real situations, the growth of protein-doped matrix crystals falls between these two extreme cases. The first case, the protein doping level is controlled by the value of $k(P_i)$, allowing for doping enrichment with a particular protein, depending on its adsorption characteristics. This sort of discrimination is known to happen. Sinapic acid, an excellent matrix for protein ion sources, only includes peptides of approximately 30 amino acids or larger, regardless of the rate of crystal growth. Smaller peptides do not appear to adsorb to the growing crystal faces either strongly enough or for long enough to be included.

It is possible to strongly affect the relative densities of protein and peptides species present in a crystal by manipulating $k(P)$. The most commonly used method is to change the solvent composition of the mother liquor, altering the amount of adsorption for a particular protein species as well as the residence time for a particular molecule on the crystal face. Solvents containing significant concentrations of formic acid, such as 1:2:3 formic acid:2-propanol:water (v/v/v), have been shown to produce less discrimination between different peptide species than solvents based on trifluoroacetic acid:acetonitrile:water mixtures [14]. The presence of nonionic detergents, such an N-octylglucoside, may also decrease discrimination effects. Presumably the values for $k(P_i)$ show less dependence on the characteristics of a particular protein (i.e., P_i) in these solvent systems.

Information is not currently available regarding the mechanism of protein binding to matrix crystals. It is known that the binding is enhanced at pH < 4, but these results are complicated by the fact that the matrices commonly used become negatively charged at pH > 4, forming salts that crystallize differently than the free acid form of the molecule. The possible binding mechanisms are hydrogen bonding, precipitation at the crystal's surface driven by the rearrangement of water around the protein-crystal composite ("hydrophobic" bonding), or ion-dipole and ion-ion interactions between the surface and the protein ("ion exchange" bonding). There have been some indications that low temperatures enhance binding, but no systematic studies have been performed.

It is clear, however, that a maximum doping level exists that a particular crystal can support without alteration of the crystal's habit. If the protein concentration is too high during crystal formation (e.g., 0.1 millimolar), growth along crystal axes that incorporate protein will be suppressed by protein molecules coating those surfaces that bind proteins preferentially, blocking the addition of further matrix molecules to those faces. The crystal growth, therefore, occurs along axes represented by faces that do not bind proteins. The resulting crystals contain a lower average protein-doping level than crystals grown in solutions containing much less protein.[4]

9.2.2 Practical Methods for Forming Matrix Crystal Deposits

A number of different methods have been described in the literature for growing protein-doped matrix crystals, aimed at solving particular practical problems. The original method described by Karas and Hillenkamp has been

---

[4] The inhibitory effect of high protein concentrations explains the phenomenon observed by anyone who has tried to prepare "good" matrix crystals: "It is better to use too little rather than too much protein." The corollary of this statement is often ignored: "Adding more protein won't help."

named the "dried-droplet" method. It entails drying a droplet of a solution containing the matrix (1–10 millimolar) and the protein (1–10 micromolar) . A simple variant on the "dried droplet" method was proposed by Vorm et al. [15]. They first placed a thin layer of the matrix compound onto a metal surface and then put a droplet of the protein containing solution on top of the layer of matrix compound. As the droplet dries, it dissolves some of the matrix, which crystallizes onto the substrate matrix layer when the droplet dries completely. The crystallized matrix layer is very thin, allowing for more reproducible mass measurements when the laser-ablated ions are analyzed with a reflectron mass spectrometer. Because the layer of protein-doped matrix is very thin, it only produces ions for <10 laser shots on a spot.

It is possible to grow large, protein-doped matrix crystals under near-equilibrium conditions, rather than in a rapidly drying droplet [16]. Supersaturated matrix solutions containing protein will form crystals that can be used directly in an ion source. Supersaturation can be achieved either by heating and cooling or by slow evaporation. These protein-doped crystals can be cleaved to expose fresh, well-defined faces to the laser beam, which are required for fundamental studies of the emission process.

The slow crystallization method produces higher protein doping levels than drying droplet crystallization, especially if the crystallization chamber is gently agitated during the crystal growth. Agitation in solutions with low protein concentrations has the effect of bringing relatively scarce, slowly diffusing protein molecules in contact with the adsorptive growing crystal by moving the solvent relative to the crystal. The total density of protein molecules adsorbed to the crystal can thus be made much higher than it would be if only diffusion were relied on to bring protein molecules to the crystal's surface. It is also possible to grow protein-doped crystals in the face of contaminants that normally inhibit crystal formation in the dried droplet technique, such as glycerol or β-mercaptoethanol. Contaminants that prevent the adsorption of proteins onto the growing crystals, such as strong detergents (most notably sodium dodecyl-sulfate), have a strong negative effect on protein doping levels.

A recently developed technique of producing crystals for use in MALDI ion sources involves growing thick films of matrix crystals [17]. The substrate for the crystals (usually a flat piece of metal) is first covered with the matrix material by rapidly drying a solution of the matrix dissolved in an organic solvent, such as 2-propanol. The small matrix crystals that cover the surface are then crushed and smeared over the substrate surface, resulting in a well-adhered layer of stressed, crystalline matrix. A saturated solution containing the matrix and protein is then placed onto the surface and allowed to dry slightly. The stressed crystalline material on the substrate acts to seed crystal formation at many sites on the surface, resulting in the rapid growth of a rather uniform polycrystalline film of protein-doped matrix crystals over the surface. The liquid drop can then be removed by blotting. The resulting film is very strongly adhered to the substrate, and it can be washed thoroughly to remove any contaminants that were present in the protein solution.

The crystals that make up this polycrystalline film are frequently oriented in the same direction because of the underlying symmetry of the matrix crystal structure: almost all known matrices belong to the closely related family of space groups $P2_1/a$, $P2_1/c$, or $P2_1/n$ [16]. Their structures are made up of relatively flat planes of hydrogen bonded matrix molecules that are stacked up on top of each other, with van der Waals forces holding the planes together. Crushing the crystals onto the substrate has the effect of pushing the planar layers of these crystal structures down onto the surface, analogous to smearing out a deck of cards. The resulting stressed surface is, therefore, covered predominantly with these planar layers facing up from the substrate, producing an oriented template for matrix crystal growth. The use of molecular templates for controlling the growth and orientation of protein-doped matrix crystals promises to be a very useful method for tailoring the physico-chemical properties of these crystals as protein ion sources.

9.3 Protein Ion Sources

Ion sources that use matrix-assisted laser desorption to generate protein ions consist of two subsystems. One subsystem is an optical apparatus that produces light with a laser and delivers that light to the surface of the protein-doped matrix crystals inside of the ion source housing. The other subsystem is a set of electrodes, voltage supplies, and switches that are used to extract ions from the region of space above the illuminated portion of the matrix crystals and condition the ion pulse produced by the laser into an ion beam that can be used in a subsequent apparatus, such as a mass spectrometer (see Fig. 2). The ion extraction subsystems most frequently used can be divided into two general types:

1. Sources where ions are generated in a field-free region and ions are extracted some time later.
2. Sources where ions are generated in a static electric field that extracts the ions immediately on production.

Most of the commercial ion sources available currently are of the second type. The presence or absence of an electric field has a profound effect on the ions that are emitted by the source, so the two types of sources will be discussed separately.

Note that although MALDI ion sources have been historically associated with time-of-flight mass analyzers (see [18] for a comprehensive review of time-

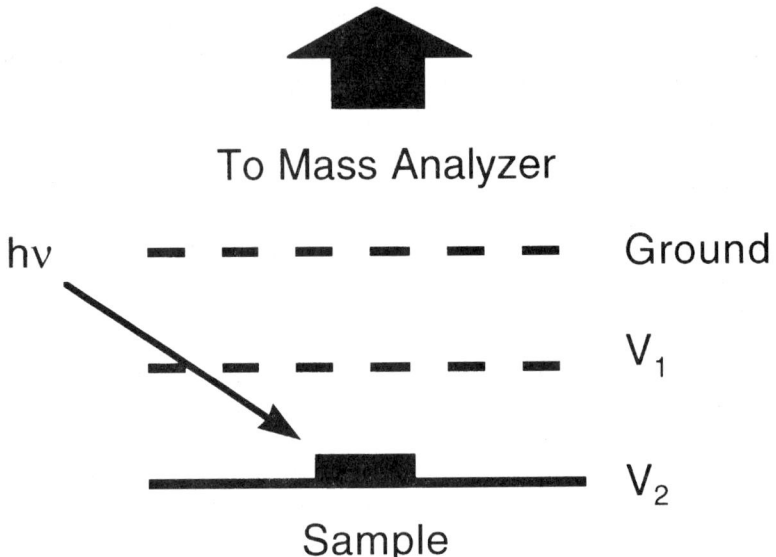

FIG. 2. Schematic of a simple MALDI ion source. V_1 and V_2 are the potentials of the corresponding electrodes. The dashed lines indicate the presence of a grid. The prepared protein-doped matrix crystals are placed at the îsampleî position and illuminated with the laser pulse (hv), emitting protein ions. The protein ions are accelerated by the fields produced by the electrodes and emerge from the grounded grid as a pulse.

of-flight techniques), there is no fundamental reason limiting the use of MALDI ion sources with other types of analyzers. Double-focusing magnet sector analyzers [19], quadrupole ion traps [20], and ion cyclotron resonance analyzers [21] have all been used to mass analyze the ions produced by MALDI sources, resulting in extremely high mass resolutions for peptide molecular ions.

9.3.1 Interaction of Photons with Matrix Crystals

Both types of ion source share a common element:ions are generated by illuminating the protein-doped matrix crystals with laser light. Ion production has no strong wavelength dependence. For α-cyano-4-hydroxycinnamic acid and 3,5-dimethoxy-hydroxycinnamic acid, protein ions have been produced over the wavelength range 266 - 420 nm, with the light emitted by a variety of lasers [22, 23]. Emission of protein ions using infrared lasers has also been demonstrated [24]. The only criterion for protein ion production appears to be that the laser light must be absorbed strongly by the matrix crystals. Minimizing the absorption of laser light by the protein reduces the amount of energy transferred to the protein during the emission process, resulting in ions that do not fragment during the ablation process. Polypeptide molecules absorb light up to approximately 280 nm because of the aromatic side chain groups on tyrosine and tryptophan. For this reason, the 266 nm light from a frequency quadrupled, Q-switched neodymium:yttrium aluminum garnet laser (Nd:YAG) originally used by Karas and Hillenkamp has been supplanted by the 337 nm light from a nitrogen laser or the 355 nm frequency-tripled light from a Q-switched Nd:YAG laser.

The fact that lasers, over a wide range of wavelengths, all produce essentially the same effect suggests that the initial method of matrix excitation is unimportant to the mechanism resulting in biopolymer ion emission. Ultraviolet lasers transfer energy to the matrix by $\pi-\pi^*$ electronic transitions, whereas infrared lasers excite specific molecular stretching modes such as O-H or C=O. A plausible hypothesis would be that the laser-induced excitations are rapidly converted into thermal motions of the matrix molecules in some volume of the matrix crystal adjacent to the surface. These motions would have to be sufficiently violent to disrupt the intermolecular forces holding the matrix crystal together, resulting in the ablation of a portion of the crystal's surface. The matrix materials used in MALDI have relatively low vacuum sublimation temperatures (200–300°C), suggesting that the binding energy of a matrix molecule in a crystal is quite low. However, the details of the energy flow among electronic, vibrational, and translational degrees of freedom in matrix crystals has not been examined in depth and any detailed mechanisms involving these energy flows would be purely speculation.

The properties required for a laser illumination system to produce protein ions have been revised several times since the discovery of the MALDI effect. The original papers and communications suggested that a laser power density in excess of 100 MW/cm^2 of 266 nm light in a 10 ns pulse was required to produce a significant current of protein ions [10]. Subsequent experimentation has shown that the early measurements were in error and that good protein ion current can be obtained with laser power densities of approximately 2 MW/cm^2 of 266 nm or 355 nm light in a 10 ns pulse [23, 25]. Studies comparing the ion emission behavior of crystals illuminated with nanosecond and picosecond pulsed lasers have shown that it is not the power density that is the most important parameter, it is the total energy in the laser pulse at a given wavelength [26]. Therefore, the power density required to produce significant protein ion current would be more properly written as an energy flux of 20 mJ/cm^2.

The dimensions of the laser beam used to produce ions from protein-doped matrix crystals has also been the subject of some study and debate. The earliest papers and communications on MALDI suggested that a tightly focused laser beam was necessary to produce protein ions. The beam diameters suggested for ion production were less than 10 micrometers. Subsequent investigation has determined that a variety of optical systems can be used to produce protein ions, using spot sizes ranging from 0.005–0.2 millimeters, so long as sufficient energy can be deposited in the matrix crystal. Recent studies have suggested that there is a nonlinear energy dependence of laser spot size with protein ion emission current: very small spots require higher laser energy fluxes to produce the same ion current per unit area from an emitting crystal [27]. These results suggest that additional mechanisms of energy dissipation may become important when dealing with very small illuminated areas, however, further work is necessary to confirm the behavior of this effect (only one experimental point was taken at very small spot sizes in [27]).

A particularly useful innovation has been the introduction of fiber optics into MALDI illumination systems, making the use of different lasers and ion source orientations much simpler in practice. In this type of system, light is focused into the fiber optic at the laser and the light emerging from the other end of the optic is imaged onto the matrix crystals. Use of long fiber optic cables produces very uniform beam intensity profiles at the output end of the cable, allowing for very uniform energy deposition on the matrix crystals.

9.3.2 Protein Ion Extraction Phenomena

The word "ablation" (rather than "desorption") is used in this section to describe the process that leads to protein ion emission. Desorption strictly refers to the removal of individual molecules from a surface; that is, it is the opposite of adsorption. Ablation refers to any process that simultaneously removes a large number of molecules from a surface, usually removing many molecular layers at the same time. The removal of a bulk portion of the surface results in strong collective effects that determine the final state of any particular molecule. In MALDI, it would be reasonable to argue that individual protein molecules do not interact with each other during the emission process. However, because protein molecules are much larger that matrix molecules (100:1–10000:1, protein:matrix molecular volumes), many layers of matrix molecules must be emitted before a single protein molecule can be freed from the protein-doped matrix crystals. Therefore, even though the protein ion emission process may have some "desorption-like" character, the underlying physical phenomenon is the ablation of a large volume of matrix molecules on every laser shot. The development of a dense plume of ablated matrix molecules, which the protein ions must pass through, has very significant consequences with respect to the design and operation of MALDI ion sources.

The physical processes involved in the protein-ion-emitting ablation event affect the behavior of those ions in the ion extraction system. These effects can be seen most clearly in the type of ion extraction systems with no electric field present in the vicinity of the matrix crystal during the ablation event (in Fig. 2, $V_2 + V_1$). The simplest of these systems consists of a flat electrode bearing the matrix crystals, a field free region above the crystals, followed by a region with a strong electric field to extract the ions. The sequence of events in an ablation experiment in this type of ion source is as follows:

1. The laser fires.
2. Ions are ablated from the matrix crystals.
3. The ions drift across the field-free region.
4. The ions enter the high field region and are extracted from the source.

This type of source is well suited to measuring the velocity of ions produced by the matrix crystals. A series of measurements performed by several groups has confirmed that all of the ions ablated from the crystals have approximately the same velocity distribution, with an average velocity of approximately 500 m/s and a distribution half-width of approximately 100 m/s [24–26]. The distribution is only weakly affected by the molecular mass of the ion ablated: to first order, all of the ions have the same velocity. This result strongly suggests that the ions are ablated from the surface as part of an expanding cloud of evaporated matrix molecules generated by the laser. Collisions with the dense expanding gas accelerate the large protein molecules included in the evaporating crystal region up to the same speed as the gas. At some point in this expansion process, protons (hydrogen ions) are exchanged between the protein and matrix molecules, causing the formation of positively and negatively charged protein ions. The expansion process continues until the density of the matrix gas becomes low enough that the protein ion can be considered a free particle, moving with the velocity imparted to it by the expansion.

Some "lagging" has been observed for protein molecules relative to small peptides, i.e., the average velocity distribution for very heavy molecules have been measured to be slightly slower than light molecules. This "lagging" effect is of interest when detailed comparisons are made between calculated velocity distributions of expanding gases and the energy transfer of these gases to different shapes of proteins; however, it is very difficult to measure lags precisely because of electric field leakage from the high field extraction region into the field-free region of the ion source. Small electric fields affect light ions much more than heavy ions, leading to a difficulty in comparing heavy ion lagging results from one apparatus to another.

The observation that all of the ions emitted during an ablation event have similar velocities has serious implications for a protein ion source. In typical gas phase sources, ions are produced with similar kinetic energies and they can therefore be transported and focused using optics that are tuned to a particular kinetic energy distribution. In a field-free MALDI ion source (in Fig. 2, $V_2=V_1$), to first order the average kinetic energy of a molecule (E_{av}) is directly dependent on its mass (M),

$$E_{av}(M) = M(v_{av}^2/2) \tag{9.4}$$

where v_{av} is the mass-independent average velocity of the expanding matrix gas. Therefore, special care must be taken to use ion optics that will transmit and focus ions with a wide range of kinetic energies. Ion trapping mass spectrometers were initially unable to trap MALDI-produced protein ions because the kinetic energy of the ions was much higher than expected.

The simple model of MALDI protein ion production suggested by the velocity distribution measurements can also be used to explain the measured protein ion production as a function of laser energy applied to the crystals [28]. Ion production exhibits a strongly nonlinear behavior with respect to laser energy. Below a certain critical value of the laser pulse energy (frequently referred to as the "threshold" for ion production), the number of ions emitted during a laser pulse is zero. Above that critical value, the number of ions produced increases very rapidly as a function of increasing laser pulse energy. The curve eventually flattens out when the laser pulse energy is several times the threshold value. If the laser pulse is causing a solid-gas phase change in a thin volume of matrix crystal, the shape of this curve can be easily explained. Below the "threshold" energy, there is insufficient energy in a pulse to overcome the intermolecular forces within a volume and the dissipative effects of heat conduction and electronic-vibronic energy conversions that remove energy from the illuminated volume. Above the threshold, increasing the laser pulse energy increases the volume of crystal that undergoes the phase transition, leading to a rapid increase in the protein ion current. High laser energies begin to produce secondary effects that decrease the detected current. These secondary effects could be detector saturation, Coulomb explosions and local cooling caused by the expansion of relatively large amounts of matrix "gas."

An alternate model proposes that there is no ion production threshold at all; it is an artifact of the measurement process [29]. When the number of ions produced per layer pulse becomes less than one, the ion current becomes difficult to detect. The alternate model proposes that protein ions are produced from ejected clusters of matrix and protein, from which the matrix molecules evaporate as the clusters fly away from the surface. The protein ions produced by such a mechanism would have the same average kinetic energy as the cluster that was originally emitted. In this model, it is the rate and intensity of gas explosions that eject the clusters that result in the ion production versus laser energy curve. Any debate between this model and the gas expansion model given above rapidly becomes a semantic argument revolving around the difference between rapidly expanding a turbulent thick gas and a collection of rapidly evaporating clusters. Further investigation is required to distinguish between the two mechanisms. Practically, both mechanisms occur with very similar time and energy scales, so the net effect on ion production in a macroscopic ion source is the same, regardless of which model is more correct in detail.

When a strong static electric field is applied in the region where protein ions are produced, protein ions extracted from the ion source have lower kinetic energies than expected from the applied fields [30, 31]. Using the ion source in Figure 2 as an example, the situation corresponds to $V_2 - V_1 \gg 1$. The phenomenon of ions being emitted with lower than expected kinetic energies is referred to as the ions having an "energy deficit." The energy deficit has been shown to increase with increasing laser pulse energy and emitted ion current. This observation appears to be incompatible with the observations at zero-field (i.e., $V_2 = V_1$). The ions produced should have an excess of energy because of their high initial kinetic energies. However, consideration of the consequences of the gas expansion model in the case of an extracting electric field shows that this result is to be expected, rather than contradictory. As the plume of ejected gas and protein leaves the crystal's surface, the protein ions will encounter resistance to acceleration relative to the expanding matrix plume because of collisions with the matrix molecules. Therefore, the presence of the electric field only affects the velocity of the protein ion once the gas plume has expanded sufficiently to allow the protein ion to accelerate without sufficient collisions to slow it down again. The work done on the protein ion by the electric field up until that point is dissipated to the gas plume as heat. The net effect of this dissipation is to reduce the kinetic energy of the protein ion as it exits the ion source by the total amount of energy dissipated as heat. If the electric field in the ion production region is high, the energy loss caused by this mechanism can easily exceed the kinetic energy given to the protein ion by the expanding gas, resulting in a protein ion with less kinetic energy than would be expected from nondissipative source. The period of time that the protein ion would be trapped in the expanding cloud of gas would be proportional to the size of the cloud, so increasing the laser pulse energy would be expected to produce a larger energy deficit.

Consideration of this dissipative energy loss mechanism suggests why linear time-of-flight mass spectrometers behave unexpectedly poorly with MALDI ion sources. The resolution and mass accuracy of a linear time-of-flight analyzer depends strongly on the stability of the energy distribution of ions emerging from the ion source. The mechanism described above for dissipative energy loss implies that unless the gas plumes generated by the laser pulse are the same for each pulse, differences in the amount of energy dissipated in the different plumes will result in significant time-of-flight shifts from shot to shot. This behavior is seen in practice. With a linear analyzer, it is necessary to carefully control the laser pulse energy being used to generate ions so that each expansion event will be as similar as possible to the previous one. It is also best to keep the laser pulse energy near the ion production threshold value, where the gas pulses are as small as possible. Careful control of the laser pulse energy minimizes the dispersion in protein ion velocities caused by the difference between the dissipative energy loss and the expansion-induced kinetic energy gain.

The use of an "energy-focusing" reflectron time-of-flight analyzer will correct for this dispersion, making reflectron instruments easier to calibrate and making laser pulse energy control less important [32–35]. Reflectron instruments are sensitive to post-ion source decay of ions, however. The dissipative energy loss mechanism has the effect of increasing the vibrational energy stored in an ion. Once extracted, that energy remains in the molecule and can cause unimolecular chemical reactions, which usually result in the loss of small neutral molecules from a protein. The loss of small neutral molecules (such as ammonia and water) results in a cluster of signals near the intact mass of a protein, creating uncertainty about the true mass of the parent ion. For this reason, reflectron analyzers have not been very useful in studying proteins, although they have been used extensively for peptides with molecular masses <5000 Daltons (see Section 9.4).

The implications of the static field-free experiments and the difficulties described for reflectron mass spectrometers have led to the development of an ion source in which the space above the crystal is field-free for some period of time after the ions are ejected, but then a field is switched on in this region [36, 37]. Keeping the electric field at zero until the ejected plume has dissipated reduces the amount of vibrational excitation that is added to the molecule during the extraction step. This type of ion source, called a "pulsed extraction" ion source, can be used to great effect if it is used in conjunction with a linear time-of-flight mass analyzer. By selecting the length of the analyzer and the geometry of the ion source electrodes correctly, it is possible to "time-focus" ions at the plane of the mass analyzer's detector, i.e., all protein ions of the same mass-to-charge ratio will arrive at the detector at nearly the same time, regardless of their initial kinetic energy distribution. This scheme only works properly for a limited range of kinetic energies at once, but the range of allowed energies is sufficiently broad to significantly improve the mass resolution of a linear time-of-flight analyzer with a MALDI ion source. The idea behind this ion source is not new; it was originally demonstrated by Wiley and McLaren [38] in 1955 using a purely gas phase, electron impact ion source.

9.4 Applications to Polymer Analysis

Matrix-assisted laser desorption/ionization ion sources have been shown to be effective for the mass spectrometric analysis of biological molecules, and the combination of ion source and mass analyzer has become a useful analytical tool in many laboratories [39]. MALDI mass spectrometry is a rapid, accurate, and sensitive technique for providing molecular weight information for a variety of compounds [40, 41]. It has been shown to be amenable to the analysis of such diverse compounds as peptides and proteins [42], oligonucleotides and nucleic acids [43], and carbohydrates [44], and other biologically important molecules. Its applications include the analysis of synthetic compounds [45, 46] as well as the characterization of natural products obtained from biological sources [47]. It can be used to analyze mixtures and is relatively tolerant of the contamination present in many biological matrixes compared to other ionization methods [48]. The utility of MALDI for the analysis of biopolymers, especially proteins, can best be illustrated by sampling the current literature, for example, [49–53]. The availability of commercial MALDI mass spectrometry instruments has made the technique available to researchers in laboratories working on a large assortment of biological problems. Combined with other bioanalytical techniques, MALDI mass spectrometry is a powerful method for the structural characterization of biopolymers.

9.4.1 Protein and Peptide Analysis

9.4.1.1 MW of Proteins and Peptides Isolated from Biological Sources. MALDI has been used to identify proteins and peptides from a variety of biological sources. Isolation and characterization of proteins from biological sources as diverse as bacteria [54, 55], invertebrates [56–59] and mammals [60–62] continue to be reported, and MALDI mass spectrometry has proven to be an invaluable tool for determining accurate molecular weight information. Cain *et al.* have demonstrated the utility of MALDI mass spectrometry to differentiate bacteria based on their protein profiles [54]. Because MALDI shows a high tolerance for biological matrixes, the bacterial proteins were analyzed with a minimum of sample preparation [54]. Lee *et al.* have isolated and characterized the protein rubredoxin from the bacterium *Heliobacillus mobilis* [55]. MALDI was used to establish the molecular weight (5671 Da), and in conjunction with enzymatic degradation, to determine the amino acid sequence of the C-terminal end of the protein [55].

An active area of research involves the characterization of peptides and proteins isolated from the human central nervous system. Brain and cerebrovascular peptides from Alzheimer's disease patients were characterized using MALDI in combination with chromatographic and sequencing methods [63]. The β-amyloid peptide, one of the main components of senile plaques in the brain tissue of Alzheimer's patients, was characterized from human cerebrospinal fluid using MALDI [64]. The ability of MALDI to tolerate organic matrixes has further been demonstrated by Li *et al.*, who have profiled the peptides from single neurons [65–69]. The neurons were placed into the matrix solution and lysed to release the neuropeptides, followed directly by mass spectrometric analysis [67]. These experiments demonstrate that MALDI is a valuable tool for the study of the synthesis and expression of bioactive peptides.

9.4.1.2 Protein and Peptide Sequencing. Gaining sequence information from the mass spectra of peptides and proteins has long been a goal of researchers [70]. MALDI mass spectra of these compounds are typically void of fragmentation, so that little or no sequence information is available directly from the mass spectrum. Strategies employed to gain sequence information generally combine chemical or enzymatic degradation methods in conjunction with MALDI analysis.

The ability of MALDI to analyze mixtures has been exploited to rapidly gain sequence information using the method of protein ladder sequencing [71, 72]. In this technique, chemical degradation is performed in solution to remove the N-terminal amino acid residue of a peptide or protein. After a short period, a mixture exists in which some of the polymers have been degraded more than others. A single analysis of this mixture shows a series of peaks, or ladder, in which the sequence can be determined from the difference in mass between adjacent peaks in the MALDI mass spectrum. All of the twenty naturally occurring amino acids—except for leucine and isoleucine, which have identical chemical composition—can be distinguished by their mass. Thiede *et al.* have demonstrated a similar method for the C-terminal ladder sequencing of peptides [73]. The enzymes carboxypeptidase Y and P, which specifically remove the C-terminal amino acid residue of peptides and proteins, were used to generate the sequence ladder. The resulting mixture was then analyzed by MALDI mass spectrometry to give the sequence [70, 73].

Tandem mass spectrometry techniques have also been employed to gain sequence information from peptide and protein ions formed using MALDI. Kaufmann *et al.* have used a reflectron time-of-flight mass spectrometer to observe the post-source decay products of peptides, gaining some sequence information [32, 74]. In this technique, the ions that undergo metastable decay after acceleration are separated from their precursor ions by applying a retarding field at a set of reflector electrodes [32]. The heavier, more energetic ions penetrate farther into the retarding field, and therefore must travel a longer distance to reach the detector than the smaller fragment ions. The fragmentation can be enhanced using photodissociation [75] or collisions with a pulsed gas [76].

9.4.1.3 Peptide Mapping. Mass spectrometry coupled with protease digestion is a powerful combination for rapidly determining the sequences of proteins [77–79]. A residue-specific protease enzyme such as trypsin, which cleaves proteins at lysine and arginine residues, generates a mixture of peptides. This mixture can be used as a "map" of the protein sequence. The molecular masses of the peptides can be rapidly determined by MALDI mass spectrometric analysis of the digest mixture [80]. To identify the protein, the peptide masses can then be compared to databases containing theoretical maps of known protein sequences [77–81]. This technique was used to generate peptide maps from a mixture of myocardial proteins following two-dimensional electrophoretic separation [82]. The digestion and MALDI analysis of the proteins were performed directly on the membrane used for the separation, demonstrating a significant savings in time and effort relative to releasing the proteins from the membrane and analyzing them separately. The proteins were then identified by comparison to peptide mass databases [82]. Without the ability to analyze the peptide mixture in a single analysis, the peptides would have to be separated chromatographically and sequenced using traditional methods, which can be a laborious and time-consuming process.

In addition to identifying unknown proteins, peptide mapping in conjunction with MALDI mass spectrometry is useful in a variety of other applications. Peptide maps generated by specific enzymatic digestion provide a convenient way to confirm tentative sequence assignments. Sequences of proteins produced by recombinant DNA techniques can be rapidly verified in this way [83]. For example, primosomal protein I from *E. coli* was produced recombinantly and the product analyzed using MALDI mass spectrometry to confirm the sequence. Figure 3(a) shows the mass spectrum of the recombinant protein obtained using a matrix of a-cyano-4-hydroxy cinnamic acid. The molecular weight of the product was determined to be 19,291 Da, lower than the molecular weight calculated from the tentative sequence of 19,469 Da. N-terminal sequencing of the intact protein was used to determine that the first residue, methionine, was missing in the recombinant product. Taking this into account, the measured mass was still lower than the expected mass by 47 Da, a difference much larger than the expected experimental error of about 0.05%. To determine the cause of the mass shift, the protein was digested with the protease trypsin, which cleaves specifically at the C-terminal side of arginine and lysine residues. There are a total of fifteen possible cleavage sites in the protein, and the masses of the expected tryptic peptides can be compared to the mass spectrum to identify any modifications. Figure 3(b) shows the resulting MALDI mass spectrum of the tryptic digest. The peaks labeled T_n show matches to the tryptic peptides that would be generated from the proposed sequence. Figure 4 shows the coverage generated from this initial analysis, in which tryptic fragments 1, 3, 9, 10, and 12 are not observed in the spectrum. The two large peaks in the mass spectrum at m/z 1336 and m/z 2258 are likely candidates for modified tryptic fragments. From the proposed sequence, it can be calculated that tryptic fragment T_3, which includes amino acid residues 27–49, should weigh 2291 Da, or 33 Da more than the observed peak. Similarly, the peak at m/z 1336 differs from the expected tryptic fragment T_9, which includes residues 134–143, by 14 Da. To verify that these are the incorrectly identified tryptic peptides, the tryptic digest was separated using liquid chromatography, and the collected fractions were identified using MALDI. The peptides corresponding to masses 1336 Da and 2258 Da

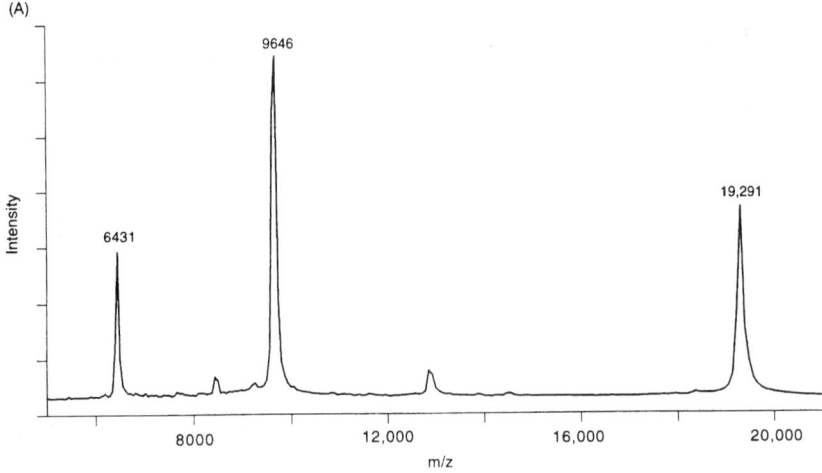

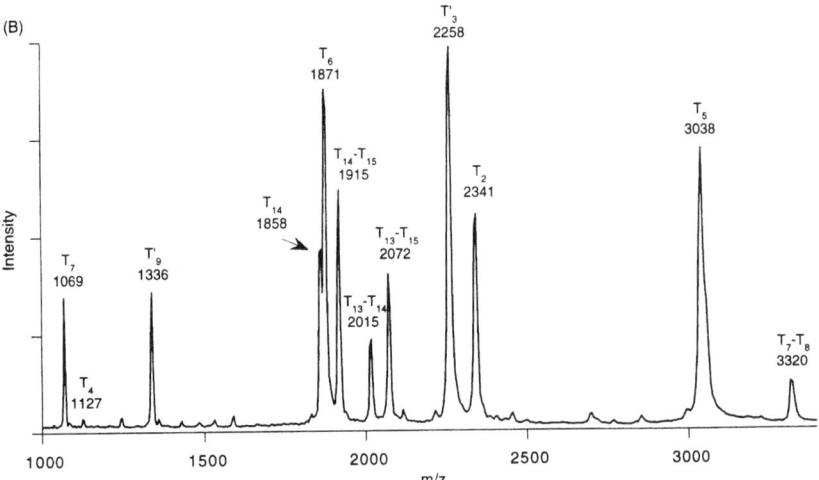

FIG. 3. (a) MALDI mass spectrum of recombinant primosomal protein I from *E. coli*. The spectrum is labeled with the mass-to-charge ratio of the peaks present. Peaks corresponding to the singly, doubly, and triply charged ions of the protein are prominent in the spectrum. (b) MALDI mass spectrum of the tryptic digest of the protein. The peaks are labeled with the mass and the number of the tryptic piece. Peaks labeled T'_n indicate those that do not match the expected mass, but correspond to modified pieces of the protein.

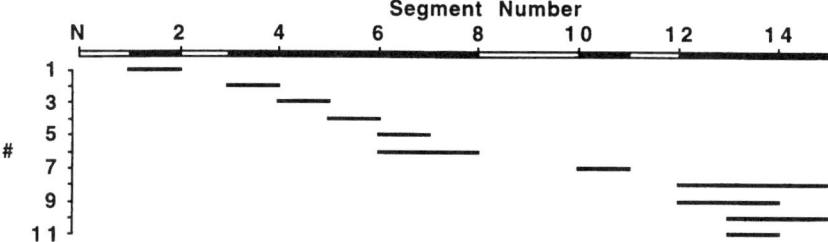

FIG. 4. Tryptic peptide mass map of recombinant primosomal protein I generated from the spectrum in Fig. 3(b). Five of the expected tryptic peptides are not observed in the spectrum. It was determined that the expected sequence differed from the actual sequence due to single amino acid substitutions on tryptic pieces T_3 and T_9.

were then sequenced on an automated sequencer. Each of these peptides was found to have a single residue different from the predicted sequence. The corrected sequence has a calculated mass of 19,290 Da, in close agreement with the measured mass. This example illustrates the utility of MALDI mass spectrometry for the rapid verification of protein and peptide sequences.

Another application of peptide mapping is to determine the sites and extent of post-translational modifications [81, 84–89]. Many proteins are chemically modified in the cell after biosynthesis. Common modifications that can be determined using mass spectrometry include glycosylation, phosphorylation, and the formation of disulfide bonds between cysteine residues. As an example, tryptic digests of recombinant glycoproteins were analyzed using MALDI mass spectrometry [83]. Because the tentative sequence was known, the sites of glycosylation were evident upon analysis of the digest mixture by detection of the corresponding glycopeptides [83]. Similarly, the sites of phosphorylation on phosphopeptides have been determined using MALDI analysis following enzymatic or chemical degradation [90, 91]. Clauser et al. used MALDI with high-energy collision induced dissociation of tryptic digests to characterize unknown proteins from human A375 melanoma cells [92]. With the use of tandem mass spectrometry, the post-translational modifications could be unambiguously identified [92]. MALDI is also useful for determining the locations of disulfide bonds in proteins [93, 94]. Many proteins contain disulfide bonds between proximal cysteine residues, which serve to stabilize their three-dimensional structure. Analysis of a protease digest mixture may show the presence disulfide-bound peptides. The disulfide bond locations may also be deduced by comparing the digest with one in which the disulfide bonds have been chemically reduced. The peptides that appear in the reduced fraction and were not present in the original fraction may have been involved in disulfide bonding.

Mass spectrometry has become useful for the investigations of protein function. The active sites of enzymes have been determined by probing the binding interactions of enzymes and their substrates. The active peptides of rat cytochrome P450 were identified after CNBr digestion and isolation of adduct-bound peptides [95]. The active-site tyrosyl residues of a class Mu glutathione transferase were identified using peptide mass mapping with MALDI analysis [96]. Hutchens *et al.* determined the binding sites of copper to histidine-rich glycoprotein by adding copper to a tryptic digest of the protein. The metal binding peptides were evident in the MALDI mass spectra of the digest mixture [97].

A method for rapidly determining protein epitopes—the sites on antigen molecules that bind to antibodies in an immunological response—involves a strategy of proteolysis followed by precipitation with the appropriate antibody. The antigenic peptides can then be rapidly analyzed by MALDI mass spectrometry [98]. Papac *et al.* determined the epitope of the gastrin-release peptide to an antibody by in situ proteolysis of the immobilized antigen-antibody complex, followed by MALDI analysis [99]. Similarly, antibodies bound to haptens—molecules that bind specifically to the antigen binding site but do not induce immune responses—can be identified using strategies involving MALDI mass spectrometry [100, 101].

9.4.2 Nucleic Acid Analysis

There have been numerous reports of investigations of MALDI for the mass spectrometric analysis of oligonucleotides [46, 102–104]. The detection of DNA as large as 500 nucleotides in length has been reported [105], and sensitivities are typically in the low picomole range. The current focus on the use of MALDI mass spectrometry for the analysis of oligonucleotides is the development of a rapid method for sequencing nucleic acids. A method using mass spectrometry would offer substantial advantages in speed over current methods for sequencing oligonucleotides, which require chromatographic separation of sequence ladders. There has been some success in the sequencing of oligonucleotides up to about 24 monomers in length using MALDI mass spectrometry [106, 107]. However, although there appears to be much promise for a rapid sequencing technique, to date the limitations of the method have prevented its general acceptance for solving problems in molecular biology.

Typical results of MALDI analysis of oligonucleotides demonstrate the problems in developing a sequencing method based on mass spectrometry. There has been limited success in analyzing mixed-base polymers large enough to make DNA sequencing feasible. The phosphate backbone is significantly more labile than the peptide backbone of proteins, making analysis of high mass polymers difficult. Zhu *et al.* and Schneider and Chait have investigated the gas phase decomposition of oligodeoxynucleotide ions and found that the fragmentation is dependent on both the base composition of the oligomer and on the matrix used for analysis [108, 109]. Homopolymers that contain only the nucleotide thymidine have been shown to be relatively stable in the gas phase [110]. Homopolymers of the other nucleotide bases and mixed polymers undergo metastable decomposition, so that the molecular ion signal is very weak for oligonucleotides longer than about 10 nucleotides in length [46]. For UV MALDI, use of the matrix 3-hydroxypicolinic acid has been shown to yield spectra with less fragmentation of the molecular ion [108, 111]. Mixed-base polymers as large as 89 nucleotides in length have been successfully analyzed using this matrix [112]. Continued optimization of the matrix conditions or chemical methods that help to decrease fragmentation such as alkylation [113] or formation of 7-deaza purine bases [114, 115] may lead to further enhancements in the ability to sequence larger oligomers.

Several studies have focused on the modifications and mutations of oligonucleotides and nucleic acids using MALDI mass spectrometry for analysis. Chang et al. detected the mutation of the cystic fibrosis gene in clinical samples, which is characterized by the deletion of three nucleotides that code for the amino acid phenylalanine [116]. This work demonstrates the utility of the method for the rapid screening of DNA mutations in clinical samples. Also, MALDI has been used for the investigations of Pt-DNA adducts, whose formation may be important in the anti-cancer activity of some platinum complexes [117, 118].

9.4.3 Oligosaccharide Analysis

With the use of appropriate matrixes, oligosaccharide ions can be formed using MALDI ions sources [119–121]. There are several examples of the mass spectrometric characterization of saccharides obtained from biological sources. The oligosaccharides from human milk showed saccharides with masses ranging up to 8000 Da [122]. Costello et al. have used MALDI to characterize glycosphingolipids, which are oligosaccharide-lipid conjugates found on the surfaces of cells [123, 124]. Quantitative analysis of oligosaccharides using MALDI has been the focus of some research as well [125, 126].

Many of the applications of MALDI for the analysis of oligosaccharides have focused on determining the type of glycosylation present in glycoproteins [127–133]. The presence of glycosylation on a protein can often have deleterious effects on the ability to analyze the sample using mass spectrometry. The presence of microheterogeneity, in which a protein may contain several closely related saccharide structures on separate protein molecules, leads to broadening of peaks and a loss of sensitivity. A typical strategy is to digest the glycoprotein with an appropriate endoprotease such as trypsin to generate a peptide map, as described previously. The glycopeptides generated can be analyzed using MALDI to show which peptide fragments contain the saccharide moiety [131]. Determining the structure of the saccharides, however, usually involves a strategy combining chromatography, chemical derivatization of the saccharides such as acetylation or methylation of the free hydroxyl groups, and tandem mass spectrometry. The utility of tandem mass spectrometry using MALDI with post-source decay analysis on a reflectron time-of-flight instrument has been demonstrated both for the structural analysis of isolated oligosaccharides and for glycopeptide mixtures [131, 132].

9.4.4 Analysis of Synthetic Polymers and Fuels

MALDI has several advantages over other techniques for the analysis of synthetic polymers. It allows for the generation of intact, high mass polymer ions, which can be detected with minimal mass discrimination using time-of-flight (TOF) mass analyzers. Synthetic polymers with molecular weights up to 70,000 Da have been recorded [134]. MALDI-TOF offers a rapid, sensitive, and accurate (0.01% using internal calibration) means of obtaining the molecular weight distributions of polymers. Compared to gel permeation chromatography, a common method used for the analysis of polymers, MALDI compared favorably in determining the molecular weight distribution of ((R)-3-hydroxybutanoates) [135]. Mainly singly charged ions are formed during MALDI, so that the spectra are easily interpretable [134]. Synthetic polymers such as polyesters, polymethylmethacrylates, polyvinyl alcohol, and polyvinyl chloride have been analyzed using MALDI-TOF [136–138]. Montaudo *et al.* demonstrated a self-calibrating procedure for poly(caprolactone), poly(Bisphenol A carbonate), Nylon 6, and poly(butyleneadipate), which uses a best-fit minimization algorithm to obtain the absolute mass values [139, 140]. This technique was shown to be useful for identifying the end groups of these polymers [139].

With the use of Fourier transform mass spectrometry (FTMS), the limitations on mass accuracy and mass resolution of TOF mass analyzers can be overcome. Resolution up to 92,000 (full width at half height, FWHH) and mass accuracy as high as 10 ppm have been obtained using MALDI-FTMS for ions of bovine insulin (MW = 5735.3 Da) [141–143]. The increased resolution allows for the separation of the repeating monomer at higher molecular weights, so that the molecular weight distribution of the polymer can be easily established. The high sensitivity of MALDI combined with the high resolution and mass accuracy of FTMS has been used for the characterization of synthetic polymers with molecular weights as large as 10,000 Da [144].

High molecular weight hydrocarbons form ions from these sources as well. MALDI has been proven useful for the formation of hydrocarbon ions from samples such as kerogens and coal extracts [145–147]. John *et al.* reported the detection of ions as large as 270,000 Da from coal samples [148, 149]. Also, synthetically prepared hydrocarbon dendrimers with mass 40,000 Da were measured with 0.01% mass accuracy [150].

References

In addition to the references given below, we have prepared a comprehensive list of papers involving matrix-assisted laser desorption. The list is in a standard spreadsheet format (Microsoft Excel) and it is available at the universal resource locator "http://128.127.10.5/preprints/contents.htm" by selecting the "literature database" entry corresponding to this chapter.

1. Loo, R. R. O., Winger, B. E., and Smith, R. D. Proton transfer reaction studies of multiply charged proteins in a high mass-to-charge ratio quadrupole mass spectrometer. *J. Am. Soc. Mass Spectrom.* **5**, 1064–1071 (1994).
2. Macfarlane, R. D., and Torgesson, D. F. Californium-252 plasma desorption mass spectrometry. *Science* **191**, 920–925 (1976).
3. Schmitter, J.-M. Performances and limits of plasma desorption mass spectrometry in the primary structure determination of proteins. *J. Chromatogr.* **557**, 359–368 (1991).
4. Benninghoven, A., and Sichtermann, W. K. Detection, identification and structural investigation of biologically important compounds by secondary ion mass spectrometry. *Anal. Chem.* **50**, 1180–1184 (1978).
5. Barber, M., Bordoli, R. S., Sedgwick, R. D., and Tyler, A. N. Fast atom bombardment of solids (F.A.B.): A new source for mass spectrometry. *J. Chem. Soc. Chem. Commun.* 325–329 (1981).
6. Andrews, P. C. and Dixon, J. E. Application of fast atom bombardment mass spectrometry to posttranslational modifications of neuropeptides, in *Hormone Action Part K: Neuroendocrine Peptides*, P. M. Conn (ed.), Vol. 168, pp. 72–103. Methods in Enzymology, Academic Press, New York, 1989.
7. Hillenkamp, F. Laser-induced ion formation from organic solids, in *Ion Formation from Organic Solids*, A. Benninghoven (ed.), Vol. 25, pp.190–205. Springer Series in Chemical Physics, Springer-Verlag, Berlin, 1983.
8. Tanaka, K., Waki, H., Ido, Y., Akita, S., Yoshida, Y., and Yoshida, T. Protein and polymer analysis up to m/z 100,000 by laser ionization time-of-flight mass spectrometry. *Rapid Commun. Mass Spectrom.* **2**, 151–153 (1988).
9. Karas, M., Bachmann, D., Bahr, U., and Hillenkamp, F. Matrix-assisted ultraviolet laser desorption of non-volatile compounds. *Int. J. Mass Spectrom. Ion Processes* **78**, 53–68 (1987).
10. Karas, M., and Hillenkamp, F. Laser desorption ionization of proteins with molecular masses exceeding 10,000 daltons. *Anal. Chem.* **60**, 2299–2301 (1988).
11. Beavis, R. C., and Chait, B. T. Cinnamic acid derivatives as matrices for ultraviolet laser desorption mass spectrometry of proteins. *Rapid Commun. Mass Spectrom.* **3**, 432–435 (1989).
12. Strupat, K., Karas, M., and Hillenkamp, F. 2,5-Dihydroxybenzoic acid: a new matrix for laser desorption-ionization mass spectrometry. *Int. J. Mass Spectrom. Ion Processes* **111**, 89–102 (1991).
13. Beavis, R. C., Chaudhary, T., and Chait, B. T. α-Cyano-4-hydroxycinnamic acid as a matrix for matrix-assisted laser desorption mass spectrometry. *Org. Mass Spectrom.* **27**, 156–158 (1992).
14. Cohen, S. L. and Chait, B. T. Influence of matrix solution conditions on the MALDI-MS analysis of peptides and proteins. *Anal. Chem.* **68**, 31–37 (1996).
15. Vorm, O., Roepstorff, P., and Mann, M. Improved resolution and very high sensitivity in MALDI TOF of matrix surfaces made by fast evaporation. *Anal. Chem.* **66**, 3281–3287 (1994).

16. Xiang, F., and Beavis, R. C. Growing protein-doped sinapic acid crystals for laser desorption: An alternative preparation method for difficult samples. *Org. Mass Spectrom.* **28**, 1424–1429 (1993).
17. Xiang, F., and Beavis, R. C. A method to increase contaminant tolerance in protein matrix-assisted laser desorption/ionization by the fabrication of thin protein-doped polycrystalline films. *Rapid Commun. Mass Spectrom.* **8**, 199–204 (1994).
18. Cotter, R. J., Time-of-flight mass spectrometry. *ACS Symposium Series* 549, Washington, DC, 1994.
19. Hill, J. A., Annan, R. S., and Biemann, K., Matrix-assisted laser desorption ionization with a magnetic mass spectrometer. *Rapid Commun. Mass Spectrom.* **5**, 395–399 (1991).
20. Chambers, D. M., Goeringer, D. E., McCluckey, S. A., and Glish, G. L., Matrix-assisted laser desorption of biological molecules in the quadrupole ion trap mass spectrometer. *Anal. Chem.* **65**, 14–20 (1993).
21. McIver, R. T., Jr., Li, Y., and Hunter, R. L. High-resolution laser desorption mass spectrometry of peptides and small proteins. *Proc. Natl. Acad. Sci. USA* **91**, 4801–4805 (1994).
22. Hillenkamp, F., Karas, M., Beavis, R. C. and Chait, B. T. Matrix-assisted laser desorption/ionization mass spectrometry of biopolymers. *Anal. Chem.* **63**, 1193A–1203A (1991).
23. Salehpour, M., Perera, I., Kyellberg, J., Hedin, A., Islamian, M. A., Håkansson, P., and Sundqvist, B. U. R. Laser-induced desorption of proteins. *Rapid Commun. Mass Spectrom.* **3**, 259–263 (1989).
24. Overberg, A., Karas, M., Bahr, U., Kaufmann, R., and Hillenkamp, F. Matrix-assisted infrared-laser (2.94 um) desorption/ionization mass spectrometry of large biomolecules. *Rapid Commun. Mass Spectrom.* **4**, 293–296 (1990).
25. Beavis, R. C., and Chait, B. T. Factors affecting the ultraviolet laser desorption of proteins. *Rapid Commun. Mass Spectrom.* **3**, 233–237 (1989).
26. Demirev, P., Westman, A., Reimann, C. T., Håkansson, P., Barofsky, D., Sundqvist, B. U. R., Cheng, Y. D., Seibt, W., and Siegbahn, K. Matrix-assisted laser desorption with ultra-short laser pulses. *Rapid Commun. Mass Spectrom.* **6**, 187–191 (1992).
27. Dreisewerd, K., Schurenberg, M., Karas, M., and Hillenkamp, F. Influence of the laser intensity and spot size on the desorption of molecules and ions in matrix-assisted laser desorption/ionization with a uniform beam profile. *Int. J. Mass Spectrom. Ion Processes* **141**, 127–148 (1995).
28. Beavis, R. C. and Chait, B. T. Velocity distributions of intact high mass polypeptide molecule ions produced by matrix assisted laser desorption. *Chem. Phys. Lett.* **181**, 479–484 (1991).
29. Ens, W., Mao, Y., Mayer, F., and Standing, K. G. Properties of matrix-assisted laser desorption. measurements with a time-to-digital converter. *Rapid Commun. Mass Spectrom.* **5**, 117–123 (1991).
30. Bokelmann, V., Spengler, B., and Kaufmann, R. Dynamical parameters of ion ejection and ion formation in matrix-assisted laser desorption/ionization. *Eur. Mass Spectrom.* **1**, 81–93 (1995).
31. Zhou, J., Ens, W., Standing, K. G., and Verentchikov, A. Kinetic energy measurements of molecular ions ejected into an electric field by matrix-assisted laser desorption. *Rapid Commun. Mass Spectrom.* **6**, 671–678 (1992).
32. Kaufmann, R., Spengler, B., and Lutzenkirchen, F. Mass spectrometric sequencing of linear peptides by production analysis in a reflection time-of-flight mass

spectrometer using matrix-assisted laser desorption ionization. *Rapid Commun. Mass Spectrom.* **7**, 902–910 (1993).
33. Cornish, T. J., and Cotter, R. J. A curved field reflectron time-of-flight mass spectrometer for the simultaneous focusing of metastable product ions. *Rapid Commun. Mass Spectrom.* **8**, 781–785 (1994).
34. Yu, W., Vath, J. E., Huberty, M. C., and Martin, S. A. Identification of the gas-phase cleavage of the asp-pro and asp-xxx peptide bonds in matrix-assisted laser desorption. *Anal. Chem.* **65**, 3015–3023 (1993).
35. Vorm, O. and Mann, M. Improved mass accuracy in matrix-assisted laser desorption/ionization time-of-flight mass spectrometry of peptides. *J. Am. Soc. Mass Spectrom.* **5**, 955–958 (1994).
36. Brown, R. S., and Lennon, J. J. Mass resolution improvement by incorporation of pulsed ion extraction in a matrix-assisted laser desorption/ionization linear time-of-flight mass spectrometer. *Anal. Chem.* **67**, 1998–2003 (1995).
37. Vestal, M. L., Juhasz, P., and Martin, S. A. Delayed extraction matrix-assisted laser desorption time-of-flight mass spectrometry. *Rapid Commun. Mass Spectrom.* **9**, 1044–1050 (1995).
38. Wiley, W. C., and McLaren, I. H. Time-of-flight mass spectrometer with improved resolution. *Rev. Sci. Instrum.* **26**, 1150–1157 (1955).
39. Siuzdak, G. The emergence of mass spectrometry in biochemical research. *Proc. Natl. Acad. Sci. USA* **91**, 11290–11297 (1994).
40. Karas, M., Ingendoh, A., Bahr, U., and Hillenkamp, F. Ultraviolet-laser desorption/ionization mass spectrometry of femtomolar amounts of large proteins. *Biomed. Environ. Mass Spectrom.* **18**, 841–843 (1989).
41. Arnott, D., Shabanowitz, J., and Hunt, D. F. Mass spectrometry of proteins and peptides: sensitive and accurate mass measurement and sequence analysis. *Clin. Chem.* **39**, 2005–2010 (1993).
42. Zaluzec, E. J., Gage, D. A., and Watson, J. T. Matrix-assisted laser desorption ionization mass spectrometry—applications in peptide and protein characterization. *Prot. Exp. Purif.* **6**, 109–123 (1995).
43. Fitzgerald, M. C. and Smith, L. M. Mass spectrometry of nucleic acids: The promise of matrix-assisted laser desorption-ionization (MALDI) mass spectrometry. *Ann. Rev. Biophys. Biomolec. Struct.* **24**, 117–140 (1995).
44. Harvey, D. J. Matrix-assisted laser desorption/ionization mass spectrometry of oligosaccharides. *Am. Lab.* **26**, 22–28 (1994).
45. Steiner, V., Bornsen, K. O., Schar, M., Gassmann, E., Hoffstetter-Kuhn, S., Rink, H., and Mutter, M. Analysis of synthetic peptides using matrix-assisted laser desorption ionization mass spectrometry. *Peptide Res.* **5**, 25–29 (1992).
46. Parr, G. R., Fitzgerald, M. C., and Smith, L. M. Matrix-assisted laser desorption/ionization mass spectrometry of synthetic oligodeoxyribonucleotides. *Rapid Commun. Mass Spectrom.* **6**, 369–372 (1992).
47. Bahr, U., Karas, M. and Hillenkamp, F. Analysis of biopolymers by matrix-assisted laser desorption/ionization mass spectrometry. *Fresenius J. Anal. Chem.* **348**, 783–791 (1994).
48. Beavis, R. C., and Chait, B. T. Rapid, sensitive analysis of protein mixtures by mass spectrometry. *Proc. Natl. Acad. Sci. USA* **87**, 6873–6877 (1990).
49. Karas, M., Bahr, U., and Hillenkamp, F. UV laser matrix desorption/ionization mass spectrometry of proteins in the 100,000 dalton range. *Int. J. Mass Spectrom. Ion Processes* **92**, 231–242 (1989).

50. Beavis, R. C., and Chait, B. T. High-accuracy molecular mass determination of proteins using matrix-assisted laser desorption mass spectrometry. *Anal. Chem.* **62**, 1836–1840 (1990).
51. Smith, E. T., Cornett, D. S., Amster, I. J., and Adams, M. W. W. Protein molecular weight determinations by MALD mass spectrometry: A superior alternative to gel filtration. *Anal. Biochem.* **209**, 379–380 (1993).
52. Overberg, A., Karas, M., and Hillenkamp, F. Matrix-assisted laser desorption of large biomolecules with a TEA-CO_2-laser. *Rapid Commun. Mass Spectrom.* **5**, 128–131 (1991).
53. Castoro, J. A., and Wilkins, C. L. Ultrahigh resolution matrix-assisted laser desorption/ionization of small proteins by fourier transform mass spectrometry. *Anal. Chem.* **65**, 2621–2627 (1993).
54. Cain, T. C., Lubman, D. M., and Weber, Jr., W. J. Differentiation of bacteria using protein profiles from matrix-assisted laser desorption/ionization time-of-flight mass spectrometry. *Rapid Commun. Mass Spectrom.* **8**, 1026–1030 (1994).
55. Lee, W. Y., Brune, D. C., Lobrutto, R., and Blankenship, R. E. Isolation, characterization, and primary structure of rubredoxin from the photosynthetic bacterium, *Heliobacillus mobilis*. *Arch. Biochem. Biophys.* **318**, 80–88 (1995).
56. Ownby, D. W., Zhu, H., Schneider, K., Beavis, R. C., Chait, B. T., and Riggs, A. F. The extracellular hemoglobin of the earthworm, *Lumbricus terrestris*. Determination of subunit stoichiometry. *J. Biol. Chem.* **268**, 13539–13547 (1993).
57. Yasuhara, Y., Koizumi, Y., Katagiri, C., and Ashida, M. Reexamination of properties of prophenoloxidase isolated from larval hemolymph of the silkworm *Bombyx mori*. *Arch. Biochem. Biophys.* **320**, 14–23 (1995).
58. Gade, G. Isolation and identification of AKH/RPCH family peptides in blister beetles (meloidae). *Physiol. Entomol.* **20**, 45–51 (1995).
59. Haebel, S., Jensen, C., Andersen, S. O., and Roepstorff, P. Isoforms of a cuticular protein from larvae of the meal beetle, *Tenebrio molitor*, studied by mass spectrometry in combination with Edman degradation and two-dimensional polyacrylamide gel electrophoresis. *Prot. Sci.* **4**, 394–404 (1995).
60. Vendeland, S. C., Beilstein, M. A., Chen, C. L., Jensen, O. N., Barofsky, E., and Whanger, P. D. Purification and properties of selenoprotein W from rat muscle. *J. Biol. Chem.* **268**, 17103–17107 (1993).
61. Wolff, E. C., Lee, Y. B., Chung, S. I., Folk, J. E., and Park, M. H. Deoxyhypusine synthase from rat testis—Purification and characterization. *J. Biol. Chem.* **270**, 8660–8666 (1995).
62. Denu, J. M., Zhou, G. C., Wu, L., Zhao, R., Yuvaniyama, J. D., Saper, M. A., and Dixon, J. E. The purification and characterization of a human dual-specific protein tyrosine phosphatase. *J. Biol. Chem.* **270**, 3796–3803 (1995).
63. Miller, D. L., Papayannopoulos, I. A., Styles, J., Bobin, S. A., Lin, Y. Y., Biemann, K., and Iqbal, K. Peptide compositions of the cerebrovascular and senile plaque core amyloid deposits of Alzheimer's disease. *Arch. Biochem. Biophys.* **301**, 41–52 (1993).
64. Vigo-Pelfrey, C., Lee, D., Keim, P., Lieberburg, L., and Schenk, D. B. Characterization of beta-amyloid peptide from human cerebrospinal fluid. *J. Neurochem.* **61**, 1965–1968 (1993).
65. Li, K. W., van Golen, F. A., van Minnen, J., van Veelen, P. A., van der Greef, J., and Geraerts, W. P. Structural identification, neuronal synthesis, and role in male copulation of myomodulin-A of Lymnaea: A study involving direct peptide profiling of nervous tissue by mass spectrometry. *Brain Res.* **25**, 355–358 (1994).

66. Li, K. W., Hoek, R. M., Smith, F., Jimenez, C. R., van der Schors, R. C., van Veelen, P. A., Chen, S., van der Greef, J., Parish, D. C., Benjamin, P. R., and Geraerts, W. P. M. Direct peptide profiling by mass spectrometry of single identified neurons reveals complex neuropeptide-processing pattern. *J. Biol. Chem.* **269**, 30288–30292 (1994).
67. Jimenez, C. R., van Veelen, P. A., Li, K. W., Wildering, W. C., Geraerts, W. P. M., Tjaden, U. R., and van der Greef, J. Neuropeptide expression and processing as revealed by direct matrix-assisted laser desorption ionization mass spectrometry of single neurons. *J. Neurochem.* **62**, 404–407 (1994).
68. Li, K. W., Vanminnen, J., Vandergreef, J., and Geraerts, W. P. M. Direct mass spectrometry of buccal ganglia and nerves reveals the processing and targeting of peptide messengers involved in feeding behaviour of lymnaea. *Neth. J. Zoo.* **44**, 432–438 (1994).
69. van Veelen, P. A., Jimenez, C. R., Li, K. W., Wildering, W. C., Geraerts, W. P. M., Tjaden, U. R., and van der Greef, J. Direct peptide profiling of single neurons by matrix-assisted laser desorption—Ionization mass spectrometry. *Org. Mass Spectrom.* **28**, 1542–1546 (1993).
70. Schar, M., Bornsen, K. O., and Gassmann, E. Fast protein sequence determination with matrix-assisted laser desorption and ionization mass spectrometry. *Rapid Commun. Mass Spectrom.* **5**, 319–326 (1991).
71. Chait, B. T., Wang, R., Beavis, R. C., and Kent, S. B. H. Protein ladder sequencing. *Science* **262**, 89–92 (1993).
72. Bartlet-Jones, M., Jeffery, W. A., Hansen, H. F., and Pappin, D. J. C. Peptide ladder sequencing by mass spectrometry using a novel, volatile degradation reagent. *Rapid Commun. Mass Spectrom.* **8**, 737–742 (1994).
73. Thiede, B., Wittmann-Liebold, B., Bienert, M., and Krause, E. MALDI-MS for C-terminal sequence determination of peptides and proteins degraded by carboxypeptidase Y and P. *FEBS Lett.* **357**, 65–69 (1995).
74. Spengler, B., Kirsch, D., Kaufmann, R., and Jaeger, E. Peptide sequencing by matrix-assisted laser-desorption mass spectrometry. *Rapid Commun. Mass Spectrom.* **6**, 105–108 (1992).
75. Beussman, D. J., Vlasak, P. R., McLane, R. D. Seeterlin, M. A., and Enke, C. G. Tandem reflectron time-of-flight mass spectrometer utilizing photodissociation. *Anal. Chem.* **67**, 3952–3957 (1995).
76. Cornish, T. and Cotter, R. J. A compact time-of-flight mass spectrometer for the structural analysis of biological molecules using laser desorption. *Rapid Commun. Mass Spectrom.* **6**, 242–248 (1992).
77. Yates III, J. R., Speicher, S., Griffin, P. R., and Hunkapiller, T. Peptide mass maps: A highly informative approach to protein identification. *Anal. Biochem.* **214**, 397–408 (1993).
78. Henzel, W. J., Billeci, T. M., Stults, J. T., Wong, S. C., Grimley, C., and Watanabe, C. Identifying proteins from two-dimensional gels by molecular mass searching of peptide fragments in protein sequence databases. *Proc. Natl. Acad. Sci. USA* **90**, 5011–5015 (1993).
79. Cottrell, J. S. Protein identification by peptide mass fingerprinting. *Peptide Res.* **7**, 115–124 (1994).
80. Mortz, E., Vorm, O., Mann, M., and Roepstorff, P. Identification of proteins in polyacrylamide gels by mass spectrometric peptide mapping combined with database search. *Biol. Mass Spectrom.* **23**, 249–261 (1994).
81. Rasmussen, H. H., Mortz, E., Mann, M., Roepstorff, P., and Celis, J. E.

Identification of transformation sensitive proteins recorded in human two-dimensional gel protein databases by mass spectrometric peptide mapping alone and in combination with microsequencing. *Electrophoresis* **15**, 406–416 (1994).
82. Sutton, C. W., Pemberton, K. S., Cottrell, J. S., Corbett, J. M., Wheeler, C. H., Dunn, M. J., and Pappin, D. J. Identification of myocardial proteins from two-dimensional gels by peptide mass fingerprinting. *Electrophoresis* **16**, 308–316 (1995).
83. Tsarbopoulos, A., Karas, M., Strupat, K., Pramanik, B. N., Nagabhushan, T. L., and Hillenkamp, F. Comparative mapping of recombinant proteins and glycoproteins by plasma desorption and matrix-assisted laser desorption/ionization mass spectrometry. *Anal. Chem.* **66**, 2062–2070 (1994).
84. Billeci, T. M. and Stults, J. T. Tryptic Mapping of recombinant proteins by matrix-assisted laser desorption/ionization mass spectrometry. *Anal. Chem.* **65**, 1709–1716 (1993).
85. Treuheit, M. J., Costello, C. E., and Halsall, H. B. Analysis of the five glycosylation sites of human alpha 1-acid glycoprotein. *Biochem. J.* **283**, 105–112 (1992).
86. Andersen, J. S., Sogaard, M., Svensson, B., and Roepstorff, P. Localization of an O-glycosylated site in the recombinant barley alpha-amylase 1 produced in yeast and correction of the amino acid sequence using matrix-assisted laser desorption/ionization mass spectrometry of peptide mixtures. *Biol. Mass Spectrom.* **23**, 547–554 (1994).
87. Manneberg, M., Friedlein, A., Kurth, H., Lahm, H. W., and Fountoulakis, M. Structural analysis and localization of the carbohydrate moieties of a soluble human interferon gamma receptor produced in baculovirus-infected insect cells. *Prot. Sci.* **3**, 30–38 (1994).
88. James, D. C., Freedman, R. B., Hoare, M., Ogonah, O. W., Rooney, B. C., Larionov, O. A., Dobrovolsky, V. N., Lagutin, O. V., and Jenkins, N. N-glycosylation of recombinant human interferon-gamma produced in different animal expression systems. *Bio-Technology* **13**, 592–596 (1995).
89. Kieliszewski, M. J., O'Neill, M., Leykam, J., and Orlando, R. Tandem mass spectrometry and structural elucidation of glycopeptides from a hydroxyproline-rich plant cell wall glycoprotein indicate that contiguous hydroxyproline residues are the major sites of hydroxyproline O-arabinosylation. *J. Biol. Chem.* **270**, 2541–2549 (1995).
90. Yip, T. T. and Hutchens, T. W. Mapping and sequence-specific identification of phosphopeptides in unfractionated protein digest mixtures by matrix-assisted laser desorption/ionization time-of-flight mass spectrometry. *FEBS Lett.* **308**, 149–153 (1992).
91. Zhang, W., Czernik, A. J., Yungwirth, T., Aebersold, R., and Chait, B. T. Matrix-assisted laser desorption mass spectrometric peptide mapping of proteins separated by two-dimensional gel electrophoresis: Determination of phosphorylation in synapsin I. *Prot. Sci.* **3**, 677–686 (1994).
92. Clauser, K. R., Hall, S. C., Smith, D. M., Webb, J. W., Andrews, L. E., Tran, H. M., Epstein, L. B., and Burlingame, A. L. Rapid mass spectrometric peptide sequencing and mass matching for characterization of human melanoma proteins isolated by two-dimensional PAGE. *Proc. Natl. Acad. Sci. USA* **92**, 5072–5076 (1995).
93. Robertson, J. G., Adams, G. W., Medzihradszky, K. F., Burlingame, A. L., and Villafranca, J. J. Complete assignment of disulfide bonds in bovine dopamine beta-hydroxylase. *Biochemistry* **33**, 11563–11575 (1994).

94. Crimmins, D. L., Saylor, M., Rush, J., and Thoma, R. S. Facile in situ matrix-assisted laser desorption/ionization mass spectrometry analysis and assignment of disulfide pairings in heteropeptide molecules. *Anal. Biochem.* **226**, 355–361 (1995).
95. Roberts, E. S., Hopkins, N. E., Zaluzec, E. J., Gage, D. A., Alworth, W. L., and Hollenberg, P. F. Identification of active-site peptides from ^{3}H-labeled 2-ethynylnaphthalene-inactivated P450 2B1 and 2B4 using amino acid sequencing and mass spectrometry. *Biochemistry* **33**, 3766–3771 (1994).
96. Ploemen, J. H., Johnson, W. W., Jespersen, S., Vanderwall, D., van Ommen, B., van der Greef, J., van Bladeren, P. J., and Armstrong, R. N. Active-site tyrosyl residues are targets in the irreversible inhibition of a class Mu glutathione transferase by 2-(S-glutathionyl)-3,5,6-trichloro-1,4-benzoquinone. *J. Biol. Chem.* **269**, 26890–26897 (1994).
97. Hutchens, T. W., Nelson, R. W., and Yip, T. T. Recognition of transition metal ions by peptides. Identification of specific metal-binding peptides in proteolytic digest maps by UV laser desorption time-of-flight mass spectrometry. *FEBS Lett.* **296**, 99–102 (1992).
98. Zhao, Y. and Chait, B.T. Protein Epitope Mapping by Mass Spectrometry. Anal. Chem., 66, 3723-3726 (1994).
99. Papac, D. I., Hoyes, J., and Tomer, K. B. Epitope mapping of the gastrin-releasing peptide/anti-bombesin monoclonal antibody complex by proteolysis followed by matrix-assisted laser desorption ionization mass spectrometry. *Prot. Sci.* **3**, 1485–1492 (1994).
100. Wengatz, I., Schmid, R. D., Kreibig, S., Wittmann, C., Hock, B., Ingendoh, A., and Hillenkamp, F. Determination of the hapten density of immunoconjugates by matrix-assisted UV laser desorption/ionization mass spectrometry. *Anal. Lett.* **25**, 1983–1997 (1992).
101. Adamczyk, M., Buko, A., Chen, Y.-Y., Fishpaugh, J. R., Gebler, J. C., and Johnson, D. D. Characterization of protein-hapten conjugates. 1. Matrix-assisted laser desorption Ionization mass spectrometry of immuno BSA-hapten conjugates and comparison with other characterization methods. *Bioconjugate Chem.* **5**, 631–635 (1994).
102. Spengler, B., Pan, Y., Cotter, R. J., and Kan, L.-S. Molecular weight determination of underivatized oligodeoxyribonucleotides by positive-ion matrix-assisted ultraviolet laser-desorption mass spectrometry. *Rapid Commun. Mass Spectrom.* **4**, 99–102 (1990).
103. Hettich, R. L., and Buchanan, M. V. Matrix-assisted laser desorption fourier transform mass spectrometry for the structural examination of modified nucleic acid constituents. *Int. J. Mass Spectrom. Ion Processes* **111**, 365–380 (1991).
104. Hettich, R. L., and Buchanan, M. V. Structural characterization of normal and modified oligonucleotides by matrix-assisted laser desorption fourier transform mass spectrometry. *J. Am. Soc. Mass Spectrom.* **2**, 402–412 (1991).
105. Tang, K., Taranenko, N. I., Allman, S. L., Chang, L. Y., and Chen, C. H. Detection of 500-nucleotide DNA by laser desorption mass spectrometry. *Rapid Commun. Mass Spectrom.* **8**, 727–730 (1994).
106. Nordhoff, E., Karas, M., Cramer, R., Hahner, S., Hillenkamp, F., Kirpekar, F., Lezius, A., Muth, J., Meier, C., and Engels, J. W. Direct mass spectrometric sequencing of low-picomole amounts of oligodeoxynucleotides with up to 21 bases by matrix-assisted laser desorption/ionization mass spectrometry. *J. Mass Spectrom.* **30**, 99–112 (1995).

107. Fitzgerald, M. C., Zhu, L., and Smith, L. M. The analysis of mock dna sequencing reactions using matrix-assisted laser desorption/ionization mass spectrometry. *Rapid Commun. Mass Spectrom.* **7**, 895–897 (1993).
108. Zhu, L., Parr, G., Fitzgerald, M. C., Nelson, C. M., and Smith, L. M. Oligodeoxynucleotide fragmentation in MALDI/TOF mass spectrometry using 355-nm radiation. *J. Am. Chem. Soc.* **117**, 6048–6056 (1995).
109. Schneider, K., and Chait, B. T. Matrix-assisted laser desorption mass spectrometry of homopolymer oligodeoxyribonucleotides. Influence of base composition on the mass spectrometric response. *Org. Mass Spectrom.* **28**, 1353–1361 (1993).
110. Tang, K., Allman, S. L., Jones, R. B., Chen, C. H., and Araghi, S. Laser mass spectrometry of polydeoxyribothymidylic acid mixtures. *Rapid Commun. Mass Spectrom.* **7**, 63–66 (1993).
111. Wu, K. J., Steding, A., and Becker, C. H. Utility of matrix-assisted laser desorption/ionization time-of-flight mass spectrometry of oligonucleotides using 3-hydroxypicolinic acid as an ultraviolet sensitive matrix. *Rapid Commun. Mass Spectrom.* **7**, 142–146 (1993).
112. Wu, K. J., Shaler, T. A., and Becker, C. H. Time-of-flight mass spectrometry of underivatized single-stranded DNA oligomers by matrix-assisted laser desorption. *Anal. Chem.* **66**, 1637–1645 (1994).
113. Gut, I. G. and Beck, S. A procedure for selective DNA alkylation and detection by mass spectrometry. *Nucleic Acids Res.* **23**, 1367–1373 (1995).
114. Kirpekar, F., Nordhoff, E., Kristiansen, K., Roepstorff, P., Hahner, S., and Hillenkamp, F. 7-deaza purine bases offer a higher ion stability in the analysis of DNA by matrix-assisted laser desorption ionization mass spectrometry. *Rapid Commun. Mass Spectrom.* **9**, 525–531 (1995).
115. Schneider, K., and Chait, B. T. Increased stability of nucleic acids containing 7-deaza-guanosine and 7-deaza-adenosine may enable rapid DNA sequencing by matrix-assisted laser desorption mass spectrometry. *Nucleic Acids Res.* **23**, 1570–1575 (1995).
116. Chang, L. Y., Tang, K., Schell, M., Ringelberg, C., Matteson, K. J., Allman, S. L., and Chen, C. H. Detection of delta-f508 mutation of the cystic fibrosis gene by matrix-assisted laser desorption ionization mass spectrometry. *Rapid Commun. Mass Spectrom.* **9**, 772–774 (1995).
117. Guittard, J., Pacifico, C., Blais, J. C., Bolbach, G., Chottard J. C., and Spassky, A. Matrix-assisted laser desorption ionization time-of-flight mass spectrometry of DNA-Pt(II) complexes. *Rapid Commun. Mass Spectrom.* **9**, 33–36 (1995).
118. Costello, C. E., Nordhoff, E., and Hillenkamp, F. Matrix-assisted UV and IR laser desorption-ionization time-of-flight mass spectrometry of diamminoplatinum(II) oligodeoxyribonucleotide adducts and their unplatinated analogs. *Int. J. Mass Spectrom. Ion Processes* **132**, 239–249 (1994).
119. Egge, H., Peter-Katalinic, J., Karas, M., and Stahl, B. The use of fast atom bombardment and laser desorption mass spectrometry in the analysis of complex carbohydrates. *Pure Appl. Chem.* **63**, 491–498 (1991).
120. Stahl, B., Steup, M., Karas, M., and Hillenkamp, F. Analysis of neutral oligosaccharides by matrix-assisted laser desorption/ionization mass spectrometry. *Anal. Chem.* **63**, 1463–1466 (1991).
121. Mock, K. K., Davey, M., and Cottrell, J. S. The analysis of underivatized oligosaccharides by matrix assisted laser desorption mass spectrometry. *Biochem. Biophys. Res. Commun.* **177**, 644–651 (1991).
122. Stahl, B., Thurl, S., Zeng, J., Karas, M., Hillenkamp, F., Steup, M., and

Sawatzki, G. Oligosaccharides from human milk as revealed by matrix-assisted laser desorption/ionization mass spectrometry. *Anal. Biochem.* **223**, 218–226 (1994).
123. Perreault, H. and Costello, C. E. Liquid secondary ionization, tandem and matrix-assisted laser desorption/ionization time-of-flight spectrometric characterization of glycosphingolipid derivatives. *Org. Mass Spectrom.* **29**, 720–735 (1994).
124. Juhasz, P. and Costello, C. E. Matrix-assisted laser desorption ionization time-of-flight mass spectrometry of underivatized and permethylated gangliosides. *J. Am. Soc. Mass Spectrom.* **3**, 785–796 (1992).
125. Harvey, D. J. Quantitative aspects of the matrix-assisted laser desorption mass spectrometry of complex oligosaccharides. *Rapid Commun. Mass Spectrom.* **7**, 614–619 (1993).
126. Garozzo, D., Spina, E., Sturiate, L., Montaudo, G., and Rizzo, R. Quantitative determination of B(1-2) cyclic glucans by matrix-assisted laser desorption mass spectrometry. *Rapid Commun. Mass Spectrom.* **8**, 358–360 (1994).
127. Sutton, C. W., O'Neill, J. A., and Cottrell, J. S. Site-specific characterization of glycoprotein carbohydrates by exoglycosidase digestion and laser desorption mass spectrometry. *Anal. Biochem.* **218**, 34–46 (1994).
128. Treuheit, M. J., Costello, C. E., and Kirley, T. L. Structures of the complex glycans found on the beta-subunit of (Na,K)-ATPase. *J. Biol. Chem.* **268**, 13914–13919 (1993).
129. Bock, K., Schuster-Kolbe, J., Altman, E., Allmaier, G., Stahl, B., Christian, R., Sleytr, U. B., and Messner, P. Primary structure of the O-glycosidically linked glycan chain of the crystalline surface layer glycoprotein of *Thermoanaerobacter thermohydrosulfuricus* L111-69. Galactosyl tyrosine as a novel linkage unit. *J. Biol. Chem.* **269**, 7137–7144 (1994).
130. Stahl, B., Klabunde, T., Witzel, H., Krebs, B., Steup, M., Karas, M., and Hillenkamp, F. The oligosaccharides of the Fe(III)-Zn(III) purple acid phosphatase of the red kidney bean—Determination of the structure by a combination of matrix-assisted laser desorption/ionization mass spectrometry and selective enzymic degradation. *Eur. J. Biochem.* **220**, 321–330 (1994).
131. Huberty, M. C., Vath, J. E., Yu, W., and Martin, S. A. Site-specific carbohydrate identification in recombinant proteins using MALD-TOFMS. *Anal. Chem.* **65**, 2791–2800 (1993).
132. Ashton, D. S., Beddell, C. R., Cooper, D. J., and Lines, A. C. Determination of carbohydrate heterogeneity in the humanised antibody CAMPATH 1H by liquid chromatography and matrix-assisted laser desorption ionisation mass spectrometry. *Anal. Chim. Acta* **306**, 43–48 (1995).
133. Lochnit, G. and Geyer, R. Carbohydrate structure analysis of batroxobin, A thrombin-like serine protease from *Bothrops* Moojeni venom. *Eur. J. Biochem.* **228**, 805–816 (1995).
134. Bahr, U., Deppe, A., Karas, M., Hillenkamp, F., and Giessmann, U. Mass Spectrometry of synthetic polymers by UV-matrix-assisted laser desorption/ionization. *Anal. Chem.* **64**, 2866–2869 (1992).
135. Burger, H. M., Muller, H.-M., Seebach, D., Bornsen, K. O., Schar, M., and Widmer, H. M. Matrix-assisted laser desorption and ionization as a mass spectrometric tool for the analysis of poly[(R)-3-hydroxybutanoates]. *Macromolecules* **26**, 4783–4790 (1993).
136. Blais, J. C., Tessier, M., Bolbach, G., Renaud, B., Rozes, L., Guittard, J., Brunot, A., Marechal, E., and Tabet, J. C. Matrix-assisted laser desorption ionisation time-

of-flight mass spectrometry of synthetic polyesters. *Int. J. Mass Spectrom. Ion Processes* **144**, 131–138 (1995).
137. Danis, P. O., Karr, D. E., Simonsick, Jr., W. J., and Wu, D. T. Matrix-assisted laser desorption/ionization time-of-flight mass spectrometry characterization of poly(butyl methacrylate) synthesized by group-transfer polymerization. *Macromolecules* **28**, 1229–1232 (1995).
138. Danis, P. O., and Karr, D. E. A facile sample preparation for the analysis of synthetic organic polymers by matrix-assisted laser desorption/ionization. *Org. Mass Spectrom.* **28**, 923–925 (1993).
139. Montaudo, G., Montaudo, M. S., Puglisi, C., and Samperi, F. Determination of absolute mass values in MALDI-TOF of polymeric materials by a self-calibration of the spectra. *Anal. Chem.* **66**, 4366–4369 (1994).
140. Montaudo, G., Montaudo, M. S., Puglisi, C., and Samperi, F. Self-calibrating property of matrix-assisted laser desorption/ionization time-of-flight spectra of polymeric materials. *Rapid Commun. Mass Spectrom.* **8**, 981–984 (1994).
141. McIver, Jr., R. T., Li, Y., and Hunter, R. L. FTMS method for high resolution matrix-assisted laser desorption. *Int. J. Mass Spectrom. Ion Processes* **132**, L1–L7 (1994).
142. Pastor, S. J., Castoro, J. A., and Wilkins, C. L. High-mass analysis using quadrupolar excitation/ion cooling in a fourier transform mass spectrometer. *J. Am. Chem. Soc.* **67**, 379–384 (1995).
143. Li, Y., McIver, Jr., R. T., and Hunter, R. L. High-accuracy molecular mass determination for peptides and proteins by fourier transform mass spectrometry. *Anal. Chem.* **66**, 2077–2083 (1994).
144. Dey, M., Castoro, J. A., and Wilkins, C. L. Determination of molecular weight distributions of polymers by MALDI-FTMS. *Anal. Chem.* **67**, 1575–1579 (1995).
145. Herod, A. A., Li, C.-Z., Parker, J. E., John, P., Johnson, C. A. F., Smith, G. P., Humphrey, P., Chapman, J. R., and Kandiyoti, R. Characterization of coal by matrix-assisted laser desorption ionization mass spectrometry. I. The Argonne coal samples. *Rapid Commun. Mass Spectrom.* **8**, 808–814 (1994).
146. Herod, A. A., Li, C.-Z., Xu, B., Johnson, C. A. F., John, P., Smith, G. P., Humphrey, P., Chapman, J. R., and Kandiyoti, R. Characterization of coal by matrix-assisted laser desorption ionization mass spectrometry. II. Pyrolysis tars and liquefaction extracts from the Argonne coal samples. *Rapid Commun. Mass Spectrom.* **8**, 815–822 (1994).
147. Li, C.-Z., Herod, A. A., John, P., Johnson, C. A. F., Parker, J. E., Smith, G. P., Humphrey, P., Chapman, J. R., Rahman, M., Kinghorn, R. R., and Kandiyoti, R. Characterization of kerogens by matrix-assisted laser desorption ionization mass spectrometry. *Rapid Commun. Mass Spectrom.* **8**, 823–828 (1994).
148. John, P., Johnson, C. A. F., Parker, J. E., Smith, G. P., Herod, A. A., Li, C.-Z., Humphrey, P., Chapman, J. R., and Kandiyoti, R. Molecular masses up to 270 000 u in coal and coal-derived products by matrix assisted laser desorption ionization mass spectrometry (MALDI-m.s.). *Fuel* **73**, 1606–1616 (1994).
149. John, P., Johnson, C. A. F., Parker, J. E., Smith, G. P., Herod, A. A., Li, C.-Z., and Kandiyoti, R. Identification of molecular mass up to 270,000 u in coal and coal-derived products by laser desorption mass spectrometry. *Rapid Commun. Mass Spectrom.* **7**, 795–799 (1993).
150. Kawaguchi T., Walker K. L., Wilkins, C. L., and Moore, J. S. Double exponential dendrimer growth. *J. Am. Chem. Soc.* **117**, 2159–2165 (1995).

10. PHYSICAL MECHANISMS GOVERNING THE ABLATION OF BIOLOGICAL TISSUE

Glenn Edwards

Department of Physics and Astronomy
and
W. M. Keck Foundation Free-Electron Laser Center
Vanderbilt University
Nashville, Tennessee

10.1 Introduction

At first glance it may appear that ablating biological tissue has much in common with efforts to ablate more traditional non-metallic materials. This is true in part. Both are dielectrics; however, the post-exposure biological response is one key difference. In other words, semiconductors neither scar nor heal. This chapter aims to summarize the status of the subfield of tissue ablation for physical scientists and other interested readers. Although this presentation may prove instructive for biomedical investigators, the target audience is physical scientists.

The following discussion emphasizes the physical mechanisms governing laser ablation of tissue. Note at the outset, however, that the application of lasers to medicine and dentistry is an active area of experimental research where past efforts have led to successful, laser-based clinical protocols that are now the standard of care [1]. Much of this experimental research has progressed successfully despite only partial understanding of the underlying mechanisms. The benefits of this experimental approach are clear; however, it seems equally clear that improved understanding of the fundamental mechanisms promises more general and effective clinical applications of tissue ablation. In particular, recent experimental results demonstrate the importance of molecular structure and dynamics in optimizing tissue removal while limiting collateral damage to the surrounding tissue, as will be described.

This chapter concentrates on the ablation of what at times has been referred to as soft tissue, where hard tissue refers to bone or teeth. The emphasis is on recent results from investigations of infrared tissue ablation; however, ultraviolet and visible laser ablation are also summarized. Mechanisms for laser-tissue interactions are outlined in Section 10.3. Section 10.4 summarizes medical applications in the ultraviolet and visible, and Section 10.5 addresses infrared laser ablation. Section 10.6 briefly addresses clinical acceptance, and Section 10.7 comments on future research.

10.2 Gedanken Experiments

In an attempt to motivate the reader, the following "thought" experiments are presented to summarize recent advances in laser ablation of tissue. View this section as an advertisement, albeit imprecise and incomplete. The following sections will provide a more scholarly discussion.

For the first series of experiments, imagine focusing a laser beam to a spot with a diameter of 100 μm at the surface of tissue. View the tissue as a slab of material with known permittivity and mechanical properties. For subablative fluences, a thermoelastic wave is generated in the tissue. As you increase the fluence, the ablative threshold is reached. To account for these results, a mechanism based on heat deposition and a change in state can be formulated: the model views tissue as a slab without concern for internal structure. There is some clinical relevance; however, it is limited by collateral damage surrounding the ablated site.

The second series of experiments aims to reduce the zone of collateral damage by confining energy in the ablated material. Thermal confinement can be approximated by limiting the pulse duration to be less than the thermal diffusion time. Stress confinement can be approximated by limiting the pulse duration to less than the transit time for sound in the exposed volume. The result is a reduction in collateral damage.

The third series of experiments aims to limit further the collateral damage by explicitly taking into account the mechanism for mechanical failure of the tissue. Choose a wavelength that simultaneously couples energy into both structural polymers and water: "micro-heating" proteins compromises the structural integrity of tissue, whereas vaporization provides the explosive force for ablation. The result is a further reduction in collateral damage.

10.3 Laser-Tissue Interactions

It is relatively easy to ablate tissue with a focused laser beam. It has proven more difficult to ablate tissue in a manner that achieves the type of biological response that results in clinical relevance. The undesirable "side effects" include both physical effects—photochemical, photomechanical, and photothermal damage—and subsequent biological effects, such as the sloughing off of additional tissue and scarring. Not all of these effects are negative; for example, a veneer of thermally denatured tissue can serve a useful hemostatic purpose.

Different wavelength and fluence regimes access multiple absorption mechanisms. Classification schemes have been proposed by other investigators to lend some order to the wide range of experimental results [2]. Experimental evidence makes a convincing case that modifying the exposure parameters can limit the undesirable physical effects in such a manner as to avoid the onset of the undesirable biological effects. For example, photochemical effects, a serious concern for cellular tissue, can be limited by shifting to longer wavelengths and collateral mechanical or thermal damage can be limited by modifying the pulse structure. Pharmacological agents may contribute to limiting some of the post-exposure biological effects.

10.3.1 Tissue Properties

The goal of this section is to introduce some basic knowledge of tissue properties to provide context for subsequent discussions. The principle components of soft tissue are water, proteins, lipids, sugars, nucleic acids, and their complexes.

Tissue is quite efficient in performing a complex set of biological functions and is highly organized. It is misleading to view tissue as a collection of water, lipids, and globular proteins without order. Cornea is an interesting case in point, although a somewhat extreme case of ordered tissue. At the length scale of 10 nanometers, cornea comprises triple helices of polypeptides, where every third amino acid is glycine, forming a highly regular structure. At the length scale of hundreds of nanometers, the triple helices pack in a regular array of microfibrils, which themselves pack into a regular array known as a collagen fibril. These fibrils form ordered sheets, which are separated by aqueous layers: this total assembly is corneal stroma, with a thickness of millimeters and a surface area of square centimeters. The stroma has layers of cells on both surfaces. The transparency of the cornea to visible light is directly attributable to this highly ordered assembly of protein. Note that the stroma is acellular, an important point with regard to ultraviolet laser ablation, which will be discussed below. Furthermore, the highly ordered array of collagen found in stroma lends itself to theoretical modeling, benefiting from a large body of knowledge regarding its structural and thermodynamic properties [3].

A survey of a few tissue types indicates some of the similarities and distinctions. Sclera (the "whites" of the eyes) and especially cornea are highly organized, collagenous tissues. Dermis is comprised of collagen, proteoglycans, and elastin in a more complex fibril architecture. Neural tissue is a non-collagenous, cellular tissue. Cellular proteins exhibit both symmetric and asymmetric (globular) forms. The principal component of tissue is water; ocular and neural tissues have about the same water content, about 70% water, whereas dermis is slightly less hydrated.

Harvesting tissue for investigational purposes is not trivial and generally requires collaboration or at least cooperation from medical or biomedical investigators. In addition, approval procedures are in place at research institutions to ensure proper handling and judicious use. Consequently model systems for tissue are often used; however, these systems lack the organization of biological tissue. Gelatin serves as a model system for investigation of laser ablation of cornea and, when doped with carbon and isopropyl alcohol, as a model system for ultrasound imaging. Relative to collagen, however, gelatin is a disordered system and has markedly different optical and mechanical properties. Milk has served as a model system for imaging studies in soft tissue, as have agar-saline mixtures. The challenge is to come up with a predominantly aqueous phantom material, where some combination of solvents organizes the phantom so that it mimics the optical, mechanical, and thermal properties of tissue and will not rapidly degrade under laboratory conditions. The result is generally a compromise and to gain credibility with the medical community. Eventually tissue needs to be investigated. In practice, we use model systems during the time-consuming process of instrumentation development, but use harvested tissue at other times to gain credibility with our medical collaborators.

The literature on tissue-equivalent phantoms concentrates on radiological imaging [4]. Although there are protocols for the preparation of phantoms and commercially available materials, they are tissue- and technique-specific and will not be included here. The interested physical scientist is encouraged to seek the collaboration of a medical or biomedical scientist.

10.3.2 Optical Properties and Dielectric Theory

Whether investigating tissue or model systems, the sample may be viewed as a lossy dielectric with or without structural symmetries. The reflection and refraction of light at a dielectric interface are familiar phenomena [5], well described by the complex refractive index κ, the reflectivity r, and E_i and E_r, the amplitudes of the incident and reflected electric fields:

$$r = \frac{|E_r|^2}{|E_i|^2} = \left|\frac{1-\kappa}{1+\kappa}\right| = \frac{(1-n)^2 + k^2}{(1+n)^2 + k^2}$$

$$\kappa = n - ik$$

n and k are linked by causality, and are related by the Kramers-Kronig relations,

$$n(\omega) = 1 + P\int_{-\infty}^{\infty} \frac{d\omega'}{\pi} \frac{k(\omega')}{\omega' - \omega}$$

$$k(\omega) = -P\int_{-\infty}^{\infty} \frac{d\omega'}{\pi} \frac{n(\omega) - 1}{\omega' - \omega}$$

providing the capability to separate the contribution $n(\omega)$ and $k(\omega)$, where P indicates the principal value integral.

This formalism implicitly emphasizes the electromagnetic radiation and is experimentally convenient; however, investigations of material failure require an explicit treatment of the lossy dielectric. An alternative, less familiar formalism is expressed in terms of the complex permittivity ε^* or the complex susceptibility χ^* [6]. The relationships between these quantities are summarized by:

$$\varepsilon^* = \varepsilon' - i\varepsilon''$$

$$\chi^* = \chi' - i\chi''$$

$$\varepsilon^* = 1 + 4\pi\chi^*$$

$$\kappa = \sqrt{\varepsilon^*}$$

Permittivities are additive, in contrast to refractive indices, a distinct advantage when casting a theoretical model for heterogeneous materials and/or multiple mechanisms for absorption. In general, there are two classes of phenomenological mechanisms to account for dielectric loss. Resonant absorption can be modeled in terms of a damped harmonic oscillator with a charge distribution leading to a net dipole, resulting in the expression:

$$\varepsilon^*_{res}(\omega) = 4\pi Ne^2 \sum_i \frac{f_i/m_i}{\omega_i^2 - \omega^2 - i\omega\gamma_i}$$

where N indicates the number of molecules per unit volume with Z electrons per molecule. Each molecule has f_i electrons with frequency of oscillation ω_i and damping coefficient γ_i ($\sum_i f_i = Z$). Energy from the electromagnetic field couples directly to the oscillating dipole, which then relaxes via the damping mechanism to the surround.

Dielectric relaxation, on the other hand, can be modeled in terms of an overdamped oscillator with a charge distribution leading to a net dipole, resulting in the expressions:

$$\varepsilon^*_{rel}(\omega) - \varepsilon_\infty = \frac{\varepsilon_s - \varepsilon_\infty}{1 - i\omega\tau}$$

$$\varepsilon'_{rel}(\omega) - \varepsilon_\infty = \frac{\varepsilon_s - \varepsilon_\infty}{1 + \omega^2\tau^2}$$

$$\varepsilon''_{rel}(\omega) = \frac{(\varepsilon_s - \varepsilon_\infty)\omega\tau}{1 + \omega^2\tau^2}$$

where τ is the characteristic time of the relaxation back to equilibrium and ε_s and ε_∞ are the static and high-frequency limits for the permittivity. The last two expressions are known as Debye's equations. For a Debye relaxation, energy is essentially transferred via the damping mechanism from the electromagnetic field to the surround.

To offer some intuition for comparison, the model of the oscillator corresponds to a resonance, but with increasing γ, the system eventually becomes overdamped and is then described as a Debye relaxation process. Consequently the most general expression for a lossy dielectric is given by:

$$\varepsilon^*(\omega) = \varepsilon_\infty + 4\pi Ne^2 \sum_i \frac{f_i/m_i}{\omega_i^2 - \omega^2 - i\omega\gamma_i} + \sum_i \frac{\varepsilon_{s,i} + \varepsilon_\infty}{1 - i\omega\tau_i} + \frac{4\pi i}{\omega} \sum_i \sigma_i$$

where σ_i is the conductivity of the ith species.

A goal of theoretical investigations of the dielectric properties of materials is to account for the phenomenological parameters f_i, m_i, γ_i, ω_i, $\varepsilon_{s,i}$, ε_∞, τ_i, and σ_i in terms of detailed mechanisms. Clearly the mechanisms will be frequency dependent, with electronic processes at relatively high frequencies, molecular processes at relatively intermediate frequencies, and ionic processes at relatively low frequencies. To account for ablation, nonlinear processes must be taken into account in these theoretical investigations. Consequently models tend to rely on semi-empirical, thermodynamic, and/or computational approaches and must ultimately be justified by their success in accounting for experimental results, e.g., subablative threshold measurements of reflection coefficient r.

10.3.3 Mechanisms

The operating parameters of modern lasers now form an enormous parameter space, as described in Chapter 4 of this book. A wide range of available wavelengths from the ultraviolet through the visible and into the infrared are available, with continuous output or pulse structures as short as femtoseconds, and in some cases, high peak and/or average powers. These lasers have the potential to access many mechanisms. It is common to recognize four general mechanisms that are not mutually exclusive: dielectric breakdown and photochemical, photothermal, and photomechanical processes.

10.3.3.1 Dielectric Breakdown. For sufficient fluences, the electric field strength is sufficient to cause dielectric breakdown of the tissue followed by shock-wave formation, cavitation, and tissue disruption [7]. This high-intensity regime accesses nonlinear mechanisms; detailed theoretical models have yet to be firmly established [8]. The thresholds for breakdown can be wavelength, pulse duration, and material dependent; for example, Zysset *et al.* [9] have determined breakdown thresholds in tissue for two pulse durations of Nd:YAG radiation: 1 TW cm^{-2} for 40 ps pulses and 0.3 TW cm^{-2} for 10 ns pulses. For comparison purposes, Vogel *et al.* [10] have investigated the threshold for water exposed to Nd:YAG radiation: 0.3 GW cm^{-2} for 30 ps pulses and 0.03 GW cm^{-2} for 6 ns pulses. Furthermore, in the water experiments, the subsequent shock waves were transient with peak pressure amplitudes of 17 kbar for the picosecond pulses and 21 kbar for the nanosecond pulses.

In general, dielectric breakdown is not a subtle mechanism for tissue ablation. For longer pulses, the plasma absorbs and/or reflects the tail end of the pulse; the consequences of the collapsing plasma are prohibitive for many biological applications. However, the duration of the plasma at threshold is shorter for picosecond pulses relative to nanosecond pulses. Femtosecond lasers promise finer control of the onset of collateral damage via photothermal and photomechanical mechanisms.

10.3.3.2 Photochemical Effects. Ultraviolet excimer lasers have been used to ablate polymers and biological tissue [11] with clinical relevance, especially in the case of cornea. The short wavelength contributes to exquisite precision. The underlying mechanism is generally attributed to the disruption of chemical bonds, with the subsequent release of both chemical and mechanical energy (the collagen in cornea is under tension) resulting in minimal collateral damage. This process has come to be known as "ablative photodecomposition." The use of ultraviolet photons has raised some concerns about the potential for photochemistry with adverse biological consequences; however, cornea is an acellular tissue.

Recently Hahn *et al.* [12] have investigated the dynamics of the ablation plume resulting from ArF excimer irradiation at 193 nm. Monitoring the Raman spectra of the plume, they observe nonevolving water spherules, where the quantity of water observed in the plume correlates closely with the original tissue water. These results support the model of ablative photodecomposition.

Venugopalan *et al.* have investigated the stress transient generated in tissue exposed to excimer radiation at 193 and 248 nm, accounting for their results in terms of photodecomposition of nonaqueous components of tissue and dielectric breakdown [13]. They conclude that ultraviolet ablation of soft tissue is a surface-mediated process without an explosive contribution.

10.3.3.3 Photothermal Mechanisms. For a range of fluences below the breakdown threshold, heat deposition can lead to changes in state in the tissue. One example is the photocoagulation of tissue in retinal surgery, where the laser-induced denaturation of tissue is a structural transition of the protein. Another example is the laser deposition of heat driving the liquid-vapor transition as the driving force for explosive vaporization. Zweig and Weber [14] account for CO_2 tissue ablation in terms of the thermodynamics of tissue water when the pulse duration is longer than a fraction of a microsecond. Similarly, Schomacker *et al.* [15] investigated the 1.81 to 2.14 micron range with 100 μs pulses from a Co:MgF$_2$ laser, demonstrating that the ablation threshold tracks the absorption due to water.

In 1984 Wolbarsht [16] recognized the similarity between industrial and medical applications of lasers, proposing that collateral damage could be minimized by using pulse durations shorter than the thermal diffusion time. The benefits of thermal confinement for tissue ablation were investigated experimentally by Walsh et al. [17], demonstrating that for Er:YAG radiation of various tissues 200 μs pulses result in 10–50 microns of collateral damage while 90 ns pulses result in 5–10 microns of collateral damage.

10.3.3.4 Photomechanical Effects. At still lower fluences, the deposition of heat by laser irradiation results in thermoelastic waves. The tissue is relatively incompressible and thus resilient to the compressive phase of the wave. It is the rarefaction phase (tensile stress) that can take the tissue beyond its yield point, leading to mechanical failure and spallation. In addition, under the appropriate conditions, shock waves can develop, which may result in tearing of the tissue.

Stress confinement is achieved when the pulse duration is less than the time it takes the sound wave to propagate out of the irradiated volume. Jacques et al. [18] have used laser flash photography to demonstrate the transition from thermoelastic waves to laser-induced spallation in indocyanine green, which absorbs at 755 nm, exposed to 140 ns pulses from an alexandrite laser at 755 nm. At still higher fluences, explosive vaporization was observed.

10.4 Ultraviolet and Visible Laser Ablation

In this section, we discuss the status of medical applications of laser ablation using ultraviolet and visible lasers. For a more medically oriented discussion, a recent series of special issues in the journal *Lasers in Surgery and Medicine* [1] summarizes new developments in various medical specialties, including laser ablation of both soft and hard tissue. Medical investigators frequently benefit from the aggressive pursuit of experimental research, not being hindered by theoretical concerns. This leads to the rapid implementation of new laser technology.

Recognizing that the eye is a light-gathering organ, one of the first applications of visible lasers was to perform retinal surgery [19]. The initial medical application was to target a visible laser through the eye, focusing on the retina, to produce localized regions of photocoagulation. The underlying mechanism is attributed to linear absorption by tissue chromophores, e.g., melanin or hemoglobin. These ophthalmic procedures are now standard practice.

The application of excimer lasers to corneal sculpting is quite mature, where photorefractive surgery has been underway outside of the U.S. for several years. This application has grown out of the enabling findings of Srinivasan [11] as discussed in the previous section and brought into application by a number of investigators [20]. During the past year, approval has been granted by the Food and Drug Administration to perform excimer-based laser procedures in the United States.

Recently the possibility of controlled dielectric breakdown or "microplasmas" has been investigated with short and ultrashort pulsed lasers; this effort is not limited to the ultraviolet [21]. In addition, active research programs are investigating photochemical applications based on photosensitive dyes, commonly referred to as photodynamic therapy [22].

10.5 IR-Laser Ablation

An advantage of the infrared is the reduced likelihood of photochemistry due to the low photon energy. The ablation characteristics of numerous infrared lasers have been investigated, where the linear spectrum of water was viewed as the first-order model to account for these results in terms of thermal confinement, explosive vaporization, and dielectric breakdown. The collateral damage observed with CO_2 and Nd:YAG laser ablation serves a useful hemostatic purpose for some tissues, however, the thermal damage is prohibitive when near vital structures or for other tissues such as cornea and neural tissues.

Itzkan et al. [23] have investigated the thermoelastic expansion of the surface of tissue at fluences leading up to the threshold for ablation using an interferometric technique, concluding that the tensile stresses are sufficient for photomechanical ablation. These investigations emphasize the importance of characterizing the onset of ablation to better our understanding of the dynamics of tissue ablation.

During the past several years we have been using a free-electron laser (FEL) to investigate infrared ablation of tissue [24]. In this section, we go into some detail to describe our research that demonstrates a significant reduction in collateral damage when the FEL is tuned to the 6.0–6.5 micron range. In addition, we summarize our ongoing effort to explore the potential for clinical applications of FEL-based tissue ablation.

10.5.1 Free-Electron Lasers

FELs are unlike conventional lasers [25–36]. The electrons are free in the sense that they are not confined to an atomic or molecular bound state. The emission of photons is a consequence of bending the path of the electrons with magnetic fields: the electrons are sent through a series of alternating magnets, known as the wiggler. The wavelength is tuned by varying the energy of the electron beam and/or the properties of the magnetic fields. In general, FELs can be constructed to operate from the far-infrared to the far-ultraviolet. The

Vanderbilt Mark-III FEL is tunable in the mid-infrared from 2 to 10 microns. The mode structure of the emitted IR beam is TEM_{00}, and the pulse structure traces back to that of the electron beam. In the Mark-III FEL, the electrons are extracted from a thermionic cathode by a microwave field in a resonant cavity electron gun. This produces a train of micropulses, which, after compression, are a few picoseconds long and separated by approximately 350 ps. This train extends for about a 5-microsecond macropulse and is repeated at up to 30 Hz, with plans for 60 Hz operation in the near future. The electrons are then accelerated to about 43 MeV by an RF linac, steered through the wiggler, and the emitted photons are captured in an optical cavity. FELs are continuously tunable over broad spectral ranges. A schematic representation of the Vanderbilt Mark-III FEL in use is shown in Fig. 1, the operating parameters are summarized in Table I, and the pulse structure is summarized in Fig. 2.

The Mark-III FEL was designed for both materials and medical research. The pulse structure was motivated by the belief that the picosecond micropulse accesses an effective dynamic regime, while the high average power, achieved by packing a train of micropulses in a macropulse of several microseconds duration, allows a reasonable rate of material modification. Vanderbilt's FEL program is a broadly multidisciplinary program in materials science, biological physics, molecular and cellular biology, biomedical research, and applications to laser surgery [37–42]. The Mark-III is being upgraded to meet the reliability and performance criteria required for a medical laser [43]. In addition to the infrared capability of the Mark-III FEL, a monochromatic X-ray project is under development at Vanderbilt [44].

In addition to the Vanderbilt program, other FELs in the United States have complementary operating parameters. Mid-IR FELs are operational at Duke University [25], Stanford University [26, 36], Los Alamos National Laboratory [32], and during the summer of 1996, the Grumman-Princeton FEL became operational [35]. Far-infrared FELs are operational at the University of California at Santa Barbara [28] and at Stanford University [33]. A storage ring FEL, producing both UV radiation and gamma rays, became operational at Duke University during the fall of 1996 [45]. There are numerous foreign FEL facilities, including FELIX in the Netherlands [27], CLIO in France [29], and FELI in Japan [34]. Investigators have produced a wide range of FEL applications research [24, 37–42, 46–54] using these FEL facilities.

10.5.2 "Photo-Thermo-Mechanical" Mechanism

Using the FEL, a team of investigators at the FEL Center [24] have investigated tissue ablation in ophthalmic, neural, and dermal tissue. Figure 3 displays infrared spectra of three tissues. Although there is some variability from tissue to tissue, it is useful to note several spectral characteristics that generalize to all

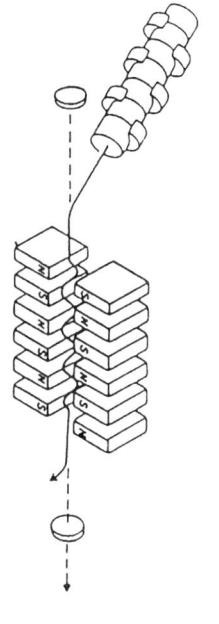

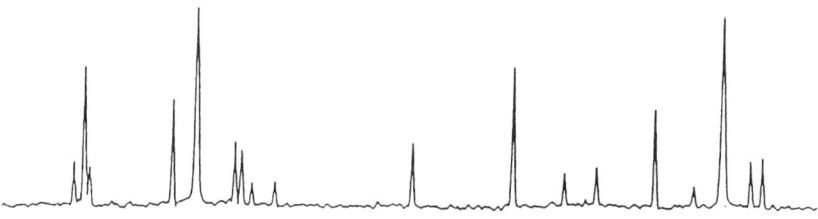

FIG. 1. A schematic representation of the Mark-III FEL tuned to a spectral feature of a generic absoption spectrum. A linear accelerator (represented as a cylinder in the upper right corner) produces electrons with energies near 43 MeV that then pass through alternating magnetic fields, known as a wiggler. The solid line represents the path of the electrons. The cavity mirrors, represented as discs, partially confine the emitted photons, whose path is represented by the dashed line.

soft tissues. The dominant feature near 3300 cm^{-1} (3 microns) is attributable to the OH-stretch mode of water. The bending mode of water is at 1640 cm^{-1} (6.1 microns) in a partially resolved spectral feature. The amide I band near 1665 cm^{-1} (6.0 microns) and amide II band near 1550 cm^{-1} (6.45 microns) also contribute to this feature. Wavelengths within this spectral feature lead to

TABLE I. Parameters of the Vanderbilt Free-Electron Laser

	Best	Typical
Wavelength	2.0–9.8 μm	2.2–9.0 μm
Linewidth	0.7%	0.85%
Jitter	0.1%	0.4%
Macropulse energy	360 mJ	50–150 mJ
Stability	5%	10%
Macropulse length	6 μs	5 μs
Repetition rate	30 Hz	20 Hz
User beam time		70 hours/week

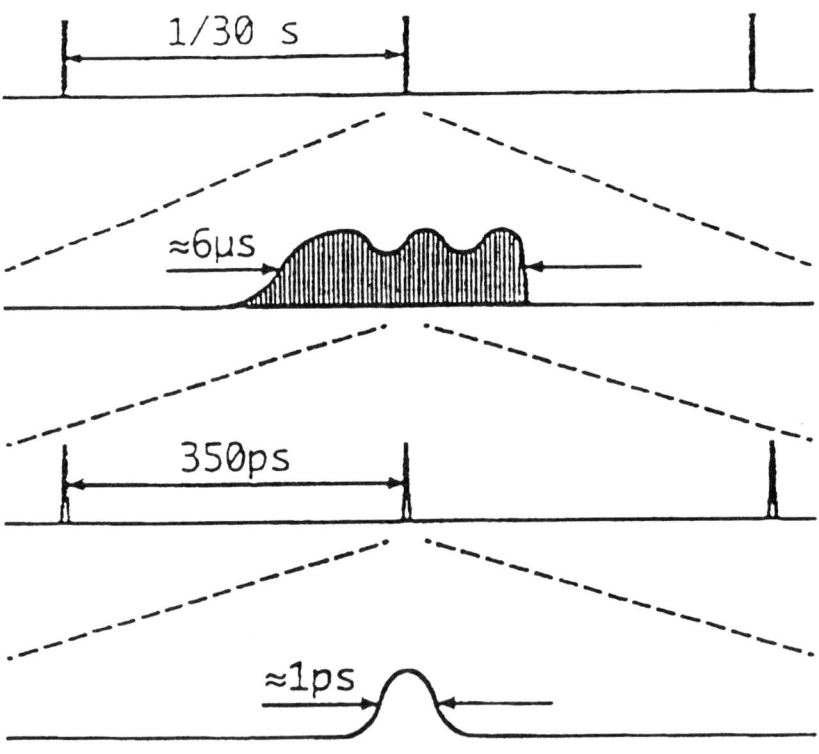

FIG. 2. Pulse structure for the Vanderbilt FEL. From top to bottom, a train of three macropulses, a single macropulse, three micropulses, and a single micropulse (from [60]).

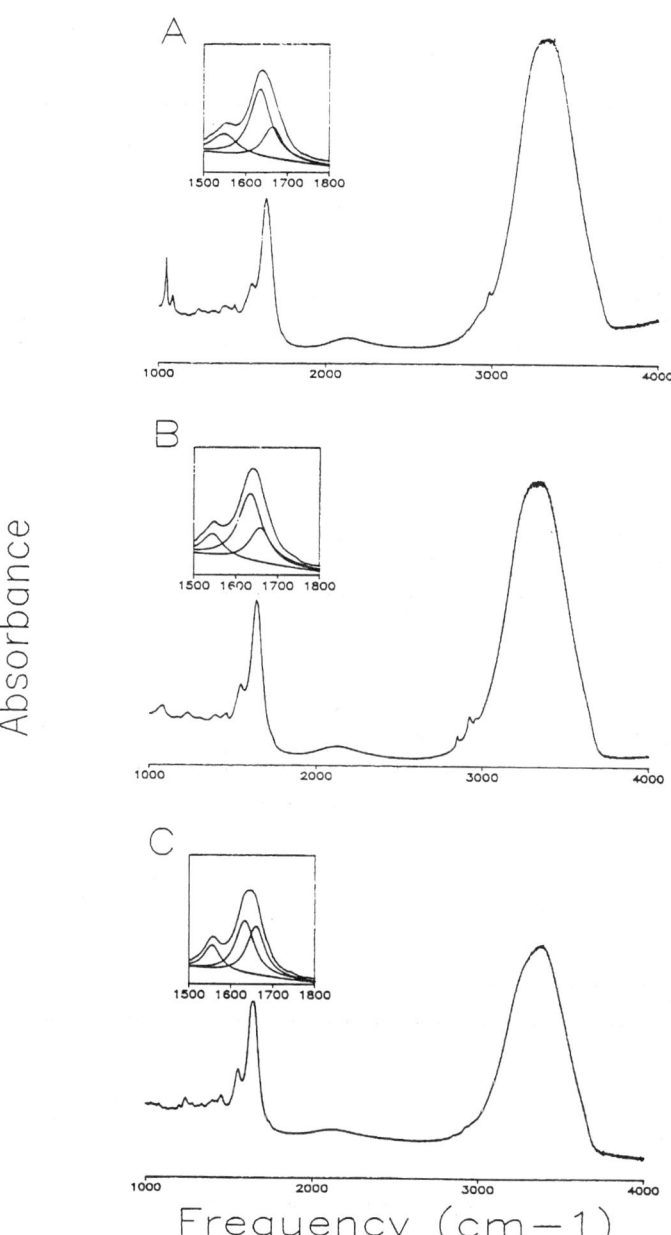

FIG. 3. Infrared spectra for a) corneal stroma, b) neural tissue, and c) dermal tissue taken with an attenuated reflectance cell and a Bruker IFS-113v Fourier-transform infrared spectrometer.

improved ablation, as determined by both ablation yield and/or the amount of collateral damage, depending on the tissue.

These observations were accounted for by a detailed consideration of corneal ablation, which exhibits order at multiple length scales as discussed previously. Although it is certainly clear that the dominant spectral feature of tissue is the OH-stretch mode of water and, as such, is the most efficient wavelength for absorbing infrared radiation in the 2–10 micron range, it does not necessarily follow that this will also be the most efficient wavelength for tissue ablation if there is more to the phenomena than the explosive vaporization of water. To account for our results in the 6.0–6.5 micron range relative to the 3 micron range, we recognized that the energy will be resonantly absorbed by both the tissue water via the shoulder of the bending mode and the tissue protein via the amide I and II modes. We proposed that the energy absorbed by the protein leads to a localized or micro-heating to drive the change in state, a structural phase transition between collagen and gelatin. Collagen is a resilient protein conformation at tension in corneal stroma. Gelatin is a disordered conformation with a marked decrease in mechanical strength. Apparently the key features are that the mechanical integrity of the tissue is compromised, then the explosive force of the expanding vapor leads to tissue removal. The net result is a significant rate of tissue removal and a reduction in collateral damage. The model was justified on thermodynamic grounds and could be generalized in a straightforward way to account for neural and dermal tissue ablation. More specifically, energy absorbed by proteins both compromises the connectivity of cellular tissue and disrupts the cell membranes, while water, once again, is the explosive component.

The reduction in collateral damage was most marked in the ablation of neural tissue at 6.45 microns, where the zone of collateral damage is less than the detectable limit (one cell width) on the sides and about 4 microns at the bottom of the ablated crater. No additional damage was observed in survival studies where the animals were sacrificed two weeks after irradiation, in contrast to a substantial onset of collateral damage that develops postexposure during survival studies using a cw CO_2 laser, the current "gold standard" for laser neurosurgery.

While the least amount of collateral damage was observed at 6.0 microns for corneal tissue, at 6.45 microns the pattern of ablation took on a flat bottomed, trapezoidal shape suggestive of layer-by-layer removal. These results indicate that both photothermal and photomechanical processes contribute to ablation in the 6.0–6.5 micron region, highlighting the importance of considering how the tissue fails in addition to the generation of explosive force in infrared tissue ablation.

Recognizing that the nonaqueous components form the structural framework of tissue, an analogy has been drawn to a hand grenade. Compare the nonaqueous components of tissue to the casing and water to the gun powder. The danger of hand grenades comes from the shrapnel that is propelled outward. "Melting" the structure prior to explosive vaporization is analogous to removing the shell casing prior to exploding the gun powder.

Our investigation of tissue ablation, as just summarized, was accounted for in terms of a thermodynamic model in the absence of dynamical information. Peterson [55] have measured the lifetime of the amide I mode with the Stanford FEL (micropulse-to-micropulse duration of 80 ns) by measuring the relaxation from saturation. The observed lifetime, about a picosecond, was short relative to 350 ps, the duration between the micropulses in the Mark-III FEL. Recently Schwettman and co-workers [56] extended these measurements to the amide II region, observing two characteristic times, one at a picosecond and the other lasting much longer. These measurements bound the longer characteristic time between 3 ns and 85 ns. Presently, the interpretation of these data is unclear. However, governing lifetimes of the order of picoseconds suggest that a single micropulse deposits sufficient energy to convert collagen to gelatin in a stromal layer. Alternatively, governing lifetimes of the order of nanoseconds or longer suggests that the vibrational energy resides in the biopolymers for a long duration relative to the Mark-III FEL micropulse spacing.

To gain additional dynamic information about the ablation process, Tribble and co-workers [42] have used the Vanderbilt FEL to investigate the onset of ablation. In these experiments, gelatin has served as a model system. A cw HeNe beam, aimed parallel to the gelatin surface, monitors the onset of ablation, while a piezoelectric detector simultaneously monitors acoustic signals propagating into the sample. Thermoelastic waves were observed at subablative fluences. Two patterns of ablation were observed, one associated with wavelengths absorbed by water only (3 microns) and the other associated with IR wavelengths absorbed by both water and protein. A fast ablative event was observed at all wavelengths. At wavelengths observed by both protein and water, a second, relatively slow, ablative event followed the fast event. Because the onset of tissue removal is characterized by the speed of sound of the material, these observations are consistent with the view that tuning to a protein band leads to a "softening" of the tissue. A detailed dynamical explanation requires further research. Currently we are extending this investigation to corneal tissue.

10.5.3 From Biophysical Research to Clinical Applications

The underlying mechanism for FEL tissue ablation has been sketched out by the thermodynamic model described in the previous section. Although the details are the subject of ongoing research, clinical applications are under investigation. There are many challenges to taking a basic experiment and completing the research and development necessary to produce a clinical application. The following sections summarize our efforts to take a basic finding from the spectroscopy laboratory to a medical application.

10.5.3.1 Research to Medical Laser. To date, the Vanderbilt FEL has performed acceptably as a research-grade laser, as summarized in Fig. 4. Through June 1996, 4,616 hours have been delivered to users, including 1,466 hours in 1995 and 785 hours during the first six months of 1996, even while FEL upgrades are being implemented. Additional improvements in performance and reliability are imperative, however, if the FEL is to emerge as a medical laser. A brief summary of our efforts is included as an example of cultural challenges associated with medical care and how they translate to technological requirements.

For medical applications, shutdowns planned weeks in advance are not critical if appropriately scheduled. However, for human surgery unscheduled downtime is a cause of concern for patient welfare and can be an expensive and nonproductive use of resources. During surgery, self-correcting failures and failures that can be fixed in a few minutes are mostly annoyances, but need to be minimized. Failures that require an operation to be abandoned, though, are very serious.

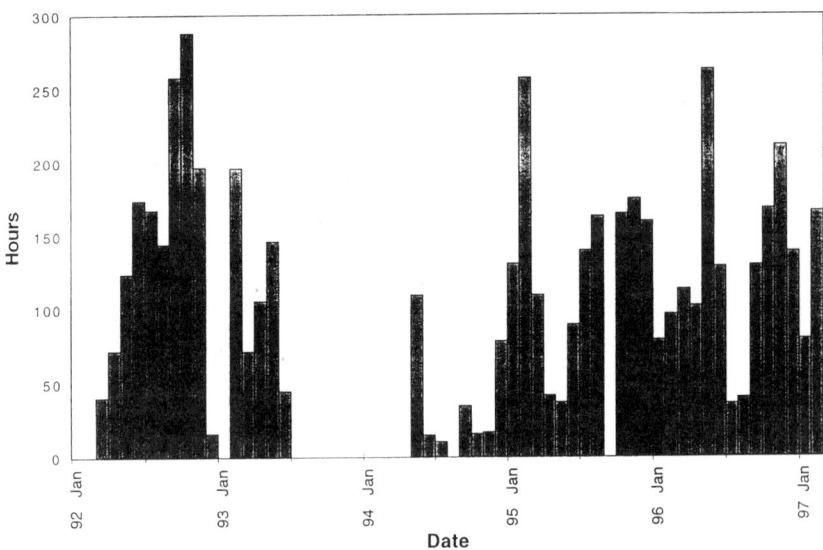

FIG. 4. Hours delivered to the users community for FEL applications research. From mid-1993 to mid-1994 access to the Center was restricted due to building construction. The klystron, which drives the linear accelerator for the Mark-III FEL, failed in September, 1995.

Recently we have assessed the performance of the Mark-III FEL and outlined a plan to upgrade it to the status of a medical laser. Although the Vanderbilt FEL is sound by design, it was found lacking in that many of its original electrical components were under-specified and there was little redundancy. Consequently we are upgrading these components and, wherever possible, implementing redundant systems. These efforts have already resulted in an overall improvement in the quality of the electron beam, improved micropulse-to-micropulse stability during the evolution of a macropulse, and extended the wavelength range.

During periods of historically best operation, the probability of a day having more than a 10-minute interruption once the FEL started in the morning was less than 5%, resulting in very few cases where an experiment was shut down once it was started. For human surgery, we need to improve even further on this last number. The subsystem that has been most prone to sudden failures has been the RF drive system for the linac. We are working to replace and upgrade many components of that system, including the thyratron and its driver, parts of the modulator, and the control electronics. Most of these problems can be relatively easily repaired. The subsystem that contributes to many long-term failures and general poor performance has been the klystron itself. Although it is not prone to sudden failure, it can destabilize the FEL to the extent that surgical applications are not practical. With very long lead times to acquire, high costs, long installation times, and high probability of catastrophic failure within the first few weeks of operation, the klystron has been the primary cause of extended shutdowns of the system for the past year. The subsystem that can lead to a slow degradation in output power and most often causes the machine suddenly to produce insufficient infrared power, or to be very difficult to start up, is the optical resonator and beam delivery system. Better alignment tools and better diagnostics of damage to the optical components are being developed to quickly identify, replace, and realign optical components as they begin to degrade. With age, the current computer control system of the FEL has become quite susceptible to noise, probably due to degradation of edge connectors. It also lacks the computational power needed to implement improved diagnostics of the FEL. It is being replaced by a VXI data acquisition system, controlled by a PowerPC-based workstation. The aim is to take advantage of faster diagnosis of FEL failures, more immunity to electrical noise, and more reliable operation of the control system itself.

10.5.3.2 Multi-Disciplinary Research. The FEL program at Vanderbilt is broadly multi-disciplinary by design. The success of the research program relies upon the combined efforts of physical and medical scientists and engineers carrying out the research and development to take these applications into the clinic. The building was designed to facilitate this activity, as shown in Fig. 5, and serves as an interesting mixture of research cultures. With regard to

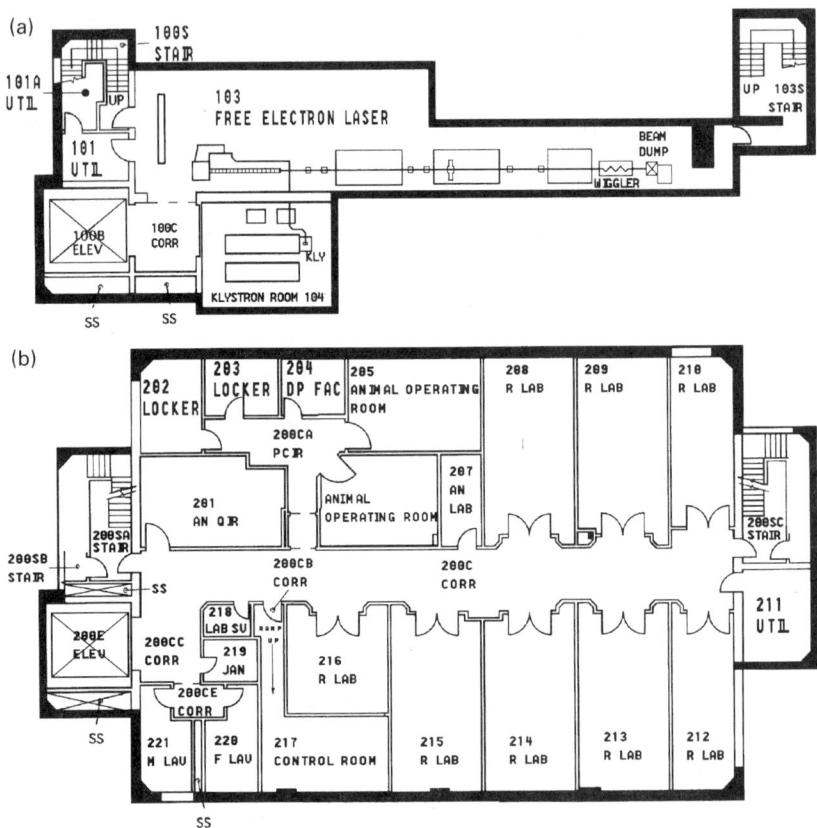

FIG. 5. The Vanderbilt FEL is housed in a special-purpose laboratory facility central to the main campus and adjacent to the Medical Center. The design of the FEL Center encourages informal interactions and promotes collaborations among Vanderbilt and visiting scientists. (a) The first floor consists of the FEL vault of approximately 1400 sq. ft., with a klystron and electronic equipment room of about 350 sq. ft. (b) The second floor, shielded from the vault by 2 m of concrete, includes the FEL control and diagnostics rooms and about 6000 sq. ft. of laboratory space (5 target rooms and 2 animal surgical suites). Each target room is equipped with conditioned power and cooling water, and a periscope to bring the FEL beam to the experiment. The second floor also includes two surgical suites, surgeons' lockers, approved animal care facilities, and a computer workroom. These facilities are supported by a tissue culture laboratory and a wet laboratory for biomedical experiments, and an electronics shop. (c) The third floor houses a library, offices, and two large "bull-pen" laboratories designed for biomedical and materials research. (d) The fourth floor is dedicated to human surgical applications of the FEL. The west (patient) entrance has ready access to a driveway for ambulances and other vehicles delivering patients, routing them to the fourth floor via an elevator. Central to the floor are two state-of-the art surgical suites and a "procedure room" reserved for clinical applications motivated by future FEL research. Adjoining the surgical suites is the "laser contingency" room which houses an optical system to distribute the IR FEL

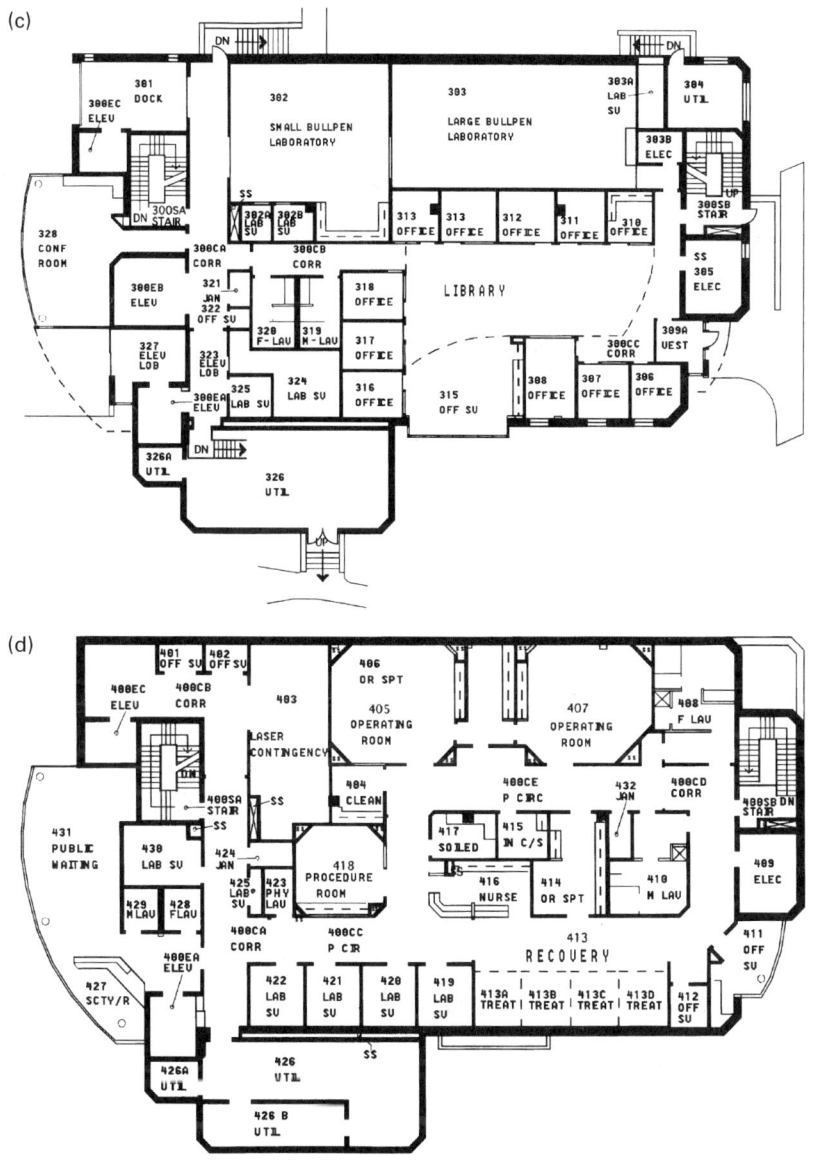

FIG. 5. Continued

beam to the surgical suites and, eventually, the procedure room, along with other equipment, including conventional lasers. Supporting these medical facilities are prep rooms, recovery areas, nursing station, anesthesia work room, locker rooms, break area, and storage and utility rooms. The fourth floor is equipped with both general and specialty surgical equipment.

medical applications on the fourth floor, it is important to recognize that the FEL and the technology in the surgical suites must function as a reliable, seamless system and the FEL must meet the perfomance standards of a medical laser.

10.5.3.3 Medical Beam Delivery. The need for medical delivery systems has been recognized at the Center for many years. An interactive image-guided stereotactic delivery system has been developed at the Center for neurosurgical applications [57]. In addition, a Computer-Assisted Surgical Technique (CAST) and delivery systems for animal surgery have also been developed [58]. Furthermore, a system based on hollow waveguides [59] is being implemented to deliver IR radiation in an aqueous environment.

We are addressing the needs of medical beam delivery in a step-wise fashion. Initially we have focused our attention on the requirements for animal experiments. Versatile, high-throughput delivery systems have been installed for animal studies in dermatology, neurosurgery, ophthalmology, and otolaryngology. Ongoing research and development is underway to upgrade this system for human surgery in the near term. The importance of biomedical engineering in making the connection from basic research to clinical application should not be underestimated.

10.6 Toward Clinical Applications

The process has started seeking approval from the U.S. Food and Drug Administration for a FEL-based clinical procedure based on infrared ablation of tissue. Previous medical applications based on conventional lasers have not met with immediate acceptance by the medical community. This is particularly interesting in light of the transition to managed care in the U.S.

10.7 Future Research

The biophysical merits of tuning to the 6.0–6.5 µm range have been demonstrated. The pursuit of clinical applications is underway. However, the infrared spectrum has not been fully investigated; in particular, still longer wavelengths that correspond to ablative mechanisms not dependent on explosive vaporization. More specifically, the potential for tissue modification by tuning to vibrational modes of nonaqueous components of tissue at longer wavelengths that do not overlap with the bending mode of water are currently being investigated at the Vanderbilt FEL Center.

We anticipate that, upon demonstration of relevance for an FEL-based clinical procedure, an alternative source of pulsed, infrared radiation based on conventional laser technology will be developed and marketed [60]. The Vanderbilt FEL Center, and the Mark-III FEL itself, will have served one of its many purposes in pioneering the application and specifying the operating parameters for a conventional IR laser to ablate tissue with clinical/surgical relevance. The current gap in conventional laser technology represents a potentially rewarding challenge for the optical science community.

A more speculative observation relates to the role of pulse structure in the future of medical applications of infrared lasers. Our understanding of the dynamical processes that govern ablation is clearly incomplete. Much of the past research has been based on concepts that do not embrace the nonlinear processes that govern ablation. This is for good reasons: much of the successful research has been driven by experimental trial and error and conceptual models traced out on the "backs of envelopes." The next frontier likely lies in the regime of pulse structure and pulse shaping to gain further control of dynamical processes. The history of nuclear magnetic resonance serves as a case in point; clever pulse schemes that take advantage of the governing mechanisms to extract out structural information have led to revolutionary applications in medical imaging. The pulse structure of the Mark-III FEL is complex and, although we routinely carry out experiments with macropulse durations between 80 nanoseconds and several microseconds [61], our understanding of the role of dynamics at the picosecond and fraction of a nanosecond timescales is naive.

10.8 Concluding Remarks

During the past decade there has been significant progress in understanding the role of photochemical, photothermal, and photomechanical processes in laser ablation of soft tissue. In many cases the advances of more conventional materials research has transferred into biomedical investigations of tissue ablation. The FEL offers a versatile tool for investigating infrared ablation; however, this research is still at the stage of basic and applied research in collaboration with clinical scientists.

Ultraviolet and visible lasers have found applications in clinical medicine. Infrared lasers also have applications where the targeted tissue is not near vital structures and the collateral damage that accompanies irradiation often serves a hemostatic purpose. The pursuit of infrared lasers for more eloquent surgical applications is underway, guided by insights into the underlying mechanisms that govern tissue ablation. The FEL has emerged as a potential medical laser.

Acknowledgments

Support provided by the U.S. DoD MFEL program, the W. M. Keck Foundation, and Vanderbilt University. I would like to acknowledge collaborations, for some aspects of the work summarized in this chapter, with Mike Copeland, Jeff Davidson, Dale Evertson,

William Gabella, Terry King, Regan Logan, Marcus Mendenhall, Lou Reinisch, Jin Shen, Roy Shores, Scott Storms, Robert Traeger, and Jerri Tribble.

References

1. For example, see Puliafito, C. A., ed. Special issue: New developments in ophthalmic lasers. *Lasers in Surgery and Medicine* **15**, 1–111 (1994); Krisnamurthey, S., and Powers, S. K., Lasers in neurosurgery, *Lasers in Surgery and Medicine* **15**, 126–167 (1994); Ossoff, R. H., Coleman, J. A., Convey, M. S., Duncavage, J. A., Werkhaven, J. A., and Reinisch, L., Clinical applications of lasers in otolaryngology—Head and neck surgery, *Lasers in Surgery and Medicine* **15**, 217–248 (1994); Deckelbaum, L. I., Cardiovascular applications of laser technology, *Lasers in Surgery and Medicine* **15**, 315–341 (1994); Wigdor, H. A., Walsh, Jr., J. T., Featherstone, J. D. B., Visuri, S. R., Fried, D., and Waldvogel, J. L., Lasers in dentistry, *Lasers in Surgery and Medicine* **16**, 103–133 (1995); Krauss, M., and Puliafito, C. A., Lasers in ophthalmology, *Lasers in Surgery and Medicine* **17**, 102–159 (1995); Reid, R. and Absten, G. T., Lasers in gynecology: Why pragmatic surgeons have not abandoned this valuable technology, *Lasers in Surgery and Medicine* **17**, 201–301 (1995).
2. Alfano, R. R., and Doukas, A. G., eds. Special issue on lasers in biology and medicine, *IEEE Journal on Quantum Electronics* **QE-20**, 1342–1532 (1984).
3. Joly, M., *A Physico-Chemical Approach to the Denaturation of Proteins*, p. 153. Academic Press, London, 1965; Flory, P. J., *Statistical Mechanics of Chain Molecules*, Ch. 7, Sect. 6. Hanser, New York, 1969.
4. *Tissue Substitutes in Radiation Dosimetry and Measurement*, International Commission on Radiological Units and Measurements, Report 44 (1989); *Phantoms and Computational Models in Therapy, Diagnosis and Protection*, International Commission on Radiological Units and Measurements, Report 48 (1992).
5. Jackson, J. D., *Classical Electrodynamics*, 2nd ed. John Wiley & Sons, 1962; Ashcroft, N. W., and Mermin, N. D., *Solid State Physics*. Saunders, 1976.
6. Frohlich, H., *Theory of Dielectrics*, 2nd ed. Oxford Science Publications, 1958.
7. Barnes, P. A., and Rieckhoff, K. E. Laser-induced underwater sparks, *Applied Physics Letters* **13**, 282–284 (1968).
8. Sigrist, M. W., and Kneubuhl, F. K. Laser-generated stress waves in liquids. *Journal of the Acoustic Society of America* **64**, 1652–1663 (1978).
9. Zysset, B., Fujimoto, J. G., Puliafito, C. A., Birngruber, R., and Deutsch, T. F. Picosecond optical breakdown: tissue effects and reduction of collateral damage. *Lasers in Surgery and Medicine* **9**, 193–204 (1989).
10. Vogel, A., Busch, S., Jungnickel, K., and Birngruber, R. Mechanisms of intraocular photodisruption with picosecond and nanosecond laser pulses. *Lasers in Surgery and Medicine* **15**, 32–43 (1994).
11. Srinivasan, R. Ablation of polymers and biological tissue by ultraviolet lasers. *Science* **234**, 559–565 (1986).
12. Hahn, D. W., Ediger, M. N., and Pettit, G. H. Dynamics of ablation plume particles generated during excimer laser corneal ablation. *Lasers in Surgery and Medicine* **16**, 384–389 (1995).
13. Venugopalan, V., Nishioka, N. S., and Mikic, B. B. Thermodynamic response of soft biological tissues to pulsed infrared-laser irradiation. *Biophysical Journal* **70**, 2981–2993 (1996).

14. Zweig, A. D., and Weber, H. P., Mechanical and thermal parameters in pulsed laser cutting of tissue. *IEEE Journal of Quantum Electronics* **QE-23**, 1787–1793 (1987).
15. Schomacker, K. T., Domankevitz, Y., Flotte, T. J., and Deutsch, T. F., Co:MgF2 laser ablation of tissue: Effect of wavelength on ablation threshold and thermal damage. *Lasers in Surgery and Medicine* **11**, 141–151 (1991).
16. Wolbarsht, M. L. Laser surgery: CO_2 or HF. *IEEE Journal of Quantum Electronics* **QE-20**, 1427–1432 (1984).
17. Walsh, Jr., J. T., Flotte, T. J., and Deutsch, T. F. Er:YAG laser ablation of tissue: Effect of pulse duration and tissue type on thermal damage. *Lasers in Surgery and Medicine* **9**, 314–326 (1989).
18. Jacques, S. L., Gofstein, G., and Dingus, R. S. Laser-flash photography of laser-induced spallation in liquid media. Invited paper. *SPIE Laser Tissue Interaction III* **1646**, 284–294 (1992).
19. Koester, C. J., Snitzer, E., Campbell, C. J., and Rittler, M. C. Experimental laser retina photocoagulation. *Journal of the Optical Society of America* **52**, 607 (1962).
20. Trokel, S. Evolution of excimer laser corneal surgery. *Journal of Cataract Refractive Surgery* **15**, 373–383 (1989).
21. Puliafito, C. A., Birngruber, R., Deutsch, T. F., Fujimoto, J. G., Stern, D., and Zysset B. Laser-tissue interactions in the nanosecond, picosecond, and femtosecond time domains, in *Photoacoustic and Photothermal Phenomena II*, J. C. Murphy, J. W. Maclachlan Spicer, L. C. Aamodt, and B. S. H. Royce (eds.), pp. 420–427 (1989).
22. Fisher, A. M. R., Murphree, A. L., and Gamon, C. J. Clinical and preclinical photodynamic therapy. *Lasers in Surgery and Medicine* **17**, 2–31 (1995).
23. Itzkan, I., Albagli, D., Dark, M. L., Perelman, L. T., von Rosenberg, C., and Feld, M. Thermoelastic basis of short pulsed laser ablation of biological tissue. *PNAS USA* **92**, 1960–1964 (1995).
24. Edwards, G., Logan, R., Copeland, M., Reinisch, L., Davidson, J., Johnson, B. Maciunas, R., Mendenhall, M., Ossoff, R., Tribble, J., Werkhaven, J., and O'Day, D. Tissue ablation by a free-electron laser tuned to the amide II band. *Nature* **371**, 416–419 (1994).
25. Madey, J. M. J. Stimulated emission of bremsstrahlung in a periodic magnetic field. *Journal of Applied Physics* **42**, 1906–1930 (1971); Brau, C. A. Free electron lasers. *Science* **239**, 1115–1121 (1988).
26. Smith, T. I. *et al.* Status of the SCA-FEL. *Nuclear Instruments and Methods in Physics Research* **A296**, 33–36 (1990).
27. van Amersfoot, P. W. *et al.* First lasing with FELIX. *Nuclear Instruments and Methods in Physics Research* **A318**, 42–46 (1992).
28. Ramian, G. The new UCSB free-electron lasers. *Nuclear Instruments and Methods in Physics Research* **A318**, 225–229 (1992).
29. Glotin, F. *et al.* First Lasing of the CLIO FEL, contributed to the 3rd European Particle Accelerator Conference (EPAC), Berlin, Germany (March 1992).
30. Brau, C. A., and Mendenhall, M. H. Medical and materials research at the Vanderbilt University Free-electron Laser Center, Nuclear Instruments and Methods in Physics Research A341, ABS21-ABS22 (1994).
31. Shea, P. O. *et al.* Accelerator Archeology—The Resurrection of the Stanford Mark III Electron Linac at Duke, in Proceedings of the 1995 Particle Accelerator Conference, 1090–1092, Dallas, Texas.
32. Nguyen, D. C. *et al.* Recent progress of the compact AFEL at Los Alamos. *Nuclear Instruments and Methods in Physics Research* **A358**, 27–30 (1995).

33. Berryman, K. W., and Smith, T. I. First lasing, capabilities, and flexibility of FIREFLY. *Nuclear Instruments and Methods in Physics Research* **A375**, 6–9 (1996).
34. Kobayashi, A. *et al.* Optical Properties of the Infrared-FELs at the FELI. *Nuclear Instruments and Methods in Physics Research* **A375**, 317–321 (1996).
35. Lehrman, I. S. *et al.* First Lasing of a Compact IR FEL, Presented at the 1996 FEL Conference, Rome, Italy. Proceedings in press.
36. Schwettman, H. A. Challenges at FEL Facilities: The Stanford Picosecond FEL Center. *Nuclear Instruments and Methods in Physics Research* **A375**, 632–638 (1996).
37. Tuncel, E. *et al.* Free-electron laser studies of direct and indirect two-photon absorption in germanium. *PRL* **70**, 4146–4149 (1993).
38. Cramer, R., Hillenkamp, F., and Haglund, R. F. Infrared matrix-assisted laser desorption and ionization using a tunable free-electron laser. *Journal of the American Society for Mass Spectrometry* (in press; 1997).
39. Park, H. K., and Haglund, R. F. Laser ablation and desorption from calcite from ultraviolet to mid-infrared wavelengths. *Applied Physics* **A** (in press; 1997).
40. Kozub, J. Photocrosslinking and photodamage in protein-nucleic acid systems resulting from UV and IR radiation. Ph.D. Dissertation (Vanderbilt University, 1995); Kozub, J., and Edwards, G., IR-UV photochemistry of protein nucleic acid systems, abstract, 2nd International FEL Users Workshop (NYC, August 1995).
41. Ueda, A. *et al.* Wavelength-selective laser ablation of diamond using hydrogen-impurity vibration modes. *Nuclear Instruments and Methods in Physics Research* **B100**, 427–430 (1994).
42. Tribble, J. A., Lamb, D. C., Reinisch, L., and Edwards, G. Dynamics of gelatin ablation due to free-electron laser irradiation. *Physical Review* **E55**, 73 85–73 89 (1997); Tribble, J. A. Role of wavelength and pulse structure in the ablation of tissue. Ph.D. Dissertation, Vanderbilt University (1996); Tribble, J. A., Lamb, D.A., Reinisch, L., and Edwards, G. Wavelength dependent delay in the onset of FEL tissue ablation, 2nd International FEL Users Workshop (NYC, August 1995).
43. Edwards, G. *et al.* Free-electron lasers: performance, reliability, and beam delivery. *Journal of Special Topics in Quantum Electronics* (in press).
44. Dong, W. W. *et al.* Current status of the VU MFEL Compton X-Ray Program. *Journal of X-ray Science and Technology* **4**, 346–352 (1994).
45. Madey, J. M. J., personal communication.
46. Austin, R. H., Roberson, M. W., and Mansky, P. Far-infrared perturbation of reaction rates in myoglobin at low temperatures, *PRL* **62**, 1912–1915 (1989).
47. Fann, W. *et al.* Dynamical test of Davydov-type solitons in acetanilide using a picosecond free-electron laser. *PRL* **64**, 607–610 (1990).
48. Zimdars, D. *et al.* Picosecond infrared vibrational photon echoes in a liquid and glass using a free-electron laser. *PRL* **70**, 2718–2721 (1993).
49. Peremans, A. *et al.* Adsorbate vibrational spectroscopy by IR-visible sum-frequency generation using CLIO-FEL: CO from CH_3OH electrochemical decomposition on Pt. *Nuclear Instruments and Methods in Physics Research* **A341**, 146–151 (1994).
50. Hill, J. R. *et al.* Vibrational dynamics of carbon monoxide at the active site of myoglobin: Picosecond infrared free-electron laser pump-probe experiments. *Journal of Physical Chemistry* **98**, 11213 (1994).
51. Heyman, J. N. *et al.* Temperature- and intensity-dependence for intersubband relaxation rates from optical rectification. *PRL* **74**, 2682–2685 (1995).

52. Tokmakoff, A. et al. Vibrational dynamics in condensed matter probed with linac based FELs. *Nuclear Instruments and Methods in Physics Research* **A358**, 540 (1995).
53. Keay, B. J. et al. Photon-assisted electric field domains and multiphoton-assisted tunneling in semiconductor superlattices. *PRL* **75**, 4098–4101 (1995).
54. Keay, B. J. et al. Dynamic localization, absolute negative conductance, and stimulated, multiphoton emission in sequential resonant tunneling semiconductor superlattices. *PRL* **75**(22), 4102–4105 (1995).
55. Peterson, K. Vibrational lifetimes of protein amide modes. 2nd International FEL Users Workshop (NYC, August 1995).
56. Schwettman, H. A., personal communication.
57. Galloway, R. L., Maciunas, R. J., and Edwards, C. A., II. Interactive image guided neurosurgery. *IEEE Transactions in Biomedical Engineering* **39**(12), 1226–1231 (1992).
58. Reinisch, L., Mendenhall, M., Charous, S., and Ossoff, R. Computer-assisted surgical techniques using the Vanderbilt free electron laser. *Laryngoscope* **104**, 1323–1329 (1994).
59. Abel, T., Hirsch, J., and Harrington, J. A. Hollow glass waveguides for broad band infrared transmission. *Optics Letters* **19**(14), 1034–1036 (1994).
60. Edwards, G. Biomedical and potential clinical applications for pulsed lasers operating near 6.45 μm. *Optical Engineering* **34**, 1524–1525 (1995).
61. Becker, K., Johnson, J. B. and Edwards, G. Broadband pockels cell and driver for a mark III type free-electron laser. *Reviews of Scientific Instruments* **65**, 1496–1501 (1994).

11. GROWTH AND DOPING OF COMPOUND SEMICONDUCTOR FILMS BY PULSED LASER ABLATION

Douglas H. Lowndes

Solid State Division
Oak Ridge National Laboratory
Oak Ridge, Tennessee
and
Dept. of Materials Science and Engineering
The University of Tennessee, Knoxville

11.1 Introduction

11.1.1 Purpose and Scope

The past decade has seen a dramatic increase in thin film growth by pulsed laser deposition (PLD), as well as intense study of the physical and chemical mechanisms responsible for the pulsed laser ablation (PLA) process. Worldwide interest in PLD was triggered by demonstrations during 1987–1989 that high quality, epitaxial, high-temperature, superconductor films can be grown [1–2]. Since then, interest has been sustained and broadened by several factors, including the concurrent development of high-power pulsed ultraviolet (UV) excimer lasers that are reliable and convenient to use and the recognition of the advantages that PLA offers for the growth of chemically complex thin-film materials at relatively low temperatures. The two most important advantages (see Section 11.2.1) are ablation's ability to transfer faithfully (stoichiometrically) the composition of a complex, multi-element material from a polycrystalline target to an epitaxial film and its capability for reactive deposition in a low-pressure ambient gas.

This chapter has two purposes. The first is to provide an introductory overview of the advantages and limitations of PLD for epitaxial film growth and of recent progress to overcome its limitations, independent of the choice of material. Basic principles underlying the solutions to problems are presented and references are given where the experimental details can be found. The second purpose is to review selectively the current status of compound semiconductor epitaxial film growth and doping by PLD. Despite the publication of books, conference proceedings, and review articles on PLA/PLD [3–17], this is the first review of compound semiconductor film growth by laser ablation. Section 11.3 deliberately is restricted to compound semiconductor materials for which PLA's characteristics may provide inherent advantages and for which film growth and in situ diagnostic data are sufficient to draw conclusions regarding key materials issues. Thus, the references given in this chapter provide a partial bibliography of compound semiconductor growth by PLD [18]; selected references to earlier compound semiconductor PLD also can be found in the bibliography by K. L. Saenger [19].

11.1.2 Pulsed Laser Deposition of Thin Films

Worldwide scientific interest in PLD is evident in the publication of more than 2,000 scientific papers (for all materials) during 1990–95 [20]. Several excellent references attest to the growing number and variety of PLD applications. These include a comprehensive textbook on PLD of thin films [3], four Materials Research Society (MRS) symposia [4–7], and a dedicated issue of *MRS Bulletin* [8]. The series of international conferences on laser ablation (COLA) [9–11] have focused on understanding desorption and ablation processes; consequently they contain much information that is relevant to epitaxial growth by PLD, as do the proceedings of a European MRS summer school [12]. European MRS annual meetings also have included laser ablation topics [13], and a number of papers on PLD can be found in MRS symposia devoted to specific materials. Finally, several authoritative overviews of pioneering PLD work also include limited descriptions of compound semiconductor film growth [14–17].

11.1.3 Why Grow Compound Semiconductors by PLD?

Recent National Research Council [21] and Dept. of Energy [22] reports have highlighted the need for research on optoelectronic semiconductors other than the group IV and III–V materials. Current device technologies are based on a tiny number of only ~10 traditional semiconductor materials. Consequently only a limited set of relevant properties (bandgaps, lattice constants, effective masses and carrier mobilities) is available [23]. The increasing importance of mesoscopic epitaxial semiconductor structures, in which epilayer thicknesses on the 1–10 nm scale are required to exploit purely quantum mechanical effects, also has caused research into synthesis and processing methods to be redirected almost exclusively toward low-temperature processing. This trend has placed great emphasis on using laser, plasma, or other energetic, beam-enhanced, low-temperature growth and processing methods, and on understanding and controlling growth mechanisms and defect structures at near the atomic level. Thus the challenge is to prepare new, chemically more complex materials as high-quality, epitaxial thin films.

There are several reasons why PLD may provide attractive opportunities for compound semiconductor growth, especially for epitaxial films and heterostructures of complex, multi-element materials. The stoichiometric transfer property means that it may be possible to grow quaternary or even more complex epitaxial semiconductors by ablating a fixed-composition target. In epilayer form, underutilized semiconductor materials such as the direct bandgap ternary and multinary I-III-VI chalcopyrites (prototype: CuInSe$_2$) could provide a rich variety of energy bandgaps and lattice constants for photovoltaic and optoelectronic device structures [24]. However, semiconductor growth for electronic applications places much more stringent requirements on stoichiometry, and for maintenance of stoichiometry locally, than did earlier PLD of oxide ceramics. PLA's ability to deposit films in reactive environments (ambient gas and laser-generated plasma) is intriguing for growth of epitaxial column-III nitrides and their alloys [25, 26]. The energetic ablation beam can induce reactions with adsorbed or ambient gases such as H$_2$S and N$_2$, to control the composition [27] or doping and electrical properties [28, 29] of compound semiconductor films. Dopants also can be incorporated directly in the ablation target and transferred to a film [30, 31]. However, the incorporation of unwanted species from the gas phase (e.g., hydrogen) or impurities from the ablation target may compensate desired dopants. Finally, PLD is attractive for superlattice and quantum well growth because it is easily possible to deposit only 0.1–1 Å per laser pulse and to "instantly" change layer composition simply by switching ablation targets; sharp compositional interfaces can be combined with attractive film-growth rates at laser repetition rates of 10–40 Hz. Multinary targets mounted in a multitarget carousel could be used to precisely match lattice constants or band edge offsets in adjacent epilayers and to increase the range of bandgaps and lattice constants available. Thus PLA may make possible multinary-based epitaxial structures with compositional and doping profiles that are not easily achieved by conventional growth methods [27, 32], for which the maintenance and controlled variation of multiple independent elemental fluxes may be difficult. However, there is another caveat: The possibility of interlayer mixing due to energetic species in the ablation beam is an open question, as is the possibility that the most energetic ablated ions or atoms will produce electrically active defects.

The need for high-quality films of chemically complex compound semiconductors places a high priority on understanding these issues, and on developing suitable new film-growth methods that perhaps will be hybrids of PLD with less energetic semiconductor-growth methods such as MBE and CBE.

11.2 Characteristics of Laser Ablation Important for Film Growth

A large part of PLD's attractiveness is that it is conceptually and experimentally simple, though the physics of the PLA process is quite complex. Figure 1 shows the essential features of a PLD film-growth system. An excimer laser beam (typical pulse duration 10–40 ns) is focused with energy density $E_d \sim 1$–5 J/cm^2

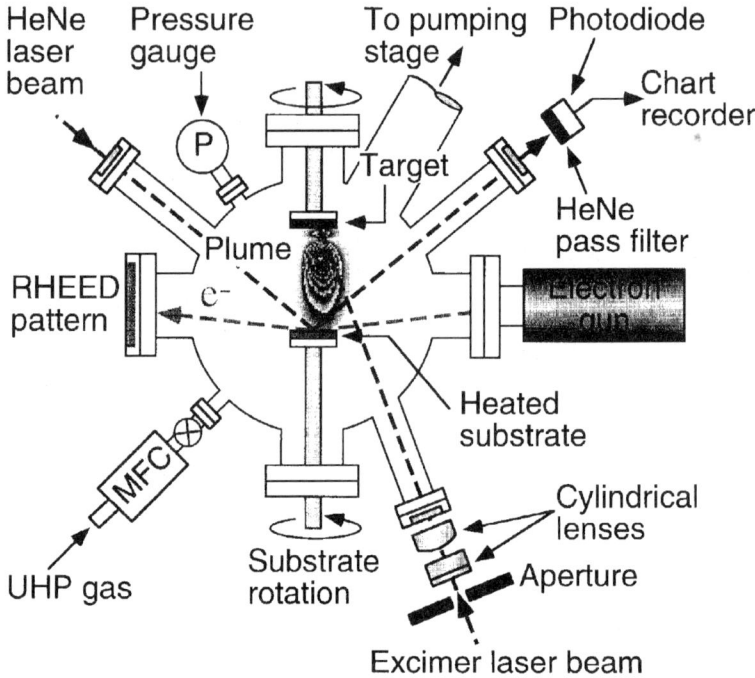

FIG. 1. Schematic diagram of a pulsed laser deposition system. (C. M. Rouleau)

through a UV-transmitting window onto a rotating polycrystalline target. Absorption of the focused laser energy creates a rapidly expanding plasma that contains both ground- and excited-state neutral atoms and ions, as well as electrons. These undergo collisions in the high-density plasma region near the target (the Knudsen layer), resulting in a highly directional flow normal to the target surface with initial velocities $\geq 10^6$ cm/s. If ablation is done in a reactive atmosphere such as oxygen, very simple oxide molecules also are formed in the expanding ablation beam. In this case, a shock front also forms due to collisions between the expanding plasma and the gas molecules; this front propagates with gradually decreasing velocity toward the substrate heater, which typically is located 4–10 cm away. [33] A biasing voltage also can be introduced into the target-substrate space to enrich the ablation plasma's content of selected species, in order to modify and control thin-film morphology and properties [34, 35].

11.2.1 Advantages of Pulsed Laser Deposition

Pulsed laser ablation has three characteristics that give it advantages for low-temperature growth of chemically complex (multi-element) thin-film materials.

These are:

1. *Congruent* (*stoichiometric*) *transfer* of material from target to film occurs over at least a limited range of deposition angles in the forward direction [36] when the focused laser energy density, E_d, and its spot size and shape are chosen properly. In most cases, all of the constituents of a complex, multi-component material can be deposited simultaneously from a single target, so that the only "starting material" requirement for PLD film growth is fabrication of a sufficiently high-purity polycrystalline target. Congruent material transfer results from the highly *nonthermal* target erosion by a laser-generated plasma (see Section 11.2.3).
2. PLA film growth takes place from an energetic beam and in a laser-generated plasma. Species in the ablation "plume" (both ground- and excited-state atoms and ions, together with electrons) possess both kinetic and potential (internal excitation) energy, which can be be used to increase sticking coefficients [37, 38] and adatom surface mobilities [39] and to enhance film nucleation via energy-enhanced chemical reactions. Consequently, epitaxial growth sometimes is possible at reduced substrate temperatures. In situ substrate cleaning and surface preparation also occur for some materials (e.g., oxide ceramics).
3. Because of the absence of electron beams or hot filaments in the deposition chamber, PLA has an enormously valuable natural capability for *reactive deposition in ambient gases*. Moreover, in the presence of plasma electrons gas-phase reaction rates increase rapidly with the kinetic energy of ablated species [40] so that simple compounds (oxides, nitrides, hydrides) are readily formed by ablation into an appropriate ambient gas. This capacity for reactive deposition into low-pressure O_2, O_3, NO_2, N_2O, or H_2O was largely responsible for PLA's great success in growing previously difficult-to-fabricate multicomponent ferroelectric, ferrite, and biocompatible oxide ceramic materials as epitaxial thin films [4].

11.2.1.1 Other Practical Advantages of PLD. PLA has other characteristics that are useful for film growth. Collisions with ambient gas molecules can be used to moderate the ablation beam's kinetic energy [28, 29, 33, 41–43]. This may prove important to avoid point defects that are produced in semiconductors at room temperature, or extended defects at elevated temperatures, when incident ion energies exceed approximately 30 eV [44]. Because most of the ablation flux is strongly forward-directed and because the laser beam can be focused to nearly a point, PLD may be superior to all other deposition methods in efficiently using rare or costly materials, or in confining the spread of toxic materials. Film growth is possible from a very small target, e.g., to grow thin films of extremely expensive or isotopically pure materials [15]. For films grown in a reactive environment, PLD growth chamber design can be simple and inexpensive; a bell jar-based system is adequate for some materials, and the transparent cover is convenient for viewing the plasma plume. Finally, quite high deposition rates can be obtained using commercial excimer lasers. For example, epitaxial $YBa_2Cu_3O_{7-x}$ films with high critical current density have been grown at 145 Å/s using a 30 W KrF (248 nm) laser running at 100 Hz [45]. Optimized deposition rates of ~600 Å-cm^2/s, which translate into >1 μm film thickness over 200 cm^2 in one hour, are possible.

11.2.1.2 Multilayered Epitaxial Structures and Laser-MBE. PLD is an especially versatile and simple method for growth of multilayered epitaxial heterostructures or superlattices in which each epilayer can be a chemically complex material. A separate, stoichiometric target can be used to grow each layer, with a multitarget "carousel" for rapid target exchange. Growth is inherently "digital" because each layer's thickness can be controlled precisely simply by calibrating the deposition rate per laser pulse and counting pulses. When a stoichiometric multi-element target is used, all elements are simultaneously present on the growing surface and so the smallest structural unit at which growth can be controlled is the chemical unit cell. Automation to grow complex multilayered structures under computer control is straightforward if the deposition conditions for alternating layers are similar. Using multi-element targets, epitaxial $YBa_2Cu_3O_{7-x}/PrBa_2Cu_3O_{7-x}$ superconducting superlattices have been grown by PLD with layer thicknesses as thin as a single unit cell [46, 47], as well as a 15-layer epitaxial integrated flux transformer and SQUID [48].

When PLA growth is combined with an ultra-high vacuum (UHV) environment and in situ monitoring tools such as RHEED, AES, and time-of-flight mass spectrometry (which require near-vacuum for operation), the result is "Laser-MBE" (L-MBE) film growth. At its ultimate level of refinement, the multi-element targets of ordinary PLD can be replaced by elemental or simple compound (e.g., metal-oxide) targets and used to deposit sequentially the subunit cell "building blocks" of a complex crystalline structure. In this way, growth can be controlled at the level of the smallest stable subunit of the unit cell (the atomic-layer level for some semiconductors), and PLD becomes "laser-sourced" MBE [49–52]. In its fullest implementation, using elemental sources in UHV, L-MBE trades away the advantages of stoichiometric transfer and reactive deposition in return for in situ monitoring and subunit cell control. In this way, new and completely artificial phases ("superlattice compounds") can be epitaxially stabilized and grown for studies of their properties [53].

11.2.2 Limitations of Pulsed Laser Deposition

PLA has two characteristics that have limited its application for film growth, especially for semiconductors and other electronic materials: particulates and the angular distribution of ablated species.

11.2.2.1 Particulates. Particulates having diameters from <0.1 μm to ~10 μm, with their size and areal density strongly dependent on the laser E_d and wavelength, usually are present on PLD films. These large particulates originate in two ways, either as irregular fragments that are blown off the target, or as round, solidified droplets that result from subsurface superheating and melting of the target. Molten droplets can be ejected from the target either by subsurface boiling (microexplosions) or by the ablation-induced recoil pressure exerted on the liquid, either of which results in "splashing" of the target. Much smaller particulates, with diameters of 1–100 nm, also can be produced by condensation from ablated vapor species; condensation of these smaller particles becomes especially noticeable for ambient gas pressures above 1 torr [54, 55a] or when a long-pulse laser [56] is used.

Small-diameter particulates cause no problem for many thin-film applications, but the larger particulates are a concern if films are to be lithographically patterned on the micron-or-smaller scale. Use of a relatively low laser E_d can significantly reduce the particulate count, but low E_d reduces the deposition rate, and too-low E_d results in nonstoichiometric transfer (see Section 11.2.3).

Both "passive" and "active" experimental techniques have been developed to minimize the number and size of particulates. By far the most important passive techniques are, first, the use of a short laser wavelength (e.g., deep UV excimer) and, second, target "conditioning" (target rotation combined with laser beam scanning) to maintain a smooth target surface. The most effective active technique is the use of a rotating-vane "velocity filter" to sweep aside the massive and relatively slow-moving particulates. However, a quite recent advance involving shallow, *transient* pulsed CO_2 laser melting of a target and pulsed excimer laser ablation of the melt promises to eliminate large particulates for all practical purposes [57, 58]. The underlying principles of these and other approaches to particulate minimization are described in Section 11.2.7.

11.2.2.2 Angular Distribution of Ablated Species. The pulsed laser beam must be focused to achieve the E_d and spot shape needed for stoichiometric transfer (Section 11.2.3). However, a near-point ablation source produces an angular distribution of ablated flux, $I[\vartheta]$, that is strongly peaked in the forward direction (typically $I[\vartheta] \sim [\cos \vartheta]^n$, with $n \sim 5\text{--}25$). Consequently uniform-thickness films would be produced only in a narrow angular range. A number of practical solutions to this problem have been developed based on producing relative motion of the ablation plume and the substrate ("painting" with the plume). For example, the laser beam can be raster-scanned over the target, or the substrate can be rotated, or these motions combined, to obtain uniform-thickness deposition. Various methods to obtain uniform film thickness are summarized in Section 11.2.8.

However, it must be pointed out that the ablation "plume" actually consists of two components: the high-intensity, forward-peaked, stoichiometric part, plus a low-intensity *evaporative* part (varying approximately as cos ϑ) that dominates at large deposition angles in vacuum [36]. Consequently a small fraction of the deposited material is expected to be nonstoichiometric if beam scanning is used. The pragmatic solutions to this problem have been either to operate at sufficiently high E_d that the evaporative component is negligible or to use films grown mostly from the central plume region. However, high E_d may exacerbate the particulate problem. Thus, obtaining thickness and compositional uniformity in PLD films, though an "intrinsic" problem, is not as serious as the particulates problem, but the two are coupled through the choice of E_d, at least for solid targets.

11.2.3 The Ablation Regime: Plasma Formation and Film Stoichiometry

For film growth, the "ablation" regime must be defined as the laser energy density range for which congruent transfer of material is obtained. The reason for this definition is that the process of pulsed emission of atoms/ions/molecules from a surface actually involves two successive E_d thresholds. The first (lower) threshold is that for the onset of laser etching, accompanied by particle emission. This pulsed atomic/molecular emission process is highly nonlinear in laser fluence and also may be highly selective for a particular chemical species [59]. The second threshold is for simultaneous stoichiometric transfer of all chemical species from target to substrate; for a multi-element target material, this occurs only at higher E_d. For example, for $YBa_2Cu_3O_{7-x}$ the KrF (248 nm) etching threshold is near 0.11 J/cm^2 [60], but stoichiometric film deposition is not obtained until $E_d \sim 0.4$–0.9 J/cm^2 [36, 61].

The reason for the two different thresholds is that atomic/molecular emission at low E_d results either from pulsed heating (and therefore really is laser thermal evaporation), or from selective desorption from particular near-surface sites (defects, cracks, impurities, etc.) as is described elsewhere in this book. If the primary low-E_d process is pulsed evaporation, variations in melting points and vapor pressures among the constituent elements of a polycrystalline target generally will result in noncongruent transfer. Desorption also is inherently chemically selective through its reliance on electronic transitions. In contrast, the stoichiometric transfer regime is reached only when E_d is sufficiently high to form a laser-generated plasma; congruent transfer, and the ablation regime for film growth, results from target erosion *mediated by this laser-generated plasma* [36, 61]. Consequently desorption studies at low laser fluence usually are not directly relevant to film growth in the useful ablation regime.

For film growth, a rule of thumb is that the onset of stoichiometric deposition coincides quite closely with the threshold fluence for formation of a dense laser-generated plasma just above the target surface [61]. The plasma's formation and expansion is described in the next section; for nanosecond laser pulses the plasma is self-sustaining throughout the laser pulse, being continuously fed material removed from the target by plasma erosion, and absorbing incident laser energy. Visible light emission from the plasma is continuously present during this period [61]. Consequently the very bright emission from near the target surface is a useful indicator of the onset of the plasma-mediated laser-target interaction that preserves stoichiometry in film deposition.

11.2.4 Overview: Physics of the Ablation Process

A qualitative understanding of how the energetic ablation beam is formed and propagates is essential in order to control and optimize it for film growth. The "capsule" overview of the ablation process given in this section is based on a longer summary published elsewhere [17] and, originally, on extensive model calculations [62–65] and experiments [59, 66–71] by many investigators.

The PLA process using nanosecond laser pulses can be thought of as occurring in three sequential time regimes [62, 63]:

1. The initial interaction of the laser beam with the target, resulting in evaporation of surface layers early in the pulse.
2. Absorption of most of the laser energy by this continuously evaporating material, resulting in formation of a sustained plasma and isothermal plasma expansion.
3. A highly anisotropic adiabatic expansion of the plasma.

The first two regimes exist during the duration of the laser pulse. The adiabatic expansion regime begins when the pulse terminates.

The initial stage of laser heating results in thermal emission of atoms, ions, and electrons from the target surface [72]; other nonthermal processes probably also contribute [66, 67, 73]. Gross omissions and simplifications in the treatment of this initial step are justified only by the fact that they do not alter the most important characteristics of ablation beams for film growth, which are determined in the two subsequent steps. Thus initial vaporization of the target can be treated as thermal, but absorption of the laser energy by the evaporated material, and the subsequent plasma expansion, are responsible for the highly nonthermal characteristics of the energetic beam of ablated species.

The mechanisms involved in laser heating of vaporized species have been extensively studied in experiments and calculations [62, 72]. A plasma is formed when initially vaporized species absorb laser energy in the presence of seed electrons, by the inverse bremsstrahlung process. It is believed that the resulting plasma effectively screens the target from the remainder of the laser pulse, continuing to absorb laser energy by inverse bremsstrahlung [74]. The plasma absorption coefficient for this process is

$$\alpha_p = 3.69 \times 10^8 (Z^3 n_i^2 / T^{0.5} v^3)[1 - \exp(-h v / k_B T)], \qquad (11.1)$$

where Z, n_i, and T are the average charge, ion density, and temperature, respectively, of the plasma, and v is the laser frequency [62].

Equation (11.1) reveals two important features of the plasma absorption process. First, because $\alpha_p \sim n_i^2$, laser energy is absorbed only close to the target where ion densities are high. Second, it is important that the laser frequency, v, be higher than the plasma frequency, v_p, or else the incident laser energy simply would be reflected by the plasma. For a KrF excimer laser, $v = 1.21 \times 10^{15}$ Hz, while the electron density required for $v_p = v$ is $1.8 \times 10^{22}/\text{cm}^3$. This very high (near-metallic) electron concentration ensures that reflection losses are negligible at excimer laser frequencies.

The plasma absorption coefficient, Eq. (11.1), controls the heating of evaporated species. However, its complex dependence on laser parameters, plasma n_i and T, and target material properties led Singh to develop an approximate method of calculation in which each parameter was considered to be a function of temperature and E_d only [62]. Consequently, the plasma temperature T is controlled by the laser E_d in his model calculations.

High expansion velocities ($>10^6$ cm/s) of the plasma into vacuum result from the large density (and, therefore, pressure) gradients that are initially present at the outer (vacuum) edge of the plasma. The electron and ion densities decrease rapidly with distance from the target, and the plasma is effectively transparent to the laser beam at large distances from the target. Consequently energy is absorbed during the laser pulse only within a thin region (the Knudsen layer) near the target surface. Light emission from this region can be seen directly in the temporally and spatially resolved plasma-imaging studies of Geohegan et al. [33, 70, 71], shown in Fig. 2. For an initial plasma expansion velocity of 10^6 cm/s, the spatial extent of the plasma at the end of a 30 ns laser pulse is ≤ 300 μm. Because the thermal diffusion time over these dimensions in a dense plasma is significantly less than its expansion time, the assumption of a uniform plasma temperature is justified [62, 72, 75]. Because the plasma volume remains small during the laser pulse, radiative (T^4) losses are small at the relatively low-temperatures characteristic of PLA plasmas; for similar reasons, electron-ion collisions are able to produce a common plasma temperature under these conditions.

Thus the overall picture is that a dynamic steady state exists during the laser pulse, in which a thin region of plasma just above the target surface continuously absorbs laser energy, while being fed evaporating material at its inner surface. The absorbed energy is converted into the thermal energy used for evaporation and the kinetic energy of the simultaneous outward plasma expansion.

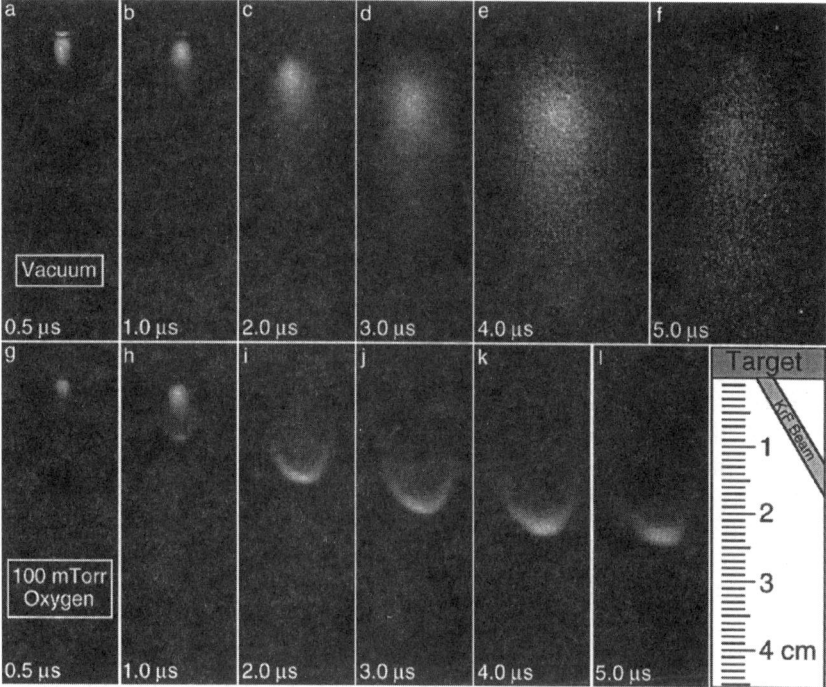

FIG. 2. ICCD photographs of the visible plasma emission (exposure times 20 ns) following 1.0 J/cm^2 KrF laser ablation of YBCO into 1×10^{-6} torr [(a)–(f)] and into 100 mtorr oxygen [(g)–(l)], at the indicated delay times after the laser pulse. The ablation geometry for each panel, with the target at the top and laser beam incident from below, is shown at the lower right. Light emission from the dense plasma (Knudsen layer) close to the target surface can be seen clearly in the first two panels of the vacuum sequence [70, with permission].

11.2.5 Angular Distribution of Ablated Species

Saenger has extensively reviewed current knowledge of the angular distribution of ablated material in vacuum and the modifications due to scattering by ambient gas molecules [76]. The reader is referred to her article for a detailed summary and analysis of methods of measurement, experimental results, and model calculations. In this section we briefly summarize those experimental results that are well established, together with selected results of modeling that provide insight into the experiments.

11.2.5.1 Measurements of Angular Distribution.

Two angular distributions are associated with PLD: the source angular distribution, $f(\theta)$, and the film thickness profile, $D(\theta)$, where θ is the angle of ejection of material, measured from the normal to the target surface. In the case of a film deposited on a flat substrate placed parallel to the ablation target and a distance h away, a point ablation source having an angular distribution $f(\theta) = (\cos \theta)^p$ will produce a film-thickness profile $D(\theta) = (\cos \theta)^{p+3}$ [77]. The most common method of determining the angular distribution is simply to measure the film thickness on a substrate as a function of the distance, x, from the projected ablation center line, i.e., measure $D(x)$ and $\theta = \tan^{-1}(x/h)$ to yield $D(\theta)$. The only assumptions in this method are (1) that all ablated species have unity sticking coefficient and (2) that no resputtering of deposited material occurs. If either assumption is violated, an undercounting of ablated material will result and $D(\theta) \ne f(\theta)$. In the case of resputtering, the outgoing and incoming fluxes also can interact, possibly further modifying the angular distribution [78].

11.2.5.2 Model Calculation of the Source Angular Distribution. For ablation in vacuum, the angular distribution of ablated species is strongly forward-directed, $f(\theta) \sim (\cos \theta)^p$ with $p \gg 1$. This forward-peaking is produced by the collisions of species in the plume with each other; Kelly pointed out that only a few collisions per atom or ion are needed to produce a strongly forward-directed beam [59]. In Singh's model [62] for the PLA plasma expansion (see Section 11.2.4), the initial plasma density is $\sim 10^{19}$–10^{20} cm^{-3}. Consequently mean free paths are short, many collisions occur between different species, and the plasma can be modeled as a fluid using the equations of gas dynamics to simulate its expansion [62]. The relationship between plasma density and pressure was assumed to be the ideal gas law, $P = nk_BT$; a linear increase of velocity with distance from the target also was assumed in order to maintain a Gaussian density profile. By combining the equations for pressure and velocity with the continuity equation and the equation of motion, Singh and Narayan arrived at an expression for the plasma dimensions during the laser heating period ($t \le \tau$) [62]:

$$X(t)\{(1/t)\,dX/dt + d^2X/dt^2\} = Y(t)\{(1/t)\,dY/dt + d^2Y/dt^2\}$$
$$= Z(t)\{(1/t)\,dZ/dt + d^2Z/dt^2\}$$
$$= k_BT/M \qquad (t \le \tau) \qquad (11.2)$$

where X, Y, and Z are the Gaussian widths of the expanding plasma, T is the plasma temperature, and M is an average mass.

The main value of Eq. (11.2) is that it explains the highly anisotropic and strongly forward-directed plasma expansion that is mediated by intraplume collisions. Early in the expansion, when the expansion velocities (dX/dt, etc.) are low, the acceleration is inversely proportional to the initial plasma dimensions. The transverse dimensions are given by the laser spot size (typically 2–3 mm), but in the direction perpendicular to the target, the initial plasma dimensions are only tens of micrometers. Consequently the initial acceleration is very high perpendicular to the substrate, decreasing as the velocity increases, and producing the characteristic elongated PLA plasma shape.

The final, adiabatic plasma expansion stage occurs after the laser pulse terminates. In Singh's model, the plasma dimensions and temperature (assumed not to vary spatially) then are related through the adiabatic equation of state,

$$T[X(t)Y(t)Z(t)]^{\gamma-1} = \text{constant} \qquad (11.3)$$

where $\gamma = C_p/C_v$. Thus thermal energy is converted into kinetic energy as the expansion velocity increases. By combining the previous equations with the adiabatic equation of state and the equation of temperature, Singh arrived at the equation governing the adiabatic expansion stage:

$$X(t)\{d^2X/dt^2\} = Y(t)\{d^2Y/dt^2\} = Z(t)\{d^2Z/dt^2\}$$
$$= (k_B T/M)\{X_o Y_o Z_0/X(t)Y(t)Z(t)\}^{\gamma-1} \qquad (t > \tau) \quad (11.4)$$

in which X_o, Y_o, Z_o are the plasma's Gaussian dimensions at termination of the laser pulse ($t = \tau$). Equation (11.4) shows that the plasma's acceleration varies inversely with its dimensions. Consequently the highest velocities are obtained in the direction perpendicular to the substrate surface.

Equation (11.4) also explains one of PLD's most striking characteristics—the relationship between the shape of the laser spot focused on the target and the shape of the resulting film deposit. If the focused laser spot is initially longer in the horizontal direction, the film will be elongated in the vertical direction, i.e., the plasma becomes elongated in what is initially its shorter dimension. An elliptical plasma with its major axis horizontal experiences larger acceleration along the minor axis (vertical) direction, resulting in a deposited film with its major axis vertical. Consequently, if a pulsed laser beam is brought to a horizontal line focus by a cylindrical lens, the resulting film deposit will be only a little wider than the line focus but will be expanded vertically into a long stripe [79].

11.2.5.3 Angular Distribution: Effect of Laser Spot Size and Other Laser Parameters.

A number of studies have been carried out in which the focused laser spot size, d, was systematically varied while keeping the focused laser energy density (J/cm^2) constant. A well-established but initially counterintuitive result is that the film thickness profile, $D(\theta)$, becomes *more sharply forward-peaked* as the spot size *increases*, so long as $d/h \ll 1$ (where h is the target-substrate separation). A consequence of this behavior, pointed out by Muenchausen and co-workers [80], is that a large-diameter laser spot and a smaller laser spot scanned over the same area give *different* angular distributions. This is a significant difference between PLD and effusive sources; for the latter, the typical cos θ angular distribution is independent of the source diameter, assuming again that $d/h \ll 1$.

For PLD the effect of laser spot size has been investigated both by varying the diameter of a round spot and by using a rectangular spot and measuring the angular variations in the two directions along the spot's long and short axes. Film thickness profiles $D(\theta)$ varying as $(\cos\theta)^{p+3}$ have been observed with p varying from ~1 to ~70 [81]. Even for a single material, large variations in $D(\theta)$ can result simply from changing the spot size. For example, $p \sim 1$ for Pb and W when using a tightly focused spot of ≤200 μm diameter, but a much narrower angular distribution, with $p \sim 7$ (Pb) or $p \sim 23$ (W), resulted when the spot size was increased to ~2 mm [78].

The increasingly strong forward-peaking of the plasma plume as the laser spot size is increased can be understood qualitatively in either of two ways. As indicated above, the forward-peaking results from the collisions of ablated species with each other as they leave the target, forming a stream that flows perpendicular to the target surface. Thus, the degree of forward-peaking increases with the number of intraplume collisions that ablated species have; this number is given approximately by the ratio of the relevant spot dimension to the species mean free path. An alternative way of viewing this, using Eq. (11.4), is that the lateral components of acceleration out of the collision region will be relatively small if the laser spot's lateral dimensions are large, corresponding to a strongly forward-directed distribution.

The effects of laser fluence, wavelength, and pulse duration on the plume's angular distribution have been examined but, according to Saenger, few generalizations are possible even for the effect of fluence, the best studied of these [76].

11.2.5.4 Angular Distribution: Effects of Ambient Gases. As was just described, the plume's angular distribution in vacuum is determined by collisions of the ablated species with each other as they leave the target. In an ambient gas, additional collisions occur with the gas molecules, scattering the ablated species and broadening their angular distribution. Intuitively, the effect of collisions on the angular distribution would be expected to be the greatest for a low mass species that is ablated into a high-mass background gas, and least in the opposite case. The effect of ambient oxygen pressure in broadening $D(\theta)$ is well documented for PLD YBCO films [82, 83], and Saenger has tabulated a number of these results [81].

For a target-substrate separation, h, the onset of collisional broadening of the plume might be expected at the pressure for which the mean free path of ablated species, Λ, first becomes smaller than h. For higher pressures, collisional broadening of the angular distribution would be expected to scale as the pressure-distance product, $P \times h$ [84]. Somewhat surprisingly, a number of experiments show that for target-substrate separations in the cm range, significant collisional broadening of the plume occurs only when the ambient gas pressure is somewhat higher than expected by this criterion, i.e., well into the mtorr range. Thus, as pointed out by Saenger, the $\Lambda \sim h$ threshold for the appearance of a collisionally broadened angular distribution apparently overestimates the effect of collisions

[85]. For example, angular distributions of carbon deposits collected at $h = 2.5$ cm were significantly broadened only for pressures >22 mtorr [86]. Similarly, recent time-resolved ion probe measurements by Geohegan *et al.* during ZnTe ablation into ambient nitrogen with $h = 10$ cm revealed a significant interaction of the ions with nitrogen molecules (resulting in the ion current splitting into scattered and unscattered components) only for nitrogen pressures ≥20 mtorr [41–43]. (See Sections 11.2.3.4–5.) These results show that a relatively large number of collisions, corresponding approximately to $h/\Lambda \sim 10$, is required to broaden or otherwise significantly alter the distribution of energetic species in typical ablation plumes.

11.2.5.5 Angular Distribution: Nonuniformities in Film Composition. Although PLD is widely described as producing films with the same composition (stoichiometry) as the ablation target, some detailed measurements have revealed variations in film composition as a function of the deposition angle θ (measured from the forward direction). Even a small compositional nonuniformity is of considerable interest, and a potentially serious problem, for semiconductor films, for which antisite defects or vacancies can compensate deliberately introduced dopant atoms and prevent their electrical activation. Compositional variations can arise in principle from three causes:

1. If the basic assumptions in using film-based measurements to infer $f(\theta)$ are correct (unity sticking coefficient and no resputtering), these measurements must reveal an actual angular variation of plume composition, i.e., intrinsically different angular distributions for different chemical species. Such differences could arise due to differences in the charge of species (neutrals and multiple ionization states) or their mass (light or heavy atoms; molecules or reaction products formed during plume transport).
2. Even if $f(\theta)$ is the same for all species, compositional nonuniformities could arise due to resputtering or sticking coefficient effects. In the case of sticking coefficients, Saenger has pointed out that merely to have an angle-dependent sticking coefficient is not sufficient; the sticking coefficients for different species must have different angular dependences for compositional nonuniformity to result [76]. An angular distribution that is the same for all species still could produce a deposit with varying composition, if resputtering occurs that is selective for a particular species. Compositional nonuniformity can result from resputtering if one chemical species has a sputtering yield that is angle-dependent, or if different species have different angular dependences for resputtering. Saenger has noted [76, p. 224] that even if all resputtering yields are constant (with different values), compositional nonuniformity still could develop if the incident flux of the energetic species responsible for resputtering has a different angular distribution from the deposited material. The latter resputtering mechanism may account for the

strongly laser fluence-dependent angular distributions of Pb, Zr, and Ti in $PbZr_xTi_{1-x}O_3$ films and for the development of a bimodal Pb distribution at high laser fluences (e.g., due to resputtering by strongly forward-directed energetic ions) [87].

3. The background gas-scattering effects mentioned in the previous section are a possible cause of angular compositional nonuniformity. The angular distribution of low-mass species would be expected to broaden, whereas high-mass species would remain mainly forward-directed, due to their relatively small and large velocity changes, respectively, for a given momentum change in collisions with ambient gas molecules. Although several groups have obtained data to describe the angular variation of composition in PLD films under vacuum deposition conditions (see below), and model calculations also have been carried out (also described below), the effect of ambient gases on the deposition thickness profile and the angular variation of film composition simply is not clear at this time (e.g., [82]), despite the great practical importance of reactive deposition. Additional experiments and modeling are needed to understand the effect of collisions between ablated species and ambient gas molecules on the deposition profile and the angular variation of composition.

With regard to intrinsically different angular distributions for different chemical elements (e.g., due to mass or charge differences), the limited experimental evidence is fairly clear but is not straightforward to interpret theoretically. Three recent experimental studies of YBCO films indicate that in vacuum and for moderate E_d values, low-mass species (e.g., Cu) are enriched in the forward direction, close to $\theta = 0$ [36, 63, 82]. There also is experimental evidence that this enrichment disappears with increasing E_d [36, 63, 88]. The tendency toward spatial (or angular) uniformity of composition with increasing E_d is supported by at least two recent model calculations [62, 65]. Saenger has noted that these PLD results contrast sharply with both the *mass-independent* angular distributions that are typical of *collision-free* effusive sources, and with the mass distributions for highly supersonic gas mixtures, which are more strongly forward-peaked for the *higher* mass components [65]. This observation provides the important clue that collisions can markedly change the angular distribution in a mass-dependent way. However, the apparent experimental observation of an intrinsically mass-dependent angular distribution and its theoretical interpretation become further complicated when we recognize that atomic masses may not be representative of the actual masses with which different species are transported from target to substrate, e.g., because the reactive formation of molecules or small clusters causes "transport masses" to differ from atomic masses.

Nevertheless, Saenger developed a model for the origin of intrinsically mass-dependent spatial nonuniformities in the composition of PLD films [65]. This model exploits the similarity between a beam of ablated material and a supersonic gas jet (free jet expansion). It is based on the idea that the extent of forward-peaking for each mass species in the ablation plume depends on the ratio of two mass-dependent velocities: the flow velocity, u, which characterizes the directed forward motion, and the mean random thermal velocity, $\mathsf{v}_\perp$, which characterizes the transverse motion. Intraplume collisions are responsible for development of the flow velocity; without collisions, $u = 0$ and an effusive angular distribution, $f(\theta) = \cos\theta$, results. However, as noted earlier, only a few collisions per particle are needed to produce a strongly forward-directed beam [59]. For a single-component gas, the usual Mach number is defined as the ratio of the flow velocity, u, to the local velocity of sound, $v_s = (\gamma k_B T/m)^{1/2}$,

$$M = u/v_s = u/(\gamma k_B T/m)^{1/2} \tag{11.5}$$

In a multicomponent gas, the Gaussian halfwidth of the transverse velocity distribution for species with mass m_i is $\mathsf{v}_{i\perp} = (2k_B T/m_i)^{1/2}$. The pseudo-Mach number, M, for a multicomponent gas expansion then is defined as $(2/\gamma)^{1/2}$ times the ratio of u and $\mathsf{v}_{i\perp}$,

$$\mathsf{M} = (2/\gamma)^{1/2}(u/\mathsf{v}_{i\perp}) = (2/\gamma)^{1/2}\{u/(2k_B T/m_i)^{1/2}\} \tag{11.6}$$

For a single-component gas this reduces to $\mathsf{M} = M$, but in a multicomponent gas the sound speed depends on the mixture's average mass, $v_s = (\gamma kT/m_{av})^{1/2}$, so that $\mathsf{M} \neq M$ and the pseudo-Mach number M is different for each species with a different mass m_i.

In Saenger's model, [65] compositional nonuniformities arise when different species have different pseudo-Mach numbers: this can occur due to the mass dependence of either u or $\mathsf{v}_{i\perp}$, as shown by Eq. (11.6). Saenger calculates the species-dependent pseudo-Mach number by regarding the ablation beam as analogous to a steady-state (continuous) adiabatic supersonic gas expansion. An expression for M is derived (Eq. 20 of [65]) from which its mass dependence can be evaluated in different expansion regimes. Within the model proposed by Saenger, the mass dependence of M turns out to be controlled by two opposing factors:

1. In highly supersonic gas expansions, all species have nearly the same flow velocity [89]. Consequently, heavy species have higher Mach numbers due to their lower transverse thermal velocities, $\mathsf{v}_{i\perp} \sim m_i^{-1.2}$. This should result in on-axis *enrichment* for *heavy* species if a uniform flow velocity for all species is achieved at high laser E_d values.
2. In weak expansions produced by low laser E_d, lighter mass species have a higher relative speed than heavy-mass species and hence have a higher collision rate. In addition, for a given momentum change (e.g., in center-of-momentum coordinates), the lighter species suffer the larger velocity change. This factor is called the "collision rate/effectiveness effect." It leaves

the heavier species with a higher effective temperature and lower flow velocity because collisions effectively cease sooner for them in the expanding plasma than for the lighter species. Lighter species continue to cool through collisions and achieve a higher flow velocity. Consequently, lighter species have higher Mach numbers under weak-expansion conditions and should be enriched on-axis for low laser E_d.

Saenger attributes the experimentally observed improvement in compositional uniformity of PLD films at higher laser E_d to an incomplete transition from the weak-expansion to the strong-expansion regime [65]. She suggests that the strong expansion limit may not be achievable, even at very high E_d, due to absorption in the ablation plume that limits the efficiency of material removal. Although the analogy between an ablation beam and a supersonic expansion may be imperfect, it does seem to account for observed spatial variations in composition and to provide physical insight into their origin.

11.2.6 The Role of Energetic Species in Pulsed Laser Deposition

11.2.6.1 Velocities and Kinetic Energies in Vacuum.
A number of investigators have measured the velocities of ablated species in vacuum. One of the most widely used techniques is optical time-of-flight measurements of the time-dependent emission from atoms and ions in excited states. Zheng *et al.* found that the temporal evolution of fluorescence from neutral Y, Ba, Cu, and O excited states, at various distances from a YBCO target, could be accounted for if their velocity distribution was similar to that in a supersonic free-jet expansion, distributed about a most probable (or stream-flow) velocity, u [64]. Most-probable velocities were in the 10^6 cm/s range. However, the distribution of velocities was much narrower than Maxwell-Boltzmann, consistent with the cooling of translational degrees of freedom that occurs when a high-pressure gas expands into a lower pressure region in a supersonic expansion [64, 90, 91]. Kelly and Dreyfus obtained similar velocity distributions, involving a stream-flow velocity common to all species, from a theoretical description based on the formation of a Knudsen layer (thin collisional layer) just above the target surface [67]. Emission from the stationary Knudsen layer was observed directly by Geohegan et al. and can be seen in the first two panels of Fig. 2 [33, 71]. Marine *et al.* also analyzed optical time-of-flight spectra following YBCO ablation and obtained velocities of $1-5 \times 10^6$ cm/s; for nanosecond-laser ablation the velocity distribution also was characterized by a common stream-flow velocity for different-mass species [92]. Lubben *et al.* measured velocities of $1.0-1.6 \times 10^6$ cm/s for neutral atoms and ions emitted during KrF (248 nm) ablation of Ge targets [93].

Geohegan and Mashburn were the first to use transient optical absorption measurements to obtain information about the motion of nonemitting *ground-*

state species, with measurements for Y, Ba, and Cu neutrals and Ba$^+$ ions during YBCO ablation [94]. A high-pressure Xe arc lamp provided a narrow pulsed (~500 ns) beam of structured continuum emission passing perpendicular to the plasma expansion. The observation of absorption in these experiments was significant because narrow atomic linewidths normally prohibit absorption spectroscopy. However, collisions in the ablation plasma can broaden spectral features to well above spectrometer resolution. By tuning the spectrometer to the center of an absorption line and varying the laser-pulsed lamp time delay, Geohegan and Mashburn were able to map out the temporal profile of absorbing species as they moved through the region being probed. The velocity distributions inferred from absorption were broader and peaked at a lower velocity than the distributions obtained from fluorescence. Furthermore, they revealed a second low-velocity component not seen by fluorescence. The velocity profiles were not Maxwellian, but consisted of a "fast" component (matching the fluorescence profile) and a "slow" component that appeared as a low-velocity shoulder [94].

Murakami and co-workers used three different temporally and spatially resolved methods to measure the velocities of laser-ablated species, including streak camera measurements of light emission, a HeNe laser deflection technique, and laser plasma X-ray absorption spectroscopy [95, 96]. Two luminous components were observed at low E_d, merging into one with a velocity ~2×10^6 cm/s with increasing E_d. In addition, two slower moving, nonluminous components were detected, the slowest of which was identified as coming from the massive particulates emitted from the target at times much later than the ablation plasma [95]. (See Section 11.2.7.2 for other measurements of particulate velocities.)

Geohegan and Mashburn's work [94] implies that the ablation plasma is only weakly ionized for laser fluences typical of film growth; most atoms and ions are in ground states. Their absorption measurements reveal that neutral species are present for quite long times. However, the ground-state atoms and ions in the "fast" component have velocities very similar to fluorescent excited-state species, typically $1-2 \times 10^6$ cm/s. These measurements, together with those of Murakami and co-workers, also reveal a significant low-velocity ablation component that cannot be seen by emission. This component indirectly confirms the importance of gas-phase collisions in the plume because absorption could not be seen except for the line-broadening due to collisions [94].

A velocity of 10^6 cm/s corresponds to a kinetic energy of 0.52 eV for a mass of 1 amu. Hence, for an ablated atom or ion with a mass of 100 amu, the typical velocity in vacuum of $1-2 \times 10^6$ cm/s corresponds to a kinetic energy of ~50–200 eV. This is a significant energy input to the growing film surface that gives PLD the potential to assist crystalline phase formation or to promote near-surface chemical reactions, but also to inflict lattice displacement damage in a growing film. We now consider these possibilities in more detail.

11.2.6.2 Energetic Beam-Assisted Growth.

Crystalline Phase Formation. In conventional vapor-phase growth, the formation of a high-quality crystalline film requires that adatoms have sufficient mobility on the growing film surface that they can sample a large number of sites during the time required to grow a monolayer [97, 98]. Too rapid deposition coupled with low surface mobility produces poorly ordered or amorphous films. Gilmer and Roland have pointed out that noble metals and noble gases have high rates of surface diffusion and crystallize readily from a vapor source, even at low substrate temperatures; in contrast, silicon and most other semiconductors have slower rates of surface diffusion, so crystalline films are more difficult to grow [99]. Both molecular dynamics simulations and actual silicon MBE growth experiments show that amorphous Si is formed at low temperatures, but epitaxial Si films can be grown on (001) substrates at temperatures above ~350 K (experiment [100, 101]) to 450 K (simulations [102, 103]). Aside from the lower surface diffusion rates, another factor complicating the growth of crystalline semiconductor films is that surfaces such as Si(001) can reconstruct to form strong dimer bonds with surface adatoms. These bonds must be broken to form the correct crystalline structure. If the dimer bonds remain intact, amorphous deposits are formed [104]. However, the dimer bonds can be broken and rearranged by energetic atoms impinging on the growing surface [105].

Motivated by these observations, Gilmer and Roland recently used molecular dynamics methods to simulate MBE growth from an energetic Si atom beam on both (001) and (111) Si surfaces [99]. To test the efficacy of strictly local heating for crystallization, they simulated MBE growth from a thermal beam of atoms with 0.17 eV kinetic energy and compared the resulting deposits with those grown from a beam of Si atoms with 5 eV kinetic energy. They found that the 5 eV atoms released about twice the energy on impact as the thermal atoms and had a dramatic effect on the crystalline order of the deposits. For both (001) and (111) surfaces, the 5 eV beam produced a crystalline film at less than half the absolute temperature required for the thermal beam. (The crystallization temperature for the (001) surface was ≤ 250 K and was much higher for (111), in agreement with experiment.) According to the authors, the effect of localized heating by nonthermal Si atoms is similar to annealing; it provides the excitation energy needed to break dimer bonds, as well as transient motion that facilitates bond rearrangements into a crystalline structure, while maintaining a low substrate temperature.

Lattice Displacement Damage. Significant momentum and energy transfer can occur in collisions of energetic incident atoms with atoms of similar mass. Such collisions may eject previously deposited atoms out of crystal lattice sites in the top few atomic layers of a growing film, resulting in crystalline defects such as interstitials and vacancies. These point defects are problematic for compound semiconductor doping because they compensate dopant atoms, resulting in electrically insulating or poorly conducting films. For bulk semiconductor materials, the energy *transfer* required to displace a host atom from its lattice position is about 15–20 eV per atom [106]. A recent simulation also showed that argon ions with energies less than 20 eV can displace surface Si atoms but do not cause bulk damage [107]. On the other hand, epitaxial semiconductor films grown by ion beam deposition at only slightly higher incident ion energies of 40–60 eV contained extensive extended defects [44, 108].

11.2.6.3 Control of Kinetic Energy via Gas-Phase Collisions.

From the discussion in the preceding section, for the growth of doped, epitaxial semiconductor films, the kinetic energies of ablated species should be sufficient to break bonds in the surface layer to facilitate doping, alloying, and the formation of a crystalline structure at low substrate temperatures. However, higher energy species that produce deeper lattice displacement damage must be eliminated. The optimum kinetic energy to achieve both is not known exactly, but it appears to be ≤20 eV.

Because PLD can be carried out at moderate ambient gas pressures (generally ≤500 mtorr), it is natural to consider using a gas to control film composition or doping via beam-assisted reactions, and simultaneously to moderate the ablation beam's kinetic energy via gas-phase collisions. At sufficiently high pressures, gas-phase collisions completely thermalize the initial kinetic energy of ablated atoms/ions so that it is not delivered to the growing film surface. Consequently, for epitaxial film growth, it is important to understand how the velocity distribution of ablated atoms and ions is modified by collisions with background gas molecules, and how many collisions are required to thermalize their kinetic energy. The answer determines the optimum gas pressure for reactive PLD at a given target-substrate separation, D_{ts}.

Geohegan [70] and Geohegan and Puretzky [41] recently used gated ICCD-array camera and ion probe measurements for time-resolved and spatially resolved studies of the propagation of ablated species in ambient gases. A detailed analysis of ICCD-array measurements such as those shown in Fig. 2 revealed four principal effects of increasing the background gas pressure [33]:

1. An increase in fluorescence from all species due to collisions on the expansion front and subsequent interplume collisions.
2. A sharpening of the plume boundary, indicative of formation of a shock front.
3. A slowing of the plume relative to its propagation in vacuum.
4. Spatial confinement of the plume.

Figure 3 shows position versus time (R vs. t) for the leading edge of the plasma emission in Fig. 2. It confirms that in 100 mtorr oxygen, the plume expansion is unaffected for times ≤1 μs, but then slows progressively. Geohegan found that a classical viscous "drag" model agrees well with the R-t data for early times and low pressures, during which time the mass of ejectants in the plume exceeds the mass of gas displaced. However, a blast wave (shock front) model fitted the data best for long times and higher background pressures, as shown in Fig. 3. The drag model predicts that the plume eventually will come to rest. In contrast, the shock model predicts continued propagation with $R \sim t^{0.4}$ and the range of the ablation plume is limited simply by attenuation [33].

During the transition from drag to shock front formation, the plasma has two propagating components, one moving faster, one slower (see Fig. 2). Ion probe measurements have revealed that this two-component character is typical of plasma expansions into background gases at lower pressures. As shown in Fig. 4(b), a single-component, ion-pulse shape was obtained for YBCO ablation into oxygen only for pressures >100 mtorr [33]. In fact, Geohegan and Puretzky have shown that this "plume splitting" phenomenon occurs quite generally for ablation of various target materials into low-pressure ambient gases [41]. Their

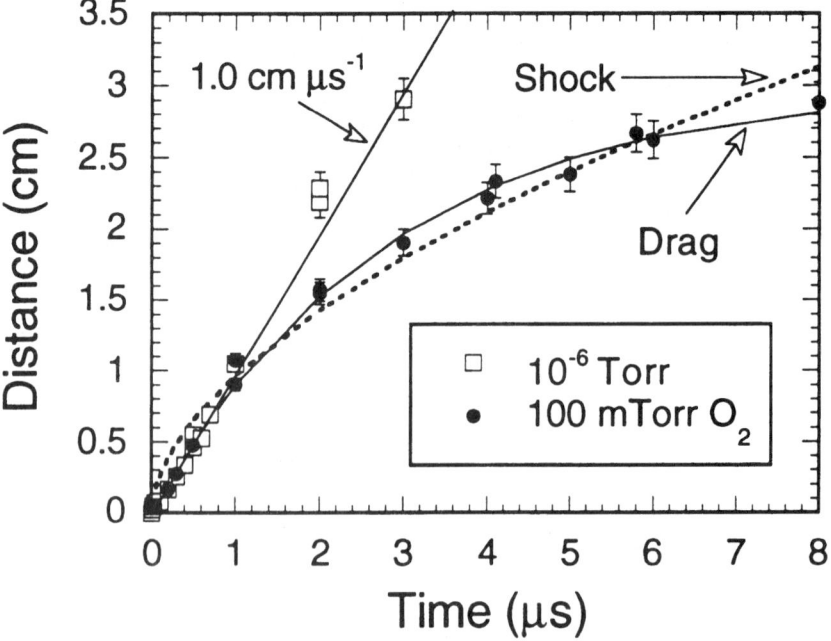

FIG. 3. Position vs. time plots for the leading edge of the luminous plasma plume (see Fig. 2), measured along the normal to the YBCO target, in vacuum and 100 mtorr oxygen. Two fits to the 100 mtorr data are shown, for a viscous drag model and a shock model. The error bars represent uncertainties in locating the position of the expansion front boundary [70, with permission].

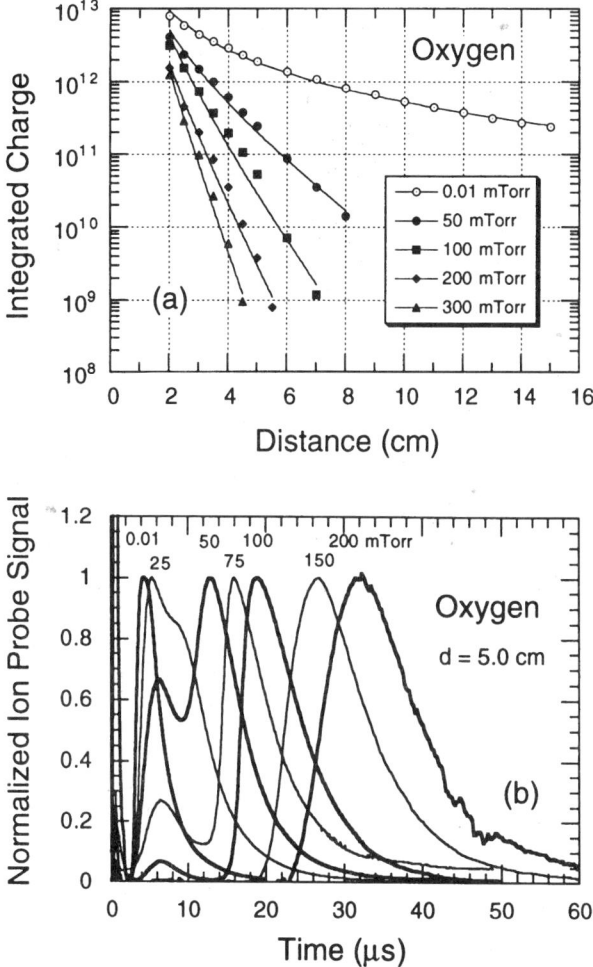

FIG. 4. (a) Integrated positive ion charge transmitted through oxygen and measured by a fast ion probe along the normal to a YBCO target irradiated at 248 nm and 3.0 J/cm^2. (b) Normalized ion-probe current waveforms measured at D_{IE} = 5 cm along the normal to the YBCO target. [From chap. 5 of [3], with permission]

experiments indicate that the fast component consists of ions that have had no collisions with ambient gas molecules, whereas the slower component consists of ions that have been scattered. Thus ions in the fast component have essentially the same relatively high kinetic energy as they do in vacuum, even though a gas is present. As is described in Section 11.3.2.4–5, epitaxial PLD ZnTe films exhibit the highest hole mobility when the nitrogen pressure is increased enough to eliminate the fast component, suggesting that fast ions do indeed create lattice defects. (Note that the ion probe is a flux-sensitive detector [flux $F(t) = N(t)v(t)$] that is biased to collect positive ions. Thus the waveforms in Fig. 4(b) are representative of moving positive charge, not stopped vapor.)

Recent Monte Carlo calculations have successfully simulated a fast transmitted flux through a background gas [109]. However, alternative explanations exist for the simultaneous presence of both fast and slow components, including ionization and snowplowing of background gas or species-dependent scattering [110–112]. Recently Wood et al. formulated an elastic-collision model for the multicomponent particle flux that is observed during ablation into low-pressure background gases [113]. Key ingredients of the model include elastic collisions with a constant cross-section, hydrodynamic transport of both plume and background species, and interpenetration of ablated species in the background gas. Numerical calculations using this model have successfully reproduced the experimental ion flux for silicon ablation into argon and helium, and for graphite ablation into argon. In particular, the division of the ablation flux into well-separated translational kinetic energy distributions centered near ~50 eV (the "vacuum" distribution) and ~1 eV (scattered species) was observed [113].

Figure 4(a) shows the integrated total transmitted positive charge versus distance as a function of background oxygen pressure. The combined attenuation and slowing of the laser plasma by background oxygen is clear. Geohegan found that the attenuation was described by $I = aI_o \exp(-bd)$, where d is the distance from the target [33]. The vacuum ion flux, I_o, varied nearly as d^{-2}, and the attenuation coefficient, b, varied linearly with background pressure, so that I varies exponentially with pressure. The blast wave (shock front) model also was found to accurately describe the entire shape of the positive ion waveforms, once single-component pulses were formed at the higher oxygen pressures.

Figures 2–4 show clearly that gas-dynamic effects play the dominant role in forming the main body of ejectants, and in controlling their propagation, under typical reactive PLD conditions. Figure 4 demonstrates an ~eight-fold velocity reduction at the ion current peak, corresponding to ion kinetic energies that are less than 2% of their values in vacuum. However, these measurements also reveal a low-pressure range in which the ablation plume has a two-component nature, with both fast (collisionless) and slower (scattered) species present. Figure 4(b) shows that the number of fast ions decreases rapidly with increasing gas pressure, but the presence of even a small number of energetic, unscattered ions means that increasing the gas pressure does not produce a simple, monotonic reduction of kinetic energy in the two-component (or "plume splitting") low-pressure range.

11.2.6.4 Pressure-Distance Scaling Laws.
Several investigators have noted that when films are grown in an ambient gas *at the lowest possible substrate temperature*, an optimal target-substrate separation, D_{ts}, exists for each

deposition pressure. For example, for YBCO films grown on MgO(001) substrates at a fixed, low substrate temperature of 680°C, Kim and Kwok measured both the zero-resistance superconducting transition temperature and the normal-state resistance ratio R(100 K)/R(300 K) and found that the best films (by both criteria) were obtained when the ambient oxygen pressure, P, was related to D_{ts} through a scaling law of the form $PD_{ts}^2 = $ constant [114]. For indium tin oxide films grown at room temperature with D_{ts} fixed at 7 cm, Zheng and Kwok found that the films' electrical resistivity was very sensitive to oxygen pressure; low resistivity films were obtained only in a narrow pressure range near 15 mtorr [115]. Similarly, Shen and Kwok made X-ray rocking curve measurements for a series of epitaxial CdS films grown on InP(001) substrates at 300°C and in Ar gas pressures of 8, 100, and 200 mtorr [116]. For each pressure they found that there was an optimum D_{ts} value to minimize the rocking curve width and the optimum D_{ts} increased with decreasing ambient gas pressure (see Section 11.3.2.5). Lowndes and Rouleau grew p-ZnTe:N films in ambient N_2 at 320°C and $D_{ts} = 10$ cm and observed a sharp peak in the hole mobility at a N_2 pressure of 50 mtorr (see Sections 11.3.2.4–5) [28, 29, 42, 43].

The qualitative explanation for most of these results is that when films are grown at a very low substrate temperature, there is an optimal velocity (kinetic energy) of incident species to assist formation of the crystalline phase. Higher energy ions and atoms can cause damage, e.g., if the pressure is too low or the substrate too close to the target. On the other hand, if the substrate is too far away or the pressure too high, the incident atoms and ions are fully thermalized by collisions and do not contribute sufficiently to surface bond-breaking or to activate diffusion. However, Lowndes and Rouleau have shown that the mechanism responsible for degraded film properties on the high-pressure side is more complicated than simply thermalization of kinetic energy [42, 43]. (See Section 11.3.2.5.)

At high growth temperatures, the quality of PLD films becomes much less sensitive to the choice of pressure and D_{ts} values. This is reasonable because bonds can be broken and incident atoms can diffuse on the growing surface using only thermal energy. Thus high-quality YBCO films can be grown at temperatures of 750–800°C for a broad range of oxygen pressures, and the electrical resistivity of PLD indium tin oxide films is nearly independent of oxygen pressure [115] at deposition temperatures of 200–300°C. However, damage due to high-energy atoms and ions still can occur. Consequently, at high deposition temperatures, a range of (P, D_{ts}) values are expected to produce optimal film properties with properties degrading only when the PD_{ts} product is too small.

11.2.6.5 Electronic Excitation Energy in PLD.

Atoms and ions are formed and transported in electronically excited states within the ablation plume. Collisions with a background gas transform initially high kinetic energies into internal (electronic) energy. This "shock heating" raises the effective temperature of the slowed species and produces a "two-temperature" pulse. Similar temperature increases can be induced even for PLD in vacuum, if some of the faster plume species are reflected from the substrate/heater surface and collide with slower incoming species above the growing film surface. These kinetic-to-internal energy conversion processes are revealed by fluorescence from high-energy Rydberg states as well as from lower atomic and ionic energy levels.

Leuchtner has pointed out that species such as Rydberg-state atoms have large internal energies, essentially equal to their ionization potentials [117]. If such species can survive transport to the substrate, this internal energy can be deposited. This raises an interesting question as to the relative importance of electronic excitation (potential) energy and kinetic (ballistic impact) energy in promoting surface diffusion and crystalline phase formation during PLD. In particular, it is important to know if there is a regime of gas pressure and other ablation conditions for which ions and atoms with initially high kinetic energies are thermalized sufficiently to avoid displacement damage, but for which large numbers of Rydberg-state atoms (or other excited-state species) still can reach the substrate and deposit their energy [117].

Leuchtner has estimated the minimum number of collisions suffered by both neutral atoms and ions in the target-substrate space. For ions, the number of collisions is energy-dependent, with the *highest* kinetic energy ions suffering *fewer* collisions than low-energy ions [117]. To estimate the absolute number of collisions needed to completely thermalize their kinetic energy, Leuchtner invoked spectroscopic measurements of the collisional thermalization of sputtered species by Park *et al.* [118], who found that 95% (99%) of the kinetic energy was thermalized when the product of pressure, P, and target-substrate spacing, D_{ts}, was ≥ 0.05 (0.1) torr-cm. The higher value corresponds to ~ 10–20 collisions, regardless of kinetic energy [117].

For the target-substrate separation of 5.5 cm used by Leuchtner for ZnO film growth, both neutrals and ions are expected to be thermalized by collisions at an oxygen pressure of 100 mtorr, but high-energy ions and neutrals should not be thermalized at 10 mtorr. Thus, at 100 mtorr oxygen pressure, the only nonthermal source of energy at the growing surface was assumed to be electronically excited species such as Rydberg atoms and ions. Although Leuchtner has noted that strong luminescence engulfing the substrate during ZnO growth is consistent with the presence of Rydberg atoms, the luminescence could equally well result from the plume-splitting phenomenon if an unscattered fast-ion component was present.

Leuchtner also compared ZnO films grown in oxygen by ablation of both Zn and ZnO targets. Using the Zn target, films of comparable quality were grown at substrate temperatures $\sim 100°C$ lower than for the ZnO target. This was attributed to the greater proportion of electronically excited species (both ions and neutrals) that were found in the Zn ablation plume from time-of-flight quadrupole mass spectrometer measurements [117].

11.2.7 Generation and Minimization of Particulates

The principal drawback of PLD is the presence of particulates embedded in the films. The particulates range in size from ~1 nm to ~1 μm. The size, the shape, and the number density of particulates depend strongly on the pulsed laser conditions, especially the laser wavelength and E_d. The target material and its near-surface morphology also are important factors.

11.2.7.1 Classification and Origins of Particulates. Observations of particulate morphology under different experimental conditions show that, aside from the effect of gas bubbles trapped in the target [15], there are three principal mechanisms for particulate production, corresponding to whether the material forming the particulate was in the solid, liquid, or vapor phase when it was ejected from the target [55]. Particulates with quite different sizes and formed by different mechanisms can be present simultaneously in PLD films.

Condensation from the Vapor Phase. Particulates formed from the vapor phase are much smaller, typically in the nanometer range, than the ~0.1 μm to 10 μm particulates produced by the target fragmentation and melting processes discussed below. The size of these ultrafine particulates can be controlled by the ambient gas pressure, whereas the large particulates are essentially the same in vacuum or a gaseous ambient. Chen pointed out the reason for this difference: At a pressure of 1 mtorr, the mean free path of ablated species is ~5 cm; but at 100 mtorr, the mean free path drops to ~0.05 cm [55a]. Consequently, in a typical ablation experiment with a target-substrate separation of a few centimeters, vapor species can undergo many collisions with gas molecules, leading to nucleation and growth of fine particulates before they arrive at the substrate.

Matsunawa *et al.* studied the formation of ultrafine metal powders in He and Ar atmospheres [54]; they found spherical or polygonal metal particulates with strongly pressure-dependent size distributions. As shown in Fig. 5, for Fe ablation into Ar, the mean particulate size at 1 torr pressure was <5 nm, but this increased to ~40 nm at 2 atmospheres pressure. The size distribution also was much narrower at low pressure. The increase in particulate size with ambient gas pressure is evidence that the ultrafine particulates form in the vapor phase, not as liquid droplets in the target (see below). If growth occurs by a collision process, the particulate size should be governed by residence time in the vapor. Indeed, Matsunawa *et al.* collected larger particulates at larger target-to-substrate separations [54]. Further details are given in the review by Chen [55].

Gaponov also has suggested the possibility of condensation from ablated vapor when long-pulse laser irradiation is used [56]. Cheung used in situ mass

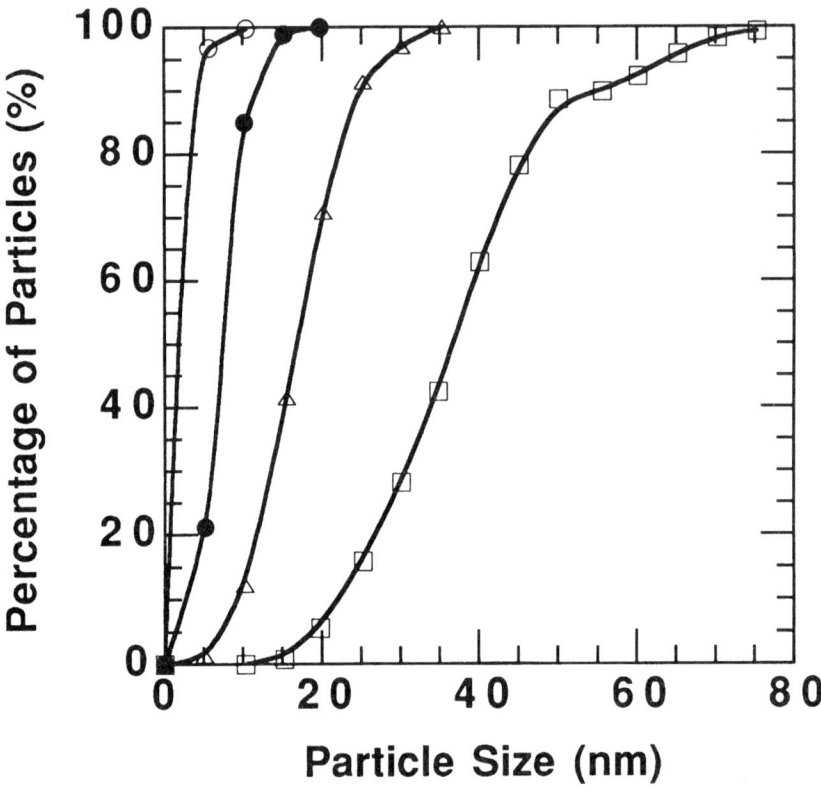

FIG. 5. Cumulative size distributions of ultrafine Fe particulates made under Ar pressures of 0.1, 1.3, 13.3, and 200 kPa [55a, with permission].

spectroscopy to correlate the mass spectrum of evaporants with the particle density on films of $Hg_{0.7}Cd_{0.3}Te$ that were evaporated with a pulsed Nd:YAG laser. With increasing power density, mass spectra showed that molecular clusters were present in addition to the expected elemental species, suggesting that the clusters have a common origin with at least some of the particulates [119].

In general, these ultrafine (nanoscale) particulates are not the "particulate problem" referred to for PLD films. However, ultrafine particulates should be expected whenever the ambient gas pressure is sufficiently high that many collisions of vapor species can occur between target and substrate. Indeed, nanoscale "cluster formation" and deposition is implicated in a change of the morphology and growth mode of p-type ZnTe:N films that was observed by Lowndes et al. [42] and Rouleau et al. [43] near 100 mtorr N_2 pressure. Moreover, PLD in low-pressure gases is a method for depositing highly nanocrystalline films [120].

Target Fragmentation. A rough target surface contains numerous microcracks and loosely attached fragments; cumulative exposure to intense pulsed laser radiation also can produce thin crater edges, columnar structures, and deep voids. These features are mechanically weak; they are easily dislodged and entrained in the explosive expansion that accompanies ablation. If these fragments are melted by subsequent absorption of laser radiation as they are ejected, circular particulates are deposited on films; incomplete disintegration and failure to melt results in irregularly shaped fragments. Such "fragmentary" particulates are prevalent at low laser power densities.

Subsurface Melting and Droplet Ejection. Surface evaporation can only extract heat to a depth $\sim(Dt)^{1/2}$ in a time t. Consequently, if laser heating of a subsurface layer takes place faster than the material closer to the surface can be removed by evaporation, superheating of the subsurface layer can occur. Under conditions of high-E_d irradiation and distributed power absorption (finite absorption coefficient, moderate absorption length), formation of a subsurface liquid layer that is heated to a higher temperature than the surface is expected to occur [121, 122].

Two mechanisms for liquid droplet ejection from the molten layer then are possible:

1. One possibility is that the liquid will be superheated, meaning that its local pressure is determined by a surface vapor pressure that is lower than the equilibrium pressure that would exist at its actual temperature. When some maximum superheating temperature is reached, the metastable superheated liquid no longer can sustain tensile forces, vapor bubbles nucleate (in order to extract heat by vapor bubble growth), and explosive subsurface boiling and "splashing" of molten material out of the target occurs [68, 121, 122]. This leads to the view that material can be ablated by a volume—not surface—process of growth of the vapor phase and subsurface boiling, resulting in rapid expulsion of material.
2. Another mechanism for liquid ejection is that, regardless of superheating and boiling, a molten subsurface liquid layer is subjected to a large force by the recoil pressure generated by the shock wave of the laser ablation plume. This recoil force also can produce splashing of similar-sized molten droplets and will be indistinguishable in effect from subsurface boiling although, as noted by Cheung [123a], the force that causes expulsion in this case is generated from above the liquid layer, not within it.

Under the moderate-to-high E_d conditions needed for stoichiometric transfer and high deposition rates, particulates are circular with smooth surfaces (sometimes even with a surrounding "splash" pattern), indicating the impact of once-molten droplets that could have been produced by either the "subsurface boiling" or "recoil expulsion" mechanisms. Stoichiometric transfer is expected because both of these mechanisms are volume processes, provided that the laser fluence is sufficiently far above the threshold fluence for liquid-layer formation that the material originating as ejected liquid dominates that from subsequent (nonstoichiometric) thermal evaporation as the surface cools.

Calculations and experiments have provided some guidance in distinguishing between these two liquid-ejection mechanisms. Schwartz and Tourtellotte [124] compared the laser-photon absorption depth, the rate of photon energy conversion to heat, and the rate of heat removal by conduction, in order to estimate the power density needed for subsurface boiling to occur. Using their results, the required power density is in the 100–1000 MW/cm^2 range, comparable to or slightly larger than in typical PLD experiments using excimer lasers. (For example, $E_d \sim 2$ J/cm^2 and 20 nsec pulse duration gives $P \sim 100$ MW/cm^2.) Cheung has suggested that subsurface boiling, referred to as "true splashing" by Ready [125], is not likely for dielectric targets but may be dominant for metals [123b].

Dyer et al. used a fast photoacoustic technique to directly record stress waves generated in a $YBa_2Cu_3O_{7-x}$ target by laser ablation; from this they obtained both the magnitude and time evolution of the surface pressure [68]. They found that a ~20 ns (FWHM) KrF laser pulse creates large transient pressures of ~400 atm at $E_d \sim 2$ J/cm^2; pressure pulses were detected for E_d as low as ~50 mJ/cm^2. Model calculations revealed that this threshold pressure pulse corresponds to a surface temperature of ~1300 K, consistent with the onset of melting (noncongruent $T_m \sim 1280$ K). Dyer et al. also related the material removal rate to the surface pressure and combined their pressure data with etch-threshold measurements by Inam et al. [60] to deduce an average mass for ablating species of ~1000–2000 amu [68]; this corresponds to a cluster size of ~1.5–3.0 $YBa_2Cu_3O_7$ units. Because evaporation of such large clusters is improbable, this work provides strong support for "explosive" material-removal models that have been proposed to explain stoichiometric transfer, and for the recoil-force mechanism [36, 121, 122].

Dijkkamp et al. [1] and Dyer et al. [68] suggested that if several $YBa_2Cu_3O_{7-x}$ molecules ablate initially as a unit, their optical properties should be similar to bulk material, and they should continue to absorb energy from the laser beam and be further fragmented. This suggests that the energetic atoms and ions that are detected spectroscopically in the ablation plume are generated from material that is initially at the front of the plasma expansion, whereas material ablated near the end of the laser pulse may remain intact, traveling slower than the main plasma expansion. Indeed, high-speed photography, optical absorption, and ion probe measurements have revealed both "fast" and "slow" components in the ablation plume (Section 11.2.6.1) and Geohegan et al. have observed the blackbody emission from slow-moving particulates [33, 71, 94].

11.2.7.2 Techniques to Minimize Particulates. To capitalize on PLA's advantages, it is essential to develop techniques to minimize the number and size of particulates in PLD films. Particulates can be avoided either by preventing their generation or, once formed, by preventing their deposition on the substrate. Various "active" and "passive" [126] particulate minimization techniques have been developed.

Short-Wavelength Laser Radiation. Particulate minimization begins with the choice of laser wavelength. Numerous experiments have shown that, by far the most effective passive approach is to use short-wavelength laser radiation, typically from a KrF (248 nm) or ArF (193 nm) excimer laser [127]. Figure 6 shows the striking decrease in particulate density and size when films of the compound semiconductor ZnO were deposited with 1064 nm, 532 nm, and 248 nm radiation [128]; similar behavior has been documented extensively for $YBa_2Cu_3O_{7-x}$ films [17, 127]. Koren *et al.* pointed out that more than one mechanism is responsible for the improved film morphology at UV wavelengths [127]. First, the photon absorption length, α^{-1}, often is shorter in the UV, which results in a thinner layer being ablated and a hotter plasma plume, as discussed below. For other materials (including $YBa_2Cu_3O_{7-x}$), α^{-1} varies only slightly from the near-UV to the near-IR. In these cases, the decreased particulate density for UV-laser ablation results from the strong absorption of UV by the particulates as they are being ejected, resulting in their further fragmentation down to vapor or to a very small size [127].

Direct evidence that large (~ 0.1–10 μm) particulates result from subsurface melting and microexplosions is provided by SEM images and model calculations for a ZnO target irradiated at various laser wavelengths and E_d values [129]. Craciun *et al.* calculated ZnO target temperature profiles for 248 nm (KrF) irradiation at $E_d = 0.7$, 2.1, and 3.5 J/cm^2 and for 532 nm (doubled Nd:YAG) irradiation at 2.1 J/cm^2, with results shown in Fig. 7. The surface of the ZnO target is at the boiling point in all the cases shown; however, for KrF radiation, the subsurface superheating is shallow (<40 nm) and becomes significant ($\sim 300°C$) only for $E_d = 3.5$ J/cm^2. In contrast, very substantial ($\sim 900°C$) and extensive (~ 150 nm) subsurface superheating of ZnO was found for doubled Nd:YAG radiation. Areas of the ZnO target surface where microexplosions occurred, resulting in the ejection of molten material, can be seen clearly in Fig. 8 for the highest KrF fluence. This explains the degradation of ZnO film quality that was observed at high E_d and resulted in an optimum KrF $E_d \sim 2$ J/cm^2 for ZnO film growth. For doubled Nd:YAG radiation, significant subsurface superheating occurred for all fluences above the ablation threshold, producing the heavily damaged target surface shown in Fig. 8(d) and preventing growth of high-quality films at the 532 nm wavelength [129]. The model calculations suggest that the microexplosions result from boiling; however, as noted above, liquid droplet ejection can equally well result from the recoil pressure exerted by the shock wave that accompanies ablation [68, 69]. These results emphasize the importance of a short laser

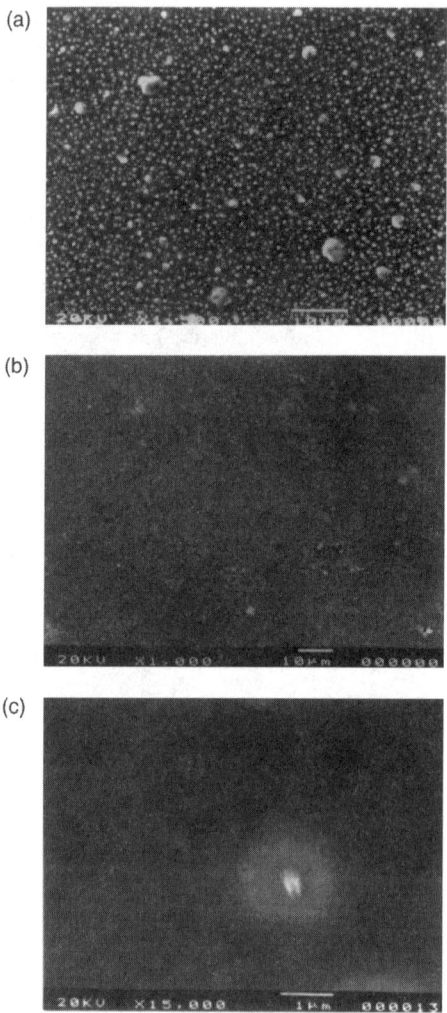

FIG. 6. SEM images of PLD ZnO films using wavelengths of (a) 1064 nm, (b) 532 nm, and (c) 248 nm [128, with permission].

absorption length, α^{-1}, and moderate E_d to minimize the splashing of particulates.

Target Preparation. A second passive technique to minimize particulates is to use a high-density target with a smooth surface. Foltyn *et al.* found that the film deposition rate is only weakly influenced by target density [130]. However, increased particulate production definitely results from the use of a low-density

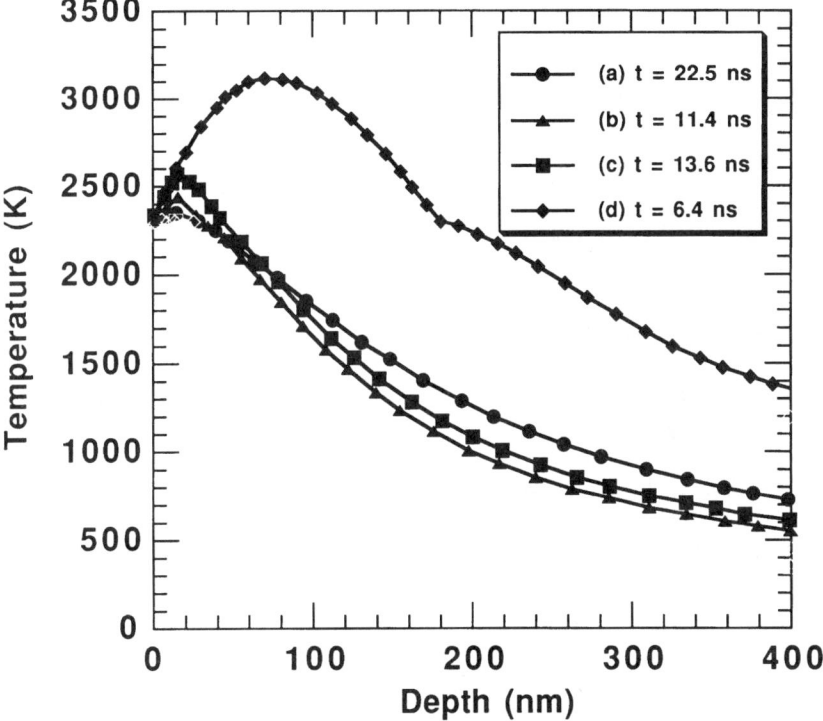

FIG. 7. Calculated near-surface temperature profiles for a ZnO target after KrF (248 nm) laser irradiation at fluences of (a) 0.7 J/cm^2, (b) 2.1 J/cm^2, and (c) 3.5 J/cm^2, and (d) for frequency-double Nd:YAG (532 nm) irradiation at 2.1 J/cm^2. The profiles were calculated at the times for which either 2 nm (case (a)) or 7 nm (cases (b)–(d)) of target material was ablated [129, with permission].

target. For compound semiconductor growth, target fabrication by mixing high-purity powders is of considerable interest in order to change the composition, energy bandgaps, and lattice constants of thin-film layers. Figure 9 compares the morphologies of ZnTe films grown from isostatically hot-pressed and cold-pressed mixtures of finely ground and well-mixed powders [42]. The principal difference is the high areal density of ~µm-sized particulates from the cold-pressed target. Most of the particulates are irregular in shape, indicating that they were produced when loosely bonded material fragmented off the target during ablation. In contrast, far fewer particulates, almost exclusively round solidified droplets, were produced from the hot-pressed target, apparently by splashing. The relatively high thermal conductivity of semiconductor materials helps to minimize splashing from a high-density hot-pressed target; for ZnTe the area covered by droplets was reduced to <0.1% [42].

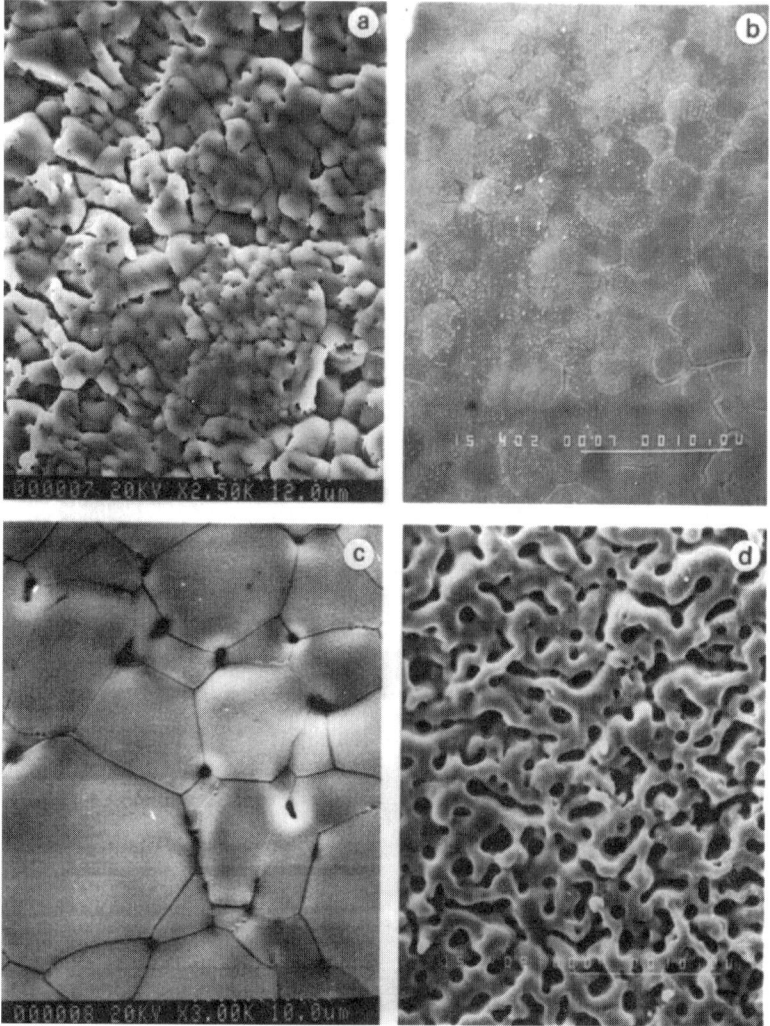

FIG. 8. SEM images of a ZnO target surface after KrF (248 nm) laser irradiation at fluences of (a) 0.7 J/cm^2, (b) 2.1 J/cm^2, and (c) 3.5 J/cm^2, and (d) frequency-double Nd:YAG (532 nm) irradiation at 2.1 J/cm^2 [129, with permission].

Cone Formation and Target Conditioning. Several changes occur in PLD if an initially smooth target surface is repeatedly pulsed laser–irradiated at a fixed angle of incidence:

1. The plume intensity and deposition rate gradually decrease with the increasing number of laser shots.

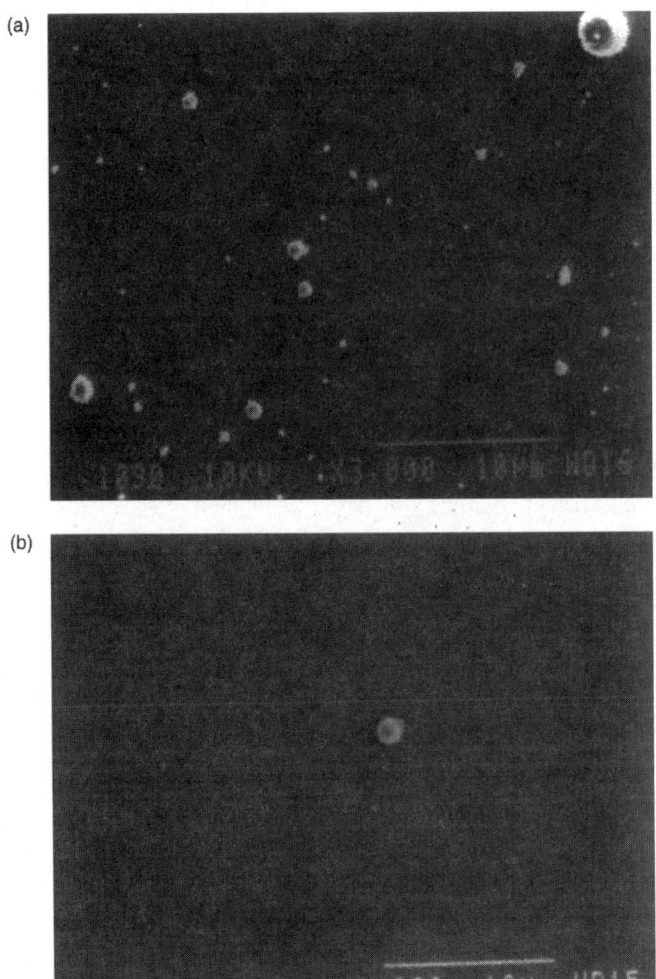

FIG. 9. Particulates on films deposited at a target-substrate separation of 10 cm, using (a) a cold-pressed $Zn_{0.99}Mn_{0.01}Te$ target ($E_d = 2.6$ J/cm^2) and (b) a hot isostatically pressed ZnTe target ($E_d = 1.9$ J/cm^2). Many comparable areas of the latter film were free of particulates [42, with permission].

2. The plume direction gradually tilts away from the surface normal, toward the direction of the incident laser beam.
3. An increasing number of particulates are deposited per unit area of film.

These effects are closely associated with texturing of the target surface by a laser beam that impinges at a large angle of incidence (measured from the normal). As the target is repeatedly irradiated, tall columnar structures with conical tips ("cones") are formed, with the column axes aligned along the incident laser beam direction. According to Foltyn et al., these result from initial erosion combined with a shadowing effect that amplifies initial features on later shots [130]. With increasing laser exposure, the deposition rate eventually reaches a steady-state value. Foltyn et al. found that the asympototic deposition rate increased linearly with increasing laser E_d, but the steady state was reached in the smallest number of shots for low fluence [131]. Target resurfacing (by sanding or polishing) produced a return to the initial deposition rate, which then decayed again with further laser exposure.

The gradual "bending back" of the ablation plume toward the incident laser beam direction apparently is most pronounced and troublesome at large laser beam angles of incidence, since it was observed by Lowndes et al. [47] and Venkatesan et al. [126] at ~60° incidence, but not by Foltyn et al. at 45° incidence.

Many groups routinely polish the target to obtain a fresh smooth surface, then "pre-ablate" it in vacuum for a fixed number of shots (with the ablation plume blocked from the substrate by a shutter) to get into a steady state. This overcomes the problems of initially varying deposition rate and plume direction.

Cone formation also is implicated in an increase in the number of particulates on PLD films [47, 126]. Cone formation can be largely suppressed by periodically reversing the laser beam's angle of incidence so that surface irregularities combined with the shadowing effect are not allowed to develop into columnar structures. With beam reversal, a nearly smooth target surface can be maintained. One convenient method for producing continual incidence-angle reversals is to use a long focal length cylindrical lens to focus the laser beam to a horizontal line that is centered symmetrically on a rotating target. With each 180° rotation of the target, the incidence direction is reversed with respect to any surface features and development of cones is prevented [17, 47, 79]. The lens also can be scanned vertically to reduce the rate at which a depression develops in the center of the target; an added benefit of vertical scanning is that it improves film thickness uniformity in the vertical direction on the substrate [47, 79]. If the laser beam is focused to only a short line or a spot on the target, cone formation still can be reduced by directing the beam to a radial distance r from the target center and periodically switching it to $-r$ (180° away) with the target rotating continuously [132].

Unequal target rotation and laser beam scan rates also can be used to prevent cone formation in the "ring source" geometry (see the discussion of uniform-thickness films in Section 11.2.8.1) [133]. Or, a true raster scan of the laser beam can be coordinated with controlled target rotation to avoid forming either a central depression or cones in the target [132]. Finally, two laser entrance windows, on opposite sides of the chamber, can be used to prevent cone formation by switching the laser beam between them as the target rotates [134].

Particulate-Free Films from a Liquid Target. Sankur and Cheung pointed out that the ultimate smooth, featureless target is a liquid [15]. For Ge, perhaps Si, a few metals such as Al, Ga, and In, and for a few dielectrics, the equilibrium vapor pressure is sufficiently low at the melting point that essentially all of the deposition rate will be due to ablation of a molten target, with a negligible contribution from thermal evaporation. Because the target is liquid, its surface should remain smooth, deposition conditions will not vary, and there should be no particulates. Sankur grew particulate-free Ge films from a molten Ge target [15], and Xiao recently grew particulate-free diamond-like carbon films [135] by ablating an organic liquid target.

Dual-Laser, Liquid-Target Ablation for Particulate-Free, Large-Area Films. Witanachchi *et al.* reported a new dual-laser ablation technique [57, 58] that provides additional insight into the physics of laser-generated plasmas and has potentially important consequences for growth of particulate-free, large-area films. In their dual-laser ablation experiments, they irradiated a polycrystalline target by both a pulsed CO_2 (10.6 μm) laser and a slightly (~75 nsec) time-delayed pulsed excimer (UV) laser. Target heating by the CO_2 laser produced a thin, transient molten layer, from which the excimer laser subsequently initiated the formation of the energetic ablation plasma plume. The time delay was chosen so that the excimer laser pulse arrived at the target immediately after the onset of melting, before any CO_2 ablation could occur, so that ablation of the thin molten volume was due to the excimer laser. However, the remainder of the CO_2 laser pulse then was absorbed by the dense excimer-generated plasma, producing plasma heating and excitation *several orders of magnitude more efficient than was possible using the excimer laser alone*. Moreover, the efficient plasma heating at the CO_2 wavelength produced an ablation plume with a greatly broadened angular distribution [57, 58]. Equation (11.1) shows that the greater excitation and expansion of the ablation plume result directly from the ~thousand-fold more efficient absorption of long-wavelength CO_2 radiation by the KrF laser-generated plasma via the inverse bremsstrahlung process.

In situ ion probe measurements by Witanachchi *et al.* revealed a six-fold increase in ablation plume kinetic energy for dual-laser ablation, in comparison with single KrF laser ablation, at a fixed KrF fluence of 1 J/cm^2. They also obtained *a thousand-fold reduction* in particulate density for KrF (248 nm) laser ablation of Y_2O_3 at 1 J/cm^2 [57]. The particulate reduction apparently was due mainly to ablation from a (transient) liquid, but the plasma's high IR absorption also may have assisted evaporation of any remaining particulates or droplets that were ejected from the target.

The significance of this dual-laser process is that it addresses the two principal problems previously associated with single-wavelength laser ablation, namely, the presence of particulates and the strongly forward-directed expansion profile that limits the area of uniform film thickness. As noted earlier, the absence of particulates when ablation takes place from a liquid target has been known for nearly a decade [15, 16]. However, until Witanachchi *et al.*'s discovery, there was no practical way to produce a liquid target of an arbitrary, multicomponent polycrystalline material. Shallow, *transient* CO_2 laser melting apparently avoids the problem caused by the high vapor pressures and/or decomposition of most molten materials, for which steady-state melting would produce a nonstoichiometric film. The expansion of the more highly excited plasma plume also helps to obtain uniform, large-area film growth. Whether the plasma's increased kinetic and excitation energies can be used to assist growth (e.g., at lower temperatures) remains to be seen. However, the apparent generality of dual-laser ablation should permit its extension to other materials. Because CO_2 lasers are efficient, inexpensive, and already in industrial use, dual-laser ablation may be another step toward PLD's use in manufacturing.

Mechanical Velocity Filter. A number of experiments have shown that the velocities of particulates vary inversely with their mass, e.g., Lubben *et al.* [93]. Dupendant *et al.* [136] used a stationary slit and rotating substrate to measure the velocity distribution in vacuum of micrometer-sized particulates produced by Nd:YAG laser ablation of several metals and $YBa_2Cu_3O_{7-x}$. They found most probable velocities of $1-2 \times 10^4$ cm/s in all cases, with a high-velocity tail extending up to 10^5 cm/s in some cases. Other measurements by Gagliano and Paek for ruby laser ablation of Al_2O_3 [254], and by Geohegan for KrF laser ablation of YBCO and BN [255], have determined particulate velocities in the range of $0.2–1.2 \times 10^4$ cm/s. Thus particulate velocities are a factor of 10 to 1000 slower than the $\geq 10^6$ cm/s velocities of ablated ions and atoms in vacuum (see Section 11.2.6.1). This difference provides a basis for very substantially reducing the density of massive particulates in films by velocity-filtering.

Lubben *et al.* [93] used a shutter to prevent particulate deposition on Ge films and to measure the number and size of particulates as a function of their arrival time. However, a simple shutter is not practical for film growth because the fastest particles cannot be eliminated without significantly reducing the useful atom/ion flux.

On the other hand, a rapidly rotating, multivaned wheel works well [137, 138]. A multivaned wheel with its rotation axis parallel to the ablation plume only transmits species whose velocities exceed NfL, where N is the number of vanes, f is the rotation frequency in Hz, and L is the vane length. Approximately one order of magnitude reduction in the particulate density can be obtained [126, 138]. Further significant reductions in particle density are expected by using higher speed vacuum-compatible motors [126]. The only experimental limitations from using a vane-type velocity filter are its requirements for (1) sufficient space between the target and substrate and (2) compatibility with the heat radiated by the substrate heater nearby.

Crossed Beams. Several investigators have removed slow-moving particulates by intersecting the ablation beam with a second, crossed and delayed, beam that either destroys or pushes aside the particles. Koren *et al.* demonstrated a reduction in both the number and size of particulates in films grown by Nd:YAG (1.06 μm, 10 ns) ablation of YBCO, when they irradiated the Nd:YAG ablation plume with a 1 μs-delayed KrF (248 nm, 30 ns) laser beam parallel to and 1.25–2.75 mm above the target surface [139]. This experiment confirmed an earlier hypothesis [127] that absorption of UV radiation produced further fragmentation of particles within the plume and is responsible for the smoother films and lower particle densities that are obtained using UV lasers. They also noted an approximate doubling in size of the visible plume when the second laser beam was present—clear evidence of coupling the second laser's energy into the plume. The superconducting properties of YBCO films grown by pulsed Nd:YAG (1.064 μm) laser ablation at a (nonoptimally low) temperature of 650°C also were improved when ablated particles were irradiated with a second, crossed and delayed, fourth-harmonic (266 nm) laser beam at distances of 0–1 mm from the target [95, 140, 141]. A crossed, pulsed oxygen jet also significantly reduced particle density and improved YBCO films [95, 141].

However, a limitation in using a crossed UV laser beam to vaporize particulates is that it must be present for all of the time during which slow-moving particles are passing a given point if really significant particulate reduction is to result. Consequently the lowest particulate densities obtained using a 1064 nm or 532 nm laser in combination with a crossed excimer beam still are much higher than those obtained with a single excimer ablation laser [55b].

Off-Axis Ablation. Kennedy [142] demonstrated a novel, simple method to completely eliminate particulates: off-axis laser ablation, analogous to off-axis sputtering. In off-axis ablation, the pulsed laser beam strikes an upward-facing target at near-normal incidence. The substrate heater is mounted above the target but also facing upward, *away from the target*, and to one side of the center of the plume. The massive particulates travel in nearly straight line trajectories, unaffected by collisions in the plume or with ambient gas molecules. Consequently, they either pass by the substrate heater or are blocked by it. Kennedy found that very smooth films and superlattices (±5 nm over 200 μm lateral distance) could be grown by this method; remarkably, the deposition rate was reduced by only a factor of three [142]. The only drawbacks of off-axis ablation are the reduced deposition rate and nonuniform film thickness; however, improved thickness uniformity probably can be obtained by rotating the substrate heater (see Section 11.2.8.1).

Gas-Dynamic Separation. Gaponov et al. demonstrated two gas-dynamic methods to separate macroscopic particulates from the laser-generated plasma plume. The underlying idea is to establish a collision region that deflects or reflects toward the substrate the evaporant used for film growth, while the undeflected particulates travel straight ahead. Both a heated reflecting screen [143, 144] and two colliding beams of ablated material [56] have been used. Although very elegant, these methods do not seem to have been exploited, probably because of the likelihood of plume contamination from the reflecting screen and/or the complexity of the colliding-plumes geometry. Again, to obtain uniform-thickness films probably would require translating or rotating the substrate.

11.2.8 Uniform-Thickness Films and Scaleup

11.2.8.1 "Painting" with the Plume.
The $(\cos\theta)^p$ angular distribution of ablated material initially was regarded as a disadvantage of PLD because uniform-thickness films were obtained only in a small region in the forward direction. More recently, several groups have demonstrated that the strongly forward-peaked distribution can be exploited by "painting" the substrate with the ablation plume to produce a highly uniform deposit.

One such approach is to rotate the substrate heater and the ablation target about parallel axes normal to their surfaces with the laser spot offset from the heater's rotation axis by an amount ρ [145–148]. The deposition rate is greatest at the plume's center, $\sim\rho$ away from the substrate heater's centerline, but this is compensated by the short residence time of points at radius ρ under the the plume. By adjusting the ratio ρ/Δ, where Δ is the target-substrate separation, the deposition thickness profile can be varied continuously from a central peak to a central minimum (saddle shape). This "offset plume" geometry produced epitaxial ZnS films that were completely uniform ($\pm 1\%$) in thickness over ~ 75 mm^2, with good thickness uniformity ($\pm 3\%$) over ~ 300 mm^2 [147, 148]. Although the small size of the ablation chamber limited the maximum deposition area in these experiments, modeling shows that the "offset" geometry can be scaled up almost arbitrarily, as long as deposition takes place in vacuum or a sufficiently low-pressure gas that collisions do not stop the ablation beam [147, 148]. Cone formation can be reduced by using a line-focused laser spot that is centered on the target, with the target's rotation axis also offset by "ρ" from the heater's axis. The plume will be elongated in the direction perpendicular to the line focus, and adjustment of ρ/Δ for uniform film thickness can be done empirically.

A second "geometrical" approach to uniform film thickness involves scanning a tightly focused laser beam around a nearly circular path near the outer edge of a disc-shaped or annular target, thus simulating a "ring source" of ablated material [149]. Such a scan can be achieved by reflecting the pulsed laser beam from a rotating mirror mounted on a finely adjustable tilt stage, or by using a pair of orthogonal beam-scanning mirrors [147]. By using a long focal length lens, the variation in laser E_d due to scanning can be made negligible; and by adjusting the ratio ρ/Δ (where ρ now is the radius of the ring source), the deposition thickness profile can be adjusted to produce a slightly saddle-shaped thickness profile. A large laser beam angle of incidence (50°–60°) may be necessary to avoid hitting the edge of the substrate heater for a given choice of ρ/Δ. Because the optimum value of ρ increases with Δ, an annular ("racetrack shaped") ablation target, rather than a circular disk, may be sufficient and is economical for large-scale applications of the ring-source geometry. Cone formation on the target can be reduced in the ring-source geometry by choosing the target rotation and laser scan rates such that the target has rotated 180° when the laser returns to any given spot [147].

As an example of the thickness uniformity that can be obtained, Gaponov and co-workers have used PLD to fabricate multilayer mirrors and other selective optical elements for soft X-rays. Optimized multilayer mirrors used "high Z/low Z" pairs of materials such as W-C, Ag-C, Au-C, or Cr-C [150]. Both flat and spherical multilayer mirrors were fabricated for use with X-ray wavelengths of 30–200 Å, though few details of experimental methods are given [151].

Mai, Dietsch, and Pompe developed a different laser ablation geometry to deposit uniform-thickness Ni/C and W/C multilayers for use as soft X-ray mirrors [152–154]. Their method is based on the observation that the ablation plume always is directed close to the target's surface normal. By using a half-cylindrical target (convex side facing up) and displacing it horizontally (perpendicular to the cylinder axis), the plume can be directed through a significant range of solid angles, "painting" a substrate suspended above. Simultaneous vertical displacement of the target compensates for the change in lens-to-target separation and keeps the origin of the ablation plume fixed in space. The decrease in deposition rate at the substrate with increasing plume angle can be compensated by scanning the target sinusoidally, rather than linearly, in the horizontal direction, thus varying the plume's dwell time. This method produced steep compositional gradients at interfaces and precise layer-thickness control over moderate areas, while still maintaining a small target-substrate separation so that deposition rate was not sacrificed. High reflectivity of grazing incidence X-rays was obtained from the resulting "high Z/low Z" structures. TEM and ellipsometric measurements on 20 × (Ni-32.5 Å/C-25 Å) multilayers fabricated by Nd:YAG laser ablation revealed local thickness variations ~1% over a 7 cm path, and interfacial roughness <2 Å. The maximum area of uniform deposition that can be obtained by this method appears to be limited by the total laser-irradiated length along the target's cylindrical axis and by the uniformity of irradiation along this direction; nevertheless, this geometry is well suited to fabricate long, narrow multilayers.

11.2.8.2 Scaleup. PLD's unique characteristics have created interest in scaleup for industrial applications. The principal requirements are:

1. Uniform substrate temperature over a large substrate area.
2. Uniform film thickness and composition over this area.
3. High deposition rate.

Scaleup meeting these objectives was demonstrated recently for several different geometries and materials [146, 155–160].

Many laboratory-scale PLD experiments have relied on substrate heating by conduction with the substrate bonded directly to the heater face with silver paste (or similar). This is unacceptable for large-area substrates at elevated growth temperatures because the bond often fails catastrophically due to thermal expansion differences. Other problems are the difficulty of obtaining uniform thermal contact over a large area, and increased outgassing. Both Foltyn *et al.* [146] and Greer and co-workers [155–158] developed purely radiative approaches to substrate heating by partially or fully enclosing the substrate in a vacuum oven. (Also see Estler [160] for details.) Foltyn's PLD system uses the offset plume geometry described above, whereas in Greer's system the focused laser beam is raster-scanned with longer dwell time at larger substrate radii to obtain good film thickness uniformity. Both arrangements keep the substrate's backside clean and free of contaminants during frontside deposition, so that epitaxial films can be grown on both sides of the substrate with little effect on film properties [146, 156, 157]. The scanned laser beam has several advantages:

1. A fixed laser beam slowly etches a groove in a rotating target, altering the direction of the ablation plume as cones are formed.
2. Compositional changes may occur in the plume as the etched pit forms.
3. Target material is used efficiently.

Greer has considered the scaleup problem in detail, including several different deposition geometries for substrates of at least 100 mm diameter; efficient target utilization; methods to keep the excimer laser entry window clean; schemes for large-area substrate heating and particulate-reduction; deposition-rate monitoring; and PLD on nonplanar surfaces [157]. According to Greer, laser beam rastering over a large-diameter target is the best approach to obtain large-area films with predictable and reproducible growth rates and film qualities [158]. Using this approach, highly uniform oxide ceramic films were deposited over 150 mm (6 inch) diameter substrates with variations of only $\pm 2.3\%$ and $\pm 0.5\%$ for thickness and composition, respectively [158]. The reader is referred to several recent publications for details [157–159].

A useful figure-of-merit when considering scaleup is the product of film area and deposition rate (e.g., Å-cm^2/s) for a laser of given output power. Wu et al. [161] obtained ~520 Å-cm^2/s using a 30 W (300 mJ, 100 Hz) XeCl laser; Foltyn et al. [146] estimated that ~600 Å-cm^2/s could be obtained for the same laser. Greer's films were deposited at >100 Å-cm^2/s (~1 Å/s for 30 Hz and 1 J/cm^2). The 600 Å-cm^2/s figure-of-merit corresponds to 20 Å-cm^2/s per Watt of laser power. If it is assumed that epitaxial semiconductor film-growth rates must be limited to ~20 Å/s in order to maintain high crystalline quality at low-to-moderate substrate temperatures, 1 Watt of laser power is required per cm^2 of substrate. Consequently a 100 W laser should be sufficient to grow a 1 μm thick epitaxial film over a 100 mm (4 inch) diameter substrate in about 7 minutes. Because laser manufacturers find that increasing the repetition rate of an excimer laser is less costly than increasing its pulse energy beyond 1 Joule, the optimum excimer laser for PLD probably will operate in the neighborhood of 100–300 Hz. However, actual deposition occupies only a small fraction of the time required to produce an epitaxial film; further cost savings and leveraging result from time-sharing a high-power laser among several deposition chambers. This mode of operation is used in many laboratories.

The results and projections described in this section suggest that PLD can be scaled up to grow large-area, high-quality epitaxial films of complex, multi-component materials at projected growth rates that are competitive with existing deposition techniques. The recent evolution of high-power excimer lasers from laboratory instruments to tools used for precision machining, UV lithography, and corneal reshaping suggests that use of an excimer laser is not a barrier to industrial PLD of high value-added films.

11.3 Growth of Compound Semiconductor Films by Pulsed Laser Ablation

11.3.1 Materials Issues for Compound Semiconductor Growth and Doping

Although an astonishing number and variety of thin films have been grown by PLD during the past nine years, it has seen little use to grow *doped, epitaxial* compound (or elemental) semiconductor films until quite recently. Current semiconductor technology is based on a small set of materials for which highly specialized, successful, and (in most cases) high-throughput growth methods were developed earlier. The particulates problem and the strongly forward-directed ablation beam made it difficult for PLD to compete with these dedicated growth methods with regard to film quality and uniformity over large areas. However, recent advances in the use of liquid targets, beam-scanning methods, and the demonstration of high deposition rates (Sections 11.2.7–8) suggest that

PLD is becoming competitive, so its other advantages for growth of chemically complex films now may come into play.

A second problem is that the requirement to maintain stoichiometry is much more stringent for electronic semiconductors than for the oxide ceramics for which PLD has been so successful. Point defect concentrations of only 1 part in 10^4 can prevent electrical activation of dopant atoms in semiconductors. The energetic ablation beam may produce lattice displacement damage during film growth, resulting in point defects or defect complexes that can compensate deliberately introduced dopants in semiconductors.

Thus a number of questions must be answered in order to discover the best way to grow doped epitaxial compound semiconductor films by laser ablation. These include:

1. **How should multinary, polycrystalline ablation targets be prepared?** Powders (if used) must have high purity to avoid accidental compensation, or accidental doping, and must be well mixed. Low-density, cold-pressed targets are relatively easy to make with high purity, but ablation produces high areal densities of particulates. Hot pressing to obtain higher density targets complicates the purity problem. Cold pressing combined with dual-laser ablation (to produce a transient-liquid target) may be the answer.
2. **Can PLA maintain stoichiometry to the required level of 1 part in 10^4 or better? Does the energetic ablation beam cause lattice displacement damage? If so, can it be controlled?** RBS and EDX are able to reliably measure departures from stoichiometry only if they exceed ~1%. Experiments designed to separately measure deviations from stoichiometry and beam-induced defect concentrations at the 0.01% level are needed to address the stoichiometry and beam-damage issues in PLD films. Thus the identification of defect electronic energy levels in PLD films is an important research area.
3. **Are ambient gases detrimental to epitaxial film quality?** Collisions with ambient gas molecules can be used to reduce the mean kinetic energy of ablated species to a few eV, which is below the threshold for lattice displacement damage and point defect generation in most semiconductors. This energy range also is generally considered ideal for energetic beam-assisted chemical reactions of ablated species with adsorbed or ambient gas molecules, to modify film composition and/or doping. However, Rouleau et al. [43] and Lowndes et al. [42] recently found that the electrical properties of epitaxial p-type ZnTe:N films were degraded for N_2 gas pressures ≥ 100 mtorr, apparently due to the onset of significant cluster formation in gas-phase collisions (see Sections 11.2.7.1 and 11.3.2.5). Experimental methods are needed to control ablation beam kinetic energy—or at least kinetic energy transfer to the film—independent of gas-phase collisions. The

effect on epitaxial film quality of gas atoms adsorbed on the growing film surface also needs further study [43].
4. **What is the best method to dope PLA semiconductor films?** Dopant atoms can be incorporated in the ablation target, but the relatively low concentration needed for even a heavily doped semiconductor (~ 1 part in 10^3) complicates target synthesis and compositional homogeneity requirements. A low-pressure ambient gas is an alternative, but only for a limited number of dopants. Pulsed delivery of gas molecules, timed either to coincide with the ablation pulse or to precede it and produce a $\sim$sub-monolayer adsorbed dopant layer may be a useful approach. Highly reactive atomic beams generated by rf or dc plasma sources can be used to supplement the ablation flux, but only in high vacuum. Considerations of efficiency, convenience, and film quality are involved in choosing between these approaches.

Several groups recently have carried out experiments that begin to address these issues in order to determine whether PLA can be used to grow epitaxial compound semiconductor films that are structurally of high quality *and highly doped*.

The selective review that follows is focused on examples of *epitaxial* growth *combined with doping* to control electrical properties. There are several reasons for this restriction. First, measurements of structural and electrical properties permit stringent comparisons to be made with the best epitaxial compound semiconductor films grown previously by MBE, CBE, and MOCVD; polycrystallinity often prevents meaningful comparisons. Second, electrical activation of dopant atoms not only is a stringent test of semiconductor quality, but it is also a formidable scientific problem for most compound semiconductor families, regardless of growth method. Thus, electrical properties measurements provide insights that can be obtained in no other way. Third, the achievement of low defect concentrations, and of high carrier concentrations and carrier mobilities, is crucial for many device applications of compound semiconductors in order to form ohmic contacts and p-n junctions.

Two significant omissions result from restricting this article to doped, epitaxial PLD films. First, Cheung and co-workers pioneered the growth of undoped, but very high-quality HgCdTe films and superlattices, using a laser-MBE method that is a hybrid of PLD with conventional MBE. That work is the subject of several reviews [14, 15, 32, 162]. Second, Dubowski *et al.* used PLD to grow epitaxial CdTe and $Cd_{1-x}Mn_xTe$ films and superlattices and have studied their optical and structural properties extensively [163–165], but deliberate doping was attempted in only one case (CdTe:In, in [166–168]) and no measurements of electrical properties were made [169]. Some of this work also has been reviewed in several other articles [16, 163–165].

11.3.2 II-VI Compounds

11.3.2.1 Epitaxial Orientation and Crystalline Phases.
The II-VI semiconductors ZnS, ZnSe, CdS, CdSe, and CdTe were grown epitaxially on (111) and (001) GaAs and InP substrates by Shen and Kwok [116, 170]. Ablation targets were obtained by pressing high-purity powders at $\sim$50,000 psi. The optimum growth temperature was near 300°C. Films grown above 400°C had poor surface morphology, and EDX measurements using the starting powder as a standard indicated that they were deficient in the column-VI element.

All of the films were fully epitaxial and in-plane aligned with mirror-like surface morphology. On (001)-oriented substrates, all of the films grew in the cubic phase (zinc blende structure). On (111)-oriented substrates, ZnS and ZnSe films were cubic, whereas CdS and CdSe were hexagonal (wurtzite structure) and CdTe grew in a mixture of phases. In general, the best lattice-matched film-substrate combinations gave the smallest X-ray rocking curve widths, which were $\geq 0.2°$ [116, 170].

Lowndes, Rouleau, and co-workers grew electrically insulating epitaxial ZnTe films by ablation through Ar gas and highly hole-doped p-type ZnTe:N films by ablation through ambient N_2 and N_2/Ar gas mixtures [28, 29, 41–43] These films were grown on semi-insulating GaAs(001) substrates at 320°C, at a target-substrate separation of 10 cm, using both high (2.7 J/cm^2) and low (0.66 J/cm^2) laser E_d. High-resolution X-ray diffraction measurements, including both $\theta - 2\theta$ plots and ϕ-scans of the ZnTe(404) reflection to assess in-plane epitaxy, revealed that films grown at relatively low laser E_d were fully epitaxial in three dimensions in the "cube on cube" alignment, with the film and substrate $\langle 00\ell \rangle$ axes precisely aligned. Films grown at the highest E_d contained only very small amounts (<1%) of the ZnTe (311) and (111) orientations in addition to the dominant in-plane-aligned (001) orientation [28, 42].

11.3.2.2 Film Stoichiometry.
Rouleau, Lowndes, and co-workers used Rutherford backscattering spectrometry (RBS) measurements to evaluate the Zn/Te ratio in $\sim$104 nm thick ZnTe films that were deposited both in vacuum and in 100 mtorr N_2 on 4 cm long Si substrates mounted radially on a rotating substrate heater, using a laser E_d of 0.66 J/cm^2 [29, 42]. A Si substrate was used instead of GaAs for RBS measurements in order to avoid interferences of the Ga and As RBS peaks with the Zn peak. The ZnTe film deposited in 100 mtorr N_2 was stoichiometric over most of its length, within the $\sim$1% resolution of RBS, whereas the film deposited in vacuum was Zn-deficient by a statistically significant amount near the center of the (rotating) heater face. The observation of a Zn deficiency under vacuum deposition conditions is consistent with MBE (near-vacuum) ZnTe growth experiments by Feldman *et al.* in which Zn/Te incident flux ratios of 2 to 3 were required to grow films with the best electrical

properties [171]. In the PLD experiments, gas-phase collisions apparently restore near-stoichiometry either by suppressing Zn evaporation or by altering the angular distribution of ablated species (see Section 11.2.5.5). These RBS data, in combination with the p-ZnTe:N electrical properties measurements described in Section 11.3.2.4–5, show that films with nearly ideal 1:1 ZnTe stoichiometry can be grown by PLA of a pure ZnTe target into ambient N_2 gas. Lowndes *et al.* found that Zn/Se ratios measured by RBS also were sensitive to both the ambient gas pressure and laser E_d [42].

11.3.2.3 Doping from the Ablation Target.
Group-III and group-VII elements are n-type dopants in a II-VI host while group-I and group-V elements are p-type dopants. Shen and Kwok reported [170] both n- and p-type doping of (001)-oriented cubic (zinc blende) CdS films that were grown at 300–400°C on semi-insulating (001) GaAs substrates. The CdS films were grown in 8 mtorr and 100 mtorr Ar using target-substrate separations of 10 cm and 5 cm, respectively. Li_3N or In powders were added to pressed-powder CdS targets, for p-type and n-type doping, respectively. A Li concentration as low as 1% in the target produced a p-type CdS film. However, because film resistivities generally were high, Hall effect measurements were possible only for higher dopant concentrations. A target with 11 mol% Li produced the lowest resistivity CdS:Li film, which had a hole concentration p $\sim 1.1 \times 10^{17}$ cm^{-3} and mobility $\mu_p \sim 9.8$ cm^2/V-s. For n-type doping, the lowest CdS:In film resistivity of 0.0017 Ω-cm, corresponding to an electron concentration in the mid-10^{20} cm^{-3} range, was obtained from a target that contained 9 mole% In. The highest Hall mobilities obtained for n- and p-CdS films were 230 and 22 cm^2/V-s, about 2/3 and 1/2, respectively, of the values expected for single-crystal CdS [172].

Shen and Kwok also reported p-type doping of ZnS, ZnSe, and CdSe using Li-doped targets [173]. These films apparently were grown using $D_{ts} = 5$ cm, in 100 mtorr Ar, at $E_d = 1.5$ J/cm^2. Hot probe (thermopower) measurements of the thermal emf generated by a temperature gradient revealed conversion from high-resistivity n-type behavior to p-type conduction for Li concentrations as low as 0.5 mole %. SEM studies showed that the films' surface morphology degraded for Li concentrations in the target greater than 6 mole % for ZnS and ZnSe, or greater than 3 mole % for CdS and CdSe. The lowest resistivities obtained were ~102 Ω-cm for ZnS:Li (9 mole % Li in the target) and ~5.8 Ω-cm for CdS:Li (11 mole %) [173]. Hall effect measurements were not made for ZnS and ZnSe because of difficulties in making ohmic contacts.

The concentrations of In and Li atoms in the targets used by Shen and Kwok were much higher than the carrier concentrations in their films. The relatively smooth film morphologies that were produced using a CdS target with 9 mole % In, or using a ZnS target with 6 mole % Li, also suggest much lower levels of dopant incorporation in the films than in the targets. The p-type doping achieved using Li prompted Shen and Kwok to suggest that PLA at least partially overcomes the formation of Li interstitials, which act as compensating donors [173]. However, their electrical measurements also suggest that low-melting point dopants such as In (156.6°C) and Li (180.6°C) may have low sticking coefficients, contrary to the stoichiometric deposition normally expected of pulsed laser deposition [173]. Direct measurements of dopant incorporation and carrier concentration are needed to resolve this question and to determine the electrically active Li fraction.

Though polycrystalline semiconductors are not the subject of this article, we note that Compaan and co-workers [174, 175] recently grew polycrystalline p-type ZnTe films on glass substrates at 300°C by mixing metallic Cu and ZnTe powders in a pressed target. Very little change of film resistivity was observed up to 0.3 wt.% Cu in the target, but the film resistivity then dropped by five orders of magnitude (to ~ 1 Ω-cm) when the target's Cu concentration was increased only another factor of three to ~ 1 wt.% Cu. A hole concentration in the 10^{18}–10^{19} cm^{-3} range was estimated, though the measured hole mobility was low because of the polycrystallinity of the films [175].

11.3.2.4 Doping from an Ambient Gas. As described in Sections 11.3.2.1–2, Rouleau, Lowndes, and co-workers grew highly hole-doped p-ZnTe:N films by ablating a pure ZnTe target through ambient N_2 gas and through N_2/Ar mixtures [28, 29, 42, 43]. Hall effect measurements at room temperature and 77 K were used to determine the free hole concentration, p, and hole mobility, μ_p, versus the ambient N_2 partial pressure. As shown in Fig. 10, the hole concentration was nearly independent of the Hall measurement temperature, indicating heavy doping and formation of an impurity conduction band. ZnTe films grown in pure Ar gas were electrically insulating, whereas ZnTe films grown in N_2/Ar mixtures had intermediate hole concentrations, consistent with doping that is controlled by the N_2 partial pressure. By slightly changing the growth conditions, a maximum hole concentration $>1.1 \times 10^{20}$ cm^{-3} was achieved.

The p-ZnTe:N film with the highest hole concentration and mobility in these experiments had nearly identical surface-normal and in-plane lattice parameters of 6.0755 (± 0.002) Å and 6.078 Å, respectively, both $\sim 0.4\%$ *smaller* than the bulk ZnTe lattice constant of 6.100 Å. A reduced in-plane lattice parameter also was found for another film grown in N_2, but not for a film grown in Ar. These results suggest that substitution of the small nitrogen ion on Te sites is responsible for the lattice contraction in PLA ZnTe:N films, consistent with their high p-type conductivity.

Geohegan *et al.* used in situ spectroscopic measurements [33] to obtain information about the mechanism for nitrogen doping in p-ZnTe:N. Optical emission measurements at a distance of 7 cm from the ZnTe target during PLA revealed no emission from either excited atomic nitrogen or excited molecular nitrogen [42]. The lack of atomic N emission from the PLA plume contrasts

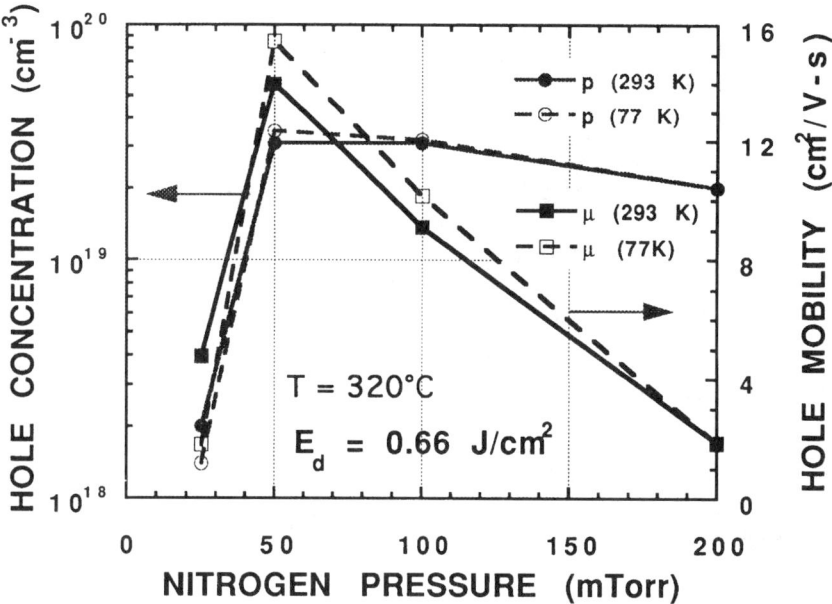

FIG. 10. Hole concentration and mobility vs. N_2 pressure (at 1 sccm flow) for p-ZnTe:N films grown at $T = 320°C$ and $E_d = 0.66$ J/cm^2 [29, with permission].

with observations of strong near-IR atomic N lines when a nitrogen RF plasma source was used for N-doping during MBE growth of ZnSe [176]. Lowndes suggested that atomic N is absent in the PLA experiments because it is very short-lived at the 25–100 mtorr pressures used, and so is unlikely to be involved in doping. Instead, the likely nitrogen-doping mechanism during PLD was considered to be kinetic energy-enhanced reactions of Zn atoms with N_2, either by forming Zn-N compounds in the incident flux (e.g., Zn_3N_2) or by dissociating N_2 molecules that were adsorbed on the growing film surface (e.g., in the metastable $A^3 \Sigma_u^+$ state as suggested by Nakao and Uenoyama [177, 178]) [42].

11.3.2.5 Crystalline Quality: Effects of Ambient Gas.

Rouleau et al. made RHEED measurements in two azimuths before and after each of a series of ZnTe film depositions [43]. In all cases, the GaAs or GaSb substrates gave streaky (2-D), high-contrast RHEED patterns prior to growth (after oxide desorption). However, as the N_2 or Ar gas pressure used during growth was increased, there were marked changes in the post-growth RHEED patterns, as shown in Fig. 11. At 25 mtorr N_2, the intensity along the sharp, high-contrast streaks became slightly modulated; at 50 mtorr N_2, a sharp, high-contrast but spotty, fish-net pattern was obtained; and at 100 mtorr N_2, only spotty, diffuse lower contrast patterns were observed, with multiple diffractions during

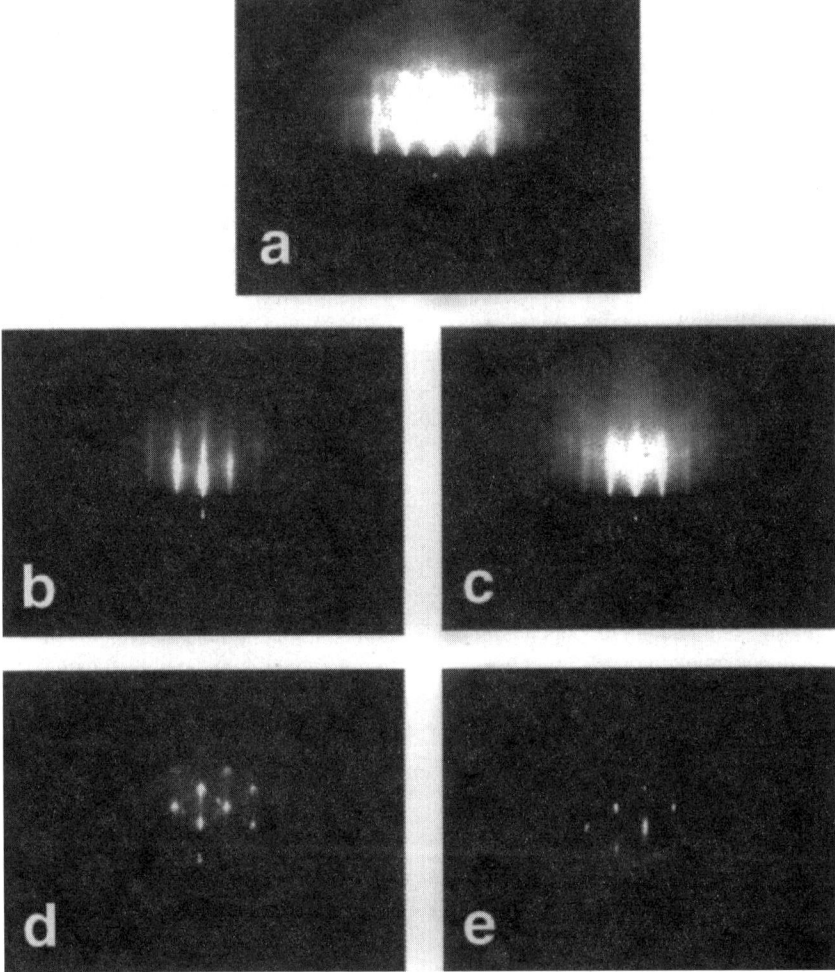

FIG. 11. RHEED patterns observed after PLD of p-ZnTe:N films as a function of the ambient N_2 pressure during growth.

azimuthal rotation. These changes reveal a gradual departure with increasing N_2 pressure from a highly ordered, 2-D surface to one that was less ordered and more 3-D, and at 100 mtorr to surface features that were misoriented [43].

Measurements of X-ray rocking curve widths were used similarly by Shen and Kwok to investigate the influence of ambient gas pressure and target-substrate separation, D_{ts}, on the crystalline quality of epitaxial CdS films grown on InP(001) substrates [116]. Other deposition parameters were fixed at $E = 30$ mJ,

$E_d = 1.5$ J/cm^2, T = 300°C, and 10 Hz rep rate. The average deposition rate and the deposition rate per laser pulse both were kept low (with the latter ≥ 0.05 Å/pulse) in order to obtain the highest quality films. Their X-ray rocking curve measurements (Fig. 12) for a series of CdS/InP(001) films grown at Ar gas pressures of 8, 100, and 200 mtorr revealed that (1) for each pressure there was an optimum target-substrate separation to minimize the XRD rocking curve width, and (2) the optimum target-substrate separation increased with decreasing ambient gas pressure. The authors interpretation of this pressure-distance scaling law (see Section 11.2.6.4) was that there is an optimal kinetic energy of incident ablated species to obtain the best film growth, and that collisions of the ablation plasma plume with Ar gas atoms can be used to reduce the plume's kinetic energy to the optimum value, for any particular value of D_{ts}. Films grown at 8 mtorr also had better surface morphology and thickness uniformity [170], in agreement with the RHEED measurements of Rouleau et al. [43].

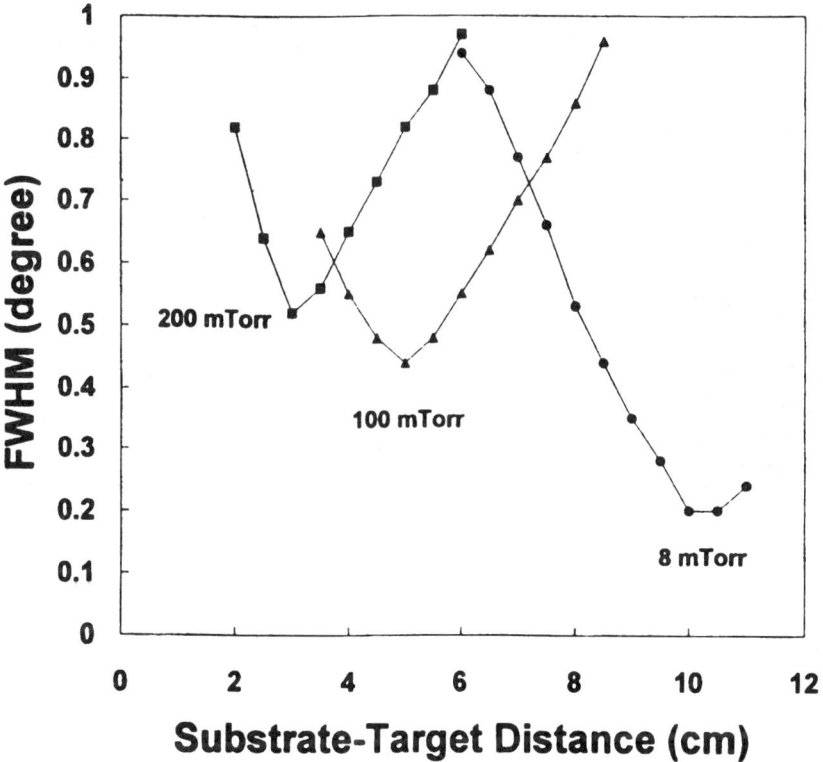

FIG. 12. X-ray rocking curve FWHM of CdS/InP(100) under different deposition conditions [116, with permission].

Shen and Kwok's results for CdS (Fig. 12) show that the *absolute minimum* value of the rocking curve width was obtained at the lowest pressure and apparently would have been obtained in vacuum for sufficiently large D_{ts}. Indeed, for ZnS films they did obtain the best deposition in vacuum at $D_{ts} = 10$ cm, by using a lower $E_d = 1.2$ J/cm^2. (Recall that Lowndes and Rouleau also obtained their best ZnTe films at a low E_d of 0.66 J/cm^2 [42, 43].)

The importance of these results for PLA growth of compound semiconductor films is that although a gas can be used to tune the kinetic energy of ablated species, the higher gas pressures apparently are deleterious to film microstructure as revealed by both RHEED [43] and X-ray rocking curve [116] measurements. This has serious implications for attempts to dope wide bandgap compound semiconductor films by ablating through a gas containing the dopant species, because if one is restricted to low gas pressures, point defects may be produced by the impact of energetic species compensating and preventing electrical activation of dopant atoms, unless a large target-substrate separation is used.

Experiments of Rouleau, Lowndes *et al.* [42, 43] and of Geohegan *et al.* [41] have focused on understanding the mechanism for degradation of epitaxial film quality with increasing ambient gas pressure, and on learning what background gas pressures can be tolerated. Time-resolved ion current measurements during ablation of various materials have revealed that, over a limited range of ambient gas pressures, the ablation plume splits into two or more components that travel at different average velocities and kinetic energies (see Section 11.2.6.3) [41]. Geohegan *et al.* interpreted the fastest pulse as ions that have undergone no collisions with gas molecules before arriving at the substrate. Figure 13 shows that for ZnTe ablation into N_2 gas under the film-growth conditions used by Lowndes, Rouleau *et al.*, the fast pulse remained significant until the N_2 pressure was increased to ~50 mtorr, which was the pressure that also gave the highest hole mobility in their ZnTe films (Fig. 10). According to Lowndes *et al.* [42], the increase of hole mobility with N_2 pressure is due to the exponential attenuation of the fast pulse of unscattered ions (and atoms), which have sufficient kinetic energy to produce lattice displacement damage and point defects. Rouleau et al quantified this idea by fitting either two or three exponentially attenuated Gaussian distributions to time-resolved ion current measurements at each pressure, and deconvoluted the ion flux into constituent "modes" of incident species with well-defined velocities and amplitudes. For the collision-free ion pulse ("mode 1" in Fig. 13), they found that 50% of incident Zn (Te) atoms had kinetic energies ≥ 19 eV (≥ 37 eV), sufficient to cause lattice damage. Rouleau's analysis also revealed two transitions as a function of N_2 pressure, from mode 1 dominance to mode 2 at ~15 mtorr, and from mode 2 dominance to mode 3 at ~70 mtorr [42, 43]. The hole mobility maximum of Fig. 10 lies in the region of mode 2 dominance and corresponds to Zn or Te kinetic energies of only a few eV. Thus the best p-ZnTe:N films were grown predominantly from an incident flux that had relatively low kinetic energy.

But Fig. 13 shows that a third peak developed and dominated the ion current distribution for N_2 pressures ≥70 mtorr, whereas Fig. 10 shows a steadily declining hole mobility for N_2 pressures of 70–200 mtorr. According to Rouleau et al., the transition to mode 3 growth at ~70 mtorr marks the onset of significant cluster formation in gas-phase collisions and subsequent cluster deposition. Cross-section TEM images of ZnTe films reveal a transition from highly microtwinned films at low pressures to a columnar grain structure with no microtwins at 100 mtorr [42]. Moreover, at 100 mtorr, both the film surface and near-surface intergranular regions were decorated with spherical particles having diameters of ~9.5 to 17 nm, apparently grown from the clusters of "mode 3" (Fig. 13).

In summary, the pronounced peak in hole mobility (Fig. 10) apparently results from the competition between two film-degradation measurements that operate

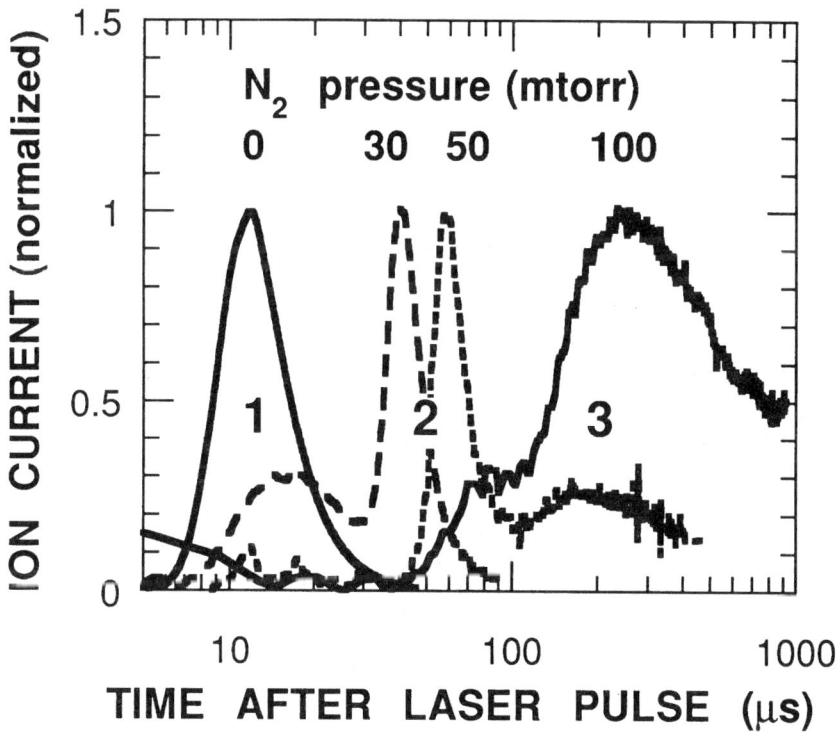

FIG. 13. Normalized ion currents measured at D_{ts} = 10 cm following ablation of ZnTe into N_2 at various pressures. The fast (mode 1) ion peak actually is exponentially attenuated with increasing D_{ts} or N_2 pressure [42, with permission].

in nearly overlapping pressure regimes for $D_{ts} = 10$ cm: defect production caused by the ablation beam's kinetic energy at low N_2 pressures, and the onset of significant cluster deposition accompanied by a change in film microstructure for N_2 pressures ~100 mtorr. Changes in X-ray rocking curve widths with increasing N_2 or Ar pressure also have been observed and interpreted according to this model [28, 42].

11.3.3 CuInSe$_2$ and Other I-III-VI$_2$ Compounds

CuInSe$_2$ (abbreviated CIS) is a direct bandgap I-III-VI$_2$ semiconductor ($E_g = 1.04$ eV) that exhibits strong optical absorption (2×10^4 cm$^{-1} < \alpha < 10^5$ cm^{-1} for $E_g < h\nu < 1.5$ eV), good resistance to radiation damage [179], and both p- and n-type conductivity [24]. These properties make CIS attractive for thin-film solar cells and infrared radiation detectors. In fact, p-type CIS and its wider bandgap alloy, CuIn$_{1-x}$Ga$_x$Se$_2$ (CIGS), are the photon-absorbing layers in photovoltaic cells that currently hold the "world record" for solar energy conversion efficiency by a polycrystalline thin-film device [180–184]. For photovoltaic applications, CIS and CIGS films are deposited on Mo-coated soda-lime glass substrates by physical vapor deposition with variable Cu, In, Ga, and Se fluxes under computer control. One of the keys to high conversion efficiency is that deposition take place at 500–550°C (uncomfortably near the softening point of soda-lime glass) so that the CIS/CIGS layer is formed by a partial-melting process. This produces polycrystalline grains (2–5 μm diam) with a columnar morphology (normal to the substrate) resulting in efficient front-to-back tranport of minority carriers, largely unimpeded by grain boundaries. p-n junction formation occurs because solidification from the melt is accompanied by a change from Cu-rich (p-type) to Cu-poor (n-type) material; n-type Cu-poor impurity phases other than CIS/CIGS also are present in the last-to-solidify material.

Improved control of this process now permits large, but still only very gradual, compositional changes to be made throughout the film thickness in order to vary the energy bandgap, control dopant type, and create a back surface field [185]. Nevertheless, CIGS films produced this way have a number of imperfections, including numerous grain boundaries, surface roughness, pinholes, and an uncertain, multiphase near-surface composition. The effects of these are minimized in high efficiency solar cells by using an n-type CdS layer to "encapsulate" the CIGS free surface [24, 180–184]. However, it is recognized that the complex chemistry and processing of the polycrystalline CIGS used in solar cells prevents its use in fundamental materials studies. Materials properties can be understood only in a highly averaged way from standard photovoltaic measurements on polycrystalline CIGS films, and there are at present few useful correlations between well-characterized single crystal I-III-VI$_2$ films or structures and their polycrystalline counterparts.

High solar cell efficiencies combined with poor understanding of basic materials properties have created a strong incentive for fundamental studies of CIS, CIGS, and other I-III-VI$_2$ compounds using epitaxial films. Potential advantages of epitaxial growth include single-phase specimens and reduced growth temperatures, as well as interfaces, compositional changes, and p-n junctions that are abrupt rather than graded. Both PLD [180–202] and MBE [203–210] have been used to initate such studies during the past several years. The strongest motivations for the PLD studies were to evaluate the congruent-transfer property by depositing CIS and CIGS from a stoichiometric target, to easily control composition by ablating a single target, and to explore growth at reduced temperatures. However, most PLD films were polycrystalline because their growth was motivated also by photovoltaic cell or IR detector interests, so amorphous (glass) or poorly lattice-matched (e.g., crystalline Si) substrates were used. As described below, the presence of polycrystalline grain boundaries in an already electronically complex material complicates the interpretation of electrical properties measurements. Consequently there remains a strong need for well-characterized epitaxial I-III-VI$_2$ thin films, both to evaluate PLD's potential for growth of complex compound semiconductors and to better understand these materials themselves.

11.3.3.1 Crystalline and Defect Structure. Semiconductors of the I-III-VI$_2$ family can be thought of as isoelectronic with binary II-VI compounds that have the zinc blende (sphalerite) structure, i.e., electronically they are II-VI compounds "on average." I-III-VI$_2$ semiconductors have the chalcopyrite structure shown in Fig. 14, in which the anions (VI) and cations (I and III) occupy different sublattices, just as in zinc blende. However, each I-III-VI$_2$ formula unit has four (not two) atoms because in the ideal chalcopyrite structure the group I and III atoms are ordered on the cation sublattice, resulting in a chalcopyrite unit cell that is approximately double the height of the zinc blende unit cell. The actual chalcopyrite unit cell often is tetragonally distorted and the c/a ratio may differ from 2 by as much as ~10%; for CIS, c/a ≈ 2.01.

This seemingly simple difference in composition and structure actually masks a substantial increase in intrinsic structural and electronic complexity of chalcopyrite semiconductors relative to zinc blende. For example, because each group-VI anion has two group-I and two group-III cations as nearest neighbors, the different I-VI and III-VI bond strengths cause the anions to be displaced from what would be the normal zinc blende anion sites, resulting in unequal I-VI and III-VI bond lengths. Furthermore, twelve intrinsic point defects can be formed in the ideal chalcopyrite structure, double the number for zinc blende. These include vacancies (V_I, V_{III}, V_{VI}), interstitials (I_i, III_i, VI_i), and a total of six antisite defects. The antisite defects are of two different types because the chalcopyrite structure permits exchanges of anions and cations with each other (resulting in I_{VI}, III_{VI}, VI_I, and VI_{III}) just as for zinc blende, but also allows exchanges between cations on the same sublattice (to produce III_I and I_{III}).

This richness in point defects is highly relevant for attempts to dope I-III-VI$_2$ compounds. Their electrical properties are believed to be dominated by their intrinsic point defect populations, which in turn are controlled by relatively

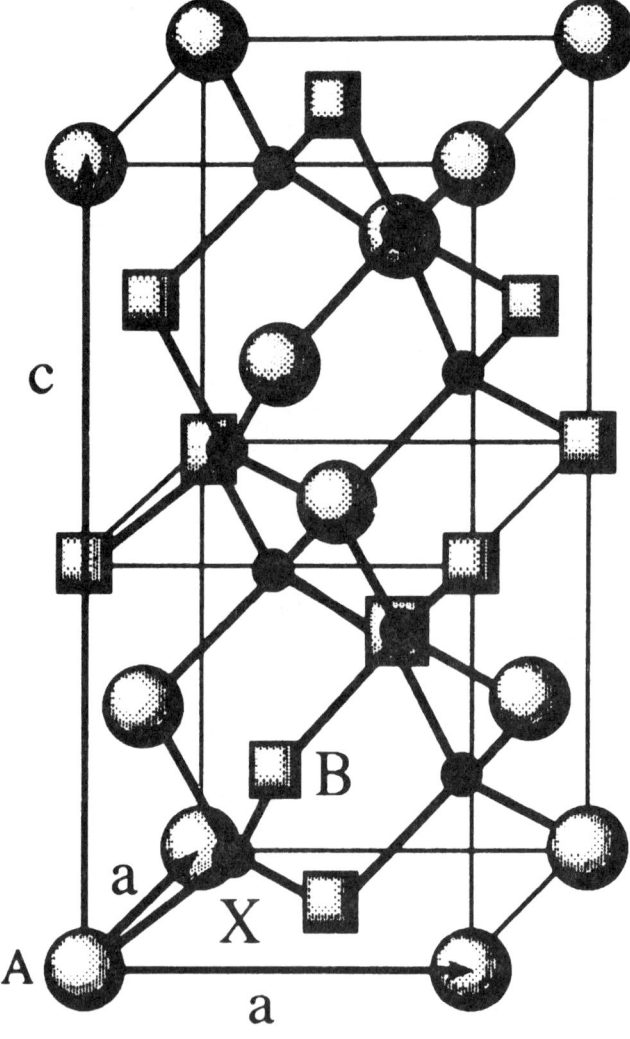

FIG. 14. The conventional tetragonal unit cell of the I-III-VI$_2$ chalcopyrite structure. The cations (I, III) are represented by the large open circles and squares, and the anions (VI) by small filled circles [24, with permission].

small deviations from the ideal 1:1:2 stoichiometry. Formation energies for all of the intrinsic defects have been calculated for CIS under various models and assumptions, as described in the next section. The result for single-crystal CIS is that the antisite defects In_{Cu} and Cu_{In} have the lowest formation energies and hence are likely to be dominant electrically, followed by vacancies [211–213]. However, the situation changes for polycrystalline CIS, because the structural disorder at grain boundaries makes it easier to form vacancies there, so they are nearly as likely as In_{Cu}. The reader is referred to recent publications by Möller for more detailed discussions of the structural and electrical complexity of I-III-VI$_2$ compounds [24, 212].

11.3.3.2 Electronic Properties of CuInSe$_2$.

Single-crystal CIS. Monocrystalline CIS can be doped n- or p-type by controlled changes of stoichiometry. Experimental and calculated results for small deviations from the ideal 1:1:2 stoichiometry are shown in Fig. 15(a). Temperature-dependent Hall mobility measurements [214] have been analyzed to yield estimated ionized defect concentrations of $N_a \sim 8 \times 10^{18}$ cm^{-3} (in p-type material) and $N_d \sim 8 \times 10^{17}$ cm^{-3} (in n-type material). Both values were significantly larger than the corresponding free carrier concentrations, which were in the 10^{16}–10^{17} cm^{-3} range, thereby illustrating the high level of compensation that is typical even of monocrystalline CIS.

Electron density calculations for CIS show that the Cu-Se and In-Se bonds are strongly polarized toward the Se (anion) site (more strongly for the In-Se bond)

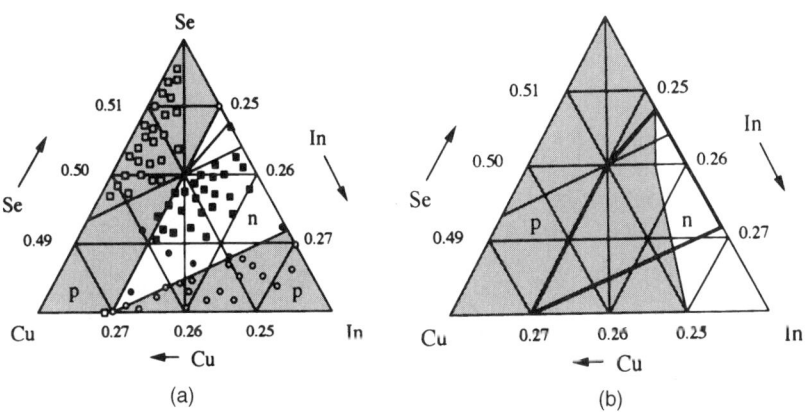

FIG. 15. (a) Calculated regions of n-type (unshaded) and p-type (shaded) conductivity, compared with experimental observations of n-type (filled symbols) and p-type (open symbols) conductivity, for single-crystal CIS as a function of stoichiometry. (b) Experimental results for regions of n-type (unshaded) and p-type (shaded) conductivity for polycrystalline CIS films (see text and Fig. 16). The boundaries of the n-type region for single-crystal CIS also are shown by the heavy lines [24, with permission].

[215]. Thus both the anions and cations in I-III-VI$_2$ materials are expected to carry significant polarization charges. For this reason, an ionic model has been used to calculate the electronic energy levels associated with the 12 different types of intrinsic point defects [211–213]. This model assumes that the point defects may be charged and that they will form in order to compensate for accidental or deliberate deviations from stoichiometry, resulting in shallow donor or acceptor electronic levels in the energy bandgap. Formation energies for all of the point defects have been calculated approximately for single-crystal CIS. Under the assumption that the formation entropies are not very different, the defects with the lowest formation enthalpies were the cation-exchange antisite defects, In$_{Cu}$ and Cu$_{In}$, and the vacancies, V$_{In}$, V$_{Cu}$, and V$_{Se}$ [211–213]. Using similar assumptions, the concentrations of all the native point defects were calculated numerically, with the result that for small deviations from stoichiometry, the electrical properties in a given composition range are expected to be controlled by only two or three defects [216]. Consequently, for material that is free of impurities, the electrical properties should be determined by the occupancy of the electronic energy levels of the intrinsic point defects that are most relevant to the composition range of interest.

Photoluminescence and temperature-dependent conductivity measurements (see Section 11.3.3.3) suggest that defects such as V$_{Se}$ introduce relatively shallow levels; thus it is plausible that these control doping. Table I shows experimental results for some shallow defect energy levels and their assignments to particular intrinsic point defects, whereas Fig. 15(a) shows that the calculated and experimental regions of n- and p-type conductivity are substantially in agreement. In Möller's words, "a rather consistent picture has emerged now for the defect chemistry in moncrystalline CuInSe$_2$" [216].

Polycrystalline CIS. Figure 16 shows experimental results of Noufi *et al.* for the carrier type and concentration in polycrystalline CuInSe$_2$ films that were co-evaporated from three elemental sources [217]. The Cu/In ratio was controlled by varying the Cu deposition rate while maintaining constant In and Se deposition rates. The carrier type changes from n to p for Cu/In ~0.9, and the

TABLE I. Experimental energy levels for shallow defects and their assignments to intrinsic point defects in CuInSe$_2$ [216].

Defect	Donor levels (meV)		Acceptor levels (meV)	
V$_{Cu}$				70
V$_{In}$			50	80
V$_{Se}$	60	80		
In$_{Cu}$	20	50		
Cu$_{In}$			50	100

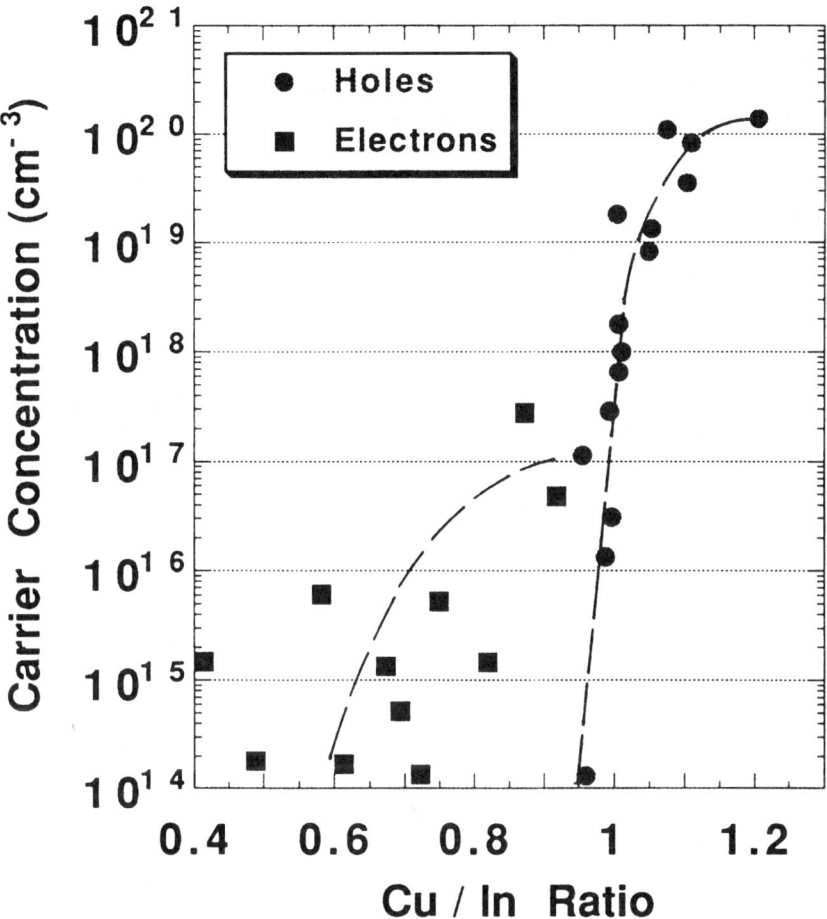

FIG. 16. Carrier type and concentration vs. Cu/In ratio for $CuInSe_2$ films with approximately constant Se concentration [217, with permission].

hole concentration increases by approximately six orders of magnitude over the narrow compositional range $0.95 \leq Cu/In \leq 1.05$. Although it is tempting to attribute the p- and n-type conductivity regions simply to excess Cu ions occupying In sites and vice versa, the situation clearly is more complicated—the conductivity type change occurs at Cu/In ~0.9, not 1.0, and the carrier concentrations on either side have quite different magnitudes. The experimental results relating conductivity type and composition for polycrystalline CIS are summarized in Fig. 15(b), which shows that polycrystalline CIS films always are p-type near the ideal stoichiometry; a substantial Cu deficiency is required to convert to n-type behavior, in agreement with Fig. 16.

Möller has noted [216] that large variations in the electrical conductivity of polycrystalline Si and Ge films, similar to that shown in Fig. 16, also have been seen over narrow ranges of dopant-atom concentration. For Si and Ge, the rapid variation occurs as traps at grain boundary sites are filled and the carrier concentration is reduced. The difference for CIS (and presumably other I-III-VI$_2$ materials) is that doping is controlled via composition, not by introducing foreign atoms.

A comparison of the two parts of Fig. 15 shows that the correlation between conductivity type and composition is different for polycrystalline and moncrystalline CIS, particularly for Se-deficient polycrystalline films that remain p-type. This difference is believed to be due to the electrical activity of the grain boundaries. The major complication that occurs for CIS is that grain boundaries are natural places to form particular types of intrinsic point defects (see Chapter 5 of [24] for examples) and, in sufficiently fine-grained material, the ratio of interfacial-to-volume sites can be significant. Consequently the correlation between electrical properties and composition is expected to be altered if the defects most likely to form at grain boundaries are different than in bulk. In that case, the formation energies calculated for the principal intrinsic point defects in single-crystal CIS are not relevant. Approximate model calculations by Möller [212] show that the defects with the lowest formation energies at grain boundaries in CIS are In_{Cu}(1.1 eV, donor), V_{In}(1.2 eV, acceptor), V_{Se}(1.3 eV, donor in the ionic model), and Cu_{In}(1.6 eV, acceptor), whereas in single-crystal CIS, they are In_{Cu}(1.6 eV, donor), Cu_{In}(1.9 eV, acceptor), V_{Se}(2.2 eV, donor in the ionic model) and V_{In}(2.4 eV, acceptor). Thus In_{Cu} has the lowest formation energy in both cases, but the populations of other intrinsic point defects are expected to be quite different in polycrystalline and monocrystalline CIS.

11.3.3.3 Pulsed Laser Deposition of CIS and CIGS.
Extensive investigations of PLD CIS and CIGS films were carried out during the past four years by three groups, in Germany [186–188], an English-Finnish collaboration [189–194], and in Belarus [195–199]. Most of the films were polycrystalline but significant results were obtained regarding structure, stoichiometry, and electrical properties. A few results also have been reported for CIS and CIGS grown epitaxially on Si(001) and Si(111) [189–192] and on GaAs(001) [42].

Target Fabrication. The electrical properties of CIS are controlled by deliberate, small deviations from the ideal stoichiometry, which in turn control native point defect populations. Consequently the preparation of targets with well-defined off-stoichiometric compositions is essential. However, the only detailed study to date of deliberately off-stoichiometric CIS/CIGS target fabrication was carried out in a collaboration of three Stuttgart groups who investigated both uniaxial hot pressing and isostatic cold pressing of powder mixtures [187].

Gremenok et al. prepared $CuIn_{1-x}Ga_xSe_2$ targets by melting the pure (5N) elements in quartz ampoules that were evacuated to 5×10^{-5} torr and held in a vertical furnace at 1050–1100°C for 24 hours [195]. The Ga content was varied from $x = 0$ to 1 and the resulting ingots were used as ablation targets. Details of the synthesis procedure were described much earlier by Bodnar and Bologa [218]. The presence of X-ray superlattice reflections and splitting of the (220)-(204) and (116)-(312) doublets (for $c/a \neq 2$) confirmed that the targets had the chalcopyrite structure. X-ray measurements as a function of Ga content also showed close agreement between the "c" and "a" lattice constants of the targets and of the films grown from them over the full compositional range ($0 \leq x \leq 1$).

Dittrich et al. fabricated off-stoichiometric CIS targets by isostatically cold pressing or uniaxially hot pressing mixtures of various high-purity (5N) starting materials, including presynthesized $CuInSe_2$, In_2Se_3, $CuSe$, and Cu_2Se [187]. Targets (25 mm in diameter and 3–6 mm thick) with the best homogeneity and highest density (~96% of theoretical) were obtained by reactive uniaxial hot pressing of $CuInSe_2$ and In_2Se_3 powder mixtures in a rough vacuum of ~1 Pa (~8 mtorr), using a commercial hot-pressing system. The optimum temperature and pressure were ~630°C and ~9 kN/cm^2 (~13 kpsi), respectively, for times up to 1 hour [186, 187]. Both quartz glass and Al_2O_3 ceramic were used as pressing tools; both were sufficiently inert that hot pressing did not produce measureable contamination of CIS [187]. By mixing $CuInSe_2$ and In_2Se_3 powders, off-stoichiometric (Cu-poor) targets were fabricated by Dittrich et al. [187]. These targets contained measureable amounts of a thiogallate Cu-In-Se phase very similar in structure to chalcopyrite CIS, but with an indicated composition of $CuIn_2Se_4$ (a Cu-deficient compound containing ordered Cu vacancies). These targets were considered best to grow photovoltaic films with a Cu/In ratio <1. Particulates produced by target ablation were present in the films and were considered to limit solar cell efficiencies, but cone formation and particulates were reduced by target rotation [187].

Hill et al. reported preparing CIS and CIGS targets from polycrystalline and Bridgman-grown single-crystal materials, but no details were given [189]. Lowndes et al. prepared 2.5 cm diameter targets by cold-pressing 99.999% purity CIS powder at 60,000 psi [42].

Stoichiometry of PLD CIS and CIGS Films. Hill, Leppävuori, and co-workers used energy dispersive X-ray analysis (EDX) and Rutherford backscattering spectrometry (RBS) to measure the composition of $CuIn_{1-x}Ga_xSe_2$ films with compositions $x = 0$, 0.05, 0.1, 0.15, 0.2, and 0.25. The films were deposited at a rate of 0.07 Å/pulse (1.8 Å/s) on fused silica and Si(111) substrates at 350°C using a XeCl (308 nm) excimer laser and $D_{ts} = 3$ cm [189]. EDX revealed a strong Se deficiency (32–45%) in the films, relative to the target material. However, if the analyzed compositions are corrected to Se = 50%, the

Cu, In, and Ga contents of the films are nearly the same as in the targets, and the In/Ga ratio of the targets was reproduced in the films up to $x = 0.2$ [189]. On the other hand, RUMP analyses of the raw RBS data gave an average Se content of 49 (± 2) % for four films, not significantly below stoichiometric. In later work by these two groups, a $CuIn_{0.75}Ga_{0.25}Se_2$ target produced films with a Se content of 46% (± 4%), and Cu and Ga contents that were stoichiometric within measurement error [191]. In their most recent work, these groups carried out EDX measurements for CIS and CIGS ($x = 0.25$) films that were deposited on fused silica, Si(001), and Si(111) substrates at temperatures of 0–550°C, using $D_{ts} = 3$ cm and deposition rates of 0.08 Å/pulse (2 Å/s) for CIS and 1 Å/pulse (10 Å/s) for CIGS [192]. The EDX film composition measurements were normalized to the composition of the target. For CIS and CIGS deposited on fused silica, the film-to-target concentration ratios were stoichiometric up to ~400°C and ~350°C, respectively, though the precision of the combined measurements probably was not better than a few percent. For CIS on Si substrates, a chemical reaction occurred above 450°C (between 350°C and 390°C for CIGS), leaving only a discontinuous Cu-Si film [192].

Gremenok et al. used X-ray microanalysis to study the composition of PLD CIS films, as well as CIGS films covering the complete $0 \leq x \leq 1$ composition range [195, 196]. A near-infrared (1.06 µm) Nd:YAG laser with a relatively long (~1 msec) pulse duration and unusually high pulse energy (150–180 J) was used for PLD. The films were deposited on glass and quartz substrates at $D_{ts} = 6$–8 cm, using temperatures of 320–380°C for CIGS and 50–400°C for CIS. For CIS, the optimum temperature was 300–350°C. All of the films were reported to have Cu, Ga, In, and Se contents near the stoichiometric composition to within experimental error (quoted as ± 0.5%) [195, 196].

Schäffler, Dittrich and co-workers used quadrupled (266 nm, 6 ns pulse duration) Nd:YAG radiation to ablate their off-stoichiometric (Cu-poor) uniaxially hot-pressed CIS targets in vacuum [186, 188]. CIS films were deposited onto soda-lime float glass substrates (some coated with Mo) at $D_{ts} = 6.5$–12 cm, using substrate temperatures in the 520–580°C range [188]. SEM observations of surface morphology were used together with X-ray diffraction and electrical measurements to characterize the CIS films. Because of the high substrate temperature (which apparently was chosen to correspond to growth conditions for co-evaporated and melt-recrystallized CIS solar cell layers), all of their as-deposited films were Se-deficient and exhibited metallic behavior. Therefore, excess Se was provided by two approaches, either in a post-deposition selenization step or by thermal evaporation of additional Se during PLD. Both selenization approaches produced single-phase CIS, but secondary phases appeared if additional Se was not provided [188]. X-ray analysis confirmed that Se-poor films were multiphase, with InSe (the Se-poor In-Se compound) and Cu-In intermetallics appearing. On the other hand, single-phase chalcopyrite films were formed by post-deposition selenization treatment, and films deposited with in situ excess Se were single-phase chalcopyrite CIS.

SEM observations also revealed significant growth-related morphological differences. Layers grown under Se-deficient conditions displayed grains with smooth surfaces that were unrelated to crystallographic boundaries, whereas under Se-saturated conditions, the film surface was made up of 0.3 μm to 1 μm grains with well-developed (112) lattice planes (equivalent to the (111) planes in zincblende) [186]. The authors suggest that there is low surface mobility of adatoms in the absence of Se, but that well-developed lateral (2-D) growth occurs under Se-saturated conditions [186].

Structure of PLD CIS and CIGS Films. Levoska, Leppävuori et al. found that CIS and CIGS films deposited on fused silica were crystalline and were oriented with the chalcopyrite (112) planes parallel to the substrate surface even at temperatures of 50–150°C [192]. However, the breadth and low intensity of XRD peaks did not allow the chalcopyrite and sphalerite structures to be resolved clearly. Above 300°C, the decreasing width and increasing intensity of XRD peaks allowed the chalcopyrite structure to be identified, and (112)-oriented films persisted to the highest temperatures (~515°C for CIS, 430°C for CIGS). In fact, for CIGS films grown at 350°C, the ratio of (112) and (220, 204) peak intensities was >1000, higher than in the best CIS films and also exceeding values of this ratio for co-evaporated or sputtered CIS and CIGS films [192].

On Si(001) substrates, Levoska, Leppävuori et al. found that CIS and CIGS films also were (112)-oriented up to 300°C [192]. However, above 300°C, CIS began to develop an epitaxial (001) orientation that became stronger with increasing temperature. The in-plane epitaxial alignment of CIS on Si(001) was confirmed by a ϕ-scan through (112) poles. However (as mentioned above), chalcopyrite CIS began to disappear for deposition temperatures >450°C on Si(001), leaving only a discontinuous Cu-Si film. For CIGS, no development of a preferred (001) orientation was observed at intermediate temperatures, and the chalcopyrite phase began to be replaced by Cu-Si at temperatures between 350°C and 390°C.

Gremenok et al. obtained single-phase chalcopyrite, (112)-oriented $CuIn_{1-x}Ga_xSe_2$ films ($0 \leq x \leq 1$) for all temperatures in the 320–380°C range [195]. The film crystallinity improved, and the (112) orientation was enhanced by increasing the substrate temperature. Systematic worsening of several structural parameters was observed with increasing Ga content: the FWHM of the (112) diffraction peak increased from ~0.2 deg to ~0.5 deg, the mosaic spread in (112) orientations increased from ~2.5 deg to ~6 deg, and the polycrystalline grain size decreased from ~1 μm to ~0.4 μm, in going from $CuInSe_2$ to $CuGaSe_2$. Gremenok et al. have tabulated experimental energy gaps and lattice constants for $CuIn_{1-x}Ga_xSe_2$ films for $0 \leq x \leq 1$ [195].

As was described in the previous section, Schäffler, Dittrich et al. deposited CIS at temperatures ≥500°C and obtained single-phase, (112)-oriented chalcopyrite CIS when additional Se was provided [186, 188].

Lowndes et al. grew highly oriented, n-type epitaxial CIS films on semi-insulating GaAs(Cr) substrates [42]. These films were grown in a 40 mtorr argon ambient in order to reduce the kinetic energy of ablated species into the range for which the best electrical properties were obtained for p-ZnTe (Sections 11.3.2.4–5). For the best CIS film (grown at 350°C with $E = 0.66$ J/cm^2 and $D_{ts} = 10$ cm), a $\vartheta - 2\vartheta$ scan revealed strong (00ℓ) diffraction peaks of the chalcopyrite phase, with no peaks indicating a more complex ordering (e.g., of vacancies). The measured lattice constants were close to those of bulk CIS, contracted slightly in-plane and expanded along the c-axis as expected for growth on GaAs. X-ray scans also were made to assess the extent of in-plane alignment, with the result that the CIS film was fully aligned epitaxially in the "cube on cube" orientation [42].

Electrical Properties of PLD CIS and CIGS Films. Schäffler et al. carried out a detailed study of the electrical properties of slightly In-rich (In/[Cu + In] = 0.53) CIS films that were deposited on soda-lime float glass at temperatures of 520–580°C (see description above) [188]. The goal in their work was to form "solar grade" material, defined as CIS with p-type conductivity $0.002 < \sigma([\Omega\text{-cm}]^{-1}) < 0.1$ and a room temperature thermopower $0.6 < \alpha(\text{mV/K}) < 0.9$. Such material lies near the middle of the range of rapidly varying hole concentration shown in Fig. 16; hence, precise compositional control is necessary. As noted earlier, addition of Se was required to produce p-type films at such high deposition temperatures. A model for the electrical behavior of CIS that they observed as a function of increasing Se deficiency is as follows: fully selenized films were p-type with $0.005 < \sigma([\Omega\text{-cm}]^{-1}) < 0.05$ and $\alpha = 0.6$–0.85 mV/K; for a slight Se deficiency, compensation (probably Se vacancies) produced lower p-type conductivity, $5 \times 10^{-6} < \sigma([\Omega\text{-cm}]^{-1}) < 5 \times 10^{-3}$ and reduced thermopower; for greater Se deficiency, the samples became n-type and the thermopower ranged from high to low values as segregation of Cu-In intermetallics began; and for strongly Se-deficient films, metallic conducting and thermoelectric behaviors were obtained, with $\sigma > 0.1$ [Ω-cm]$^{-1}$ and $\alpha \sim 30$ µV/K, as segregation of Cu-In intermetallics dominated the electrical properties [188]. The authors note that this behavior is similar to that of Se-deficient CIS prepared by other processes.

Solar grade CIS films were produced both by PLD in excess Se vapor and by post-selenization of PLD films deposited in vacuum. These films were used to fabricate the first PLD CIS solar cells, which consisted of a CIS absorber layer, a CdS encapsulation layer, an n-type ZnO:Al window, and an Al grid front contact, with a photovoltaic conversion efficiency of 8.5% [186, 188]. According to the authors, the major problems remaining to be solved for PLD CIS solar cells are a pronounced falloff in quantum efficiency in the near infrared (not understood) and the presence of μm-size particulates that resulted in inhomogeneous film growth [186, 188].

CIS films deposited by Gremenok et al. in vacuum at temperatures of 280–350°C (see above) were p-type with conductivities of 0.01–0.1 $[\Omega\text{-cm}]^{-1}$. Their CIGS films became more resistive with increasing Ga content, reaching $\sigma \sim 10^{-5}$ $[\Omega\text{-cm}]^{-1}$ for $CuGaSe_2$ [195, 196]. Zaretaskaya et al. noted that a polycrystalline target produced higher conductivity p-type CIS films ($\sigma \sim 3$ $[\Omega\text{-cm}]^{-1}$) than a single-crystal target ($0.02 < \sigma([\Omega\text{-cm}]^{-1}) < 0.4$), although no Hall effect was observed in these films because of either their low hole mobility or their high hole concentration, which was estimated to be $>10^{21}$ cm^{-3} [197].

Zaretaskaya et al. also discovered that p-type CIS films with still higher conductivity ($\sigma \sim 25$ $[\Omega\text{-cm}]^{-1}$) were produced by ablating CIS into a 100 mtorr oxygen ambient. These films displayed increasing conductivity with decreasing temperature in the 78–300 K temperature range, which the authors noted was similar to the $\sigma(T)$ behavior for CIS films deposited by other methods [219]; they attributed this behavior to impurity conductivity typical of a highly compensated semiconductor. However, these conductivity values are well above the range of "solar grade" material defined by Schäffler et al. [188] and correspond more closely to the values that Schäffler et al. attributed to Cu-In intermetallics. However, Zaretskaya et al. reported that their films were near-stoichiometric, single-phase chalcopyrite CIS.

Ahmed et al. have described the electrical properties of $CuIn_{0.75}Ga_{0.25}Se$ films deposited by the English-Finnish collaboration [191]. As-deposited films were both n- and p-type with conductivity values in the range $10^{-3} < \sigma([\Omega\text{-cm}]^{-1}) < 10$. An annealing study was carried out to investigate the reason for the n-type conductivity. Annealing in vacuum or a reducing atmosphere had no effect on conductivity, but annealing in a Se atmosphere converted the conductivity from n-type to p-type [191]. This was regarded as evidence that the n-type behavior of as-grown films was due to Se vacancies, consistent with EDX measurements that revealed a Se deficiency (as described above) [191].

Hall effect measurements by Lowndes et al. for an epitaxial CIS film grown on GaAs(Cr) showed n-type conductivity with $n = 6 \times 10^{17}$ cm^{-3} and $\mu_n = 55$ $cm^2/V\text{-s}$. Their X-ray measurements revealed a series of eight weak diffraction peaks coming from a high density of particulates (observed by SEM) on this film, due to their use of a cold-pressed target. Six of the eight peaks were indexed to InSe, the Se-deficient In-Se phase. Thus both the X-ray and electrical measurements were consistent with n-type doping due to Se vacancies. No post-selenization studies were carried out to attempt to change the conductivity type. The results of Lowndes et al. are significant because the 350°C growth temperature for a fully epitaxial CIS film is ~175°C lower than both the temperature range in which CIS films with columnar grains are grown for photovoltaic cells [180–184] and the ~520–580°C range used by Schäffler et al. [188].

Optimum PLD Conditions for CIS and CIGS. The measurements of film stoichiometry and structure summarized above show that the optimum deposition temperature for CIS is in the 320–400°C range and is slightly lower for CIGS. The temperatures >500°C used by Schäffler et al. result in excessive Se loss, although it remains to be established that "solar grade" semiconducting CIS or CIGS can be grown in situ at temperatures below 400°C. Unfortunately the limited absolute accuracy of EDX and RBS measurements, and the great sensitivity of film electrical properties to small deviations from stoichiometry, make it difficult to meaningfully relate the two. If a small, but electrically significant, Se deficiency persists even for temperatures ~350°C (as indicated by the many of the results described above), PLD into Se vapor or a low-pressure oxygen ambient may be the only way to grow electrically high quality as-deposited films. A combination of photoluminescence (PL) with temperature-dependent Hall effect and resistivity measurements will provide the most stringent test of the quality of thin films whose properties are controlled by intrinsic point defects. Such measurements still are needed for epitaxial PLD CIS and CIGS films.

11.3.3.4 Pulsed Laser Deposition of Other I-III-VI2 Materials.

Uchiki and co-workers grew polycrystalline [200] and epitaxial [201] films of $AgGaS_2$ (E_g = 2.68 eV) on quartz and GaAs(001) substrates, respectively. Ablation targets were formed by mixing and pressing powders of two phases, Ag_2S and Ga_2S_3, in Ag_2S-to-Ga_2S_3 molar ratios of M = 1, 1.25, 1.5, and 1.75. Films were grown in vacuum using D_{ts} = 2 cm, substrate temperatures of 400–650°C and deposition rates approaching 10 Å/pulse (at 0.5–2 Hz).

X-ray analysis of polycrystalline films grown from a stoichiometric target (M = 1) revealed highly (112)-oriented chalcopyrite $AgGaS_2$ together with Ag islands and some unidentified peaks; temperatures $\geq$600°C were required to inhibit Ag island formation. X-ray microanalysis showed that these films were nonstoichiometric, with composition $Ag_{0.7}GaS_{1.5}$, but with an optical absorption edge at 2.65 eV, close to the expected energy bandgap. PL spectra at 77 K revealed no excitonic emission and were dominated by a broad green emission centered at 2.46 eV, possibly due to sulphur vacancies [200].

Near-stoichiometric epitaxial films, with composition $Ag_{0.9}GaS_2$, were grown from a Ag-rich target (M = 1.25) at 600°C in vacuum; Ag islands were present, and formed in larger numbers in films grown from targets with larger M. PL measurements at 77 K for the M = 1.25 film showed the broad green emission that was observed for polycrystalline films, but weak emission also was present at 2.70 eV, i.e., near E_g. The temporal behavior of this emission was studied using 70 ps, 355 nm light pulses for excitation and a streak camera (80 ps resolution) for PL detection. The PL decay time was shorter than the streak camera resolution, indicating a strong interaction of carriers with defects and/or impurities, in contrast to the ~nanosecond decay times for good-quality films or bulk crystals [201].

The short PL decay times and the presence of Ag islands demonstrate that it is difficult to precisely control the composition and quality of films deposited by ablation of an unreacted, two-phase target. A lower deposition rate per pulse should improve film quality and perhaps allow the substrate temperature also to be reduced.

11.3.4 Column-III Nitrides

AlN, GaN, and InN have direct energy bandgaps of 6.2, 3.4, and 1.9 eV, respectively, in the wurtzite structure. Column-III nitride films are of interest for opto-electronic devices such as light-emitting diodes and lasers that could make possible full-color optical displays, as well as for photodetectors. Alloys based on GaN and AlN are especially interesting for blue and near-UV lasers for use in very high-density optical data storage if they can be doped reliably. If bipolar doping can be achieved in (Al, Ga)N alloy films, their wide bandgap, high stability, and relative lack of chemical reactivity also would make possible transistors and other electronic devices capable of operating at much higher temperatures and under more adverse conditions than present devices based on silicon and conventional III-V compounds.

Feiler, Williams, and collaborators recently demonstrated PLA growth and doping of column-III nitride epitaxial films [25, 26]. AlN, GaN, and InN were grown epitaxially on sapphire(0001) substrates at temperatures of 460°C–520°C, much lower than the temperatures reported for epitaxial nitride films grown by MOCVD (and many grown by MBE). The most intriguing discovery by Feiler et al. was that their GaN films were doped (accidentally) either n-type or p-type, depending on whether the ambient atmosphere was high vacuum or ~10 mtorr nitrogen (N_2), respectively [25, 26]. Electrically conducting p-type InN films also were grown, while AlN films were insulating [25].

Nitride films ~90–400 nm thick were deposited by 1 Hz KrF laser irradiation of targets prepared by cold-pressing column-III nitride powders. The sapphire(0001) substrates were annealed at 800°C prior to deposition until sharp LEED images were obtained. The focused laser energy density was 1.5–3.5 J/cm^2 and the target-substrate separation was only 3 cm, so that the substrate and growing film were immersed in the fluorescent energetic part of the ablation plume [400].

AlN and GaN films were single crystalline in the same orientation, with the film (0001) || sapphire(0001) and the film [1$\bar{1}$00] || sapphire[11$\bar{2}$0], i.e., the films were rotated by 90° in-plane with respect to the sapphire substrate [25]. This orientation places the films in compression in the in-plane direction and minimizes their lattice mismatches to 13.2% and 16.0% for AlN and GaN, respectively. Feiler et al. pointed out that if the films were not rotated, they would have been under tension in-plane, with much larger mismatches of 34.6% (33.1%) for AlN (GaN) [25]. The InN films also were highly oriented, with the film and substrate (0001) planes parallel. The dominant in-plane alignment was InN[1$\bar{1}$00] || sapphire[10$\bar{1}$0], i.e., the InN film was mainly rotated 60° with respect to the substrate. However, InN XRD ϕ-scans contained broad peaks and high background scattering between peaks, showing that a large number of small crystallites also were present with random in-plane orientations. This was consistent with a high density of particulates, easily visible in SEM images, due to the use of a cold-pressed target. The dominant 60°-rotated InN orientation also minimizes the lattice mismatch of InN to sapphire but places InN under tensile strain along the in-plane direction.

GaN films grown in high vacuum to thicknesses of 90 nm and 190 nm were n-type with an electron concentration and mobility of 1×10^{18} cm^{-3} and 300 cm^2/V-s, respectively [25, 26]. In contrast, a GaN film grown in 10 mtorr of nitrogen (N$_2$) was p-type with a hole concentration and mobility of 7.5×10^{18} cm^{-3} and 150 cm^2/V-s, respectively. In both cases, Feiler et al. attributed the conductivity to unintentional impurity doping. Silicon, at ~100 ppm, was the major impurity in the GaN powder used for target fabrication and is an n-type dopant in GaN. It is thought that ~1 ppm carbon monoxide (CO) in the N$_2$ ambient gas used during ablation may have resulted in carbon incorporation in the films grown in N$_2$, and a net p-type doping that compensated the n-type doping by Si impurities. For both n-type and p-type PLD GaN films, the carrier mobility values are considerably higher than expected from baseline data for GaN films grown earlier by other methods [401, 402]. A 190 nm thick GaN film grown by PLD in vacuum had an X-ray rocking curve width of 280 arc sec for the GaN(0002) peak, which is very good for a film of this thickness. Feiler et al. also experimented with growth of a thick GaN overlayer at 700–800°C, but these films were inferior in quality, possibly due to N loss at elevated temperatures [26].

The InN films prepared by Feiler et al. were strongly p-type both when grown in vacuum ($p = 6.5 \times 10^{20}$ cm^{-3} and $\mu_p = 30$ cm^2/V-s) and in a N$_2$ ambient ($p = 4.7 \times 10^{19}$ cm^{-3} and $\mu_p = 240$ cm^2/V-s). This contrasts with the n-type doping of InN that is obtained by conventional growth techniques, for which the electron doping is believed to be due to nitrogen vacancies. The source of hole doping in the PLD InN films is not known [25]. The distinctly different doping behavior in PLD emphasizes the need to understand compound semiconductor growth and doping in the energetic PLD environment.

AFM images showed that the electrically insulating AlN films were free of particulates. However, SEM and XRD revealed Ga (In) droplets on the surfaces of GaN (InN) films that were grown in a few mtorr of ambient N$_2$. For growth in vacuum, there were either no droplets (GaN) or fewer, larger droplets (InN). Feiler et al. point out that these results are consistent with the decrease in melting points and thermodynamic stability in going from AlN to GaN to InN [25].

The ~500°C temperature at which doped GaN and InN films were grown on sapphire(0001) by PLA, without use of a buffer layer, is much lower than the temperatures at which almost all epitaxial nitride films have been grown by MOCVD. The PLA films were grown at a target-substrate separation of only 3 cm and so were immersed in the excited laser ablation plasma. These conditions produced GaN films that contained particulates but whose crystallinity and electrical properties rival those of films grown at much higher temperatures by other methods, especially when the thinness of the PLD films and the absence of any strain-relieving buffer layer is taken into account [25, 26].

An implication of these experiments is that growth directly in the ablation plume may be an important factor permitting high-quality GaN films to form at low temperatures, although it is not known whether the kinetic or the electronic excitation energy thereby made available is most beneficial. Indeed, direct comparisons of the microstructure of MOCVD and PLD GaN films, as a function of film thickness, support the idea that PLD and MOCVD growth mechanisms for GaN must differ [25, 26]. The growth of strongly p-type InN films also suggests distinctly different growth conditions and doping behavior during PLD than with conventional growth methods [25]. The occurrence of accidental impurity doping in the GaN experiments, both in the ablation target and in the gas phase, promises considerable versatility and convenience for future PLA growth and doping of these materials.

11.3.5 Transparent Conducting Oxides

Transparent conducting oxide (TCO) thin films have been used for many years in a wide variety of applications, some of which currently include solar cells, liquid crystal and electroluminescent displays, heat-reflecting mirrors, gas sensors, and thin-film resistors. Three doped wide bandgap semiconducting oxide materials have had the most extensive use: tin oxide (SnO_2:F), indium tin oxide (In_2O_3:Sn), and zinc oxide (ZnO·Al). Among these, indium tin oxide [220–222] has become the "TCO of choice" [223] for many applications because it combines a low sheet resistance of 10 $\Omega/\square$ (i.e., a high conductivity $\sigma \sim 2.5\text{--}5 \times 10^3$ [Ω-cm]$^{-1}$) with low optical absorption throughout much of the visible region. Recently, however, a need has developed for new TCO materials with a sheet resistance approaching 1 $\Omega/\square$, as well as better transparency in the blue-green region [223, 224].

PLD has been used to deposit all three of the traditional TCO materials as well as to synthesize new conducting oxides such as $GaInO_3$ [223] and zinc indium oxide [224], which have better visible transmission than ITO. The most extensive PLD studies of the growth and doping of a TCO were carried out for crystalline ZnO. These results are reviewed in the next section, followed by an overview of recent PLD studies of other TCO materials.

11.3.5.1 Zinc Oxide. ZnO is a wide bandgap (E_g = 3.26 eV), n-type semiconductor that crystallizes in the wurtzite structure (c = 5.2 Å, a = 3.26 Å, c/a = 1.60) and is optically transparent throughout the visible region (λ_g = 3804 Å) [225]. Its n-type electrical conductivity can be varied over several orders of magnitude by doping with column-III (e.g., Al) or -IV elements [225–228]. Conduction in ZnO also is controlled by interstitial Zn atoms and O vacancies [225, 229, 230] and by Zn vacancies [227]. The (0001) faces of ZnO have the lowest surface free energy; consequently, ZnO films exhibit a strong preference for growth with the [0001] (c-axis) direction oriented perpendicular to either an amorphous or crystalline substrate. ZnO also exhibits good piezoelectric and piezo-optical properties. For ZnO films the largest piezoelectric coefficient is obtained when the c-axis is perpendicular to the substrate [225, 226]. This combination of properties has made ZnO films attractive for a variety of applications, including transparent conducting electrical contacts, gas sensors, optically coupled devices, varistors, and microwave acoustic resonators. Consequently c-axis–oriented crystalline ZnO films have been grown by most of the common deposition techniques, at temperatures generally in the 200–450°C range [231].

PLD first was used to grow ZnO films by Sankur and Cheung (with a pulsed CO_2 laser) [232] and by Nakayama (with pulsed nitrogen and dye lasers) [233]. Sankur and Cheung observed c-axis–perpendicular growth of ZnO on glass, Si, GaAs, SiO_2, Au, Ti, and sapphire substrates. No evidence of any preferred in-plane alignment was reported (and none is expected for the amorphous substrates) [232].

There has been considerable interest in growing highly c-axis–oriented crystalline ZnO films at the lowest possible substrate temperature, both for device applications that require compatibility with other heat-sensitive materials, and also to coat difficult-to-heat curved surfaces or long lengths of fiber [128, 129]. Several groups recently grew highly c-axis–oriented ZnO films by PLD at low temperatures; both ZnO and metallic Zn targets were used, in vacuum or low-pressure oxygen ambients [128, 129, 234–240]

Srikant et al. also recently reported detailed studies of the epitaxial growth relationships, defect equilibria, and electrical properties for epitaxial ZnO films grown on four different crystallographic orientations of sapphire [227, 241] These studies were motivated by needs for fully in-plane aligned ZnO films in order to study anisotropic properties of ZnO; oriented ZnO buffer layers for epitaxial heterostructure growth; and bicrystal films with controlled misorientations for studies of the effect of piezoelectricity on ZnO varistor behavior [241].

Low Temperature Growth of Highly Textured ZnO. Striking results were obtained by Ianno et al. [128] who studied how the crystalline orientation and stoichiometry of ZnO films varied with the ablating laser's wavelength and fluence. A wavelength/fluence combination was discovered that permits growth of stoichiometric and highly (002)-textured ZnO films on Si(001) substrates even at room temperature. Nd:YAG (both 1064 nm and 532 nm) and KrF (248 nm) lasers (10 ns pulse width) were operated at 10 Hz to ablate a rotating polycrystalline ZnO sputtering target in vacuum ($\sim 10^{-5}$ torr). The Si(001) substrate size and the film thickness were kept nearly constant so that meaningful comparisons of films could be made from XRD and AES data.

For deposition at room temperature with a fluence of 1.0 J/cm^2, Ianno et al. found that 1064 nm radiation produced a dull gray film containing (101)-oriented metallic zinc. In contrast, 532 nm radiation produced a transparent film containing (101)-oriented ZnO (the dominant powder diffraction peak of ZnO). KrF excimer (248 nm) irradiation produced a transparent film containing the desired (002)-oriented ZnO, together with a significant fraction of randomly oriented or amorphous material [128]. For deposition at room temperature, they also found clear correlations between the presence of Zn$^+$ in the ablation plume (revealed by optical emission spectra), the laser wavelength, and the orientation and oxygen content of ZnO films (measured by XRD and AES depth profiling, respectively). Strong Zn$^+$ ion emission and highly oxygen-deficient films (O/Zn $\sim$ 0.49) were found for 1064 nm irradiation, with weaker Zn$^+$ emission and a smaller oxygen deficiency (O/Zn $\sim$ 0.80) being found for 532 nm irradiation. In contrast, ZnO films deposited using 248 nm radiation were stoichiometric (O/Zn $\sim$ 0.98) and no Zn$^+$ emission was seen.

These striking differences result strictly from the strong dependence of the ZnO target's absorption coefficient on photon energy because the laser fluences and pulse durations were identical [128]. ZnO's absorption length, α^{-1}, is large for below-gap radiation (1064 nm and 532 nm), but it is much shorter at 248 nm (hν = 5.0 eV). These differences produce the wavelength-dependent changes in the energy deposition profile and the ablation process that were discussed in connection with Figs. 6–8. The high particulate density for 1064 nm ablation of ZnO was accompanied by a high deposition rate of 3.0 Å per laser pulse, whereas at 248 nm there were few particulates and a deposition rate of only 1.0 Å/pulse. Thus the strong wavelength-dependence of ZnO film stoichiometry and crystalline orientation at room temperature clearly results from the large differences in the ablated material arriving at the substrate [128].

Ianno et al. also found that the oxygen content increased with increasing substrate temperature for films deposited using 1064 nm and 532 nm radiation [128]. ZnO films deposited using these longer wavelengths also became stoichiometric and (002)-oriented for T $\geq$ 300°C.

For 248 nm ablation at 1.0 J/cm^2, XRD revealed a rapid increase in the amount of (002)-oriented material as the temperature was raised above room temperature and the FWHM of the ZnO(002) diffraction peak also narrowed. Highly (002)-textured films and a nearly constant FWHM of 0.477° (in 2ϑ scans) were obtained for temperatures of 200°C to 300°C.

However, Ianno et al. also discovered that for room-temperature deposition of ZnO at 248 nm, the (002) FWHM actually narrowed with decreasing laser fluence and the amount of (002)-oriented material simultaneously increased dramatically [128]. No oriented material was obtained at 2.0 J/cm^2, but this changed to ~80 (002) counts/sec and FWHM (2ϑ scan) = 0.52° at 1.0 J/cm^2, and to ~2000 counts/sec and FWHM = 0.31° at 0.5 J/cm^2. The latter is an even smaller FWHM than was obtained at elevated temperatures using 248 nm radiation at 1.0 J/cm^2. Thus, for 248 nm ablation, the *energy fluence* controlled *the amount of ZnO film texture at room temperature.*

Ianno et al. carried out similar experiments using reduced fluences at 532 nm and 1064 nm in order to determine whether the improved 248 nm texture with decreasing E_d resulted simply from the reduced deposition rate (which decreased linearly with fluence). However, reducing the deposition rate had no effect on film quality at the longer laser wavelengths [128]. Thus these experiments demonstrate that a low energy fluence and the characteristics of ablation plumes produced by short-wavelength radiation are necessary to grow highly textured ZnO films at room temperature.

Optimum ZnO Growth Conditions. Craciun et al. grew ZnO films on Si and Corning glass substrates using a polycrystalline ZnO target [129, 235, 236]. In agreement with Ianno et al., they found that short-wavelength (UV excimer) laser irradiation always produced better quality films than did the longer Nd:YAG wavelengths. Craciun et al. also investigated the effect of substrate temperature and ambient oxygen pressure on the position and FWHM of the ZnO (002) XRD peak in 2ϑ scans. Using a KrF laser at 2.1 J/cm^2 (near their optimal E_d value; see Fig. 8), they found that the best ZnO films were deposited at temperatures of 300°C to 375°C and in oxygen pressures of a few mtorr [129, 235]. These conditions are similar to those used by Zheng and Kwok to grow conducting indium tin oxide (ITO) films by PLD [242]. For ITO growth, it had been suggested that an oxygen pressure of a few mtorr was beneficial for film structure because it produced a more uniform velocity/kinetic energy distribution for ablated species, while still providing sufficient kinetic energy to enhance surface mobility and assist crystallization [114, 242]. However, ambient oxygen also strongly affects the films" electrical conductivity, as discussed below.

Textured ZnO Growth on Silica. Hu et al. used PLD of a metallic Zn target in oxygen to deposit highly c-axis–oriented ZnO films on fused silica substrates [234]. Their work was motivated by interest in integrating optical waveguiding films onto a Si motherboard, which requires growth of the highly textured waveguiding film on a planar silica (SiO_2) coating. The pronounced tendency of hexagonal ZnO to grow c-axis perpendicular, together with its interesting piezoelectric properties, makes it a candidate waveguiding material [243].

A metallic Zn target, considered purer and easier to fabricate than ZnO, was irradiated by a KrF laser at 3 Hz and 1.0 J/cm^2 using a target-substrate separation of 3.5 cm. A reactive oxygen ambient pressure of 750 mtorr, much higher than for the experiments described above using ZnO targets, was necessary to obtain pure ZnO films. Lower oxygen pressures produced a metallic Zn impurity, whereas higher pressure caused the ablation plume to shorten and film uniformity to deteriorate. ZnO(002) FWHM XRD measurements were used to locate an optimum temperature of 375°C–425°C for deposition on fused silica substrates. However, note that all of these ZnO films also were post-annealed in 0.5 atm oxygen for 30 minutes at the deposition temperature. SEM images showed that the ZnO films were composed of a network of c-axis–oriented grains with in-plane grain dimensions of ~0.3 µm. The refractive index of 1.2 µm thick films was ~5% less than for bulk ZnO; some optical waveguiding capability was demonstrated [234].

Craciun *et al.* also reported PLD of textured ZnO films on GaAs(001), both with and without a thin SiO_2 buffer layer [236]. Earlier work using sputtered ZnO films had established that a SiO_2 buffer relieves stress and promotes growth of highly (002)-oriented textured ZnO films [244–246]. A comparison of ZnO films grown on GaAs (lattice constant $a = 5.653$ Å) under the optimum deposition conditions of Craciun *et al.* (see above) showed that with an SiO_2 buffer layer the ZnO(002)/GaAs(002) XRD intensity ratio increased by more than a factor of four, the ZnO(002) rocking curve FWHM decreased from 3.97° to 2.19°, and the ZnO c-axis strain decreased from 1.03% to 0.26%. Thus the SiO_2 buffer layer significantly improved the crystallinity and reduced the strain in PLD ZnO films.

Electrical Conductivity and Stress in PLD ZnO Films. The n-type electrical conductivity of ZnO can be modified over several orders of magnitude by doping with column-III and -IV elements [225–227]. Conduction in ZnO also is controlled by interstitial Zn atoms and O vacancies [225, 229, 230] and in heavily Al-doped ZnO by Zn vacancy diffusion [227].

Using the optimized conditions described above, Craciun *et al.* grew ZnO films on glass, Si, and SiO_2/GaAs substrates for which the ZnO(002) FWHM was ~0.15° and the average optical transmittance in the visible region was ~85%. The films' room temperature electrical conductivity varied with the oxygen pressure used during growth, from ~40 $(\Omega\text{-cm})^{-1}$ for 1.3 mtorr of O_2 to ~100 $(\Omega\text{-cm})^{-1}$ in 2×10^{-5} torr of oxygen [129, 235].

Cheung and co-workers recently investigated PLD ZnO films for two different applications, as thin-film acoustic resonators used in RF filters [237] and as gas sensors to be used in life-support systems on long-duration space missions [238, 239]. For the former application, it is necessary to avoid the compressive stress typically found in sputtered piezoelectric films; whereas for gas sensors, long life, fast response, and high sensitivity to gas composition are required.

Cheung et al. used a low-pressure (15–20 mtorr) O_2 ambient to grow highly c-axis–oriented ZnO films on a variety of single-crystal, polycrystalline and amorphous substrates. XRD spectra showed that films grown at $E_d = 0.8$ J/cm^2 and T ~ 150°C were c-axis oriented, unstressed, and remained undistorted when released from the substrate, whereas sputtered films contained ~0.1% compressive strain and bowed when released [237]. By introducing ammonia directly into the ablation plume, they were able to dope normally n-type ZnO films with N atoms, compensating the intrinsic donors (O vacancies) and reducing their films' conductivity. Because of difficulties in forming ohmic contacts, AC measurements (10^4–10^7 Hz) were used to characterize film conductivity. Films grown in pure O_2 were quite lossy, but those doped with N at ~5 × 10^{18} cm^{-3} produced capacitive structures in which the ZnO dielectric loss tangent was substantially <0.1 throughout the 10^4 Hz to 10^7 Hz frequency range.

For gas sensors, Cheung and Johnson used $E_d = 2$ J/cm^2 to grow transparent, c-axis–oriented ZnO films even at room temperature [238]. Films grown in ~20 mtorr O_2 were n-type (n~10^{19}–10^{20} cm^{-3}) due to O vacancies. Nitrogen doping, using ammonia, reduced the electron density and produced films with resistivities 3 to 6 orders of magnitude higher. Cheung and Johnson found that the fastest response times for oxygen-gas detection were obtained using these very thin and highly resistive (N-doped) ZnO films. The explanation given is that the rate-limiting step in detector response is oxygen diffusion along grain boundaries; the substitutional replacement of oxygen atoms by nitrogen in ZnO causes internal stress that is believed to enhance the oxygen diffusion rate [238].

These applications demonstrate that PLD is able to produce low-stress, highly c-axis–oriented, high resistivity ZnO films at low deposition temperatures. The high resistivity results from using nitrogen doping to compensate intrinsic oxygen-vacancy donors and film properties are stable over time [237, 238].

Epitaxial ZnO Films on Sapphire: In-plane Alignment. Srikant et al. recently reported PLD of fully oriented (in-plane aligned) ZnO films on sapphire substrates of four different orientations (C-, A-, R-, and M-plane) [227, 241]. A tripled Nd:YAG laser (355 nm wavelength) was used to ablate a stoichiometric ZnO target doped with 1 at.% Al. Operating at 10 Hz, $E_d = 6$ J/cm^2 and $D_{ts} = 5.5$ cm produced a deposition rate of ~2 nm/s (~2 Å/pulse). The sapphire substrates were chemically cleaned and annealed in flowing oxygen at 1400°C for 4 hours prior to growth.

ZnO epitaxial growth was investigated for temperatures of 250–800°C and in oxygen pressures ranging of 10^{-5} torr to 10 mtorr. In striking contrast to the low-temperature growth experiments that were summarized above, the highest quality films (as determined by narrowing of X-ray ω-scan line widths) were obtained at growth temperatures near 750°C in an oxygen pressure of 10 mtorr. For C- and A-plane sapphire, ZnO grew with the c-axis perpendicular to the substrate under all growth conditions. For the M- and R-orientations, ZnO films grew with the c-axis either parallel or perpendicular to the substrate, depending on the substrate temperature and oxygen pressure. However, in all cases except the M-plane at high temperatures, a single unique in-plane relationship always was found between the sapphire substrate and the ZnO film.

All of the films, though epitaxial, contained a slight mosaic structure consisting of oriented grains separated by low-angle ($\sim 1°$) grain or sub-grain boundaries. This characteristic is important to understanding their electrical properties, as described in the next section.

Epitaxial ZnO Films on Sapphire: Electrical Properties. Numerous device applications have spurred research aimed at optimizing the electrical and optical properties of Al-doped n-ZnO films deposited on glass substrates [228, 247]. However, little information was available regarding the electrical properties of fully epitaxial Al-doped ZnO films grown on sapphire until the work by Srikant et al. [227]. These investigators carried out Hall effect measurements of the electron concentration and mobility for ~ 0.12 μm thick ZnO films that were grown at 600°C and in 10 mtorr oxygen on sapphire substrates of four different crystal orientations, i.e., not far from the optimum conditions determined by the structural studies described in the preceding section.

A key discovery by Srikant et al. was that, *because of the high energies* of ablated species, their as-grown ZnO films *initially contained defect concentrations, and hence carrier concentrations, that greatly exceeded thermal equilibrium values* [227]. To carry out a *near-equilibrium* analysis of carrier concentration and mobility data, their films were annealed for 5 hours at 750°C in various oxygen pressures. Annealing produced large decreases in carrier concentration (e.g., by an order of magnitude or more) but the resulting equilibrium carrier concentrations still were approximately two orders of magnitude larger than those obtained in undoped ZnO at the same oxygen partial pressures [248], i.e., the post-annealing carrier concentrations were dominated by the Al donors [227].

As shown in Fig. 17, equilibrium carrier concentrations were the same in ZnO films grown on C-, A-, and M-plane sapphire, but were lower for growth on R-plane sapphire. Srikant et al. attributed this difference to a lower incorporation of Al atoms during growth on the R-plane. They also studied the carrier mobility of ZnO films as a function of the oxygen partial pressure used during annealing at 750°C. As shown in Fig. 17, the slope of the carrier mobility was "single crystal"-like for carrier concentrations greater than 2×10^{19} cm^{-3}, but the electron mobility decreased dramatically for lower concentrations.

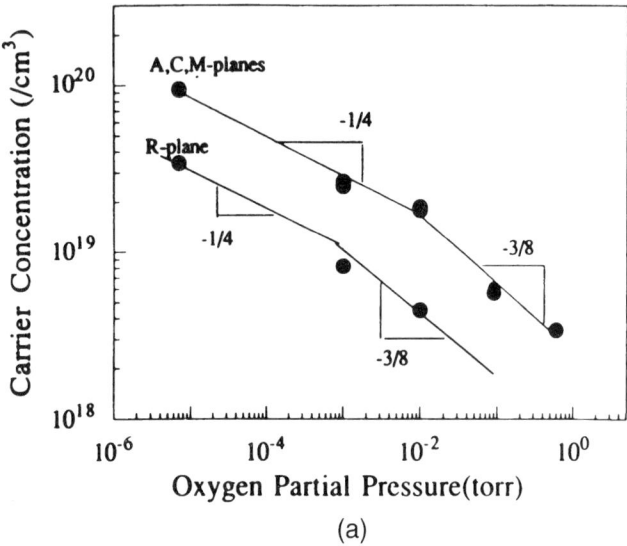

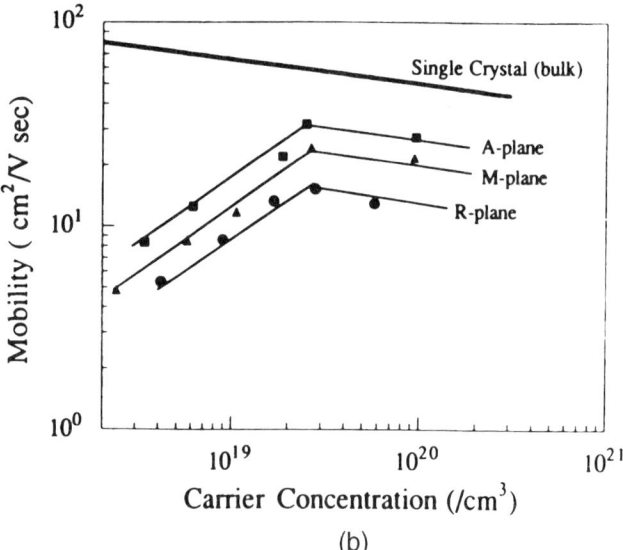

FIG. 17. (a) Equilibrium electron concentration vs. oxygen partial pressure after annealing as-grown ZnO films at 750°C. (b) Electron mobility vs. electron concentration in ZnO films annealed at 750°C [227, with permission].

To model the oxygen pressure-dependent carrier concentrations (Fig. 17(a)), Srikant et al. assumed that the mosaic structure of epitaxial ZnO films was the result of small-angle grain boundaries that behave as potential barriers to in-plane electrical transport [227]. By considering the various possible defect reactions that can occur during oxygen annealing and applying electroneutrality conditions to them, Srikant et al. were able to infer that the observed $p^{-1/4}$ and $p^{-3/8}$ oxygen partial pressure dependences (Fig. 17(a)) are due to Zn vacancy diffusion as the rate-controlling reaction step [227]. The overall picture is that when the oxygen partial pressure is changed, transport along grain boundaries is rapid and equilibration then is controlled by the diffusion of singly and doubly ionized Zn vacancies from the free surface and from the grain boundaries into the grains. Their model results in a crossover of the mobility from "single crystal" to "polycrystalline" behavior, depending on whether the potential barrier height at the grain boundaries is smaller or larger than a critical value [227].

Srikant et al. also argue that the absolute differences in the electron mobility of ZnO films grown on different crystal planes of sapphire (Fig. 17(b)) cannot be due to the grain boundaries, but must result from differences within the grains themselves. They attribute the mobility differences to different electron scattering rates by different dislocation densities associated with the different lattice mismatches of ZnO to A-, M-, and R-plane sapphire.

11.3.5.2 Other Transparent Conducting Oxides.

Indium Tin Oxide (ITO). Zheng and Kwok grew highly conducting ITO films by ArF (193 nm) laser ablation of a sintered ceramic In_2O_3(90 wt.%)/SnO_2(10 wt.%) target at $E_d = 1$ J/cm^2 in a low-pressure oxygen ambient [115, 249]. Under optimum deposition conditions, they obtained films with conductivities of 7.1×10^3 (Ω-cm)$^{-1}$ and 1.8×10^3 (Ω-cm)$^{-1}$ at 310°C and room temperature, respectively. When a surface scattering contribution to the resistivity was subtracted, the equivalent bulk conductivity for the room-temperature ITO film was $\sim 3.6 \times 10^3$ (Ω-cm)$^{-1}$, the highest conductivity value reported for any TCO film at room temperature without post-annealing [249]. Optical transmittance in the visible region was >90% for films 150 nm thick and was nearly 100% in the 600–800 nm region [115, 249].

A striking result for the ITO films grown at room temperature was that the highest electrical conductivity was obtained only in a narrow oxygen pressure range near 15 mtorr, for a target-substrate separation of 7 cm. The authors attributed this to the pressure-distance scaling law discussed in Section 11.2.6.4, through which gas-phase collisions reduce the kinetic energies of incident species to values that are optimal to assist low-temperature film growth. In contrast, Zheng and Kwok found that the ITO film conductivity was independent of oxygen pressure at high deposition temperatures [115, 249]. This is consistent with the idea that, at high temperatures, the thermal energies of adatoms are sufficient for diffusion on the growing film surface and for stoichiometric film formation without any energetic-beam "assist." Zheng and Kwok also suggested that for low-temperature film growth, the oxygen content of ITO films must have come mainly from the ablation target because energetic ablated oxygen atoms and ions are much more reactive than ambient O_2 [115]. Jia et al. also grew ITO films on InP using similar PLD methods [250].

Tin Oxide. Dai, Su, and Chuu used a pulsed Nd:glass (1.06 μm) laser operating at 1–10 Hz and focused to a high fluence of 20–150 J/cm^2 to deposit transparent, conducting SnO_2 films onto Pyrex glass and polished aluminum substrates, in vacuum and at room temperature [251]. Undoped SnO_2 pressed-powder targets were used at D_{ts} = 2–3 cm. The high laser fluence and close spacing produced very high deposition rates of 24 (34) Å/pulse on Pyrex (aluminum); a total of 500 laser shots were used per film. Despite the high deposition rate per pulse, the films had a columnar cross-sectional morphology and (002) texture, but they were decorated with numerous small (~0.2 μm) grains at the surface. The films' resistivity and visible transmittance were 3.0×10^{-3} Ω-cm and ~75%, respectively, slightly worse than for rf-sputtered SnO_2 films deposited on unheated substrates ($2–3 \times 10^{-3}$ Ω-cm, 80–85%). Auger electron spectroscopy indicated that the PLD SnO_2 films were nearly stoichiometric, which the authors attributed to the high laser fluence. Perhaps the most interesting aspect of this work is its demonstration that PLD is able to produce SnO_2 films with reasonably good optical transmittance and low resistivity without substrate heating or a post-annealing treatment.

GaInO$_3$. Phillips et al. recently reported the properties of $GaInO_3$ films deposited on glass and quartz substrates by both PLD and dc reactive sputtering [223]. $GaInO_3$ ablation targets were prepared by standard ceramic processing methods; doped targets were produced by substituting up to 10% Sn for In or Ge for Ga. A KrF excimer laser was operated at 10 Hz and 2 J/cm^2. The PLD films had a slightly granular microstructure but no evidence of precipitates for dopant concentrations ≤10%; sputtered films were smoother but contained precipitates for 15% Sn doping. The index of refraction of $GaInO_3$ was found to be $n = 1.65 \pm 0.1$, which is considerably closer to the index of glass ($n \sim 1.5$) than ITO ($n = 2.0$). The absorption coefficient of $GaInO_3$ was much lower over the entire visible range than the absorption coefficients of state-of-the-art TCOs such as commercial ITO, ZnO:Al or SnO_2:F. $GaInO_3$ films ~1 μm thick had transmissions as high as 90% throughout the visible region.

$GaInO_3$ films with the highest conductivity (~345 [Ω-cm]$^{-1}$) and nearly the best transparency were grown at 250°C in an oxygen pressure of 1 mtorr [223]. This conductivity is still 3 to 10 times less than the conductivity of the best TCO films (see below). Temperature-dependent resistivity together with room-temperature Hall, magnetoresistance, and thermoelectric measurements showed that $GaInO_3$ is an electron conductor with a metallic temperature coefficient of resistance (~1.5 μΩ-cm/K) near room temperature that reverses sign below about 120 K. Carrier densities in the $1 \times 10^{19} - 4 \times 10^{20}$ cm^{-3} range were inferred, which are similar to or slightly lower than for other TCOs. A post-growth H$_2$ anneal produced an optimized oxygen content and reduced resistivity. The removal of oxygen had a larger effect on conductivity than changing the concentration of the cation dopant, from which it was inferred that doping of GaInO$_3$ is due to both oxygen vacancies and cation substitution. Phillips *et al.* point out that because the optical transmission characteristics of GaInO$_3$ are significantly better than other TCOs, somewhat thicker films can be used to compensate for its higher resistivity [223].

Zinc-Indium-Oxide. Motivated by the need for a material with transparency similar to that of GaInO$_3$ but with higher electrical conductivity, Phillips *et al.* also investigated zinc-indium-oxide [224]. They found that PLD zinc-indium-oxide films have electrical conductivity similar to that of ITO, but better transparency in both the visible and infrared. Greatly superior transmission in the 1.0–1.5 μm range makes zinc-indium-oxide especially attractive for transparent electrodes on infrared devices [224].

PLD was carried out using ceramic Zn$_x$In$_{2-y}$Sn$_y$O$_{3+x+0.5y-\delta}$ targets in which Sn is a dopant and $1.4 \leq x \leq 4$ and $0 \leq y \leq 0.4$. Al, Ga, and Ge also were used as dopants. As for GaInO$_3$, the best zinc-indium-oxide films were grown at 250°C in 1 mtorr of oxygen. X-ray diffraction revealed that these zinc-indium-oxide films (and those deposited at room temperature) were amorphous with no long-range order greater than ~2 nm [224].

Phillips *et al.* also found that zinc-indium-oxide films deposited by either PLD or reactive sputtering contained less Zn than did the targets, except for films deposited at room temperature. For PLD, a Zn$_{2.5}$In$_{1.9}$Sn$_{0.1}$O$_{5.55-\delta}$ target (with Zn/[In + Sn] = 1.25) produced a film with a Zn/[In + Sn] ratio of only 0.6, for deposition at 250°C in 1 mtorr O$_2$. A slightly lower ratio of 0.5 was obtained for sputter deposition at 450°C in 0.5 mtorr O$_2$. In both cases, the films' conductivities were ≥ 2500 (Ω-cm)$^{-1}$. The significance of this departure from stoichiometric deposition is that the conductivity of zinc-indium-oxide films was found to depend strongly on the Zn/[In + Sn] ratio, and was sharply peaked at ~2500 (Ω-cm)$^{-1}$ near a value of 0.5–0.6 for this ratio.

Zinc-indium-oxide films exhibited metallic resistivity behavior below room temperature with an increase in resistivity below 100 K, very similar to the behavior of GaInO$_3$. Because high conductivities were observed even in films containing no Sn, the authors inferred that oxygen vacancies are donors in zinc-indium-oxide, as they are in GaInO$_3$ and ITO. Because the optical transmission characteristics of zinc-indium-oxide films were significantly better than those of state-of-the-art TCOs, while their electrical conductivity approached that of ITO, Phillips *et al.* suggested that thicker zinc-indium-oxide films can be used to reduce sheet resistance while preserving high optical transmission. In particular, zinc-indium-oxide is the preferred TCO material for transparent electrodes to be used in the 1.0–1.5 µm near-infrared region [224].

Cd₂SnO₄. J. McGraw et al. recently reported the properties of PLD amorphous and crystalline Cd_2SnO_4 films that were deposited on optical grade quartz substrates by ablating a hot-pressed ceramic target into ambient oxygen or nitrogen gas at 400 mtorr pressure [252]. A KrF laser was used with E_d = 1–2 J/cm², 5 Hz repetition rate, and D_{ts} = 7 cm. For a deposition time of 15 minutes, film thicknesses ranged from ~380 nm at room temperature to ~150 nm at 550°C. Films deposited in oxygen at room temperature and in nitrogen at 500°C were amorphous, but all films grown in oxygen at $T \geq 500°C$ were crystalline with the cubic spinel structure. Crystalline films grown in oxygen in the 500–550°C range had similar carrier concentrations (4.2–5.6×10^{20} cm^{-3}), mobilities (28.7–34.9 cm²/V-s), and resistivities (3.2–4.9×10^{-4} Ω-cm). Interestingly, the amorphous film grown in nitrogen at 500°C had the highest mobility, 44.7 cm²/V-s, and a carrier concentration of 2.9×10^{20} cm^{-3}.

Optical properties of Cd_2SnO_4 films were studied because of interest in using them to filter heat (infrared) radiation in thermophotovoltaic cells [252]. Films for this application must be highly transparent from their energy bandgap (E_g ~ 3.1 eV, λ_g ~ 400 nm for crystalline Cd_2SnO_4) out to the plasma edge wavelength. The position of the plasma edge is critical to filter heat radiation and is controlled primarily by the carrier concentration. As expected, the plasma edges were steeper and occurred at shorter wavelengths for crystalline Cd_2SnO_4 films than for the amorphous films with lower carrier concentrations. A crystalline film deposited at 500°C had an IR reflectivity $\geq 80\%$ for wavelengths ≥ 3 µm and an average transmissivity $\geq 5\%$ in the visible to near-IR region, making Cd_2SnO_4 a candidate for both plasma filter and transparent electrical contact applications. The authors also compared the electrical properties of PLD Cd_2SnO_4 films with those grown by electroless deposition and sputtering and found that the higher mobility and comparable-to-better carrier concentrations and resistivity of PLD films made them superior overall [252].

11.3.6 Concluding Remarks

We close this chapter by summarizing what has been learned about the optimal conditions for PLD of epitaxial compound semiconductor films. Some suggestions also are given regarding the types of information and experiments that are needed to advance PLD of compound semiconductors.

11.3.6.1 Optimum Deposition Conditions. Methods to minimize or eliminate particulates and to obtain films with uniform thickness and composition were discussed in Sections 11.2.7.2 and 11.2.8. Here we focus on the optimal pulsed laser and ambient gas conditions.

The pulsed laser beam should be passed through an aperture before focusing it in order to remove its low-energy fringe and produce a well-defined energy density. This also helps to minimize particulate production. As described in Section 11.2.7.2 the target should be rotated to prevent cumulative local damage, and the laser beam angle of incidence should be reversed periodically to counteract cone formation.

The laser beam should be focused to only low-to-moderate energy density (generally ≥ 1 J/cm^2) in order to minimize particulate production by subsurface boiling ("spitting"). However, the acceptable upper limit on E_d is expected to scale with the thermal diffusivity of the target material, so it could be higher. The lower limit on E_d is set by the requirement that a well-defined laser-generated plasma should be formed on each pulse to ensure near-stoichiometric transfer; convenience and the possibility of pulse-to-pulse energy variations require operating well above the minimum E_d.

PLD's average deposition rate typically exceeds MBE or ALE growth rates by about an order of magnitude [253]. However, the biggest difference between PLD and continuous-deposition methods is that PLD is characterized by brief periods of high "instantaneous" flux that are followed by relatively long periods in which there is little or no incident flux (other than slow-moving particulates, if present). Depending on ambient gas conditions, the energetic part of the ablation plume is delivered to the surface in time intervals ranging from ~ 10 μsec to <1 msec. Hence, instantaneous deposition rates for PLD can range from 10^3–10^5 Å/sec, approximately four orders of magnitude higher than for MBE. Consequently, when PLD is used for low-temperature epitaxial growth, it may be necessary to keep the deposition rate per pulse fairly low (probably <0.1 Å/pulse) to allow time for atoms to diffuse to the correct crystalline lattice sites without being buried by other atoms in the same pulse. (For example, Shen and Kwok found that a low deposition rate of ~ 0.05 Å per laser pulse was necessary to obtain the best epitaxial CdS films at low substrate temperature and low ambient pressure [116].) For epitaxial growth at the lowest possible substrate temperature, the pulse repetition rate (PRR) also should be relatively low (probably ≤ 10–20 Hz) to provide adequate time for surface migration of atoms between the arrival of successive pulses. The maximum PRR depends on the substrate temperature, but a PRR of 10–20 Hz usually maintains a reasonable average deposition rate.

If it is necessary to increase the deposition rate per pulse, this can be accomplished by increasing both the pulse energy and the focused spot size. This ablates more material and also changes its angular distribution (see Section 11.2.5.3) but it maintains a constant E_d so that particulate production and plasma conditions should not change much. For this purpose, a laser with a high pulse energy may be required.

For epitaxial growth in vacuum, there is considerable evidence that the most energetic atoms/ions produce lattice displacement damage and point defects that degrade films' electrical properties [29, 42, 43, 114–116, 170]. Possible solutions to the damage problem while continuing to work in high vacuum are not established at this time but may involve (1) using a large target-substrate separation, D_{ts} [116], or (2) tilting the substrate heater face so that the ablation beam's angle of incidence is large and the momentum transfer per incident atom is below the threshold for lattice displacement damage.

For epitaxial growth in an ambient gas and at the lowest possible substrate temperature, several investigators have demonstrated that there is a pressure-distance scaling law that defines an optimal D_{ts} value for each gas pressure [29, 42, 43, 114–116, 170]. The physical basis for this relationship is simply that a sufficient number of gas-phase collisions will reduce the kinetic energies of incident species to below the threshold energy (or momentum) transfer for lattice displacement damage. On the other hand, if the substrate is too far from the target or the gas pressure is too high, the moderately energetic atoms and ions that can assist low-temperature growth and doping processes are no longer present in the incident flux. However, if the substrate temperature is increased, the presence of moderately energetic incident species is less important for crystalline phase formation. Thus increasing D_{ts} or the ambient gas pressures has little effect; the pressure-distance scaling relationship is "washed out" for large values of either parameter. However, lattice damage still can be expected for low pressures and/or short D_{ts} (see Fig. 10), so there probably is a minimum value of the pressure-D_{ts} product to avoid degradation of film electrical properties. The experiments by Rouleau et al. [43] show that it may be necessary to maintain the absolute gas pressure below a few tens of mtorr to ensure two-dimensional growth (see Fig. 11). This suggests working at large D_{ts} values, consistent with the results of Shen and Kwok [116].

11.3.6.2 Future Research Needs. The experiments discussed in this chapter demonstrate that phase-pure, fully epitaxial doped compound semiconductor films can be grown by pulsed laser ablation. The excited ablation plume provides a distinctly different semiconductor growth environment than does MBE or CVD, resulting in low growth temperatures and unusual doping behavior in some cases [25]. Doping from the gas phase can be carried out without dc or rf plasma sources, and the ablation beam's kinetic energy can be moderated by gas-phase collisions, though there are indications (see above) that ambient pressures should not exceed a few tens of mtorr [28, 29, 42, 43]. Dopants can be incorporated in ablation targets, but questions remain regarding the fraction of the dopant atoms that are incorporated in films and/or electrically activated [170, 173].

The most serious remaining question about epitaxial semiconductor growth by laser ablation comes from evidence that although ablation preserves stoichiometry to a high degree, it does not prevent compositional errors of order 1%, which in semiconductors generally produces massive numbers of native point defects (vacancies, interstitials) that control and/or compensate doping.

One of the principal needs now is to investigate in situ methods to supply the deficient species (e.g., Se in CIS or Zn in ZnTe and Zn-In-O) in order to control the electrical properties of as-grown compound semiconductor films. The difficulty of interpreting electrical properties measurements for the polycrystalline CIS and CIGS films deposited by several groups (see Section 11.2.2) emphasizes the importance of growing and studying epitaxial films. Epitaxial films may have only limited relevance to potentially low-cost polycrystalline devices, but they are extremely valuable in understanding intrinsic limitations of the PLD process, as well as the intrinsic properties of complex semiconductor materials that are free of grain boundary effects. Photoluminescence measurements of defect and dopant energy levels and temperature-dependent Hall effect and resistivity measurements are needed to fully understand doping and defect behavior.

Finally, there is a continuing need for studies of the ablation environment using in situ diagnostics with temporal, spatial, chemical, and mass resolution. These studies are needed both in the gas phase (e.g., emission and absorption spectroscopy, ion probe, ICCD-array camera, and TOFMS measurements) as well as on the growing film surface. For the latter, photon-based measurements such as time-resolved ellipsometry and time-resolved synchrotron X-ray diffraction provide new opportunities to study growth mechanisms and the evolution of crystalline structure.

The combination of in situ diagnostics with electronic properties measurements will allow us to develop a fundamental understanding of the non-equilibrium growth and doping of compound semiconductors in the energetic ablation environment.

Acknowledgments

It is a pleasure to acknowledge assistance provided by C. M. Rouleau and D. B. Geohegan in preparing the figures, and by B. J. Shoopman in preparing the reference list. This chapter was prepared at Oak Ridge National Laboratory, managed by Lockheed Martin Energy Research Corp. for the U.S. Department of Energy under contract DE-AC05-96OR22464.

The submitted manuscript has been authored by a contractor of the U.S. Government under contract DE-AC05-96OR22464. Accordingly, the U.S. Government retains a nonexclusive, royalty-free license to publish or reproduce the published form of this contribution, or allow others to do so, for U.S. Government purposes.

References

1. D. Dijkkamp, T. Venkatesan, X. D. Wu, S. A. Shaheen, N. Jisrawi, Y. H. Min-Lee, W. L. McLean, and M. Croft. Preparation of Y-Ba-Cu oxide superconductor thin

films using pulsed laser evaporation from high-T_c bulk material. *Appl. Phys. Lett.* **51**, 619 (1987).
2. X. D. Wu, D. Dijkkamp, S. B. Ogale, A. Inam, E. W. Chase, P. F. Micelli, C. C. Chang, J. M. Tarascon, and T. Venkatesan. Epitaxial ordering of oxide superconductor thin films on (100) $SrTiO_3$ prepared by pulsed laser evaporation. *Appl. Phys. Lett.* **51**, 861 (1987).
3. D. B. Chrisey and G. K. Hubler, eds. *Pulsed Laser Deposition of Thin Films*. John Wiley and Sons, New York, 1994.
4. D. C. Paine and J. C. Bravman, eds. Laser ablation for materials synthesis. *Mat. Res. Soc. Symp. Proc.* **191**, Materials Research Society, Pittsburgh, 1992.
5. B. Braren, J. J. Dubowski, and D. P. Norton, eds. Laser ablation in materials processing: Fundamentals and applications. *Mater. Res. Soc. Symp. Proc.* **285**, Materials Research Society, Pittsburgh, 1993.
6. H. A. Atwater, J. T. Dickinson, D. H. Lowndes, and A. Polman, eds. Film synthesis and growth using energetic beams. *Mater. Res. Soc. Symp. Proc.* **388**, Materials Research Society, Pittsburgh, 1995.
7. R. Singh, D. P. Norton, L. D. Laude, J. Narayan, and J. Cheung, eds. Advanced laser processing of materials—Fundamentals and applications *Mater. Res. Soc. Symp. Proc.* **397**, Materials Research Society, Pittsburgh, 1996, in press.
8. *MRS Bulletin*, vol. 17, no. 2, February, 1992.
9. J. C. Miller and R. F. Haglund, Jr., eds. *Laser Ablation Mechanisms and Applications*. Springer-Verlag, New York, 1992.
10. J. C. Miller and D. B. Geohegan, eds.Second Intl. Conf. on Laser Ablation (COLA-93). *AIP Conf. Proc.* **288**, Amer. Inst. of Physics, New York, 1993.
11. E. Fogarassy, D. B. Geohegan, and M. Stuke, eds. Laser ablation—Mechanisms and applications III (COLA-95). *European Mater. Res. Soc. Symp. Proc.* **55**, 1996, in press.
12. E. Fogarassy and S. Lazare, eds. *Laser Ablation of Electronic Materials: Basic Mechanisms And Applications*. North-Holland, Amsterdam, 1992.
13. I. W. Boyd, ed. Proc. of Symp. E: Laser surface processing and characterization. E-MRS Spring Conference, Strasbourg, France, 1991. *Appl. Surface Sci.* **54**, 1–535 (1992). Also see earlier conferences in this series, published in *Applied Surface Science*.
14. J. T. Cheung and H. Sankur. Growth of thin films by laser-induced evaporation. *CRC Crit. Rev. in Solid State and Mater. Sci.* **15**, 63 (1988).
15. H. Sankur and J. T. Cheung. Formation of dielectric and semiconductor thin films by laser-assisted evaporation. *Appl. Physics* **A47**, 271 (1988).
16. J. J. Dubowski. Pulsed laser evaporation and epitaxy of thin semiconductor films. *Chemtronics* **3**, 66 (1988).
17. D. H. Lowndes. Growth of epitaxial thin films by pulsed laser ablation. In *Modern Topics in Single Crystal Growth*, L. A. Boatner (ed.). Amer. Inst. of Physics, New York, 1996, in press.
18. The INSPEC database was used to compile a bibliography for compound semiconductor growth by PLD. Although an effort was made to obtain a comprehensive listing, omissions may have occurred if the key words used in this search did not appear in an article's title or abstract. The author apologizes in advance to uncited authors for any such unintentional omissions.
19. K. L. Saenger. Bibliography of films deposited by pulsed laser deposition [3, pp. 581–604].

20. INSPEC database, 1990–95.
21. *Materials Science and Engineering for the 1990s: Maintaining Competitiveness in the Age of Materials*, p. 98. National Research Council, National Academy Press, Washington, D.C., 1989.
22. Research assistance task force report on *Photovoltaic Materials: Innovations and Fundamental Research Opportunities* (A. Zunger, ed.). *J. Electronic Mater.* **22**, 3–72 (1993).
23. A. Zunger, S. Wagner, and P. M. Petroff. New materials and structures for photovoltaics. *J. Electronic Mater.* **22**, 3 (1993).
24. H. J. Möller. *Semiconductors for Solar Cells*, Ch. 8. Artech House, Norwood, MA, 1993.
25. D. Feiler, R. S. Williams, A. A. Talin, H. Yoon, and M. S. Goorsky. Pulsed laser deposition of epitaxial AlN, GaN, and InN thin films on sapphire(0001). *J. of Crystal Growth* **171**, 12 (1997).
26. D. Feiler, R. S. Williams, A. A. Talin, H. Yoon, K. Matney, and M. S. Goorsky. Pulsed laser deposition of epitaxial GaN on sapphire(0001). *Applied Physics Letters*, 1996, in press.
27. J. W. McCamy, D. H. Lowndes, and J. D. Budai. Pulsed-laser ablation growth of epitaxial $ZnSe_{1-x}S_x$ films and superlattices with continuously variable composition. *Appl. Phys. Lett.* **63**, 3008 (1993).
28. D. H. Lowndes, C. M. Rouleau, J. W. McCamy, J. D. Budai, D. B. Poker, D. B. Geohegan, A. A. Puretzky, and Shen Zhu. Growth of highly doped p-type ZnTe films by pulsed laser ablation in molecular nitrogen [6, p. 85].
29. C. M. Rouleau, D. H. Lowndes, J. W. McCamy, J. D. Budai, D. B. Poker, D. B. Geohegan, A. A. Puretzky, and Shen Zhu. Growth of highly doped p-type ZnTe films by pulsed laser ablation in molecular nitrogen. *Appl. Phys. Lett.* **67**, 2545 (1995).
30. W. P. Shen and H. S. Kwok. P-type compound semiconductor thin films grown by pulsed laser deposition [6, p. 91].
31. W. P. Shen and H. S. Kwok. Highly doped p-type, n-type CdS thin films and diodes. In *New Materials for Advanced Solid State Lasers*, B. H. T. Chai, S. A. Payne, T. Y. Fan, and A. Cassanho (eds.), p. 173. Materials Research Society, Pittsburgh, 1994.
32. J. T. Cheung and J. Madden. Growth of HgCdTe epilayers with any predesigned compositional profile by laser molecular beam epitaxy. *J. Vac. Sci. Technol.* **B5**, 705 (1987). See also, J. T. Cheung. Bandgap engineering of HgCdTe by pulsed laser deposition [3, pp. 519–533].
33. D. B. Geohegan. Diagnostics and characteristics of pulsed laser deposition laser plasmas [3, pp. 115–165].
34. S. Witanachchi, H. S. Kwok, X. W. Wang, and D. T. Shaw. Deposition of superconducting Y-Ba-Cu-O films at 400°C without postannealing. *Appl. Phys. Lett.* **53**, 234 (1988).
35. R. K. Singh, J. Narayan, A. K. Singh, and J. Krishnaswamy. In situ processing of epitaxial Y-Ba-Cu-O high-T_c superconducting films on (100) $SrTiO_3$ and (100) yttria-stabilized ZrO_2 Substrates at 500–650°C. *Appl. Phys. Lett.* **54**, 2271 (1989).
36. T. Venkatesan, X. D. Wu, A. Inam, and J. B. Wachtman. Observation of two distinct components during pulsed laser deposition of high-T_c superconducting films. *Appl. Phys. Lett.* **52**, 1193 (1988).
37. C. T. Rettner, L. A. DeLouise, and D. J. Auerbach. Effect of incident kinetic energy

and surface coverage on the dissociative chemisorption of oxygen on W(110). *J. Chem. Phys.* **85**, 1131 (1986).
38. A. V. Hamza, H.-P. Steinruck, and R. J. Madix. The dynamics of the dissociative adsorption of alkanes on Ir(110). *J. Chem. Phys.* **86**, 6506 (1987).
39. J. E. Greene, S. A. Barnett, J.-E. Lundgren, and A. Rockett. In *Low-Energy Ion/Surface Interactions During Film Growth*, T. Itoh (ed.), Ch. 5. Elsevier, Amsterdam, 1988.
40. C. E. Otis and R. W. Dreyfus. Laser ablation of $YBa_2Cu_3O_{7-x}$ as probed by laser-induced fluorescence spectroscopy. *Phys. Rev. Lett.* **67**, 2102 (1991).
41. D. B. Geohegan and A. A. Puretzky. Collisional effects of background gases on pulsed laser deposition plasma beams [6, p. 21].
42. D. H. Lowndes, C. M. Rouleau, D. B. Geohegan, A. A. Puretzky, M. A. Strauss, A. J. Pedraza, J. W. Park, J. D. Budai, and D. B. Poker. Pulsed laser ablation growth and doping of epitaxial compound semiconductor films [7, in press].
43. C. M. Rouleau, D. H. Lowndes, M. A. Strauss, S. Cao, A. J. Pedraza, D. B. Geohegan, A. A. Puretzky, and F. Allard. Effect of ambient gas pressure on pulsed laser ablation plume dynamics and ZnTe film growth [7, in press].
44. R. A. Zuhr, S. J. Pennycook, T. S. Noggle, N. Herbots, T. E. Haynes, and B. R. Appleton. Ion beam deposition in materials reearch. *Nucl. Instr. and Methods in Physics Res.* **B37/38**, 16 (1989).
45. X. D. Wu, R. E. Muenchausen, S. Foltyn, R. C. Estler, R. C. Dye, C. Flamme, N. S. Nogar, A. R. Garcia, J. Martin, and J. Tesmer. Effect of deposition rate on properties of YBa2Cu3O7-x superconducting thin films. *Appl. Phys. Lett.* **56**, 1481 (1990).
46. Q. Li, X. X. Xi, X. D. Wu, A. Inam, S. Vadlamannati, W. L. McLean, T. Venkatesan, R. Ramesh, D. M. Hwang, J. A. Martinez, and L. Nazar. Interlayer coupling effect in high-T_c superconductors probed by $YBa_2Cu_3O_{7-x}/PrBa_2Cu_3O_{7-x}$ superlattices. *Phys. Rev. Lett.* **64**, 3086 (1990).
47. D. H. Lowndes, D. P. Norton, J. W. McCamy, R. Feenstra, J. D. Budai, D. K. Christen, and D. B. Poker. Superconductivity in nonsymmetric epitaxial $YBa_2Cu_3O_{7-x}/PrBa_2Cu_3O_{7-x}$ superlattices: The superconducting behavior of Cu-O bilayers. *Phys. Rev. Lett.* **65**, 1160 (1990).
48. L. P. Lee, K. Char, M. S. Colclough, and G. Zaharchuk. Monolithic 77 K dc SQUID magnetometer. *Appl. Phys. Lett.* **59**, 3051 (1991).
49. J. T. Cheung, J. Bajaj, and M. Khoshnevisan. *Proc. of the Infrared Information Symp., Detector Special Group.* Boulder, 1983.
50. J. T. Cheung, G. Niizawa, J. Moyle, N. P. Ong, B. M. Paine, and T. Vreeland, Jr. HgTe and CdTe epitaxial layers and HgTe°CdTe superlattices grown by laser molecular beam eitaxy. *J. Vac. Sci. Technol.* **A4**, 2086 (1986).
51. M. Kanai, T. Kawai, S. Kawai, and H. Tabata. Low-temperature formation of multilayered Bi(Pb)-Sr-Ca-Cu-O thin films by successive deposition using laser ablation. *Appl. Phys. Lett.* **54**, 1802 (1989).
52. H. Koinuma and M. Yoshimoto. Controlled formation of oxide materials by laser molecular beam epitaxy. *Appl. Surf. Sci.* **75**, 308 (1994); see also H. Koinuma, M. Kawasaki, and M. Yoshimoto. Laser-MBE for atomically defined ceramic film growth [7, in press].
53. D. P. Norton, B. C. Chakoumakos, J. D. Budai, D. H. Lowndes, B. C. Sales, J. R. Thompson, and D. K. Christen. Superconductivity in $SrCuO_2$-$BaCuO_2$ superlattices: Formation of artificially layered superconducting materials. *Science* **265**, 2074 (1994).

54. A. Matsunawa, S. Katayama, A. Susuki, and T. Ariyasu. Laser production of ultrafine metallic and ceramic particles. In *Laser Welding, Machining, and Materials Processing*, C. K. Albright (ed.), p. 205. IFS Publications, United Kingdom, 1986. See also *Trans. J. Welding Res. Institute* **15**, 61 (1986).
55. L.-C. Chen. Particulates generated by pulsed laser ablation [3, pp. 167–198], (a) p. 187; (b) p. 182.
56. S. V. Gaponov, A. A. Gudkov, and A. A. Fraerman. Processes occurring in an erosion plasma during laser vacuum deposition. III. Condensation in gas flows during laser vaporization of materials. *Sov. Phys. Tech. Phys.* **27**, 1130 (1982).
57. S. Witanachchi, K. Ahmed, P. Sakthivel, and P. Mukherjee. Dual-laser ablation for particulate-free film growth. *Appl. Phys. Lett.* **66**, 1469 (1995).
58. S. Witanachchi and P. Mukherjee. Role of temporal delay in dual-laser ablated plumes. *J. Vac. Sci. Technol.* **A13**, 1171 (1995).
59. R. Kelly. On the dual role of the Knudsen layer and unsteady adiabatic expansion in pulse sputtering phenomena. *J. Chem. Phys.* **92**, 5047 (1990).
60. A. Inam, X. D. Wu, T. Venkatesan, S. B. Ogale, C. C. Chang, and D. Dijkkamp. Pulsed laser etching of high-T_c superconducting films. *Appl. Phys. Lett.* **51**, 1112 (1987).
61. D. B. Geohegan, D. N. Mashburn, R. J. Culbertson, S. J. Pennycook, J. D. Budai, R. E. Valiga, B. C. Sales, D. H. Lowndes, L. A. Boatner, E. Sonder, D. Eres, D. K. Christen, and W. H. Christie. Pulsed laser deposition of thin superconducting films of $HoBa_2Cu_3O_{7-x}$ and $YBa_2Cu_3O_{7-x}$. *J. Mater. Res.* **3**, 1169 (1988).
62. R. K. Singh and J. Narayan. Pulsed-laser evaporation techniques for deposition of thin films: Physics and theoretical model. *Phys. Rev.* **B41**, 8843 (1990).
63. R. K. Singh, O. W. Holland, and J. Narayan. Theoretical model for deposition of superconducting thin films using pulsed laser evaporation technique. *J. Appl. Phys.* **68**, 233 (1990).
64. J. P. Zheng, Q. Y. Ying, S. Witanachchi, Z. Q. Huang, D. T. Shaw, and H. S. Kwok. Role of the oxygen atomic beam in low-temperature growth of superconducting films by laser deposition. *Appl. Phys. Lett.* **54**, 954 (1989).
65. K. L. Saenger. On the origin of spatial nonuniformities in the composition of pulsed-laser-deposited films. *J. Appl. Phys.* **70**, 5629 (1991).
66. R. Kelly and R. W. Dreyfus. On the effect of Knudsen-layer formation on studies of evaporation, sputtering, and desorption. *Surf. Sci.* **198**, 263 (1988).
67. R. Kelly and R. W. Dreyfus. Reconsidering the mechanisms of laser sputtering with Knudsen-layer formation taken into account. *Nucl. Instr. and Meth. in Physics Res.* **B32**, 341 (1988).
68. P. E. Dyer, S. R. Farrar, and P. H. Key. Fast time-response photoacoustic studies and modeling of KrF laser ablated $YBa_2Cu_3O_7$. *Appl. Surface Sci.* **54**, 255 (1992).
69. P. E. Dyer, P. H. Key, and P. Monk. Ablation studies of Y-Ba-Cu-O in oxygen using a pulsed CO_2 laser. *Appl. Surface Sci.* **54**, 160 (1992).
70. D. B. Geohegan. Fast ICCD photography of YBCO laser ablation plume propagation in vacuum and ambient oxygen. *Appl. Phys. Lett.* **60**, 2732 (1992).
71. D. B. Geohegan. Physics and diagnostics of laser ablation plume propagation for high-T_c superconductor film growth. In *Metallurgical Coatings and Thin Films 1992, Vol. II*, B. D. Sartwell, G. E. McGuire, and S. Hofmann (eds.), p. 138. Elsevier, Amsterdam, 1992.
72. J. F. Ready. *Effects of High Power Laser Radiation*. Academic Press, Miami, 1971.

73. T. Nakayama, M. Okigawa, and N. Itoh. Laser-induced sputtering of oxides and compound semiconductors. *Nuc. Instr. Meth. in Physics Res.* **B229**, 301 (1983).
74. I. V. Nemchinov and S. V. Popov. Time of start of screening of a surface evaporating under the influence of laser radiation. *JETP Lett.* **11**, 312 (1970).
75. A. F. Haught and D. H. Polk. High-temperature plasmas produced by laser beam irradiation of single solid particles. *Phys. Fluids* **9**, 2047 (1966).
76. K. L. Saenger. Angular distribution of ablated material [3, pp. 199-227].
77. Ibid., p. 215.
78. A. D. Akhsakhalyan, S. P. Gaponov, V. I. Luchin, and A. P. Chirimanov. Angular distribution of a laser ablation plasma expanding into vacuum. *Sov. Phys. Tech. Phys.* **33**, 1146 (1988).
79. D. H. Lowndes, D. P. Norton, J. W. McCamy, R. Feenstra, J. D. Budai, D. K. Christen, and D. B. Poker. In situ growth of high quality epitaxial $YBa_2Cu_3O_{7-x}$ thin films at moderate temperatures by pulsed laser ablation. *Mat. Res. Soc. Symp. Proc.* **169**, 431 (1990).
80. R. E. Muenchausen, K. M. Hubbard, S. Foltyn, R. C. Estler, N. S. Nogar, and C. Jenkins. Effects of beam parameters on excimer laser deposition of $YBa_2Cu_3O_{7-x}$. *Appl. Phys. Lett.* **56**, 578 (1990).
81. [76, pp. 204–206].
82. M. C. Foote, B. B. Jones, B. D. Hunt, J. B. Barner, R. P. Vasquez, and L. J. Bajuk. Composition variations in pulsed laser deposited Y-Ba-Cu-O thin films as a function of deposition parameters. *Physica* **C201**, 176 (1992).
83. D. B. Chrisey, J. S. Horwitz, K. S. Grabowski, M. E. Reeves, M. S. Osofsky, and C. R. Gosset. The influence of target-substrate bias on pulsed laser deposited $YBa_2Cu_3O_{7-x}$. *Mater. Res. Soc. Symp. Proc.* **169**, 435 (1990).
84. R. E. Muenchausen, S. R. Foltyn, N. S. Nogar, R. C. Estler, E. J. Peterson, and X. D. Wu. Laser-induced target modification effects on pulsed laser deposition of Y-Ba-Cu-O superconducting thin films. *Nucl. Instr. Meth.* **A303**, 204 (1991).
85. [76, p. 211].
86. A. A. Gorbunov and V. I. Konov. Thickness profiles of films deposited from laser erosion plasmas. *Sov. Phys. Tech. Phys.* **34**, 1271 (1990).
87. S. K. Hau, K. H. Wong, P. W. Chan, C. L. Choy, and H. K. Wong. Angular distribution of XeCl laser deposition of $Pb(Zr_{0.48}Ti_{0.52})O_3$ films. *J. Mater. Sci. Lett.* **11**, 1266 (1992).
88. R. A. Neifeld, S. Gunapala, C. Liang, S. A. Shaheen, M. Croft, J. Price, D. Simons, and W. T. Hill III. Systematics of thin films formed by excimer laser ablation: Results on $SmBa_2Cu_3O_7$. *Appl. Phys. Lett.* **53**, 703 (1988).
89. J. B. Anderson. Molecular Beams. In *Molecular Beams and Low Density Gas Dynamics*, P. O. Wegener (ed.). Marcel Dekker, New York, 1974.
90. J. P. Zheng, Z. Q. Huang, D. T. Shaw, and H. S. Kwok. Generation of high-energy atomic beams in laser-superconducting target interaction. *Appl. Phys. Lett.* **54**, 280 (1989).
91. N. H. Cheng, Q. Y. Ying, J. P. Sheng, and H. S. Kwok. Time-resolved resonant absorption study of 532-nm laser-generated plumes over $YBa_2Cu_3O_7$ targets. *J. Appl. Phys.* **69**, 6349 (1991).
92. W. Marine, M. Gerri, J. M. Scotto d'Aniello, M. Sentis, Ph. Delaporte, B. Forestier, and B. Fontaine. Analysis of the plasma expansion dynamics by optical time-of-flight measurements. *Appl. Surf. Sci.* **54**, 264 (1992).
93. D. Lubben, S. A. Barnett, K. Suzuki, S. Gorbatkin, and J. E. Greene. Laser-induced

plasmas for primary ion deposition of epitaxial Ge and Si films. *J. Vac. Sci. and Technol.* **B3**, 968 (1985).
94. D. B. Geohegan and D. N. Mashburn. Characterization of ground-state neutral and ion transport during laser ablation of $YBa_2Cu_3O_{7-x}$ using transient optical absorption spectroscopy. *Appl. Phys. Lett.* **55**, 2345 (1989).
95. K. Murakami. Dynamics of laser ablation of high-T_c superconductors and semiconductors, and a new method for growth of films [12, pp. 125–140].
96. O. Eryu, K. Murakami, K. Masuda, A. Kasuya, and Y. Nishina. Dynamics of laser-ablated particles from high-T_c superconductor $YBa_2Cu_3O_y$. *Appl. Phys. Lett.* **54**, 2716 (1989).
97. J. A. Venables, G. D. T. Spiller, and M. Hanbucken. Nucleation and growth of thin films. *Rep. Prog. Phys.* **47**, 399 (1984).
98. J. Hu and R. G. Gordon. Textured aluminum-doped zinc oxide thin films from atmospheric pressure chemical-vapor deposition. *J. Appl. Phys.* **71**, 880 (1992).
99. G. H. Gilmer and C. Roland. Simulations of crystal growth: Effects of atomic beam energy. *Appl. Phys. Lett.* **65**, 824 (1994).
100. D. J. Eaglesham, H.-J. Gossman, and M. Cerullo. Limiting thickness h_{ep}: For epitaxial growth and room-temperature Si growth on Si(100). *Phys. Rev. Lett.* **65**, 1227 (1990).
101. B. E. Weir, B. S. Freer, R. L. Headrick, D. J. Eaglesham, G. H. Gilmer, J. Bevk, and L. C. Feldman. Low temperature homoepitaxy on Si (111). *Appl. Phys. Lett.* **59**, 204 (1991).
102. M. Schneider, I. K. Schuller, and A. Rahman. Epitaxial growth of silicon: A molecular-dynamics simulation. *Phys. Rev.* **B36**, 1340 (1987).
103. E. T. Gawlinski and J. D. Gunton. Molecular-dynamics simulation of molecular-beams epitaxial growth of the silicon (100) surface. *Phys. Rev.* **B36**, 4774 (1987).
104. H. J. Gossman and L. C. Feldman. The influence of reconstruction on epitaxial growth: Ge on Si(100)-(2 × 1) and Si(111)-(7 × 7). *Surf. Sci.* **155**, 413 (1985).
105. M. Kitabatake and J. E. Greene. In *Low Energy Ion Beam and Plasma Modification of Materials*, J. M. E. Harper, J. R. McNeil, and S. M. Gorbatkin (eds.), p. 9. Materials Res. Soc., Pittsburgh, 1991.
106. S. M. Sze. *Semiconductor Devices: Physics and Technology*, p. 416. Wiley, New York, 1985.
107. M. V. Ramana Murty and H. A. Atwater. Defect generation and morphology of (100) Si surfaces during low-energy Ar-iron bombardment. *Phys. Rev.* **B45**, 1507 (1992).
108. V. P. Ageev, A. D. Akhsakhalyan, S. V. Gaponov, A. A. Gorbunov, V. I. Konov, and V. I. Luchin. Influence of the wavelength of laser radiation on the energy composition of an ablation plasma. *Sov. Phys. Tech. Phys.* **33**, 562 (1988).
109. J. C. S. Kools. Monte Carlo simulations of the transport of laser-ablated atoms in a diluted gas. *J. Appl. Phys.* **74**, 6401 (1993).
110. R. R. Goforth and D. W. Koopman. Collisional generations of precursor ions by laser-produced plasma expanding in a gas. *Phys. Fluids* **17**, 698 (1974).
111. D. W. Koopman and R. R. Goforth. Collisional coupling in counterstreaming laser-produced plasmas. *Phys. Fluids* **17**, 1560 (1974).
112. D. W. Koopman. Momentum transfer interaction of a laser-produced plasma with a low-pressure background. *Phys. Fluids* **15**, 1959 (1972).
113. R. F. Wood, K.-R. Chen, C.-L. Liu, J. N. Leboeuf, J. Donato, A. A. Puretzky, and D. B. Geohegan. Elastic collision model of multi-component particle flux in laser

ablation processes. *Bull. Amer. Phys. Soc.*, in press; and, submitted for publication.
114. H. S. Kim and H. S. Kwok. Correlation between target-substrate distance and oxygen pressure in pulsed laser deposition of $YBa_2Cu_3O_7$. *Appl. Phys. Lett.* **61**, 2234 (1992).
115. J. P. Zheng and H. S. Kwok. Low resistivity indium tin oxide films by pulsed laser deposition. *Appl. Phys. Lett.* **63**, 1 (1993).
116. W. P. Shen and H. S. Kwok. Crystalline phases of II-VI compound semiconductors grown by pulsed laser deposition. *Appl. Phys. Lett.* **65**, 2162 (1994).
117. R. E. Leuchtner. Pulsed laser deposition of ZnO: Energetic Rydberg state atoms and their impact on film growth [7, in press].
118. W. Z. Park, T. Eguchi, C. Honda, K. Muraoka, Y. Yamagata, B. W. James, M. Maeda, and M. Akazaki. Investigation of the thermalization of sputtered atoms in a magnetron discharge using laser-induced fluorescence. *Appl. Phys. Lett.* **58**, 2564–6 (1991).
119. J. T. Cheung. Semi-Annual Rept. No. 4. DARPA contract no. MDA903-79-C-0188, 1981. See also [14].
120. T. Yoshida, S. Takeyama, Y. Yamada, and K. Mutoh. Nanometer-sized silicon crystallites prepared by excimer laser ablation in constant pressure inert gas. *Appl. Phys. Lett.* **68**, 1772 (1996).
121. S. Otsubo, T. Minamikawa, Y. Yonezawa, A. Morimoto, and T. Shimizu. Thermal analysis of target surface in the Ba-Y-Cu-O film preparation by laser ablation materials. *Jap. J. Appl. Phys.* **29**, L73 (1990).
122. R. Singh, D. Bhattacharya, and J. Narayan. Relationship between thermal stress and structural properties of SrF_2 films on (100) InP. *Appl. Phys. Lett.* **57**, 2022 (1990).
123. J. T. Cheung. History and fundamentals of pulsed laser deposition [3, pp. 1–22]. (a) p. 15; (b) p. 14.
124. H. Schwartz and H. A. Tourtellotte. Vacuum deposition by high-energy laser with emphasis on barium titanate films. *J. Vac. Sci. and Technol.* **6**, 376 (1969).
125. J. F. Ready. Development of plume of material vaporized by giant-pulse laser. *Appl. Phys. Lett.* **3**, 11 (1963).
126. T. Venkatesan, X. D. Wu, R. Muenchausen, and A. Pique. Pulsed laser deposition: Future directions [8, pp. 54–58].
127. G. Koren, R. J. Baseman, A. Gupta, M. I. Lutwyche, and R. B. Laibowitz. Laser wavelength dependent properties of $YBa_2Cu_3O_{7-x}$ thin films deposited by laser ablation. *Appl. Phys. Lett.* **55**, 2450 (1989).
128. N. J. Ianno, L. McConville, N. Shaikh, S. Pittal, and P. G. Snyder. Characterization of pulsed laser deposited zinc oxide. *Thin Solid Films* **220**, 92 (1992).
129. V. Craciun, S. Amirhaghi, D. Craciun, J. Elders, J. G. E. Gardeniers, and I. W. Boyd. Effects of laser wavelength and fluence on the growth of ZnO thin films by pulsed laser depositon. *Appl. Surf. Sci.* **86**, 99 (1995).
130. S. R. Foltyn, R. C. Dye, K. C. Ott, E. Peterson, K. M. Hubbard, W. Hutchinson, R. E. Muenchausen, R. C. Estler, and X. D. Wu. Target modification in the excimer laser deposition of $YBa_2Cu_3O_{7-x}$ films. *Appl. Phys. Lett.* **59**, 594 (1991).
131. S. R. Foltyn, R. E. Muenchausen, R. C. Estler, E. Peterson, W. Hutchinson, K. C. Ott, N. S. Nogar, K. M. Hubbard, R. C. Dye and X. D. Wu. Influence of beam and target properties on the excimer laser deposition of $YBa_2Cu_3O_{7-x}$ thin films [4, p. 205].
132. C. M. Rouleau: (a) unpublished; (b) unpublished computer simulation.
133. D. H. Lowndes, unpublished results.

134. D. N. Mashburn, private communication.
135. R.-F. Xiao. Growing diamond films from an organic liquid. *Appl. Phys. Lett.* **67**, 3117 (1995).
136. H. J. P. Dupendant, J. P. Gavigan, D. Givord, A. Lienard, J. P. Rebouillat, and Y. Souche. Velocity distribution of micron-size particles in thin film laser ablation deposition (LAD) of metals and oxide superconductors. *Appl. Surf. Sci.* **43**, 369 (1989).
137. W. P. Barr. The production of low scattering dielectric mirrors using rotating vane particle filtration. *J. Phys.* **E2**, 1112 (1969).
138. R. Edwards and A. Pique (Neocera, Inc., College Park, MD), private communication.
139. G. Koren, R. J. Baseman, A. Gupta, M. I. Lutwyche, and R. B. Laibowitz. Particulates reduction in laser-ablated $YBa_2Cu_3O_{7-x}$ thin films by laser-induced plume heating. *Appl. Phys. Lett.* **56**, 2144 (1990).
140. H. Chiba, K. Murakami, O. Eryu, K. Shihoyama, T. Mochizuki, and K. Masuda. Laser excitation effects on laser ablated particles in fabrication of high-T_c superconducting thin films. *Jap. J. Appl. Phys.* **30**, L732 (1991).
141. O. Eryu, K. Yamaoka, K. Murakami, and K. Masuda. Novel methods of laser ablation of $YBa_2Cu_3O_y$. In *Extended Abstracts of the 1991 Int. Conf. on Solid State Devices and Materials*, p. 438. Bus. Center Acad. Soc. Japan, Tokyo, Japan, 1991.
142. R. J. Kennedy. A new laser ablation geometry for the production of smooth thin single-layer $YBa_2Cu_3O_{7-x}$ and multilayer $YBa_2Cu_3O_{7-x}/PrBa_2Cu_3O_{7-x}$ films. *Thin Solid Films* **214**, 223 (1992).
143. S. V. Gaponov, B. M. Luskin, and N. N. Salashchenko. Homoepitaxial superlattices with nonoriented barrier layers. *Solid State Commun.* **39**, 301 (1981).
144. S. V. Gaponov, A. A. Gudkov, B. M. Luskin, V. I. Luchin, and N. N. Salashchenko. Formation of semiconductor films from a laser erosion plasma scattered by a heated screen. *Sov. Phys. Tech. Phys.* **26**, 598 (1981).
145. K. B. Erington and N. J. Ianno. Thin films of uniform thickness by pulsed laser deposition. *Rev. Sci. Instr.* **63**, 3525 (1992); *Mat. Res. Soc. Symp. Proc.* **191**, 115 (1990).
146. S. R. Foltyn, R. E. Muenchausen, R. C. Dye, X. D. Wu, L. Luo, and D. W. Cooke. Large-area, two-sided superconducting $YBa_2Cu_3O_{7-x}$ films deposited by pulsed laser deposition. *Appl. Phys. Lett.* **59**, 1374 (1991).
147. D. H. Lowndes, J. W. McCamy, Shen Zhu, and D. B. Poker. Growth of uniform-thickness thin films by pulsed-laser ablation. In *Solid State Division Progress Report*—1992, p. 21. Report ORNL-6722, Oak Ridge National Laboratory, Oak Ridge, TN, 1992.
148. J. W. McCamy, D. H. Lowndes, J. D. Budai, B. C. Chakoumakos, and R. A. Zuhr. Growth of epitaxial ZnS films by pulsed-laser ablation. *Mater. Res. Soc. Symp. Proc.* **242**, 243 (1992).
149. V. M. Dubkov. Analysis of the condition for obtaining homogeneous optical coatings by the pulsed laser evaporation method. *Sov. J. Opt. Technol.* **49**, 168 (1982).
150. S. V. Gaponov, S. A. Gusev, Yu. Ya. Platonov, and N. N. Salashchenko. Synthetic multilayer reflectors and selectors for soft x rays. I. Choice of materials and design of multilayer mirrors. *Sov. Phys. Tech. Phys.* **29**, 442 (1984).
151. A. D. Akhsakhalyan, S. V. Gaponov, S. A. Gusev, V. I. Luchin, Yu. Ya. Platonov, and N. N. Salashchenko. Synthetic multilayer reflecting and selective elements for soft x-ray radiation. II. Fabrication of multilayer soft x-ray mirrors by laser pulse deposition. *Sov. Phys. Tech. Phys.* **29**, 448 (1984).

152. H. Mai and W. Pompe. Manufacture and characterization of soft x-ray mirrors by laser ablation. *Appl. Surf. Sci.* **54**, 215 (1992).
153. H. Mai, R. Dietsch, and W. Pompe. Preparation of soft x-ray monochromators by laser pulse vapour deposition (LPVD). In *X-ray Optics and Microanalysis*, P. B. Kenway et al. (eds.), p. 503. IOP, Bristol, United Kingdom, 1993.
154. R. Dietsch, H. Mai, W. Pompe, and S. Vollmar. A modified plasma source for controlled layer thickness synthesis in laser pulse vapour deposition (LPVD). *Adv. Materials for Optics and Electronics* **2**, 19 (1993).
155. J. A. Greer and H. J. Van Hook. Uniformity considerations for "in-situ" laser-ablated $YBa_2Cu_3O_{7-x}$ films over three inch substrates. *Mat. Res. Soc. Symp. Proc.* **169**, 463 (1990).
156. J. A. Greer. High quality YBCO films grown over large areas by pulsed laser deposition. *J. Vac. Sci. and Technol.* **A10**, 1821 (1992).
157. J. A. Greer. Commercial scale-up of pulsed laser deposition [3, pp. 293–311].
158. J. A. Greer and M. D. Tabat. Large-area pulsed laser deposition: Techniques and applications. *J. Vac. Sci. and Technol.* **A13**, 1175 (1995).
159. J. A. Greer and M. D. Tabat. On- and off-axis large-area pulsed laser deposition [6, p. 151].
160. R. C. Estler, N. S. Nogar, R. E. Muenchausen, X. D. Wu, S. Foltyn, and A. R. Garcia. A versatile substrate heater for use in highly oxidizing atmospheres. *Rev. Sci. Instr.* **62**, 437 (1991).
161. X. D. Wu, R. E. Muenchausen, S. Foltyn, R. C. Estler, R. C. Dye, A. R. Garcia, N. S. Nogar, P. England, R. Ramesh, D. M. Hwang, T. S. Ravi, C. C. Chang, T. Venkatesan, X. X. Xi, Q. Li, and A. Inam. Large critical current densities in $YBa_2Cu_3O_{7-x}$ thin films made at high deposition rates. *Appl. Phys. Lett.* **57**, 523 (1990).
162. J. T. Cheung and H. Sankur [12, p. 325].
163. J. J. Dubowski. Pulsed laser ablation: a method for deposition and processing of semiconductors at an atomic level. *Materials Science Forum* **173-174**, 73 (1994).
164. J. J. Dubowski. Structural and optical properties of semiconductor heterostructures and superlattices grown by pulsed laser evaporation and epitaxy (PLEE). *Proc. of the SPIE* **2045**, 112 (1994).
165. J. J. Dubowski, J. R. Thompson, S. J. Rolfe, and J. P. McCaffrey. Laser-induced growth of $Cd_{1-x}Mn_xTe$ and $CdTe-Cd_{1-x}Mn_xTe$ superlattices. *Superlattices and Microstructures* **9**, 327 (1991).
166. J. M. Wrobel, C. E. Moffitt, and J. J. Dubowski. Photoluminescence study of reproduction of energy levels in CdTe films grown by pulsed laser evaporation and epitaxy. *Mat. Res. Soc. Symp. Proc.* **285**, 459 (1993).
167. J. M. Wrobel, J. J. Dubowski, and P. Becla. Low temperature photoluminescence in In-doped CdMnTe/CdTe quantum wells and superlattices. *Proc. of the SPIE* **2403**, 251 (1995).
168. D. Labrie, J. B. Aufgang, J. M. Wrobel, J. J. Dubowski, and P. Becla. Photoreflectance and photoluminescence spectroscopies of In-doped CdMnTe/CdTe multiple quantum wells and superlattices. *Proc. of the SPIE* **2403**, 259 (1995).
169. J. J. Dubowski, personal communication.
170. W. P. Shen and H. S. Kwok. In *New Materials for Advanced Solid State Lasers*, B. H. T. Chai et al. (eds.), p. 173. Mat. Res. Soc., Pittsburgh, 1994.
171. R. D. Feldman, R. F. Austin, P. M. Bridenbaugh, A. M. Johnson, and W. M. Simpson. Effects on Zn to Te ratio on the molecular-beam epitaxial growth of ZnTe on GaAs. *J. Appl. Phys.* **64**, 1191 (1988).

172. S. M. Sze. *Physics of Semiconductor Devices*, 2nd ed. John Wiley and Sons, 1981.
173. W. P. Shen and H. S. Kwok. P-type II-VI compound semiconductor thin films grown by pulsed laser deposition [6, p. 91].
174. A. Compaan, A. Bhat, C. Tabory, S. Liu, Y. Li, M. E. Savage, M. Shao, L. Tsien, and R. G. Bohn. Polycrystalline CdTe solar cells by pulsed laser deposition. In *Proc. 22nd IEEE Photovoltaic Specialists Conf.*, p. 957. IEEE, 1991.
175. A. Compaan and A. Bhat. Laser-driven physical vapor deposition for thin-film CdTe solar cells. *Int. J. Solar Energy* **12**, 155 (1992).
176. R. P. Vaudo, J. W. Cook, Jr., and J. F. Schetzina. Atomic nitrogen production in nitrogen-plasma sources used for the growth of ZnSe:N and related alloys by molecular-beam epitaxy. *J. Cryst. Growth* **138**, 430 (1994).
177. T. Nakao and T. Uenoyama. Adsorption and dissociation mechanism of excited N_2 on ZnSe surface. *Jpn. J. Appl. Phys.* **32**, 660 (1993).
178. T. Uenoyama, T. Nakao, and M. Suzuki. Interaction between N_2 and stabilized ZnSe surface. *Cryst. Growth* **138**, 301 (1994).
179. G. A. Landis and A. F. Hepp. Thin film photovoltaics: status and applications to space power. In *European Space Power Conf.*, p. 517. (Florence, Italy, 1991).
180. J. R. Tuttle et al. 17.1% efficient $Cu(In, Ga)Se_2$-based thin film solar cell. *Progress in Photovoltaics* **3**, 235 (1995). M. A. Contreras, A. M. Gabor, A. L. Tennant, S. Asher, J. Tuttle, and R. Noufi. 16.4% total-area conversion efficiency thin-film polycrystalline $MgF_2/ZnO/CdS/Cu(In, Ga)Se_2/Mo$ solar cell. *Prog. in Photovoltaics* **2**, 287 (1994).
181. A. M. Gabor et al. High efficiency polycrystalline $Cu(In, Ga)Se_2$-based solar cells. 12th NREL Photovoltaic Program Review, *AIP Conf. Proc.* **303**, 59 (1994).
182. H. W. Schock. High efficiency polycrystalline thin film $CuInSe_2$ and $CuInS_2$ based solar cells. *Optoelectronics-Devices and Technologies* **9**, 511 (1994). H. W. Schock. Solar-cells based on $CuInSe_2$ and related compounds—recent progress in Europe. *Solar Energy Materials and Solar Cells*, **34**, 19 (1994).
183. T. Walter and H. W. Schock. Fundamental studies and development of technologies for $CuInSe_2$ based thin film solar cells in the EUROCIS program. 12th NREL PV Program Review. *AIP Conf. Proc.* **303**, 67 (1994).
184. NREL's CIGS Cell Establishes New Record 17.7% Conversion Efficiency. *Photovoltaic Insider's Report* **XV**, no. 4 (April, 1996).
185. The author would like to acknowledge R. Noufi and M. Contreras for helpful discussions of current CIGS solar cell fabrication methods and materials problems.
186. H. Dittrich, M. Klose, M. Brieger, R. Schäffler, and H. W. Schock. $CuInSe_2$ thin film solar cells by pulsed laser deposition, p. 617 in *Proc. of 23rd IEEE Photovoltaic Specialists Conf.*, IEEE, 1993.
187. H. Dittrich, M. Klose, M. Brieger, R. Schäffler, and H. W. Schock. Complex semiconductor compound targets for pulsed laser deposition, *Materials Sci. Forum* **173-174**, 129 (1995).
188. R. Schäffler, M. Klose, M. Brieger, H. Dittrich, and H. W. Schock. Pulsed laser deposition and characterization of $CuInSe_2$ thin films for solar cell applications. *Materials Sci. Forum* **173-174**, 135 (1995).
189. A. E. Hill, S. Leppävuori, R. D. Tomlinson, J. Levoska, E. Ahmed, and J. Frantti. Laser ablation of $CuInSe_2$ and $CuIn/GaSe_2$ alloys for solar cell applications. *Mater. Res. Soc. Symp. Proc.* **285**, 483 (1993).
190. J. Levoska, A. E. Hill, S. Leppävuori, O. Kusmartseva, R. D. Tomlinson, and R. D. Pilkington. Laser ablation deposition of $CuInSe_2$ thin films on silicon and fused silica. *Jpn. J. Appl. Phys.* **32**, Suppl. 32-3, 43 (1993).

191. E. Ahmed, A. E. Hill, S. Leppävuori, R. D. Pilkington, R. D. Tomlinson, J. Levoska, and O. Kusmartseva. A comparative study of pulsed laser deposition and flash evaporation of $CuIn_{0.75}Ga_{0.25}Se_2$ films, *Adv. Materials for Optics and Electronics* **4**, 423 (1994).
192. J. Levoska, S. Leppävuori, F. Wang, O. Kusmartseva, A. E. Hill, E. Ahmed, R. D. Tomlinson, and R. D. Pilkington. Pulsed laser ablation deposition of $CuInSe_2$ and $CuIn_{1-x}Ga_xSe_2$ thin films. *Physica Scripta* **T54**, 244 (1994).
193. S. Leppävuori, J. Levoska, A. E. Hill, R. D. Tomlinson, J. Frantti, O. Kusmartseva, H. Moilanen, and R. D. Pilkington. Laser ablation deposition as a preparation method for sensor materials. *Sensors and Actuators* **A41-42**, 145 (1994).
194. S. Leppävuori, A. E. Hill, J. Levoska, E. Ahmed, R. D. Pilkington, and R. D. Tomlinson. Copper indium diselenide single-crystal and thin-film infrared sensors. *Sensors and Actuators* **A46-47**, 395 (1995).
195. V. F. Gremenok, E. P. Zaretskaya, I. V. Bodnar, and I. A. Victorov. Chalcopyrite $CuGa_xIn_{1-x}Se_2$ thin films produced by laser-assisted evaporation. *Jpn. J. Appl. Phys.* **32** Suppl. 32-3, 90 (1993).
196. V. F. Gremenok, E. P. Zaretskaya, V. V. Kindyak, A. S. Kindyak, I. V. Bodnar, and I. A. Victorov. Production of highly oriented thin $CuInSe_2$ films by pulsed laser vaporization. *Tech. Phys. Lett.* **20**, 24 (1994).
197. E. P. Zaretskaya, V. F. Gremenok, I. V. Bodnar, Yu. V. Rud', N. I. Konstantinova, and I. K. Polushina. Photoelectric properties of laser-deposited p-$CuInSe_2$ layers and structures based on them. *J. of Appl. Spectros.* **60**, 265 (1994).
198. V. F. Gremenok, V. V. Kindyak, E. P. Zaretskaya, A. S. Kindyak, I. A. Victorov, and I. V. Bodnar. Preparation of chalcopyrite $CuGa_xIn_{1-x}Se_2$ films by pulsed laser evaporation. *Semiconductors* **29**, 881 (1995).
199. V. V. Kindyak, A. S. Kindyak, V. F. Gremenok, I. V. Bodnar, and Ya. I. Latushko. On the structure of the edge absorption in laser-deposited $CuInTe_2$ thin films. *Tech. Phys. Lett.* **21**, 268 (1995).
200. H. Uchiki, O. Machida, A. Tanaka, and H. Hirasawa. Preparation of $AgGaS_2$ films by excimer laser desorption. *Jpn. J. Appl. Phys.* **32**, L765 (1993).
201. H. Uchiki, H. Hirasawa, and I. Hasegawa. Epitaxial growth of $AgGaS_2$ on (100) GaAs by excimer laser deposition. *Jpn. J. Appl. Phys.* **33**, L983 (1994).
202. I. Taguchi, H. Ezumi, S. Keitoku, T. Tamaru, and H. Osono. Stress-induced raman frequency shifts in $CuInSe_2$ thin films prepared by laser ablation. *Jpn. J. Appl. Phys.* **34**, L135 (1995).
203. S. Niki, H. Shibata, P. J. Fons, A. Yamada, A. Obara, Y. Makita, T. Kurafuji, S. Chichibu, and H. Nakanishi. Excitonic emissions from $CuInSe_2$ on GaAs(001) grown by molecular beam epitaxy. *Appl. Phys. Lett.* **67**, 1289 (1995).
204. R. Shioda, Y. Okada, H. Oyanagi, S. Niki, A. Yamada, and Y. Makita. Characterization of molecular beam epitaxy grown $CuInSe_2$ on GaAs(001). *J. of Crystal Growth* **150**, 1196 (1995).
205. S. Niki, Y. Makita, A. Yamada, O. Hellman, P. J. Fons, A. Obara, Y. Okada, R. Shioda, H. Oyanagi, T. Kurafuji, S. Chichibu, and H. Nakanishi. Heteroepitaxy and characterization of $CuInSe_2$ on GaAs(001). *J. of Crystal Growth* **150**, 1201 (1995).
206. A. Yamada, Y. Makita, S. Niki, A. Obara, P. Fons, and H. Shibata. Growth of $CuGaSe_2$ film by molecular beam epitaxy. *Microelectronics J.* **27**, 53 (1996).
207. A. Yamada, Y. Makita, S. Niki, A. Obara, P. Fons, H. Shibata, M. Kawai, S. Chichibu, and H. Nakanishi. Band-edge photoluminescence of $CuGaSe_2$ films grown by molecular beam epitaxy. *J. Appl. Phys.* **79**, 1 (1996).
208. B.-H. Tseng, S.-B. Lin, K.-C. Hsieh, and H.-L. Hwang. Stoichiometric control of

CuInSe$_2$ films using a molecular beam epitaxy technique. *J. of Crystal Growth* **150**, 1206 (1995).
209. A. N. Tiwari, S. Blunier, K. Kessler, V. Zelezny, and H. Zogg. Direct growth of heteroepitaxial CuInSe$_2$ layers on Si substrates. *Appl. Phys. Lett.* **65**, 2299 (1994).
210. A. N. Tiwari, S. Blunier, M. Filzmoser, H. Zogg, D. Schmid, and H. W. Schock. Characterization of heteroepitaxial CuIn$_3$Se$_5$ and CuInSe$_2$ layers on Si substrates. *Appl. Phys. Lett.* **65**, 3347 (1994).
211. H. Neumann. *Verbindungshalbleiter*, p. 392. Akadem. Verlag., Leipzig, 1986.
212. H. J. Möller. Structure and defect chemistry of grain boundaries in CuInSe$_2$. *Solar Cells* **31**, 77 (1991).
213. G. K−hn and H. Neumann. A^{I}B^{III}C$_2^{VI}$-Halbleiter mit Chalcopyritstruktur. *Z. Chem.* **27**, 197 (1987).
214. T. Irie, S. Endo, and S. Kimura. Electrical properties of p- and n-type CuInSe single crystal. *Jpn. J. Appl. Phys.* **18**, 1303 (1979).
215. J. E. Jaffe and A. Zunger, Electronic structure of the ternary chalcopyrite semiconductors CuAlS$_2$. *Phys. Rev.* **B28**, 5822 (1983).
216. See Möller [24, pp. 300–307], for a more complete discussion of these calculations.
217. R. Noufi, R. Axton, C. Herrington, and S. K. Deb. Electronic properties versus composition of thin films of CuInSe$_2$. *Appl. Phys. Lett.* **45**, 668 (1984).
218. I. V. Bodnar and A. P. Bologa. Investigation of the CuGa$_x$In$_{1-x}$Se$_2$ solid solutions. *Cryst. Res. and Technol.* **17**, 339 (1982).
219. S. Wagner. *Copper Indium Diselenide for Photovoltaic Applications*. Amsterdam, 1986.
220. G. Haacke. Transparent conducting coatings. *Ann. Rev. Mater. Sci.* **7**, 73 (1977).
221. Z. M. Jarzebski. Preparation and physical properties of transparent conducting oxide films. *Phys. Status Solidi* **A71**, 13 (1982).
222. C. G. Granqvist. Solar energy materials: Overview and some examples. *Appl. Phys.* **A52**, 83 (1991).
223. J. M. Phillips, J. Kwo, G. A. Thomas, S. A. Carter, R. J. Cava, S. Y. Hou, J. J. Krajewski, J. H. Marshall, W. F. Peck, D. H. Rapkine, and R. B. van Dover. Transparent conducting thin films of GaInO$_3$. *Appl. Phys. Lett.* **65**, 115 (1994).
224. J. M. Phillips, R. J. Cava, G. A. Thomas, S. A. Carter, J. Kwo, T. Siegrist, J. J. Krajewski, J. H. Marshall, W. F. Peck, and D. H. Rapkine. Zinc-indium-oxide: A high conductivity transparent conducting oxide. *Appl. Phys. Lett.* **67**, 2246 (1995).
225. O. Madelung, ed. *Landolt-Bernstein New Series*, Vol. 17b. Springer-Verlag, Berlin, 1982.
226. E. Kaldis, ed. *Current Topics in Materials Science*, Vol. 7. North-Holland, Amsterdam, 1981.
227. V. Srikant, V. Sergo, and D. R. Clarke. Epitaxial aluminum-doped zinc oxide thin films on sapphire: II. Defect equilibria and electrical properties. *J. Amer. Ceram. Soc.* **78**, 1935 (1995).
228. Z. C. Jin, I. Hamberg, and C. G. Granqvist. Optical properties of sputter-deposited ZnO:Al thin films. *J. Appl. Phys.* **64**, 5117 (1988).
229. Y. Natsume, H. Sakata, T. Hirayama, and H. Yanagida. Low-temperature conductivity of ZnO films prepared by chemical vapor deposition. *J. Appl. Phys.* **72**, 4203 (1992).
230. A. P. Roth and D. F. Williams. Properties of zinc oxide films prepared by the oxidation of diethyl zinc. *J. Appl. Phys.* **52**, 6685 (1981).

231. See [128, 129, 227, 228, 234–236, and 247] for references to device applications of ZnO films and their growth by various methods.
232. H. Sankur and J. T. Cheung. Highly oriented ZnO films grown by laser evaporation. *J. Vac. Sci. Technol.* **A1**, 1806 (1983).
233. T. Nakayama. Laser-induced sputtering of ZnO, TiO_2, CdSe, and GaP near threshold laser fluence. *Surf. Sci.* **133**, 101 (1983).
234. W. S. Hu, Z. G. Liu, X. L. Guo, C. Lin, S. N. Zhu, and D. Feng. Preparation of c-axis oriented ZnO optical waveguiding films on fused silica by pulsed laser reactive ablation. *Materials Lett.* **25**, 5 (1995).
235. V. Craciun, J. Elders, J. G. E. Gardeniers, and I. W. Boyd. Characteristics of high quality ZnO thin films deposited by pulsed laser deposition. *Appl. Phys. Lett.* **65**, 2963 (1994).
236. V. Craciun, J. Elders, J. G. E. Gardeniers, J. Geretovsky, and I. W. Boyd. Growth of ZnO thin films on GaAs by pulsed laser deposition. *Thin Solid Films* **259**, 1 (1995).
237. C. W. Seabury, J. T. Cheung, P. H. Kobrin, R. Addison, and D. P. Havens. High performance microwave air-bridge resonators. *Proc. of IEEE Ultrasonic Symposium*, Seattle, Nov., 1995, in press.
238. J. T. Cheung and S. R. Johnson. Solid state oxygen sensor development. *24th Int. Conf. on Environmental Systems* and *5th European Symp. on Space Environmental Control Systems*, Friedrichshafen, Germany, June 20–23, 1994. Soc. Aero. Eng. technical paper 941267, SAE Publications Group, Warrendale, PA.
239. J. T. Cheung, U. S. patent no. 5,448,906, Sept. 12, 1995.
240. R. E. Leuchtner [7, in press].
241. V. Srikant, V. Sergo, and D. R. Clarke. Epitaxial aluminum-doped zinc oxide thin films on sapphire: I. Effect of substrate orientation. *J. Amer. Ceram. Soc.* **78**, 1931 (1995).
242. J. P. Zheng and H. S. Kwok. Preparation of indium thin oxide films at room temperature by pulsed laser depositon. *Thin Solid Films* **232**, 99 (1993).
243. S. N. Venkatesh, E. S. Ramakrishnan, and K. C. Jungling. Studies of MZS and MZOS structures with zinc oxide deposited by conventional RF diode and magnetron sputtering techniques, J. D. Dow and I. K. Schuller (eds.). *Mater. Res. Soc. Symp. Proc.* **77**, 289 (1987).
244. W.-C. Shih and M.-S. Wu. Growth of ZnO films on GaAs substrates with a SiO_2 buffer layer by RF planar magnetron sputtering for surface acoustics ware applications. *J. Cryst. Growth* **137**, 319 (1994).
245. Y. Kim, W. D. Hunt, F. S. Hickernell, and R. J. Higgins. Surface acoustic wave properties of ZnO films on (001)-cut(110)-propagating GaAs substrates. *J. Appl. Phys.* **75**, 7299 (1994).
246. H. K. Kim, W. Kleemeier, Y. Li, D. W. Langer, D. T. Cassidy, and D. M. Bruce. Thin-film induced stress in GaAs ridge-waveguide structure integrated with sputter-deposited ZnO films. *J. Vac. Sci. Technol.* **B12**, 1328 (1994).
247. F. C. M. Van de Pol. Thin-film ZnO-properties and applications. *Amer. Ceram. Soc. Bull.* **69** 1959 (1990).
248. J. S. Choi and C. H. Yo. *J. Chem. Phys.* **37**, 1149 (1976).
249. J. P. Zheng and H. S. Kwok. Preparation of indium tin oxide films at room temperature by pulsed laser deposition. *Thin Solid Films* **232**, 99 (1993).
250. Q. X. Jia, J. P. Zheng, H. S. Kwok, and W. A. Anderson. Indium tin oxide on InP by pulsed laser deposition. *Thin Solid Films* **258**, 260 (1995).

251. C. M. Dai , C. S. Su, and D. S. Chuu. Growth of highly oriented tin oxide thin films by laser evaporation deposition. *Appl. Phys. Lett.* **57**, 1879 (1990).
252. J. M. McGraw, P. A. Parilla, D. L. Schultz, J. Alleman, X. Wu, W. P. Mulligan, D. S. Ginley, and T. J. Coutts. Pulsed laser deposition of cadmium stannate, a spinel-type tranparent conducting oxide [6, p. 51].
253. G. K. Hubler. Comparison of Vacuum Deposition Techniques [3, Ch. 13, p. 327].
254. F. P. Gagliano and U. C. Paek. *Applied Optics* **13**, 274 (1974).
255. D. B. Geohegan. Fast-ICCD photography and gated photon counting measurements of blackbody emission from particulates generated in the KrF-laser abalation of BN and YBCO [5, p. 27].

12. LASER ABLATION IN OPTICAL COMPONENTS AND THIN FILMS

Michael Reichling

Fachbereich Physik
Freie Universität Berlin, Germany

12.1 Introduction

Laser ablation from optical materials is one of the most challenging topics in ablation physics because interactions of laser light with nominally transparent materials have to be considered. Thus, at moderate intensities, virtually no interaction of the laser light with the material can be detected and there is a distinct threshold for a strong—often called catastrophic—interaction accompanied by ablation or at least a highly nonlinear dependence of the ablation yield on incident laser intensity. Although ablation is a massive effect, it is often initiated by absorption processes with extremely small cross section. Therefore, for optical materials, an understanding of the primary absorption process is the key to understanding desorption and ablation. This chapter deals to a large extent with the description of absorption mechanisms in bulk and thin film optical materials.

Optical materials have been of great practical importance since scientists started fabricating optical components in order to intentionally manipulate light. Optical components are used to reflect, transmit, focus, or guide light beams or to change their polarization and spectral characteristics. In recent years, nonlinear optical properties providing new possibilities of manipulating light pulses have gained much attention. Since the seventeenth century, optical components have been made from bulk optical materials and the main techniques for shaping and surface preparation have been developed. The advent of vacuum and evaporation techniques in the nineteenth century opened up the whole new field of thin film optics that are today found in all domains of conventional as well as novel optical applications. For conventional applications, the quality of optical components is mainly judged in terms of their optical performance and the long-term stability of their specifications. However, since the advent of high-power laser systems, the problem of *laser damage*, i.e., the degradation or destruction of optical components by the laser light, appeared to be one of the most relevant limitations for the use of optical components. Therefore, the *laser damage threshold* (or inversely, the *laser resistivity*) has become the most important figure-of-merit for high-power optics. To understand the extremely high demands on the laser resistivity of optical components in advanced applications, we should keep in mind that a 10 mW HeNe-laser when tightly focused provides an average intensity of not more than 100 kW/cm^2, commercially available nanosecond Nd:YAG or excimer lasers have a peak irradiance of typically 1 GW/cm^2, and pulses of an amplified Ti:sapphire system may have more than 1 TW/cm^2 at a pulse length of some hundred femtoseconds. This describes an increase of a factor of 10^7 in available peak power density that the optics must handle. One of the most demanding applications for optics is its use in ultra-high power lasers of tremendous dimensions designed for nuclear fusion experiments [1]. Here the problem is handling light pulses containing megajoules of pulse energy, and this is an extreme challenge for coating manufacturing because it requires excellent coating quality for deposition on optical components of a typical diameter of one meter and more (see Fig. 1). Therefore, many production and characterization techniques for advanced coatings are a spin-off from such fusion laser programs and these programs have been a constant stimulus for research on laser damage mechanisms in optics.

The questions of laser damage [2] and laser ablation are closely related [3, 4], and, therefore, the laser damage threshold is often simply identified with the threshold for laser ablation. In fact, laser damage of optical materials is most

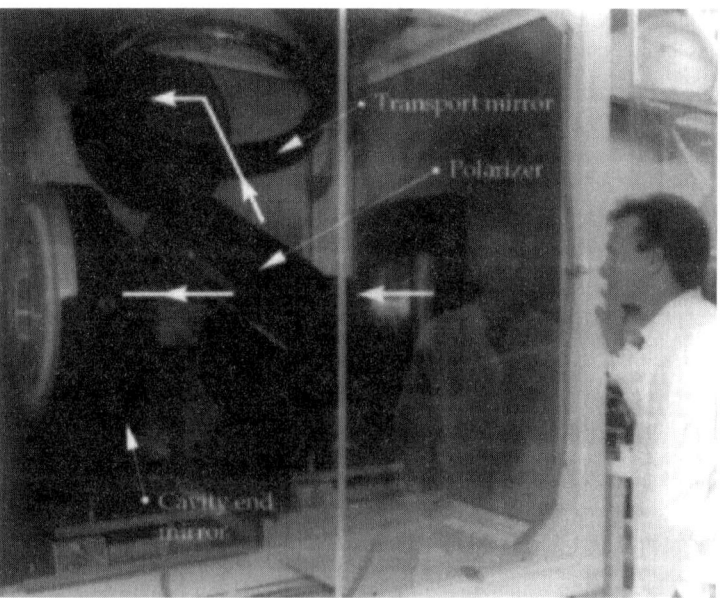

FIG. 1. Mirrors and polarizing optics for the 1.8 MJ laser at the Lawrence-Livermore National Laboratory designed for the Inertial Confinement Fusion program. The project involves a large number of optical components with diameters of up to 1m that have to withstand 500TW laser pulses. (From Ref. 1, reprinted with permission.)

often accompanied by the emission of material. However, the interaction of laser light of increasing intensity with optical materials also results in a variety of near-threshold effects, and it is a difficult question of definition and measurement sensitivity to derive a damage threshold. On the other hand, desorption or even ablation from thin films can be used for *laser conditioning*, i.e., an improvement of the overall laser resistivity of the optical system.

Because this book is concerned with ablation phenomena, the use of the term "threshold" will always imply the *ablation threshold* and not the damage threshold. *Desorption* describes particle emission below the ablation threshold and mostly cannot be easily characterized by a threshold for the materials discussed here.

Research on ablation from optical materials is, however, not only motivated by the laser damage problem, but also by an increasing interest in intentional ablation, i.e., in the laser structuring of optical materials for incorporation into optoelectronic devices. This is also a field of applied research, and much of its efforts are focused on finding laser process conditions for an ablation of the optical material in as controlled a manner as possible.

This chapter starts with a review of experimental techniques used for desorption and ablation measurements from dielectrics in Section 12.2 and then the organization follows a division of the content according to the two main types of optical materials, i.e., bulk dielectrics and dielectric thin film systems addressed in Sections 12.3 and 12.4, respectively. Section 12.3 starts with a brief description of the respective materials, their use for optical components and some of their properties relevant either for their optical performance or the understanding of desorption and ablation phenomena in these materials (Section 12.3.1). This introduction is followed by a discussion of the main sources and types of absorption, i.e., the processes determining the primary energy transfer from the laser light into the material (Section 12.3.2). Then ablation phenomena are described and their basic physical concepts are discussed (Section 12.3.3). Although the discussion of laser ablation from bulk optical materials can be exemplified by a few characteristic materials and phenomena, optical thin films cover such a broad range of applications, materials, designs and ablation phenomena that a comprehensive description is beyond the scope of this chapter. Thin film systems are mostly well characterized with respect to their absorption properties and damage thresholds, however, thin film ablation phenomena are much less well studied. Therefore, Section 12.4 is a very condensed summary of optical thin film applications and properties, followed by the description of selected ablation phenomena related to thin film defects.

In the past, the main stimulus for the investigation of damage in optical materials came from the development of lasers that provide more and more pulse energy. Today the major developments in laser technology are clearly toward shorter pulses and increased peak power. Desorption and ablation with ultra-short laser pulses involves new types of interactions and requires physical concepts that often differ considerably from those used for longer pulses. Because this will be the future of ablation also with respect to optical materials, new concepts have to be developed to describe ablation with pulses shorter than 1 ps. However, this field has not proceeded very far for optical materials. So far, studies on the interaction of ultra-short laser pulses are mostly restricted to basic interaction mechanisms [5, 6], pulse length scaling laws [7–9], and damage measurements [10–12]; however, ablation measurements are scarce [13–16]. Hence, this chapter is fully devoted to phenomena relevant in the nanosecond and picosecond time regime, and the femtosecond regime is left to a forthcoming review.

12.2 Experimental Techniques

Laser ablation studies from optical materials have been performed in a variety of experimental configurations, including measurements in air, in a well-defined gas atmosphere, and in ultra-high vacuum. The choice of the environment depends either on the application or on the technique that might require a specific atmosphere. For instance, an ultra-high vacuum is required for basic research with well-defined surface conditions, whereas an applied study of optical components under realistic working conditions is performed in air. On the other hand, techniques like time-of-flight or quadrupole mass spectroscopy used for desorption measurements require at least a high vacuum, but the prerequisite for a detection of shock waves driven by ablation is a gas atmosphere.

According to the broad spectrum of applications for optical materials, ablation studies have been done with laser light covering the entire optical spectrum from the infrared to the ultraviolet. By far, the most studies have been done with ns pulses but some also with mode-locked systems providing ps time resolution. Light sources include the free-electron laser; infrared lasers like the Nd:YAG laser and often their frequency doubled, tripled, or quadrupled radiation in the visible and ultraviolet spectral range; and typical UV lasers like nitrogen and excimer lasers. Only few experiments involving dye lasers have been reported. The reason for this is threefold:

1. This is done for experimental simplicity because most ablation mechanisms are not very sensitive to wavelength.
2. Some ablation measurements require fluences that cannot be provided by dye-lasers.
3. Many ablation studies have been designed to gain information for an improvement of the damage resistance of optics for standard laser systems and are, therefore, performed at the relevant wavelength.

Generally the techniques used for ablation studies on optical materials do not differ considerably from those used for other materials. Therefore, the main experimental approaches are only briefly introduced here and more details are found in other chapters of this book. The only exception is the photoacoustic shock-wave detection that is a key instrument for ablation threshold determination of bulk as well as thin film optical materials and will be described in some detail here. Although the primary interest of this chapter is ablation and not damage in optical materials, the last section is devoted to the investigation of the damaged surface because the damage morphology often provides important insight into the ablation process and because damage inspection is most important for the practical purpose of optical device testing.

12.2.1 Mass and Energy Analysis

The analysis of laser desorption and ablation from optical materials involves the detection of electrons, ions, neutrals, and even small clusters emitted from the surface. The simplest device to measure these species is a collector plate or a linear flight tube with a channel plate at the end [17]. Such an arrangement covers a large solid angle and therefore is most sensitive. The polarity of ions detected can be chosen by potentials of grids in the flight path. Gating of these grids allows a discrimination between fast electrons and slow ions and, in a long flight tube, also a time-of-flight energy analysis (TOF). A mass separation, however, is only possible in rare cases where the energy spectra of different masses do not overlap. The ionization of neutrals is mostly done by a conventional filament technique or by photo-ionization in a strong UV laser pulse. The latter method again can be used for gating to obtain time-of-flight information where the utmost sensitivity in the detection of single desorption events can be obtained with the resonance ionization technique [18]. An extension of the flight tube is the reflectron geometry providing both energy analysis and mass analysis at the same time [19]. In fact, this is one of the most advantageous techniques because it combines a high transmission with the capability of recording a rather complete piece of information for single laser shots. This is especially important for studies of highly nonlinear processes that strongly depend on surface properties where results may vary from shot to shot and averaging does not yield consistent results. In some cases, however, a very precise separation of masses is required, and a quadrupole mass filter is used for ion and neutral analysis [20]. At a further cost in transmitted flux, energy resolution can be obtained by a Bessel box or hemispherical energy analyzer.

12.2.2 Spectroscopic Techniques

Other very sensitive and element-specific methods for the detection of particle emission that have been used for optical materials are based on the detection of fluorescence. The fluorescence either is a result of excitation during the ablation process or occurs when particles desorbed in the ground state are excited in a laser beam. If the latter laser-induced fluorescence (LIF) technique is used with a pulsed laser, again a TOF analysis can be applied [21]. With a precise spectral analysis in combination with TOF, it is possible to obtain translational, vibrational, and even rotational temperatures of desorbed species [22]. Very common is the detection of fluorescence light by a simple arrangement based on a monochromator and photomultiplier while also more sophisticated arrangements like optical multichannel analysis (OMA) are used. Such CCD-camera-based techniques would also allow an analysis of the plasma plume during ablation. However, plasma diagnostics have been used to investigate optical materials in very rare cases only.

12.2.3 Photoacoustic Shock Wave Detection

The measurement of ablation by a detection of particle-driven pressure waves in the adjacent gas is a simple but also sensitive and precise way to obtain information about surface heating and the ablation threshold of optical materials under normal atmospheric conditions.

The technique is based on the detection of a transient pressure wave that is created by the sudden compression of the gas in front of the sample [23]. If the sample is laser heated, rapid thermal expansion of the gas and/or surface launches an acoustic wave (*thermal piston*) propagating with sound velocity. In the case of ablation, however, a shock wave is driven by material ejected with supersonic velocity. Such a shock wave is characterized by a high initial amplitude and a propagation velocity that rapidly decays to the parameters of the acoustic wave propagation limit as a function of distance from ignition. The process of shock-wave generation is well understood and, therefore, with a carefully calibrated experimental setup, the energy content of acoustic and shock waves can be measured. Thus the technique can provide quantitative data for the energy transfer of the laser light to the ablation region [24].

In a typical experiment, the amplitude of the pressure pulse is recorded as a function of the incident laser fluence. As schematically depicted in Fig. 2, three regions can be identified for fluences above the detection threshold F_D. First, a linear increase in amplitude indicates surface heating and the propagation of an acoustic pulse (*pulsed photoacoustic effect*). Above the ablation threshold F_A, a shock wave with hypersonic speed is emitted and the amplitude rises steeply as a function of fluence. At fluences well above the threshold, the saturation of the signal amplitude indicates absorption of laser energy by the plasma [25].

Pressure pulses are normally detected by the mirage effect, i.e., the deflection of a probe laser beam in front of the sample surface due to the transient gradient in the index of refraction of the gas [26]. The principal scheme for the mirage

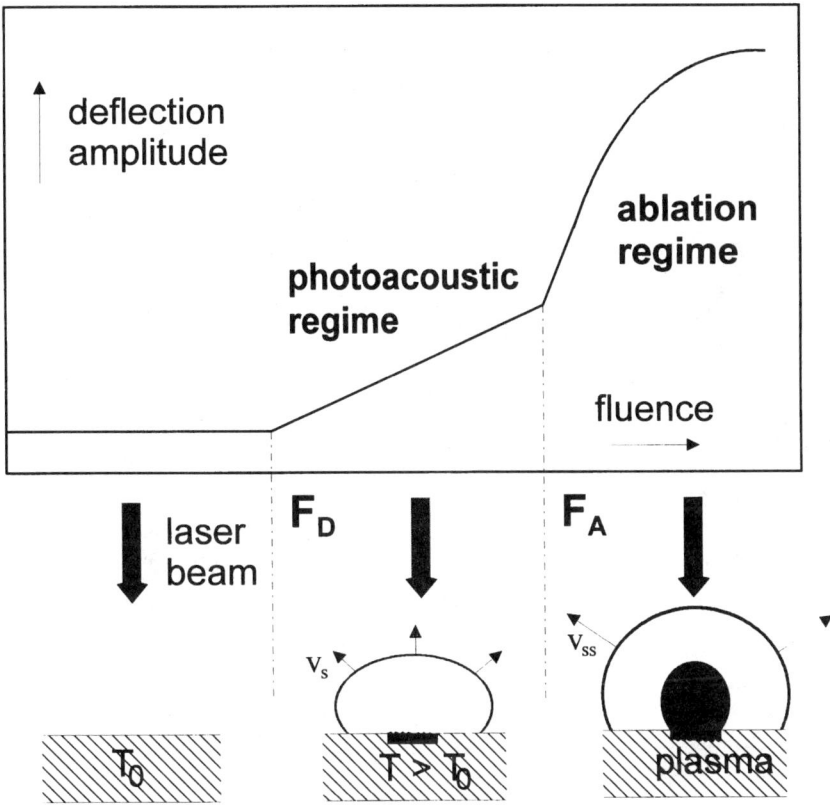

FIG. 2. Scheme of the laser fluence dependent amplitude in photoacoustic mirage deflection experiments. The slowly rising acoustic signal above the detection threshold (F_D) propagating with sound velocity v_s is due to the photoacoustic response of the laser heated surface. Above the ablation threshold (F_A), the signal is determined by a shock wave driven by material ejected with supersonic velocity v_{ss}. Signal saturation at high fluences is due to an absorption of laser light by the plasma plume.

detection and a typical setup used for experiments described below is shown in Fig. 3. The heating or damaging laser light from the excimer laser (pump laser) passes a variable attenuator adjusting the intensity and is focused on the sample surface into a spot of some hundred micrometers. Measurement of the incident laser pulse energy with a pyroelectric detector and of the beam profile with an electronic camera system allows an accurate determination of the incident laser fluence. The transient change of the index of refraction in the ambient atmosphere at a distance of some mm in front of the surface leads to a mirage deflection of a HeNe laser beam (probe laser) aligned parallel to the sample surface. Data analysis of the mirage signal is mostly restricted to deriving the

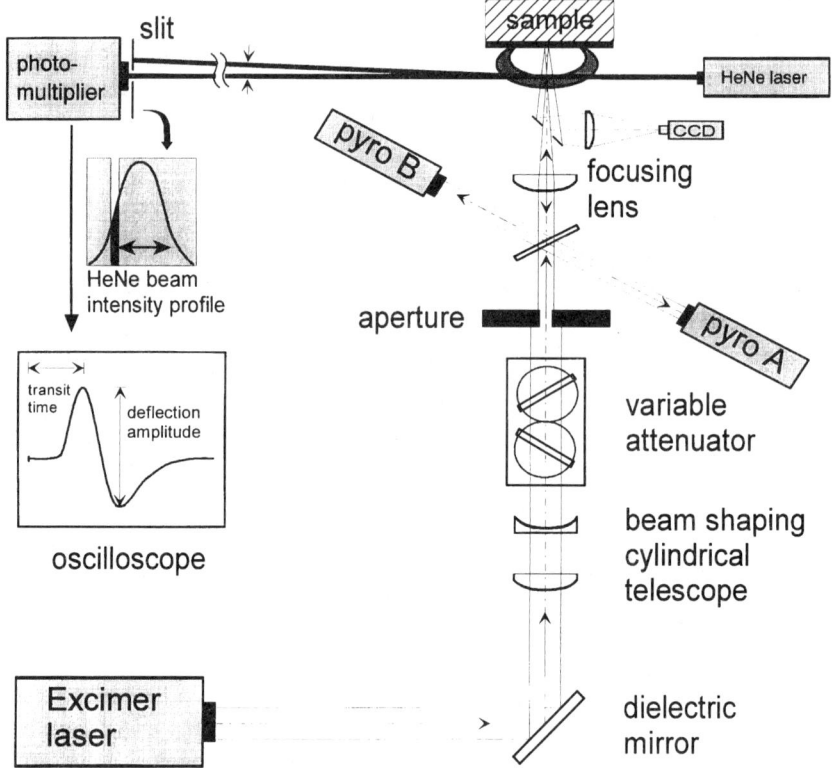

FIG. 3. Experimental setup for mirage detection of acoustic and shock waves during desorption and ablation experiments.

deflection amplitude (peak to peak) and transit time of the acoustic pulse that can easily be obtained from the trace of the recording digitizing oscilloscope. For a more complete data analysis, the signal is integrated or its form is compared to theoretical model calculations.

12.2.4 Surface Inspection Techniques

Analyzing ejecta has certainly helped to characterize desorption and ablation processes, but inspecting the damaged surface can also yield important information. Often a simple observation with an optical microscope allows a decision about the dominant surface process. Numerous more sophisticated investigations with higher lateral resolution have been performed with scanning electron microscopy (SEM) [27]. The ultimate in resolution can be obtained by scanning force microscopy (SFM) that has been used to investigate the topography of ultra-short pulse laser erosion of dielectrics.

12.2.5 Damage Detection

As already pointed out, for the practical purpose of the characterization of optical components and materials, the determination of the damage threshold is most interesting. Various schemes, and often a combination of them, are used to measure damage thresholds; however, by far the most measurements are based on optical inspection using a Nomarski microscope [28]. As will be explained below, the damage threshold is not a very well defined quantity for optical materials, but it may vary drastically as a function of location on the sample. Therefore, the threshold fluence given for a sample normally represents an averaged value. Several schemes have been proposed to obtain such averaged values; the most commonly used are sketched in Fig. 4. Most often the sample is irradiated with a large number of laser pulses with fluences covering a range around the expected threshold fluence, and whether or not damage has occurred is determined with the microscope. The average threshold is defined as the mean value between the highest fluence where no damage occurs and the lowest fluence where damage is detected. Another important parameter for damage testing is the mode of irradiation; i.e., whether single shots or multiple shots are applied. For multiple-shot experiments, the damage threshold also differs for a train of irradiations at a constant fluence or increasing fluence. The former method is significant for the study of incubation phenomena while the latter includes surface cleaning and conditioning effects [29].

12.3 Laser Ablation from Bulk Optical Materials

12.3.1 Bulk optical materials and their properties

Bulk materials in crystalline or amorphous form are the basis of any optical component. Besides the well-known glasses used for mass applications, a great choice of different materials can be selected for a specific purpose. This review is restricted to four of the most common materials, namely, CaF_2, NaCl, sapphire and fused silica; work on other materials is only briefly discussed or referred to. Laser ablation from MgO, which is another important optical material, is discussed in detail in Chapter 3. This selection covers two halide crystals, two oxide crystals, and an amorphous material and is based on practical importance and also represents a group of materials for which most material on experimental and theoretical research is available.

Calcium difluoride (CaF_2) is a crystalline material with one of the largest bandgaps (12 eV) available and, therefore, covers widespread use for windows,

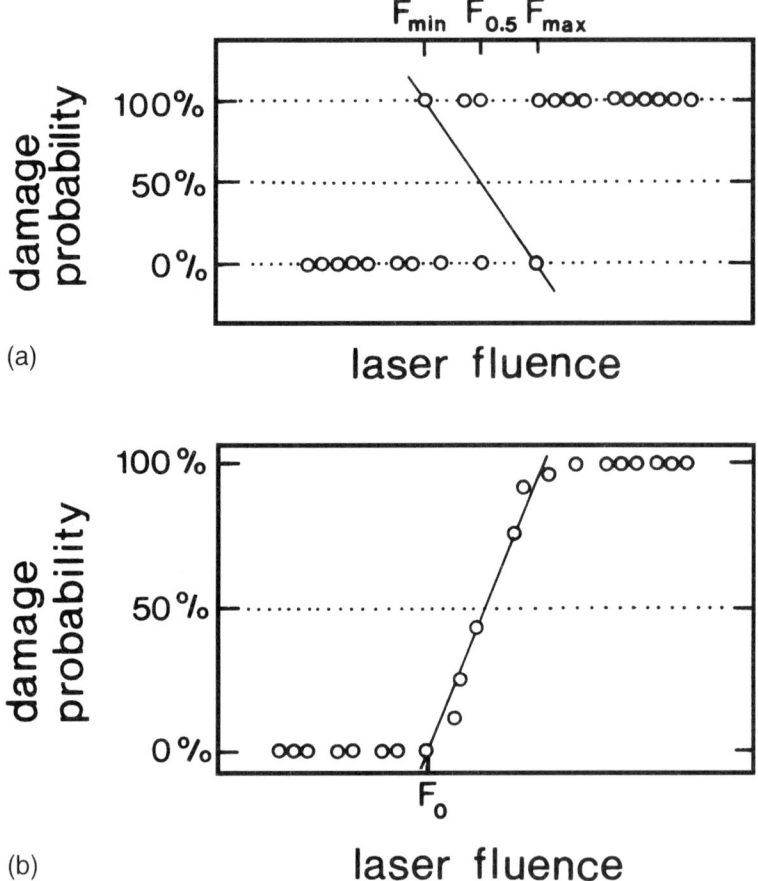

FIG. 4. Averaging schemes for damage threshold analysis. (a) The average damage threshold $F_{0.5}$ is derived from multiple shots of statistically varying fluence applied on different locations of the sample. $F_{0.5}$ is defined as the mean between the lowest fluence causing damage and the highest fluence not causing damage. (b) For each fluence, a number of irradiations is applied and the damage threshold defined as the fluence with 50% damage probability.

prisms, and lenses. In standard quality, it is well suited for most IR applications, whereas its use with UV light often requires high purity material. For example, CaF_2 will certainly be a material of choice for future projection [30] optics in excimer laser lithography [31] at 193 nm where other conventionally used materials fail. CaF_2 is not very hard compared to some of its competitors, but its surface is rather inert with respect to environmental contamination and especially insensitive to water. Sodium chloride (NaCl) has a very high transmission in the IR for wavelengths up to 15 μm and is mainly used in IR spectroscopic assemblies and for CO_2 laser optics. Its practical use is limited by its hygroscopic nature. Magnesium oxide (MgO) is outstanding with respect to the extremely high melting point of 2800°C and, therefore, very suitable as a refractory substrate. Sapphire (Al_2O_3) is the choice for any severe environment application because its hardness of 2000 Knoop is unsurpassed by comparable optical materials. Amorphous quartz (vitreous silica, SiO_2) represents the group of traditional optical glasses and has widespread use where UV transmission is required. Furthermore, it is an important substrate material for optical coatings.

12.3.2 Absorption Mechanisms

Laser ablation from optical materials in most cases involves the interaction of laser light with a nominally transparent material; i.e., at moderate laser fluences, the interaction is negligible or cannot be measured at all. The energy deposition at high fluences is mediated either by some *residual absorption* or by a *nonlinear absorption process*. Cross sections for such processes are generally rather small, and a great variety of subtle effects may contribute to absorption. The key for a successful description of ablation phenomena from optical materials is, therefore, an understanding of the basic mechanisms of energy deposition in the weakly absorbing materials. However, the cause of absorption is often very difficult to identify for a specific ablation problem. In fact, this peculiarity is a major reason for the predominance of speculative and phenomenological models for the description of ablation in optical materials despite three decades of intensive research in this field.

In the following discussion the present knowledge about weak absorption in relevant dielectrics is reviewed. This compilation of material includes experimental absorption data, a description of physical models for energy deposition, and theoretical results explaining some of the absorption phenomena. We first describe intrinsic absorption, i.e., the uptake of radiation energy by a perfect crystal at nonzero temperature. Depending on wavelength and material, this can be a linear process or at high intensities predominantly a multiphoton absorption. Such intrinsic absorption is quite well understood and can be measured in carefully designed laboratory experiments. However, for the practical purpose of optical components testing, they are mostly irrelevant because in this case defect-related processes are dominant, as will be discussed in the next section. Defects can be point defects in the perfect lattice, structural imperfections, impurities, aggregates of these species, or even a combination of them, e.g., defect or impurity aggregates localized at dislocations. A dynamic description of defects is required when they are created during laser irradiation and act as absorption centers for the remaining part of the laser pulse or following pulses.

Such a phenomenon is called *incubation* and may increase absorption by orders of magnitude. The last section is devoted to the absorption properties of surfaces and surface defects. These are most relevant for optical materials because surface finishing, always inducing defects, is a prerequisite for any optical application and the surface of the optics is exposed to environmental influences that most likely induce absorption.

12.3.2.1 Intrinsic Absorption. Optical materials are dielectrics with a bandgap of several eV; i.e., an energy gap larger than the photon energy normally applied. The intrinsic absorption mechanisms that may occur in a pure material are sketched in Fig. 5 where energies are drawn to scale for a possible

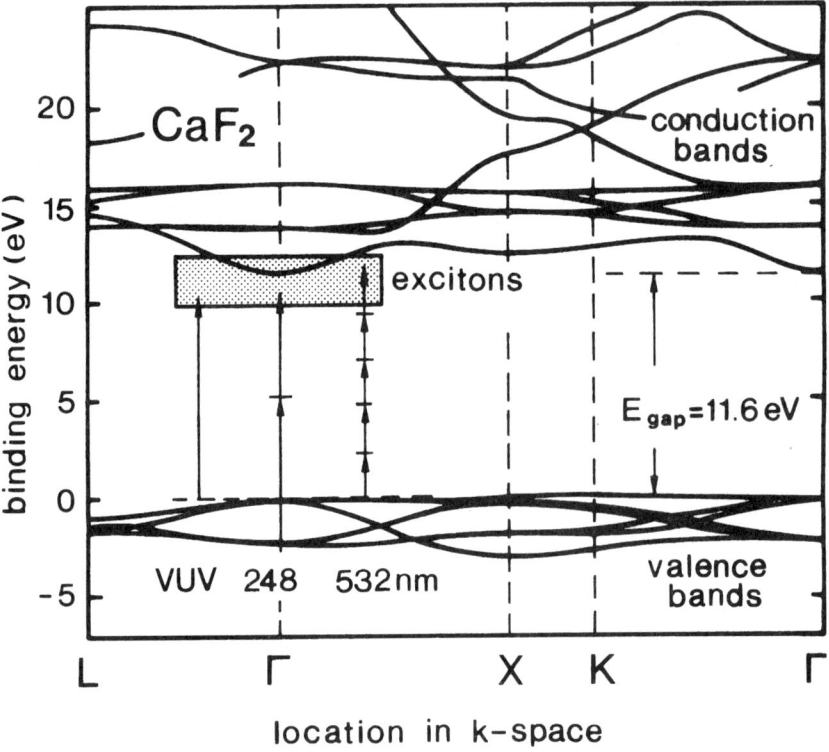

FIG. 5. Absorption scheme for intrinsic one-photon excitonic absorption and multiphoton absorption in a pure bulk crystal. Energies are drawn to scale for CaF_2 and photon energies of 2.3 eV, 5 eV and 10 eV. The scheme demonstrates that two-photon absorption into excitonic states is possible for 248 nm light of an excimer laser; however, for an interaction with frequency-doubled light at 532 nm from a Nd:YAG laser, a five-photon process would be required. The band structure is drawn according to a theoretical calculation [137]; however, the value for the bandgap in CaF_2 was adjusted to an experimental value [138].

absorption of 2.3, 5.0, and 10 eV photons by CaF_2, representing an interaction of 532 nm doubled Nd:YAG laser light, 248 nm excimer laser light, and VUV photons with the crystal.

The simplest process is a linear absorption of a VUV photon (≈ 10 eV) via an interband transition from the top of the valence band into the ground state of the exciton associated with the bottom of the conduction band or into the bottom of the conduction band itself. Such transitions have been identified as pronounced peaks in VUV reflectance spectra [32] of bulk crystals, in the energy loss structures of energetic electrons passing through thin CaF_2 films [33] and recently by spectroscopic ellipsometry [34]. In these spectra, a pronounced peak at 11.2 eV was attributed to the first excitonic interband transition at the Γ point, while the band edge is defined by an absorption shoulder at 11.6 eV. The onset of excitonic absorption is seen as a steep increase in absorptivity in a logarithmic plot of the absorption constant as a function of photon energy [35], as shown in Fig. 6. Detailed analysis of the data reveals that the dependence of the absorption coefficient on photon energy can be well described by an exponential function and the onset of absorption is temperature dependent. The temperature dependence of the exponential absorption tail is a well-known phenomenon in dielectrics and is referred to as the *Urbach rule*. The lineshape and shift with temperature of the excitonic absorption is a result of a strong electron-lattice coupling that is especially present in halide crystals [36, 37]. According to the absorption data, 10 eV is expected to be an energy with some interaction with CaF_2 at room temperature.

A linear absorption of 10 eV photons is not a relevant process for the discussion of laser-solid interactions, but two-photon absorption of 248 nm (5 eV) light from a KrF excimer laser can be expected. Highly sensitive calorimetric experiments with 25 ns pulses could not provide evidence for such a process [38], but a measurement of luminescence excited with 280 fs pulses yielded strong indications for two-photon absorption into the excitonic state [39]. However, there are also contradictory results [40] and such nonlinear measurement might also be influenced by defect-related absorption, which is discussed below. For 532 nm photons, a nonlinear process involving at least five photons would be required for an excitation. The cross-section for such a high order process is extremely low and not expected to be measurable for nanosecond pulses.

Multiphoton absorption of orders up to four have been reported for NaCl with a bandgap of 8.5 eV, i.e., much smaller than that of CaF_2. Such experiments have been carried out with picosecond pulses on ultrapure crystals reducing extrinsic effects to a minimum and consistent results obtained by several authors confidently established the nature of the observed nonlinearities as intrinsic multiphoton absorption. With a highly sensitive photoacoustic technique, the energy deposition could precisely be measured and follows the fourth power for

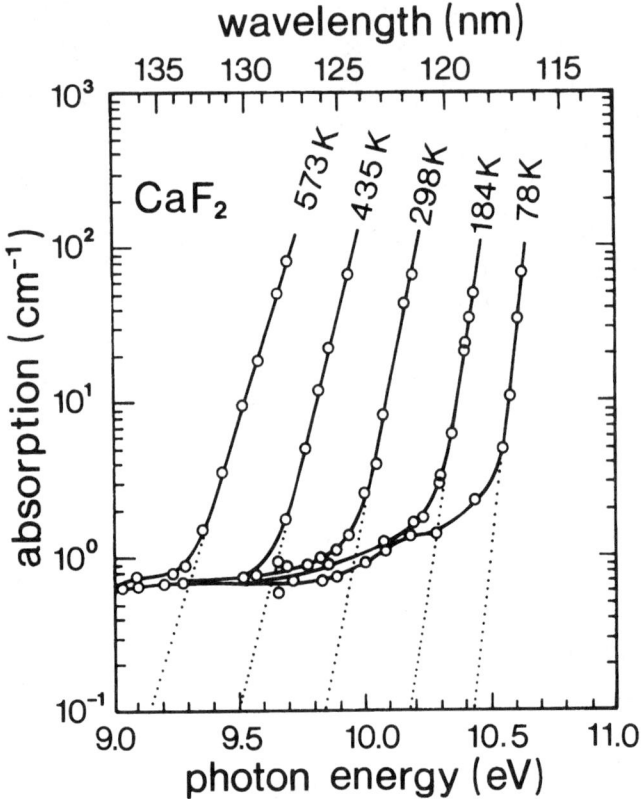

FIG. 6. Absorption spectrum of CaF$_2$ in the region of the UV absorption edge for various crystal temperatures. The temperature-dependent exponential tail extending into the bandgap region is known as the Urbach tail [36, 37] and determines the transparency region. (From Ref. 38, reprinted with permission.)

532 nm photons as shown in Fig. 7 [41, 42]. Two-photon absorption at 266 nm can directly be measured in transmission [42].

12.3.2.2 Incubation and Transient Absorption. The dynamics of two-photon excitation in NaCl has been studied with a pump-probe technique measuring the transient absorption of a 532 nm pulse after excitation with 266 nm for 30 ps pulses [43]. It was found that excitons are formed within a time shorter than the experimental resolution of this experiment. An important observation was, however, that after the formation of excitons, the absorption signal had two components, a transient component decaying within nanoseconds at room temperature and a stable absorption prevailing on a macroscopic timescale. The latter component was attributed to the formation of stable F

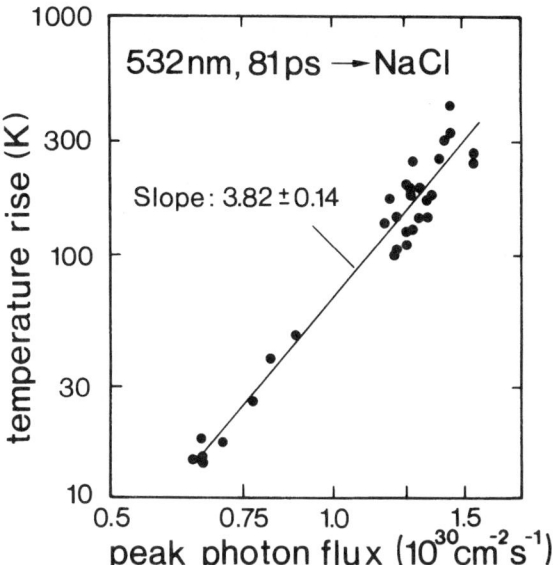

FIG. 7. Energy deposition (i.e., transient temperature rise) from 532 nm (81 ps) photons in ultrapure NaCl detected by a calibrated photoacoustic transducer. The intensity-dependent slope of the absorbed energy indicates a four-photon process as expected from the wavelength and the bandgap of NaCl. (From Ref. 41, reprinted with permission.)

centers induced by the multiphoton absorption. The formation of stable color centers following electron-hole pair creation is a well-understood process in halides and other dielectrics, and it follows a sequence of self-trapping, i.e., localization of the exciton and a consecutive separation into a stable F–H center pair [44]. Self-trapping is due to the strong electron-lattice coupling and the efficiency for the formation of stable F centers strongly depends on temperature and material [44]. F centers are readily formed in alkali halides like NaCl [45], but due to the more complicated structure for example, of CaF_2, a permanent coloration of this material without an additional excitation is much harder [46].

The formation of stable defect centers by laser irradiation [47] implies a dramatic and often irreversible change of the absorption properties of the irradiated material and is commonly referred to as *incubation*. Incubation phenomena have been observed in many materials and do not necessarily need intensities provided by lasers [48], but in the high fluence regime, they may have a significant, sometimes even dominant, influence on ablation. Strong color center formation following multiphoton absorption has, for example, been observed in fluorides [38] and silica glasses [49]. However, incubation is not only limited to the formation of point defects, but laser light can also create and stabilize larger scale imperfections, which are treated in Section 12.3.2.3 [50].

The energy transfer from laser light into the material may be enhanced not only by the described permanent laser-induced absorptivity changes, but also by a transient absorption from electrons in the conduction band. This phenomenon called *free electron heating* has been described theoretically [51] and demonstrated with picosecond [52] and femtosecond [53] laser pulses in SiO_2 and halides. The idea behind this concept is an uptake of energy by electrons from a strong laser field and, following electron impact, ionization creating more charge carriers. Although this electron heating mechanism can be initiated by a single electron in the conduction band, multiple steps of absorption and ionization will create an *electron-avalanche*, i.e., a plasma of excited electrons in the conduction band that is cooled by electron-phonon collisions. The absorption of laser light by the plasma is wavelength dependent, and the process is found to be much more efficient in the infrared spectral range than for ultraviolet light.

12.3.2.3 Impurity-Related Absorption.

Light-induced absorption centers as described in Section 12.3.2.2 are not the only cause of absorption from states in the bandgap region. If crystals are not grown with extreme care under high-purity conditions, they contain at least traces of impurities that may act as centers for linear or nonlinear absorption. Also surface preparation such as polishing introduces impurities and dislocations that may act as strong absorption centers, as will be discussed in Section 12.3.2.4. The general schemes for impurity-related absorption is sketched in Fig. 8, again illustrated for the CaF_2 band structure. Like point defects, impurities generally create occupied electronic states in the bandgap that may interact with the laser light via linear absorption or a multiphoton process. The latter may, however, be resonantly enhanced by unoccupied states in the upper part of the bandgap [54]. Generally it is found that the density of occupied states in the bandgap falls off rapidly with increasing energetic distance to the valence band and is close to the experimental detection limit at the Fermi energy (see Fig. 9 in Section 12.3.2.4 as an example for CaF_2). For nominally pure CaF_2, the Fermi level is found about 3 eV below the conduction-band minimum [55]; i.e., for light with a wavelength shorter than 400 nm, linear absorption from bandgap states into the conduction band is the most probable process.

Nonlinear absorption of 308 nm laser light with intensities below the ablation threshold has been studied in standard quality halide crystals, namely, NaCl, KBr, KCl, LiF, NaF, CaF_2, and MgF_2, with pulsed photothermal deflection. Second- and third-order dependences were found for the increase of the intensity-dependent signal amplitude, indicating that multiphoton absorption is the dominant process for energy deposition in these crystals [56]. However, this is a speculation since a clear correspondence between the photon energy of 4 eV and the bandgap, excitonic, or impurity energies could not be given. Spectral features and a second-order intensity dependence indicated multiphoton absorption by Ce^{+3} impurities in a study of linear and nonlinear photoconductivity from charge-producing defects in CaF_2 [57]. Two possible mechanisms involving charge-compensating F^- interstitials were proposed, however, these interpretations also remain speculative. At first, this appears to be surprising because trivalent and other impurities in alkaline-earth halides in general [58, 59] and in special Ce^{+3} ions [60] and F^- ions [61] have been well-studied species in earth-alkaline halides for a long time. However, there are two fundamental problems in assigning absorptions to certain defects. First, the spectroscopic features in the impurity-containing crystal are often very broad and unspecific due to a manifold of configurations and broadening mechanisms. Second, these crystals mostly contain a variety of impurities with overlapping effects. To isolate effects from one species would require a careful study of optical properties as a function of doping with this impurity in combination with other methods like electron paramagnetic resonance allowing a clear-cut identification. Such an elaborate study has never been presented and would, even if existing, only be of limited value because results are only valid for crystals exactly reproduced from the original ones.

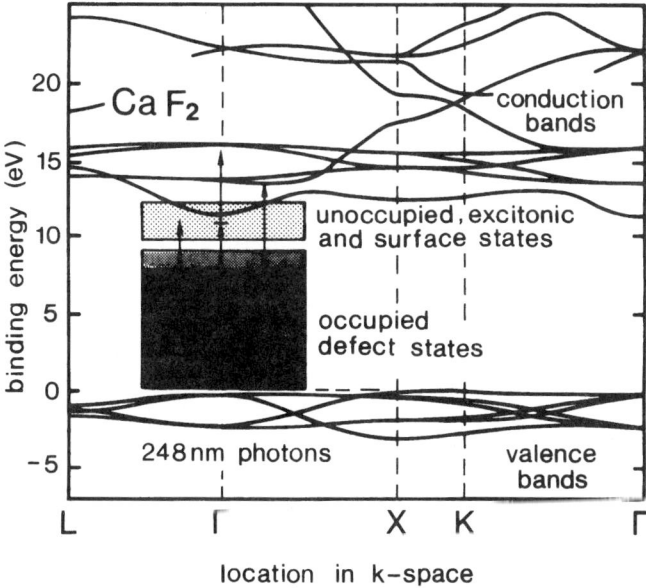

FIG. 8. Absorption scheme from impurity-related states in the bandgap of CaF_2. The lower shaded area denotes a band of occupied defect states with a density of states strongly decreasing with energetic distance to the valence band. The upper band in the gap is formed by unoccupied excitonic, surface, and defect states that are populated by optical transitions and may serve as resonant intermediate states for two-photon processes.

The description of absorption is even more phenomenological and speculative for macroscopically sized impurities in optical materials. Such impurities have been described as cracks, pores, and inclusions, and they quite frequently are limiting for the use of optical components [62]. The energy deposition at such inclusions is generally difficult to calculate. As a first approximation, they can be assumed to be spherical particles with a certain absorptivity, thereby neglecting the influence of more complicated shapes, interface effects, electrostriction, and other higher order effects [63]. More elaborate approaches include effects of self-focusing [64] and electrodynamic field instability [65]. For the latter case, it was shown that due to laser light intensity dependent changes in the refractive index of inclusion and host, the electric field can be strongly localized, leading to extremely high amplitudes at the inclusion site. This involves a threshold behavior for the absorption that cannot be anticipated from the linear absorption properties of the inclusion.

12.3.2.4 Surface Absorption. One of the most important factors for the discussion of absorption phenomena related to ablation from optical materials is surface absorption. The electronic properties and, hence, absorption of the perfect surface differ considerably from the bulk, and for any optical application, a surface preparation is needed that mostly introduces additional

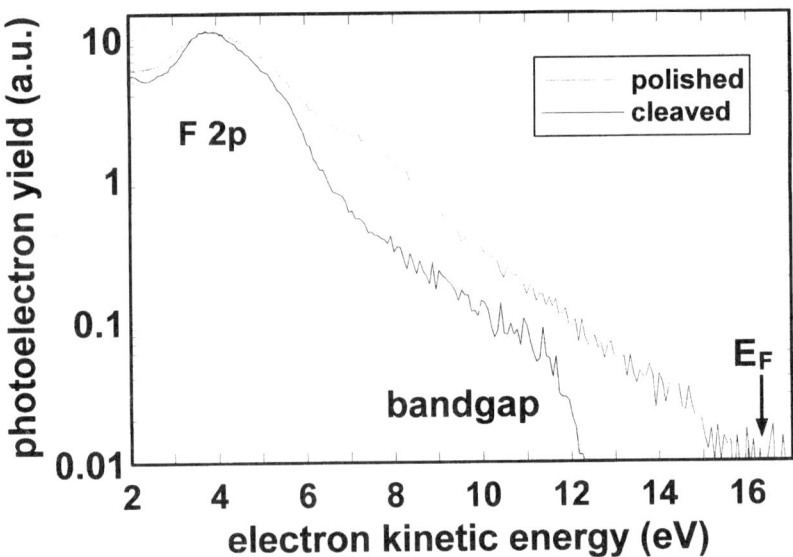

FIG. 9. Ultraviolet photoelectron spectrum obtained with 21.2 eV light from a cleaved and a polished CaF_2 surface. Polishing introduces a considerable amount of occupied states in the bandgap reaching up to the Fermi level. (from Ref. 78, reprinted with permission.)

absorption. Furthermore, the surface is directly exposed to the environment with a possibility of contamination and degradation during preparation or handling. Despite the importance of surface processes, this is the field where the least conclusive results have been presented, and if data is available, they are mostly based on phenomenology.

Surface contaminations are most relevant for infrared applications where the radiation can resonantly interact with the contaminant molecule. Absorption by water and alcohols present in cleaning agents for optical surfaces have been identified for KCl and CaF_2 at 2.8 and 3.8 μm with highly sensitive laser calorimetry [66, 67]. It was found that only strong exposure producing a thin film on the surface causes a measurable effect. Subtle effects on absorption and laser damage threshold have also been found for 10 μm during aging experiments on halide windows and dielectric coatings in a humidity chamber [68]. In this case, aging could be correlated with morphological changes of the surface and laser damage results.

In the UV spectral range, the surface electronic structure is most important for absorption of a reasonably clean surface; e.g., a sapphire surface subjected to excimer laser light exhibits linear absorption of up to 3% of the laser pulse energy [69]. The absorption decreases with increased laser wavelength from 193 to 308 nm and can be greatly reduced by chemical etching of the surface. From these observations, it was concluded that the absorption presumably arises from F and F^+ centers present at the surface.

For CaF_2, possible absorbing states of the perfect and defective surface have been investigated theoretically, and it was found that such states are positioned in the bandgap close to the bottom of the conduction band [70, 71]. Because some of them are unoccupied, surface states provide possibilities for light absorption over a broad spectral range. Absorption can either take place from surface states below the Fermi level into bulk excitonic states and the conduction band or into surface states above the Fermi level. Ultraviolet photoelectron spectroscopy (UPS) [72] and inverse photoemission [73] provided some information about occupied and unoccupied surface levels in CaF_2. However, an unambiguous identification of their nature has never been presented nor was any kind of light absorption clearly attributed to surface states.

The situation is even more complicated for real surfaces. The structure of the cleaved CaF_2 surface has been subject to investigations with various levels of lateral resolution [74–76], revealing a large number of surface features ranging from simple cleavage steps and tips to triangular hillocks that undergo transformations when heated to temperatures of about 1000°C. The first attempts to model the imperfect surface theoretically and to predict the electronic structure are restricted to the simplest features [77]. A direct measurement of occupied states in the bandgap with UPS shows that a continuum of states extends from the top of the valence band far up into the bandgap [78]. In Fig. 9, the density of occupied bandgap states of a cleaved crystal is compared to that of a polished surface. It is found that a detectable density of states extends to energies of about 6 eV above the valence band for the cleaved surface, whereas it reaches the Fermi level for the polished surface, dramatically demonstrating the impact of surface preparation on absorption. In summary, it can be said that surface absorption is essential to a complete understanding of the interaction of laser beams with optical crystals; however, an identification of specific causes of absorption is even more difficult than for the bulk.

12.3.3 Desorption and Ablation Processes

The deposition of laser light energy in the material as described in the previous section is the first step in a sequence finally resulting in material emission. In this section, the main processes following the absorption are discussed. It has to be pointed out, however, that for higher fluence levels the uptake of energy is not at all restricted to the primary process of light absorption, but some of the ablation scenarios involve a strong further interaction of the light field with the specimen or ejecta. This is one major reason why a large amount of energy can be transferred to the dielectric having only an extremely small initial absorptivity and to a large extent explains the extremely strong nonlinearity observed for some of the ablation phenomena.

The description of laser-induced material removal can roughly be divided into three regions characterized by the change in emission yield as a function of laser fluence. These regions are schematically depicted in Fig. 10(a), however, this represents a general guideline only, and the results for specific materials may differ, e.g., in a way that certain steps extend over a very narrow fluence range.

- Region I is the desorption region where the emission of single particles is initiated by the primary absorption and the emission events are all independent from each other. In this region, the yield follows the intensity dependence of the absorption mechanism and normally does not exhibit any threshold behavior; i.e., the lower limit of particle detection is limited by the sensitivity of the detection method.
- Region II is where the intensity dependence is modified by the onset of collective effects involving a nonlinear interaction of the laser light field with the material even if the primary absorption process is linear. The enhanced interaction normally yields such an increase in particle emission that the transition from region I to region II clearly defines an ablation threshold and particle energies are so high that a plasma is ignited. This is the intensity regime with the largest slope in the intensity dependence because the effects of nonlinear absorption are combined with nonlinearities from the plasma interactions. As a rule of thumb, the slope of the yield curve is high for a

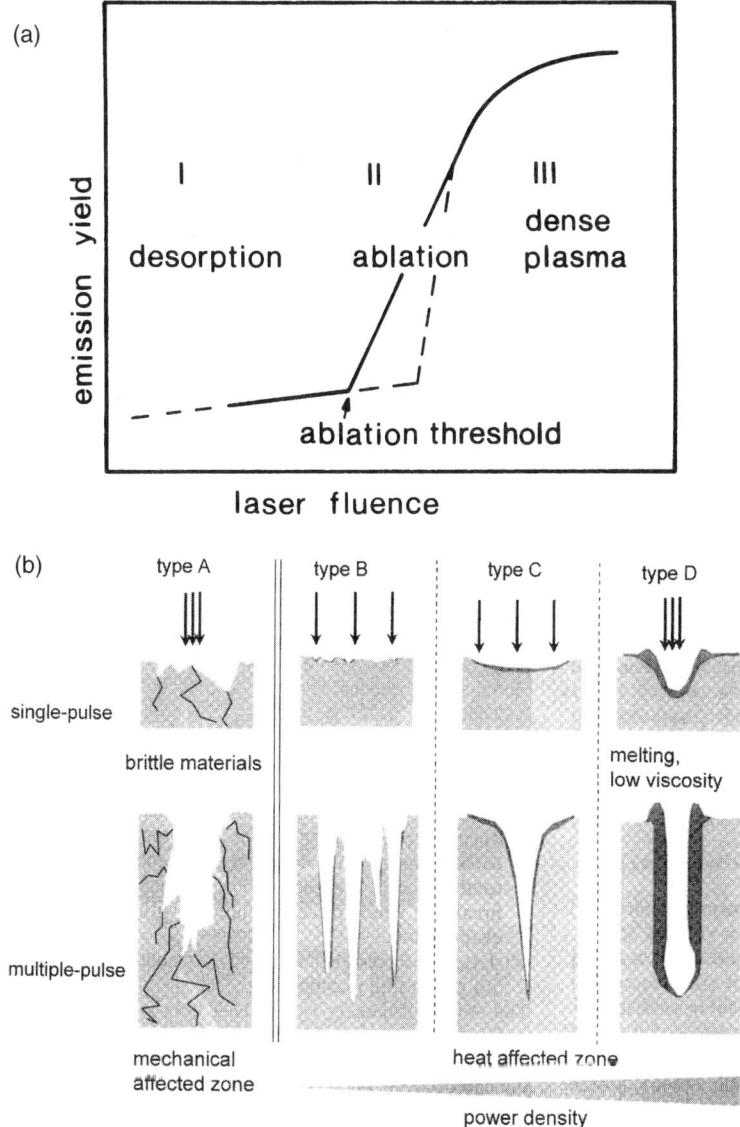

FIG. 10. (a) Scheme for the description of laser sputtering from optical materials in various fluence regions. Region I is the desorption region for low-yield single particle emission. The ablation region II is characterized by a super-linear increase of the ablation yield as a function of laser fluence. The intensity-dependent yield saturates in region III where laser light is absorbed by the dense plasma. (b) Classification of ablation processes with respect to material properties (see explanations in text). (From Ref. 79, reprinted with permission.)

highly nonlinear or impurity-related absorption, whereas the yield increases more slowly for a linear and spatially homogeneous absorption.
- Region III is only relevant for pulse durations longer than the time needed for the formation of the plasma because it describes a strong interaction of the laser pulse with the evolving plasma. For rising fluence, an increasing amount of light is absorbed by the plasma leading to plasma heating but a saturation in emission yield.

Beside the problem of laser resistivity of optical components, laser ablation from optical materials is technically most relevant in the context of laser patterning. The specific processes acting during ablation are rather complex and cannot be described by the simple scheme introduced in Fig. 10(a). A phenomenological classification of various types of ablation for single- and multiple-pulse irradiation based on material and process parameters [79] is presented in Fig. 10(b). Type A ablation is operative for brittle materials where fracture due to thermoelastic stress is able to break out whole pieces of material without melting or evaporation. Type B also yields a very inhomogeneous ablation topography, however, in this case, very localized heating due to defects or beam interference effects results in local evaporation and material removal. In type C ablation, the fluence is sufficiently high for evaporation over the beam cross-section, but the vapor pressure is not yet high enough to remove molten material. This type of ablation yields clean but conical holes for irradiation with multiple shots. The best results in terms of structuring control and precision are obtained for type D ablation where the melt is ejected. In multishot experiments with high fluence, it is possible to drill holes with a large aspect ratio by type D ablation.

In the following sections, experimental desorption and ablation results obtained on bulk optical materials with nanosecond and picosecond pulses are summarized and some of the models proposed for their interpretation are reviewed. As pointed out in the introduction, ablation from transparent dielectrics involves a manifold of very complex processes that are very difficult to measure independently and require extensive modeling. This is true also for ablation from other materials; however, in the special case of transparent dielectrics, considerable uncertainty is added due to the lack in knowledge about primary absorption. The high nonlinearities involved in ablation models makes them especially sensitive to variations in the starting parameters, i.e., exactly to the primary source of absorption. Extremely subtle changes in absorption, e.g., due to surface preparation, may not only alter results quantitatively but invoke a completely different ablation mechanism. This is the reason why many experiments published in the literature are frequently not reproducible especially in different laboratories using different materials and preparation techniques. Therefore, results presented in the following sections are a cross-section through available data and interpretations, but they cannot provide a conclusive picture of ablation from bulk optical materials.

12.3.3.1 Laser Desorption from NaCl.

Among the dielectrics discussed here, NaCl is the material where laser sputtering in the desorption regime is best understood. On the one hand, this is due to the well understood mechanism of multiphoton absorption in high-purity crystals as discussed in Section 12.3.2.1. On the other hand, together with MgO, this is the material for which the most extensive experimental and theoretical studies have been carried out.

Sodium chloride is an ionic crystal with strong electron-lattice coupling, and it is well known that electronic excitation in the bulk of such crystals results in the formation of Frenkel defect pairs. In the bulk of a pure crystal, the creation and diffusive motion of such defects occurs regardless of the nature of the excitation; thus, laser and electron sputtering yield very similar results [80]. The fundamental processes of Frenkel pair creation [81] and separation are depicted schematically in Fig. 11. The originally created unrelaxed exciton localizes within picoseconds into a self-trapped exciton (STE) that may separate into a pair of stable F- and H-center defects, i.e., into a halogen vacancy filled with an electron and a covalently bound halogen molecule having one halogen on an interstitial site. This occurs if the lattice provides enough thermal activation to overcome the potential barrier for H-center diffusion. Due to the interstitial nature of the H-center, at the surface this process can directly result in the emission of a halogen atom [82]. The concept of such a molecular ion moving in a collision cascade along the (110)-direction toward the surface has been developed as a general model for electron and laser sputtering and is often referred to as the *Pooley-Hersh-mechanism* [83]. The directional emission of halogen atoms after the creation of F-H pairs by four-photon absorption of ns-ruby laser pulses has been demonstrated for NaCl and other alkali halide crystals [84] yielding direct evidence for the collision cascade transport of halogen to the surface. Representative results of the laser intensity dependent emission yield and the direction for chlorine atoms emitted from KCl are shown in Fig. 12.

This model can only explain the emission of halogen, which, in fact, is the predominant species being sputtered, while metal was found to be emitted in much smaller quantities. Recently the laser induced Na emission from NaCl has been studied in some detail for ns pulses of 400 to 600 nm photons in a fluence regime where multiphoton excitation is excluded [85]. The main result was that for a clean stoichiometric surface, metal emission is negligible, whereas a strong emission signal was found after preparation of the surface with low-energy electrons that are known to produce F-centers. From the dependence of the emission yield from laser wavelength and fluence, crystal temperature, and a pretreatment of the surface with electrons, it was concluded that metal is emitted from regular lattice sites on the surface that are neighbored by F-centers or small

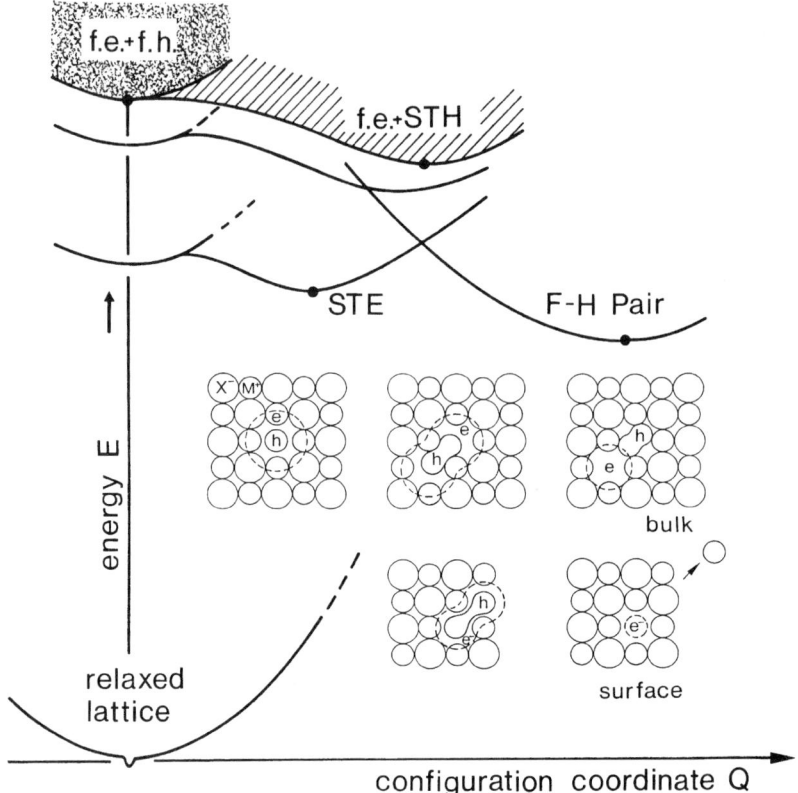

FIG. 11. Schematic diagram of the relaxation of an exciton in alkali halides. The adiabatic potential surface is drawn as a function of the main configuration coordinate; i.e., the (110)-direction in the crystal with NaCl structure. The primary electron-hole pair localizes to form a self-trapped exciton (STE). The STE that is a nearest neighbour pair of an F- and an H-center can separate into next neighbor defects and form independent species. The H-center is mobile and may propagate in a collision cascade along the (110) direction. At the surface, this process results in halogen emission. (From Ref. 81, reprinted with permission.)

F-center aggregates. The interpretation of emission of metal atoms from a slightly damaged surface was supported by theoretical calculations showing that such a process is energetically possible for metal ions in the vicinity of F, F' and F_3 centers [86]. The process of emission requires the localization of energy at the Na ion site and is demonstrated in two steps in Fig. 13. The pictures show the NaCl lattice with the calculated spin densities of the surface F-center and the nearby Na ion. It is seen that the lattice and especially the surface Cl ion closest to the F-center undergo a strong relaxation when the metal ion is separated from

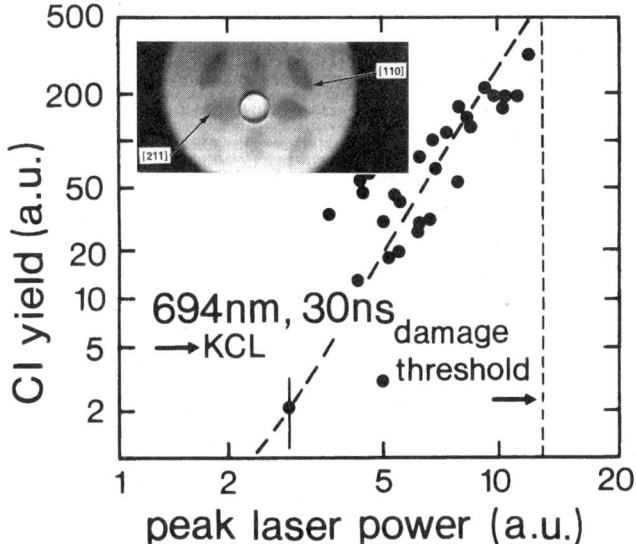

FIG. 12. Intensity dependence of neutral Cl emission from KCl after excitation with 694 nm (30 ns) ruby laser pulses. Atoms were sampled with a mass spectrometer positioned in the (110)-direction of the crystal. The inset shows an emission pattern formed on a photographic plate in front of the sample. (From. Ref. 84, reprinted with permission.)

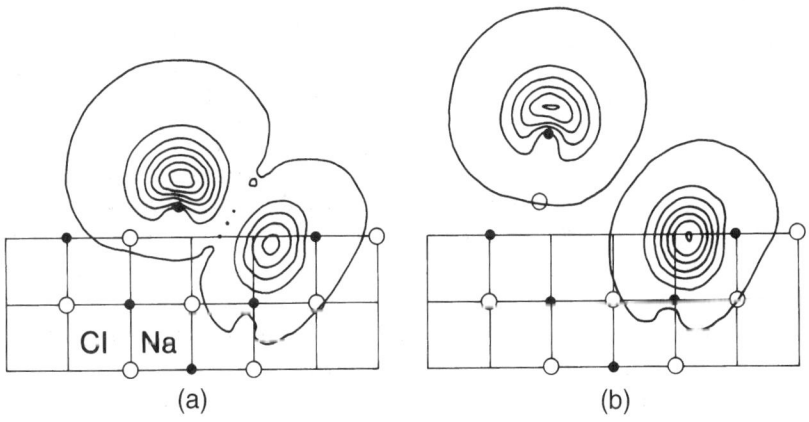

FIG. 13. Theoretical model for steps of the desorption of a Na atom from NaCl(100) via an excitonic mechanism. Contour lines describe the electron density in the vicinity of the involved species. (a) The equilibrium configuration of the 2p excited state of the F-center on the NaCl(100) surface. (b) The scenario for Na desorption and displacement of the closest Cl ion. (From Ref. 86, reprinted with permission.)

the surface. A calculation of possible trajectories for the desorption shows that emission in the (110) direction is likely as it was found for the halogen emission. Similar calculations have been performed for surface sites with low coordination such as kinks and corners, and it was found that the energies required for emission might even be lower for these configurations.

In summary, the desorption of halogen and metal from NaCl and similar crystals can be well described by excitonic processes where the predominant halogen emission is a result of a highly directed diffusion in the bulk while metal emission only occurs at preexisting surface defect sites.

12.3.3.2 Ablation and Desorption from Al_2O_3. Particle emission from Al_2O_3 in the desorption and ablation regime has been studied extensively for several wavelengths in the visible and ultraviolet region. A common goal of all studies was to find evidence for the electronic nature of the primary sputtering process in contrast to thermal evaporation following laser heating.

In first studies, sapphire targets were irradiated in vacuum with 532, 266, 248, and 193 nm, nanosecond pulses of above-threshold intensity laser light, and the fluorescence was observed during laser irradiation while the damaged surface was inspected by SEM [87]. Depending on the wavelength of the exciting laser light, the fluorescence spectra yielded evidence for F and F^+-centers, however, no correlation with any sputtering result could be found. Therefore, color center absorption was ruled out as a primary cause of sputtering. Clear differences with regard to the topography of ablation features were found between results for 532 and the UV multishot irradiation. Surface damage produced by visible light exhibited an exfoliation of flakes and evidence for melting, whereas the UV-radiation produced well-shaped craters with sharp structures reproducing features of the beam intensity profile. With a large number of laser shots, the specimen could be perforated.

As a conclusion, it was proposed that the main mechanism for ablation at 532 nm is heating and thermal stress-induced exfoliation accompanied by melting on asperities, and sputtering with UV light is due to an electronic process. A similar damage topography was found for irradiation with picosecond pulses at 266 nm in air at atmospheric and reduced pressure [88, 89]. As shown in Fig. 14, the etch depth was found to first rise moderately with the number of applied laser shots up to a certain threshold describing the onset of explosive sputtering with a much higher etch rate. In the above-threshold regime, the ablation plume appeared to be much more directional and supersonic ejection velocities have been estimated. Particles collected from the plume had the form of molten droplets indicating hydrodynamic sputtering at these high sputter rates. Typical particle velocity distributions obtained from laser-induced fluorescence studies on ejecta from sapphire irradiated with ns pulses of 248 nm light are shown in Fig. 15. Analysis of the distribution functions revealed a non-Maxwellian distribution and kinetic energies up to 20 eV, depending on laser

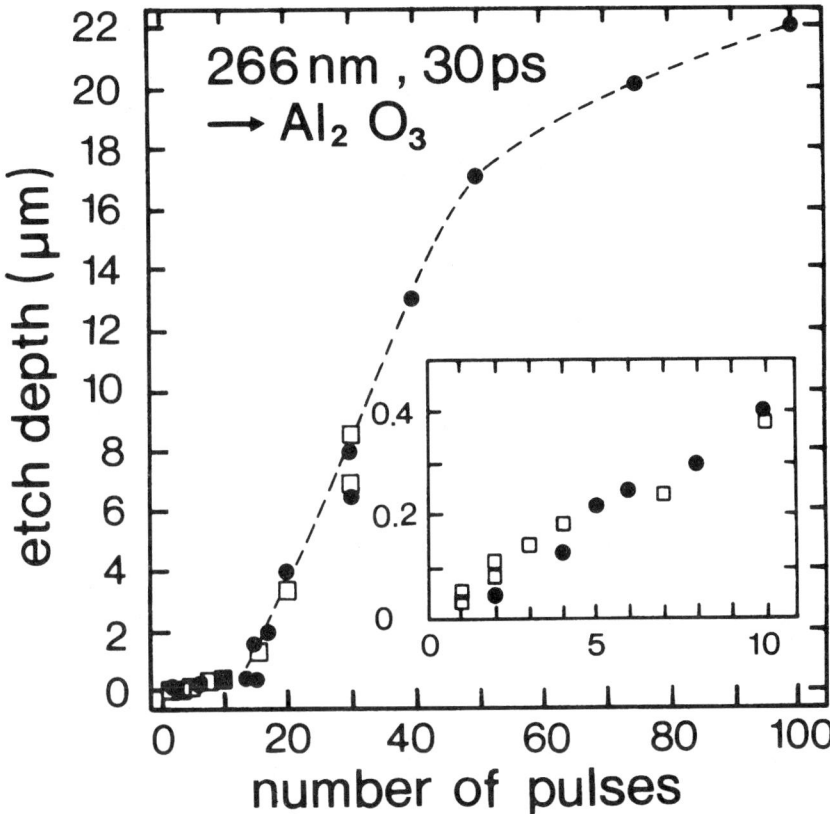

FIG. 14. Etch depth in sapphire for 266 nm (30 ps) laser light as a function of the number of applied shots. The strong rise in etch depth at 15 pulses marks the transition from a *gentle-etch* regime to the *explosive sputtering* regime characterized by massive ejection of material with supersonic velocity. (From Ref. 88, reprinted with permission.)

fluence, and the rotational energies of molecular ejecta were found to be in the order of some hundreds of Kelvin [90]. These findings strongly supported the assumption of a nonthermal origin of ablation. The efforts to clarify the interaction of UV photons with the sapphire surface have been described in Section 12.3.2.4 and lead to the conclusion that the primary source of absorption are surface states from the dangling bonds on the relaxed surface. Pulse pair ablation results for 200 ps, 1064 nm laser light performed under UHV conditions on well-prepared (1120) surfaces showed that even at this low photon energy, the ablation is governed by a low order process ($n = 1$ or 2) and suggests that there are occupied states high in the bandgap of bulk sapphire [91]. In desorption experiments with ns light at 1064 nm and 355 nm carried out at

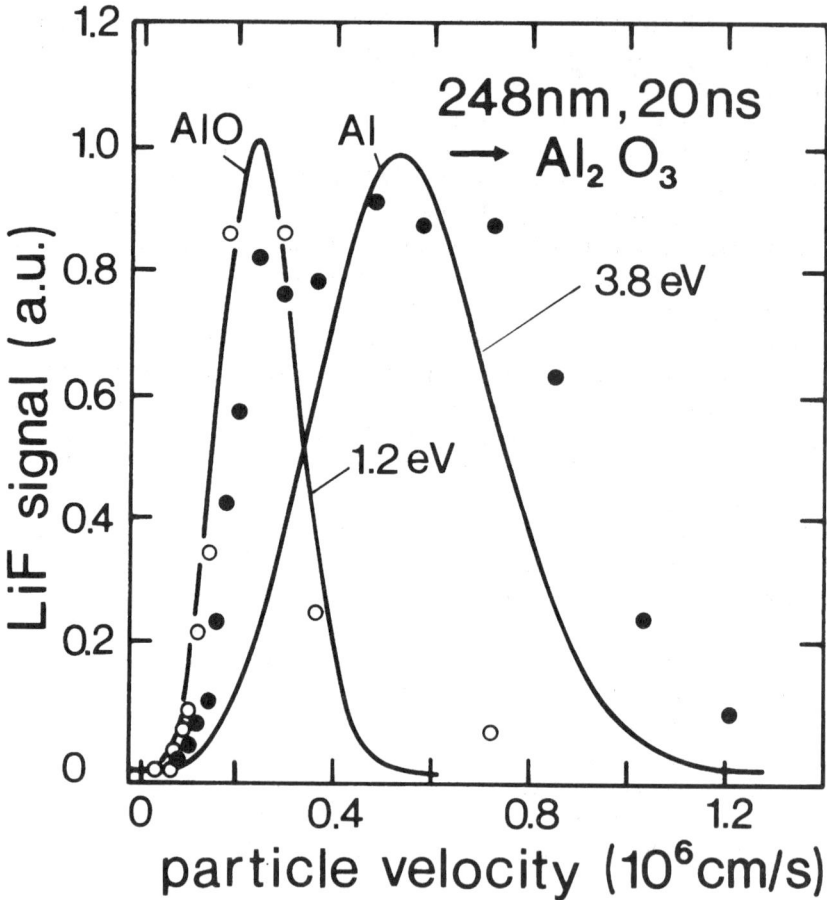

FIG. 15. Time-of-flight velocity distribution of Al atoms and AlO molecules emitted from sapphire after irradiation with 0.7 Jcm^{-2} of 248 nm (20 ns) laser light. The solid lines show Maxwell-Boltzmann distributions fitted to the data. Although these indicate very high translational energies, the rotational distributions yielded temperatures of 500 to 2000 K. (From Ref. 90, reprinted with permission.)

fluences well below the ablation threshold on such well-defined surfaces, Al$^+$ ions were found to be the predominant desorbing species [92]. The kinetic energy of the ions appeared to be about 7 eV regardless of laser light wavelength and fluence again suggesting some kind of electronic sputtering. However, none of the described experiments could yield conclusive evidence for a specific mechanism for the desorption explaining particle energies.

12.3.3.3 Ablation and Desorption from CaF$_2$.

Ablation from CaF$_2$ has been studied for most typical laser wavelengths in the infrared [93], visible [94] and ultraviolet [95]. A common observation of several authors is that desorption below the onset of massive damage is very hard to detect. It may, however, be strongly enhanced when preparing F-centers by irradiation of the crystal with X-rays [96]; i.e., most studies focused on regions II and III in the classification scheme of Fig. 10(a). The predominant species found above the ablation threshold are neutrals, namely, Ca^0 and CaF^0 and less abundant F^0, while the emission of the respective ionic species is much weaker. The CaF$_2$ molecule has never been observed, neither as a neutral nor in an ionic state.

Typical results for ablation at 532 nm (7 ns) are shown in Fig. 16, where the emission of neutral and ionic species from one surface spot is plotted against step-wise increased laser intensity and correlated with the transmission of the CaF$_2$ crystal for the exciting laser light [97]. Particles were detected with a quadrupole mass filter allowing some discrimination between ions and neutrals. It was found that the kinetic energy of the emitted particles covers a wide range, and therefore they were grouped into three categories according to the time of arrival at the detector of the quadrupole mass filter. Fast particles had energies up to several hundred eV and, therefore, were not mass selected. The upper graph also does not exhibit any significant contribution of neutrals because the signal is virtually independent from the ionizer being switched on or off. As can be seen in the middle graph, the result is markedly different for slow particles that were emitted with thermal energies. They add considerably to the total yield, and considering the low ionization probability of the order of 10^{-4}, this measurement confirms that neutrals are the predominant ejecta [93]. The lower figure shows the total yield of neutrals integrated over the 3 sec dwell time between laser pulses and therefore includes delayed emission. The two thresholds observed are due to damage at the rear and front sides of the crystal, respectively. Rear side damage, marked by the sharp decrease in transmission, occurs at lower intensities because it is favored by electric field enhancement at the exit interface. At this threshold, a peaked emission of fast ions and delayed neutrals is observed that is attributed to particles emitted from the front surface that is fractured by the shock wave launched when damage occurs at the rear side. A large yield of delayed particles is observed also above the front side ablation threshold indicating that fracto-emission plays a major role for the ablation process. Indeed, surface fracture along the natural $\{111\}$ cleavage planes of CaF$_2$ has been found to be the predominant single-shot damage result [100, 101] (see Fig. 17(b)) and other phenomena like melting occur only at very high intensities and upon application of a large number of pulses [94]. LIF results suggest that CaF^0 is emitted directly from the surface and was found to be a stable species even in a dense plasma plume excited by a 4 J/cm^2, 248 nm (20 ns) pulse in air [102]. At higher fluences

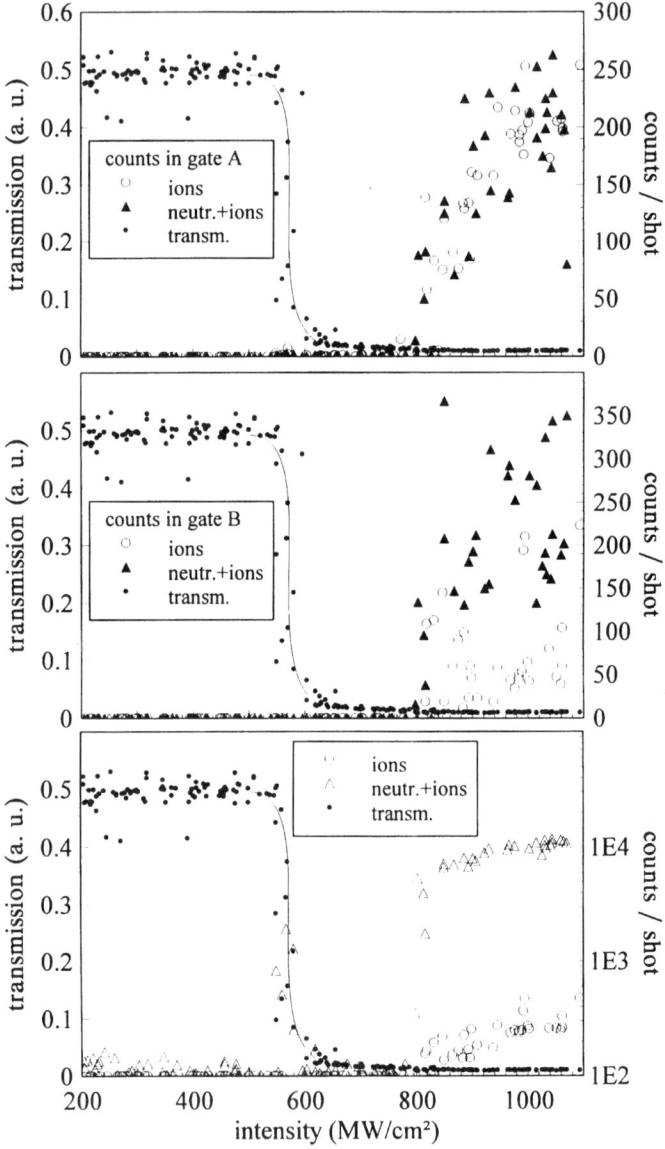

FIG. 16. Ablation products and optical transmission of CaF_2 as a function of irradiation intensity for ablation with 532 nm (7 ns) pulses. Counts in gate A and B are particles emitted with hyperthermal and thermal energies, respectively. The lowest frame shows the total yield including delayed emission. A strong delayed emission from neutrals is found at the intensity of transmission breakdown and from the plasma formed above the ablation threshold. (From Ref. 97, reprinted with permission.)

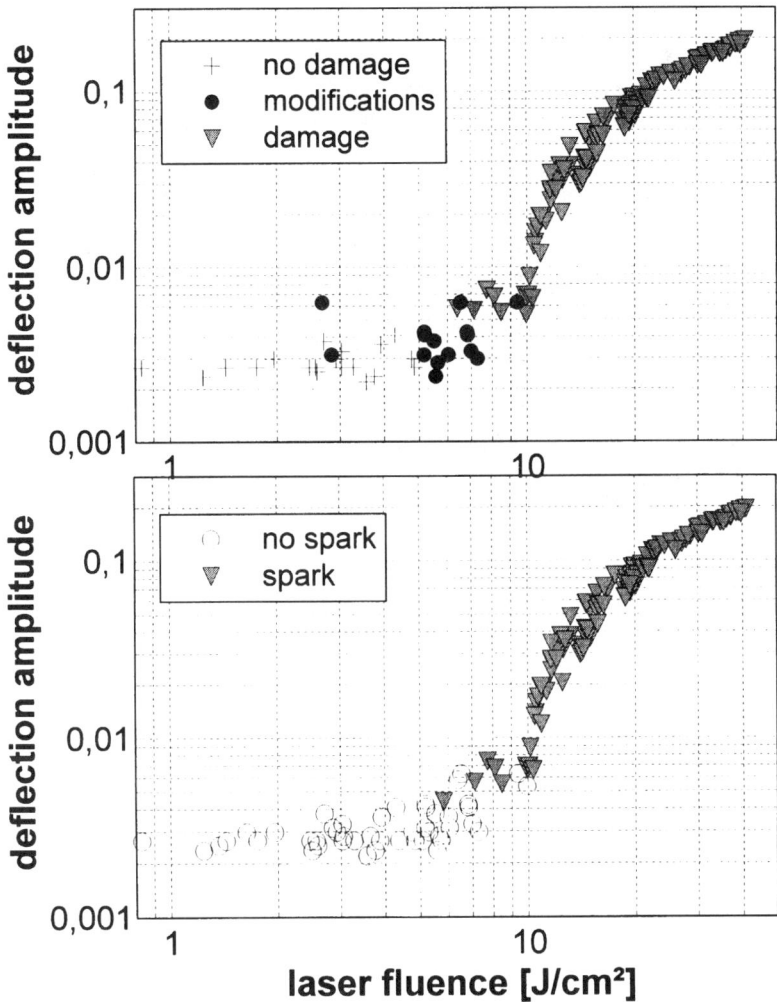

FIG. 17(a). Surface heating and ablation of CaF_2 irradiated with 248 nm (14 ns) laser pulses correlated with the observation of plasma luminescence. *Damage* denotes clearly visible cracking of the surface, while *modifications* are subtle damage effects only detectable from their charging characteristics during SEM observation. Plasma luminescence was detected by eye as *sparks*.

in region II, more ions are emitted with energies of up to several hundred electron volts [103]. This result, together with the observed sharp onset of ablation and the steep slope of intensity-dependent yield curves, strongly suggests that in this case ablation is initiated by subtle absorption discussed

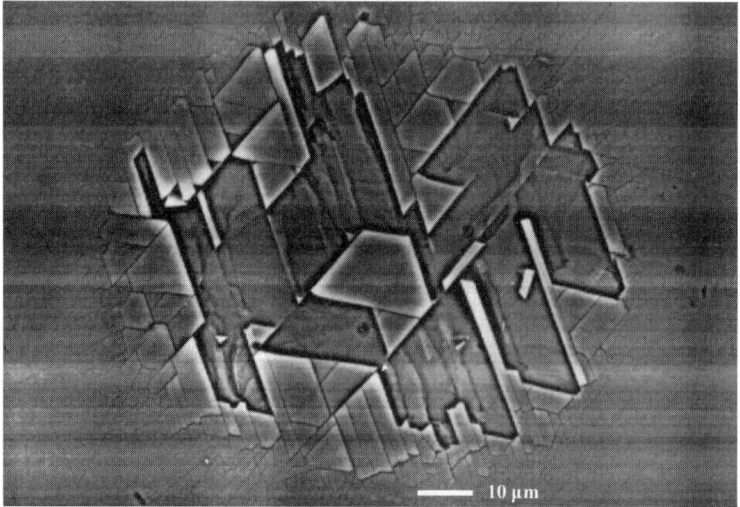

FIG. 17(b). Ablation topography after irradiation with one laser pulse with a fluence well above the ablation threshold. Ablation predominantly occurs in form of massive tiles formed by cracking due to thermoelastic stress in the surface layer. Cracks are aligned in the (111)-directions of the crystal. (From Ref. 78, reprinted with permission.)

below and then driven by the avalanche ionization process (see Section 12.3.2.2).

The onset of ablation and correlation with damage for polished $CaF_2(111)$ has been investigated with irradiation at 248 nm (14 ns) with photoacoustic mirage detection, as shown in Fig. 17(a). The linear increase of signal amplitude below the ablation threshold at 10 J/cm^2 is due to laser heating that, for some events, leads to surface fracture accompanied by light emission. At even lower fluences, subtle material modifications can be detected [78]. These modifications and the fracture are precursors of the ablation that is always accompanied by a luminescent plume. The conclusion from these findings is that the primary interaction of the laser light is linear absorption by defect states (cf. UPS results shown in Fig. 9), resulting in heating and thermoelastic stress and finally fracture, when a critical threshold is exceeded. Ablation is initiated by fracto-emission in which density and energy of ejecta is small at the onset of ablation. Above the ablation threshold, the emitted particles gain enough energy to form a plasma and massive ablation starts. As evident from Fig. 17(b), however, the major ablation products for high fluence irradiation are large tiles of fractured CaF_2 ejected by thermoelastic forces and possibly further driven by a developing plasma.

So far, only the results for ablation from surfaces that had been prepared by standard hard polishing procedures have been discussed. The importance of localized absorption for the ablation is impressively confirmed by measurements on crystals with different surface finishing. In Fig. 18, ablation results for a cleaved surface and a surface prepared by diamond turning [104] obtained under the same experimental conditions as those for the hard polished surface shown in Fig. 17 are compared. The cleaved surface does not exhibit a clear ablation

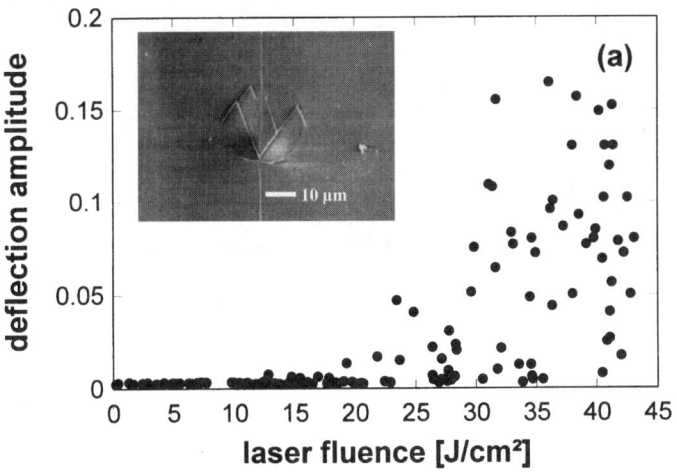

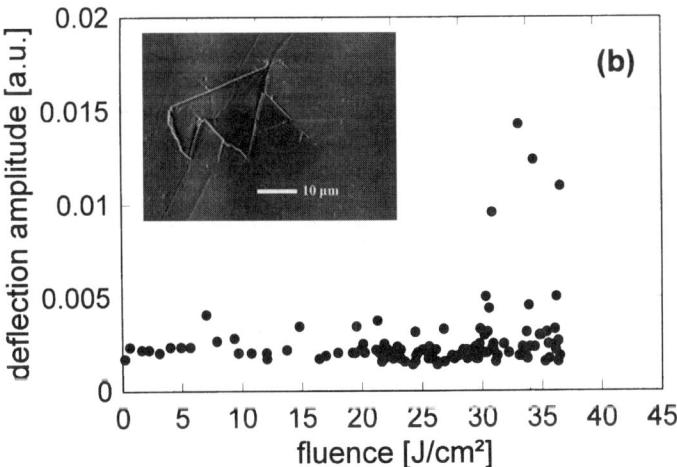

FIG. 18. Ablation from cleaved (a) and diamond turned (b) CaF_2 surfaces subject to 248 nm (14 ns) laser light. Ablation was detected by the photoacoustic mirage effect where the mirage signal amplitude is plotted as a function of incident laser fluence. Each pulse was applied to a new location on the sample. High amplitudes arise from ablation following increased absorption at step edges or residual polishing artifacts as shown in the insets (SEM micrographs). (From Ref. 105, reprinted with permission.)

threshold; however, the average laser resistance is much higher than for the mechanically polished surface. Careful inspection of the damage sites reveals that all ablation events below approximately 35 J/cm^2 can be attributed to local features like step edges as shown in the inset. Most of these absorption centers can be removed by advanced polishing techniques like diamond turning, and thus surfaces with an ablation threshold of 30 J/cm^2 can be prepared. There are strong indications that also ablation with fluences above this threshold is defect related [105], and currently the search for intrinsic mechanisms is clearly limited by the state-of-the-art in surface preparation and characterization.

12.3.3.4 Ablation from Fused Silica.
Work on laser ablation from fused silica was mainly motivated by the development of techniques for high-precision patterning of this material that is widely used in the optics and optoelectronics industries. Several studies aimed for a definition of irradiation parameters allowing clean ablation at high etch rates.

The bandgap of 8 eV for fused silica precludes ablation in the visible spectral range, and results for UV pulses strongly depend on wavelength. Ablation with nanosecond pulses at 248 nm is initiated by defects or impurities and yields a very irregular ablation topography due to thermoelastic expansion and fracture in the vicinity of localized absorptions [106]. However, sacrificing the cleanliness of ablation, it is possible to drill deep holes even at 308 nm [14]. For 193 nm (20 ns) light, a very clean ablation can be obtained (see inset in Fig. 19); however, the etch rate at this wavelength strongly depends on laser fluence and exhibits a threshold behavior as shown in Fig. 19. Ablation starts at about 0.5 J/cm^2 and rises slowly as a function of fluence up to 1 J/cm^2 where a dramatic increase in ablation efficiency is observed, with a maximum measured ablation rate of approximately 140 nm/pulse. The good results obtained for ablation at 193 nm were attributed to an enhanced, homogeneous absorption at this wavelength induced by color centers formed during the impact of the laser pulse [106]. Apparently the ablation behavior of fused silica is similar to that of organic polymers (see Chapter 6), and concepts of laser-induced defect creation and accumulation developed for this class of materials, can with some modifications be transferred to the silica ablation problem.

The idea of an ablation promoted by laser-induced absorption was pursued by multi-wavelength ablation experiments using 266 nm (8 ns) laser light and various anti-Stokes Raman-lines generated in a H$_2$ Raman scattering cell [109]. The shortest usable wavelength is the ninth order pulse at 133 nm generated with approximately 10^{-4} of the fundamental intensity. Despite this low efficiency, the additional absorption induced by this short wavelength irradiation enhances the ablation rate for the 266 nm fundamental light from 0 to 13 nm/pulse at a fluence of 2.1 J/cm^2. This technique allows very precise micro-patterning of fused silica as shown in Fig. 20, where $24 \times 24 \times 0.7$ μm^3 trenches with a flat bottom and only little distortion by debris at the edges have

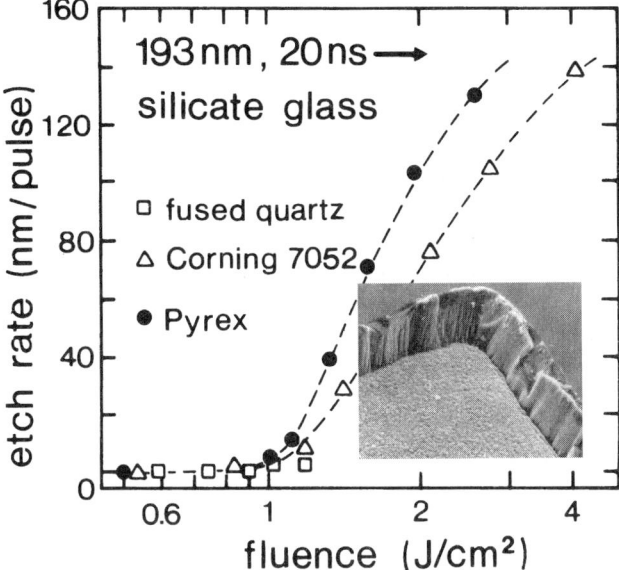

FIG. 19. Etch depth per pulse as a function of laser fluence at 193 nm (20 ns) for etching different silica glasses. The inset (SEM micrograph) shows clean ablation with 450 pulses of 1.57 J/cm² fluence (From Ref. 106, reprinted with permission.)

been ablated. Two effects contribute to the laser-enhanced ablation, a transient increase in absorption by excited states that is only effective during the impact of vacuum ultraviolet (VUV) radiation and a stationary effect that has been attributed to the photodissociation of Si-O bonds. From time-resolved absorption measurements, the lifetime of the transient component was estimated to be 1.7 ns and a 60% absorption increase per pulse was found [110]. Evidence for photodissociation was found by XPS-analysis and the absorption from the reaction products was stable for several days. However, a wavelength-selective ablation study revealed that the bond-breaking mechanism only works for spectral components with a wavelength below 180 nm [111].

12.4 Laser Ablation From Thin Film Optical Materials

12.4.1 Optical Thin Film Materials and Their Properties

Today virtually no optical component of practical relevance is purely based on a bulk optical material. Optical thin film systems are indispensable parts of the optical systems and in many cases not only improve the performance of

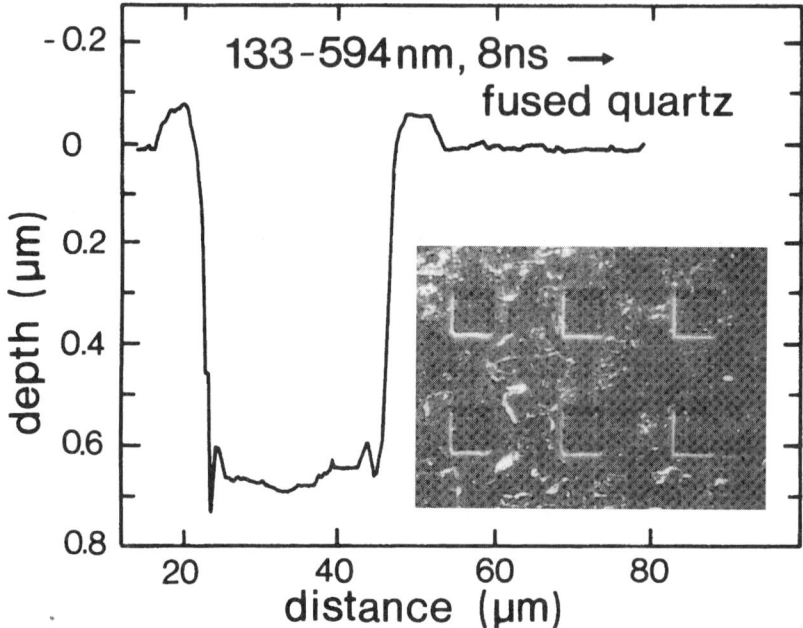

FIG. 20. Surface profile and SEM micrograph (inset) of square trenches in a fused silica substrate ablated at 2.1 J/cm^2 with 60 pulses (1–8 ns) of multi-wavelength light from a Raman cell pumped with frequency-quadrupled light from a Nd:YAG laser. The multi-wavelength light included UV radiation transiently creating absorption centers for visible components. (from Ref. 109, reprinted with permission.)

bulk components, e.g., as anti-reflection (AR) coatings avoiding stray light in optical lens systems but often provide the basic functionality of the optical system as, for example, in the case of many lasers that cannot be operated without mirrors based on high-reflection (HR) coatings.

Although the materials incorporated in thin film systems are mostly the same oxides and fluorides used for bulk optical components, the physical properties, and hence the ablation behavior of thin films and systems, strongly differ from those of bulk systems as a result of coating preparation and design. The latter may, for example, introduce an electric field enhancement initiating breakdown at a specific depth in the layer stack. Thin-film preparation and processing impacts absorption properties in a variety of ways. One of the most important degradation problems is the incorporation of contaminants during deposition as will be discussed in detail below.

Deposition and characterization techniques for optical thin films as well as the coating design are highly developed and their structure is well investigated.

However, much of the available technology and knowledge is based on "alchemy," and thin film research is still far away from a complete and quantitative understanding of the physical phenomena determining the structural, mechanical, and optical properties of thin films on a microscopic scale [112, 113]. Naturally thin film optical properties are best studied in view of the applications, and this includes the problem of laser damage where detailed investigations have been performed for almost all relevant systems [2]. Surprisingly, only very few systematic studies have been devoted to the question of thin film laser ablation that is closely related to the damage problem. This chapter covers only the specific case of ablation induced by thin film defect centers located at the interface between the layer stack and the substrate and thus only a small section of possible ablation processes.

The conventional way of thin film deposition is by electron beam evaporation or plasma sputtering techniques. In recent years, *laser sputtering* techniques have been developed that use laser light for the evaporation of thin film starting materials and in some cases allow the preparation of films with unique properties. Laser sputtering is another application of laser ablation from bulk materials treated in Section 12.3 and has developed into an active field of research that has been reviewed elsewhere [114, 115].

12.4.2 Absorption in Thin Film Systems

The increased interaction of light with a thin film system compared to bulk material has several causes. First, it contains a high number of interfaces introducing occupied electronic states in the bandgap of the constituent materials that allow optical transitions into the unoccupied conduction band. This does not only denote the interfaces between thin film layers or the substrate, but also the much larger interface area created by the growth of many films in a columnar microstructure [116]. Second, even when utmost care is taken to avoid contamination during thin film deposition, evaporation or sputtering always bears a great risk of introducing unwanted absorbing components into the film system or onto the substrate prior to deposition. Third, the intensity of light inside the film system does not decay exponentially as in a bulk material but is often an oscillating function with maxima at locations of minimum film stability, e.g., at interfaces [117]. Additionally thin film materials have to be selected by their refractive index required for the specific coating design, and it is often not possible to choose the material of lowest absorption.

A variety of methods has been developed to measure absorption in thin films [118]. Because the task for thin film absorption measurements is the detection of minute residual absorptions, often in the parts per million (ppm) range [119], transmission measurements are excluded and most methods applied are to some extent related to calorimetry, i.e., the measurement of a sample temperature rise induced by heating with a focused laser beam. This principle is also exploited in photothermal techniques involving a modulated laser beam [120]. The photothermal detection combines utmost sensitivity with the capability of detecting local absorptions with a lateral resolution down to the diffraction limit of the involved laser beams, typically 1 µm. The two most commonly applied photothermal detection schemes applied for thin film absorption measurements are the photothermal *mirage technique* [121, 122] and the *surface displacement technique* [123, 124] respectively. Both techniques are based on a noncontact measurement of the pump-laser-induced temperature rise in the sample by a probe laser beam. The mirage detection scheme records the deflection of the probe beam by the thermally induced local change in refractive index in the sample or the atmosphere above its surface. For surface displacement measurements, the local thermoelastic deformation of the surface is recorded via the change in angle of reflection of a probe beam focused onto the surface. Both techniques are well suited for thin film measurements where the former is somewhat more advantageous in terms of sensitivity and ease in absolute calibration, whereas the latter is best for measurements requiring highest lateral resolution.

Thin film deposition techniques and coating design have been highly developed, but a major problem is the presence of small defects incorporated into the film system that may severely degrade coating quality and are most detrimental for thin film–based high-power laser optics [125]. Photothermal microscopy is able to localize such thin film defects introduced during film deposition [126, 127] or handling and cleaning of the sample [128], and it is also possible to quantify their absorption. For the best presently available oxide multilayer coatings [129] with an average ablation threshold of up to 20 J/cm^2, it has, in fact, been shown that the local threshold for ablation may be much lower than the average and is solely determined by defect absorption as identified by photothermal displacement microscopy [130].

Figure 21 is an illustrative example for a quantitative measurement of defect absorption with mirage microscopy on a fused silica substrate coated with a TiO_2 thin film [131]. The upper frame is a scan over the uncoated substrate surface exhibiting a predominant absorption close to the detection limit but also localized absorptions of up to 400 ppm resulting from substrate contaminants present even after applying elaborate cleaning procedures. The lower frame shows the same area scanned after deposition of the TiO_2 film under the same conditions. The substrate contaminants are still visible, but there are also additional defects incorporated during film deposition. Furthermore, the increased mean absorption level due to the film can be deduced from this measurement. It will be demonstrated in the following section that especially the defects present at the interface between the film stack and the substrate in multilayer systems are centers for the onset of ablation in high-quality coatings.

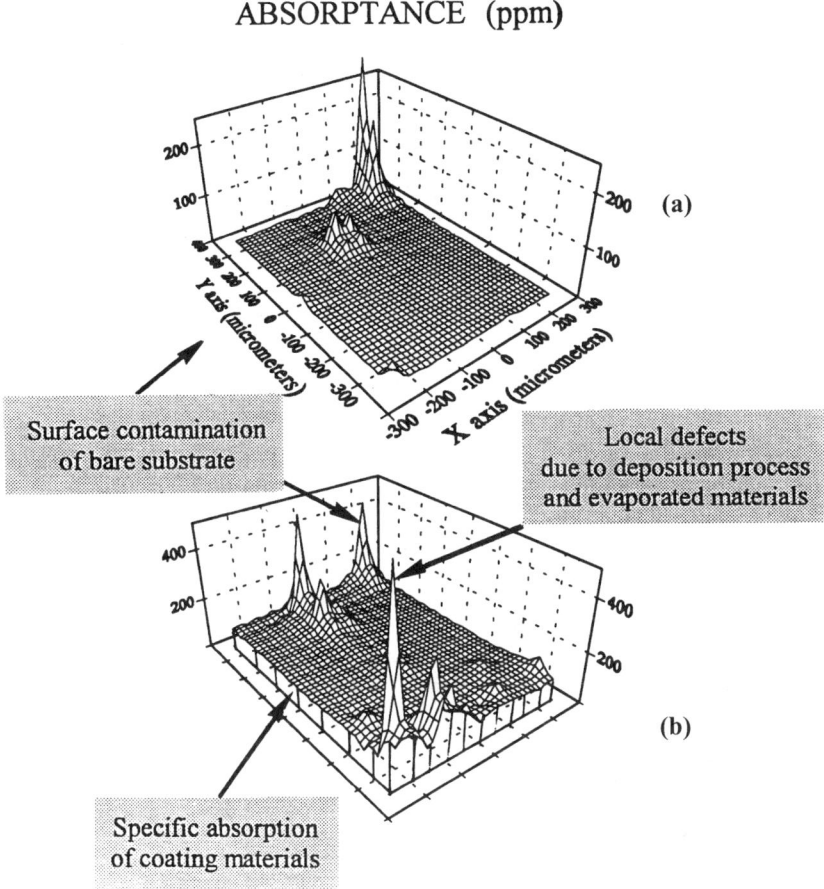

FIG. 21. Photothermal mirage scans on a fused silica substrate before (a) and after (b) deposition of eight $\lambda/4$ layers ($\lambda = 600$ nm) TiO_2 by an ion-assisted deposition technique. Absorption of substrate surface contaminations is present in both measurements, while additional absorption centers are created during deposition. Deposition also increases the background absorption level. (From Ref. 131, reprinted with permission.)

12.4.3 Thin Film Ablation Mechanisms

In singe-layer oxide films, the ablation threshold is well correlated with the overall absorption expected from the band structure of the respective materials. Figure 22 shows photoacoustic mirage results for TiO_2, ZrO_2, and HfO_2 films irradiated with 248 nm (14 ns) laser light. The lowest threshold is found for TiO_2 with a bandgap energy below the photon energy (5 eV), whereas the other oxides with larger bandgaps exhibit considerably higher ablation thresholds

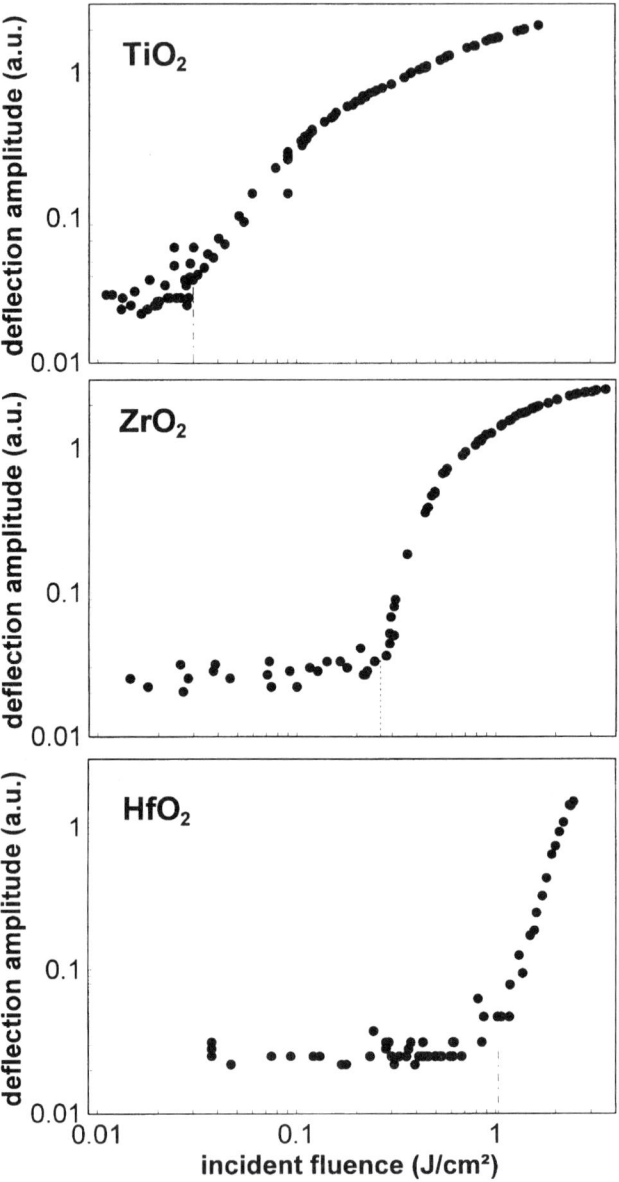

FIG. 22. Photoacoustic mirage data on ablation with 248 nm (14 ns) laser light from oxide films with λ ($\lambda = 248$ nm) thickness deposited on fused silica. Bandgap energies of ZrO_2 and HfO_2 are larger than the 5 eV laser photon energy while the bandgap of TiO_2 is smaller. This changes both the ablation threshold as well as slope of intensity-dependent signal rise. (From Ref. 132, reprinted with permission.)

[132]. The curves, however, also differ qualitatively with respect to the slope of the intensity-dependent deflection signal above threshold. The slow rise found for TiO_2 is typical for linear, homogeneous heating as the driving force for ablation, whereas the steep slopes for the other materials are indicative of highly nonlinear processes related to defect states in the bandgap.

In highly reflective multilayer coatings, the threshold for ablation is essentially determined by such defect states. For Al_2O_3/SiO_2 multilayer systems with a threshold of 19 J/cm^2, a defect-related local threshold of only 9 J/cm^2 was found [130]. A direct correlation between absorption sites and ablation events can be established by comparing data from photothermal absorption microscopy performed before pulse laser irradiation with an optical inspection of ablation craters after irradiation. This is demonstrated in Fig. 23 showing a measurement where the sample has first been inspected by photothermal microscopy (Fig. 23(a)) and then irradiated with 248 nm (14 ns) laser pulses applied in successive raster scans with increasing fluence. As soon as first indications for ablation were detected, the irradiation was stopped and the sample inspected by raster reflection microscopy (Fig. 23(b)). The correlation of defects and ablation sites is evident from the comparison of the micrographs.

The ablation morphology found in such experiments strongly depends on the thermomechanical properties of the film system under investigation [130, 133]. The ablation mechanism is difficult to deduce in detail if material melting is the predominant effect. However, if ablation is due to a thermomechanical process,

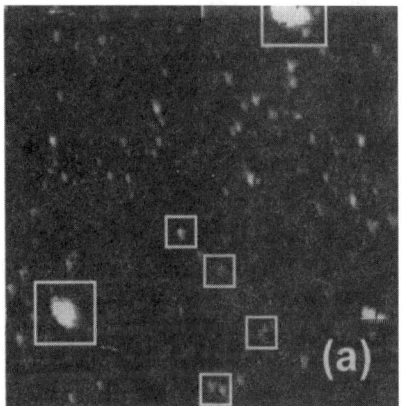
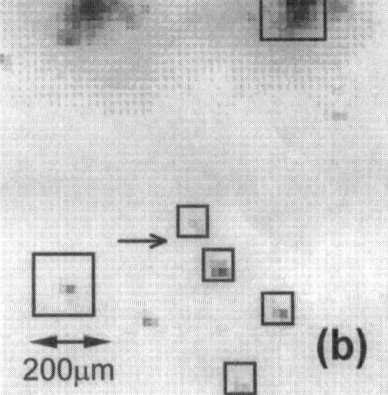

FIG. 23. Correlation of defects detected with photothermal displacement microscopy in an Al_2O_3/SiO_2 HR multilayer coating with damage sites found after irradiation with 9 J/cm^2 of 248 nm (14 ns) laser pulses. (a) A map of photothermal displacement amplitudes exhibiting thin film defects of various absorption strengths. (b) The intensity of a probe beam reflected from the surface is recorded. Correlated defects and damage sites are marked with squares. (From Ref. 130, reprinted with permission.)

the ablation morphology allows a conclusion about how material is ejected. The ablation crater shown for such an example in Fig. 24 is characterized by a perfectly circular shape and a structure of terraces sharply cut out from the stack. Signs of melting are only found at the crater bottom that is the substrate surface. The small dot found in the center marks the location of the absorbing defect initiating the ablation. The rapid laser-induced evaporation of this defect causes elastic stress, delaminating the first layer from the substrate and thereby bulging the whole layer stack. This in turn results in tensile stress, leading to mechanical failure in the first layer. When the resulting crack reaches the bottom of the second layer, it changes direction and propagates between the first two layers, which is the plane of least surface energy. The next change in direction occurs when the tensile stress in the second layer reaches a critical level and breaks. In this way, the crack propagates from the bottom of the stack to the surface, and finally a conical cap is ejected.

A similar thermomechanical ablation mechanism is found for the ejection of nodule defects [134]. Nodular defects [116] are created during thin film deposition and initiated by the deposition of clustered starting material, forming a seed for a defective thin film growth. Figure 25 is a cross-section through a

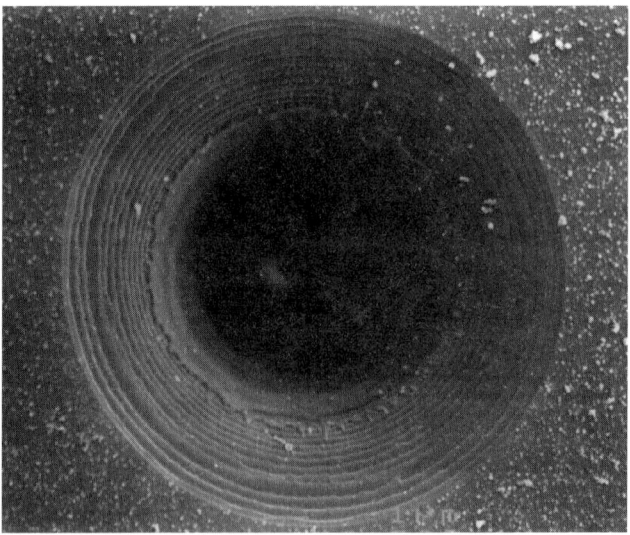

FIG. 24. Damage crater (13 μm outer diameter) in an Al_2O_3/SiO_2 HR multilayer coating initiated by a sub-μm-sized defect located at the interface between substrate and layer stack. Evaporation of the defect results in a delamination of the stack from the substrate and tensile stress in the layers. When the stress exceeds a critical value, the thin film material cracks and a conical cap is ablated. The melting temperature is only reached at the interface in the vicinity of the defect.

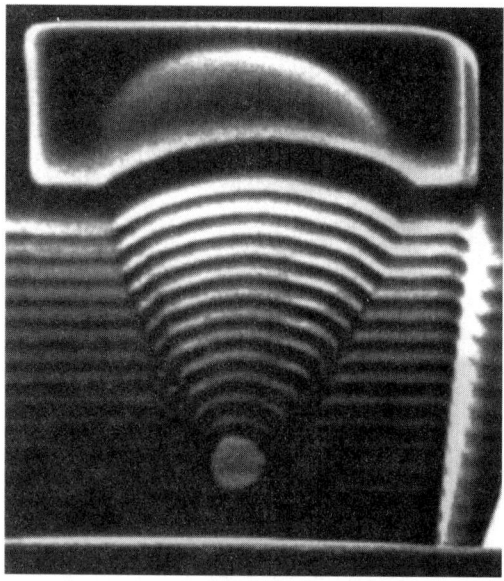

FIG. 25. Nodular defect in a HfO$_2$/SiO$_2$ HR multilayer coating. The electron micrograph displays a cross-section through the film stack showing the circular seeding defect in the third layer. (From Ref. 135, reprinted with permission.)

HfO$_2$/SiO$_2$ multilayer system and shows a cluster directly deposited on the third layer. This defect is overgrown by the following layers and continued deposition leads to a broadening of the feature and a conical structure in the cross-section [135]. A top-view image of a fully developed nodule is shown in Fig. 26(a), and its reaction to laser irradiation is shown in frames (b) and (c), respectively. With a 3.6 J/cm^2 laser pulse, i.e., a fluence well below the ablation threshold of the unperturbed film system, the nodule is completely ejected by a thermomechanical mechanism similar to that described above. Obviously the area of maximum susceptibility to fracture is the surface of the cone defined by the overgrown nodule (see the cross-section in Fig. 25)), and this mechanical instability leads to the very clean ablation of the cone, as can be seen in frame (b). Frame (c) demonstrates that such an ablation event increases the overall stability of the coating because further irradiation with laser pulses does not result in more ablation, even at a fluence level close to the threshold for the unperturbed coating. Effectively this type of ablation leads to a laser conditioning; i.e., an improvement in laser resistivity by subthreshold laser light irradiation, and is exploited to improve the performance of large area laser mirrors for ultra-high power lasers [136].

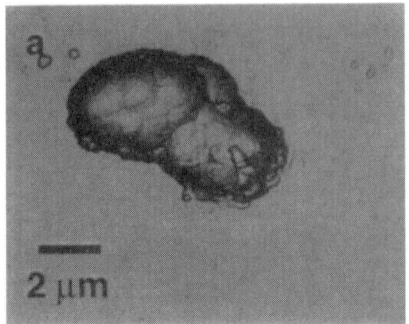
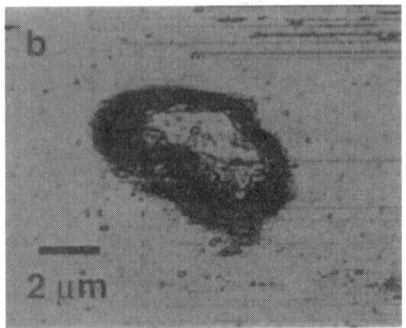
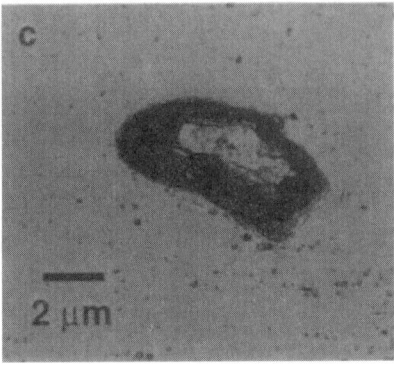

FIG. 26. Demonstration of laser conditioning in a HfO_2/SiO_2 HR multilayer coating by ablation of a nodular defect. (a) Top-view STM image of nodule as grown. (b) Nodule has been ablated by irradiation with a 1.06 μm (11 ns) laser pulse at $3.6/cm^2$ fluence. (c) Irradiation of the same area with a subsequent 17.8 J/cm^2 pulse does not cause further ablation. (From Ref. 136, reprinted with permission.)

References

1. Campbell, J. H., Atherton, L. J., DeYoreo, J. J., Kozlowski, M. R., Maney, R. T., Montesanti, R. C., Sheenan, L. M., and Barker, C. E. Large-aperture, high-damage threshold optics for beamlet. *LLNL Inertial Confinement Fusion Annual Report 1995* (URCL-LR-105821-95-1).
2. Woods, R. M., ed. *Laser Damage in Optical Materials*. SPIE Milestone Series, Vol. MS 24, Bellingham, 1990. A source for more recent results are volumes of *Laser-Induced Damage in Optical Materials*, published annually by the Society of Photo-Optical Instrumentation Engineers, Bellingham, WA.
3. Matthias, E., and Dreyfus, R. W. From laser-induced desorption to surface damage. In *Photoacoustic, Photothermal and Photochemical Processes at Surfaces and in Thin Films*, P. Hess (ed.), pp. 89–129. Topics in Current Physics, Vol. 47, Springer-Verlag, Heidelberg, 1989.

4. Chase, L. L., Laser ablation and optical surface damage. In *Laser Ablation*, J. C. Miller (ed.), Springer Series in Materials Science, Vol 28, pp. 53–84. Springer-Verlag, Berlin, 1994.
5. Tanimura, K., Fujiwara, H., and Suzuki, T. Time resolved spectroscopy for radiation damage processes induced by electronic excitation in insulators. *Nucl. Instrum. Meth.* **B116**(1-4), 26–32 (1996).
6. Audebert, P., Daguzan, Ph., Dos Santos, A., Gauthier, J. C., Geindre, J. P., Guizard, S., Hamoniaux, G., Krastev, K., Martin, P., Petite, G., and Antonetti, A. Space-time observation of an electron gas in SiO_2. *Phys. Rev. Lett.* **73**(14), 1990–1993 (1994).
7. Hooker, C. J., Lister, J. M. D., Osvay, K., Sheerin, D. T., Emmony, D. C., and Cowell, R. L. J. Pulse-length scaling of laser damage at 249 nm in oxide and fluoride multilayer coatings. *Opt. Lett.* **18**(12), 944–946 (1993).
8. Du, D., Liu, X., Korn, G., Squier, J., and Mourou, G. Laser-induced breakdown by impact ionization in SiO_2 with pulse widths from 7 ns to 150 fs. *Appl. Phys. Lett.* **64**(23), 3071–3073 (1994).
9. Stuart, B. C., Feit, M. D., Rubenchik, A. M., Shore, B. W., and Perry, M. D. Laser-induced damage in dielectrics with nanosecond to subpicosecond pulses. *Phys. Rev. Lett.* **74**(12), 2248–2251 (1995).
10. Mann, K., Pfeiffer, G., and Reisse, G. Damage threshold measurements using femtosecond excimer laser. In *Laser-Induced Damage in Optical Materials: 1992*, H. E. Bennett, L. L. Chase, A. H. Guenther, B. Newnam, and M. J. Soileau (eds.), pp. 415–423. SPIE Proc. 1848, Bellingham, 1993.
11. von der Linde, D., and Schüler, H. Breakdown threshold and plasma formation in femtosecond laser-solid interaction. *J. Opt. Soc. Am.* **B13**(1), 216–222 (1996).
12. Stuart, B. C., Feit, M. D., Herman, S., Rubenchik, A. M., Shore, B. W., and Perry, M. D. Nanosecond-to-femtosecond laser-induced breakdown in dielectrics. *Phys. Rev.* **B53**(4), 1749–1761 (1996).
13. Küper, S., and Stuke, M. Ablation of UV-transparent materials with femtosecond UV excimer laser pulses. *Microlectron. Engin.* **9**, 475–480 (1989).
14. Ihlemann, J., Wolff, B., and Simon, P. Nanosecond and femtosecond excimer laser ablation of fused silica. *Appl. Phys.* **A54**, 363–368 (1992).
15. Ashkenasi, D., Varel, H., Rosenfeld, A., Noack, F., and Campbell, E. E. B. Pulse-width influence on the laser-induced structuring of CaF_2 (111). *Appl. Phys.* **A63**, 103–107 (1996).
16. Kautek, W., Krüger, J., Lenzner, M., Sartania, S., Spielmann, C., Krausz, and F. Laser ablation of dielectrics with pulse durations between 20 fs and 3 ps. *Appl. Phys. Lett.* **69**(21), 3146–3148 (1996).
17. Matthias, E., Nielsen, H. B., Reif, J., Rosén, A., and Westin, E. Multiphoton-induced desorption of positive ions from barium fluoride. *J. Vac. Sci. Technol.* **B5**(5), 1415–1422 (1987).
18. Estler, R. C., Apel, E. C., and Nogar, N. S. Laser mass-spectrometric studies of optical damage in CaF_2. *Opt. Soc. Am.* **B4**(2), 281–286 (1987).
19. Kreitschitz, O., Husinsky, W., Betz, G., and Tolk. N. H. Time-of-flight investigation of the intensity dependence of laser-desorbed positive ions from CaF_2. *Appl. Phys.* **A58**(1-4), 563–571 (1994).
20. Nakayama, T. Laser-induced sputtering of ZnO, TiO_2, CdSe, and GaP near the threshold laser fluence. *Surf. Sci.* **133**, 101–113 (1983).
21. Arlinghaus, H. F., Calaway, W. F., Young, C. E., Pellin, M. J., Gruen, D. M., and Chase, L. L. High-resolution multiphoton laser-induced fluoescence spectroscopy

of zinc atoms ejected from laser-irradiated ZnS crystals. *J. Appl. Phys.* **65**(1), 281–289 (1989).
22. Dreyfus, R. W., Kelly, R., and Walkup, R. E. Laser-induced fluorescence studies of excimer laser ablation of Al_2O_3. *Appl. Phys. Lett.* **49**(21), 1478–1480 (1986).
23. Petzoldt, S., Elg, A. P., Reichling, M., Reif, J., and Matthias, E. Surface laser damage thresholds determined by photoacoustic deflection. *Appl. Phys. Lett.* **53**(21), 2005–2007 (1988).
24. Petzoldt, S., Reif, J., and Matthias, E. Laser plasma threshold of metals. *Appl. Surf. Sci.* **96–98**, 199–204 (1996).
25. Matthias, E., Siegel, J., Petzoldt, S., Reichling, M., Skurk, H., Käding, O., and Neske, E. In-situ investigation of laser ablation of thin films. *Thin Solid Films* **254**, 139–146 (1995).
26. Azzeer, A. M., Al-Dwayyan, A. S., Al-Salhi, M. S., Kamal, A. M., and Harith, M. A. Optical probing of laser-induced shock waves in air. *Appl. Phys.* **B63**, 307–310 (1996).
27. Kelly, R., Cuomo, J. J., Leary, P. A., Rothenberg, J. E., Braren, B. E., and Aliotta, C. F. Laser sputtering. Part I. On the existence of rapid laser sputtering at 193 nm. *Nucl. Instr. Meth.* **B9**, 329–340 (1985).
28. Mann, K., and Gerhardt, H. Setup of a damage testing facility for excimer laser radiation. In *Excimer Lasers and Applications*, D. Basting (ed.), pp. 136–140. SPIE Proc. 1023, Bellingham, 1989.
29. Kozlowski, M. R. Damage-resistant laser coatings. In *Thin Films for Optical Systems*, F. R. Flory (ed.), pp. 521–549. Marcel Dekker Inc., New York, 1995.
30. Dressler, L., Rauch, R., and Reimann, R. On the inhomogeneity of refractive index of CaF_2 crystals for high performance optics. *Cryst. Res. Technol.* **27**(3), 413–420 (1992).
31. Jain, K. *Excimer Laser Lithography*. SPIE Press, Bellingham, 1990.
32. Rubloff, G. W. Far-ultraviolet reflectance spectra and the electronic structure of ionic crystals,. *Phys. Rev.* **B5**(2), 662–684 (1972).
33. Frandon, J., Lahaye, B., and Pradal, F. Spectra of electronic excitations in CaF_2, SrF_2 and BaF_2 in the 8 to 150 eV range. *Phys. Stat. Sol.* **B53**, 565–575 (1972).
34. Barth, J., Johnson, R. L., Cardona, M., Fuchs, D., and Bradshaw, A. M. Dielectric function of CaF_2 between 10 and 35 eV. *Phys. Rev.* **41**(5), 3291–3294 (1990).
35. Tomiki, T., and Miyata, T. Optical studies of alkali fluorides and alkaline earth fluorides in VUV region. *J. Phys. Soc. Jap.* **27**(3), 658–678 (1969).
36. Keil, T. H. Theory of the Urbach rule. *Phys. Rev.* **144**(2), 582–587 (1966).
37. Dexter, D. L. Interpretation of Urbach's rule. *Phys. Rev. Lett.* **19**(24), 1383–1385 (1967).
38. Eva, E., and Mann, K. Calorimetric measurement of two-photon absorption and color center formation in UV-window materials. *Appl. Phys.* **A62**(2), 143–149 (1996).
39. Hata, K., Watanabe, M., and Watanabe, S. Nonlinear processes in UV optical materials at 248 nm. *Appl. Phys.* **B50**, 55–59 (1990).
40. Taylor, A. J., Gibson, R. B., and Roberts, J. P. Two-photon absorption at 248 nm in ultraviolet window materials. *Opt. Lett.* **13**, 814–816 (1988).
41. Jones, S. C., Shen, X. A., Braunlich, P. F., Kelly, P., and Epifanov, A. S. Mechanism of prebreakdown nonlinear energy deposition from intense photon fields at 532 nm in NaCl. *Phys. Rev.* **B35**(2), 894–897 (1987).

42. Jones, S. C., Fischer, A. H., Braunlich, P., and Kelly, P. Prebreakdown energy absorption from intense laser pulses at 532 nm in NaCl. *Phys. Rev.* **B37**(2), 755–770 (1988).
43. Williams, R. T., Bradford, J. N., and Faust, W. L. Short-pulse optical studies of exciton relaxation and *F*-center formation in NaCl, KCl, and NaBr. *Phys. Rev.* **B18**(12), 7038–7057 (1978).
44. Song, K. S., and Williams, R. T. *Self-Trapped Excitons.* Springer-Verlag, Berlin, 1993.
45. Suzuki, Y., Abe, H., and Hirai, M. Double laser excitation spectroscopy on picosecond photochemical reactions in alkali halide crystals. In *Ultrafast Phenomena VI*, T. Yajima, K. Yoshihara, C. B. Harris, and S. Shionoya (eds.), Springer Series in Chemical Physics, Vol. 48. Springer-Verlag, Berlin, 1988.
46. Tanimura, K., Katoh, T., and Itoh, N. Lattice relaxation of highly excited self-trapped excitons in CaF_2. *Phys. Phys. Rev.* **B40**(2), 1282–1287 (1984).
47. Williams, R. T. Optically generated lattice defects in halide crystals. *Opt. Engin.* **28**(10), 1024–1033 (1989).
48. Dexter, D. L. Optical properties of solids. *Nuovo Cimento* **VII**(X), 245–286 (1958).
49. Efimov, O. M., and Mekryukov, A. M. Investigation of energy structure of lead silicate glasses by non-linear absorption spectroscopy technique. *J. of Non-Crystalline Solids* **191**, 94–100 (1995).
50. Wu, S.-T., and Bass, M. Laser-induced irreversible absorption changes in alkali halides at 10.6 μm. *Appl. Phys. Lett.* **39**(12), 948–950 (1981).
51. Epifanov, A. S. Theory of electron-avalanche ionization induced in solids by electromagnetic waves. *IEEE J. Ouant. Electron.* **QE-17**(10), 2018–2022 (1981).
52. Shen, X. A., Jones, S. C., and Braunlich, P. Laser heating of free electrons in wide-gap optical materials at 1064 nm. *Phys. Rev. Lett.* **62**(23), 2711–2713 (1989).
53. Daguzan, Ph., Guizard, S., Krastev, K., Martin, P., Petite, G., Dos Santos, A., and Antonetti, A. Direct observation of multiple photon absorption by free electrons in a wide bandgap insulator under strong laser irradiation. *Phys. Rev. Lett.* **73**(17), 2352–2355 (1994).
54. Matthias, E, and Green, T. A. Laser-induced desorption. In *Desorption Induced by Electronic Transitions DIET IV*, G. Betz and P. Varga (eds.), Springer Series in Surface Sciences, Vol. 19, pp. 112–127. Springer-Verlag, Heidelberg, 1990.
55. Huisinga, M., Reichling, M., and Matthias, E. Ultraviolet-photoelectron spectroscopy and photoconductivity. *Phys. Rev.* **B55**(12), 7600–7605 (1997).
56. Petrocelli, G., Scudieri, F., and Martellucci, S. Nonlinear absorption in ionic crystals determined by pulsed photothermal deflection. *Appl. Phys.* **B52**, 123–128 (1991).
57. O'Connell, R. M, and Marrs, C. D. Linear and nonlinear photoconductivity from charge-producing defects in CaF_2. *J. Appl. Phys.* **70**(4), 2313–2321 (1991).
58. Görlich, P., Karras, H., and Lehmann, R. Über die optischen Eigenschaften der Erdalkalihalogenide vom Flußspat-Typ (1). *Phys. Stat. Sol.* **1**, 389–440 (1961).
59. Hayes, W., ed. *Crystals with the fluorite structure.* Clarendon Press, Oxford, 1974.
60. Visser, R., Dorenbos, P., van Eijk, C. W. E., Meijerink, A., Blasse, G., and den Hartog, H. W. Energy transfer processes involving different luminescence centers in BaF_2:Ce. *J. Phys. Condens. Matter* **5**, 1659–1680 (1993).
61. Twidell, J. W. Radiation induced movement of charge compensating ions in CaF_2. *J. Phys. Chem. Sol.* **31**, 299–305 (1971).

62. Bloembergen, N. Role of cracks, pores and absorbing inclusions on laser-induced damage thresholds at surfaces of transparent dielectrics. *Appl. Opt.* **12**, 661–664 (1973).
63. Bennett, H. S. Absorbing centers in laser materials. *J. Appl. Phys.* **42**(2), 619–630 (1971).
64. Kelly, P. J., Ritchie, D. S., Bräunlich, P. F., Schmid, A., and Bryant, G. W. Deformation of intense laser beams tightly focused inside NaCl: A comparison of the multiphoton-polaron and avalanche models of optical breakdown. *IEEE J. of Quant.. Electr.* **QE-17**(10), 2027–2033 (1981).
65. Gruzdev, V. E., and Libenson, M. N. Electrodynamics instability of dielectric nonlinear sphere in transparent linear surrounding medium. In *Laser Damage in Optical Materials: 1994*, H. E. Bennett, A. H. Guenther, M. R. Kozlowski, B. E. Newnam, and M. J. Soileau (eds.), pp. 553–558. SPIE Proc. 2428, Bellingham, 1995.
66. Harrington, J. A. Surface absorption in KCl and CaF_2 at 2.8 and 3.8 μm. *Appl. Opt.* **18**(15), 2534–2536 (1979).
67. Gallant, D. J., Law, M., and Pond, B. Effect of cleaning on the optical absorption of calcium fluoride and fused silica at 351 nm. *Nist spec. Publ.* **752**, 159–167 (1988).
68. Kennedy, M., Plaß, W., Ristau, D., and Giesen, A. Environmental stability of CO_2 laser optics. In *Laser-Induced Damage in Optical Materials: 1995*, H. E. Bennett, A. H. Guenther, M. R. Kozlowski, B. E. Newnam, and M. J. Soileau (eds.), pp. 145–155 (1996).
69. Dreyfus, R. W., McDonald, F. A., and von Gutfeld, R. J. Energy deposition at insulator surfaces below the ultraviolet photoablation threshold. *J. Vac. Sci. Technol.* **B5**(5), 1521–1527 (1987).
70. Westin, E., Rosen, A, and Matthias, E. Molecular cluster calculations of the electronic structure of the (111) surface of CaF_2. In *Desorption Induced by Electronic Transitions DIET IV*, G. Betz and P. Varga (eds.), Springer Series in Surface Sciences, Vol. 19, pp. 316–321. Springer-Verlag, Heidelberg, 1990.
71. Stankiewicz, B., and Modrak, P. Electronic structure of CaF_2-LCAO calculations for slabs with the (111) free surface. *Vacuum* **45**(2/3), 205–208 (1994).
72. Poole, R. T., Szajman, J., Leckey, R. C. G., Jenkin, J. G., and Liesegang, J. Electronic structure of the alkaline-earth fluoride studied by photoelectron spectroscopy, *Phys. Rev.* **B12**(12), 5872–5877 (1975).
73. Bouzidi, S., Angot, T., Langlais, V., Debever, J.-M., Sporken, R., Longueville, J. L., and Thiry, P. A. Inverse-photoemission spectroscopy of electron irradiated epitaxial CaF_2 on Si(111). *Surf. Sci.* **307**, 1038–1044 (1994).
74. Bezerianos, N., and Vook, R. W. Thermal vaporization from the (111) CaF_2 face. *J. Appl. Phys.* **43**(4), 1417–1422 (1972).
75. Overney, R. M., Haefke, H., Meyer, E., and Güntherodt, H.-J. Cleavage faces of alkaline earth fluorides studied by atomic force microscopy. *Surf. Sci. Lett.* **277**, L29–L33 (1992).
76. König., G., Lehmann, A., and Rieder, K. H. He atom scattering studies of CaF_2(111). *Surf. Sci.* **331–333**, 1430–1434 (1995).
77. Stankiewicz, B., and Kisiel, W. Influence of fluorine desorption on electronic structure of CaF_2-LCAO slab calculations. *Vacuum* **45**(2/3), 209–210 (1994).
78. Reichling, M., Gogoll, S., Stenzel, E., Johansen, H., Huisinga, M., and Matthias E. Laser-damage processes in cleaved and polished CaF_2 at 248 nm. In *Laser-Induced Damage in Optical Materials: 1995*, H. E. Bennett, A. H. Guenther, M. Kozlowski,

B. E. Newnam, and M. J. Soileau (eds.), pp. 260–271. SPIE Proc. 2714, Bellingham, 1996.
79. Körner, C., Mayerhofer, R., Hartmann, M., and Bergmann, H. W. Physical and material aspects in using visible laser pulses of nanosecond duration for ablation. *Appl. Phys.* **A63**, 123–131 (1996).
80. Itoh, N. Laser sputtering in the electronic excitation regime: Comparison with electron and ion sputtering. *Nucl. Instr. Meth.* **B27**, 155–166 (1987).
81. Tanimura, K., and Itoh, N. Mechanisms of atomic processes induced by electronic excitations in solids. *Nucl. Instr. Meth.* **B33**, 815–819 (1988).
82. Itoh, N., and Tanimura, K. Formation of interstitial-vacancy pairs by electronic excitation in pure ionic crystals. *J. Phys. Chem. Sol.* **51**(7), 717–735 (1990).
83. Townsend, P. D., Browning, R., Garlant, D. J., Kelly, J. C., Mahjoobi, A., Michael, A. J., and Saidoh, M. Sputtering patterns and defect formation in alkali halides. *Rad. Eff.* **30**, 55–60 (1976).
84. Schmid, A., Bräunlich, P., and Rol, P. K. Multiphoton-induced directional emission of halogen atoms from alkali halides. *Phys. Rev. Lett.* **35**(20), 1382–1385 (1975).
85. Kubo, T., Okano, A., Kanasaki, J., Ishikawa, K., Nakai, Y., and Itoh, N. Emission of Na atoms from undamaged and slightly damaged NaCl(100) surfaces by electronic excitation. *Phys. Rev.* **B49**(7), 4931–4937 (1994).
86. Puchin, V., Shluger, A., Nakai, Y., and Itoh, N. Theoretical study of Na-atom emission from NaCl(100) surfaces. *Phys. Rev.* **B49**(16), 11364–11373 (1995).
87. Rothenberg, J. E., and Kelly, R. Laser sputtering. Part II. The mechanism of the sputtering of Al_2O_3. *Nucl. Instr. Meth.* **B1**, 291–300 (1984).
88. Tam, A. C., Brand, J. L., Cheng, D. C., and Zapka, W. Picosecond laser sputtering of sapphire at 266 nm. *Appl. Phys. Lett.* **55**(20), 2045–2047 (1989).
89. Brand, J. L, and Tam, A. C. Mechanism of picosecond ultraviolet laser sputtering of sapphire at 266 nm. *Appl. Phys. Lett.* **56**(10), 883–885 (1990).
90. Dreyfus, R. W., Kelly, R., and Walkup, R. E. Laser-induced fluorescence studies of excimer laser ablation of Al_2O_3. *Appl. Phys. Lett.* **49**(21), 1478–1480 (1986).
91. Hamza, A. V., Hughes, R. S., Chase, L. L., and Lee, H. W. H. Investigation of the laser-Al_2O_3 (1120) surface interaction using excitation by pairs of picosecond laser pulses. *J. Vac. Sci. Technol.* **B10**(1), 228–230 (1992).
92. Schildbach, M. A., and Hamza, A. V. Sapphire (1120) surface: Structure and laser-induced desorption of aluminium. *Phys. Rev.* **B43**(11), 6197–6206 (1992).
93. Estler, R. C., Apel, E. C., and Nogar, N. S. Laser mass-spectrometric studies of optical damage in CaF_2. *J. Opt. Soc. Am.* **B4**(2), 281–286 (1987).
94. Reichling, M., Johansen, H., Gogoll, S., Stenzel, E., and Matthias, E. Laser-stimulated desorption and damage at polished CaF_2 surfaces irradiaited with 532 nm laser light. *Nucl. Instr. Meth.* **B91**, 628–633 (1994).
95. Kreitschitz, O., Husinsky, W., Betz, G., and Tolk, N. H., Laser-induced desorption of positive ions from wide band gap insulators. *Nucl. Instr. Meth.* **B78**, 327–332 (1993).
96. Stenzel, E., Bouchaala, N., Gogoll, S., Klotzbücher, Th., Reichling, M., and Matthias E. Defect formation in CaF_2 due to Al Kα x-ray irradiation. In *Defects in Insulating Materials ICDIM 96*, G. E. Matthews and R. T. Williams (eds.), pp. 591–594. Materials Science Forum, Trans Tech Publications, Zürich, 1997.
97. Matthias, E., Gogoll, S., Stenzel, E., and Reichling, M. Laser-stimulated desorption from CaF_2 crystals. *Rad. Eff. Def. Sol.* **128**, 67–78 (1994).
98. Boling, N. L., Crisp, M. D., and Dubé, G. Laser induced surface damage. *Appl. Opt.* **12**(4), 650–660 (1973).

99. Dickinson, J. T., Jensen, L. C., Langford, S. C., and Hirth, J. P. Atomic and molecular emission following fracture of alkali halides: A dislocation driven process. *J. Mater. Res.* **6**(1), 112–125 (1991).
100. Gogoll, S., Stenzel, E., Reichling, M., Johansen, H., and Matthias, E. Laser damage of CaF$_2$(111) surfaces at 248 nm. *Appl. Surf. Sci.* **96–98**, 332–340 (1996).
101. Gogoll, S., Stenzel, E., Johansen, H., Reichling, M., and Matthias, E. Laser damage of cleaved and polished CaF$_2$(111) surfaces at 248 nm. *Appl. Surf. Sci.* **96–98**, 332–340 (1996).
102. Mitzner, R., Rosenfeld, A., and König, R. Time-resolved absorption studies of excimer laser ablation of CaF$_2$. *Appl. Surf. Sci.* **69**, 180–184 (1993).
103. Cronberg, H., Muydermann, W., Nielsen, H. B., Matthias, E., and Tolk, N. H. Intensity dependence of laser-induced emission of particles from CaF$_2$(111) near the plasma threshold. In *Desorption Induced by Electronic Transitions DIET IV*, G. Betz and P. Varga (eds.), Springer Series in Surface Sciences, Vol. 19, pp. 157–162. Springer-Verlag, Heidelberg, 1990.
104. Puttick, K., and Gee, T. Brittle materials yield to plastic deformation. *Opto & Laser Europe* **9**, 25–28 (1994).
105. Stenzel, E., Gogoll, S., Sils, J., Huisinga, M., Johansen, H., Reichling, M., and Matthias, E. Laser damage of alkaline-earth fluorides at 248 nm and the influence of polishing grades. *Appl. Surf. Sci.* **109/110**, 162–166 (1997).
106. Braren, B., and Srinivasan, R. Controlled etching of silicate glasses by pulsed ultraviolet laser radiation. *J. Vac. Sci. Technol.* **B6**(2), 537–541 (1988).
107. Davis, G. M., and Gower, M. C. Time-resolved transmission studies of poly(methyl methacrylate) films during ultraviolet laser ablative photodecomposition. *J. Appl. Phys.* **61**(5), 2090–2092 (1987).
108. Sauerbrey, R., and Pettit, G. H. Theory for the etching of organic materials by ultraviolet laser pulses. *Appl. Phys. Lett.* **55**(5), 421–423 (1989).
109. Sugioka, K., Wada, S., Tsunemi, A., Sakai, T., Takai, H., Moriwaki, H., Nakamura, A., Tashiro, H., and Toyoda, K. Micropatterning of quartz substrates by multiwavelength vacuum-ultraviolet laser ablation. *Jpn. J. Appl. Phys.* **32**, 6185–6189 (1993).
110. Sugioka, K., Wada, S., Tashiro, H., Toyoda, K., Ohnuma, Y., and Nakamura, A. Multiwavelength excitation by vacuum-ultraviolet beams coupled with fourth harmonic of a Q-switched Nd:YAG laser for high-quality ablation of fused quartz. *Appl. Phys. Lett.* **67**(19), 2789–2791 (1995).
111. Sugioka, K., Wada, S., Tashiro, H., Toyoda, K., and Nakamura, A., Novel ablation of fused quartz by preirradiation of vacuum-ultraviolet laser beams followed by fourth harmonics irradiation of Nd:YAG laser. *Appl. Phys. Lett.* **65**(2), 1510–1512 (1994).
112. Hummel, R. E., and Guenther, K. H., eds. *Thin Films for Optical Coatings*, Handbook of Optical Properties, Vol. 1. CRC Press, Boca Raton, 1995.
113. Flory, F. R., ed. *Thin Films for Optical Systems*. Marcel Dekker, New York, 1995.
114. Cheung, J. T., and Sankur, H. Growth of thin films by laser-induced evaporation. *CRC Crit. Rev. Sol. State Mat. Sci.* **15**(1), 63–109 (1988).
115. Afonso, C. N. Pulsed laser deposition of films for optical applications. In *Insulating Materials For Optoelectronics: New Developments*, F. Agullo-Lopez (ed.). World Scientific Publishers, London, 1995.
116. Guenther, K. H. Microstructure of vapor-deposited optical coatings. *Appl. Opt.* **23**(21), 3806–3816 (1984).

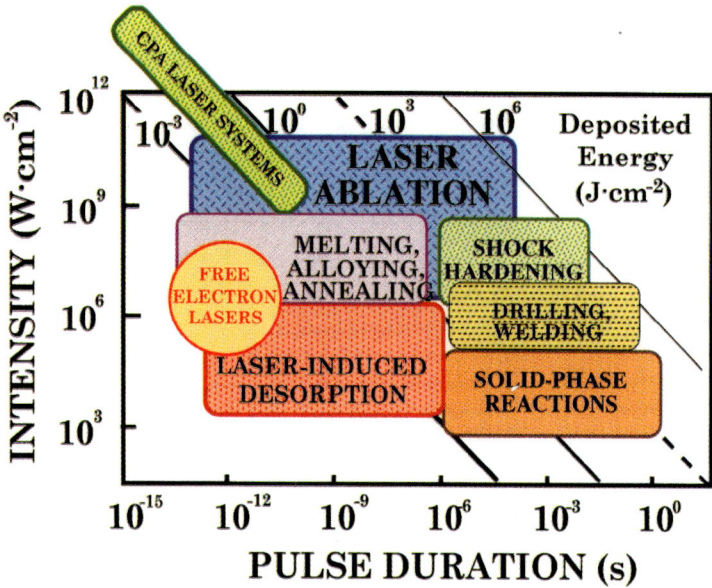

Plate 1(a). Materials processing using lasers, located with respect to laser-pulse duration and laser intensity. The diagonal lines are lines of constant total energy.

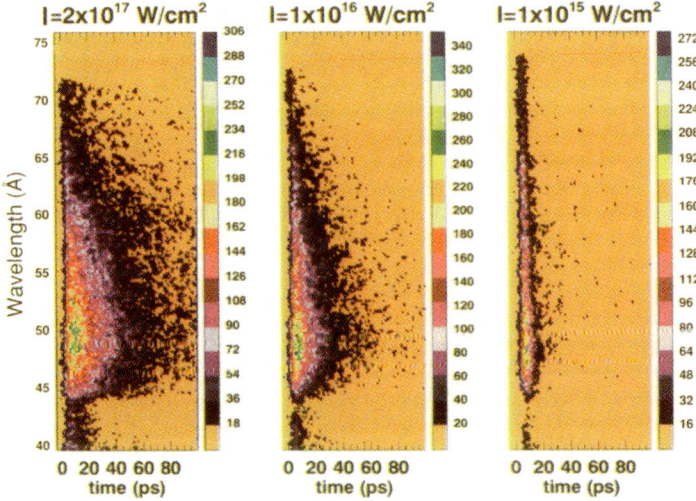

Plate 1(b). Streak-camera time-and-wavelength-resolved record of X-rays produced by laser ablation of an Au target, for three different intensities. Adapted from Refs. [6, 7].

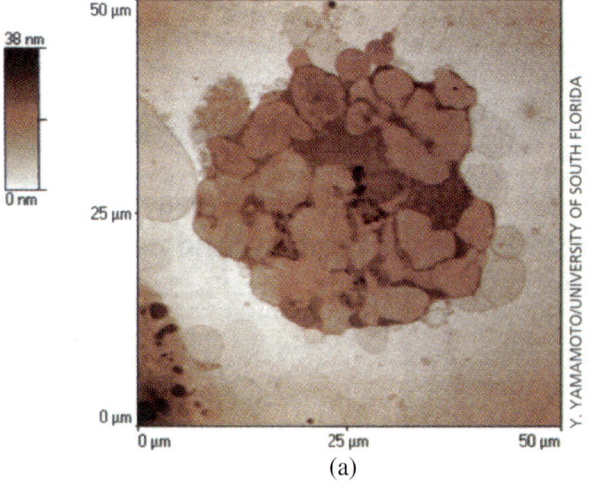

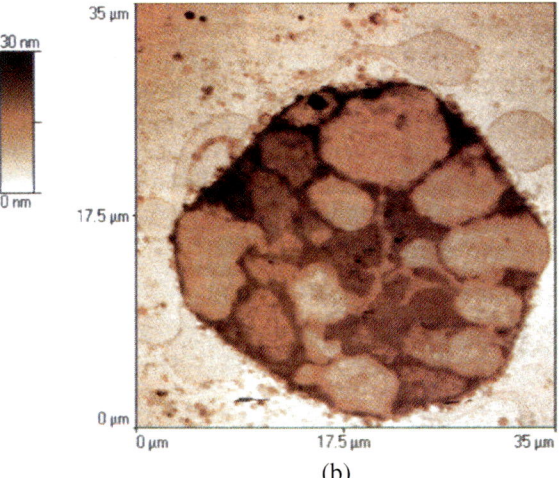

Plate 2. Micrographs of mouse macrophages made with a laser-produced X-ray source at high temporal and spatial resolution, at two different resolutions. From Ref. [9].

Plate 3. Photograph of a laser marking system in operation in an electrical-components assembly line (top), and a comparison of the printed (bottom) and laser-marked (top) components. From Ref. [12].

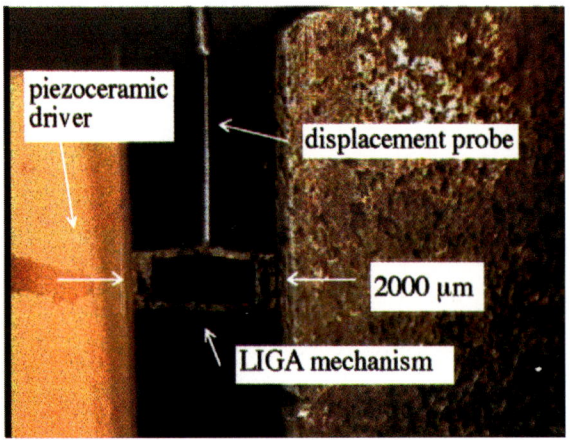

(a)

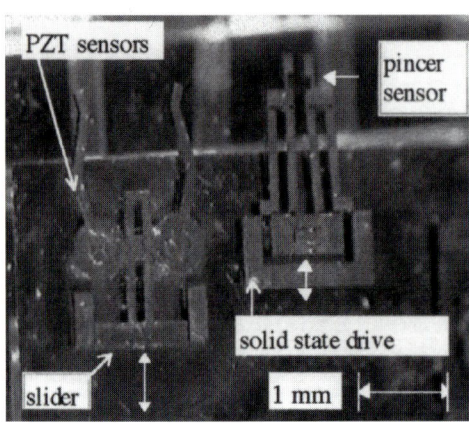

(b)

Plate 4. Microgripper assemblies fabricated by the LIGA process. These structures are typical of what one would like to fabricate in ceramics and oxide materials by laser ablation for use in smart (e.g., sensor-actuator) adaptive microscale devices. From Ref. [19].

117. Amon, O., and Baumeister, P. Electric field dictribution and the reduction of laser damage in multilayers. *Appl. Opt.* **19**(11), 1853–1855 (1980).
118. Welsch, E. Absorption measurements. In Ref. 112.
119. Loriette, V., Roger, J. P., Boccara, A. C., and Gleyzes, P. Probing of low loss materials at 1.064 µm for interferometric gravitational waves detection. *J. de Phys. IV (Colloque)* **4**(C7), C7/631–C7634 (1994).
120. Almond, D. P., and Patel, P. M. *Photothermal Science and Techniques*. Chapman & Hall, London, 1996.
121. Boccara, A. C., Fournier, D., Jackson, W., and Amer, N. M. Sensitive photothermal deflection technique for measuring absorption in optically thin media. *Opt. Lett.* **5**(9), 377–379 (1980).
122. Jackson, W., Amer, N. M., Boccara, A. C., and Fournier, D. Photothermal deflection spectroscopy and detection. *Appl. Opt.* **20**(8), 1333–1344 (1981).
123. Olmstead, M. A., Amer, N. M., Kohn, S., Fournier, D., and Boccara, A. C. Photothermal displacement spectroscopy: an optical probe for solid and surfaces. *Appl. Phys.* **A32**, 141–154 (1983).
124. Welsch, E. Photothermal surface deformation technique—A goal for nondestructive evaluation in thin-film optics. *J. Mod. Opt.* **38**(11), 2159–2176 (1991).
125. Stolz, C. J., and Kozlowski, M. R. Reduce defects to increase laser damage threshold. *Laser Focus World*, November issue, 83–88 (1994).
126. Welsch, E., and Reichling, M. Micrometer resolved photothermal displacement inspection of optical coatings. *J. Mod. Opt.* **40**(8), 1455–1475 (1993).
127. Reichling, M., Welsch, E., Duparré, A., and Matthias, E. Photothermal absorption microscopy on ZrO_2 and MgF_2 films. *Opt. Eng.* **33**(4), 1334–1342 (1994).
128. Commandré, M., and Roche, P. Characterization of optical coatings by photothermal deflection. *Applied Optics* **35**(25), 5021–5034 (1996).
129. Kaiser, N., Uhlig, H., Schallenberg, U. B., Anton, B., Kaiser, U., Mann, K., and Eva, E. High damage threshold Al_2O_3/SiO_2 dielectric coatings for excimer lasers. *Thin Solid Films* **260**, 86–92 (1995).
130. Reichling, M., Bodemann, A., and Kaiser N. A new insight into defect-induced damage in UV multilayer coatings. In *Laser-Induced Damage in Optical Materials 1994*, H. E. Bennett, A. H. Guenther, M. Kozlowski, B. E. Newnam, and M. J. Soileau (eds.), pp. 307–316, SPIE Vol. 2428, Bellingham, 1995,
131. Roche, P., Commandré, M., Escoubas, L., Borgogno, J. P., Albrand, G., and Lazarides, B. Substrate effects on absorption of coated surfaces. *Appl. Opt.* **35**(25), 5059–5066 (1996).
132. Reichling, M., Siegel, J., Matthias, E., Lauth, H., and Hacker E. Photoacoustic studies of laser damage in oxide thin films. *Thin Solid Films* **253**(1, 2), 333–338 (1994).
133. Welsch, E., Ettrich, K., Blaschke, H., Thomsen-Schmidt, P., Schäfer, D., and Kaiser, N. Investigation of the absorption induced damage in ultraviolet dielectric thin films. *Opt. Eng.* **36**(2), 504–514 (1997).
134. Tench, R., Kozlowski, M., and Chow, R. Defect geometries and laser-induced damage in multilayer coatings. In *Optical Thin Films IV: New Developments*, J. D. Rancourt (ed.), pp. 60–66. SPIE Proc. 2262, Bellingham, 1994.
135. Tench, R. J., Kozlowski, M. R., and Chow, R. Investigation of the microstructure of coatings for high power lasers by non-optical techniques. In *Optical Interference Coatings*, F. Abelés (ed.), pp. 596–602. SPIE Proc. 2253, Bellingham, 1994.
136. Tench, R., Kozlowski, M., Cohen, J., Chow, R. Laser damage and conditioning at defects in optical coatings. In *Optical Interference Coatings Vol. 17*, 1995 OSA

Technical Digest Series, pp. 214–216. Optical Society of America, Washington DC, 1995.
137. Gan, F., Xu, Y.-N., Huang, M.-Z., Ching, W. Y., and Harrison, J. G. Optical properties of a CaF_2 crystal. *Phys. Rev.* **B45**(15), 8248–8255 (1992).
138. Tomiki, T., and Miyata, T. Optical studies of alkali fluorides and alkaline earth fluorides in VUV region. *J. Phys. Soc. Jap.* **27**(3), 658–678 (1969).

13. THE FUTURE OF LASER ABLATION

Richard F. Haglund, Jr.

Department of Physics and Astronomy
and W. M. Keck Foundation Free-Electron Laser Center
Vanderbilt University
Nashville, Tennessee

13.1 Introduction

The preceding chapters of this book have covered a broad spectrum of both fundamental science and applications. Impressive as these applications are, the last words remain to be written in the expanding story of laser ablation and its scientific, industrial, and medical uses. Among the clearest indicators of this fact are the annual market surveys of the laser industry carried out by the laser industries popular magazines *Laser Focus World* and *Photonics Spectrum*. The surveys for 1997, by laser type and by industrial application, indicate a market for laser processing systems that is rapidly approaching the billion-dollar-per-year mark. Even discounting the medical market where lasers made some of the earliest inroads, a substantial fraction of those sales are in materials processing areas where laser ablation plays a role.

In this chapter, we consider a variety of applications—some already implemented, some in prospect, and some only beginning to be realized. The intent is to highlight a selection of applications that illustrate the versatility of laser ablation at different wavelengths, pulse durations, and pulse repetition rates in industrial processes. These applications are frequently driven by special capabilities of particular lasers or the requirements of specialized industrial end uses. In addition to the ability simply to carry out a particular task, lasers for applications must often satisfy a wide variety of other constraints, including cost, ease of operation and maintenance, safety, and adaptability for changing manufacturing processes. Some major areas of current and future application are listed in Table I; many others can be found in [1].

Plate 1(a) shows the processing operations that can be carried out with lasers at varying pulse durations, wavelengths, and pulse-repetition frequencies. Aside from illustrating the great range of applications for laser ablation, Plate 1(a) also illustrates the special opportunities created by the development of ultrashort pulse, chirped-pulse amplifier lasers and free-electron lasers. These two newcomers to the laser field are opening previously unexplored parameter space in materials processing and other industrial applications. Free-electron lasers, with their unique combination of high average power, high peak power, broad

TABLE I. Representative Industrial Uses of Laser Ablation [2]

Level of maturity	Polymer surfaces	Micromachining	Materials or electronics processes information storage	Semiconductor or metal surfaces
Commercializable • Demonstrated • Major market • Cost/capacity barrier	• Surface texturing • Surface conductivity • Surface amorphization	• Fuel injectors • Spinnerets • Slitting coated films • Marking • Ink jets	• Flat panel displays • Large-area photovoltaics • Large-area diamond coatings	• Laser glazing and annealing • Carburizing and nitriding • Elemental analysis • Marking
Developable • Demonstrated • Development needed • Large market potential	• Antimicrobial nylon • Pulsed laser deposition • Enhamced adhesion	• Subthreshold ablative cutting • Ceramics machining	• Pulsed laser deposition • MEMS/ASIMs/MEOMs	• Adhesive bond pretreatment • Solvent-free cleaning • Removal of corrosives
Prospective • Preliminary results • Research, development needed • Enabling facilities needed	• Surface activation of carbon fibers for composites	• Smart materials deposition • Large-area drilling (air foils)	• Enhanced multi-level CDs • Embedded layer processing	• Thermal barriers, coatings • Metglass coatings • Shape-memory alloys

tunability, and high pulse repetition frequency have some especially attractive potential industrial applications [2].

This chart as it stands gives no hint of the qualitative changes that the use of lasers can bring to industrial applications. Industry is being pulled in two somewhat competing directions simultaneously: first, to more meet increasingly sophisticated and differentiated consumer demands for customized or individualized products (which means smaller batch sizes); and second, to do so with reduced production costs, higher product quality, and shorter throughput times. The laser makes possible higher production speeds, reduced tool wear, greater flexibility in changing procedures, and (if laser parameters are properly chosen) smaller collateral damage zones in the laser-processed materials. Laser ablation, and other laser applications in materials processing and process monitoring, are likely to play an even more important role in the future of both high- and low-technology industries.

13.2 Material Cutting and Joining

Laser cutting, drilling, and welding drives much of the industrial interest in lasers; typically this segment of the laser applications market is almost half the total value of laser-manufactured parts, with 90% of that amount going to general two- and three-dimensional cutting. The range of applications covers everything from the manufacture of micrometer-thick precision parts to the cutting of thick, complex shapes in the shipbuilding industry. The unusual properties of the laser that make these applications attractive include the fact that it is contactless; that it can work in almost any direction, and in vacuum, gas or fluid environments; and particularly that some operations, such as cutting and joining (e.g., welding), can be accomplished with the same tool simply by varying the machine parameters. In typical applications, the cutting laser may be used in a variety of modes, including cutting in a gas or fluid jet to remove material efficiently and maintain material throughput. Although almost every extant laser type has been used for cutting and drilling, the most popular types being the high-average-power Nd:YAG and CO_2 lasers.

The most publicized laser cutting and drilling operations have been in applications requiring micrometer-level precision. The copper-vapor laser (CVL), for example, has recently shown significant promise for drilling high-aspect ratio holes in stainless steel, opening up possible niches in microdrilling of inkjets for computer printers (diameter ~30 μm), fuel-injector nozzles, and other similarly demanding applications. The combination of high pulse repetition rates, high power (up to 500 kW), and small focal spots by virtue of the visible laser light, makes lasers attractive for these applications, provided that the lasers can be made stable enough—a requirement that the laser industry increasingly fills very well. An electron micrograph of such a drilled hole is shown in Fig. 1.

The ability of the laser to drill and cut precisely will undoubtedly assure it a significant future in this application. Other significant potential applications remain to be explored—such as the drilling of microholes to influence airfoil properties—that require not only precision, but the ability to perform the operation over very large areas in a short period of time. The realization and reduction to practice of such large-area processing techniques will demand that we acquire a much deeper understanding of the relationship between materials properties and optimum laser pulse duration, repetition frequency, and wavelength.

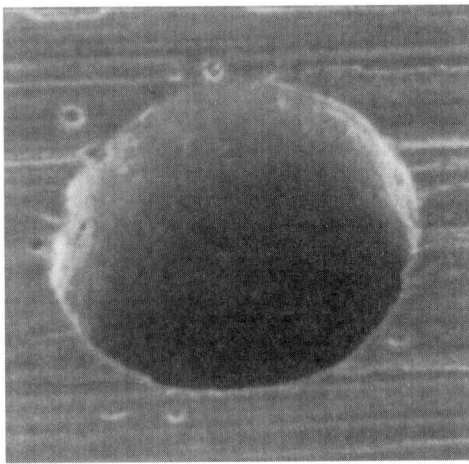

FIG. 1. A 30 μm-diameter hole in copper, made with a copper vapor laser [7].

13.3 Cluster-Beam Generation by Laser Ablation

An important dream in both chemistry and physics has been the production of large quantities of mass-selected clusters for catalytic, electronic, and optical applications. The awarding of the Nobel Prize in Chemistry for 1996 signaled scientific recognition for the discovery of C_{60} in a laser ablation cluster-beam source, as well as for the initial exploitation of Buckminsterfullerene's unique properties. Equally significant, in the long run, have been the many other studies in cluster physics and chemistry that have resulted from the rapid deployment of laser-ablation sources all over the world. At this point, clusters of everything from alkali halides (used in a laser-ablation experiment described in Chapter 2) to semiconductors to novel mixed compounds such as the metallocarbohydrenes [3] are being produced in beam sources of this type.

The original laser-ablation cluster source of Smalley *et al.* has largely been supplanted by the design of Milani and DeHeer [4]. In typical source designs of this type (Fig. 2), a laser pulse, usually from an excimer laser but sometimes using harmonics of a Nd:YAG laser, is delivered to a rod of target material. Simultaneously a burst of He or Ar gas is delivered to the target region by a fast-pulsed gas jet, entraining the laser-ablated material from the rod. The rapid cooling of the ablation plume in an adjacent expansion chamber causes clusters to nucleate with high efficiency; those clusters with primarily forward momentum are able to pass through a skimmer nozzle into the experimental region where they are now routinely interrogated by a wide variety of optical and electronic spectroscopies.

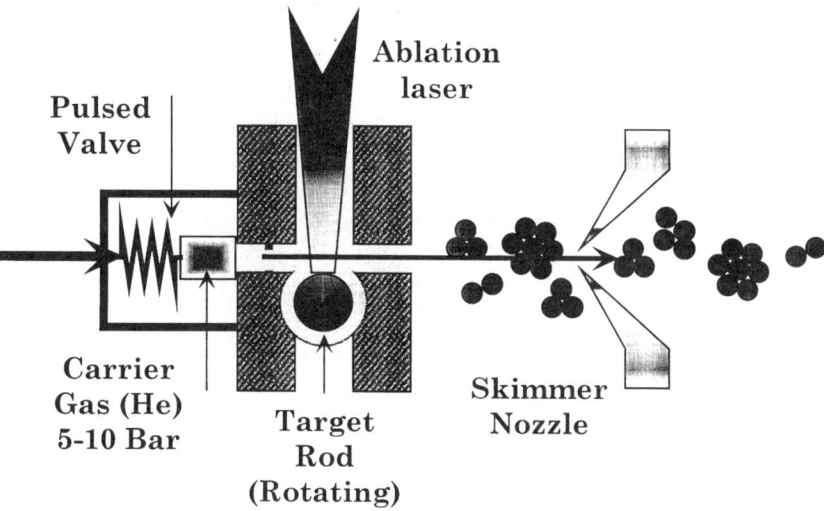

FIG. 2. Cluster-beam source of the Milani-deHeer type, showing the ablation region and the expansion chamber where the clusters are formed.

In addition to the many fundamental studies of the electronic, optical, and chemical properties of individual clusters that have been made possible by these sources, they are now beginning to be used to provide the clusters for materials synthesis. To cite only one example: There is a long-standing interest in the linear and nonlinear optical properties of metal and semiconductor clusters embedded in dielectric host materials. The production of suitably uniform clusters is, however, a major challenge when pursued by conventional means, such as melt-glass, sol-gel, and ion implantation. All of these fabrication techniques share the common defect that the thermodynamics of the nanocluster formation is not fully under control. In melt glass fabrications, for example, the rate of formation of the clusters, as well as their mean size and size distribution, is governed by the kinetics of the glass formation process; set the temperature too high, or put in too much metal, and nanocluster formation is inhibited. The metal or semiconductor material simply precipitates out of the glass. On the other hand, with cluster-beam fabrication, it is possible to generate mass-selected clusters, deposit them on a substrate, and then use other techniques—for example, electron-beam evaporation, or physical vapor deposition, or even pulsed-laser deposition—to cover the nanoclusters as desired. This produces unprecedented control over the size of the clusters. Such sources have been constructed at more than one laboratory [5]. Figure 3 shows a supported antimony cluster made by one such source.

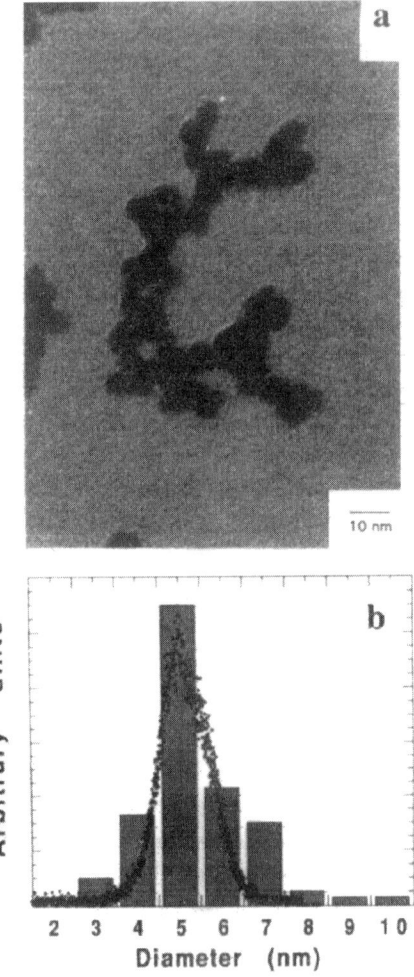

FIG. 3. (a) High resolution electron micrograph of antimony clusters supported on HOPG graphite substrate. (b) Size distribution of the antimony particles. From Ref. [5].

13.4 Laser Plasma X-ray Generation

X-ray microscopy has long been a dream of researchers in the biological sciences. Currently major efforts are underway to build large X-ray absorption facilities at the national synchrotron laboratories. However, both the temporal resolution and spectral brightness of laser-produced plasmas can exceed by substantial margins what is available at synchrotron light sources; in one recently reported experiment [6], the source was six orders of magnitude brighter and three orders of magnitude shorter in duration than soft X-rays from typical synchrotron sources.

Plate 1(b) shows the streak-camera records of three X-ray bursts from an Au target generated by the 1.06 µm output of an 0.5 ps Nd:YAG laser in a chirped-pulse amplifier configuration, focused to a spot about 30 µm in diameter. The streak record with its charge-coupled-device camera recorded the intensity as a function of wavelength as a function of time following the impact of the laser pulse; for the lowest intensities, very short pulses are observed. Dispersive optics (e.g., a Bragg crystal) or multilayer X-ray optics can be used to select a narrow-band portion of the spectrum. This particular source, at the University of Michigan's Center for Ultrafast Optical Science, has a pulse repetition frequency of 10 Hz, and has an efficiency of about 10% for conversion of laser light to total X-ray emission.

Lasers with short pulse duration and high brightness, make it possible to image biological systems on time scales short compared to the time in which these delicate materials can be damaged by the intense bursts of X-rays. Although Ångstrom-resolution electron microscopy offers better resolution than X-rays (typically 100 Å), the fact that electron-microscope samples are "dead," that is, killed, sectioned, and stained, inherently limits the useful information available. Plate 2 shows an image of a mouse macrophage made using a nanosecond laser ablation X-ray source and a gold target [8]. The false-color scale is keyed to the wavelength of the source, thus providing selective information about the X-ray absorption in the macrophage. Although the spatial resolution is not as good as one would see in an electron microscope, it is much better than that found in visible microscopy. Moreover, coupled to available information from X-ray spectroscopy, such an image makes possible the visualization of specific elemental information not obtainable by visual microscopy [9].

For now, these systems remain limited to some degree by cost, complexity, and the difficulty of fabricating the necessary X-ray optics. However, intense efforts underway worldwide are likely to see changes in this situation before long. Moreover, the cost and complexity of the lasers needed to produce these plasmas are within limits such that one could imagine them ultimately being no more expensive than many other items of standard analytical equipment routinely used in hospitals and university laboratories.

13.5 Lasers in the Semiconductor Industry

Lasers find an astonishing number of uses in the semiconductor industry, from marking components to drilling vias, from breaking circuit connections and component marking to repairing lithography masks. The chart in Fig. 4 shows where in the processing of typical semiconductor fabrication steps lasers are

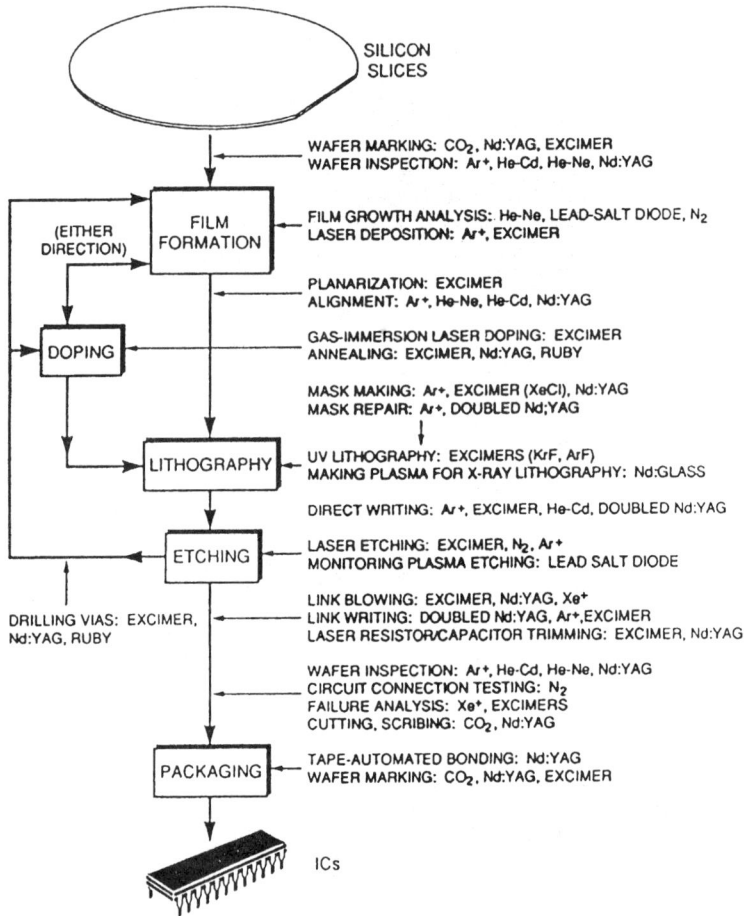

FIG. 4. Chart of semiconductor fabrication processes involving lasers, showing the types of lasers used. Laser ablation is involved in many of these processes.

used, listing also the types of lasers. In several of these processes, laser ablation is one of the key elements.

Producers of circuit boards use several different lasers to drill the vias—small holes that are used to lead from one level to another in a multilayer circuit board—because the dry photochemical etch provided by the laser is environmentally clean and inherently flexible (Fig. 5(a)). The inherent ability to alter the hole size by changing the focal spot size of the laser also makes this manufacturing process flexible. Interestingly, it turns out that there are wavelength differences, even though the process appears to be fundamentally thermal

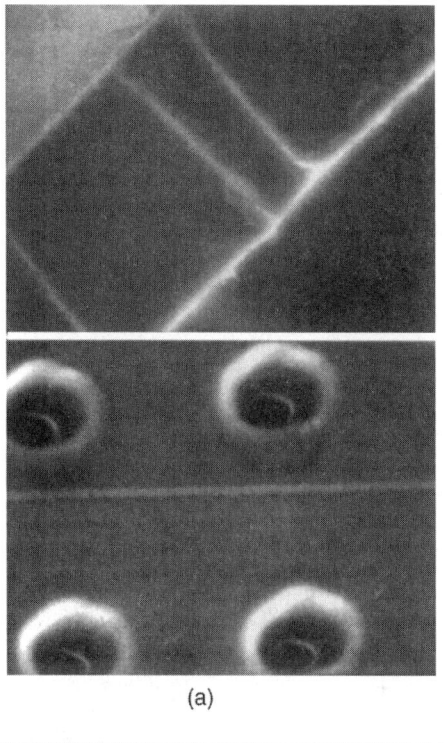

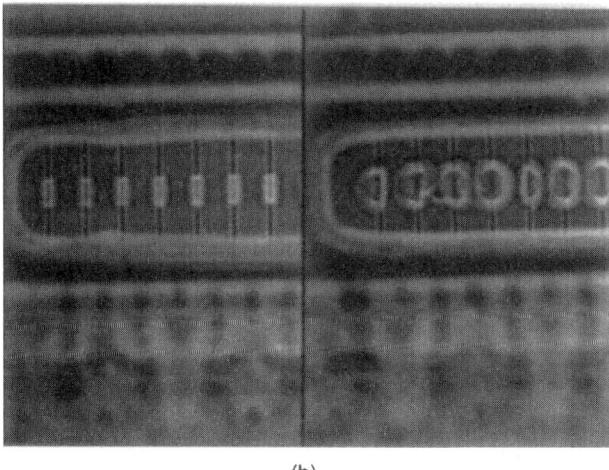

FIG. 5. (a) Precision via drilling using the third harmonic (355 nm) from a Nd:YAG laser [9]. (b) Circuit repairs in a Si "microchip" made using two different lasers, operating at 1.045 μm (right) and 1.341 μm (left) [10].

in the polyimide insulating layer between multilayer boards. Ablation at 355 nm, for example, is actually more efficient than at 266 nm. Thus electronic absorption and thermomechanical processes are clearly involved in the ablation mechanism [10].

Lasers are also now being used to repair high-density 16- to 256-Mbit memory chips by severing faulty connections, where the challenge is to ablate the offending connection without damaging the surrounding Si material [11]. In earlier processing schemes, Q-switched Nd:YAG lasers operating at 1.047 μm were used for this task, but as the storage densities of chips have increased, it has become increasingly difficult to cut circuit links without damaging neighboring circuit elements. By switching to the Nd:YLF laser operating at 1.321 μm, the laser process was adapted to the more demanding requirements. The absorption coefficient of Si is approximately 10^2 lower at 1.321 μm than at 1.047 μm, thus dramatically reducing the heat flow in the vicinity of the repair spot. The Nd:YLF operates at pulse repetition frequencies up to 4.6 kHz with as much as 6 mJ per pulse (average power 25 W). A comparison of a section of similar circuits with laser-ruptured metal circuit links, showing the improvement in circuit morphology that results from the longer wavelength laser, is shown in Fig. 5(b). Similar results can be achieved by continuous-wave lasers, by using a halogen gas in the work area to induce photochemical ablation.

Another common application in electronics (and elsewhere, in settings as diverse as the automobile and hand-tool manufacturing industries) is in *laser marking* [12]. Tiny, distinctive, and legible markings are indispensable for many electronic components; in many other applications, laser marking provides an image that is difficult to counterfeit and that is not as easily rubbed away as a printed or painted image. In general, the technique may involve either melting or vaporization of a substrate overlaid on a color-contrasting base material. Plate 3 compares CO_2 laser marking of capacitor cases with the former printed image. The contrast is striking but does not tell the entire story of the advantages of laser marking: In this particular instance, because of the varying speed of the assembly line, it was necessary to find a way to measure the speed at which the part was moving simultaneously with marking it. In an unintended shakedown test of one such commercial system during its development, the assembly line stopped—and the marking system continued to mark the capacitors perfectly as the line slowed down.

13.6 High-Speed Color Printing

A novel application of laser ablation has recently been developed by a collaboration between the University of Illinois and Graphics Technology International using focused laser beams to achieve high-resolution transfer of

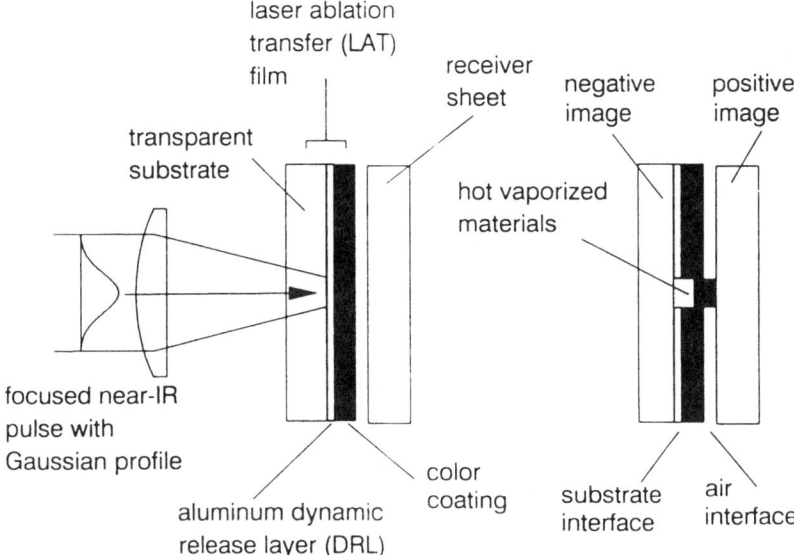

FIG. 6. Schematic of the laser-ablation transfer (LAT) printing process. The laser is focused through a transparent substrate onto an aluminum release layer covered with a color coating, which is ejected by ablation from the DRL and retained as a positive image on the receiver sheet. From Ref. [13].

pigmented material to a substrate. This scheme, called laser ablation transfer (LAT), is depicted schematically in Fig. 6. A Gaussian, near-infrared laser beam (Nd:YAG) is focused through a transparent substrate onto a thin aluminum film which acts as a dynamic release layer (DRL) for the color-coating material. The coating, which absorbs at the Nd:YAG wavelength (1.064 µm), can be loaded with a variety of black-and-white and color pigments to suit the requirements of the printing application.

Studies with an ultrafast laser microscope and ultrafast shadowgraphy show that extremely rapid heating of the DRL by a nanosecond laser pulse, at rates approaching 10^9 deg·s^{-1}, leads to the efficient ablation and thermal decomposition of a small fraction of the color-coating material [13]. Time-resolved optical thermometry shows that the temperature of the ablated material approaches several hundred degrees during the ablation process. The ablated color coating, moving at near sonic velocities (see Fig. 7), is rapidly transferred to the receiver sheet, where its cooling time is estimated to be of order several microseconds. The process produces a negative in the color-coating material and a positive image on the receiver sheet.

Among the key issues for the effective use of this technology, of course, is the extreme accuracy and reproducibility in the amount and dimensions of the

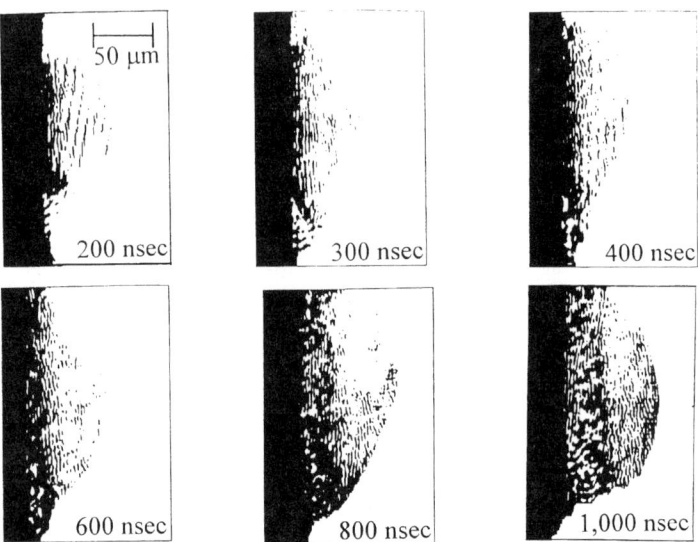

FIG. 7. Time-resolved micrographs of LAT film during picosecond laser ablation. The air-coating interface is the transition zone between the dark region on the left and the lighter region on the right. Initial velocity of the ablated material is approximately Mach 1. From Ref. [14].

ablated material. Satisfactory performance can be achieved with nanosecond lasers. However, *mode-locked* picosecond lasers could, in principle, generate images at a rate of megapixels per second, although these lasers presently are not considered suitable for an industrial environment. Recent experiments [14] have shown that picosecond pulses are even more efficient in ablation efficiency than nanosecond pulses, with ablation thresholds an order of magnitude smaller than those measured with nanosecond lasers (roughly 15 vs. 150 mJ·cm^{-2}). The deployment of mode-locked picosecond lasers running at MHz pulse repetition frequencies thus could open the possibility for both high-speed printing and proofing of color images.

Future deployment of this application will undoubtedly capitalize on diode-pumped Nd:YAG laser systems as they are reduced in size and as mode-locked systems are made more reliable.

13.7 Laser Ablation in Microfabrication

Laser ablation can play a significant role in the production of micro-electro-mechanical and micro-opto-electro-mechanical systems (MEMS and MOEMS).

These devices—ranging from microsensors to microactuators and controllers—have captured the imagination of design engineers and are now being fabricated in industrial quantities for a wide variety of technological and medical end users. While silicon has been the material of choice in MEMS systems up to the present, it is increasingly clear that metals and oxides will have to be added to the MEMS-MOEMS toolkit to make some of the most interesting systems. Laser ablation can be effectively used in a variety of unusual microfabrication situations, such as:

- Rapid prototyping of MEMS or MOEMS devices, especially in oxide materials.
- Achieving site-specific activitation or chemistry in conjunction with material removal.
- Carrying out processing on or modification of embedded interfaces.

The laser has been used, often in conjunction with chemical processing, to produce microstructures with increasing efficiency, both by microablation and by pulsed-laser deposition in conjunction with advanced LIGA techniques.

Excimer lasers were among the early leaders in this field [15] because their ultraviolet wavelengths permitted extremely small focal spots. In a now famous photomicrograph from Lambda Physik, a human hair has been drilled with micron-scale resolution to produce an appropriate advertising slogan for the excimer laser (Fig. 8). In their most recent configurations, excimer lasers have become extremely reliable from the standpoint of high-voltage engineering, requiring substantially less maintenance than formerly and yielding much longer gas-fill lifetimes. Moreover, the development of high repetition rate excimers is making it possible for these lasers to function more effectively on assembly lines in microelectronics and elsewhere.

The particular interest in using excimers and other lasers for microfabrication stems from the same features that makes them attractive generically in cutting and joining applications, but there are additional advantages to be gained because of the complex materials requirements for fabricating piezoelectric and ferroelectric materials in MEMS applications. Thus far in the development of micro-electro-mechanical systems, researchers have focused on layered silicon technology, producing actuators, sensors, and even micro-instrumentation in polysilicon [16]. In spite of all that has been accomplished, however, it is generally recognized that piezoelectric sensor-actuator systems will be required to provide the control bridge to macroscopic systems [17].

The development of the LIGA (*LIthographie, Galvanoformung, Abformung*) process opened new avenues for the development of solid-state devices [18]. LIGA fabrication begins with the exposure of a thick photoresist (PMMA) to X-ray radiation through a mask; the developed photoresist yields a mold that is electroplated with, for example, nickel, to form high-aspect-ratio components up

FIG. 8. Microscopic ablation of trenches in a human hair, carried out with a finely focused excimer laser. Photograph courtesy of Lambda Physik. From Ref. [15].

to 2000 μm in length and over 200 μm in thickness. Plate 4 shows two examples of microgrippers, both constructed using the LIGA process [19]. The assembled gripper (top) comprises various LIGA components, and the other (bottom) is made from a single component using solid-state kinematic links. These grippers are designed to be the final stage of a complex haptic interface for MEMS assembly of structures and will allow the manipulation of the MEMS scale environment with force-reflective feedback at the macroscopic level. Such devices are currently under intensive investigation [20, 21] for applications such as microsurgery. By using micro-pulsed-laser-deposition to coat the the metal pincers with a smart material such as PZT, it would become possible to fabricate integrated sensors and thus obtain vital, high-sensitivity sensory data to enhance the quality of the man-machine interface.

13.8 Conclusions

For some, the current boom in industrial applications of the laser—including but certainly not limited to laser ablation—is the final answer to the old joke about the laser as "a solution in search of a problem." But it may in fact be that the really big breakthroughs in using laser ablation will be like the photocopier and automatic teller machines: we will not really know that we had to have them until someone invents them. The current economic drivers in the industrial sector are pushing manufacturing techniques that effectively match component design and fabrication processes to specific materials and processing characteristics. This leads, as one observer has noted, to "the development of highly economical alternative processes instead of simply using lasers as a substitute for traditional tools or methods" [22]. The fact that laser manufacturing processes can bring to large areas and mass production runs a precision once reserved only for serial-production items of the highest quality is certain to lead to many more applications in the future.

It is almost obligatory in a summary chapter of this type to end with optimistic forecasts about the future. In this case, however, substantial optimism seems to be justified. Rapid developments in new laser technology are driving costs down and efficiency up, and the improved performance of high-repetition-frequency, high-average-power laser systems is making it reasonable to regard laser ablation as a "high throughput" process. The range of available laser frequencies, pulse duration, peak power, and fluence is expanding, so that matching laser characteristics to applications is more readily accomplished. Increased fundamental understanding of the laser-materials interaction is making it possible to select lasers for specialized fabrication processes, such as ablation, and for an increasingly wide range of target materials, with more understanding and less guesswork. The marriage of the laser with inexpensive, powerful computer controls is certain to make industrial applications at both large and small scale attractive to industry. Finally, the growing demand for higher quality and more individualized products is precisely the kind that can be satisfied by the high-precision application of laser ablation.

References

1. See, for example, the recent *Optical and Quantum Electronics* **27**, special issue (December 1995) devoted to industrial applications of the laser.
2. Kelley, M. J. FEL for the polymer processing industries. *SPIE Proceedings* **2988**, Paper 2988-29. San Jose, CA, Feb. 13–14, 1997. See also Shinn, M. D. Jefferson Laboratory IR demo project. *SPIE Proceedings* **2988**, Paper 2988-21. San Jose, Feb. 13–14, 1997.
3. Guo, B. C., Wei, S., Purnell, J., Buzza, S., and Castleman, A. W. Metallo-Carbohedrenes [$M_8C_{12}^+$ (M = V, Zr, Hf and Ti)]: A class of stable molecular cluster ions. *Science* **256**, 515–516 (1992).
4. Milani, P., and deHeer, W. A. Improved pulsed laser vaporization source for production of intense beams of neutral and ionized clusters. *Rev. Sci. Instrum.* **61**(7), 1835–1838 (1990).
5. For a review, see Melinon, P., Paillard, V., Dupuis, V., Perez, A., Jensen, P., Hoareau, A., Perez, J. P., Tuaillon, J., Broyer, M., Vialle, J. L., Pellarin, M.,

Baguenard, B., and Lerme, J. From free clusters to cluster-assembled materials. *Int. J. Mod. Phys. B* **9**(4-5), 339–397 (1995).

6. Workman, J., Maksimchuk, A., Liu, X., Ellenberger, U., Coe, J. S., Chien, C.-Y., and Umstadter, D. Picosecond soft-x-ray source from subpicosecond laser-produced plasmas. *J. Opt. Soc. Am. B* **13**, 125–131 (1996).
7. Messenger, H. W. Terawatt lasers generate x-ray continuum from gold target. *Laser Focus World* **31**(2), 15–16 (February 1995).
8. Kado, M., Richardson, M. C., Gabel, K., Torres, D., Rajyaguru, J., and Muszynski, M. J. Ultrastructural imaging and molecular modeling of live bacteria using soft x-ray contact microscopy with nanosecond laser plasma radiation. *Proc. SPIE* **2523**, 194–201 (1995).
9. DeMeis, R. Laser plasma "x-rays" living organisms. *Laser Focus World* **32**(6), 37–38 (June 1996). The X-ray images were prepared by Y. Yamamoto at the University of South Florida.
10. Anderson, S, G. Solid-state uv lasers drill vias effectively. *Laser Focus World* **32**(2), 31–32 (February 1996).
11. DeMeis, R. Longer-wavelength laser repairs increase chip yield. *Laser Focus World*, 40–41 (August 1996).
12. Hayes, O. Marking applications now encompass many materials. *Laser Focus World* **33**(2), 153–160 (1997).
13. Tolbert, W.A., Lee, I-Yin S., Doxtader, M. M., Ellis, E. W., and Dlott, D. D. High-speed color imaging by laser ablation transfer with a dynamic release layer: Fundamental mechanisms. *J. Imaging Sci. Tech.* **37**(4), 411–421 (1993).
14. Tolbert, W. A., Lee, I-Yin S., Wen, X., Dlott, D. D., Doxtager, M. M., and Ellis, E. W. Laser ablation transfer imaging using picosecond optical pulses: Ultra-high speed, lower threshold and high resolution. *J. Imaging Sci. Tech.* **37**(5), 485–490 (1993).
15. See, for example, Rebhan, U., Endert, H., and Zaal, G. Micromanufacturing benefits from excimer-laser development. *Laser Focus World*, 91–96 (November 1994).
16. Gabriel, K. J. Engineering microscopic machines. *Scientific American* **260**, 118–121 (1995).
17. Fujita, H. Future of actuators and microsystems. *Sensors and Actuators* **A56**, 105–111 (1996).
18. Mohr, Bleg, Wallrabe. Microactuators fabricated by the LIGA process. *Journal of Microelectromachanical Systems* **2**(4) (December 1992).
19. Photographs courtesy of Prof. Ephraim Garcia, Director, Vanderbilt University Smart Structures Laboratory. Pokines, B. J., Garcia, E. "Forging Solid-State Micro-actuators through the Merging of Smart Material and Microelectromechanical Systems," *Proc. SPIE* Conference on Smart Structures and Materials, San Diego, CA (1997).
20. Dario, P., Valleggi, R., Carrozza, M., Montesi, M., and Cocco, M., Microactuators for microrobots: A critical survey. *J. Micromech. Microeng.* **2**, 141–157 (1992).
21. Goldfarb, M., and Celanovic, N. Minimum surface-effect microgripper design for force-reflective telemanipulation of microscopic environment. *Proceedings of the ASME International Mechanical Engineering Conference and Exposition*, November, 1996.
22. Reinhart, G., Milberg, J., Lindl, H., and Trunzer, W. Laser systems in modern production environments. *Opt. and Quant. Electronics* **17**(12), 1103–1125 (1995).

Index

A

Ablation
 definition of, 15
 mechanisms, 78–106, 485–486
 mid-infrared, 102–104
 near-infrared, 102
 off-axis, in PLD, 39
 threshold, 186, 428, 584
 ultraviolet, 90–91, 94–95, 103
 visible, 102
Abrasion, effect of on desorption, 151, 156
Absorption, soft-tissue, 461–462
Absorption depth, 335–339, 583–590
Acoustic emission, 391
Adiabatic potential energy surface (APES), 49–56
Adsorbates, in desorption, 76
Alkali halides, 72–75
α-cyano-f-hydroxycinnamic acid, 418
Alumina, 76–77, 100, 232, 598–600
Aluminum (Al), 113–114, 404
Aluminum nitride (AlN), 245–258, 284
Ambient gas, 245–247, 383–387
Amorphic diamond, 189
Angular distribution
 in ablation, 481–482, 485–488
 ambient-gas effects, 488–491
 spot-size effects, 487–488
Anthracene, 110–112
Atomic emission spectroscopy (ICP-AES), 385
Atomic force microscopy (AFM), 291, 292, 299–306
Auger electron spectroscopy (AES), 291, 292, 318, 323–326

B

Band structure, 23, 584, 589
Beam deflection probes, 217, 218
Beam delivery, 191–201
Beam profile, 200, 201
Beam shaping, 198–200
Blackbody temperature, 206
Boiling, 186, 229, 234, 281, 398
Boltzmann distribution, 33–35, 124–126, 209, 228, 241, 493, 600
 of ejecta, 123, 125
Bond-orbital calculations, 120
Bond orbitals, 28–30
Breakdown, dielectric, 463
Brushite ($CaHPO_4 \cdot 2H_2O$), 142

C

Cadmium tin oxide (Cd_2SnO_4), 554
Calcium carbonate ($CaCO_3$), 104–106, 146, 151, 160
Calcium fluoride (CaF_2), 94–95, 98–99, 585–588, 601–606
Charge screening, 49
Cinnamic acid, 418
Cleaning, laser-assisted, 357–359
Clusters, 2, 212, 428
 desorption from, 74–75
 generation of beams, 4, 401–402, 628–629
 model calculations, 116
CO_2 laser, 334
Coherent anti-Stokes Raman scattering (CARS), 116–118
Cohesive energy, 30
Collective effects, 592
Collector plate, 599
Collisions, in ablation plume, 494–499
Concentric hemispherical analyzer (CHA), 319
Conductivity, laser modification of, 360–361
Cone formation, in laser ablation, 508–511
Configuration coordinate, 40, 49–52, 59
Congruent transfer, 479
Contact front, 245
Cooling transitions, 50–51
Crossed beams, for particulate control, 513
Cu plume, 259–261, 270, 285
Cylindrical mirror analyzer (CMA), 320

642 INDEX

D

Damage threshold, 573, 574
Defect-induced desorption, 16, 56, 75–78
Defect structure, 529–531
Depth of focus, 191
Desorption
  laser-induced, 15, 58–78
  molecules, 62–65
  neutral atoms, 140, 157–158
  photon-stimulated (PSD), 72
  positive ions, 142–149
Dexter-Klick-Russell processes, 52
Dielectric breakdown, 454
Dielectric constant
  optical, 22
  static, 22
Dielectric function
  metal, 21
  lossy insulator, 457–458
Diffusion, laser assisted, 365
Doping in PLD, 519, 521–523
Double-pulse experiments, 100–101
Dried-droplet method, 422
Droplets, 187

E

Effusion, 242–243
Electron avalanche, 87, 95, 588
Electron energy loss spectroscopy (EELS), 76, 291
Electronic excitation, in ablation plume, 499–501
Electronic process, 18, 176, 225, 232–234, 272, 282, 413
Electronic sputtering, 176
Electron-hole pairs, 24, 26
Electron-lattice coupling, 30–32
Electron-phonon coupling, 81
Electron spectroscopy for chemical analysis (ESCA), 318
Electron traps, 141, 148, 149, 150–152
Elemental analysis, 224
Energetic-beam-assisted growth, 494–495
Energy deficit, 429
Energy spectra
  of atoms in plume, 523
  of electron emission, 14–16

  of positive ions, 155–164
Enthalpy function, 181
Equation of state, 115
Etching, chemical, 355–357
Excitons, 34–36, 584
Exfoliation, exfoliational sputtering, 102, 176, 236
Explosive boiling, see Phase explosion
Explosive vaporization, 89

F

Fast-atom bombardment (FAB), 415
Film
  doping in PLD, 475–572
  growth by PLD, 475–572
  stoichiometry in PLD, 482–483, 520–521
Fluence, 174–175
Fluence dependence
  of ablation yields, 616
  of atom emission, 158
  of ion emission, 145
Fluorescence contours, 250–251
Formic acid, 421
Fourier-transform mass spectrometry (FTMS), 438
Fracture, 92–98, 102–104, 592–594, 604
Franck-Condon excitation, 58
Free-carrier heating, 119, 588
Fused silica (a-SiO_2), 93–96, 606–607

G

Gallium arsenide (GaAs), 70–71, 293
Gallium indium oxide ($GaInO_3$), 552–553
Gallium phosphide (GaP), 69, 77–78, 159
Gas dynamics, 225, 230–231
Gaussian beams, 173
Gaussian laser beams, 179–180
Gentisic acid, 418
Gold (Au), 83–85
Green's function, 349

H

Heat conduction, 178–179, 239
Heat transfer, 177–178

Hertz-Knudsen equation, 226
High-T_c superconductors, 3
Hole formation, 189
Holography, 112–115
Homogeneous vapor nucleation, 235
Homogenization, 198–200
Hydrodynamic sputtering, ablation, 237
Hydrodynamical ablation, 176

I

Impact collision ISS (ICISS), 213
Impurity absorption, 588
Inclusions, 421
Incubation effect, 579, 586–590
Indium tin oxide (ITO), 551–552
Inductively coupled plasma (ICP), 384–393
Industrial applications, 626
 of PLD, 516–517
Insulators
 ablation of, 92–106
 band structure, 23, 139, 140, 144, 149, 584, 589
 defects, 144, 148–149
 desorption from, 71–75
Intensity, 24–25
Interferometry, 112–115, 354–355
Inverse Bremsstrahlung, 187, 391, 396
Ion emission, 150, 155, 156
Ion scattering spectroscopy (ISS), 293, 306–315
Ion threshold, 428
Irradiance, 24
Island growth, 308

K

Kapton films, 188
Kinetic energy of ablation products, 492–493
Knotek-Feibelman model, 60–61
Knudsen layer, 226–228, 239–242, 271, 281, 492

L

Ladder sequencing, 432
Lagging, 427

Langmuir probe measurements, 160–163
Laser
 chirped-pulse amplifier (CPA), 93–99
 excimer, 334–335
 femtosecond, 93–99, 190–191
 free-electron (FEL), 17, 102, 190, 457–467
 nanosecond, 188–190
 Nd:YAG, 102, 334
 picosecond, 93–94, 96–97, 190
 ruby, 2, 72–73
 Ti:sapphire, 335
Laser ablation, 15, 78–106
Laser ablation transfer (LAT) printing, 634–636
Laser beam
 angle of incidence, 195–197
 effects of optical elements, 197–198
 focal spot properties, 191
 Gaussian, 173, 179–180
 homogenization, 198–200, 367, 426
 polarization effects of, 195–197
 spatial profile, 174
 temporal profile, 174–175
Laser drilling, 627–628
Laser fusion, 2
Laser-induced desorption, 58–78, 139, 145, 154, 162
Laser-induced periodic surface structures (LIPSS), 341–342
Laser-induced plasmas, 376–405
Laser intensity, 24, *see also* Irradiance
Laser surgery, 3, 461–465
Latent interface, 182
Lattice relaxation, 45–49
Lithium niobate (LiNbO$_3$), 100–101
Lithography, excimer laser, 367–369
Localized modes of excitation, 38–42, 45–48
Local thermodynamic equilibrium (LTE), 176
Local transient laser fluence, 174
Low EIS (LEIS), 306
Low-energy electron diffraction (LEED), 292
Low-energy electron microscopy (LEEM), 297
Low-energy ion scattering (LEIS), 310–312

M

Magnesium oxide (MgO), 144, 150
Marking, laser, 636
Mass ablation rate, 387–389

Mass spectrometry
  quadrupole (QMS), 210–211
  resonance ionization (RIMS), 210–217
  time-of-flight (TOF-MS), 2, 143, 208,
    429–430, 577
Matrix-assisted laser desorption/ionization
    (MALDI), 3, 413–439
Matrix crystallization, 417–425
Matrix effect, 316
Maxwell-Boltzmann distribution, *see*
    Boltzmann distribution
Medium-energy ion scattering (MEIS), 291,
    306
Melting, 180
  of semiconductors, 120
  pulsed laser, 180–183, 347–350
Melting threshold, 176
Melt interface, 182, 183
Menzel-Gomer-Redhead model, 60–63
Mesoscale properties, 78–79
Metallicity, 38–40, 42–44
Metallization, during ablation, 87–89
Metal-matrix ceramics, 188
Metals
  ablation, 79–86
  desorption, 62–67
  surface modification of, 363–364
Microfabrication, 636–638
Micromachining, 194
Microscale properties, 54–62
Molecular beam epitaxy (MBE), 291
  laser, 480
Molecular dynamics simulations, 120–121
Morphology, modification of surface,
    353–359
Multilayer structures, epitaxial, 480
Multiphoton absorption, 25–27, 338–339
Multiphoton excitation, 585
Multiphoton ionization, 69, 186
Multiphoton transitions, 24–26

N

Near-field scanning optical microscopy,
    110–111
Negative-U interaction, 48–49
Nickel oxide (NiO), 125
Nicotinic acid, 417, 418
Nomarski microscope, 581

Nonlinear index of refraction, 37–38
Nucleic acid analysis, 436, 437

O

Off-axis ablation in PLD, 513
Oligonucleotides, 436
Oligosaccharide, 437–438
Optical damage, 581–615
Outflow
  from reservoir, 243–245
  into ambient gas, 245–247
Oxide thin films, 611–616

P

Particles, 376, 381–393
Particulate generation, in PLD, 501–502
Particulates, in PLD, 481
Patterning, 354–355
PBSCCO plume, 258–261, 285
Peak surface temperature, 179
Penetration depth, optical, 18, 20, 24–25, 176
Peptide mapping, 436
Peptide sequencing, 432
Phase explosion, 226, 235, 245, 275–278, 271,
    281
Phase transformation, 103, 462, 463
Phase transition, 335, 346–350
Phonon kicking, 53, 54
Phonons, 27, 28, 30
Photoacoustic deflection (PAD), 217–218
Photoacoustic spectroscopy, 463, 578–580
Photochemical ablation, 189
Photochemical effects, 455
Photochemical process, 176, 339
Photodesorption, *see* Desorption
Photoelectrons, 2, 140
Photoelectron spectroscopy (PES), 292
Photoemission, *see* Photoelectrons
Photoionization, 166–167
Photolysis, 352–353
Photomechanical effects, 456
Photothermal ablation, 175, 189, 462–463
Photothermal effects, 105–106, 343–347,
    455–456
Picosecond lasers, 93–94, 96–97, 190
Plasma absorption, 186, 187

INDEX 645

Plasma desorption, 414, 415
Plasma emission, 376–379
Plasma formation, 165
  in PLD, 482, 483
Plasma heating, 163
Plasma shielding, 49, 100, 390, 393, 396
Plasma temperature, 380, 381
Plasmon polariton, 33
Plasmon resonance spectrometry, 107–108
Plasmons, 177
PLD, *see* Pulsed laser deposition
Plume, 225
Plume emission, 104–106, 154–166
  from MgO, 154–158
  from AlN, 249–252, 255–258
Plume emission spectroscopy, 377–381
Plume expansion, 382–384
Plume formation, 152–168, 232–234, 376–383
Plume interferometry, 98–100
Polarity, 38–40, 44–46
Polarization dependence, 195–197
Polarons, 33, 34
Poly(dimethylsiloxane) (PDMS), 124, 125
Poly(ethylene) (PE), 285
Poly(ethylene terephthalate) (PET), 285
Poly(imide) (PI), 274–278, 285, 338, 360–361
Polymer analysis, 430, 431
Polymers, 336
Poly(methyl methacrylate) (PMMA), 285
Poly(styrene) (PS), 285
Poly(tetrafluoroethylene) (PTFE), 361–362
Positive ion emission, 142–149
Post-translation modifications, 435
Preferential ablation, 398–400
Primary mechanism of ablation, 283
Projection machining, 192–195
Protein ions, 417, 423–440
Protein sequencing, 432
Protoplasma, 157
Pulsed extraction ion source, 430
Pulsed laser deposition (PLD), 3, 189, 399
  apparatus, 477–478
  characteristics, 478–480
  from liquid target, 511–512
  optimum conditions for, 554–556
  scaleup in, 514–517
  stoichiometry in, 518
  target fragmentation in, 503–504
  target preparation for, 506–508, 518

Pyrolysis, 351–352
Pyrometry, 206

Q

Quantitation, 397–400

R

Rate equation models, 337–338
  of free-carrier heating, 177–120
  of ion emission, 150–152
  of lattice heating, 117–120
  of photoemission, 164–166
  of positive-ion emission, 150–152
Rayleigh length, 191
Reaction coordinate, 40, 59
Recombination, 35–36
  non-radiative, 51–56
  radiative, 50
Recondensation, 278–281
Reflection high-energy electron diffraction (RHEED), 294–297
Reflectron, 430, 432
Retarding field analyzer (RFA), 319
Rotational temperature, 123–124, 598–600
Rotation effect, 258–261
Rutherford backscattering (RBS), 291, 306, 309–310

S

Sapphire (Al_2O_3), *see* Alumina
Scanning microprobe, 324
Scanning transmission electron microscopy (STEM), 292
Scanning tunneling microscope (STM), 111–113, 291–292, 299–303
Secondary ion mass spectrometry (SIMS), 292, 315–318, 415
Secondary mechanism of ablation, 284
Secondary neutral mass spectrometry (SNMS), 316–318
Self-focusing, 37–38, 177

Self-trapped exciton recombination
    luminescence, 117–120
Self-trapped excitons, 56–57
Semiconductor(s)
    ablation of, 86–92
    applications, 630–632
    bandstructure of, 69–70
    CIGS, 544–550
    CIS, 544–550
    desorption from, 67–71, 78
    doped epitaxial, by PLD, 517–519
    laser annealing of, 364
    I-III-VI$_2$, 524–537
    II-VI, 520–528
    III-nitrides, 541–543
    surface modification of, 362–364
Sensible heat, 182
Shadow-cone effect, 312
Shear constant, 89–90
Shifted Maxwellian distribution, 241
Shock wave, 335, 350–351, 390–391
Silicon (Si), 91
Silicon dioxide (SiO$_2$), 152
Sinapic acid, 420
Skin depth of metals, 21
Sodium chloride (NaCl), 586, 595–598
Sodium nitrate (NaNO$_3$), 142–143, 146–147, 150–152
Spallation, 106
Spatial profile, 174
Speckle, laser, 341–342
Spot-shape effects, 259–261
Sputtering, electronic, 176, 232, 282
Sputtering, exfoliational, *see* Exfoliation
Sputtering, hydrodynamic, *see* Hydrodynamic sputtering
Sputtering, thermal-spike, *see* Thermal-spike sputtering
Stanski-Krastanov growth, 308
Stefan problem, 180
Stress, thermal, 236
Subsurface heating, 238–239
Subsurface melting, 503–506
Superconducting thin films, 401
Surface absorption, 77–78, 590–592
Surface analysis methods, 292, 327
Surface breakdown, 157, 163
Surface oxides, 177
Surface patterning, 189
Surface plasmon
    ablation mediated by, 65–67
    desorption mediated by, 65–68, 212
    resonance spectroscopy, 107–109
Surface reflectivity, 178, 201–205
Surface roughness, 83–86
Surface state, desorption induced by, 57–58, 68–69
Surface temperature, 179
Surface vaporization, 183–186
Synthetic polymers, 438

T

Tandem mass spectrometry, 432
Target conditioning, 508
Target preparation, for PLD, 506–507
Target temperature, 201–218
Teflon AF, 361
Thermal ablation, 175
Thermal conductivity, 344
Thermal diffusion length, 344
Thermal diffusion penetration depth, 178
Thermal lensing, 179
Thermal processes, 340–341
Thermal processing, 343–347
Thermal-spike sputtering, 238, 281
Thermal sputtering, 176
Thermionic emission, 186
Thermistors, 205, 206
Thermodynamic critical temperature, 226
Thickness uniformity in thin films, *see* Uniformity
Thin-film coating, 607–616
Thin films, 2
Time-resolved holography, 112
Tin oxide (SnO$_2$), 552
Ti:sapphire laser, 335
Tissue
    ablation of soft, 449–452
    dielectric properties of, 452–454
    structural properties of, 451–452
Translational temperature, 83–85, 106–107, 158–160, 598–600
Transparent-conducting oxides (TCO), 543–554
Transverse charge, 22, 30
Transverse electromagnetic modes (TEM), 173
Two-hole localization, 47–50

U

Uniformity, of film thickness in PLD, 514–516
Unsteady adiabatic expansion, 226–227
Urbach rule, 584

V

Vacuum-isolated ions, 413, 414
Van der Merwe growth, 297, 308
Vaporization, 234, 239
Velocity distribution, 427, 428
Velocity filtering, in PLD, 512
Vibrational processes, 18, 103–104, 413

Volmer-Weber growth, 308

X

X-ray generation, by laser ablation, 629–631
X-ray microscopy, 631
X-ray photoelectron spectroscopy (XPS), 318–322

Z

Zinc-indium-oxide ($ZInO_3$), 553–554
Zinc oxide (ZnO), 544–551

ISBN 0-12-475975-0